Sustainable Waste Management Challenges in Developing Countries

Agamuthu Pariatamby
Sunway University, Malaysia

Fauziah Shahul Hamid
University of Malaya, Malaysia

Mehran Sanam Bhatti
University of Malaya, Malaysia

A volume in the Advances in Environmental Engineering and Green Technologies (AEEGT) Book Series

IGI Global
DISSEMINATOR OF KNOWLEDGE

Published in the United States of America by
IGI Global
Engineering Science Reference (an imprint of IGI Global)
701 E. Chocolate Avenue
Hershey PA, USA 17033
Tel: 717-533-8845
Fax: 717-533-8661
E-mail: cust@igi-global.com
Web site: http://www.igi-global.com

Library of Congress Cataloging-in-Publication Data

Names: Pariatamby, Agamuthu, editor. | Shahul Hamid, Fauziah, editor. | Bhatti, Mehran Sanam, editor.
Title: Sustainable waste management challenges in developing countries / Agamuthu Pariatamby, Fauziah Shahul Hamid, and Mehran Sanam Bhatti, editors.
Description: Hershey, PA : Engineering Science Reference, [2020] | Includes bibliographic references. | Summary: "This book examines waste management practices in developing countries such as waste generation and waste composition. It also explores the roadblocks and potential opportunities involved in achieving integrated waste management in developing countries"-- Provided by publisher.
Identifiers: LCCN 2019018306 | ISBN 9781799801986 (hardcover) | ISBN 9781799802006 (ebook) | ISBN 9781799801993 (softcover)
Subjects: LCSH: Refuse and refuse disposal--Environmental aspects--Developing countries. | Sustainability. | Developing countries--Environmental conditions.
Classification: LCC TD790 .S87 2020 | DDC 628.4/45091724--dc23
LC record available at https://lccn.loc.gov/2019018306

This book is published in the IGI Global book series Advances in Environmental Engineering and Green Technologies (AEEGT) (ISSN: 2326-9162; eISSN: 2326-9170)

British Cataloguing in Publication Data
A Cataloguing in Publication record for this book is available from the British Library.

List of Reviewers

Asit Aich, *Municipal Administration Department, Government of West Bengal, India*
Arlen Ancheta, *Mother Earth Foundation, Philippines*
Md Ashikuzzaman, *Khulna University, Bangladesh*
Hamidi Abdul Aziz, *Universiti Sains Malaysia, Malaysia*
Benedict Basnayake, *University of Peradeniya, Sri Lanka*
Adem Basturk, *Yildiz Technical University, Turkey*
Goran Bošković, *University of Kragujevac, Serbia*
Emenike Chijioke, *Hezekiah University, Nigeria*
Alejandro Cittadino, *CAECE University, Argentina*
Ahmet Demir, *Yildiz Technical University, Turkey*
İbrahim Demir, *İstanbul Technical University, Turkey*
D. M. S. H. Disanayaka, *Rajarata University of Sri Lanka, Sri Lanka*
Renalda El Samra, *Rafik Hariri University, Lebanon*
Premakumara Dickella Gamaralalage, *IGES Centre, Japan*
Sunil Herat, *Griffith University, Australia*
Yasuhiko Hotta, *Institute of Global Environmental Strategies, Japan*
Aleksandar Jovović, *University of Belgrade, Serbia*
Muhammad Safar Korai, *Mehran University of Engineering and Technology, Pakistan*
Rania Maroun, *Sustainable Alternatives, Lebanon*
Edison Muzenda, *Botswana International University of Science and Technology, Botswana*
Suzan Oelofse, *Council for Scientific and Industrial Research, South Africa*
Saim Özdemir, *Sakarya University, Turkey*
Andante Hadi Pandyaswargo, *Waseda University, Japan*
Tusar Kanti Roy, *Khulna University of Engineering and Technology, Bangladesh*
Atilio Armando Savino, *Association for Solid Waste Studies, Argentina*
Hul Seingheng, *Institute of Technology of Cambodia, Cambodia*
Maria Francesca O. Tan, *University of the Philippines, Philippines*
Yulinah Trihadiningrum, *Institut Teknologi Sepuluh Nopember, Indonesia*
Yahaya Yakubu, *Africa Environmental Sanitation Consult Limited, Ghana*
Mi Yan, *Zhejiang University of Technology, China*
Lou Ziyang, *School of Environmental Science and Engineering, China*

Table of Contents

Detailed Table of Contents

waste management have been exemplified in this chapter. Finally, the authors apprised the issues and challenges of sustainable solid waste management practices and proposed the way forward for Bangladesh to have sustainable solid waste management.

Driven by economic development, population growth, change in life style, and consumption patterns, Cambodia is faced with equally rapid increase of solid waste, with MSW disposal amount attaining 1,709,379 tons/year in 2018. Various policy instruments and legislations have been developed over the years in response to this long-existing crisis, which effectively translated into tangible improvements on the ground. But municipalities continue to suffer from weak waste management system including collection, transportation, treatment, and disposal. Resource recovery almost entirely depends on informal sector while efforts for reduce and reuse are still weak, while presenting vast opportunity if effective policies are employed and implemented. The ongoing governance reform involving functional decentralization of waste management service is being implemented although at a slow pace.

This chapter focuses on the generation, treatment, and management policy, but excluding reduction and recycling information, of municipal solid waste (MSW) in China. Cities in China generated 215 million tons of MSW in 2017. It presents a big challenge for sustainable development of cities. Currently, sanitary landfill is the dominant method for MSW treatment, treating 57.2% of total waste. MSW composting has seen a decrease in application, only used to treat less than 2.5% of total MSW. Meanwhile, waste to energy in the form of incineration for energy production has developed significantly in the last decade. The percentage of MSW amount treated by incineration is 40.2% from the total amount of treated MSW in 2017, a sharp increase from only 4.9% in 2003, due to China's commitment to WtE. The content of this chapter shall include comprehensive data from government and industry, including plant capacity, quantity, technology, emission, policy and regulation promoting, and guiding WtE in China.

The demands of the circular economy and the sustainable development goals call for a critical appraisal of the solid waste management systems of developing economies to divert decision-making towards the development of sustainable strategies and support systems for purposes of modernization. This chapter uses a combination of literature review, network queries to key experts within municipalities, and the authors' experiences and outputs of an ongoing collaborative research to evaluate and present the realities of solid waste management and recycling in Ghana. The goal is to chart a locally appropriate and sustainable pathway to drive system improvement. The analysis and evidence suggest that structural and coordinated cooperation between researchers of higher educational and research institutions and municipal authorities creates an enabling platform to build human resource capacities, to bridge data gaps, to identify what works, and to drive decision-makers' commitment towards the development of realistic action plans and continuous policy strategies.

Chapter 6
Waste Management Under the Legal Framework in India.. 144
Sadhan Kumar Ghosh, Jadavpur University, India

India, the second biggest country in the world, has nearly 1.25 million people living in 29 states and seven union territories covering an area of 3,287,000 sq. km. India's economy grew at an impressive 8.2% in the first quarter of 2018-19. Traditionally, India has the habit of reuse and recycling the materials wherever possible. As the city agglomeration is increasing the waste generation is increasing. The number of towns/cities have increased from 5,161 in 2001 to 7,935 in 2011, whereas the number of metropolitan cities having million plus population has increased from 35 to 53 number as per 2011 census. It is projected that half of India's population will live in cities by 2050. Waste management in India has been experiencing a paradigm shift through the establishment of Swachh Bharat Mission in urban and rural India in 2014 and the revision and establishment of waste management rules in six types of wastes including transboundary movement in 2016. This study presents the overall waste management scenario and the legal framework in India.

Chapter 7
Challenges of Waste Management in a Developing Context: Lessons From Lebanon 166
Mutasem El-Fadel, American University of Beirut, Lebanon
Amani Maalouf, American University of Beirut, Lebanon

Over the past years, several efforts were undertaken in an attempt to improve solid waste management in Lebanon. However, with the lack of a well-established national plan and defined policies and legislation, all attempts achieved little or no progress. Several challenges continue to hinder the sector's successful development at various levels: politically, legally and institutionally, socially, land use, and last but not least, technically. This chapter provides an overview of waste management practices in Lebanon such as waste generation, waste composition, current status of waste management practices, policy, and legislation, while highlighting the main challenges for implementing an integrated solid waste management towards a sustainable system.

Chapter 8
Waste Management Challenges in Malaysia... 186
Agamuthu Pariatamby, Sunway University, Malaysia
Mehran Sanam Bhatti, University of Malaya, Malaysia

Malaysia, an upper-middle class country, populated with approximately 31 million people generated nearly 13.9 million tonnes of municipal solid waste (MSW) at per capita of 1.23 kilogram per day in 2016. Landfills and open dumps, being the absolute opposite of sustainable waste management, received about 80% of this generated MSW. Waste collection is on a par with developed nations, and almost all urban MSW is being collected for disposal. However, illegal dumping still occurs sporadically, and it can account for 10% of the total MSW generated. Hence, Malaysia is facing a stiff challenge in reducing the amount of waste sent to landfills and adopting sustainable waste management. National Solid Waste Management Department of Malaysia targets to divert 40% of MSW from landfills by 2020. There are total 296 landfills in Malaysia, and only 165 landfills are currently functional whereas the number of sanitary landfills is 8 out of 165 operating landfills. The national recycling rate of MSW was 17% in 2016 and the national recycling target in 11th Malaysian Plan is 22% by 2020.

Myanmar, the largest country in mainland Southeast Asia, has been facing considerable challenges with the management of solid waste in the recent past because of increasing income and consumption patterns, urban growth, and lack of effective waste management policies, treatment, and disposal methods. Waste management is also a crosscutting issue that touches on many aspects of social and economic development, and as such is widely associated with a range of global challenges including public health, climate change, poverty reduction, food security, resource efficiency, and sustainable production and consumption. This country chapter therefore presents an overview of the current waste management in Myanmar, discusses key challenges and opportunities, and identifies some policy recommendations towards its improvement.

This chapter covers different aspects of solid waste management (SWM) in Pakistan. Being a developing country, Pakistan is facing serious environmental issue as a result of improper and limited SWM practices. The current status of waste generation is far high than waste management operation working in the country. Detailed information has been provided on account of waste generation per capita, waste composition and types, current status of waste management practices, policy and legislations on solid waste management. Furthermore, particular challenges in local setups for achieving waste management goals have been highlighted. There is urgent need to look into the various options to deal with proper management of solid waste in the country. In this regard, various projects by governmental and private sector have been initiated. Particularly, projects related to waste to energy (WTE) have been planned and operational in certain parts of the country.

In response to the growing problems on waste management in the country, the Philippines' Republic Act 9003 (RA 9003), also known as the Ecological Solid Waste Management Act of 2000, was enacted in January 26, 2001. This chapter hopes to provide the brief background of the country particularly on the issues related to waste management such as population, economic situation, urbanization, and modernization, among others. It will also discuss the definition, classification, and generation of waste both in urban and rural areas. In addition, it will also review the existing policies and the current waste management practices, and the lessons learned based on case studies of both successful and failed experiences. Furthermore, this will also provide the latest updates on the compliance of the local government units to RA 9003, identify the challenges, opportunities, and the proposed recommendations on how the waste management in the country can possibly be improved and become more sustainable.

This chapter aims to describe the status and challenges of waste management (WM) in Rwanda. Currently, waste is managed by the Ministry of Local Government, with the participation of private companies which are only in charge of waste collection. In the city of Kigali, waste is managed by the city of Kigali whereas in other four remaining provinces waste is managed by the districts. Implementation of waste management policy is carried out by a government-owned company called Water and Sanitation Corporation (WASAC) Ltd. The per capita solid waste generation rate in the city of Kigali is equal to 0.57 kg/person/day. A lot of legislations and regulations on WM are in place but their enforcement is weak. The Government of Rwanda should do more in terms of enforcing WM legislations and regulations.

In most of the countries of South East Europe, the current situation in the field of municipal waste management is characterized by unreliable and incomplete data about the physical characteristics of generated municipal waste. The first and the second part of the chapter contain the description of specific project activities that are related to measuring quantity and composition of municipal waste in municipalities in the Republic of Serbia. The methodology description for the determination of the quantity and composition of waste is analyzed in this chapter. The potential of implementing the circular economy concept in developing countries is reflected primarily in the optimization of production and industrial activities; the example of Serbia in the third part of chapter shows the opposite. In the fourth part, results of application of the AHP method are presented. Taking the obtained results as a basis, it can be concluded that the applied model can be efficiently used to support decision-making in this kind of problem.

Solid waste management (SWM) is a challenge in developing countries such as the Republic of South Africa (RSA). This book chapter highlights the drivers and state of SWM in RSA and suggests alternatives to make solid waste a resource. The SWM strategy of the country has a role in pushing waste up its hierarchy towards minimal generation, reuse, and recycling through extended producer responsibility and economic instruments. However, the lack of an all-inclusive planning and management has challenged the success of these initiatives. In recognition of these flaws, the private sector is teaming up with the government and individuals to bridge service and value chains in sustainable SWM by formalising waste pickers, initiating waste-to-energy initiatives, promoting recycling at all stages of the waste cycle, and adopting practices that divert wastes from landfills. These initiatives if taken up will promote better economic turnover through the production of alternative energy, environmental conservation, and creation of employment opportunities in RSA.

Sri Lanka faced many catastrophes in managing wastes because of human errors and climate change impacts. Consequently, government administrators with stakeholder participation drafted a comprehensive policy on all aspects of waste management and now requires planning processes in the national, provincial, and local authority levels. Evidenced-Based Interactive Database is under construction. It will be a platform for all actors to participate in managing all types of wastes, inclusive of quantity and quality of wastes. This chapter describes municipal solid waste management, including present status of technology applications with the emphasis on 3R to divert wastes from landfills. It contains reports on e-wastes collection programs and some processing, polythene recycling, hazardous wastes management, including healthcare wastes, disaster and construction and demolition wastes, informal sector and recyclers, economics of waste management, and case studies. Future directives are given with coupling of intended nationally determined contributions for sustainable waste management.

Resilient economic urbanisation growth in Tanzania signifies national increasing trend of solid waste generation that necessitates proper management. About 50% of Tanzanians suffer from intestinal schistosomiasis and soil transmitted helminths resulting from haphazard solid waste disposal. Various efforts to address solid waste management in Tanzania include establishment of public private partnership and major reforms in policy and regulatory frameworks. Despite these efforts, 60% of waste generated is mismanaged and normally disposed in illegal places along roadsides, sewage canals, and reserved open spaces. Cornerstone of proper solid waste management lies on separation of the waste at source. It is recommended that 'the smart garbage collection bin' be adopted at community collection points and charging higher prices for generators who do not separate the waste.

Thailand covers an area of 513,120 km2 and is located in the southern region of the continent of Asia. The total population was around 67.2 million in 2016. With increasing population growth, city expansion and rapid industrialization, generated 27.37 tonnes of waste in 2017. Thailand government has recognized problems associated with solid, hazardous, and electronic waste generation and issued policies/regulations to support the waste management system. Initiatives such as roadmap for municipal and hazardous waste, national master plan on cleaner production and cleaner technology, and national integrated e-waste management strategy have been introduced to improve the waste management system. However, there are many issues preventing the implementation of a successful waste management system in Thailand. Society has suffered from waste mismanagement because of lack of awareness and realistically applicable technology. There are also weaknesses in the processes of public participation, policy implementation, and institutional support.

Due to geographical position at the crossroads between Middle East and Asia and Europe, Turkey has a significant role for neighbor developing countries in terms of economical and social growth. Since solid waste management system mainly depends on these factors, Turkey has been always comprehensive model for many developing countries with its management system. This chapter, consisting of eight sub-sections, evaluates the management of municipal solid waste in Turkey with current data from the Ministry of Environment and Urbanization and provides an overview of the development of waste management practices in Turkey. In addition to the future perspectives and planning, the current status, waste amount, waste characteristics, prevention studies, recycling and re-use applications, collection, treatment and disposal applications of municipal wastes, packaging wastes, and medical wastes are reported in this chapter.

This chapter will provide an overview of waste management in Vietnam, including drivers, the current waste management situation, impacts, an institutional framework, issues, and challenges. There have been improvements in the last 10 years. However, waste management in Vietnam continues to face issues and constraints related to inefficient waste reduction, a national level efforts to implement separation at the source, environmental pollution caused by informal sector recycling efforts, and 70% to 75% of municipal solid waste in landfills. These issues are caused by incomplete legislation, inadequate financial resources, weak enforcement and implementation efforts, and low stakeholder awareness. Based on the analysis, several recommendations are provided for future improvement.

This chapter summarizes and compares the current situation of waste management practices in developing countries as reported in other chapters of this book. Due to increasing rate of urbanization, population, and economic growth, generation of waste is also on the rise. It further makes the waste management situation trickier for developing countries as effective regulations and policies are either lacking or their stringent enforcement is absent. Thus, it leads to major issues of inefficient waste collection, open dumping, open burning, and over-reliance on landfilling. It was also discovered that holistic approach of all relevant stakeholders from governmental level to waste generators level is glaringly missing in most of the developing countries. At the end of the chapter, several recommendations are presented to overcome the challenges.

Foreword

As it will become unequivocal in this book, the developing countries are witnessing brisk economic growth, rapidly increasing population, and equally fast changing living standards, that have contributed to an increase in per capita generation of waste. Judging by the growing trends, waste generation is expected to grow further with an even greater rate. With the changing lifestyles, the consumption and disposal habits of users are also changing. This has given rise to highly concentrated and more detrimental forms of waste categories, especially in developing countries, e.g., excess of plastic waste or sanitary napkins, etc. Therefore, each country with its changing demographics and consumption habits of the citizens, has very specific and unique characteristics of waste. Thus, this calls for the need to devise its own solutions to tackle the ever-growing waste generation and related issues as highlighted by each country chapter in this book. The chapters then further go on to elucidate that there is no one solution to all problems. We can learn and adapt from others but we need to develop our own path for sustainable waste management which works in our country, city or community. This book further confirms the traditional solid waste management of the developing countries, which is also known as the linear waste management approach, that is heavily focused on the collection, transportation, storage and disposal of the waste. Moreover, unlike developed countries, the developing countries have to be more innovative in the formation of sustainable waste management, owing to their limited infrastructure and reasonably smaller economies. This puts extensive pressure on urban municipalities to create sustainable waste management systems.

There is a dire need to shift from the traditional waste management approach by involving proven concepts like 3R (reduce, reuse, recycle) to create a more sustainable future for waste management. A strong policy intervention and the involvement of disruptive technologies are also needed. The main aim should be focused on the reduction of waste at source, by following simple techniques like source segregation. Along with that, the promotion of solutions such as waste to energy to increase the capacities of end of the line waste treatment systems is also needed as the approach taken by China. Urban waste management can no longer be a responsibility of the governing bodies or municipalities, it has to reach the grassroots of the society. People need to understand their responsibilities towards the environment and how their actions contribute to global phenomena like global warming. With that said, this shift in urban waste management requires combined social, technical and economic inputs, which will also aid in the creation of a new job sector of sustainable urban waste management.

This publication entitled *Sustainable Waste Management Challenges in Developing Countries* essentially addresses the country-specific status reports of 19 countries from 4 continents, including the national solid waste status and the challenges involved in terms of technology and policies. Additionally, a distinct feature of this book is the synergy chapter at the end, which compares, analyzes and synthesize the information from all 19 chapters and concludes with recommendations. This book contributes to the policy development of integrated waste management, while enlisting the potential opportunities that are brought along with sustainable waste management. This book can help waste management experts of a country to overcome the roadblocks in sustainable waste management and achieve Sustainable Development Goals.

C. Visvanathan
Asian Institute of Technology, Thailand

Preface

Industrial revolutions have resulted with astounding creations that ease human civilization on the Earth. Yet, these revolutions come with a high price due to human ignorance and limited human capability to predict the future. With the creation of various materials that bring comfort and luxury to the human wellbeing, the generation of waste is increasing rapidly while the composition is becoming more complicated than ever. The earlier revolutions only required simple approach in managing wastes, which normally were organic based, where disposal of waste into disposal sites was sufficient to solve the waste issue. However, moving towards the industrial revolution 4.0, the complexity of waste has extraordinarily escalated, that mere disposal into disposal sites is no longer a sustainable option. While developed nations with advanced technologies can manage their waste in a profitable manner, turning waste into resources by adopting the concept of circular economy, developing nations are yet to arrive at the same conclusion. Many developing nations which undergo economic transition are struggling to manage their own waste. In fact, providing a proper waste management system to the whole nation can be an extremely difficult task. The most challenging sectors for waste management in developing nations are waste collection system, and sustainable disposal options.

The situation faced by many developing nations calls for the need of a manuscript that puts scientific details and case studies into the framework of resource management and sustainable strategies. This publication called Sustainable *Waste Management Challenges in Developing Countries* responds to that need. All stakeholders namely authorities, practitioners, researchers, students and general public will find this book enlightening with current updates on the situation of waste management of various developing countries in the world. It highlights the real state of affairs of waste management in each country, which may not be easily obtained from other resources.

IMPORTANT FEATURES OF THIS BOOK

The topics discussed in this book fit the waste management world today. This is extremely relevant to the developing nations that are harassed and threatened with the impacts from improper waste management. Successful case studies of one country can be beneficial for other countries for planning the improvement in the waste management, on the basis of experience sharing concept. Since chapters are contributed by developing countries' experts, the information and experiences shared have very similar foundation from the issues and challenges perspectives. Therefore, the solutions and strategies provided in the case studies are more applicable to suit the needs of the other developing economies. This book deliberates the current waste management practices in developing countries, whereas exceptional case studies including successful and failed scenarios are also reported.

To elucidate the current scenario of waste management in developing countries while increasing its user-friendly features, illustrations, graphs and diagrams are also included. Additionally, selected references are included at the end of each country chapter.

TOPICS IN THIS BOOK

A total of 20 chapters are presented in this book. The chapters form a comprehensive and coherent investigation in solid waste management, including definitions used, waste generation, collection, transportation, treatment and disposal and 3R (reuse, recycle, recover) activities. Several case studies are included to exemplify the real situation experienced in Asia, Africa, Latin America and East Europe. It is greatly envisaged to form an excellent source of reference on MSW management in developing countries for both developing and developed nations.

The first chapter represents a developing country in South America, namely Argentina. This chapter on waste management in Argentina discussed the obligations of the country to improve its waste management system by refining waste collection strategies and its disposals, particularly for household and domestic wastes. It further highlights the role of Ecological Coordination Society of the State Metropolitan Area (CEAMSE) and waste pickers for improving the waste management system. The chapter concludes that integrated and improved waste management is essential for a Sustainable Development Agenda and can be achieved by closing open dumpsites, by providing and improving waste collection services, and by integrating appropriate waste management strategies for all waste streams. Additionally, it stresses the importance of promoting circular economy, appropriate organic waste management, ensuring adequate control and treatment of hazardous waste, promoting a suitable set of policies, regulations and economic instruments, selection of the best available technologies, and the formalization of the informal sector as an integral part of recycling sector.

The second chapter, a chapter on solid waste management in Bangladesh highlights the alarming increase in waste generation which will exceed 21 million tons by 2025, which is a true representation of waste scenario in developing countries where waste generation is projected to increase in coming years. The issues pertaining to waste management, particularly the policies and regulations on waste management in Bangladesh are delicately discussed. The chapter also contains emerging issues of agricultural waste, industrial waste, hazardous waste, construction waste and E-wastes. Additionally, it elaborates the existing practices of 3R in Bangladesh and its impacts on the waste management system in Bangladesh. However, the chapter reports that weak enforcement of 3R strategies by the local authorities has hindered 3R practices in the country. This Bangladesh chapter is further strengthened by case studies of community-based approach in waste management, waste recycling sector and medical waste management.

The third chapter of the book discusses the solid waste management scenario in Cambodia. This chapter highlights that institutional drivers such as policy instruments and legislations have been strengthened over the years, thus they played a significant role in improving the current state of waste management in the country. Nevertheless, it is reported that many local authorities still fail to provide sufficient collection services. In addition to that, vehicles for waste transportation are yet to improve to a satisfactory standard while waste treatment are almost nonexistence that waste is send indiscriminately to disposal sites without pre-treatment. In Cambodia, resource recovery is highly dependent on informal sector while reduction and reutilizing efforts are insignificant. The chapter concludes with the discussion on the implementation of functional decentralization of waste management service in the country.

The fourth chapter elucidates waste management in China. It focuses on the trend of generation, composition and treatment of municipal solid waste, resulting from the progressive development and economy growth in China. The chapter highlights the challenges for sustainable development of cities due to excessive increase in waste generation. This chapter also discusses the issues related to the China's ban on the importation of waste since 2018. The policy on waste management in the country was also deliberated highlighting the diversion of waste stream from landfilling to waste to energy plants. The strength of the chapter is the detailed information on China's shift from landfills to waste to energy and comprehensive data.

The fifth chapter discusses the current scenario of waste management in one of the developing countries from the African continent, Ghana. This chapter presents the current scenario of waste management and recycling in the country by incorporating literature review, interviews with stakeholders in waste sector and author's research findings. It highlights the generation and subsequent management of municipal solid waste (MSW), E-waste, agricultural waste, medical waste, construction and demolition waste among others. The chapter reasons and presents that framework and strategies for a country can be formulated by stakeholders including researchers, research institutions and municipalities working in coordinated and unison manner.

The sixth chapter is on India, the second biggest populous country in the world. The chapter describes the waste management scenario in India from the perspective of urbanization and country's economy which is projected to grow rapidly. As a result, higher amount of waste is being generated despite the common practice of reuse and recycling among the Indians. Furthermore, the chapter highlights the modifications of waste regulations and introduction of new policies and strategies that are shaping the waste management system in India.

The seventh chapter provides the situation of waste management in Lebanon, highlighting that despite the initiatives taken for improvement, the waste management system is deeply affected by lack of explicit policies, national plans and legislations. Additionally, it discusses the challenges including political, social, legal and institutional that are impeding the successful development of an integrated waste management in the country. This chapter explains the current status of waste management practices in Lebanon from waste generation, composition and management to policy and legislation.

The eighth chapter on Malaysia presents a comprehensive overview of all types of solid waste generated in the country. While the chapter explains the legal instruments, national targets and management of generated solid waste, it further proposes recommendations on achieving integrated waste management. Furthermore, the generation of marine plastic debris from Malaysia is also elucidated in the chapter.

The ninth chapter provides an overview of waste management system in Myanmar, the largest country in mainland Southeast Asia. The chapter also highlights the challenges in solid waste management that faces the country due to increasing income and changes in consumption patterns, urbanization, and absence of effective waste management policies, treatment and disposal methods. This country chapter concludes with important policy recommendations for the improvement of waste management in Myanmar.

The tenth chapter deliberates the practices of waste management in Pakistan. The chapter highlights serious environmental issues faced by the country as a result of improper and ineffective solid waste management. The current status of waste generation is provided with detailed information on the account of waste generation per capita, waste composition and waste types. The chapter also discusses policies and legislations on the solid waste management in Pakistan. Furthermore, this chapter elucidates the challenges faced by local setups for achieving waste management goals. While an urgency to deal with proper waste management in the country is stressed, the chapter states that some projects i.e. waste-to-energy, have been envisaged and introduced in some parts of the country.

The eleventh chapter of the book on the Philippines, presents the background of the country along with the issues of waste management. It highlights that a monumental Philippines' Republic Act 9003 (RA 9003) or also known as the Ecological Solid Waste Management Act of 2000 was ratified to deal with increasing waste management problems in the country. The chapter also gives the definition, classification and generation of waste, both in urban and rural areas. In addition, a review of the existing policies, the current waste management practices, and the role of informal sector in waste management are discussed. The current situation of waste sector is supported by both successful and unsuccessful case studies and elucidates the lessons learned. Lastly, the chapter examines the compliance of the local government units on RA 9003, recognizes the challenges and opportunities faced and summarizes the recommendations for moving towards a sustainable waste management.

The twelfth chapter describes the status and challenges of waste management in Rwanda. The responsibility of waste management lies with Ministry of Local Government and the waste collection services are provided by private companies. This chapter also elaborates on the implementation of waste management policy by a government-owned company called Water and Sanitation Corporation (WASAC) Ltd. The chapter states rules and regulations pertaining to waste management and also reasons the weak enforcement of respective rules and regulations by the authorities. Therefore, the chapter concludes that the Government of Rwanda must take more participation in order to improve the current waste management state in the country.

The thirteenth chapter focuses on Republic of Serbia. The chapter lays the foundation that the present state of affairs in the management of municipal solid waste in Southeast Europe is marked with unreliable and incomplete data on waste generation and composition. Thus, the chapter presents the waste generation and composition data of the municipalities in the Republic of Serbia, while describing relevant project activities that produced the aforementioned data. The chapter also highlights that Republic of Serbia fails to implement the circular economy in the form of optimization of production and industrial activities. This chapter proposes analytic hierarchy process (AHP) method for decision making in waste management sector. The chapter concludes with supporting data and elaboration on the results from the implementation of AHP method.

The fourteenth chapter reflects on the solid waste management and its challenges in the Republic of South Africa (RSA). This chapter highlights the drivers and state of solid waste management in RSA and suggests alternatives to make solid waste a resource. It describes national strategy of implementing the waste management hierarchy towards waste prevention, waste reduction, waste reutilization, and material recycling through extended producer responsibility and other economic instruments but states that the absence of an all-inclusive planning and management in the waste sector has hindered the execution of an integrated waste management. Furthermore, the chapter establishes that in order to improve the current situation of waste management, private sector is building a partnership with government and individuals for bridging service and value chains in sustainable solid waste management. In order to do so, it aims to integrate waste pickers into a formal waste management system, initiate waste-to-energy options, promote recycling at all stages of the waste cycle and reduce landfilling.

The fifteenth chapter demonstrates the current waste management scenario of various waste streams such as municipal solid waste, E-waste, hazardous waste, healthcare waste, disaster waste, and construction and demolition waste in Sri Lanka. This chapter establishes that due to calamities related to waste sector either because of human error or impact of climate change has directed government administrators and relevant stakeholders to formulate a comprehensive policy on waste management and now necessitates planning processes in the national, provincial and local authority levels. This chapter also elaborates the issues and challenges pertaining to 3R, informal sector, while also explaining the waste economic of the country. The country status of waste management is supported by case studies.

The sixteenth chapter covers the waste management issues and challenges in Tanzania. This chapter further elaborates on the haphazard and improper solid waste disposal in the country that has led to the deterioration of the public health. Efforts to address solid waste management in Tanzania which include establishment of Public Private Partnership and major reforms in policy and regulatory frameworks are also thoroughly discussed. The chapter concludes with some highlights on the practice of various strategies at community level in order to improve the current waste management problem in Tanzania.

The seventeenth chapter includes the waste management scenario in Thailand. The chapter elucidates on the impact of increasing population, expansion of cities, and rapid industrialization on the waste generation in the country. It also elaborates on the recognition of problems associated with generation of solid waste, hazardous waste and E-waste by the government in Thailand. The existing policies and regulations to support the waste management system in the country are also highlighted together with initiatives such as Roadmap for municipal and hazardous waste, National master plan on cleaner production and cleaner technology, and National Integrated e-waste management strategy. Furthermore, this chapter discusses the issues that thwart the implementation of a successful waste management system in Thailand, including the mismanagement of waste due to lack of awareness and current waste technology, lack of public participation, ineffective policy implementation, and insufficient institutional support in waste management sector.

The eighteenth chapter highlights the waste management system in Turkey, which has a very unique geographical location. This chapter examines the management of municipal solid waste in Turkey based on the latest waste data from the Ministry of Environment and Urbanization of Turkey. Additionally, the chapter elaborates the current status of waste management by delineating the amount of waste generation, waste composition, waste collection and treatment, practices of recycling and reuse, and the methods of final disposal. It also focuses on packaging wastes and medical wastes. Lastly, the future perspectives and planning of waste sector are discussed.

The second last chapter of the book presents the waste management scenario in Viet Nam, by providing the current waste management situation, and its impacts, the institutional framework, the issues and challenges faced by the country in waste management. The chapter then goes on to explain the issues and constraints in detail by discussing inefficient reduction of waste, source separation, waste collection, environmental pollution resulting from informal recycling and overreliance on landfilling for waste disposal. The chapter reasons that the issues arise due to incomplete legislation, inadequate financial resources, weak enforcement and implementation efforts, and low stakeholder awareness. In the end, the chapter gives some recommendations achieving the sustainable waste management in Viet Nam.

The last chapter synthesizes a compilation and comparison of the current situation of waste management practices in the 19 developing countries covered in this book. The chapter explains that the rapid urbanization, fast growing population and economic boost are contributing towards the generation of waste in developing countries and the increase in waste is expected to continue. Then, the chapter discusses the elements of sustainable waste management by focusing on implementation of 3R and adoption of extended producer responsibility, which are not as widely practiced as in developed countries. Furthermore, the chapter highlights the major issues such as weak enforcement of legislations, poor waste collection in majority of developing countries, the practice of open dumping, and inclination towards landfilling and the negligence of role of informal sector by the waste administrators. The chapter concludes with the recommendations to address the issues and challenges in waste management in developing countries.

EXPECTED IMPACTS TO WASTE MANAGEMENT IN DEVELOPING COUNTRIES

This book elaborates the real situation of waste management among aforementioned developing nations that are currently struggling to implement an integrated and effective solution to their waste problem. The current scenario presented in this book would provide a clear picture to all the relevant stakeholders of waste management sector. Successful case studies can be adopted with minimal modification since all developing countries are undergoing similar economic transition and are facing similar challenges. On the other hand, case studies that highlight the failures, can be used as lessons to prevent the repetition of decision-making mistakes when faced with similar challenges. It is hoped that this book will benefit all developing nations by learning from each other's success and failures in order to create a sustainable waste management for their country. Furthermore, developed nations could also utilize this book as an authoritative guide for the waste management data from developing countries.

Agamuthu Pariatamby
Fauziah Shahul Hamid
Mehran Sanam Bhatti
September 2019

Acknowledgment

We would like to express our appreciation to all the authors of the chapters who diligently worked towards realizing this book; without their contributions it would not have been possible to complete the book on schedule.

Editors would like to thank our publisher, IGI Global, for giving us the opportunity to edit a book on one of major issues facing today's world - sustainable waste management. We would also extend our gratitude towards The Book Development Team of IGI Global including Ms Courtney Tychinski and Ms Halle N. Frisco for assisting us throughout this book project.

We are also grateful to Prof. Chettiyappan Visvanathan for endorsing our book and contributing the Foreword.

Lastly, we are indebted to all the reviewers who evaluated the chapters and gave critical comments to improve the contents and made sure that the chapters are of high quality. We appreciate their invaluable contribution to the book.

Agamuthu Pariatamby

Fauziah Shahul Hamid

Mehran Sanam Bhatti

Chapter 1
Sustainable Waste Management Challenges in Argentina

Atilio Armando Savino
Association for Solid Waste Studies (ARS), Argentina

Ernesto de Titto
ISALUD University, Argentina

ABSTRACT

A general picture of waste management in Argentina is presented, taking into consideration the obligations of the state to guarantee the collection of household waste and to dispose of it in authorized sites and the generator's responsibility on the management of hazardous waste. Separate boxes are presented to present the activities performed by organized waste pickers and by CEAMSE, a state-owned company responsible for waste disposition in the metropolitan area of Buenos Aires. Waste management is essential for a sustainable development agenda. As witnessed in Argentina, this requires (1) closing open dumpsites, (2) a reliable and complete waste collection service, and (3) proper treatment of all waste streams. It is thus clear the need to (1) promote a circular economy; (2) tackle the management of organic waste; (3) ensure the adequate control and treatment of hazardous waste; (4) promote a suitable set of policies, regulations, and economic instruments; (5) select the best available technologies; and (6) formalize the informal phase of recycling.

OVERVIEW

In Argentina, local governments are legally obliged to guarantee the collection of household waste and to dispose it in authorized sites. This can be done by the State itself or by hiring third parties.

The situation is different for hazardous waste whose management is under the responsibility of the generator (industries, stores, universities, healthcare centers, etc.), who must perform source separation, temporary storing at adequate storage and employ the authorized transporters and operators for transporting, treating and final disposal.

DOI: 10.4018/978-1-7998-0198-6.ch001

Several locations adopted regionalization models for municipal waste management that has reaped benefits for all the involved parties from the environmental and economic perspective. However, this model deserves further development especially since different technologies for treatment of different waste streams are required which are costly and necessitates a certain scale to be sustainable.

The recovery of recyclable materials from garbage must also be considered. Waste recycling is a growing practice nowadays. However, it is not rare meeting unorganized families collecting and selling materials that other people have considered waste, as a way of life. Thus, people of different age, including kids, without personal protection elements, handle waste daily without even taking into account the implication this activity has on their health.

Besides, although legally forbidden it is not rare abandoning waste in public sites, especially in the periphery of largest cities. It is known that uncontrolled burning and open dumpsites with the expected consequences of environmental impacts on the air, soil and surface and underground water, associated with several health risks, should be avoided.

Solid waste treatment plants and sanitary and secure landfills, such as any work project or activity with entity to cause an impact in the environment are required by the environmental legislation to carry out an Environmental Impact Assessment's procedure prior to its habilitation. Nevertheless, the so-called NIMBY effect[1], especially in connection with the selection of sites intended for sanitary landfill still is a relevant issue in Argentina. Communities' rejection to have a sanitary landfill nearby is frequent, affecting the possibility of having sites suitable for the final disposal of waste.

In these troubled situations, a number of faults are immediately noticed that derive basically from pending issues of governments that go beyond the site itself, such as: 1) most cities have grown in a disorganized and unplanned manner, being it difficult to find examples of consistent works that pertain environmental management of the territory; 2) the weak control of the premises by the authorities and the scarce information regarding design, operation and monitoring of the sites; 3) the lack of timely inclusion of the citizens in the decision making process, and/or in choosing the opportunity to provide information to the community and 4) the low impact of communitarian opinions. In many conflicting cases the governments make huge efforts to explain facts and supply information when the conflict has already been unleashed.

COUNTRY PROFILE

The Argentine Republic extends over the south of the American continent, the islands of the South Atlantic and part of Antarctica. The surface of its continental region is 2,791.810 Km^2 and that of Antarctica and the islands of the South Atlantic of about a million Km^2 (969.464 Km^2).

The country spans in latitude from the Tropic of Capricorn to the South Pole, and in longitude from the highest peaks of America in the Andes to sea level on the Atlantic coasts, which determines a wide variety of climates.

The wetlands in the East of its territory comprise forests and subtropical mountains and the Pampas Plain in which the main agricultural and livestock activities of the country are developed. The arid and semiarid zones occupy two thirds of the surface of Argentina. They have a high variability in their landscape with xenophiles' forests, areas with deserts, mountain and spinal, in the North and central region. In the South there is the Patagonia, almost all arid and with strong winds.

Figure 1. Map of Argentina

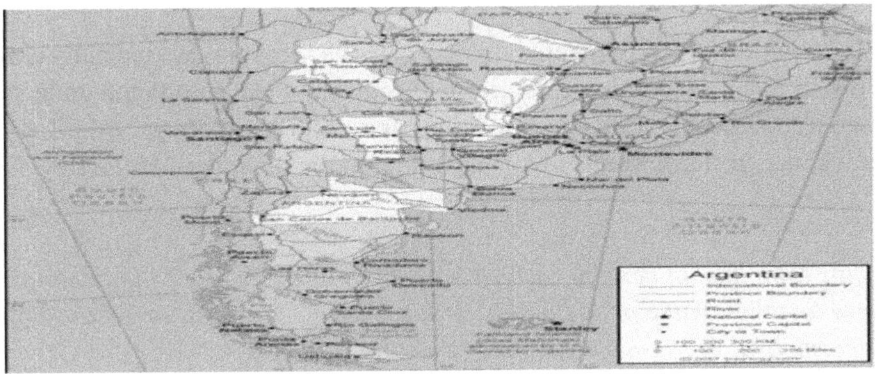

The population of the country at the end of 2015 was over 43 million inhabitants with a proportion of the urban population of about 91%. The Argentine population has grown in the period 2000-2015 at an average annual rate of 10.7 per thousand. The average density is 15 inhabitants per Km^2. The distribution is very uneven and most of the population is concentrated in the largest urban centers.

According to official data (PAHO-Argentine Ministry of Health, 2018), Argentina's crude birth rate is 18 per thousand, while mortality is 7.7 per thousand. Infant mortality of live births is 9.7 per thousand, having declined since 1980, when it rose to 33.2 per thousand. Life expectancy is about 75 years, having increased in two years, compared to that recorded in the early nineties. In the case of women life expectancy is 79 years and in men of 72 years.

There is a growing demographic ageing: while at the beginning of the nineties the percentage of older adults (over 65 years old) was 8.9%, currently it is 10.4% of the population. At the same time, the number of young people (up to 14 years) in the total population has been reduced from 30.6% in 1991 to 25.1%.

Argentina is a Republican, representative and federal state; it has a decentralized political-administrative organization formed by 2253 municipalities in 23 provinces and the Autonomous City of Buenos Aires (CABA) as presented in Fig. 1. In Environmental Matters and according to the basic principles of the National Constitution, each province enjoys the dominion and the administration of its natural resources. The Nation has among its faculties to dictate the norms with the minimum budgets of environmental protection (Law 25,675).

With a Gross Domestic Product (GDP) of more than US$500 billion, Argentina is one of the largest economies in Latin America.

Argentina has vast natural resources in energy and agriculture. Within its 2.8 million square kilometers of territory, Argentina is endowed with extraordinary fertile lands and has great potential for renewable energy. It is a leading food producer with large-scale agricultural and livestock industries. In addition, Argentina has significant opportunities in some manufacturing subsectors, and innovative services in high tech industries.

Argentina has a mixed economic system which includes a variety of private freedom, combined with centralized economic planning and government regulation. Argentina is a member of the Common Market of the South (Mercosur).

DEFINITION OF WASTE

The terms "waste" (residuos) "scraps" (desechos) or "garbage" (basura) are typically used interchangeably, as synonyms, in Spanish-speaking countries to designate, under a common understanding, all the products or materials that people decide to discard because they are no longer useful.

This chapter will use the term waste, which according to the dictionary of the Royal Spanish Academy means

things or materials from human activities which have been intentionally or unintentionally discarded because they have lost their value or usefulness, efficiency, effectiveness, or because of their overuse, destruction or decomposition caused by humans or nature,

since it is an all-encompassing and broad term for the purposes and objectives of this document.

Under this definition, waste can be in liquid, gaseous or solid state, and it will finally be received in the natural sinks: water, air and soil.

In order to be more precise, this definition matches the definition of "waste" established by the BASEL CONVENTION ON THE CONTROL OF TRANSBOUNDARY MOVEMENTS OF HAZARDOUS WASTES AND THEIR DISPOSAL, ratified by Argentina in June 1991 through the Law 23.922, even though the official translation of the Basel Convention into Spanish uses the term "desecho", "'wastes are substances or objects which are disposed of or are intended to be disposed of or are required to be disposed of by the provisions of national law".

To narrow the context even further, this chapter will refer to solid waste generated as a result of human activities and/or nature, and the eventual consequences of its treatment or lack of treatment for soil, air or water quality, and land-use planning.

WASTE MANAGEMENT

Evolution

In Argentina, as in most countries, from the end of the 19th century to the end of the 1960s, the biggest concern was how to make waste disappear from the sight of the population. Services were called Waste Collection and Urban Cleaning and institutions, at the national level, were under the Health Ministry, along with the provision of water and the treatment of excreta. The most common practice was uncontrolled disposal and burning.

The environmentalist movement that became public in the 1970s and which started drawing attention to the pollution levels of the different natural receiving media and to the limits of an uncontrolled economic growth (the Club of Rome), also noted the importance and the role of waste for sustainable development.

Then, a considerable body of law and institutions started to be built and to establish obligations, controls and standards, especially in developed countries. Their implementation together with the technological development and the understanding that all processes needed to be environmentally sound, socially accepted and economically viable, started to bring the system to its current complexity.

Argentina followed that path, especially from 1992 when the Agenda 21 was adopted at the Earth Summit in Rio de Janeiro, Brazil, with enormous international support.

It should be recalled that chapter 21 of said Agenda is dedicated to laying the strategic foundations of a sound waste management system. Another result of that meeting was the United Nations Convention on Climate Change, soon recognized as a significant driver for the sector.

In short, the initial drivers, both in Argentina as in most developed countries, started attending to the public health concerns and to the protection and care of the local environment up to its global impact, complementing the transition from a waste management system to a resource management system (UNEP, 2017).

Overview

Providing a national overview of the different waste streams in the country is a complex and difficult task. There is a large amount of relevant information regarding Municipal Solid Waste (MSW) for most of the provinces, but when it comes to other streams such as hazardous waste, waste from health care facilities, construction and demolition, food, etc., information is scarce and it is hardly possible to say coverage is national.

MSW Generation

Even though there is information relative to the generation of MSW available for the provinces in Argentina, data for each province often varies depending on the source used. Although standardized methods are usually applied to obtaining generation rates at the source; these rates are often obtained by indirect methods; for instance, from the records of the volume of waste received at a final disposal site. This reflects the volume of collected waste but not necessarily the volume generated at the source, limiting the interpretation of data, since collection coverage rates can vary widely between the provinces, but also between the cities of one province, as well as between different sectors in the same city. Moreover, between the source where waste is generated and the final disposal site, there can be a reduction due to the segregation of some materials during transport and the intervention of waste pickers (WP, "cartoneros" in Argentina) recovering from waste informally.

It was estimated that 45.000 tons of MSW were generated in the country per day in 2015.

As it can be seen on Figure 2 the amount of MSW per capita per day has been relatively constant between 2001 and 2009 to increase significantly since then, with the limitations described above.

Figure 2. MSW generation in Argentina

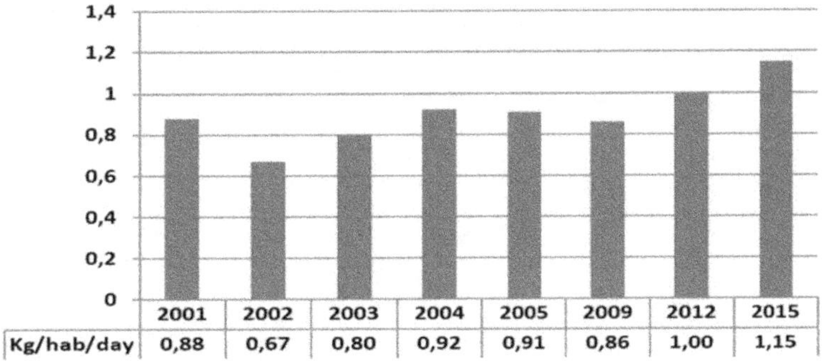

Evolution of daily MSW generated per capita over the last 15 years. Source: Sustainable Development Indicators System 8va Ed., SAyDS, 2015.

Few studies have been done on MSW generation in Argentina. Although it may vary considerably in time and space, recent data from the CABA revealed a close correlation between generation rates and the income level of the population, paralleling findings in other countries. As shown in Table 1, people within high and medium socioeconomic level produces almost 50% more MSW than people in the lower socioeconomic level.

MSW Composition and Properties

MSW composition is also a dynamic indicator that can vary considerably in time and space, as it was mentioned for generation. It is influenced by such factors as the introduction of new materials in products and change in people's consumption habits, among others. Under this precaution the results of evaluation of samples of MSW composition in the CABA from 1972 to 2015 (Instituto de Ingeniería Sanitaria, Facultad de Ingeniería, Universidad de Buenos Aires- CEAMSE, 2015) are presented on Figure 3.

The composition of MSW in the CABA changed significantly through the last 40 years. Food waste keeps first with a clear trend of decline from 63% of the total in 1972 to 41-44% since 2005. The papers and cartons fraction maintain constant between 15% and 20% of the total with a peak of 24% in 2001 but returning to its historical values in studies since then. The plastics component that was negligible in 1972 (2 to 3%) went on to represent the 15-20% of the total weight of MSW from 2005 to the present, although it has a marked seasonality with lower values during winter season. The glass maintains a constant presence with a sustained value in the order of 3 to 6% in weight, as well as disposable diapers and dressings that have a constant presence since 2001 with 4 to 5%, being the fourth components in importance.

These changes can be partly explained by changes in consumption habits, for example plastic packaging expansion, and in waste collection, such as the fact that construction waste collection has been organized more recently.

Table 1. Per capita MSW production according to socioeconomic level

Socioeconomic Level	Kg/hab. x day
High and medium high	0,97
Medium	0,89
Medium low	0,82
Low	0,68
Volumetric weight (Tn/m^3) 0,364	

Per capita MSW weight production parallels population´s socioeconomic level. (Source: Study on the quality of MSW in the CABA, IIS-FIUBA, 2015.)

Figure 3. MSW composition in CABA
Evolution of MSW composition in CABA over the last 43 years
Source: Study on the quality of MSW in the CABA, IIS-FIUBA, 2015

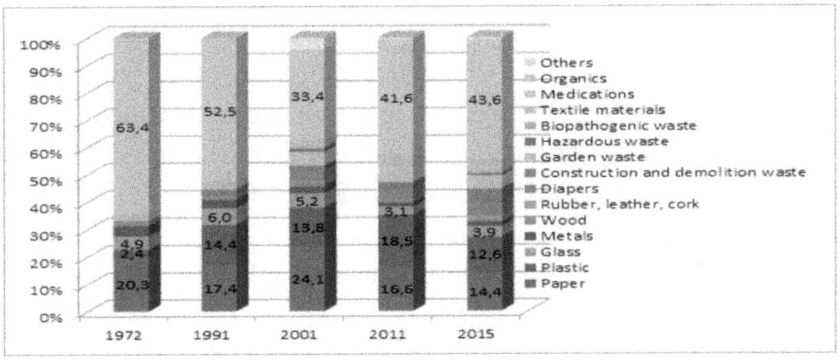

Trends in MSW Generation

As it was mentioned above the population in Argentina is mostly urban and there is no much room for this proportion to grow. Therefore, the estimation of projected MSW generation is essentially based on population growing.

This projection has not considered a change in the per capita MSW generation rate because there are no robust data available for the country. Moreover, the moderation in national economic growth expectations in the recent years makes the definition of this variable even more difficult. As shown in Figure 4 it can be estimated an increase of some 3 Tn/day.

General Status Of Msw Management

This section examines the status of two significant steps in the provision of services in the country: firstly, MSW collection and then final disposal.

Collection Coverage

MSW collection is a complex activity and, in economic terms, it is the most resource-intensive one of the urban services. In general, there has been an increase in both service coverage and quality with time.

Figure 4. Expected trend in MSW generation in Argentina
Expected evolution of MSW generation in Argentina.
Source:Regional Waste Management Outlook in Latin America and the Caribbean, 2015.

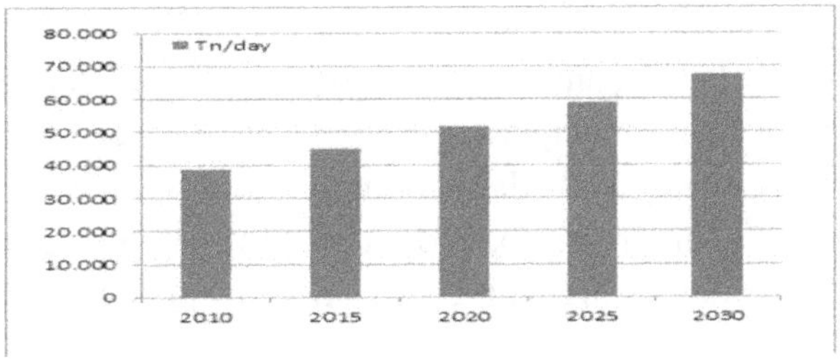

Traditionally, the highest-income areas in a city, as well as the highest-income cities in a country, have high levels of collection coverage, leaving less efficient services, usually consisting of containers that are not looked after as often as necessary, to marginal areas.

At country level, the Evaluation 2010 (IDB-AIDIS-PAHO, 2011) reported over 90 per cent coverage average, numbers comparable to the ones that characterize high-income countries. There are, however, marginalized areas and neighborhoods in large cities, or municipalities located in rural areas with values below 70 percent. This value is fitting, especially when compared to the data of other regions in the world reported by UNEP (UNEP-ISWA, 2015): Africa (25% to 70%); Asia (50% to 90%); Europe (80% to 100%) and North America (100%).

A complete picture of MSW in Argentina is presented in Table 2.

Controlled Disposal

Final disposal of MSW has improved significantly in recent decades in the country, so much that a significant share of the volume of waste collected is currently disposed of in sanitary landfills, especially in middle and large cities where the percentage of coverage reaches 81.1% (Table 2). Unfortunately, several provinces still dispose some fraction of their waste in places that meet the conditions of a sanitary landfill only partially or not at all. This practice has led, in some provinces to the use of the term "controlled landfill", a facility which generally consists of old dumpsites ("basurales" in Spanish) in which some control measures such as fencing, signage, access control, scale (or some even more advanced measures, such as coverings membranes and waste compaction, gas collection, among others), have been put in place in order to have better operating conditions.

Despite the significant progress made, there still is a lot left to do. While significant amounts of waste are still disposed of without any precautions (close to 49%, Table 2), being burned or used as animal feed, with the health risks these practices entail, and a considerable number of dumpsites are in operation in the country, energy valorization of MSW in some final disposal sites is now being implemented.

Resource Recovery

Accomplishing an efficient resource recovery from MSW depends critically on the implementation of segregation at source and separation (differentiated) collection program. In this way materials will reach valorization facilities with a better quality and at higher quantities, regardless of the methodology used.

Table 2 shows that 21 out of the 24 provinces have mechanized separation plants with an installed capacity to treat 17.7% of the waste generated, although most of them work below their capacity. There is here room for improving.

The Importance of Segregation

Good segregation at source is the key for any valorization program. In practically all provinces segregation is carried out in all stages of waste management: at the source, sweeping, collection, transfer and, of course, at final disposal sites, mainly by WP. Many municipalities have implemented separate collection programs (basically organic/inorganic fraction), under different schemes and with greater or lesser success, laying the foundations for achieving higher recycling rates by providing less contaminated materials.

Table 2. MSW in Argentina

Regular Service of Waste Recollection		Data Source
Homes[1]	89,9%	National Census 2010
Urban homes[1]	94,8%	
Urban homes in 31 main cities[1]	95,4%	Permanent Home Inquiry 2014
Collection vehicles/10.000 inhabitants	1,34	World Bank 2015
Collection vehicles older than 10 years	45,0%	
Collection vehicles with compacting equipment	70,5%	
Municipalities collecting construction and demolition waste	76,7%	
Municipalities collecting in markets	61,0%	
Level of satisfaction of consumers		GIRSU Project 2014
Very good	21%	
Good	55%	
Bad	17%	
Very bad	7%	
Municipal Sweeping of MSW		World Bank 2015
Manual	62,7%	
Mechanical	18,9%	
Total	81,6%	
Municipalities maintaining public parks and gardens	89,5%	
Municipalities cleaning fairs and expositions	61,8%	
MSW transfer stations		IDB-AIDIS-PAHO 2011
MSW transferred[2]	20,0%	
Population served	16,9%	
Provinces with transfer stations	9/24	
Mechanical plants for MSW treatment		World Bank 2015
In the country	150	
MSW treated[3]	17,7%	
Plants capable to treat over 100 Tons/day	9	
Provinces with mechanical plants	21/24	
MSW disposition		
In sanitary landfills	51,2%	
Population served	55,2%	
Population from 31 main cities served	81,1%	
Provinces with at least one sanitary filling	18/24	
In controlled landfills	9,9%	
In open dumps	24,6%	
Burned or disposed using other inadequate practises	0,8%	

[1]at least twice a week;[2] 76.3% of it in CEAMSE stations; [3] 44.0% in CEAMSE plants;

It is important to stress that the difference between waste sorting and recycling is often unclear, so it is necessary to note that the terms can refer, interchangeably, to: (i) the percentage of waste actually separated for being then reused, recycled or subject to another activity that extends its life cycle; (ii) the percentage of waste removed from the waste stream for possible recycling; (iii) the percentage of waste actually incorporated into recycling processes and (iv) a combination of all of the options mentioned.

The efforts of the generators who segregate the waste produced (segregation at source) requires continuity through selective or separate collection in order to warrant that the different waste streams can go on through differentiated management, separating the materials which can be subjected to some treatment that will allow for valorization and/or recovery, from waste that has to be finally disposed of given its characteristics and/or lack of a market for them.

There are different mechanisms for differential collection, with the following two being the most widespread, both in use in several municipalities:

a) Different types of waste are collected on pre-set days of the week; the widespread practice is usually to separate MSW into organic and inorganic, or wet and dry fractions. This basic classification can be expanded to incorporate specific components (glass, paper, etc.) as the efficient implementation of separate collection schemes progresses.
b) Use of collection vehicles with different compartments that enable the selective collection of the different waste streams separated at their point of origin.

Technologies for Material Recovery

The technologies used for recovering materials from MSW have improved and diversified in recent years, so that currently there is a wide array of possibilities. However, it is advisable to raise awareness about the advantages and disadvantages of these technologies since not all of them are necessarily suitable for developing countries. The lack of knowledge and qualified staff in municipalities has often favored the acquisition of inappropriate –and almost always very expensive– technologies, which have resulted in failures.

The most significant part of MSW is the organic fraction. It is therefore highly recommendable to adopt technologies for its management, which requires consideration for variables such as volume, moisture content, where it is generated, etc. The lack of an adequate infrastructure for processing organic waste is still a challenge for Argentina, where a high percentage of this waste is still disposed of in sanitary landfills –at best– which additionally contributes to considerable problems, such as the formation of leachates and greenhouse gas (GHG) emissions, among others.

Even though the application of technologies to the recovery and treatment of waste on a large scale is still limited, the significant presence of organic matter in final disposal sites has promoted the use of biogas from sanitary landfills (when feasible), followed by composting (when organic matter is properly separated). Large cities facing space constrains together with the expiration of the useful life of current final disposal sites, have then started shifting to the development and/or application of technologies based on the principle that waste is a resource, both as a material and for generating energy.

Moreover, there are commitments to reduce GHG emissions and, given the waste sector's considerable methane contribution, the search for alternatives to reduce methane emissions becomes more important.

The search for alternatives is pushed by other important factors, like the energy sector. Moreover, there are commitments to reduce GHG emissions and, given the waste sector's considerable methane contribution, the search for alternatives to reduce methane emissions becomes more important.

This updated vision of waste management reinforces the relevance of regionalization proposals, a tool that should help municipalities to overcome their weakness to access technologies and financing.

Examples of technologies used and operational in Argentina aimed at generating energy or recovering materials from waste are:

Mechanical Sorting: The most basic and simple technology for recovering resources is made up by material separation facilities, usually including a conveyor belt on rolling pins where recyclable materials from MSW is manually separated by the staff. This technology can be located in urban areas or at final disposal sites, and its dimensions and capacities range from a few to several thousand tons a day.

As mentioned, there are 150 mechanical separation plants in Argentina (Table 2). Although there are no precise statistical, it is assumed that there are also about 30 plants with manual separation.

Energy Generation from Biogas in Sanitary Landfills: Another option for recovering energy from waste is the combustion of methane from the biogas generated in sanitary landfills (or in large decommissioned dumpsites), as well as its use for producing heat or electricity. An example of this will be presented in CEAMSE box.

Additional advantages of this technology are, aside from reducing GHG emissions, that it also allows the displacement of fossil fuels, thus contributing to meeting the country commitment to climate change, and the obtention of income through carbon markets and from the sale of electricity.

Composting: This technology has been present in Argentina for a number of years, but still does not have the desired impact on MSW management. Different factors can explain the lack of success of composting projects in LAC countries (USAID, 2013). Despite this, there are successful cases of composting facilities in the country, whether government-owned, public-private, private, community-owned or owned by neighbor associations, although there is no a formal record of them.

The composting plant Environmental Complex North III, with a treatment capacity for 1,200 to 2,000 ton/month of green waste[2] was built by the CEAMSE in 2001. The compost produced is distributed among CEAMSE facilities (where it is used as covering material in sanitary landfills), municipalities, educational organizations, hospitals and NGOs when the quality allows. Another example is in the city of Rosario, which has the composting plant Bella Vista, with a design capacity of 200 tons per day, employing 12 previously informal WP that make up 30 per cent of the plant's staff[3].

Anaerobic Digestion (AD): AD is an alternative for the treatment of organic waste which has garnered interest in recent years. Although it has been used for many years for treating agricultural waste, it is still not widespread for the treatment of the organic fraction of MSW, and its use on a commercial scale is still budding and the process is under evaluation.

Mechanical Biological Treatment (MBT): MBT consists of a combination of mechanical processes with biological reactors, with the singularity that they are located in the same plant, and it can comprise several treatment technologies. The biological reactor can be bio-drying, composting or AD[4].

Argentina has a MBT plant in the Environmental Complex North III-CEAMSE[4], with capacity to treat 1,000 ton/day of MSW. Up to 60 per cent of it, about 420 tons of organic waste and about 180 tons of potentially recyclable waste (plastic, paper, glass and metals), can be recovered. The biological stage includes loading of food scraps/yard waste to bioreactors, where they are encapsulated and sealed for 21 days under moisture control. The product of this process is used as a coverage layer in the same

Sanitary Landfill. CEAMSE built and operates a second plant, located in Ensenada in the province of Buenos Aires, close to the provincial capital La Plata, which started operative in 2017 processing around 1000 tons/day.

Incineration: "Many countries in Latin America are beginning to explore the use of waste-to-energy plants as part of their integrated solid waste management plans" (UN Habitat, 2010). Feasibility studies carried out by the IDB in Latin American and Caribbean municipalities concluded that high investment, operating and maintenance costs are a significant barrier to the development of this technology in the Region, especially if they are to compete with the current costs of final disposal (IDB, 2013). A considerable portion of the investment costs corresponds to the need for the installation and operation of equipment to prevent emissions to the atmosphere, such as persistent organic pollutants and heavy metals, among others. Investors promoting MSW incineration are faced with the difficulty municipalities have to pay rates that are commensurate with capital and operating costs, which can mean longer periods until return on investment.

Besides, when addressing any project proposal on this subject, it is also essential to consider the engagement of stakeholders and local communities, among other reasons, to inform how the potential impacts on environment and health will be prevented during the operational phase of the project. In conclusion, the operation of waste to energy equipment requires a high level of specialization, social approval and will be economically viable only with enough financial and governmental support.

The CABA built years ago a legal framework that includes waste incineration as a possible technology to be used but did not apply it yet.

Legislation for Resource Recovery, Waste Prevention and Sustainable Consumption and Production (SCP)

Regulating the recovery of resources or materials contained in waste implies a direct contribution to the main principles of circular economy, based on the conservation and improvement of the natural capital and optimization of the use of natural resources. Besides, it is a useful tool to reduce the amount and hazard of waste intended for sanitary landfills, which is a valuable goal in Argentina, where there is the problem of lack, in various cities, of controlled sites for the final disposal.

Although rules in Argentina have references to resource recovery and to the concept of prevention, meaning actions aimed at reducing the generation of waste and minimizing or eliminating its hazard[5], these provisions are generally expressed as part of the principles or objectives in the laws, but not establishing specific measures on them. There is still a long way ahead when it comes to specifically forcing certain fractions of waste to receive a management that implies the reincorporation of materials to the economic circuit, and to the regulation of sustainable consumption and production, today more a criterion or course to follow than a requirement of the legal system.

To note are the significative advances made by the CABA[6], and the strategy on SCP and cooperation approved by MERCOSUR in July 2007, including an action plan to implement the strategy.

However, the problem is not solved with the creation of perfect regulations but with their specific ability of being implemented and enforced. In that sense, other variables come up, including the governments' ability to make the prescriptions of the law be verified later in reality. That requires setting goals, preparation, human resources and materials, and the development of a series of tools which efficiently cooperate in the achievement of said goals and desired changes. This is a pending challenge for Argentinean authorities.

OTHER WASTE STREAMS AND EMERGING ISSUES

Besides MSW there are other waste streams –such as industrial toxic, explosive, reactive, or those generated in health-care facilities– which deserve our attention for many reasons.

Hazardous Waste (HW)

Argentina installed the concern for HW more than 25 years ago, with the Law No. 24.051. This law and its regulatory Decree (No. 831/93) have the main objective of providing a regulatory framework for the management of HW of industrial origin, including waste from health care establishments.

Given the federal character of Argentina this national standard requires for its validity the accession of the provinces, so its dictation in 1992 helped to install the subject and to promote a wide diversity of provincial norms (40 laws and 22 decrees so far). To date 15 provinces adhered to the law, making it in force in its territory and 9 provinces opted to develop their own laws on the subject (so in Argentina there are 10 different laws). In addition, among the 15 jurisdictions that adhere to the national law, 7 also adopt their statutory decree and constitute a single normative framework for all of them. The remaining 8 provinces adhere to national law, but have developed their own statutory decree, generating 8 different normative frameworks. Therefore, there are 18 different legal frameworks for HW in the country.

In spite of that heterogeneity, all Provinces adopt the categories of HW subject to control that are part of the Annex I to Law 24.051, inspired by those of the Basel Convention, which Argentina adhered to by the National Law No. 23.922.

Another relevant fact for HW management comes from the fact that there are different criteria regarding the transportation of HW through the country. While several provinces authorize under certain and defined conditions this movement, others do not let it under any condition. This restriction should be faced with the map of existing treatment plants which are not available in all provinces. Among other problems this makes that HW must sometimes travel more than 500 Km to be properly treated, with the economic charge and transportation risks implied. As it can be seen in Fig. 5 only 5 out of the 24 provinces authorize the transportation of HW through their frontiers while 9 do not let it at all. Partial authorizations are in place in the remaining 10 provinces.

Figure 5. HW transportation restrictions
HW transportation through provincial frontiers is authorized in 5 provinces (green), forbidden in 9 provinces (red) and partially restricted in 10 provinces (yellow).
Source: Final Report Demonstrating and Promoting Best Techniques and Practices for Reducing Health Care Waste to Avoid Environmental Releases of Dioxins and Furans and mercury reduction, GEF-UNDP Project. 2013

There is no specific regulation regarding prohibitions to the export of HW. However, the competent authority applies the principles of the Basel Convention, and its national environmental laws (Minimum transboundary movement, and Self-Sufficiency), thus limiting exports to situations where there is no local technology and/or sufficient capacity for dealing with HW in an environmental sound manner.

A case by case evaluation is considered when requesting an export for recovery, when local capacity relies only on final disposal. However, the import of present or potentially HW as well as radioactive wastes are forbidden by the Article 41 of the Argentine Constitution. Also, Argentina bans import or transit of wastes defined as hazardous according to the National Law 24.051. Moreover, Executive Decree 181/1992 prohibits import or transit of wastes listed in its Annex I.

Export of HW is regulated by Resolution 58/2017. Special procedures do exist for some specific waste streams such:

a) PCB waste (Y10) through Resolution 522/2014, with a National Plan of Reduction and Elimination of PCB, enacted by National Law 25.670 and Decree 853/2007.
b) Containers used for holding phytosanitary substances (Y4) where a National Plan considering Management and Reduction measures are established by National Law 27.279.

In addition, guidelines for specific wastes, are also developed with the major stakeholders, and formalized through regulation (i.e. – waste tires are dealt with by Resolution 523/2013 which includes a comprehensive analysis and reduction and elimination measures for this source of waste).

It must be stressed that the country is involved in various national and international programs and projects associated with the chemicals and waste management agenda which aim to prevent and minimize waste generation, its reuse, recovery and recycling, alternative fuel use and subsequent disposal. Most of these initiatives are implemented with the assistance of international funding agencies, especially the GEF, implementing agencies like UNEP, UNDP, UNIDO and the support of the Basel Convention Regional Center for Capacity Building and Transfer of Technology for the South America Region (BCRC-Argentina).

However, recovery of data regarding HW is still problematic as illustrated on Table 3.

Table 3. Total amount of hazardous wastes and other wastes generated (metric tons)

	2012	2013	2014	2015	2016	2017
Total amount of hazardous wastes generated in the years for which official data are available						130.327
If possible, total amount of hazardous wastes under Art. 1 (1)a (Annex VIII) generated	249.259	253.826	310.044			
If possible, total amount of hazardous wastes under Art. 1 (1)b generated	43.240	40.946	60.756			
If possible, total amount of other wastes generated (Annex II)	14.600.000	14.000.000	1.569.500			

Officially accepted amounts of hazardous waste generated in Argentina. Source: National Report 2018 in Electronic Reporting System of the Basel Convention, 2017.

Household Hazardous Waste (HHW)

There is a small fraction in the MSW stream consisting of domestic or household hazardous waste (HHW), which can have a widely varying composition, depending on the location and size of the municipality, the social class, the existence or not the separation practices, etc. It is made up by a wide array of products at the end of their useful life and the degree of hazard varies depending on their chemical and physical characteristics. The consumer products found most frequently in the MSW stream that can potentially be HHW are detergents and whiteners, disinfectants, drain cleaners, oven cleaners, printer toner, motor oil, batteries, brake fluid, enamels, paints, pesticides, compact lamps, as well as those related to medical care, such as needles, cotton, gauze, etc., among others.

When mixed with conventional MSW and managed together with that stream, they can pose a risk for the operators and cause various problems when reaching treatment or final disposal sites by becoming integrated into the composition of biogas and leachate from the site, thus increasing the risk of affecting the mediate and immediate environment.

A sampling carried out in 2008 in the CABA reported HHW to be 1 per cent[7] and 0.4 per cent[8] in weight. As it can be seen from these significant differences, this waste stream needs to be further studied, because HHW are likely to increase since new substances are being incorporated into household consumer products (for example, the composition of some personal care products, such as cosmetics, hygiene products and medications, among others, can include nano-materials or nanoparticles) (UNEP, 2015). The considerable range of variation in the values found suggests that it is necessary to pursue investigating in this area, while adequate HHW stockpiling and management systems are to be implemented in the municipalities.

Healthcare Waste (HCW)

Even though most of the waste generated in healthcare establishments have characteristics that can be grouped with MSW, there are also waste which can be classified as hazardous due its infectious nature (aside from waste which is hazardous due to its toxicity, corrosiveness, etc., generated during maintenance activities and in laboratories). Inadequate management of this waste can pose significant risks for the environment and human health, first for the staff responsible for handling the waste internally, and then for the operators of the waste management systems from health-care facilities. The risk is increased when this waste is mixed with MSW in places where operate informal WP, who are exposed to possible injuries from needles, glass and other materials which could transmit infectious diseases such as hepatitis, AIDS, etc.

As mentioned above HCW is included (articles 19 and 20) in the national law for HW, additionally involving the intervention of the Ministry of Health in its regulation. However, under the federal nature of Argentina, provinces retain the capacity to establish local regulations and to point out the complexity of the legal framework regarding HCW it should be noted that provincial regulations are not always contained on their legislation of HW.

A total of 15 provinces have a legal framework specific for HCW: 11 have their own legislation, while 3 (Chubut, San Juan and Neuquén) make a special reference for HCW within the legislation of HW and 1 (Jujuy) into the regulation of the General Law of the Environment.

A comparative evaluation of the legal framework, which can be seen in Figure 6, was recently published attempting to call the attention of the authorities to build more homogeneous criteria (de Titto, Montecchia, Brunstein & Chesini, 2015).

Legal diversity regulates HCW in Argentina: local regulations in 19 provinces adds up to the 5 (in red in the figure) following the nationally established previsions.

Different technologies are used to treat HCW throughout the country by several companies, as shown on Figure 7. This sector faces the problem of distances since their number is low. It has been estimated that around 17% of the health services are at 500 or more Km. of a treatment plant. On the other side costs to operate these facilities require a significative volume of HCW to operate.

National distribution of HCW treatment plants superimposed on provinces colored according to their regulations on acceptance (green, without restrictions; yellow, with some restrictions) or not (red) of HW from outside the province as shown in figure 5. To the right, the city of Buenos Aires in a higher scale. Colored dots indicate the technology employed in each plant.

Figure 6. HCW: Regulation diversity in Argentina

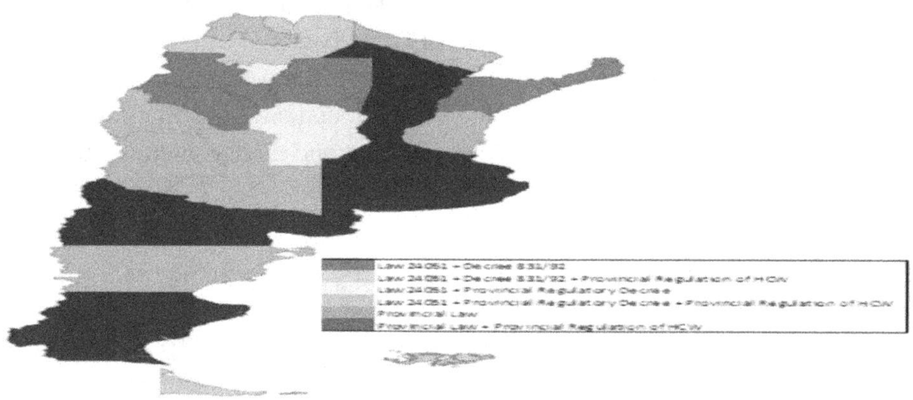

Figure 7. Available HCW treatment in Argentina
Source: Final Report DEMONSTRATING AND PROMOTING BEST TECHNIQUES AND PRACTICES FOR REDUCING HEALTH CARE WASTE TO AVOID ENVIRONMENTAL RELEASES OF DIOXINS AND FURANS AND MERCURY REDUCTION, GEF-UNDP Project. 2013

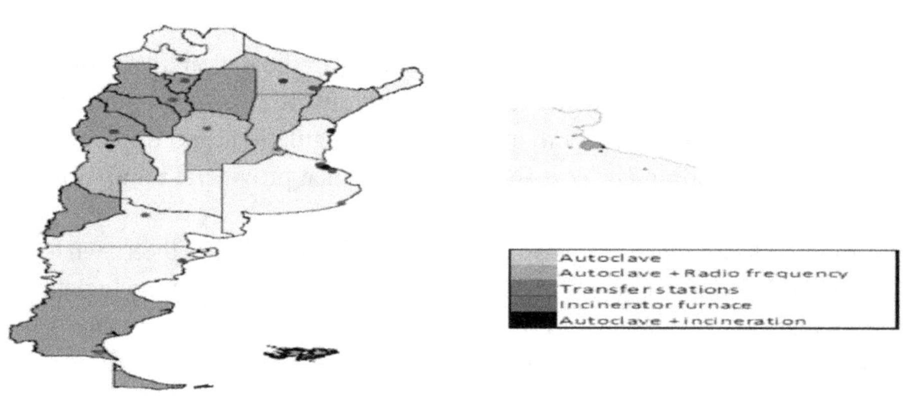

Waste Electrical And Electronic Equipment (WEEE)

The production, consumption and disposal of electronic and electrical devices have grown in Argentina, leading to a growing volume of waste from these products at the end of their useful life, just like in the rest of the world.

WEEE may contain heavy metals and rare earth metals as well as other compounds, such as flame retardants. This make WEEE an hazardous waste, posing significant environmental and health risks, because of the possible release of these compounds into groundwater through the subsoil or to the atmosphere during occasional fires (Gullett et al., 2010). Besides, WEEE in the waste are a missed opportunity for the recovery of valuable materials.

The most significant challenges in order to achieve an adequate management of WEEE are (ITU, 2015):

- Legal loopholes for WEEE management.
- Sustainable management throughout the life cycle of WEEE.
- Extended Producer Responsibility (EPR) as a State-level policy.
- Lack of awareness about WEEE management.
- Availability of recycling technologies and transference.
- Sustainable economic models that overcome financial difficulties.

Argentina has been paying attention to this subject since 2006 both nationally and in the MERCOSUR, including a specific law in the CABA (Law N° 2807/LCBA/08).

In concordance, Argentina takes part in the *Regional Latin American and Caribbean Platform for Electronic Waste* (RELAC, in Spanish), based in Santiago, Chile, which combines the individual efforts of countries in order to consolidate information that is currently available in its online portal[9], an important media for raising awareness about the regional issues in connection with this area. The United Nations University, in turn, has recently generated relevant information for the Region that mostly relates to statistical indicators and public policy. The United Nations International Telecommunication Union (ITU) has also worked with an approach more geared toward sustainable management of WEEE in the Region.

Finally, it is important to emphasize that, even though significant progress have been made in the development of legal instruments, as well as in the necessary infrastructure for adequately managing electronic waste, the same cannot be said for waste derived from electrical equipment, mainly consisting of household appliances and other types of equipment, such as beverage and food vending machines and air conditioners, among others. Unfortunately, no good statistics are available on this subject.

Flow Of Waste Associated To Transport (End-Of-Life Vehicles, Tyres, Shipbreaking)

The transport of people and cargo brings with it the generation of various types of waste, which can range from the vehicle itself as a unit that is disposed of at the end of its useful life, to the different parts that make them up and which have a varying valorization potential. Among these wastes are some which can be classified as hazardous.

End-of-life Vehicles (ELVs)

Argentina produces cars since the '50s, with a growing tendency in a market closely linked to ups and downs of the local economy and a shared production model linked to Brazil in a MERCOSUR agreement over the last 20 years. The Figure 8 illustrates the decadal evolution of the national production and the profile of the present situation.

Total number of circulating vehicles, including domestically produced and imported vehicles, has doubled since 2006, reaching 14 million vehicles by 2016, with a mean age of 11, 7 years (Figure 8). It is interesting to recall that Argentina has 3,8 hab/vehicle, the lowest ratio in the Region.

Decadal production of vehicles 1950-2018 in Argentina (left) and distribution of the fleet in 2016 (right). Source: Authors elaboration from National Observatory of Transportation Data, Universidad Tecnológica Nacional.

As it can be seen, the production and consumption of transport vehicles has increased significantly in Argentina. That poses the fact that at the end of its lifespan vehicles become hard to manage waste not only because of its volume but also because of the diversity of materials included, including some hazardous materials.

The average composition of ELVs (68% ferrous metals, 9% plastics -some chlorinated or with brominated flame retardants-, 8% non-ferrous metals, 6% glass and rubber and 9% other materials -textiles, batteries, fluids, electric components, etc.-) (SEMARNAT–IPN, 2009) points out the high potential for recycling.

The size of the vehicle fleet, with most motor vehicles concentrated in large metropolitan areas, is a challenge for those responsible to establish suitable management schemes for ELVs.

In Argentina the CABA has organized a Plan for the management of ELVs, which includes cars abandoned in a public places and vehicles sequestered by the Police. Before being compacted, the cars are subjected to a process of decontamination in which batteries, gas pipes and other elements that can be applied to specific treatment are extracted. Since installed in 2012 the Plan has processed around 7.000 vehicles.

Figure 8. Vehicles in Argentina

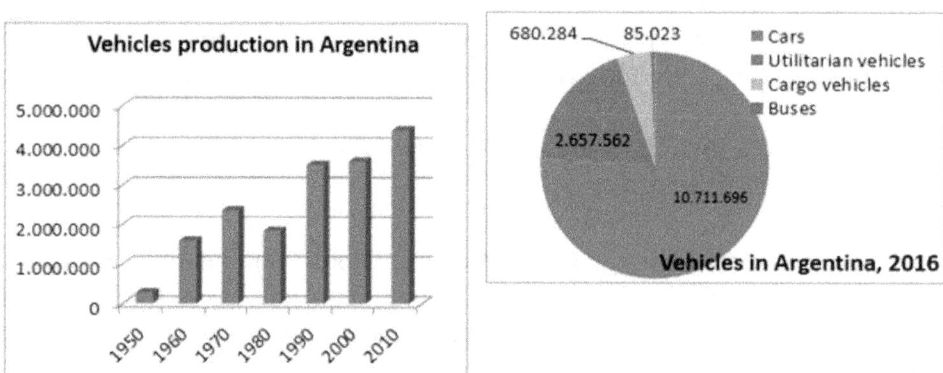

Tyres

Not being a hazardous waste *per se*, perhaps the most significant problem caused by the tyres out of use (TOU) is their potential for accumulating water, thus creating an environment favorable for the reproduction of insects' vectors of diseases such as yellow fever, dengue fever, Chikungunya and the Zika virus, which have emerged or re-emerged in recent years. A detailed and thorough overview of the necessary aspects to be considered about tyres, from the point of view of the environment and human health, can be found on the *Revised technical guidelines for the environmentally sound management of used and waste pneumatic tyres* of the Basel Convention (UNEP, 2011).

In Argentina the TOU belong to the category Special Waste of Universal Generation (REGU in Spanish) and their treatment as waste has been a cause of concern for years. In 2003 the INTI (National Institute for Industrial Technology) created a Permanent Commission of Tyre Recycling Work, coordinated by the INTI-Rubber Center and integrated by manufacturers, importers, reconstructors, business chambers and public bodies. The Commission articulates the public and private sectors for the development of a model of integral management of TOU. In 2007 an agreement was signed between the CEAMSE and the Company REGOMAX from which the first yielded a land of 2 hectares, in the San Martín municipality, for 20 years and derived the delivery of the TOU to the recycling plant. For its part, REGOMAX built the plant and invested approximately two million dollars in capital goods. REGOMAX, which started regular activities in 2010, receives at no cost the tyres in the plant, obtains the grinding of the rubber of the tyres for the manufacture of floors, sports pavements and synthetic fields, recovering the steel and the remainders of the fabric of the structure. This is achieved by replacing the importation of the rubber grind, one of the main objectives of the venture, as well as facilitating the export of synthetic fields and sports flats.

Later on, other initiatives were launched in other locations (i.e. Arroyo Seco, Berazategui, Salta) and several provinces did regulate this matter (Catamarca, -Law 5.359, Decree 1.663/2012-; La Rioja -Law 9.291; Decree 1.521/2012; Rio Negro –Law 5003/2014-; San Juan -Law 1434/2016- and Santa Cruz -Law 3316, Decree 634/2013, for example).

In Argentina it has been estimated that TOU account for 130,000 tons per annum, half of them in Buenos Aires Metropolitan Area. In 2018 REGOMAX processed around 12.000 tons[10].

Currently, the National Congress is working on a bill to establish the Minimum Budgets for Environmental Protection for the Integral Management of TOU. The application and interpretation of this law is to be based on the principles established in General Law of the Environment –N° 25.675- and the principle of Extended Liability of the Producer (EPR)[11] and in particular the principle of Individual Responsibility of the Producer (IRP).

Emerging Wastes (Nano-waste, Bio-polymers, Composite Materials, Wind Turbine Generators)

Technological development and the inclusion of new materials and chemical substances used in the manufacture of various products provide them with new properties and features, which directly affects their condition as wastes when they reach the end of their useful life. Thus, the permanent-changing offer of products poses a challenge for waste management which needs adapting to these new waste streams.

Emerging wastes are a challenge for waste management in Argentina. So far, is lacking an in-depth analysis of the behavior of these streams in conventional waste management and treatment facilities (recycling, final disposal, etc.) as well as the effects on microbial degradation (composting or biogas), their potential for penetrating watertight barriers in sanitary landfills or for entering the biogas, and any other health and environmental risks associated.

Particularly important is case of wind turbines, whose number installed and in operation has increased significantly in recent years for the crucial general goal of mitigating GHG emissions and reducing costs by lowering the prices of new technologies. Among the drawbacks of this technological alternative is the fact that a few years ago it was detected that there was an unclear space for the decommissioning phase of these equipment within their life cycle, so it is important to pay attention to this sector as a potential generator of waste.

Construction And Demolition Waste (CDW)

The construction sector is one of the most prominent in the economy of a country, because it is an important source of employment, it demands products and services, as well as the creation of infrastructure, and so it can be deemed as an indicator of the development of a country or region. However, even though the construction sector brings tangibles and unobjectionable benefits, it also generates, as a consequence, a significant volume of waste classified as CDW.

Argentina stands out as the most urbanized of the Region, with 91 per cent of the population living in urban areas. This degree of urbanization is reinforced by the fact that in the last census 47 per cent of the total population and 52 per cent of the urban population in Argentina lived in cities with more than half million inhabitants (INDEC Argentina, 2010). These numbers also indicate the relevance of the construction sector in the country. It represents around 5% of the gross domestic product (GDP) and employs almost 450.000 people (INDEC, 2018).

However, data relative to CDW are not routinely and consistently controlled and reported in Argentina, so the scarce data available must be taken with caution since the classification of this category can also vary according to the municipality.

The most developed project to deal with CDW is the one that takes place in the CABA operative since 2004 in the Transfer Station located in the Pompeya neighborhood. During 2017 this plant recovered 21.244,72 tons of so called "arides" to be reused.

Disaster Waste (Emergencies)

Because of its location in the globe, as well as its geographical characteristics, Argentina is vulnerable to numerous natural hazards, including coastal and river floods, droughts, earthquakes, landslides, and volcanoes that can reach the category of natural disaster (CEPIS/PAHO, 2003)[12]. Estimates indicate Argentineans have a higher natural disaster loss per capita than most countries in the region.

Even though the frequency of natural phenomena in the Region in the past ten years has been decreasing, there is an upward trend since 1970 where the main cause of natural disasters Argentina was flooding (UNEP, 2016). Argentina is among the ten emerging economies world-wide with the highest flood hazard exposure. Over the past two decades, estimated losses due to floods exceeded $3 billion per year, equivalent to roughly 0.7 percent of gross domestic product in 2012. Sea level rise is projected to increase

Buenos Aires Province's coastal flood risk. Climate change is also expected to cause temperature rise and higher occurrence of extreme weather events, including excess rainfall leading to river flooding[13].

Disaster risk management (DRM) is the responsibility of the Ministry of Internal Affairs. The National Civil Protection Directorate coordinates disaster relief policy, plans, and federal interventions with local actors—including damage assessments. To further advance its DRM agenda, the Government of Argentina is prioritizing:

- Enhancing the legal and institutional framework to better prioritize national and provincial actions to increase Argentina's resilience to natural hazards.
- Strengthening capacities to conduct hazard and risk assessments for improved decision-making.
- Developing a risk-informed financial risk management strategy;
- Reducing exposure to extreme flooding in the provinces of Chaco, Corrientes, and Buenos Aires.
- Supporting the city of Buenos Aires to efficiently manage flood risk and improve drainage systems in the Cildáñez, Maldonado, and Vega Basins while improving transport accessibility and reducing the impact on the poor.

All these events generate a significant volume of waste, their nature depending on the type and intensity of the phenomenon and the affected area. The lack of adequate emergency plans can make the aftermath very complicated. So far, no specific national plans for disaster waste have been developed.

WASTE AND CLIMATE CHANGE

According to the Intergovernmental Panel for Climate Change (IPCC)[14], emissions from the waste and wastewater sector accounted for 3% of total GHG emissions from different sources in 2010. Therefore, the sector's contribution to global emissions is almost marginal. Climate change, however, can have a negative impact on the waste management system through its undeniable consequences, such as:

- Infrastructure and facilities can be affected by floods or extreme events.
- Increased temperatures or rainfall may influence the system degradation or leachate generation processes.
- Collection and transport services can be disrupted and the accumulation of uncollected waste clearly poses a risk to the health of workers and the surrounding population, due to the increase in vectors and pathogens.

Since several gases are produced during the different waste management operations[15], the advances registered in the country by increasing the fraction of MSW disposed in sanitary landfills have, paradoxically, resulted in the increase of methane emissions as sanitary landfills offer improved conditions for the anaerobic degradation than open dumpsites, this not meaning that a dumpsite is preferable over a sanitary landfill.

The evolution of the sectorial contribution to GHG emissions in Argentina presented in Figure 9 shows the growing contribution of energy and changes in the use of land while the contributions of agriculture, industrial processes and waste sectors are relatively stable through the years.

Figure 9. Evolution of GHG emissions in Argentina
Source: Third National Communication on Climate Change, SAyDS, 2015.

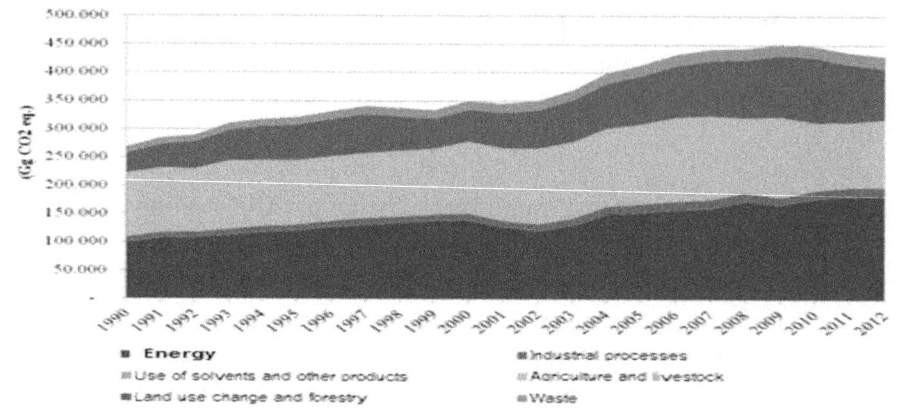

The evolution of GHG emissions with time shows that while emissions are increasing, the contribution of the waste management sector keeps relatively stable.

When contributions are break down for a given year it can be seen that the waste management sector is a poor contributor to GHG emissions, as presented in Figure 10.

Sectorial contribution to GHG emissions in Argentina.

FOOD WASTE

It has been estimated that approximately one third of the total food produced for human consumption –equivalent to 1.3 billion tons– is lost or wasted annually. Other negative impacts linked to this waste are emission of CO_2e, wasting of water and misuse of fertilizer during the production of that food (Lipinski, Hanson, lomax, Kitinoja, Waite & Searchinger, 2013). Given the relevance of agriculture in Argentina profile this item is highly relevant.

Figure 10. Waste contribution to GHG emissions in Argentina
Source: Third National Communication on Climate Change, SAyDS, 2015.

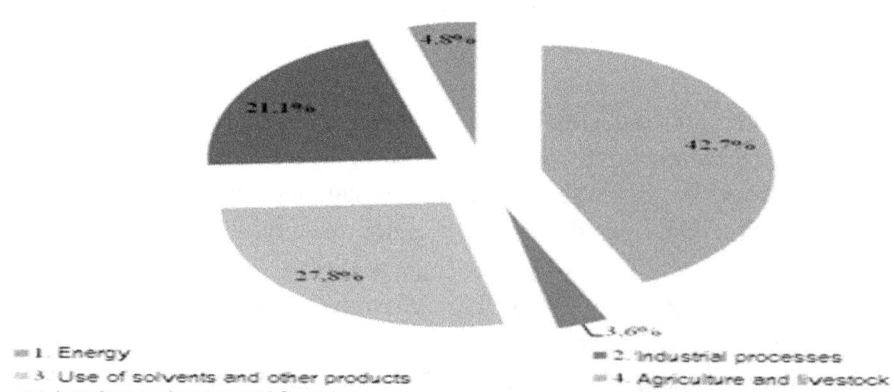

For the dual purpose of reducing food losses and facing the hunger problem, Argentina has committed to the development of food banks, which are *"not-for-profit, solidarity organizations that contribute to reducing hunger and malnutrition in the world. In order to achieve this, they receive surplus food from stores, companies or individuals, to be distributed among the population in need"*[16].

In 2003 was founded the Argentinean Network of Food Banks (Red de Bancos de Alimentos de Argentina), a Civil Association that groups 15 Food Banks and several nongovernmental organizations (NGOs). Then, in 2004 it was dictated the law 25.989 giving birth to a Special Regime for Food Donation -which was amended last October to clarify levels of responsibility for donors-, a legal framework that open up the participation of several governmental areas such as Social Development and Agriculture) strengthening the whole system.

This network works with food producers and distributors, supermarkets and Fruit and Vegetable Concentrator Markets. From its foundation till 2017 the Network has distributed over 85,000 tons of food. In 2017 they provided almost 10 million kilos of food to 2,400 community organizations reaching more than 330,000 people[17].

SANITARY LANDFILL LEACHATES

As a general rule, the volume of leachate generation must be considered and provided to the corresponding authorities in the process of authorization to build and operate a sanitary landfill, by the application of a model that allows obtaining the required data. It should be noted that the volume of leachate generation depends on the local weather conditions and the building and operational characteristics of the site while measuring the leachate flow can be a difficult task depending on the conditions and characteristics of each site.

In the CEAMSE Complexes in the metropolitan area of Buenos Aires, leachates are treated by a biological step and another physical-chemical one, before discharging them to the water bodies and once the permissible maximum limits of pollutants have been reached. CEAMSE currently has 6 treatment plants. The Northern Environmental Complex has a more advanced removal technology, which consists in an ultrafiltration and nanofiltration (UF/NF) membrane, with 85% removal efficacy of pollutants. The capacity of treatment is variable, for example in the Ensenada Environmental Complex is 6.000 m^3/month, while the capacity of treatment of Gonzalez Catan Environmental Complex is 10.000 m^3/month, with a 15.000 m^3/month expected extension[18]. The volume of leachate generation in other sanitary landfills of Argentina is not available.

ECONOMICS OF WASTE MANAGEMENT

Financing is a key issue for the sustainability of waste management schemes.

The financing model is defined by the way in which the four essential component–generators or municipalities, the operator or service provider, revenues to pay for the cost of the service, and investments required for the development of infrastructure- are organized and related to one another.

Investments in the public sector must start by analysing the costs and benefits that the implementation of a service would mean to the society, beyond the income, costs of operation and maintenance of collection, transport and final disposal, and financial indicators that are traditionally revised. This is crucial for the waste sector since their access to resources must compete with the investments that are to be made in other sectors essential for the well-being of the population such as health, poverty reduction, employment, education or infrastructure. Given that budget constraints are always significant, economic analysis of waste management becomes relevant by taking into account the positive (e.g. natural resource efficiency, green job creation and a healthy and clean living environment.) and negative impacts (externalities) on the environment, health and the economy in general.

Quantifying these costs is not an easy task especially where models directly financed by municipalities prevail and in many cases the costs of the service are scarcely recovered.

It should be noted that in Argentina home generators do not particularly pay for the waste management service. Collection, treatment and urban cleaning (cleaning the streets and pavements and public sites in general) costs, are included in a generic rate, which also covers other services - such as public lightning- without any breakdown of the items paid. In short, it is not clear how this fee is calculated and/or how much is paid for each one of the services.

Given the need for municipalities in Argentina to know the real cost of waste management, so that they can adapt their tax collection structure to the real needs for resources to cover maintenance and operation costs and new investment needs, in 2011, the GIRSU Executing Unit developed the "GIRSU Economic and Financial Matrix", with financing from the Secretary of Environment and Sustainable Development (SAyDS-Argentina, 2011).

The Matrix is a computer tool (spreadsheet) that allows the calculation of the waste management cost from a comprehensive point of view and the development of performance indicators.

The tool divides the comprehensive waste management of each municipality into phases: final disposal, sweeping and cleaning, yard and green areas waste, collection, transfer, transport, recovery and sale of materials, composting, among others; and for each phase the tool allows the calculation of costs for the different activities/programs according to their category (e.g. staff, capital and consumer goods), collates information on waste generation, population, financial-accounting, among others, and allows the development and analysis of local performance indicators (SAyDS-Argentina, 2011).

In training workshops held in the provinces, the tool is introduced to representatives from different municipal areas, such as treasury/finance, environment and public services convened in workshops where management experiences are shared, and different performance indicators are developed and analyzed to estimate local costs as well as the trend towards a "comprehensive" view on cost management to embody the notion that integrated waste management is a "cross-cutting" issue.

WASTE MANAGEMENT GOVERNANCE

The goal of a waste management governance system is to achieve the best possible management in a certain context. Its content may be different on each locality, depending on the current particular situation of each place. Thus, even though the ultimate goal (to achieve an optimum socially, environmentally and economically-balanced environmental quality and public health) should be the same for everyone, the specific situation of each community shall be essential to define its own objective for a particular period.

The system must determine the rules or principles and the tools to achieve the goal set, identifying the players that can or must be participating in that process. It is also necessary to identify the instruments to make feasible the achievement of the goal, to analyze if they suitably respond to that purpose or if it is necessary to generate new instruments or adapt the existing ones. Hence, it is current and totally relevant the application of Principle 10 of the Rio Declaration: *the best way to approach environmental issues* is that in which access to information and full participation in the public decision-making processes is guaranteed.

Building a waste management governance system is a huge challenge regarding coherence and efficacy in a federal country like Argentina where national, provincial and municipal authorities can regulate and establish its own rules and institutions regarding both waste regulation and management.

In Argentina the National Constitution recognizes the autonomy of local governments to regulate environmental matters. Thus, the national government sets a base level, the minimum requirements of environmental protection for the whole country, while the local governments can complement said requirements[19]. In this context the national environmental authority produced back in 2006 the National Strategy for Municipal Waste Management (ENGIRSU according to his denomination in Spanish) (SAyDS-Argentina, 2005).

Furthermore, there is a space to discuss and convene the national environmental policy in the Federal Environmental Council (Consejo Federal de Medio Ambiente, COFEMA) –created by an inter-province convention in the year 1990 and ratified in 2002 by the General Environmental Law 25.675-. There all provinces, the CABA and the national environmental authority, have the possibility to agree and dictate resolutions with a binding effect. However, there are still significant legal loopholes that require the work of COFEMA in that field [20] prompting this organism to fully perform its relevant legally attributed role [21]. In terms of MSW, there is a major unresolved matter regarding the management thereof, a matter on which COFEMA should work, especially in relation to the possible regionalization.

Other pending matter is the legal and practical development of the concept of Extended Liability of the Producer (EPR) Principle[27] and its applications. Some advances have been done in the near past, like in the regulation of pesticides cans but still much remains to be done in this matter.

Direct Regulations

One of the cornerstones of waste management-related governance in the legal regulations to be complied with by the generators and handlers of waste as well as the penalties applicable upon noncompliance. It is a coercive, ordering and compulsive instrument necessary for waste management organization: the regulation defines the types of waste and who are bound to perform - or not to perform - certain activities in connection with those. Among other requirements, they establish the environmental licenses required to waste handlers, the mandatory nature of the Environmental Impact Assessment (EIA) for the facilities associated with management and provisions on the selection of sites for final disposition.

The existence of open dumpsites and burning of waste - both long-standing prohibition in most legislation and the lack of comprehensive management systems in many cities show that even when direct regulation is clearly indispensable, it is also evident that legal regulations alone are not sufficient to achieve good governance in terms of waste.

Strict regulations in force for decades in developed countries show that, even making relevant progress in comprehensive waste management and taking consistent steps towards a circular economy, it is still necessary to use other type of tools that can contribute to achieve the desired changes. Thus, to discourage or prompt certain behaviors conducive to minimization, reuse, recycling and sustainable production and consumption can be important to establish economic and information-based instruments. In Argentina these instruments have been provided for in some regulations. However, although provided for, they are rarely used. The experience shows that, broadly speaking, there is a long way ahead, not only regarding effective implementation of the legal rules, but also -and especially- in the use of said instruments, which are undoubtedly complementary to direct regulation.

Case Study 1: Municipal Waste Management in the Metropolitan Area of Buenos Aires

Document prepared by Marcelo Rosso and Carlos A. Fontán.

The Metropolitan Area of Buenos Aires presents an enormous complexity from the urban point of view with regard to the provision of public services. More than 16 million people (almost 40% of the national population) live in an area smaller than 0.2% of the country's mainland. In this region, coexist several jurisdictions and the planning is almost non-existent. These characteristics make its operation and the development as a region really complicated. One key public service problem is the Municipal Solid Waste (MSW) management: more than 20 thousand tons are produced daily and they need to be collected, transported, treated and disposed in an environmentally and economically sustainable way.

To deal with this problem in 1978 was founded the state-owned enterprise CEAMSE (Coordinación Ecológica Área Metropolitana Sociedad del Estado), made up of the governments of the City and the Province of Buenos Aires. Over the years CEAMSE's responsibility grow to cover more municipalities in the region, as illustrated in Figure 11.

Maps show the geographical extension covered by CEAMSE services: 25 provincial departments were added till 2018 to the CABA and 21 provincial departments originally served in 1980. Source: CEAMSE memories.

Paralleling CEAMSE expansion of territory coverage grew the amount of MSW, as indicated in Figure 12. From around 2/2.5 million tons processed in the '80s to over 6 million tons processed actually.

Yearly amount of MSW received in CEAMSE plants between 1978 and 2018.

In terms of geographic scope, the company serves now more than 16 million inhabitants, representing more than 36% of the total Argentine population, which generates approximately to 40% of the total waste generated in the country. During two decades (80's and 90's) CEAMSE provided controlled final disposal zones (sanitary landfills) to the collected MSW. At the end of the 1990s, CEAMSE embodied

Figure 11. Departments attended by CEAMSE in 1980 and 2018

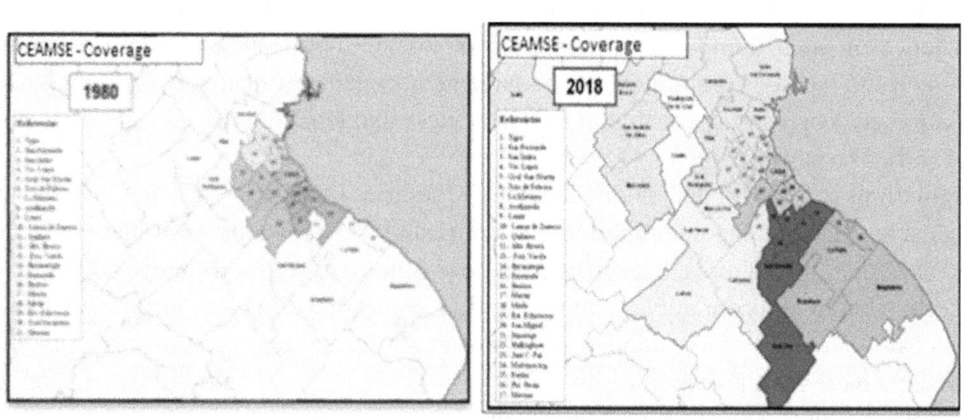

Figure 12. MSW received at CEAMSE facilities, 1978-2018
Source: CEAMSE memories.

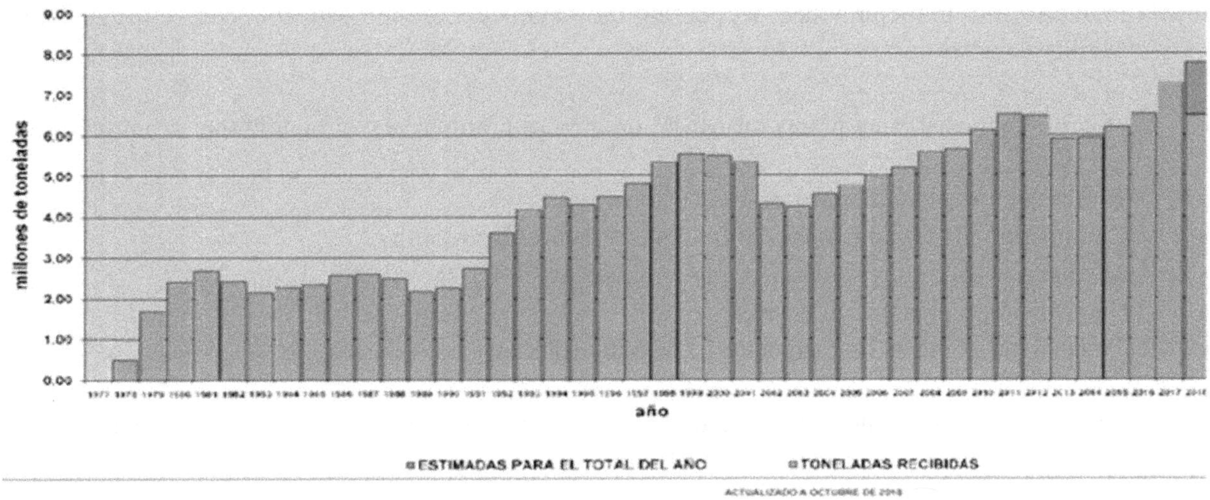

the concept of an "environmental complex" to transform sanitary landfills into "areas where garbage undergoes a set of processes for recycling the waste so that they can return to the productive circuit and where the technology allows not only to reduce the environmental impact but also to take advantage of the biogas from organic waste decomposition to generate renewable energies". Nowadays, CEAMSE has 5 transfer stations in order to achieve "economy of transportation, work and energy saving, reduction of wear and tear/or breakage of equipment costs, greater versatility, better resolution of the discharge front in the landfills and lower emissions". In addition, CEAMSE has three environmental complexes in operation and one in the post-closing stage. In such facilities waste separation and classification, biogas collection, leachate treatment and compost or organic amendment production take place (http://www.ceamse.gov.ar/).

The evolution of CEAMSE services are summarized in Figure 13.

Line of time showing the changes in MSW management during the last 20 years.

Figure 13. Time-evolution of MSW management at CEAMSE facilities

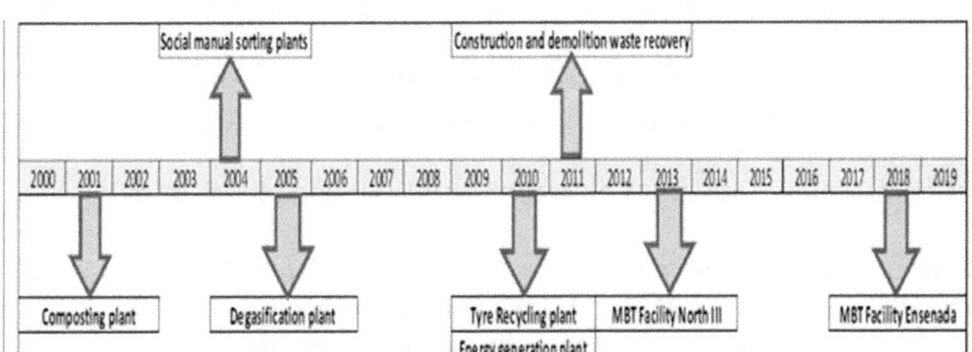

Line of time showing the changes in MSW management during the last 20 years.

As an example of the evolution of management let's quote that the Environmental Complex North III produces energy from waste. Daily, it receives 17 thousand tons of MSW generated in the CABA and 31 municipalities in the province, accounting for 90 per cent of the waste collected in the AMBA.

The facility consists of:

- 12 Social manual sorting plants for MSW, created to promote recycling and the social inclusion of recyclers.
- 1 Tyre recycling plant that processes 700 tons/month.
- 1 Composting plant that receives 800 tons of green waste/month.
- 1 Mechanical Biological Treatment plant that receives 1,100 tons/day.
- 2 Degasification plants with flares.
- 2 Degasification plants that generate 15 MW/h of energy.

The biogas produced in the Environmental Complex North III is collected by an extraction plant consisting of a network of interconnecting wells and pipes that flow into blowers and then to flares or, otherwise, to the energy generation plant.

The four degasification plants are registered in the Clean Development Mechanism, which enables emission reduction projects in developing countries to obtain Certified Emission Reductions (CERs). Between 2010, when they began operation, until 2016 degasification plants have reduced a total of 6,028,198 tons of CO_2 equivalent.

Benefits:

- CO_2 emissions are reduced approximately 1,0 million tons/year.
- 15 MW/h of energy are injected into the national electrical grid, meeting the daily electricity needs of approximately 25,000 homes.

This project also contributes to the sustainable development of the area immediately surrounding the landfill in the Province of Buenos Aires, by:

- Reducing methane emissions from the sanitary landfill.
- Eliminating gas emissions and odours, in benefit of the surrounding area.
- Reducing the risk of fire and explosion in the sanitary landfill through safe extraction of biogas.
- Positively impacting the local economy by using local labour and materials, whenever possible.

Case Study 2: Some Aspects of the Wastepicker or *Cartonero* Phenomenon in the Autonomous City of Buenos Aires, Argentina

Document prepared by Pablo J. Schamber and Francisco M. Suárez

Zabbaleens in Cairo (Egypt), *cartoneros* in Buenos Aires (Argentina), *recicladores* in Bogotá (Colombia), *catadores de lixo* in Rio de Janeiro (Brazil), *pepenadores* in Autonomous City of Mexico (Mexico), are some of the denominations used to refer to those in peripheral countries who survive from the activity linked to the collection and commercialization of materials that other people have considered waste. According to some estimates, this population living in marginal areas of large metropolitan cities represents no less than 2% of the inhabitants of their respective countries (Bartone, 1988), and they contribute with over 20% of the inputs used by industries that recycle.

These common features should not lead to avoiding the observation of the particularities that exist in each context, closely linked to waste management systems and the way in which urban waste pickers (WP) are inserted into them. In this sense, the Autonomous City of Buenos Aires has been a paradigmatic case of inclusion for some years. Although the local situation is part of a trend shared with other cities of the countries in the region and is related to the purposes pursued by international organizations such as the Regional Initiative for Inclusive Recycling (IRR, public-private partnership led by the Multilateral Fund of Investments -FOMIN- of the Inter-American Development Bank Group - IDB- which receives contributions from large companies such as Coca Cola and PepsiCo (http://reciclajeinclusivo.org), the fact that here, in the most populated city of Argentina, 12 cooperatives of *WP* have to their exclusive charge the concession of the local service of collection and treatment of the inorganic waste, is something unprecedented.

In fact, in Buenos Aires around 5,500 WP receive a fixed amount equivalent to 300 US Dollars in individual bank accounts if they comply with the total number of days of scheduled collection work, regardless of the quantity and type of materials collected. We have not been able to know that this circumstance is similar to the situation of the WP of any other place in the world. Here, since 2014, through contractual agreements signed between the local government and 12 urban recovery cooperatives, exclusive areas (covering the whole urban territory) were awarded for the collection of this type of waste, infrastructure for its treatment (classification plants known as Green Centers) and logistic support for the transfer of materials and people (trucks and vans), as well as the *incentives regime*, as the monthly assignments to the aforementioned WP are known. On their side, the organizations committed themselves to the effective fulfillment of the contractual terms and to gradually include independent WP (not belonging to any cooperative) that operate in the awarded areas.

Images of waste pickers collecting material downtown Buenos Aires (above) and manually sorting materials in a plant (below).

This recognition of the urban WP as actors in the public management of the waste as well as their effective participation in the system did not happen overnight, nor did it result from the exclusive will of a government. In order to understand the profound metamorphosis that led to the fact that from the last years of the 20th century to the first decade of the 21st century, the "*cartonero*" phenomenon went from being something ignored and invisible, it transited as an error to be corrected and repressed, until

Figure 14. Waste pickers in Buenos Aires

finally it became a question unavoidable of public policies related to the management of urban solid waste, it is necessary to address a complex variety of aspects, some of a strictly vernacular nature as well as others of a global scale. These include the significant acceleration of unemployment levels in the late 1990s, the lack of public or private initiatives that would seriously promote source separation or encourage recycling simultaneously with the social recognition that *cartoneros* did this as a worthy way of earning a living, the important increase in the demand for recyclable waste by the local industry from the great devaluation of the national currency and the recovery of consumption after 2002, the collaboration and technical assistance to strengthen the organization of *cartoneros* by nongovernmental organizations and sectors of the academy, the mobilization in defense of their interests by the organized *cartoneros*, to account for some.

REFERENCES

Bartone C. (1988). The Value in Wastes. *Decade Watch,* 3-4.

CEPIS/PAHO. (2003). Gestión de residuos sólidos en situación de desastres [Management of solid waste in disasters]. Serie salud ambiental y desastres.

de Titto, E., Montecchia, M. F., Brunstein, L., & Chesini, F. (2015). Normativas para la gestión de residuos biopatogénicos en Argentina [Regulations on Bio-Pathogenic Waste Management in Argentina]. *Rev Argent Salud Pública*, *6*(24), 7–14.

Electronic Reporting System of the Basel Convention. (2017). Retrieved from http://ers.basel.int/ERS-Extended/FeedbackServer/fsadmin.aspx?fscontrol=respondentReport&surveyid=75&voterid=49405&readonly=1&nomenu=1

Gullett, B. K., Wyrzykowska, B., Grandesso, E., Touati, A., Tabor, D. G., & Ochoa, G. S. (2010). PCDD/F, PBDD/F, and PBDE Emissions from Open Burning of a Residential Waste Dump. *Environmental Science & Technology*, *44*(1), 394–399. doi:10.1021/es902676w PMID:19950929

UN Habitat. (2010). *WTE Industry in Latin America 2010.* Author.

IDB. (2013). *Guidebook for the Application of Waste to Energy Technologies in Latin America and the Caribbean.* IDB.

IDB-AIDIS-PAHO. (2011). *Regional Evaluation on Urban Solid Waste Management in Latin America and the Caribbean - 2010 Report.* Author.

INDEC-Argentina. (2010). *Censo Nacional de Población, Hogares y Viviendas.* National Census of Population, Homes and Housing.

Instituto de Ingeniería Sanitaria, Facultad de Ingeniería, Universidad de Buenos Aires- CEAMSE. (2015). *Estudio de Calidad de los Residuos Sólidos Urbanos de la CABA* [Study on the quality of Urban Solid Waste in the CABA]. Author.

ITU. (2015). *Gestión Sostenible de Residuos de Aparatos Eléctricos y Electrónicos en América Latina* [Sustainable Management of waste from electrical and electronic equipment in Latin America]. ITU.

Lipinski, B., Hanson, C., Lomax, J., Kitinoja, L., Waite, R., & Searchinger, T. (2013). *Reducing Food Loss and Waste*. Working Paper, Installment 2 of Creating a Sustainable Food Future. Washington, DC: World Resources Institute. Available online at: http://www.worldresourcesreport.org

PAHO-Argentine Ministry of Health. (2018). *Argentina Indicadores Básicos 2017* (Argentine, Basic Indicators 2017). Author.

Secretary of Environment and Natural Resources (SEMARNAT) -IPN. (2009). *Estudio de análisis, evaluación y definición de estrategias de solución de la corriente de residuos generada por los vehículos usados al final de su vida útil*. Author.

Secretary of Environment and Sustainable Development (SAyDS), Argentina. (2005). *Estrategia Nacional para la Gestión Integral de Residuos Sólidos Urbanos* [National Strategy for Integral Municipal Solid Waste Management, ENGIRSU]. Author.

Secretary of Environment and Sustainable Development (SAyDS). (2011). *Economic and Financial Matrix for calculating the costs of the integrated municipal solid waste management in Argentina*. Argentina: Author.

The World Bank. (2015). *Diagnóstico de la Gestión Integral de Residuos Sólidos Urbanos en la Argentina*. Author.

UNEP. (2011). *Revised technical guidelines for the environmentally sound management of used and waste pneumatic tires*. Document UNEP/CHW.10/6/Add.1/Rev.1. 2011. UNEP.

UNEP. (2015). *Plastics in Cosmetics*. UNEP.

UNEP. (2016). *GEO-6 Regional Assessment for Latin America and the Caribbean*. Nairobi, Kenya: United Nations Environment Programme.

UNEP. (2017). *Regional Waste Management Outlook in Latin America and the Caribbean*. UNEP.

UNEP-ISWA (2015). *Global Waste Management Outlook*. Author.

USAID. (2013). *Experiencias internacionales en el composteo de residuos sólidos orgánicos*. USAID.

ENDNOTES

1 "Not in my back yard" connotes the unwillingness of individuals who reside in an area to accept the setup of any activity or premises that they deem might affect their quality of life and the value of their property. They generally do not object to the activity or premises in themselves but to the proximity with their homes, due to the environmental impacts that could be caused.

2 http://www.ceamse.gov.ar/reciclaje/compostaje/

3 http://www.rosario.gov.ar/web/servicios/medio-ambiente/planta-de-compostaje-bella-vista

4 http://www.ceamse.gov.ar/arranco-la-nueva-planta-de-separacion-de-residuos-en-norte-iii/#.VwRCw3phPod

5 Argentina established in 2006 through the Law Nº 26.,184 the ban to manufacture, assemble or import cylindrical or prismatic, common carbon-zinc and manganese alkaline cells and primary batteries, whose content of mercury, cadmium and lead was over 0.0005%, 0.015% and 0.200% in

weight, respectively. They It also banned the sale thereof as from three years following the enactment of the law and established that. In this context, the manufacturers, assemblers and importers must proceed to certify the composition of the batteries, by an authorized institution, as a prior requirement to sell. Additionally, the law established a requires that clear ment related to the information to be included in the composition has to accompany each battery, its shield (air-tightness) and the technical rules of duration each battery, its shield (air-tightness) and the technical data of duration they must comply with.

6 The CABA has implemented the sustainable public procurement though rules of diverse hierarchy. From the local constitution that sets forth that the city can promote the public and private initiative in the economic activity within the framework of a system which ensures the social welfare and sustainable development, by means of a Joint Resolution No 323/SSGO-APRA/12 it created a Work Group of Sustainable Public Procurement, made up of sections of the Environmental Protection Agency and the Ministry of Finance. This group is engaged in the analysis, development and implementation of the regulations in force concerning Sustainable Public Procurement. Even when the greatest advances have occurred in the field of energy efficiency - with rules related to tender specifications for the purchase of public lightning items, traffic lights and electric appliances - the Recommendation Guideline for Responsible Consumption and Purchase of Paper, which contains directions for the best choice and use, based on the purchase of certified products and the possibility of recycling them. Within this framework, Law No. 3540 related to forestry management establishes that the purchases and procurement that include timber products and byproducts shall be made to providers certified by the Forest Stewardship Council (FSC), the Program for the Endorsement of Forest Certification (PEFC) and other organizations that can guarantee the environmental sustainability of the entire productive circuit. Furthermore, by Provision No. 237/DGCYC/2014 the “Guideline of Recommendations to Progressively Improve Packaging and/or Container Sustainability in Public Procurement” was approved. For more information, see http://www.buenosaires.gob.ar/hacienda/compras/guia-para-el-consumo-y-compra-responsable-de-papel and http://www.buenosaires.gob.ar/hacienda/compras/compras-sustentables/guia

7 http://www.alihuen.org.ar/santa-rosa-recicla.-basura-cero.-la-pampa/residuos-peligrosos-generados-en-nuestras.html

8 http://basuracero.org/sitio/base.php?sec=28

9 www.residuoselectronicos.net

10 https://realidadeconomica.com.ar/regomax/

11 Extended Producer Responsibility (EPR) is a “policy approach under which producers are given a significant responsibility – financial and/or physical – for the treatment or disposal of post-consumer products. Assigning such responsibility could in principle provide incentives to prevent wastes at the source, promote product design for the environment and support the achievement of public recycling and materials management goals” (https://www.oecd.org/env/tools-evaluation/extendedproducerresponsibility.htm)

The EPR Principle is based on four main columns: the duty of preventing pollution, the Life Cycle concept, the Polluter-Pays Principle and the internalization of costs. This principle changes the traditional view on who must take charge of the generated waste management targeting the analysis to the product, including its characteristics, composition, volume and even its mass consumption instead of keeping the State’s whole responsibility for managing all waste generated from consumption. This approach changes essentially the allocation of economic and factual responsibility for waste management to the product's manufacturer, producer or importer, being the ones who dominates the value chain and who can best influence the behaviors of the remaining links, as well as who can make decisions concerning the design and composition of the products and marketing strategies.

12 PAHO defines a natural disaster as the occurrence of a natural phenomenon in a limited space and time that disrupts normal patterns of life, causing human, material, and economic loss, and environmental damage. And adds: it is an ecological event of such magnitude that requires external intervention to manage its effects.

13 https://www.gfdrr.org/argentina

14 Intergovernmental Panel for Climate Change (IPCC) http://www.ipcc.ch/home_languages_main_spanish.shtml

15 Waste Sector's contribution to GHG production is mostly (97 per cent) through methane (CH_4) emissions resulting from the anaerobic decomposition of waste in sanitary landfills. Methane is a pollutant GHG 28 times more potent than carbon dioxide (CO_2) and persists in the atmosphere for 12 years. Emissions of carbon dioxide (CO_2) resulting of some treatments (composting, incineration, sanitary landfills) or burning in dumps also count while those released in collection, transport and recycling activities are not quantified in this sector, but rather in the transport sector, Nitrous oxide is also emitted during the composting and incineration processes. Black carbon is generated by indiscriminate burning of waste, by the combustion that moves the waste transport and operation equipment, and by the equipment used in recycling.

16 https://www.foodbanking.org/food-bank-resources/global-food-bank-community

17 https://www.bancodealimentos.org.ar/

18 http://www.ceamse.gov.ar/ciclo-de-la-basura/complejo-ambiental/tratamiento-de-liquidos/

19 The National Constitution of Argentine establish in its art. 41 that in environmental maters the provinces can dictate rules complementary to the minimum requirements set by the federal government.

20 In relation to waste management, aAn unsolved mater very important aspect to deal with and that remains unresolved is the existence of different provincial criteria regarding the inter-jurisdictional transportation of wastes and the prohibitions of access, described above.

21 According to the above-mentioned convention, i.e., the Articles of Incorporation of COFEMA, it was created to "Formulate a comprehensive environmental policy, as a precautionary and corrective measure, based on the corresponding diagnosis, taking into consideration the local, provincial, regional, national and international scales." The General Environmental Law, instead, by creating the Environmental Federal System with the objective of coordinating the national environmental policy, recognizes COFEMA as an instrument to make this objective feasible and assigns specific responsibilities in environmental planning of the territory and environmental education, responsibilities that are elaborated on by other sectoral laws. That is the case in Law 25.916 that regulates the Comprehensive Household Waste Management and mentions COFEMA as the agency of inter-jurisdictional coordination, making it responsible for "a) Agreeing on policies of comprehensive management of household wastes; b) Setting technical and environmental criteria to employ in different stages of the comprehensive management; c) Agreeing, along with the Enforcement Authority, on the goals of valorization of household wastes".

APPENDIX: ABBREVIATIONS AND ACRONYMS

AIDIS Asociación Interamericana de Ingeniería Sanitaria (Interamerican Association of Sanitary and Environmental Engineering)
CEAMSE Coordinación Ecológica Área Metropolitana Sociedad del Estado
CDW Construction and demolition waste
COFEMA Consejo Federal de Medio Ambiente (Environmental Federal Council)
EPR Extended Producer Responsibility
GHG Greenhouse Gas
GIRSU Gestión Integral de Residuos Sólidos Urbanos (Integral Management of Municipal Solid Waste)
HCW Healthcare waste
HHW Household hazardous waste
HW Hazardous waste
IDB Inter-American Development Bank
INTI Instituto Nacional de Tecnología Industrial (National Institute for Industrial Technology)
MSW Municipal Solid Waste
PAHO Pan American Health Organization
SAyDS Secretaría de Ambiente y Desarrollo Sustentable (Secretary of Environment and Sustainable Development), Argentina
UNFCCC United Nations Framework Convention on Climate Change
WP Waste Pickers
WEEE Waste electrical and electronic equipment

Chapter 2
Sustainable Solid Waste Management in Bangladesh:
Issues and Challenges

Md. Ashikuzzaman
https://orcid.org/0000-0001-6603-7189
Khulna University, Bangladesh

Md. Hasan Howlader
Khulna University, Bangladesh

ABSTRACT

Solid waste management is a grave concern for Bangladesh as by 2025 waste generation per capita will be 0.75 kg/capita/day and total amount of waste will reach 21.07 million tons per year. This chapter attempts to uncover the facts regarding waste management along with the policies and regulations existing in Bangladesh by reviewing published secondary documents. The chapter also contains emerging issues of agricultural, industrial, hazardous, construction, and electronic generation and 3R practices in Bangladesh. Legal instruments for waste management in Bangladesh are also examined. It has been found that enforcing authorities lack the capacity to implement their strategies regarding 3R practices for waste management. Case studies about community-based approach, waste recycling sector, and medical waste management have been exemplified in this chapter. Finally, the authors apprised the issues and challenges of sustainable solid waste management practices and proposed the way forward for Bangladesh to have sustainable solid waste management.

DOI: 10.4018/978-1-7998-0198-6.ch002

INTRODUCTION

Waste is inevitable as it is generated by daily human activities through discarded materials of domestic chores, agriculture, business, industry, natural and man-made disasters, and medical services, etc. According to Bangladesh Environment Conservation Act, 1995, waste is, "any solid, liquid, gaseous, radioactive substance, the release, disposal, and throwing away of which may cause deleterious changes to the environment." Management of waste comprised of collection, carrying, treating or disposal of discarded materials and substances (Unnisa & Rav, 2012). Sustainable solid waste management is one of the prerequisites for sustainable environmental management because the unsafe disposal of solid waste pollutes the environment and causes human health hazards.

Bangladesh is considered as a densely populated country where 162.7 million (BBS, 2018) people live in an area of 147570 square kilometers and 1,115.62 people live in per square kilometer (Kormoker et al., 2017). The population growth rate of the country is 1.37% (BBC Monitoring, 2019), where the urban and rural populations are 23.3% and 76.7%, respectively. The urban population of the country will be 38% by the year 2020 (Population Division of the UN, 2016). This quick advancement in urbanization, industrialization, and consequently improved lifestyles generate the bulk of solid wastes. The per capita national income of Bangladeshi people reached USD 1,751, with 7.86% GDP growth in the fiscal year 2017-18 that has overcome its status from the least developed to a lower-middle-income country (BBS, 2017, 2018a, 2018b). Appreciable economic development has ushered considerable changes in the lifestyle leading towards urbanization and lavish living standard. Previous studies (Abedin & Jahiruddin, 2015; Rahman, 2017; Shams et al., 2017) showed that fast urbanization and population growth are liable for a large volume of solid waste production in Bangladesh. The volume of waste generated in 1970 was 11,00,000 tons which in 2015 amplified to 52,00,000 tons with an annual increase rate of 1,34,300 tons (Shams et al., 2017). In two separate studies, it was found that in 2013 the total amount of municipal solid waste (MSW) generated in urban areas of Bangladesh was 52,00,919 ton/year by 3,69,86,768 urban inhabitants (rate 0.35 Kg/cap/day) whereas in 2015 the total amount of urban wastes was 5,11,000 ton/year by 3,94,88,000 residents (0.32 Kg/cap/day) (Shams et al., 2017; Islam, 2017). Another source reports MSW generation of 1,47,78,497 ton of waste in 2012 with the country's population of 15,57,27,053 (Kaza et al., 2018). The average per capita MSW production in different municipal areas ranges from 0.2 to 0.56 kg/cap/day (Ahsan et al., 2014). In the urban areas of Bangladesh, the waste generation is expected to raise 0.6 Kg/cap/day (Bhuiyan, 2010) where the total generated waste volume would be 57,718 tons/day (Hoornweg & Perinaz, 2012) by 2025.

Since waste is the ever-present issue for every society, the management of solid waste is also an ancient phenomenon. Bangladesh has had a traditional waste management system since its independence in 1971. Early waste management in Bangladesh allowed indiscriminate open dumping and burning, disposal of wastes into water bodies, landfilling, and direct disposal into agricultural land in rural areas. Night soil collection was handled through Bullock Cart for solid waste, along with nighttime waste collection, house-to-house collection performed by the Water Supply and Sewerage Authority (WASA) in urban areas. With the advent of time, the country shifted its waste management system from traditional to a modern approach. Wastes are now considered resources, so Bangladesh has shifted from waste management to resource management.

Bangladesh began recycling and composting of organic wastes in 1980 and 1990. Sanitary landfill with gas recovery of the residual wastes began in 2010 (Shams, et al., 2017). With a view to complying with the existing national and international legal frameworks (BIGD, 2015), the major initiatives taken by Bangladesh in order to manage waste sustainably are illustrated in Table 1.

In spite of having tremendous efforts of GoB and Non-Governmental Organizations (NGOs), waste management is far from the desirable status in Bangladesh. Even in Dhaka, the capital city of Bangladesh has proved to be a nightmare for waste management as the amount generated per day is approximately 4124 tons, of which almost 40% goes uncollected (Kabir, 2015). Because of low awareness, lack of motivation, financial, and technological paucity, 40-60% of wastes remain uncollected and are not disposed of in a safer manner (Ahsan et al., 2014). Thus, low or inadequate collection, insufficient transport coverage, and inappropriate processing technology and treatment are accountable for substandard management of MSW that leads to environmental pollution, great health hazards and environmental risks. So solid waste management is a conundrum for Bangladesh. The aim of this chapter is to portray the waste management scenario of Bangladesh, based on secondary document analysis.

CLASSIFICATION OF WASTES IN BANGLADESH

Typology of wastes generated in Bangladesh based on sources is shown in Table 2.

WASTE FLOW

Figure 1 deals with the waste flow path from the generation source to the final dumping sites along with the treatment processes. In Bangladesh, typically waste, irrespective of the type, is usually dumped into the roadside community bins or secondary deposit sites (SDS). In most cases, it is the responsibility of

Table 1. Major collaborative initiatives of Solid waste management in Bangladesh

Sl no.	Major Initiatives	Inception Year	Funding Organization (s)
1	Recycling Training Centre	2006	Government of Bangladesh (GoB) and United Nations Development Program (UNDP)
2	Strengthening of Solid Waste Management in Dhaka City	2007	Japan International Cooperation Agency (JICA), Dhaka City Corporation (DCC) and Local Government, Rural Development and Co-operatives (LGRD&C)
3	Improvement of Solid Waste Management in Dhaka toward the low carbon society	2009	JICA partnership with DCC and LGRD&C
4	Preparation of Solid Waste Management Plan for 19 towns of Bangladesh	2009	United Nations International Children's Emergency Fund (UNICEF), GoB and Waste Concern
5	Urban Public and Environmental Health Development Project	2010	GoB, Asian Development Bank (ADB)
6	Bangladesh towards zero waste	2013	European Union (EU)
7	Value for waste (First Phase)	2013	Waste Concern, Swisscontact

Source: Authors' compilation 2019

Table 2. Typology of waste

Source	Location	Waste Type
Domestic	Households	Kitchen waste, wood, metal, plastics, etc.
Commercial	Commercial firms, restaurants, offices, shops, hotels, and banks	Paper, plastic, e-waste, metal, oil, batteries, etc.
Industrial	Small, medium and large manufacturing and processing units	Solid and liquid wastes, ashes, concrete, debris, metal, residues, etc.
Municipal	Cleaning of urban facilities and treatment plants	Plastics, dust, sludge, debris, concrete, etc.
Educational Institutions and Hospitals	Classrooms, teacher/office rooms, student hostels	Paper, plastics, useless furniture, food waste, glass, hazardous waste, metals, etc.
Agriculture	Cropping fields and farms	Crop residues, metal, plastics, dust, and other organic wastes

Source: Kumar, 2016

Figure 1. The flow of MSW in Bangladesh
(Source: Ahsan et al., 2014)

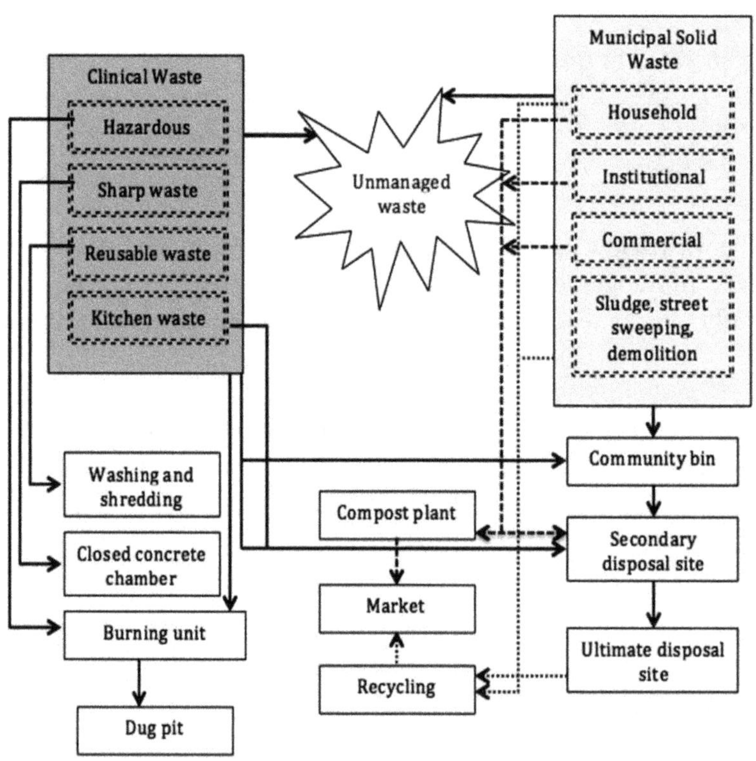

the urban authority to collect and dispose of waste in the ultimate disposing points or sites. Large portions of non-hazardous inorganic waste are reused and recycled by the inhabitants at the generating points. NGOs and Community Based Organizations (CBOs) providing door-to-door collection services also segregate the wastes and reuse them without processing. The residents store their daily waste in a plastic or metal bin within the home or on their premises to supply them to the waste collectors. During their collections, the collectors keep the waste in a single chamber van that they use to separate the useable and recyclable wastes, such as clothing, plastic products, and scrap metal, in hanging sacks behind the van, which are then sold in the informal sector recyclers. NGOs working in waste collection provide the collected wastes in composting plants to produce compost for the market sale. In recent times, government and NGOs-led initiatives have been taken to produce electricity and biogas from recycled wastes. In the case of medical wastes, there are separate bins allocated for kitchen waste (black), general waste (gray), reusable waste (yellow), and sharp waste (red). Generally, medical waste is separated by the users and nurses at the generation points. The reusable wastes after cleaning are stored for recycling. In a seven-foot-deep chamber made of concrete, the hazardous medical wastes are encapsulated. The non-hazardous medical wastes are burnt in an incinerator (Ahasan et al., 2012). Outside of this pattern, some of the residents still throw their wastes indiscriminately into the drain, onto the road, or into nearby water bodies, as these residents are reluctant to spend money for waste disposal.

WASTE GENERATION / 3R PRACTICES, AND WASTE MANAGEMENT TECHNOLOGIES

Waste Generation and 3R Practices

The estimated population of Bangladesh by 2025 would be 78.44 million and they will generate 47,000 tons of wastes daily (Waste Concern, 2014). According to Ahsan et al. (2014), households, businesses, and institutions contribute respectively 78%, 20%, and 1% of waste while the other sectors responsible for only 1%. Item-wise generation rate of MSW in Bangladesh is illustrated in Table 3.

The National 3R (Reduce, Reuse, and Recycle) strategy which was participatory in nature was introduced in 2010 in Bangladesh for waste management with the support of Japan's Ministry of Environment and Centre for Regional Development of the UN where Waste Concern provided the technical assistance

Table 3. Item-wise waste production in major urban areas of Bangladesh

Items	Weight in Percentage						
	Dhaka	Chattogram	Khulna	Rajshahi	Barisal	Sylhet	Average
Food Waste	68.3	70.5	78.9	70.0	81.1	73.5	74
Paper	10.7	4.63	9.5	9.0	7.2	8.6	8
Plastic	4.3	8.7	3.1	9.0	3.5	3.5	5
Textile and Wood	2.2	2.4	1.3	6.0	1.9	2.1	2
Leather and Rubber	1.4	5.8	0.5	1.1	0.1	0.6	2
Metal	2.0	2.65	1.1	3.0	1.2	1.1	2
Glass	0.7	1.0	0.5	1.1	0.5	0.7	1
Other	10.4	7.4	5.1	0.8	4.5	9.9	6

Source: Shams et al., 2017

in preparing it (Nasrin, 2016). To uphold this strategy, a wing has been created in the Environment and Forests ministry to implement activities along with a committee comprised of inter-ministerial personnel to coordinate the activities across different ministries. To monitor and evaluate the 3R activities of the aforesaid committee, a cell is created in the environment department (DoE, 2010).

Waste Management Technologies

In Bangladesh, the system of managing waste is unorganized. Figure 2 shows the process of managing waste in Bangladesh. From the functional point of view, the country has three (formal, informal and community) systems of waste management (Abedin & Jahiruddin, 2015). In the formal system, the municipal authorities are solely accountable for the overall management of waste-related activities. Informal waste management is done by a huge labor force mainly engaged in the trade of recycling (DoE, Waste Concern, & ITN-BUET, 2004; Ahsan et al., 2014). The initiatives of the CBOs and NGOs for collecting and managing the wastes from primary sources are known as community initiatives.

The technologies and methods used by the city authorities in Bangladesh are traditional and similar to other developing countries that are highly labor-intensive. The words "source reduction" seem to be unheard of by waste managing authorities. Informal sectors are involved in doing the recycling and reusing activities (DoE, Waste Concern, & ITN-BUET, 2004). People typically segregate valuable wastes to sell them to vendors. Pickers retrieve recyclable wastes from dumpsites, community bins or roadsides.

Figure 2. Waste management process in Bangladesh (Adapted from Abedin & Jahiruddin, 2015; DoE et al., 2004)

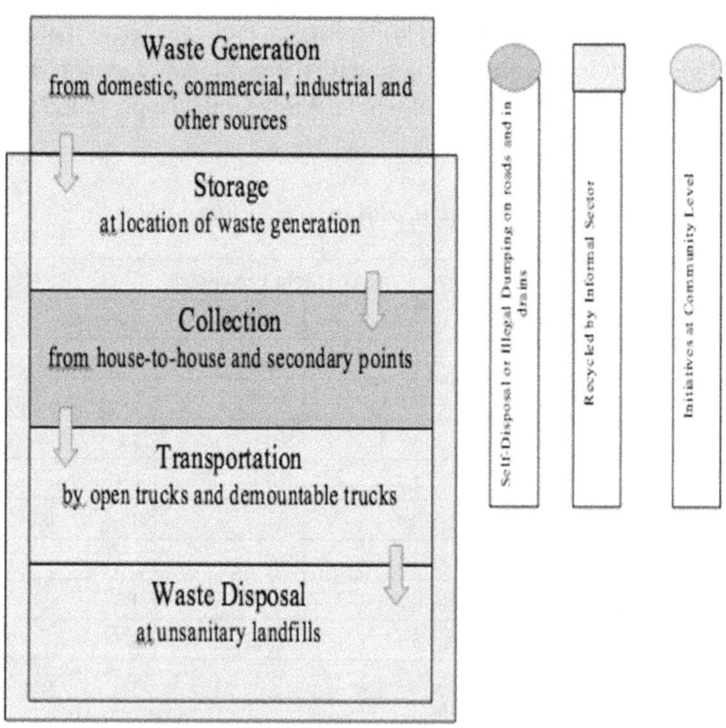

There is no provision of storage of wastes at household level resulted in dumping wastes in the community bins and designated open spaces. The city authority generally uses open trucks, tractors, demountable containers and trailers for collection and transportation of wastes. The filthy stench is a part of everyday life for city dwellers and the odor becomes excruciating during the rainy season. Incineration technologies are not common due to the high cost involved. In large cities, incinerators burn medical wastes. Finally, open dumping is adopted for landfilling (Abedin & Jahiruddin, 2015; DoE, Waste Concern & ITN-BUET, 2004).

WASTE MANAGEMENT POLICY AND REGULATIONS IN BANGLADESH

Major waste management related policies, rules, and acts are shown in Table 4.

GENERATION AND MANAGEMENT OF HAZARDOUS WASTE, AND TRANSBOUNDARY MOVEMENT

With the increasing use of chemical substances and electronic goods, the volume and number of hazardous wastes are also increasing rapidly in Bangladesh. Some people also buy the second hand or reconditioned electronic goods without considering the negative effects of these products on human health and environment. Majority of these end users dispose these goods and expose substances directly in their nearby environment (Nahar, 1998). The sources and generation of hazardous wastes are shown in Table 5.

The management of hazardous waste in Bangladesh is very much alarming. Most of the hazardous waste producers or importers are less conscious about the harmful effects of open dumping or exposing here and there. Aside of this, the legal and regulatory framework is not followed properly. Govt. of Bangladesh introduce its regulatory framework in 1996 but it could not bring any significant changes to the behavior of people regarding hazardous wastes rather some people continuing their business on importing reconditioned goods and toxic chemicals. To control the transfer of hazardous waste outside national boundaries or transboundary movement, the Basel Convention was adopted in 1989 and in 1992 it entered into force. Transboundary movement according to his convention is "any kind of movement of hazardous waste at least between two separate nations or two separate national boundaries". In spite of being a signatory of the Basel Convention, Bangladesh could not strictly follow the provisions of the convention, as there prevails illegal imports of scrap and second-hand products from developed countries, such as ship scraps and scrap metal from developed countries.

Brain-damaging mercury and toxic electronic and plastic wastes from the United States; cancer-causing asbestos from Canada; defective steel and tin plates from the EU, Australia, and the U.S.; toxic waste oil from the United Arab Emirates, Iran, and Kuwait; toxic zinc ash, residues and skimming; lead waste and scrap; used batteries; and waste and scrap of metals, such as cadmium, chromium, cobalt, antimony, hafnium, and thallium from Germany, Denmark, the Netherlands, the United Kingdom, Belgium, and Norway are all dumped on Bangladesh (ESDO, 2012).

Table 4. Chronology of regulatory framework for waste management in Bangladesh

Year	Legal Instruments	Key Features
1972	Constitution of Bangladesh (CoB) 1972	Article 18A of the CoB portrays that declares the state obligation to ensure and improve the earth and to protect and defend the characteristic assets, biodiversity, wetlands, woods, and natural life for the present and future citizens.
1992	National Environment Policy 1992	Distinguishes and manages exercises that degrade and pollute the earth.
1995	Environmental Conservation Act 1995	Provides definition of wastes and prescribes disposal standards for various kinds of waste.
	National Environmental Management Action Plan (NEMAP) 1995	3R is being advanced under the Sustainable Environment Management Program (SEMP).
1997	Environmental Conservation Rules 1997	Sets the standards for mechanical wastes and gushing release, and enterprises need to adhere with few principles to pick up a Clearance Certificate for their activity.
	Bangladesh Environmental Conservation Rules (ECR) 1997	Prescribes waste transfer benchmarks principally for industrial squanders.
1998	Urban Management Policy Statement 1998	Privatization of urban facilities for slum tenants, including arrangements of water and sanitation, and the waste transfer was prescribed.
	National Policy for Water Supply and Sanitation 1998	Guides the authorities of local government to engage private sectors for waste collection and management and taking waste recycling measures.
1999	National Agriculture Policy 1999	Advances utilization of manure and natural compost by farmers to enhance soil efficiency and food security.
2004	Private Sector Infrastructure Guideline 2004	Supports the private sector to invest in waste recycling and management activities. It has additionally distinguished the waste sector as one of the important sectors for private ventures.
	Dhaka Declaration on Waste Management by SAARC countries during 10-12, October 2004	The SAARC nations consent to support NGOs and privately owned businesses to build-up community-based composting, segregation at the source, separate collection, and resource recovery from wastes, with specific focus on composting.
2005	National Industrial Policy 2005	Suggests the cleaner production practices in industries using Environmental Management Systems (EMS).
	Draft National Solid Waste Management Rules 2005	3R standard has been embraced unequivocally.
	Environment Management Plan 2005	Prioritizes reduction and recycling of wastes.
	National CDM Strategy 2005	Encourages CDM projects that are pro-poor by taking carbon financing into account.
	Poverty Reduction Strategy Paper (PRSP) 2005	Emphasizes on source separation, recycle, reduce, and reuse of solid wastes generated from various sources.
	Dhaka Environment Management Plan 2005	EMS was promoted among businesses along with recycling and less land filling. .
	Solid Waste Management Action Plan for Eight Secondary Towns in Bangladesh 2005	4R principle i.e. reduce, reuse, recycle, and recovery of waste was introduced in this plan.
	Private Sector Housing Development Guideline 2005	Proposes for waste recycling, composting and biogas generation space in housing areas.
	National Sanitation Strategy 2005	Assets recovery and recycling as an option in contrast to dumping is distinguished as key to improving urban sanitation.
2006	Fertilizer Act 2006	Composting was encouraged and a standard for compost was developed in 2008.
	National Urban Sector Policy 2006	"User pays" principle was proposed to extend services and reduce municipal costs.
	Lead Acid Battery (LAB) Recycling and Management Rules 2006	Brought changes in collection and recycling of LAB.
2007	Fertilizer Management Rules 2007	Management and standardization of fertilizer quality is emphasized.
2008	National Renewable Energy Policy 2008	Biogas production was promoted along with other green energy.
	Biomedical Waste Management Rules 2008	Source separation along with a separate SWM system development is recommended.
	Circular to Promote Compost by the Ministry of Agriculture (MoA), on 23 April 2008	Promotes compost use, reduce environmental impacts, as well as financial burden of subsidizing fertilizer.
	National Renewable Energy Policy, 2008	Promotes production of biogas and other green energy from waste and also supports incentives for CDM to encourage green energy projects.
2009	Local Government (City Corporations) Act 2009	Inhabitants should maintain buildings under city authorities jurisdiction.
	Local Government (Municipality) Act 2009	
2010	Bangladesh Environment Conservation (Amendment) Act, 2010	Amendment of ECA 1995.
	National Solid Waste Management Handling Rules 2010	Encourages 3R principles along with public participation in SWM process.
	National 3R (Reduce, Reuse and Recycle) Strategy for Waste Management 2010	Promotes user pay principles and 3R strategies as well as implementation of proper technology.

continued on the following page

Table 4. Continued

Year	Legal Instruments	Key Features
2011	Hazardous Waste and Ship Breaking Waste Management Rules 2011	Deals with ship-breaking and hazardous waste management.
	Sixth Five-year plan (FY2011-2015)	Emphasizes source separation and 3R approach.
	Ship-Breaking and Recycling Rules 2011	Deals specifically with ship-breaking and recycling issues.
	Land Use Policy 2011	Propose a definite site for managing e-wastes in every district.
	National Science and Technology Policy 2011	Encourages research and use of green technology for reducing waste as well as adverse effects of climate change
2012	National Agricultural Policy 2012	Promotes the use of compost fertilizer to improve soil productivity.
2013	National Environment Policy 2013	Recognizes 3R principles of waste management.
	National Information and Communication Technology Act 2006 (Amendment up to 2013)	Describes the relevancy to management of discarded ICT equipment.
2014	National Policy for Water Supply and Sanitation 2014	Recycling, composting and biogas production measures were promoted by GoB
	National Sanitation Strategy 2014	Regarding improving sanitation in urban area priority has been given to recycling and recovery than disposal only.
2015	Revised National Urban Policy 2015	Emphasizes Clean Development Mechanism (CDM) and recycling.
	National Information and Communication Technology (ICT) Policy 2015	For tackling climate change and managing disasters emphasize have been given on using green technology for e-waste management.
	Import policy Order 2015-2018	Forbids the import of products like reconditioned office hardware to reduce e-waste.
	Export Policy 2015-2018	For exporting e-devices and goods that are refurbished all restriction are withdrawn.
2016	National Industry Policy 2016	Administration will provide motivations to industry/businesses for properly managing waste using 3R techniques.

Source: Authors' compilation 2019

Table 5. Sources and volumes of hazardous wastes generation in Bangladesh

Waste Sector	Solid Waste Tons/Year	Liquid Waste m^3x10^6
Textile	113,720	99.75
Hospitals	12,271	N/A
Tannery	26,250	1.30
Fertilizers	277	7.80
Pesticides	357	10,920
Oil Refining	4	0.61

Source: Green, 2010

AGRICULTURAL BIOMASS AND UTILIZATION, WASTE-TO-ENERGY

Energy crisis in Bangladesh is overwhelming as a result the per capita consumption of energy is one of the lowest in the world. In 2011 the electricity consumption in Bangladesh surpassed 39.5 billion kWh. Furthermore, 38% of inhabitants have no access to electricity, with an annual accessibility of 321 kWh per capita (Halder et al., 2014; Hossen et al., 2017).

The prospect of biomass to energy is enormous in Bangladesh and improving plant productivity and utilizing prevailing resources can achieve this. Proper technological advancement can rejuvenate bio-energy. Biomass covers 65% of total energy consumption in Bangladesh (Mamun, 2007). Major types of biomass in Bangladesh are:

- Fuel wood.
- Forestry crops and residue/Forest biomass.
- Crop biomass.
- Agricultural crop/Fruits/Vegetables.
- Animal waste and human excreta.
- Municipal and industrial solid wastes.
- Fish wastes.

The rain-fed environment of Bangladesh produces biomass resources in huge quantities (Halder et al., 2014). The existing characteristics of farm wastes as well as solid wastes make it a potential source of energy and power in this country (Hossen et al., 2017). In rural Bangladesh, firewood, leaves, branches, twigs, cow dung, rice straw and husks are used as the cooking fuels.

According to several studies (Halder et al., 2014; Hossen et al., 2017; Mamun, 2007), based on its origins, biomass resources have different energy potentialities. In fiscal year 2012-2013, the amount of residues from agricultural-crop was almost 43 million tons with a net recovery of almost 36.5 million tons. It is estimated that, 582.33 PJ (petajoules) can be recovered (Halder et al., 2014). Moreover, estimates suggest that maximum 3447 PJ, more than Bangladesh's current requirement, can be recovered from crop residue and forest resources (Hossen et al., 2017). The potentiality of biomass to energy is huge in Bangladesh. But lack of proper management makes this potential resource a waste and the government are also unable to address the potentialities as strategies for electricity generation are completely opposite of biomass utilization.

E-WASTE, INDUSTRIAL WASTE AND EXTENDED PRODUCER RESPONSIBILITY (EPR)

E-Waste and 3R Practices in Bangladesh

E-waste means electronic wastes produced from discarded electronic gadgets and appliances used in households and industries i.e. television, mobile phone sets, refrigerators, computers, air conditioners, etc. (Rode, 2012). In Bangladesh, each year, the average generated e-waste volume is 2.7 million metric tons (in Table 6). Globally, the overwhelming e-waste growth rate is 4% due to rapid technological breakthrough and short lifespan of electronic products (Lundgren, 2012). In spite of having a 3R policy regarding e-waste management, the implementation of it very poor at industrial and household level. The 3R policy was introduced to eradicate disposal of wastes into floodplains, rivers, and open dumps by 2015 along with promote waste recycling practice through separation of wastes at source and

Table 6. Item-wise volume of E-waste produced in Bangladesh

Item	Quantity (Million Metric Tons)
Ship Breaking Yards	2.5 per year
Television Sets	0.182 per year
Computers	0.0084 per 30 years
Mobile Phones	0.0006 per 10 years
CFL Bulbs	0.0001 per 6 years
Mercury Bulbs	0.0018 per 10 years
Thermometers	0.0002 per 50 years
Health care services	0.009 per 30 years
Total	2.702 per year

Sources: ESDO, 2012

financial incentives for recycling activities. Primarily, this program was implemented at the household level, but still the production sectors, such as industry and the commercial sector do not implement it. However, Bangladesh also developed 4R (Reduce, Re-use, Recycle and Recovery of waste) practices for solid waste management (Country Presentation Bangladesh, 2010). Some projects working with 4R principles are as follows:

- Source Separation of Waste at Uttara Model Town (Sector 5).
- Community Based Urban Solid Waste Management in Bangladesh.
- Plastic Waste Recycling by Informal Sector.
- Lead Acid Battery Recycling by Formal and Informal Sector.
- Battery Buy Back for Recycling by Rahimafrooz.

E-Waste Disposal Practices

The most general practice of e-waste is reusing without any treatment or processing at domestic level. Most of the old mobile phones, televisions, and computers are dumped into repair shops. If the technicians fail to repair the TV set, Mobile set, or the computer parts, then the user keep them in the repairer's shop. There is also a market for scrap electronic products like, cell phone body, batteries, charger, television, radio, etc. For business, some semiformal and informal sectors are started refurbished and recycled process in Bangladesh and it is emerging. Due to the absence of a formal e-waste processing plant, Bangladesh faces problems of managing them. Among the total generated e-waste, only 20% to 35% are recycled in Dhaka and Chattogram and the rest end up being dumped indiscriminately here and there, which creates hazards to the health of citizens and the environment (Islam & Emon, 2016).

Industrial Waste and Extended Producer Responsibility (EPR)

Generally industrial waste means any useless byproduct produced during industrial processes and activities are considered as industrial waste. The study conducted by Waste Concern and Asian Development Bank (ADB) in 2008 found that seven major industries i.e. tannery, textile, fertilizer, paper and pulp, health care, oil refinery and pesticide, produce three types of wastes including solid waste (26,884 tons/year), wastewater (109.47 million cubic meter/year) and sludge (0.113 million ton/year). On the other hand, in 2025, the estimated volume of industrial waste solid waste, wastewater and sludge would be 53,874 tons/year, 2472.07 million cubic meter/year and 2.81 million ton/year respectively (Amin, 2017). According to (ESDO, 2012) EPR is 'an environment protection strategy that makes the producer responsible for the entire life cycle of the product, especially for taking back, recycling and final disposal of the product'. But this strategy is not followed accordingly in Bangladesh as the e-waste and industrial wastes are thrown along the roadside, and into rivers and other open places indiscriminately and thoughtlessly, which destroys the ecology, and poses risks to human health and environment. Some of the companies in Bangladesh offer to replace old products for a discounted rate, but this offer is not circulated publicly, thus rural people do not receive the information regarding the replacement offer. In order to get the enforcement of the EPR policy, the Department of Environment should take the lead in compelling the industry or factory to dispose of their expired products.

CONSTRUCTION, DEMOLITION AND DISASTER WASTE

Construction and Demolition Waste

The construction sector of Bangladesh consumes a large amount of natural resources that generates huge volumes of waste. The amount of demolition waste is also noteworthy as both construction and demolition (C&D) waste constitutes the largest amount of waste of the total solid wastes. Bangladesh is continuing huge development works and mega projects, such as the Padma Bridge, which are ongoing around the country, as well as lot of construction activities that can be seen everywhere (Chowdhury et al., 2016; Fatemi, 2012). The demolition of existing structure is going on simultaneously. An astounding increase in the level of urbanization has been observed in the major cities of Bangladesh over the last few decades. Almost one third of the municipal population of this country live in Dhaka city alone. Rising need for housing facilities has been noted, which has led to an acceleration of the construction industry. This scenario continues in other cities in Bangladesh.

Research and technical resources related to the construction waste management scenario in Bangladesh is few and far between (Chowdhury et al., 2016). Bangladesh produces 2.24 million tons of construction and demolition waste e.g. bricks, tiles, concrete, plastic, aluminum, steel, timber, and glass every year (Jamal & Sobuz, 2017). In Bangladesh, manual wrecking is the predominantly used method, while mechanical wrecking is scarcely used, and explosive charge demolition is unaccustomed to builders and developers (Fatemi, 2012). Usual waste materials generated from building demolitions are listed in Table 7 along with their source.

Table 7. Demolition waste materials

Waste Material	Source of Generation
Electric Wire	Electric wiring
Plumbing Pipes	Water supply and drainage
Timber	Door and window panel
Plaster & Floor Finish	Building surface
Concrete Piece	Slabs, beams, columns, RCC walls
MS Rods	Slabs, beams, columns, RCC walls
Full Brick	Masonry works, partition walls
Half Brick	Masonry works, partition walls

Source: Modified from Saifullah, Uddin, Alam, & Quazi, 2009

The manual method is the most practiced way of demolition in Bangladesh. It is less expensive but takes time. On the other hand, though the mechanical method of demolition takes less time, it is more costly. Moreover, construction and demolition workers are accustomed to the traditional ways and safety concerns arise with new methods (Saifullah et al., 2009). Proper planning, assessment, selection of consultants and contractors, design, and operation, each step has influence on successful project output and the government should craft guidelines for companies to include a waste management plan in their proposed projects (Chowdhury et al., 2016). Researchers are also concerned about the frequency of development and redevelopment projects in major urban centers, which will only create problems if not managed properly (Fatemi, 2012).

Disaster Waste

Natural hazards are common in Bangladesh and the frequency of disasters, both natural and man-made, are also high. A disaster can create varying quantities and types of waste and debris, depending on the disaster type, intensity, location, resilience, and the preparedness of the affected community. To provide guidance and proposed procedures on how to handle and manage debris that is beyond the capacity of the affected community, supportive debris management guidelines and procedures are required (Ministry of Disaster Management and Relief, 2015). National Debris Management Guidelines (NDMG) has been developed to facilitate robust national debris management in the event of a disaster.

The government of Bangladesh has developed and implemented a disaster management approach focused on three key documents: The Disaster Management Act (2012); the National Plan for Disaster Management (2010-2015); and the Standing Orders on Disaster, which are determining the process for post-disaster management, wherein debris management is included. According to the draft NDMG, the national disaster response coordination group will preview the debris management in the event of a high magnitude disaster.

According to the drafted NDMG guidelines, the type and composition of the debris arising depends on numerous factors. The major types of disaster debris are:

- Concrete, both as cast concrete, reinforced concrete, as well as concrete blocks and elements.
- Bricks largely as masonry bricks from buildings, as well as roofing tiles of same materials.

- Rocks/stones, typically as rough-cut stones used in the walls of buildings.
- Vegetation from damaged and uprooted trees, agricultural crops, bushes, and other green waste.
- Wood from structural timbers, as well as internal paneling and furnishings, hereunder also straw from roofing.
- Asphalt from hard standing areas within the plot of the building or structure as well as roads.
- Soils, mud, and sand from landslides, soil erosions, and movements, as well as excavations in connection with cleanup and demolition works.
- Plaster from coatings of walls, often mainly of gypsum.
- Metals, such as reinforcement bars, structural steel, and metallic structures, like tanks and industrial plant.
- White goods, such as fridges, freezers, cooking ovens, etc.
- Electronic wastes, such as TVs, stereos, computers.
- Furnishings, internal general wastes, which include paper, cardboard, plastics, glass, etc.
- Personal belongings from homes, offices, and industrial sites.
- Cars, boats, and other mobile items where disasters such as flooding can move these items and deposit them nearby or on top of structures, etc.
- Industrial wastes, such as fuel, chemicals, solvents, and other hazardous materials, substances and wastes from damaged industrial and commercial plant.
- Human and animal corpses.

The composition of debris can vary from disaster to disaster, but the option for recyclability is comprehensively discussed in the drafted NDMG guidelines.

IMPACT OF WASTE MANAGEMENT ON GLOBAL WARMING / GHG

Municipal solid waste emits greenhouse gases, which are a matter of global concern, and for Bangladesh this amount was 64.46 kg/cap/year of CO_2 (Waste Concern, 2014). On the other hand, managing wastes also adds to the generation of greenhouse gases and in a scenario analysis study during 2015 found that 420.88 kg CO_2 emitted for managing one ton of waste (Islam, 2017). In 2005, an average household of 4.72 people in Bangladesh emitted 304.25 kg/year of CO_2, whereas the average household of 4.40 people emitted 380.07 kg/year of CO_2 during 2014 (Waste Concern, 2014). This information portrays the increasing trend of CO_2 emission by a single household in Bangladesh.

ECONOMICS OF WASTE MANAGEMENT

The higher the per capita GDP, the larger the waste generation volume. In 1991 and 2014, total waste generation was 6,493 and 23,688 (tons), respectively, while the per capita GDP was USD 220 and USD 1,113 (Amin, 2017). The collection, transportation, treatment activities, and processes needed in waste management require money. For every country, huge amounts of money are needed to finance labor costs for waste collection, purchasing transportation vehicles, constructing treatment plants, and in meeting the operational costs. In Bangladesh, management of wastes, 5% to 20% of the total annual budget is spent (BBS, 2010). In two major cities i.e. Dhaka and Chittagong, the operating costs for MSW management are approximately BDT (Bangladeshi Taka) 626.24 ($7.97)/ton and BDT 439 ($5.59)/ton respectively (Islam, 2016). On the other hand, sustainable waste management saves money, public and environmental health through reduction of noxious gases, global warming, and carbon footprint. The informal sectors save BDT 10,705.5M ($15.29M USD) annually through recycling 4% to 15% solid wastes produced in different urban areas (Enayetullah et al., 2005). Through recycling of 60% of used plastic wastes, Bangladesh saves $600M US that will be spent on virgin materials import (Islam & Emon, 2016). Sustainable management of waste can provide fuel and electricity if it is being composted, anaerobic digested, and thermos chemically processed. These different types of waste treatments demand different cost patterns.

ISSUES AND CHALLENGES IN WASTE MANAGEMENT

Bangladesh is developing at a fast pace with regards to population growth, especially in urban areas and economic advancement that increase the volume of MSW. Besides, rapid industrialization, expansion of construction and demolition work, and over dependence on electronic goods also adds a large amount of wastes in Bangladesh. People do not care and the attitude towards waste is merely 'not in my backyard'. This type of behavior leads to ignorance, and people start throwing garbage into open drains and ditches in the surrounding areas. Industries do the same by dumping waste into adjacent open fields or waterways that also create environmental problems. Changing people's attitudes towards waste is the first challenge in Bangladesh. Reduction in open dumping and throwing away in water bodies and roadsides will come naturally when attitudes can be changed. Existing legal instruments are conversant to some extent, but lack of execution of these rules and regulations encourage people to haphazardly manage waste.

Besides, regarding waste management Bangladesh may face the following challenges:

- To reduce the environment pollution especially air, water, and soil pollution, water-logging, greenhouse gas emission etc. as the unmanaged and untreated open burning and dumping plastic waste, abandoned electronic goods, food wastes, residues from informal recycling of organic wastes easily get mixed into water and soil, and spread bad odors in atmosphere.
- To control the public health hazards like spread of water-borne diseases, communicable diseases, and parasitic diseases through the untreated hazardous organic wastes like fecal sludge, biomedical wastes, and discharge of chemical wastes from industry etc.
- To invest a huge amount of fiscal budget for procuring vehicles, waste containers, construction of landfill sites, and processing plant etc.
- To finance for implementation of 3R or 4R projects and clean development mechanisms.
- To provide sufficient number of manpower for carrying waste management.

These challenges ultimately may affect the achievement of SDGs targets relating to life on land, good health, clean water and sanitation, climate action, sustainable cities and communities, and industry, innovation and infrastructure.

CASE STUDIES

Community Based Approach for SWM

The failure of conventional approach gave birth to community led waste management system. The conventional approach demands households to take responsibility and transport the waste to bins usually located along the roads. These are either permanent or movable bins provided by city authorities. In the next step of the conventional approach, the responsibility of waste transportation is boasted upon city authorities to dispose waste to its final destination at designated disposal site. Low motivations among people along with travel distance from home to roadside bins are the causes of conventional approach's failure. As a result, city dwellers disposed waste in open spaces, around bins, parks and drains.

In the community-based approach, at first the areas of a city are divided into smaller blocks. Usually a block consists of 500 households. A non-motorized vehicle is used to collect waste and transfer it to secondary collection points. The responsibility of the process (home to secondary points) of solid waste transfer is boasted upon community. In the next phase of the process, city authorities take the responsibility of disposing waste to open dumpsites. Implementation of a community-based management of solid waste program was done in several phases with the collaboration of community, city authorities and the collaborator, usually the ward commissioner and NGO representative. It is followed by building community organizations after defining primary collection blocks and finally stating the operation of primary collection system. People in the major cities of Bangladesh accepted this approach, but problems still remain in crowded areas (slums) where narrow roads hinder primary collection. However, in small towns, as well as in sub-urban areas, households are scattered, and primary collectors have to cover a large area to reach the threshold population of 500 households. As a result, the conventional approach is used in these areas, which leads to the related problems of this approach.

Recycling in Bangladesh

The informal sectors in Bangladesh recycled 4% to 15% solid wastes produced in different urban areas, which saved about BDT 10,705.5M ($15.29M USD) annually (Enayetullah et al., 2005). Approximately, 60% of post-use plastic waste was recycled in Bangladesh, which resulted in a saving of $600M US on the import of virgin materials (Islam & Emon, 2016). Moreover, over dependence on chemical fertilizer can be reduced if organic fertilizer is produced and used, and will save BDT 350M per year (Nasrin, 2016). Besides, it can produce energy if the wastes are treated before dumping and landfilling. Additionally, waste can provide fuel and electricity if it is being composted, anaerobic digested, and thermos chemically processed.

Medical Waste Management (MWM)

Health Care Facilities (HCFs) have been found to collect their in-house waste systematically but mostly private organizations are involved in this process. The largest HCF in Bangladesh is Dhaka medical college where in-house waste collection and segregation is in operation along with Holy Family Hospital (private) in Dhaka manages medical waste properly, although every HCFs have limitations that require further improvement. PRISM Bangladesh (a non-government organization) has developed a disposal facility for medical waste treatment and management in most divisional cities of Bangladesh. Financial support is provided by international organizations (CIDA, WSP, WHO, the Embassy of Japan and UNDP) with the collaboration of city authorities. City authorities provide dumpsites near the city limit for the final disposal of medical waste. PRISM Bangladesh takes part in the collection and transportation. Incineration facilities are available in some cities to burn hazardous waste. Other mentionable NGOs, Waste-Concern, and Prodipon are playing important roles in medical waste management. Apart from these NGOs, BASA (Bangladesh Association for Social Development) in Tongi and Shawpno in Bagura are also involved in the management of medical waste.

THE WAY FORWARD

Waste management especially MSW management in Bangladesh is regarded as a neglected issue due to lack of motivation and consciousness of residents, ineffective policy implementation, and funding constraints. In the days ahead, rate solid waste generation will continue to increase as urban dwellers, factory/industry, and consumption of electronic goods continues to grow rapidly. In such a situation, for the sustainable management of solid waste, the following issues need to be addressed:

- Awareness raising programs regarding waste disposal, and the benefits of recycling and reuse should be adopted through mass and social media.
- Budget allocation for solid waste management should be increased.
- Adopt more projects/programs on solid waste management in collaboration with NGOs, donors, and other international organizations.
- The city corporation or municipal authority should establish a definite waste disposal point in every urban community and collect the waste from that point every day.
- For industrial waste and e-waste, the extended producer responsibility (EPR) principle should be strictly enforced in line with the environmental conservation act.
- Establish sewage and industrial waste treatment plants in city or municipal area.

CONCLUSION

The state of solid waste and its management in Bangladesh is one of the major concerns. Though the country has established legitimate framework and guidelines for waste management, the deficiency in execution has directed to an intensification of waste problem driven by the ever-increasing population. A distinguishable risk to the healthiness of the populace always impedes because of the lack of planned disposal services, which leads to poor collection and open dumping. Likewise, unrestrained leachate, mixing biomedical and municipal waste, and the existence of pollutants in wastes stipulate to the growing health hazards. Despite the limitation of resources, the government of Bangladesh are committed to incorporating technologies for waste management. Essential measures, in terms of policies, plans, and programs have been taken into consideration to embrace international initiatives in environmental management. Other stakeholders, such as NGOs and CBOs are also engaged in the country's efforts to ensure proper solid waste management along with industries, environmental activists, researchers, and the public. However, transforming policies and plans into experience poses the principal challenges along with financial constraints and managerial inefficiency. Therefore, the institutional guidelines and determination of the citizens to work on a contemporary platform is needed to ensure sustainability of waste management in Bangladesh.

REFERENCES

Abedin, M. A., & Jahiruddin, M. (2015). Waste generation and management in Bangladesh: An overview. *Asian Journal of Medical and Biological Research*, *1*(1), 114–120. doi:10.3329/ajmbr.v1i1.25507

Ahsan, A., Alamgir, M., Imteaz, M., Daud, N. N., & Islam, R. (2012). Role of NGOs and CBOs in waste management. *Iranian Journal of Public Health*, *41*(6), 27. PMID:23113191

Ahsan, A., Daud, N. N. N., El-Sergany, M. M., Alamgir, M., Shams, S., & Rowshon, M. K. (2014). Assessment of municipal solid waste management system in a developing country. *Chinese Journal of Engineering,* 1–11. Retrieved from /561935 doi:10.1155/2014

Al Mamun, M. R. (2007). *Utilization pattern of biomass for rural energy supply in Bangladesh.* Bangladesh Agricultural University.

Amin, A. N. (2017). *State of the 3Rs in Asia and the Pacific country chapter: The people's republic of Bangladesh.* doi:10.1111/eth.12371

Asian Development Bank. (2018). *Asian Development Bank member fact sheet: Bangladesh.* Retrieved from https://www.adb.org/sites/default/files/publication/27753/ban-2017. pdf

Bangladesh Bureau of Statistics (BBS). (2010). Compendium of environment statistics of Bangladesh 2009. Dhaka: Bangladesh: Bureau of Statistics (BBS), Ministry of Planning, Government of Bangladesh.

Bangladesh Bureau of Statistics (BBS). (2017). *Bangladesh statistics 2017.* Statistics and Informatics Division (SID), Ministry of Planning. Dhaka: Bangladesh. Retrieved from http://bbs.portal.gov.bd/sites/default/files/files/bbs.portal.gov.bd/ page/a1d32f13_8553_ 44f1_92e6_8ff80a4ff82e/Bangladesh%20 %20Statistics-2017.pdf

Bangladesh Bureau of Statistics (BBS). (2018a). *Bangladesh economic review 2018.* Finance Division, Ministry of Finance, Government of the People's Republic of Bangladesh. Retrieved from https://mof.portal.gov.bd/sites/default/files/files/mof.portal.gov.bd/ page/e8bc0eaa_463d_4cf9_b3be_26ab70a32a47/01.%20 BER-2018_Cover%20page.pdf

Bangladesh Bureau of Statistics (BBS). (2018b). Statistical pocket book, Bangladesh 2017. Dhaka: Bangladesh Bureau of Statistics (BBS), Ministry of Planning, Government of Bangladesh.

BBC Monitoring. (2019). *Bangladesh country profile.* Retrieved from https://www.bbc.com/news/world-south-asia-12650940

Bhuiyan, S. H. (2010). A crisis in governance: Urban solid waste management in Bangladesh. *Habitat International, 34*(1), 125–133. Retrieved from . 002 doi:10.1016/j.habitatint.2009.08

BIGD. (2015). *The state of cities: Solid waste management of Dhaka city–towards decentralized governance*. Dhaka, Bangladesh: BRAC Institute of Governance and Development, BRAC University.

Chowdhury, F. H., Raihan, M. T., Islam, G. M. S., & Ramiz, F. (2016). Construction waste management practice: Bangladesh perception. *Proceedings of 3rd International Conference on Advances in Civil Engineering*.

Concern, W. (2014). *Bangladesh waste database 2014*. Retrieved from www.wasteconcern.org

CP Bangladesh, . (2010). *2nd Meeting of the regional 3R forum in Asia*. Kuala Lumpur, Malaysia: Author.

Department of Environment (DoE). (2010). *National 3R strategy for waste management*. Ministry of Environment and Forests, GOB, Dhaka. Retrieved from http://doe.portal.gov.bd/sites/default/files/files/doe.portal.gov.bd/publications/7dc258b2_4501_400d_b066_5e56d84f439f/National_3R_Strategy.pdf

DoE, Waste Concern, & ITN-BUET. (2004). *Country paper, Bangladesh*. In SAARC Workshop on Solid Waste Management, Dhaka, Bangladesh. Retrieved from http://wasteconcern.org/wp-content/uploads/2016/04/SAARC-Country-Paper_Banglaadesh.pdf

Enayetullah, I., Sinha, A. H. M. M., & Khan, S. S. A. (2005). *Urban solid waste management scenario of Bangladesh: Problem and prospects*. Dhaka, Bangladesh: Waste Concern.

Environment and Social Development Organization-ESDO. (2012). *Guidelines for e-waste management in Bangladesh*. Dhaka, Bangladesh: Author.

Fatemi, M. N. (2012). Strategies to reduce construction and demolition (C&D) waste for sustainable building design in Dhaka: Role of architects. *Proceedings of International Seminar on Architecture: Education, Practice and Research*.

Government of Bangladesh (GoB). (1995). *National environmental management action plan final report*. Dhaka, Bangladesh: Author.

Government of Bangladesh (GoB). (1998). *Urban management policy statement*. Dhaka, Bangladesh: Author.

Green, D. W. J. (2010). Technical assistance consultant's report: Managing hazardous wastes. Hong Kong, China: Academic Press.

Halder, P. K., Paul, N., & Beg, M. R. A. (2014). Assessment of biomass energy resources and related technologies practice in Bangladesh. *Renewable & Sustainable Energy Reviews*, *39*, 444–460. doi:10.1016/j.rser.2014.07.071

Hoornweg, D., & Bhada-Tata, P. (2012). *What a waste: a global review of solid waste management* (Vol. 15). World Bank.

Hossen, M. M., Rahman, A. H. M. S., Kabir, A. S., Hasan, M. M. F., & Ahmed, S. (2017). Systematic assessment of the availability and utilization potential of biomass in Bangladesh. *Renewable and Sustainable Energy Reviews*, *67*, 94–105. .2016.09.008 doi:10.1016/j.rser

Islam, K. M. (2016). Municipal solid waste to energy generation in Bangladesh: Possible scenarios to generate renewable electricity in Dhaka and Chittagong City. *Journal of Renewable Energy*.

Islam, K. M. N. (2017). Greenhouse gas footprint and the carbon flow associated with different solid waste management strategy for urban metabolism in Bangladesh. *The Science of the Total Environment, 580*, 755–769. doi:10.1016/j.scitotenv.2016.12.022 PMID:28024747

Islam, M. S., & Emon, M. M. H. (2016). *Current status with facing challenges by entrepreneurs and future prospect of plastic sector in Bangladesh.* Paper presented at the International Plastic Summit, Bangladesh Plastic Goods Manufacturers and Exporters Association (BPGMEA), Dhaka, Bangladesh.

Jamal, M. S., & Sobuz, M. H. R. (2017). A sustainable approach of demolition, reconstruction, reuses and recycling construction materials in Bangladesh. *BAUET Journal, 1*(1). Retrieved from https://bauet.ac.bd/uploads/0.57107300 1505019927Article-9.pdf

Kabir, M. R. (2015). Municipal solid waste management system: A study on Dhaka North and South city corporations. *Journal of Bangladesh Institute of Planners, 8*, 35–48.

Kaza, S., Yao, L., Bhada-Tata, P., & Woerden, F. V. (2018). *What a waste 2.0: A global snapshot of solid waste management to 2050. Urban Development Series*. Washington, DC: World Bank. doi:10.1596/978-1-4648-1329-0

Kormoker, T., Proshad, R., & Khan, M. M. (2017). Analysis of water quality in urban water supply system of Bangladesh. *Journal of Environmental & Analytical Toxicology, 7*(4), 2161–0525. doi:10.4172/2161-0525.1000492

Kumar, S. (2016). *Municipal solid waste management in developing countries.* CRC Press. doi:10.1201/9781315369457

Lundgren, K. (2012). *The global impact of e-waste: addressing the challenge*. International Labor Organization. Retrieved from https://www.ilo.org/wcmsp5/groups/public/---ed_dialogue/---sector/documents/publication/wcms_196105.pdf

Ministry of Disaster Management and Relief. (2015). *Bangladesh national debris management guidelines*. Dhaka, Bangladesh: Author.

Nahar, N. (1998). *Handling, storage, use and disposal of toxic and hazardous materials in Bangladesh.* Bangladesh University of Engineering and Technology.

Nasrin, F. (2016). Waste management in Bangladesh: Current situation and suggestions for action. *International Research Journal of Social Sciences, 5*(10), 36–42. Retrieved from http://www.isca.in/IJSS/Archive/v5/i10/7.ISCA-IRJSS-2016-113.pdf

Rahman, M. A. (2017). *E-waste management: A study on legal framework and institutional preparedness in Bangladesh.* North South University. Retrieved from http://www.northsouth. edu/newassets/files/ppg-research/PPG_5th_Batch/14._Anis_E-waste_Management. pdf

Rode, S. (2012). E-waste management in Mumbai metropolitan region: Constraints and opportunities. *Theoretical and Empirical Researches in Urban Management, 7*(2), 89-103. Retrieved from http://um.ase.ro/no72/5.pdf

Saifullah, I., Uddin, S. M. K., Alam, A., & Quazi, B. (2009). Practice of demolition waste management. *Proceedings of the International Conference on Solid Waste Management - WasteSafe*. .192510.13140/RG.2.1.4222

Shams, S., Sahu, J. N., Rahman, S. S., & Ahsan, A. (2017). Sustainable waste management policy in Bangladesh for reduction of greenhouse gases. *Sustainable Cities and Society*, *33*, 18–26. doi:10.1016/j.scs.2017.05.008

Unnisa, S. A., & Rav, S. B. (Eds.). (2012). *Sustainable solid waste management*. CRC Press. doi:10.1201/b13116

Chapter 3
Solid Waste Management in Cambodia

Ran Yagasa
Institute for Global Environmental Strategies, Japan

Rithy Uch
https://orcid.org/0000-0001-5780-8914
Cambodian Education and Waste Management Organization, Cambodia

Phalla Sam
Cambodian Education and Waste Management Organization, Cambodia

ABSTRACT

Driven by economic development, population growth, change in life style, and consumption patterns, Cambodia is faced with equally rapid increase of solid waste, with MSW disposal amount attaining 1,709,379 tons/year in 2018. Various policy instruments and legislations have been developed over the years in response to this long-existing crisis, which effectively translated into tangible improvements on the ground. But municipalities continue to suffer from weak waste management system including collection, transportation, treatment, and disposal. Resource recovery almost entirely depends on informal sector while efforts for reduce and reuse are still weak, while presenting vast opportunity if effective policies are employed and implemented. The ongoing governance reform involving functional decentralization of waste management service is being implemented although at a slow pace.

INTRODUCTION

Rapid economic development, population growth, and changes in lifestyle, consumption patterns and industrial structure are all intensifying the waste management problem in Cambodia. Increasing waste generation with diversity of waste including hazardous and toxic waste is causing negative impacts on the local environment such as air, water, and land, as well as impacting on human health. On the other hand, local governments struggle to improve the current waste management system to cope with the

DOI: 10.4018/978-1-7998-0198-6.ch003

generated amount of waste as well as different types of waste to minimize its environmental and public health consequences. Following decades of struggle, the country's waste management system is currently in the midst of a transformation with the decentralization of service responsibility from province to municipalities (district). The full-fledged results of this tectonic change are yet to be seen, but it is expected to contribute to the development of an Integrated Waste Management System and will hopefully result in improved service delivery and minimized risk to society.

WASTE DEFINITION AND CLASSIFICATION

Waste Definition

Sub-decrees No. 36 on Solid Waste Management (1999) and No.113 on Waste Management (2015) provide legal definitions of waste in general, and constitute the key waste management legislation in the modern legal system of Cambodia while also supported by many subject-specific laws and regulations.

Sub-decree No. 36 defines "Solid waste" as comprising all the waste arising from human activities, including animal waste that is discarded as useless or unwanted, while it distinguishes "solid waste" and "garbage" as follows:

1. Solid waste refers to hard objects, hard substances, products or refuse which are useless, disposed of, are intended to be disposed of, or required to be disposed of.
2. Garbage is the part of solid waste which does not contain toxins or hazardous substances, and is discarded from dwellings, public buildings, factories, markets, hotels, business buildings, restaurants, transport facilities, recreation sites, etc.

Sub-Decree No. 113 also defines "garbage", "Municipal Solid Waste (referred to as 'solid waste in towns' in the text)", "industrial waste" and "hazardous waste" in its annex as follows:

1. Garbage: referring to tools, equipment, products that were left or created by actions or from the everyday livelihoods of humans, not including toxic substances or hazardous waste.
2. Solid waste in towns: referring to solid waste that was left or created by business activity or services, not including toxic substances or hazardous waste.
3. Industrial waste: referring to solid waste that was left or created by productivity of factory or enterprise, not including toxic substances or hazardous waste.
4. Hazardous waste: referring to solid substances, liquid, gas, radioactive substances, explosive substances, flammable substances, disease-transmitting substances, or substances that cause burning and decay, antioxidants, substances that cause poisoning, cancer, or other chemical substances that cause danger to humans, animals or damage to plants, public property, resorts, public buildings, education establishments, business activities, services, crafts, factories, agricultural activities or mining business.

Waste Classification

Figure 1 illustrates the waste classification based on the key Sub-decrees above and other relevant national waste management legislations in Cambodia as of March 2019. However, literature has also discussed slightly different classifications or has pointed out the necessity to develop clear definitions and classifications (Sethy et al., 2014).

WASTE GENERATION AND COMPOSITION

Country Overview

Home to 16 million people, Cambodia is located in the central part of Mekong region with a territory of 181,035 sq.km, bordering Thailand in the west, Viet Nam in the east, Laos in the North and Gulf of Thailand in the south.

The economic development over the past two decades has pushed Cambodia into lower middle income status with its GNI per capita reaching USD1,230 in 2017, while reducing its poverty rate from 47.8% in 2007 to 13.4% in 2014 (World Bank, 2019). The population of Cambodia has been rapidly increasing in the past few decades, from 8,973,372 in 1990 to 16,005,373 in 2017, with the growth rate hovering between 1.5 and 2.0 in the past 10 years. With continuous internal migration from rural to urban areas, the urbanization rate has also increased by approximately 10 percentage points during this period, attaining 23% in 2017 (World Bank, 2019).

Figure 1. Waste classification and waste management legislations of Cambodia (Adapted from Kol Mardi et al., 2018)

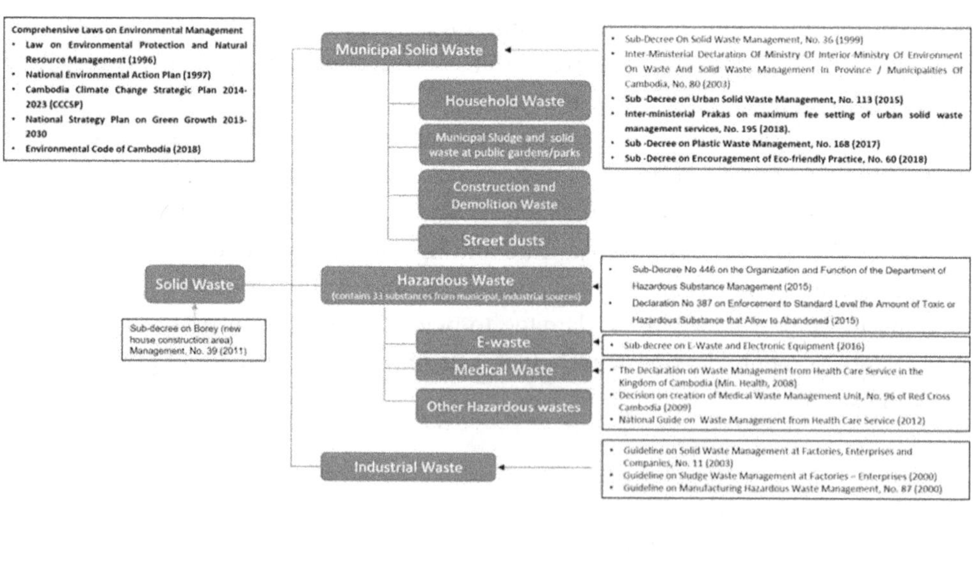

Figure 2. Map of Cambodia (CartoGIS Services, 2018)

Waste Generation

Driven by economic development, population growth, migration, coupled with changes in lifestyle and industrial structure the amount of waste generation, as well as per-capita generation, is considered to have increased, while shifting the waste composition particularly in urban areas. *Figure 3* presents the trend of total MSW generation and MSW generation per capita from 2004 to 2015, which both show dramatic increases over the past 10 years. Per-capita generation varies widely, with a notable gap between urban population centers such as Phnom Penh, Preah Sihanouk, Siem Reap, and the rural provinces such as Koh Kong, Ratanakiri, and Mondulkili thereby reflecting the differences in local economy, lifestyles and consumption patterns (*Table 1*).

The total generation is also projected to increase. Based on official MoE data in 2014, a 2018 World Bank report estimates that national waste generation will rise to 1,702,523 tons and 2,641,058 tons in 2030 and 2050 respectively, driven by the aforementioned drivers (Kaza et al., 2018).

The amount of waste collected and disposed of, which is directly measured at the final disposal sites, also increased from 317,550 tons/year in 2004 to 1,177,655 tons/year in 2015, and more recently reached 1,709,379 tons/year in 2018 (Ministry of Environment Cambodia, 2019b), with a 10% annual increase on average in the past decade. Records of disposal amount at final disposal sites in major cities also present a gradual increase. For instance, in Phnom Penh the amount almost doubled in the last decade from 343,657 tons/year in 2007 to 808,530 tons/year in 2017. However, precautions are required when interpreting data since this figure is also considered to be positively influenced by improvements to the collection operations (Mardi et al., 2018; Yagasa et al., 2018).

Figure 3. Total MSW generation and MSW generation per capita from 2004 to 2015
(Source: Liu et al., 2018)

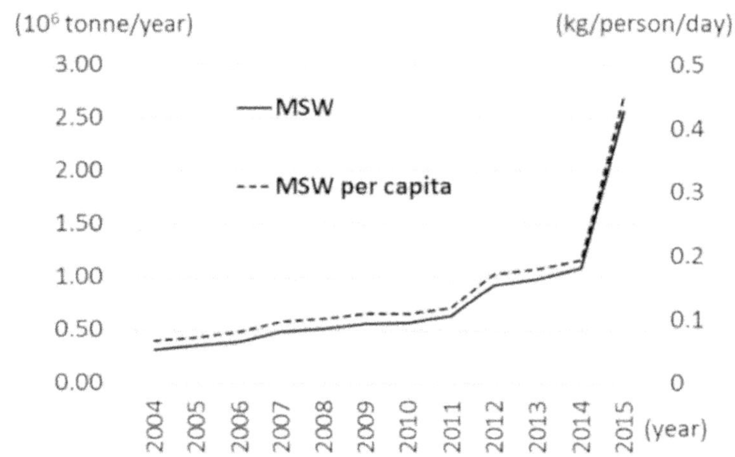

Waste Composition

Although the government has made little systematic effort to produce data and most data is related to Phnom Penh, increasing efforts have been paid in the past decade to generate waste composition data based on stand-alone projects. *Table 2* provides a summary of recent waste composition studies conducted in Phnom Penh over the past decade. The most recent conducted by Seng (2016) reporting that city's MSW comprises 51.9% organic kitchen waste, 19.5% plastic, 1.4% PET, 9.9% paper, 2.3% glass, also containing grass and wood (2.3%), textile (2.1%), glass (1.6%), cans (0.7%), rubber and leather (0.2%), ceramic (0.5%), and others (4.1%). Household sources tend to have higher proportion of kitchen waste with 57.4% compared to 39.8% for non-household sources, while higher rates were reported for paper, plastic and inorganic wastes for non-household sources in general (Seng, 2016). Such trend was also confirmed by Yim (Yim et al., 2013; Yim et al., 2014). The table also shows a slight decline in organic waste and an equally moderate increase in plastic waste between 2005 and 2016. As an attempt to produce a substitute for national figures, Sang-Arun (2012) also reported the average composition of waste in four major cities in the country – Phnom Penh, Battambang, Siem Reap, and Kampong Cham – which comprised of 64% organic, 4% paper, 10% plastic, 2% metals, 2% textiles, 3% glass, 6% wood and dry matter, and 9% others, suggesting a generally high ratio of organic waste across the country (Sang-Arun et al., 2012), which has significant implications on waste management planning.

WASTE MANAGEMENT, 3R PRACTICES AND TECHNOLOGIES

Waste Collection and Transportation

Characterized by uneven and problematic implementation, waste collection and transportation are two of the most visible public health, environmental, social and political issues related to waste management in Cambodia. Without financial resources, municipalities across Cambodia have traditionally, and are

Table 1. Waste generation, per-capita waste generation, waste collection and waste recycling in Cambodia in 2015

No.	Name of Province	Estimated Waste Generation	Per-capita Waste Generation	Waste Collection/Disposal Amount		Recyclable Waste Collected by (junkshops)	
		(t/year)	(Kg/Cap./Day)	Amount (t/year)	Rate (%)	Amount (t/year)	Rate (%)
1	Phnom Penh Capital	839,500	0.82	671,600	80.0%	50,370	6.0%
2	Kandal Province	205,495	0.45	35,040	17.1%	18,980	9.2%
3	Takeo Province	138,010	0.45	3,093	2.2%	16,359	11.9%
4	Kep Province	9,125	0.65	4,380	48.0%	1,825	20.0%
5	Kampot Province	120,450	0.51	15,513	12.9%	78,840	65.5%
6	Preah Sihanouk Province	63,989	0.85	41,975	65.6%	438	0.7%
7	Koh Kong Province	12,049	0.30	9,553	79.3%	1,905	15.8%
8	Kampong Speu Province	116,425	0.40	7,300	6.3%	3,628	3.1%
9	Kampong Chhanang Province	88,330	0.45	19,203	21.7%	18,633	21.1%
10	Pursat Province	73,109	0.45	2,318	3.2%	164	0.2%
11	Battambang Province	36,829	0.66	32,907	89.4%	1,732	4.7%
12	Paylin Province	10,950	0.43	6,570	60.0%	4,745	43.3%
13	Buteay Meanchey Province	91,615	0.75	39,603	43.2%	22,630	24.7%
14	Oudor Meanchey Province	36,698	0.45	1,768	4.8%	0	0.0%
15	Preach Vihear Province	4,380	0.55	4,015	91.7%	183	4.2%
16	Siem Reap Province	319,740	0.84	112,913	35.3%	13,162	4.1%
17	Kampong Thom Province	115,705	0.50	9,125	7.9%	33,580	29.0%
18	Kampong Cham Province	308,060	0.74	37,230	12.1%	12,410	4.0%
19	Tboung Khmom Province	110,350	0.70	6,315	5.7%	1,168	1.1%
20	Kratie Province	40,551	0.45	16,973	41.9%	17,476	43.1%
21	Steung Treng Province	12,410	0.30	8,760	70.6%	0	0.0%
22	Ratanakiri Province	22,254	0.35	7,300	32.8%	1,570	7.1%
23	Mondulkiri Province	9,223	0.35	2,300	24.9%	374	4.1%
24	Svay Rieng Province	110,918	0.50	25,349	22.9%	2,442	2.2%
25	Prey Veng Province	155,855	0.45	60,955	39.1%	21,900	14.1%
Total / Average		3,052,021	0.534	2,571,240	36.7%	324,512.70	13.6%

Source: (Ministry of Environment Cambodia, 2019b)

predominantly relying on the private sector for MSW collection service. Naturally, the business feasibility has determined whether municipalities can successfully invite private contractor to deliver collection services, creating a wide gap between densely populated economic centers with large potential revenue base, and the rest of territory where population and economic activities are less concentrated. As a result, in large cities such as the capital city, Phnom Penh, as well as Battambang and Seam Reap, waste collection services are outsourced to major private waste management companies such as CINTRI Cambodia

Table 2. Recent waste composition studies in Phnom Penh

Literature	JICA, 2005	Mongtoeun, 2013	Mongtoeun, 2014	Seng, 2016	Seng, 2016	Seng, 2016
Source	Household	Household	Commercial	Household	Commercial	Household and Commercial
Sampling Year	2003	2012	2013	n.a.	n.a.	n.a.
Organic Waste (Food waste)	63.3	66.27	51.34	57.4	39.8	51.9
Paper	6.4	4.55	14.84	6	18.6	9.9
Plastic	15.5	9.81	17.65	18.1	27	20.9
PET	-		1.6	-	-	-
Glass	1.2	1.4	4.2	1.4	2	1.6
Metal	0.6		1.04	1.1	1.2	1
Grass and wood (Garden waste)	6.8	0.77	4.57	2.5	1.8	2.3
Textile	2.5	0.76	1.15	2.6	1.2	2.1
Rubber, leather	0.1	0.38	0.23	0.2	0.2	0.2
Ceramics, stones, soil	1.5		1.4	0.6	0.3	0.5
Electronics, batteries	-		0.15	-	-	-
Others	2.1	17.41	1.96	10.2	8.1	9.5

Source: Developed by authors based on (KOKUSAI KOGYO, 2005; K. Seng, 2015; Yim et al., 2013, 2014)

Co. Ltd. (CINTRI) and GAEA Waste Management (GAEA) through service agreements, typically monitored by the municipal waste management division with an assigned mandate and corresponding budget. However, many areas, particularly rural ones, do not have access to such services. The estimated waste collection rates in 25 provinces across Cambodia in 2015 are provided in *Table 1*, suggesting an implementation gap with only 6 out of 25 provinces enjoying collection rate of above 50 percent.

According to the MoE database on the modality of waste collection service provision in 185 municipalities in the country, 53% of respondents relied on market contractors and 23% relied on private waste management companies, while 24% reported an absence of service providers. No responses were confirmed for direct service provision by local administrations (Ministry of Environment Cambodia, 2019b). Market waste – one of the major waste generation sources in many cities – is often collected by separate waste collection companies under an independent contract signed by the Ministry of Economy and Finance. In many municipalities without a household waste collection, such market waste collections are the only services offered.

In large cities, curbside collection is implemented with various types of both open and closed waste vehicles (typically trucks but also compactors in limited major cities), sometimes combined with small-scale carts which can be used to access narrow streets for primary collection. Street sweeping and cleansing of public areas are also implemented in large cities.

The areas where a waste collection service is provided often come up against challenges to service quality in terms of collection rate, service coverage, and inconsistent and inefficient service operation that often stems from road congestion and use of outdated equipment. Untimely and incomplete waste collection often result in uncollected waste piles on the street. This situation is a frequent source of complaints by residents (Singh et al., 2018; Yagasa et al., 2018).

In areas without a waste collection service, inappropriate waste management practices are commonly practiced, such as backyard burning, burying and/or illegal disposal in public spaces such as streets, vacant land, and canals and water bodies. Littering is widely observed elsewhere, resulting in contamination of soil, air and water and posing public health risks. In urban areas, canals and drainage systems attract waste piles from littering, thereby accumulating solid waste in the structure, and resulting in lower drainage capability and the risk of flood in high precipitation season (Urban Voice Cambodia, 2018). In 2015, a JICA study team investigated the weight and composition of the waste disposed of into Phnom Penh's two major drainage channels and found that 2,296.2 kg/day of MSW (consisting of plastics, organic waste, and plastic containing mud), equivalent to approximately 0.1% of the MSW transported to the city's final disposal site in Dangkor (CTI Engineering International, Nippon Koei, & Kitakyushu Water Service, 2016).

Intermediate Treatment

Intermediate treatment processes such as incineration, aerobic digestion, and anaerobic digestion for the purpose of reducing the volume of incoming waste to final disposal are almost absent in Cambodia, except for the manual extraction of recyclable materials at the disposal site or during earlier stages. Open burning of waste in public spaces or backyards is still commonly observable, resulting in emissions of GHGs and SLCPs.

Final Disposal

In the absence of intermediate treatment solutions, open landfill is the most common waste disposal method in the country. Some 185 cities and districts have a final disposal site, 79 do not have access to a landfill, while 12 have reserved land to be organized as a landfill. This suggests that even a final disposal site is not a universally accessible solution, forcing many to rely on irregular open disposal and burning. MoE (2019) reports that 1,709,379 tons of MSW were disposed of in 2018 in 106 final disposal sites across the country, all of which are operated as open dump sites[1]. The amount of waste disposed nationally tripled in the past decade from 518,053 tons/year in 2008, to 1,089,429 tons/year in 2014, and to 1,709,379 tons/year in 2018 (Ministry of Environment Cambodia, 2019b). The increasing trend of generated waste and weak efforts on the 3Rs are putting pressure on final disposal sites across the country. Moreover, preparation of land for final disposal is prohibitively expensive and residential agreements are difficult to forge. For instance, Phnom Penh's Dangkor final disposal site, which was designed to operate for 25 years at the time of opening, is approaching its maximum capacity at the time of writing in 2019, with the increase in incoming waste outpacing the initial plan.

The absence of sanitary landfill management practice and infrastructure is contributing to the deterioration of the soil, water and air in the regional environment. For instance, Seingheng et al. (2018) analyzed leachate and sediment samples from the Dangkor final disposal site in Phnom Penh and confirmed that the biochemical parameters of fresh leachate exceeded the effluent standards, posing a contamination

Figure 4. Transition of waste disposal amount in Cambodia from 2004 to 2018 (Ministry of Environment Cambodia, 2019b)

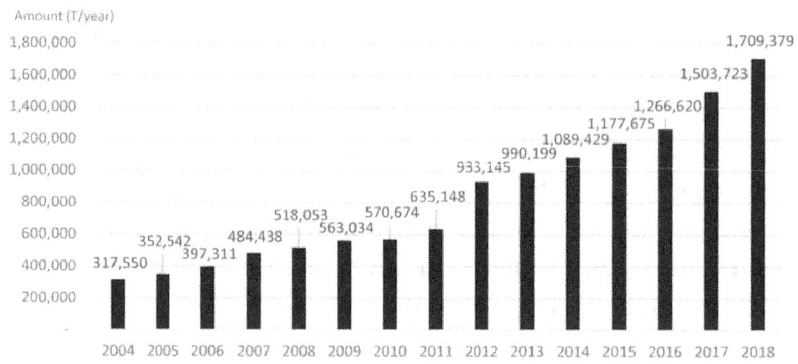

risk to surface water and ground water contamination (Seingheng et al., 2018). In addition, the absence of diversion of organic waste and frequent landfill fires induced both accidentally and intentionally are generating GHG and SLCP such as methane and black carbon, influencing global and regional climate, air quality, and human health.

Reduce, Reuse, Recycle

Overview

Information is not widely documented on waste reduction and reuse in the upper stream of the product-waste value chain, namely the production and consumption stage. However, existing data on recycling suggest the existence of strong, if not adequate, material recovery routes for inorganic waste driven by the informal sector and, in comparison, underutilized organic waste.

Source Segregation

The existing legislations and policies have placed emphasis on source segregation as a key strategy to promote the 3Rs. However, it is widely considered that source segregation of MSW is not widely practiced except for segregation of marketable recyclables, resulting in diverse waste materials being disposed of as mixed waste irrelevant of recyclability or potential hazardousness. (Bunrith et al., 2012; Ministry of Environment, 2019; PPCA, 2015; Yim et al., 2014). This is largely due to challenges of engaging general public: for instance, segregation rules have either not been developed or not well informed to the residents, failing to translate the law into operation, but also stems to a systemic problem where absence of treatment process in lower stream of the waste management chain do not allow for meaningful segregation.

Despite the absence of segregation rules, recyclable inorganic waste materials such as glass, metal, and PET are well segregated driven by economic interests with to varying degrees. Recent evidence suggests that recyclables are well segregated by households (Mongtoeun et al., 2019; Seng et al., 2013) whereas segregation is not widely practiced among commercial institutions (Yim et al., 2014) despite relatively homogenous waste composition and the high potential for resource recovery and utilization.

Segregation of organic waste is not common among households in Phnom Penh, except for households and commercial institutions that sell food waste to pig farmers as animal feed, which is practiced on a limited scale (Bunrith et al., 2012). However, studies have also found that waste generators are willing to participate in segregation if such programs are institutionalized (Bunrith et al., 2012; Yim et al., 2013), indicating a strong potential for promoting material recovery targeting both organic and inorganic waste thereby reducing the amount of waste to be bought to the final disposal site.

Recycling of Inorganic Waste

Recycling of inorganic waste, including cans, glass bottles, paper, plastics are almost entirely handled by the unregulated informal sector. In Cambodia, there is no established recycling value chain comprising of recyclers for reprocessing collected recyclables into materials, and manufacturers that utilize the recycled materials in the production process. For this reason, activities observed in Cambodia are often limited to collection with minimum processing such as shredding and bailing, whereas materials are exported as secondary resources to neighboring countries such as Viet Nam and Thailand for ultimate recycling.

According to the MoE dataset, the amount of recyclable waste purchased by 445 junkshops operating across the country in 2018 were: paper – 36,829 tons; plastic – 23,886 tons; glass – 7,269 tons; aluminum – 26,664 tons; and iron – 105,770 tons (*Figure 5*). In the same year, 8,793.31 tons of plastics were exported under the authorization of MoE, while glass (470.1 tons), aluminum (1,549.86 tons), iron (200 tons), copper (14.36 tons), and other mixed recyclables (180.02 tons) were also exported under the authorization scheme (Figure 5). While the exact recycling rate is unknown, MoE estimates that more than 90% of metal, and more than 70% of plastic generated as waste are recovered whereas the rate for paper, construction materials and e-waste remains around 50-60% (Ministry of Environment, 2019). The amount of exported recyclable waste constitutes only a fraction of the amount purchased by junkshops. Apart from the issue of data accuracy/reliability, further investigation is required to explain the gap, which may be partially explained by removal of the unmarketable component from purchased material

Figure 5. Amount of recyclable waste purchased by junkshops for recycling (Ton/year) and number of junkshops from 2010 to 2016 (Adapted from Ministry of Environment Cambodia, 2019b)

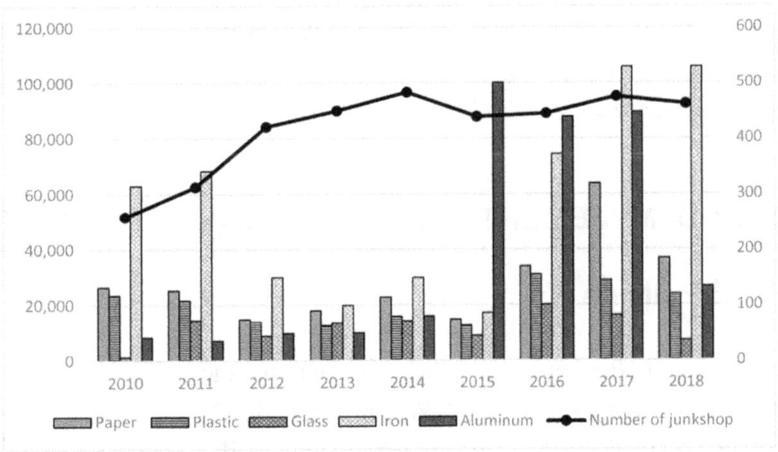

Figure 6. Amount of recyclable waste exported under the due process of MoE (ton/year) (Adapted from Ministry of Environment Cambodia, 2019b)

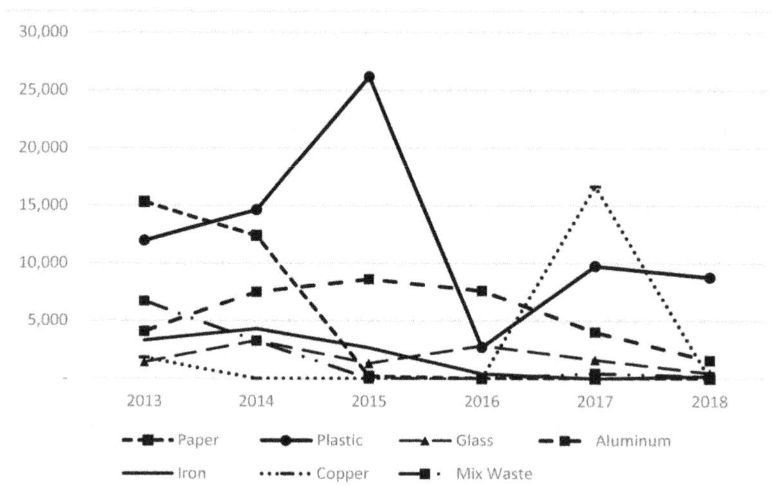

in pre-processing stage, existence of exporting route not captured by the authorization process, and emergence of domestic material recyclers. Also, the annual trend is considered to be strongly influenced by material prices driven by market demand.

Recycling of Organic Waste

On the other hand, organic waste is still not fully utilized as a resource despite a generally large proportion of organic components in typical waste composition across Cambodia. Cases of organic waste utilization are often reported while the precise scale is less documented. Utilization in animal feeding has traditionally been widely practiced, although statistics are lacking to assess its impact, while Seng (2012) reports a decline in this waste stream due to "marketable animal feed, difficulty of food waste transport and the speed of the animal production".

Industrial scale composting is also observed in large cities such as in Phnom Penh and Battambang operated by both city government and NGOs long active in the waste sector such as Cambodian Education and Waste Management Organization (COMPED) and Community Sanitation and Recycling Organization (CSARO) (COMPED, 2018; CSARO, 2018).

DEVELOPMENT AND IMPLEMENTATION OF NATIONAL POLICIES

National and Local Policies

Cambodia developed National 3R Strategy in 2008 with the assistance of UNEP. Following an attempted to develop a "National Integrated Solid Waste Management Strategy" as an internal initiative within MoE which did not come to fruition, MoE has been working with UNEP and CCET to develop a National Waste Management Strategy and Action Plan since 2016 as an attempt to upgrade the preceding national policies on waste management and the 3Rs (Ministry of Environment, 2019; Sethy, 2017). A few instances of local policy development which translate the national policies into respective localities can also be observed. Battambang city has developed its strategy with the support of Climate and Clean Air Coalition (CCAC) under its Municipal Solid Waste Initiative, which aimed at reducing GHG and SLCP from the waste sector. Under the strong leadership of the city governor at that time, some of the actions outlined by the strategy also advanced into implementation (Yagasa et al., 2018). Phnom Penh also developed its waste management strategy with the support of UNEP and CCET, which was published in October 2018 (Kol Mardi et al., 2018).

Decentralization Reform

Cambodia's waste management is currently in the midst of a transformation with the decentralization of service responsibility from province to municipalities (Khan/district). In 2015, the Inter-ministerial *prakas*, or ordinance, on the Usage of Environmental Sanitation Service Fund was jointly issued by MoE, Ministry of Interior (MoI), and Ministry of Economy and Finance (MEF), which was intended to provide municipalities with financial resources for improving waste management with the fund of USD 5 million subsequently prepared. Sub-decree on Urban Solid Waste Management (No. 113) was issued later in the same year which underlined the functional transfer from province to municipalities, clarifying the specific roles and responsibilities of relevant institutions and societal members.

The responsibilities assigned to municipalities under the Sub-decree are:

- Prepare management plans and annual action and budget plans for waste management.
- Prepare and implement any measures for waste management through legal instruments.
- Establish, control and manage cleaning, collecting and transporting services and waste management services within its territorial jurisdiction.
- Advise citizens on environmental sanitation and waste management services.
- Promote education and information dissemination on waste management, while advancing source separation, 3Rs, and use of recycled products (The Royal Government of Cambodia, 2015).

The full-fledged results of this governance reform are yet to be seen; municipal administrations will require access to adequate fiscal and human resources as well as administrative support to fulfill their newly assigned responsibility, and to design and implement their waste management strategy for orchestrating an improved service delivery.

Analyzing the inter-governmental relationship between provincial and municipal administrations, the result of a case study by Mun (2016) found a few instances where municipal plans were intervened by provincial administrations, and concluded that while the functional transfer is being implemented without major challenges, the expected efficiency gains of the functional transfer may be "negated if the centralizing tendency does not simultaneously subside" (Mun, 2016). A survey report by the National League of Local Councils also found that municipalities faced difficulties in mobilizing support from the provincial administration, Department of Economy and Finance (DEF) and MEF (Muny, 2016).

UTILIZATION OF WASTE FROM AGRICULTURE AND ANIMAL HUSBANDRY

Utilization of Agricultural Biomass

COMPED (2012) estimates the amount of agricultural biomass generated from major agricultural production in each province in Cambodia, namely, rice (husk and straw), sugar cane (bagasse and top), corn (stalk, cob, and husk), groundnut (husk, shell, and straw), and coconut (frond, husk, and shell) (*Figure 7*). As the graph suggests, there is a regional variation of agricultural waste reflecting production trends, with rice as a majority product. Although there is a significant data gap on post-harvest wastage of agricultural products, MoE estimates that approximately 5-10 of the crops and agricultural produce are being wasted before reaching consumers (Ministry of Environment, 2019).

Utilization of Waste from Animal Husbandry

Management of waste from animal husbandry presents another environmental challenge. Advances in hygiene control and production techniques in cattle, swine and poultry farming are resulting in an increase of animal waste from the animal production sector, posing public health and climate challenges. According to MAFF statistics, production of cow, swine and poultry all increased significantly between 2011 and 2015 by 173% (from 4,770 in 2011 to 13,000 in 2015), 153.6% (from 164,000 in 2011 to

Figure 7. Estimation of agricultural biomass waste generated in Cambodia (Adapted from COMPED, 2012)

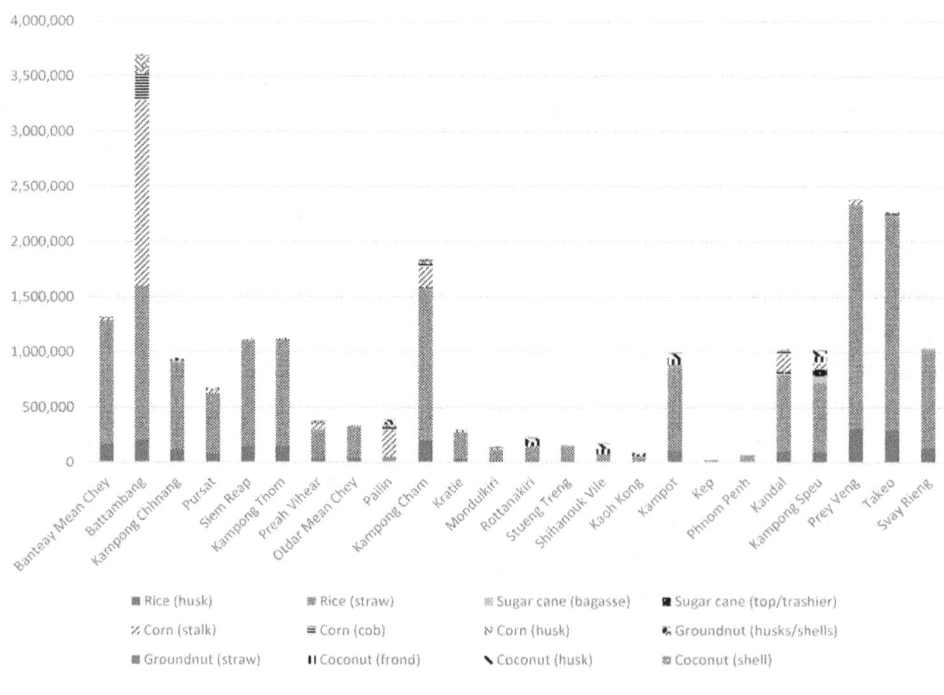

416,000 in 2015), and 69.5% (from 1.77 million in 2011 to 3 million in 2015) respectively. Family-scale production also slightly increased for swine and poultry with a growth rate of 12% (from approx. 2.09 million to 2.35 million) and 18.2% (from 22 million to 26 million) respectively during the same period, while cattle production decreased by 14.7% (from 3.4 million to 2.9 million).

With the aim of providing a decentralized sustainable solution, an anaerobic digestion system on all scales is currently being promoted by the National Bio-digester Program (NBP) under the initiative of the Ministry of Agriculture, Forestry and Fisheries (MAFF) (MAFF, 2016), which started in 2006. The program aims to advance electrification of rural farms, provide a low-cost treatment solution to generated animal waste, and mitigate GHG production. About 25,000 AD systems had been constructed by 2016, and 90% of these are still in operation (MAFF, 2016).

MARINE AND COASTAL PLASTICS

Waste Generation

Management of plastic waste has gained wider public attention in Cambodia over the past several years, in response to ever increasing plastic products. Reflecting the dramatic increase of single-use plastic bags in the country, the policy discourse and government response have focused on this area but a comprehensive policy on plastic products and waste covering various types and management stages – from production, consumption, and post-consumption stage – has not materialized.

ACRA Foundation (2015) reports that Phnom Penh consumes 10 million plastic bags every day, and an average urban Cambodian use more than 2,000 plastic bags every year (Fondazione ACRA, 2015). According to MoE, plastic bags constitute 17.3, 20.7, and 34% of waste in Phnom Penh, Siem Reap and Preah Sihanouk province respectively (Cambodia Tribune, 2018).

Legislation

In October 2017, the government of Cambodia issued Sub-decree No. 168 GNKR.BK on Management of Plastic Bags which intends to reduce waste plastic bags by directly influencing the importation, production and use of single-use plastic bags in terms of quantity and quality, through clarifying the duties of relevant authorities and waste generators, general measures in managing plastic bags, as well as penalties for non-compliance. Production and importation of biodegradable and bioplastics are also encouraged though financial incentives, and citizens are encouraged to reduce the usage of plastic bags by replacing them with other more nature-friendly alternatives (Ministry of Environment Cambodia, 2018).

Initiatives

In addition, examples of 3Rs targeting plastic wastes can also be observed:

- The NGO "Plastic Free Cambodia" has been conducting consumer education programs since its foundation in Siem Reap in 2015, with a strong focus on plastic waste reduction[2].
- Since early 2018, environmental NGOs – OEOO, Thuringian-Cambodian Society e,V., and COMPED initiated a joint project comprising of clean-up operations and environmental educa-

tion in Battambang, particularly focusing on plastic waste. A specially designed boat was donated by German plastics manufacturing companies for plastic waste clean-up operations on the local river. The project is also intended to raise awareness through wider media exposure together with environmental education programs[3].

- A Japanese plastic recycler GOMI RECYCLE 110 CO., LTD is launching a recycling plant in Svey Rien where mixed plastic waste from municipal sources is manufactured into high quality construction material and furniture for the domestic market[4].

INDUSTRIAL WASTE MANAGEMENT

Waste Generation

While cumulative composition data for industrial waste produced in Cambodia is not available, Sarom Industry reports waste textile (60%), sludge (15%), and leather (10%) make up the top three waste categories, with plastic (5%), paper (5%) and others (5%) equally sharing the rest, reflecting the city's industrial structure which consists of a major textile industry (Ministry of Environment Cambodia, 2019b).

Legislation

Management of industrial waste, defined as non-hazardous solid waste generated from production processes, is subject to the national guidelines on Solid Waste Management at factories, enterprises and companies (enacted in 2003) and on Sludge Waste management at Factories and Enterprises (enacted in 2000). A national strategy for addressing industrial waste does not exist, and there are also no measures to date for reducing waste in the upper-stream through improving resource efficiency in production processes. Green and social accounting is not widely adopted by companies and industries (Ministry of Environment, 2019).

Collection, Transportation, and Disposal

Industrial waste is collected, transported and disposed of on a contract-basis where private industrial waste management companies are active, such as Phnom Penh, Kandal, Kampong Speu and Takeo. Most of the recyclable and marketable resource materials are extracted by the waste generators for sale in the informal economy while residuals are disposed of through paid collection services. Despite the fine scheme introduced by Sub-decree No. 113 which bans the mixture of industrial, medical and hazardous waste with those from municipal sources, mixed waste is still common practice in many provinces due to absence of industrial waste collection service.

Controlled and uncontrolled dumping to final disposal sites owned by private waste management companies is the major disposal method. Accordingly MoE, the total disposal amount has slightly declined in 10 years from 201,078 tons/year in 2006 to 150,695 tons/year in 2016, experiencing a major decline in 2010 and 2011 (see *Figure 8*), which is considered to be the result of a decline in industrial production during the global economic recession (Ministry of Environment Cambodia, 2019b).

Figure 8. Trend of industrial waste disposed at industrial landfill sites in Cambodia (Adapted from Ministry of Environment Cambodia, 2019b)

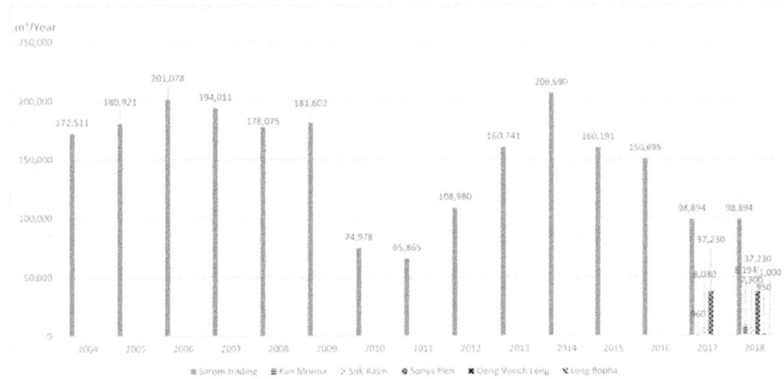

Sorom Trading Co., Ltd. has long been the sole service provider since its operation was approved in 2002 serving some 390 factories in Phnom Penh, Kandal, Kampong Speu, Sihanouk Vile and Takeo provinces. However, a number of industrial waste companies have started to emerge in the past two years, while the national total disposal amount stayed at the same level as in 2016 (145,114 tons/year in 2017 and 153,568 tons/year in 2018), suggesting new entrants are competing for a larger share of the same pie without extending services elsewhere.

Although limited in number, intermediate treatment processes are also observed for the management of waste from industrial sources. For instance, six incinerators are in operation as of 2015 treating 3,525 tons/year, five of which are equipped with WtE function and accepting waste from garment industry – one of Cambodia's major exporting industries (Liu et al., 2018).

CONSTRUCTION AND DEMOLITION WASTE

Waste Generation

Strong economic development is considered to be behind the substantial amount of construction and demolition (C&D) waste generated in Cambodia, particularly in urban areas under the construction boom. The most recent national waste projection by UNDP estimates that C&D waste accounts for more than one third of total waste generation in Cambodia, and it is expected to increase up to more than 40,000,000 tons/year by 2050 (Perrin, 2018). It also estimated that soil, stone, and concrete make up the majority of C&D waste at 75%, but the waste also contains organic waste (6%), metal (5%), glass (2%), hazardous materials (2%), and plastics (1%).

Legislations

Article 46 of Sub-decree No. 113 clarified the responsibility of generators on appropriate storage, collection, transportation and final disposal of C&D waste either on by themselves or using waste management companies, while stipulating financial penalties for non-compliance. However, there is no legislation specifically targeting C&D waste, with municipalities being tasked to design a management system (The Royal Government of Cambodia, 2015). Due to the absence of an institutionalized data collection mechanism, these is a significant information gap in terms of real measurement data for C&D waste.

Collection, Transportation, and Disposal

In practice, some generated waste is collected and transported directly to the designated final disposal site under a waste management contract. Soil, bricks, rocks and stones produced in one location are often reused and recycled at another, while metals and glass are also segregated and sold where possible. Waste with no market value or method of application is often transported and dumped illegally on open public spaces.

IMPACT OF WASTE MANAGEMENT ON GLOBAL WARMING / GHG

Overview

The renewed political commitment for addressing global climate change with the conclusion of the Paris Agreement in 2015 has accelerated mitigation efforts at all governance layers, from international to local level. The waste sector has also been given attention as one highly cost-effective sector, with UNEP (2010) estimating that approximately 3-5%of the global GHG emission originates in the waste sector, while methane constitutes 90%of GHGs from the sector (UNEP-IETC, 2010; UNEP, 2017). In addition to the reduction of CO_2, whose atmospheric lifetime is relatively longer, mitigation of Short-lived Climate Pollutants (SLCPs) – a set of powerful climate warming agents with relatively short lifetime such as black carbon (BC), methane (CH_4), tropospheric ozone (O_3), and hydrofluorocarbons (HFCs) – is also being highlighted as the target substance for reduction efforts (UNEP, 2011; UNEP & WMO, 2011).

Ithis context, mitigation has become one of the sustainability considerations for designing and establishing waste management system at national and city level. GHGs and SLCPs are emitted across the waste management value chain, from collection, transportation, to intermediate treatment, recycling and final disposal. Emissions are largely dependent upon the amount and composition of waste, process flow, technologies and methodologies employed at each stage, and the amount of waste processed by each technology option. For instance, open dumping and landfilling, particularly disposal of untreated organic waste, is considered the major emission source of methane (CH_4) from the sector, while emissions of black carbon (BC) come from open burning and incomplete combustion of fuels from obsolete vehicles. There are also emissions of GHGs (CO_2 and N_2O) from operations of machinery or waste processing that require attention and mitigation measures (CCAC, 2014; Menikpura & Premakumara, 2018).

Emission Data

Despite the importance of mitigation measures in the waste sector, emission data from the sector are not widely available, with only a few studies conducted to estimate emissions.

Based on MoE statistics in 2015, Sethy (2017) estimated that net emissions from the waste sector (both MSW and wastewater) in the year 2000 amounted to 229.24 $GgCO_2$-eq, with methane constituting 94%and N_2O accounting for 7% (Sethy, 2017) (*Table 3*).

Table 3. Estimation of GHG emissions from waste sector in 2000

GHG Source Categories	CH_4	N_2O	Total CO_2-eq
Solid waste disposal on land	9.69	-	203.46
Wastewater management	0.49	0.05	25.73
Waste incineration	-	-	0.00
Other	0.00	0.00	0.00
Total	**10.18**	**0.05**	**229.24**

Source: (Sethy, 2017)

Based on the data in 2015, Singh et al. (2018) estimated the amount of GHG and SLCP emissions from the waste management system in Phnom Penh. Given that waste collection and landfilling are the major sources of emissions in the city, the reports calculated that 3,014.34 tons/CO_2-equivalent of GHGs and SLCPs are released under current conditions (Table 4), although more detailed source information is required to improve this figure (Singh et al., 2018).

Seng (2015) also investigated the methane generation from Phnom Penh's Dangkor final disposal site based on the IPCC Waste Model, and estimated that emissions will gradually increase up to 12.0Gg in 2030, also suggesting that food and kitchen waste is the major contributor followed by paper and green waste (Seng et al., 2015).

Planning and Implementation

It is only recently that attention has been paid to the waste-climate nexus in waste management planning, with Battambang being one of the first cities to highlight the issue in its municipal strategy development. The city joined the Climate and Clean Air Coalition – Municipal Solid Waste Initiative (CCAC-MSWI)

Table 4. Summary of GHG/SLCP emissions from Phnom Penh's waste management system in 2015

Description	Technology	Unit	CH_4	BC	CO_2	N_2O
Waste Collection and transportation by city	Transportation	Kg/month (unit 'kg' used here to show the magnitude of small amount of emissions)	0	1	8,085	0
Treatment for separate waste	Composting		-	-	-	-
	Anaerobic digestion		-	-	-	-
	Recycling		-	-	-	-
Treatment for mixed waste	MBT		-	-	-	-
	Incineration		-	-	-	-
	Landfilling		903,365	8,379	4,072,921	-
Uncollected waste	Open burning/ scattered		27,396	0	0	0
Net emissions per monthly generated waste		Tons/month	931	8	4,081	0
Net BC emissions per monthly generated waste		Tons of BC/month	8.38			
Net climate impact from other GHGs per monthly generated waste		Tons of CO_2 eq/ month	30142.34			

Source: (Singh et al., 2018)

as one of its city partners in 2014, and subsequently developed a Work Plan for reducing GHG and SLCP in waste sector. Some of the actions included in the Work Plan, such as prevention of open burning in the streets and at the final disposal site, promotion of source segregation and diversion of organic waste from markets to composting facilities thereby reducing the amount of organic waste decomposed under anaerobic condition in the landfill site, were also implemented based on the plan.

In Phnom Penh, recent inter-sectoral development plans such as *Phnom Penh City Strategic Action Plan for Climate Change (2017)*, and *Phnom Penh Green City Strategic Plan 2017-2026 (2017)*, as well as its sector-specific strategy, *Phnom Penh Waste Management Strategy and Action Plan 2018-2035* have positioned projects and programs highlighting the clear linkage between waste and climate (City of Kitakyushu & Phnom Penh Capital Administration, 2017; Kol Mardi et al., 2018; Phnom Penh Capital Administration, 2017).

ISSUES AND CHALLENGES IN WASTE MANAGEMENT AND THE WAY FORWARD

Weak Waste Collection Service and Unsystematic Waste Discharging-Collection Mechanism

Achievement of complete collection by extending service coverage and enhancing service quality is considered the highest priority, particularly in the urban context, to manage public health, and reduce the economic, social and environmental impacts of generated waste (Bual, 2018; Kol Mardi et al., 2018; Ministry of Environment, 2019; Ministry of Environment Cambodia, 2019a). In addition, calls for complete waste collection are also visible in relation to disaster risk reduction, namely minimizing flood risks by preventing waste from entering water ways (Urban Voice Cambodia, 2018).

However, many cities and regions struggle to attain complete collection. This is often the result of multiple, complex, and interconnected issues, including poor collection services delivery (in terms of service coverage and collection rate) by private contractors, inappropriate discharging behavior by residents, and low institutional capacity of subnational administrations to manage public private partnership (PPP) contracts in a transparent and results-oriented manner, as well as a lack of coordination among the relevant stakeholders. In addition, collection companies often cannot extend their service to low-income areas where willingness-to-pay is small. These areas are often located on the periphery of the city at a distance which makes operation costs high, and with poor infrastructure (such as poor road conditions, lack of street lights) which makes operations difficult. The rate of user fee collection is not always high, thereby suppressing revenue on which service quality and sustainability depend. Waste discharging rules and collection timetables are often undeveloped or not widely shared where they exist, resulting in improper and unsanitary discharging practices in many cities, as well as inefficient and time-consuming collection by the servicer.

Due to these multiple issues, the waste collection problem in Cambodia requires a mixture of various interventions. For instance, in response to the behavioral issue of residents, information, education and communications (IEC) programs should be designed and implemented to build awareness and influence the behavior of waste generators. The urban-rural implementation gap may be addressed by two options. In Battambang, service coverage was extended to low-income areas by inviting a new waste collection company based on the national sanitation budget (Yagasa et al., 2018). Another intervention is the introduction of differentiated (progressive) user fee system which would allow operators to charge higher fee to mid-to-high income areas to compensate for the low-cost recovery in low-income areas.

Municipal capacity to manage PPP should also be strengthened. In the midst of ongoing decentralization in the waste sector in Cambodia, privatization of waste management services continues to be an option for municipalities, with the pressure to find cost-effective solutions to waste management on one hand, and limited financing options accessible on the other hand. In this context, controlling the quality of service provided by the waste management companies through monitoring, addressing financial and operational sustainability through improved cost recovery, and competitive and results-based selection of contractors hold the key to achieving sustainable waste management across the country. Local governments as outsourcing agents should be able to control the private waste management service companies through appropriate contracts, and impose service improvements or replace an underperforming contractor if necessary.

Weak 3R Efforts

Tapping Potential for Organic Waste Utilization

Given the generally high proportion of organic component in generated waste, there is a strong potential for promoting utilization of organic waste, such as composting, anaerobic digestion system, RDF, and animal feeding. According to the Cambodian Agronomic Soil Classification (CASC) system, soil in most lowland areas of potential rice cultivation lack basic nutrients including nitrogen, phosphorus potassium and organic matters (Matschullat, 2014). This nutrient deficiency represents an opportunity for the application of compost derived from agricultural and municipal waste. This does require the cooperation of waste generators for controlling input materials, market channels for the sale of output products, a supportive cost recovery system for sustainability of business model, administrative capacity and political will. Similarly, the country's strong demand for inexpensive sources of energy also provides opportunities for waste-to-energy systems for organic waste, such as RDF and anaerobic digestion. In addition, a two-track approach of applying a combination of decentralized and centralized systems based on the local conditions should be sought so that every opportunity is tapped.

Promoting Material Recovery and Recycling

A vibrant informal economy provides opportunities to advance material recovery and circulation which Cambodia can benefit from, if it is backed by supportive policies, programs and projects. The important role played by the informal sector in resource recovery and job creation has long been unacknowledged, with this sector being subject to various controls and regulations. A shift in perspective is required in designing waste management policy towards forming a collaborative relationship. Resources can also be mobilized for cultivating domestic resource circulation routes with a view towards creating a circular economy. This can be achieved if the domestic recycling industry is developed by creating a supportive environment for investment in this area, and bridging producers and users of recovered materials through promotion of industrial symbiosis, consumer education and green public procurement.

Exploring the Model of PPP in Value Recovery

Despite the existing potential for resource recovery and utilization, municipalities are often faced with difficulties when trying to introduce solutions. This can be due to lack of public financing, or failure to attract private investments from potential recyclers. For instance, 25 companies have conducted a feasibility study between 2011 and 2015, looking at establishing a value recovery facility in Phnom Penh, but there have been no cases of any investment (Denney, 2016). While various factors may lie behind such decisions by companies, this lack of investment is largely due to low business feasibility, particularly of organic waste utilization projects, which often face difficulty in marketing the end products to a profitable extent.

Project operators should consider promotion of a market-oriented approach to technology selection with advanced planning of how the resulting product can be applied, rather than marketing the product once it is produced. This is important for ensuring the sustainability of the project, for instance, understanding and strategizing how compost can be used by whom and where, and planning the controlling strategy for input material (organic waste).

Likewise, municipalities can also extend support, for instance, in the form of providing land where facilities can be constructed, or attracting private/development finance for investment, while also creating a supportive environment through promoting source segregation and educating users of final products (e.g. communicating the benefits of organic agriculture to farmers to enhance the compost user base) through IEC programs. In this regard, several models of PPP can be considered based on how and to what extent private entities are engaged and to what extent municipalities should invest their public funds. On one end of the spectrum, the private sector can concentrate on operational profitability with land and initial investment financed with public funds. On the other end of the spectrum, the stakes are high for private entities which must take land, initial investment and operational costs into account without any public support (Enayetullah, 2019). In any case, the tariffs for incoming waste (resources) must be set high enough to allow project operators to be profitable and to ensure business sustainability. To this end, municipalities will need to first choose the modality of PPP.

Promoting Reduce and Reuse

While improving waste collection and disposal is given higher priority in Cambodia, less attention has been paid to upstream considerations. For the production stage, policies and programs for promoting resource efficiencies and industrial symbiosis can be developed as measures for waste reduction upstream. In 2017, the government of Cambodia issued *Sub-decree No. 168 on Management of Plastic Bags*, which included an incentive for the promotion of biodegradable bags and encouraged reduction efforts by consumers, while banning non-qualified single-use plastic bags. For retail and consumption, similar policies may also be considered for other consumer products.

Public Awareness on Waste Management

Building public awareness on waste management issues is imperative to address inappropriate waste management practices such as littering and open burning, as well as to advance source segregation and the 3Rs. Efforts and activities for inducing behavioral change need to be strengthened, complementing the development of technical and managerial components of waste management. Particularly, programing and budget allocation for information, education, and communication (IEC) on waste management should be strengthened.

Currently, the importance of waste management, 3Rs, behavior of individuals and their relevance to sustainable society are not widely shared or understood by the general public, with environmental awareness raising activities not frequently conducted and often focusing on public area clean-ups and greening (Mardi et al., 2018; PPCA, 2015; Singh et al., 2018).

In this regard, a recent study on residents in Phnom Penh city by Mongtoeun, Fujiwara and Spoann (2019) found that while respondents are not aware of solid waste management issues, 82%expressed their willingness to participate in environmental programs, with 42%specifically on solid waste management, and 37% on source segregation and separation (Mongtoeun et al., 2019). These findings suggest that there is the potential for IEC programming to translate into wider behavioral change, from issues of littering and open burning to orderly disposal of waste at collection points. This can be achieved if IEC is approached with target-tailored information and PR strategy, and if other factors such as provision of effective collection service are equally satisfied.

Mainstreaming environmental education including waste management issues should also be considered, especially focusing on primary education. Currently, systematic environmental education does not exist in the Cambodian education system. There is a strong need to establish and institutionalize a systematic, continuous education program. While many NGOs and development partners are active, their activities are often not coordinated, with approaches that are too diverse and on scale too inadequate to have large scale impact. Key component includes developing quality textbooks to designing and integrating environmental education into the formal curriculum, as well as capacity development for teachers through strengthening the teacher education center.

Final Disposal

Securing a sustainable waste disposal system is also an urgent issue in Cambodia. Currently there is no final disposal site which employs a sanitary landfill structure equipped with installations to prevent environmental degradation such as impermeable base layers for preventing leachate penetration, vent pipe for landfill gas management, or a leachate collection and treatment system. With the typical waste composition high in organic content, a lack of landfill gas management results in methane build-up in the waste layer which often causes landfill fires (Sassoon & Chheng, 2018). In addition, high precipitation due to the tropical climate produces an enormous amount of leachate which is often released into the natural environment without any treatment.

Data, Planning, and Results-based Management of Policy Cycle

Publicly-available data on waste often rely on different methodologies or are sampled at different points. This poses a difficulty for comparison and mid-to-long term analysis and planning which need to be based on a consistent data set. The amount of country-wide waste generation in 2008 was estimated by Sethy el al. (2014) as 6,818,000 tons/year based on the population that year of 14 million, and a per-capita generation rate of 0.87kg/capita/day. This was the only available figure at the time (Sethy et al., 2014). The figure was also estimated in 2014 as 2,797,633 tons/year based on the population that year of 16,391,068 with an updated per capita waste generation rate of 0.498. Meanwhile the Ministry of Environment (MoE) officially reported 1,089,000 tons/year for the same year. More recently, MoE estimated the figure in 2015 to be 3,052,021 tons/year based on per capita generation rate reported for each province (Table 1). There is an urgent need to establish a national data management system capable of producing reliable and continuous data sets, and more importantly to build capacity at the municipal level to generate data on daily basis. There is also an issue of the urban-rural gap in existing data. Much of the literature has focused on the capital city of Phnom Penh, with fewer studies on other cities and municipalities. Efforts by academic and development communities should also be mindful of this urban-rural gap.

Planning, implementation and evaluation of waste management policies, programs and projects, guided by shared goals and targets, integrating appropriate indicators, and utilizing reliable data in the policy cycle is critical in identifying and upscaling workable solutions within various resource constraints, and in harnessing cooperation of various stakeholders. However, NLLC (2016) reports that only 17% of municipalities have integrated waste management as a matter to be addressed in their development plans (Muny, 2016). Only a few major municipalities such as Battambang and Phnom Penh currently have their own waste management strategy to guide their day-to-day implementation and facilitate actions by various stakeholders. Reliable data are difficult to obtain and existing data are not widely shared to track the progress. Monitoring and evaluation of the implemented programs and projects are often not conducted or shared, preventing institutional learning and development of effective planning in the future.

Public Financing

While financial information for the public sector is limited, it is understood that public finance for waste management is almost entirely limited to the national sanitation fund. The authorization and disbursement procedure is time consuming and prone to intervention by upper tiers of government, while options for financing and cost recovery for waste management are limited for municipal administrations (Denney, 2016; Mun, 2016).

Beyond the procedural aspect of national funding, a model of financially sustainable waste management should be explored and economic instruments designed from a system-wide perspective. For instance, as described in the Case Study on Phnom Penh below, disposal fees at the final disposal site which were meant to promote sustainable landfill management, are in fact functioning as a disincentive for the collection company to increase its collection rate under the unique circumstances of the city.

Building Collaborative Relationship with Stakeholders and Enhancing Public Participation

As previously described, a general lack of attention to waste management issues in municipal development planning reflects the lower political priority resulting in resources directed to this area amongst many competing development issues. Building and harnessing strong political will is critical to bring waste management to the forefront. Accordingly, promotion of environmental education in formal and informal sectors, as well as continuous awareness raising efforts targeting general consumers and various institutions are critical. For the management of waste from industrial and commercial sources, engaging industrial and commercial actors from the planning stage such as promotion of resource efficient practices in the production stage. In addition to the strategic benefit of stakeholder engagement towards improved waste management, the accountability aspect also needs to be highlighted. National and sub-national administrations must promote information disclosure and institutionalize feedback mechanism to reflect the voice of taxpayers in the public service delivery.

CASE STUDY: WASTE COLLECTION IN PHNOM PENH

Phnom Penh, the capital city of Cambodia, is one of many cities in the country that are faced with a waste collection problem. The annual waste disposal amount brought into the city's final disposal site in Dangkor has increased more than two-fold from 343,657 tons in 2007 to 808,530 tons in 2017. Depending on the accuracy of the waste generation amount, MoE (2019) estimates the collection rate to be around 80% in the city. Collection services are unevenly provided with the city center enjoying quality services and areas in the periphery forced to endure poor services or even no services (Kol Mardi et al., 2018; Ministry of Environment Cambodia, 2019b). The consequences are disproportionately felt by the urban poor – 35.7% (99 of the 277) of urban poor communities currently do not have access to collection service due to low payment capacity and poor access roads to such communities, thus facing severe health risks (Sahmakum Teang Tnaut, 2019). User satisfaction cannot be described as high among those living in the service area, with only 14%expressing satisfaction while 56%being neutral and 27%expressing dissatisfaction (Mongtoeun et al., 2019).

Such implementation problems stem from multiple issues, whose nature is beyond operational. The economic and governance aspects are analyzed below. Major factors influencing waste collection services in Phnom Penh are organized as a Causal Loop Diagram (CLD) in *Figure 9* to provide clarity.

Governance Aspect

In Phnom Penh, the city administration concluded a contract with its waste collection service provider, CINTRI, in 2002, allowing exclusive rights to provide waste management service in the city. Definitions in the language employed in the contract are reportedly vague and open to interpretation, causing a dispute between the city government and the contractor over what constitutes MSW and the responsibility of the service provider, while in reality, the terms of contracts are confidential and appear to be known to only a handful of officials in Phnom Penh Capital Authority (PPCA). This complicates matters for PPCA staff who are unable to monitor CINTRI's performance based on agreed targets and terms in the absence of contract (Denney, 2016).

In addition, the single provider contract to collect waste from its entire territory for 49 years is preventing the city from making a transition to competitive selection and thereby benefitting from the promise of privatization such as cost-efficient service, despite its desire to improve the service quality. A series of reform efforts were attempted by PPCA to revoke the CINTRI contract during 2014-2015, with growing demand both in PPCA and MOE, which subsequently led to an announcement by the Council of Ministers to review the contract (Denney, 2016). However, the Contract is said to have designated Singapore and New York for jurisdiction over any legal dispute, and this has prevented PPCA from taking decisive action towards revocation[5]. While the exact reasons are unclear, such revocation has still not taken place as of the end of 2018.

Figure 9. Causal loop diagram of waste collection system in Phnom Penh (Adapted from Denney, 2016; Mardi et al., 2018; Sahmakum Teang Tnaut, 2019; Singh et al., 2018)

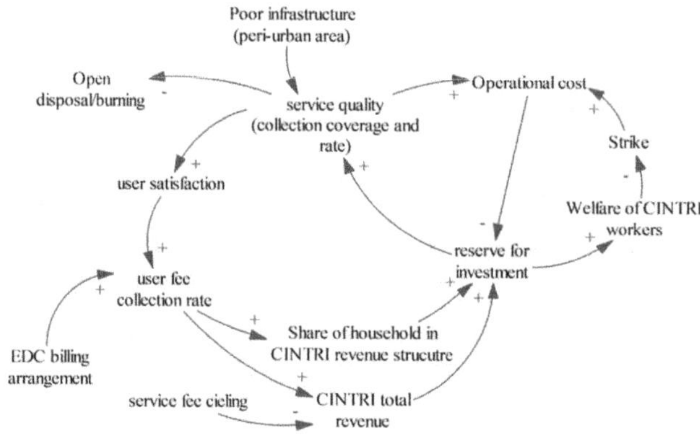

Economic Aspect

The low rate of cost recovery is also intensifying the problem. CINTRI's low service quality – untimely collection, unsanitary collection practice by some workers, and incomplete service coverage – is resulting in complaints and non-payment of waste collection fees among unsatisfied service beneficiaries in Phnom Penh, and is posing a business risk to CINTRI, suppressing the profit unless otherwise directed to investments for improving service quality. Currently, the fee collection is included in the electricity bill issued by the city's power company EDC (Electricite Du Cambodge) under its agreement between CINTRI, with the fee collection operation handled by EDC (Kol Mardi et al., 2018). While no financial statement has been submitted by CINTRI to PPCA, this agreement is considered to be providing reasonable stability to the company's fee collection rate. However, faced with increasing complaints from service users, this fee collection arrangement is also being reviewed (Spiess & Sokhorng, 2017). While no action has been taken at the time of writing, it would further undermine the company's financial and operational stability if the current arrangement is discontinued.

The deteriorating state of CINTRI's financial condition can also be observed from its refusal to pay the disposal fees (USD0.75/ton) charged by Dangkor Landfill Management Authority at the city's final disposal site. According to PPCA, fees have not been paid since January 2018 despite repeated request by the city authority. The company states that it has no financial reserves to pay. In addition, *Inter-ministerial Prakas (ordinance) on maximum fee setting of urban waste management service (No. 195 PrK PR)* was issued in this regard to set the price ceiling for chargeable waste management fees, directly intervening the pricing. Without the freedom to reflect the revenue needs into its user fees, CINTRI is unable to expand its revenue base which is the foundation of extending its service.

CONCLUSION

Waste management systems in Cambodia have evolved over the past few decades but have been outpaced by the growth of waste generation and diversification. Waste collection has been strengthened in urban areas, and has also been initiated in areas which previously could not be accessed. However, large swathes of the population are still without access to a basic sanitation service. In the absence of recycling infrastructure and institutionalized source segregation, a significant proportion of waste remains unutilized, and is directly transported and disposed of as mixed waste. A strong informal economy does exist for collection and simple processing of inorganic recyclable materials, but it is vulnerable to resource price fluctuations. Meanwhile efforts to reduce and reuse in the upper stream are still weak, and ideally require a resource perspective into waste management. Mismanaged waste across the waste management value chain from generation points to the final disposal sites is causing pollution of air, water and land, thereby exhibiting local and global risks to public health, resource management and to the climate.

National policies, laws, regulations, and funding programs have been developed over the past few decades aiming for improved solid waste management, but have had mixed results at the implementation stage. The ongoing decentralization of waste management responsibilities to municipalities through Sub-decree No. 113 and the national fund is resulting in early signs of improvement albeit at a slow pace. If further national actions such as periodical updates and implementation of a national strategy, provision of sustained funding adequate to finance the given responsibility, and development of operational capacity, are met with local efforts, for instance, in planning, engaging stakeholders, and inviting investment to the sector, such a transition may yield even higher achievements.

REFERENCES

Bual, H. (2018). *Waste Management*. Retrieved April 12, 2019, from https://www.khmertimeskh.com/548828/waste-management/

Bunrith, S., Ochiai, S., & Kaneko, H. (2012). Resident willingness of food waste separation in Phnom Penh, Cambodia. In *Proceedings of the 23rd Annual Conference of Japan Society of Material Cycles and Waste Management* (p. 615). Japan Society of Material Cycles and Waste Management. Retrieved from https://www.jstage.jst.go.jp/article/jsmcwm/23/0/23_615/_pdf/-char/ja

Cambodia Tribune. (2018). Sub-decree to Discourage Too Much Use of Plastic Bag. *Cambodian Tribune*. Retrieved April 7, 2019, from https://cambodiatribune.com/sub-decree-to-discourage-too-much-use-of-plastic-bag/

Carto G. I. S. Services. (2018). *Cambodia base*. Retrieved April 14, 2019, from http://asiapacific.anu.edu.au/mapsonline/base-maps/cambodia-base

CCAC. (2014). *The Climate and Clean Air Coalition to* (J. Maréchal, S. Bonnard, & N. BorgfordParnell, Eds.). Paris: Climate and Clean Air Coalition to Reduce Short-Lived Climate Pollutants. Retrieved from www.ccacoalition.org

City of Kitakyushu & Phnom Penh Capital Administration. (2017). *Phnom Penh Capital City Strategic Action Plan for Climate Change*. Phnom Penh: Authors.

COMPED. (2012). *Multi Fuel Policy Development for Solid Biomass Utilization and Capacity Building Plan for EEP Mekong Countries Overview of solid biomasses volumes and availability in Cambodia*. Phnom Penh: Author.

COMPED. (2018). Retrieved April 9, 2019, from https://comped-cam.org/

CSARO. (2018). *CSARO Organization Information*. Retrieved April 9, 2019, from http://www.bigpond.com.kh/users/csaro/organization_info.htm

CTI Engineering International, Nippon Koei, & Kitakyushu Water Service. (2016). *The Study on Drainage and Sewerage Improvement Project in Phnom Penh Metropolitan Area*. Authors. Retrieved from http://open_jicareport.jica.go.jp/pdf/12270294_01.pdf

Denney, L. (2016). *Reforming solid waste management in Phnom Penh* (WORKING POLITICALLY IN PRACTICE SERIES-CASE STUDY No. 8). Retrieved from https://asiafoundation.org/wp-content/uploads/2016/06/Working-Politically-and-Flexibly-to-Reform-Solid-Waste-Management-in-Phnom-Penh.pdf

Enayetullah, I. (2019). *IRRC Lessons Learned: Contribution to Urban Resource Efficiency, Circular Economy and Sustainability in the Asia Pacific Region*. Dhaka: Waste Concern.

Fondazione, A. C. R. A. (2015). *Cambodia's plastic bag ecosystem and usage summary of research report excess baggage: Reducing plastic bag waste in major cities of Cambodia*. Retrieved from www.excessbaggage-cambodia.org

Kaza, S., Yao, L., Bhada-Tata, P., & Van Woerden, F. (2018). *What a Waste 2.0: A Global Snapshot of Solid Waste Management to 2050. What a Waste 2.0: A Global Snapshot of Solid Waste Management to 2050*. World Bank. doi:10.1596/978-1-4648-1329-0

Kokusai Kogyo. (2005). *The study on solid waste management in the municipality of Phnom Penh in the Kingdom of Cambodia Final Report*. Retrieved from http://open_jicareport.jica.go.jp/pdf/11784469_01.pdf

Kol Mardi, M., Sam, N., Vathanak Thida, K., Channarith, K., Yagasa, R., Akhmedkhodjaeva, N., … Totoki, Y. (2018). *Phnom Penh Waste Management Strategy and Action Plan 2018-2035*. Phnom Penh Capital Administration. Retrieved from https://www.ccet.jp/publication/phnom_penh_waste_management_strategy_and_action_plan

Liu, C., Hotta, Y., Totoki, Y., Yagasa, R., Aoki-Suzuki, C., Hengesbaugh, M., & Inoue, M. (2018). *State of the 3Rs in Asia and the Pacific-Experts' Assessment of Progress in Ha Noi 3R Goals*. Retrieved from http://www.uncrd.or.jp/content/documents/6777[full document] State of the 3Rs in Asia and the Pacific.pdf

MAFF. (2016). *Policy On Biodigester Development in Cambodia*. MAFF.

Matschullat, J. (2014). Save Cambodia's wildlife: Atlas of Cambodia. Maps on socio-economic development and environment. *Environmental Earth Sciences*, *72*(4), 1295–1298. doi:10.100712665-014-3325-3

Menikpura, N., & Premakumara, D. G. J. (2018). *User's Manual - Emission Quantification Tool (EQT) for Estimating Short Lived Climate Pollutants (SLCPs) and Other Greenhouse Gases (GHGs) from Waste Sector Institute for Global Environmental Strategies (IGES) Prepared on behalf: Climate and Clean Air Coalition (CCAC)*. Hayama, Kanagawa: Initiative. Retrieved from https://www.ccet.jp/sites/default/files/2018-12/User Manual_EQT version II_June 2018.pdf

Ministry of Environment. (2019). *Regional 3Rs Forum in Asia and the Pacific Cambodia - Country Report*. UNCRD. Retrieved from http://www.uncrd.or.jp/?page=view&nr=1174&type=13&menu=198

Ministry of Environment Cambodia. Sub-Decree No. 168 on the management of plastic bags, Pub. L. No. No. 168 GNKR.BK, Ministry of Environment Cambodia 11. (2018). *OD Mekong Datahub*. Retrieved from https://data.opendevelopmentmekong.net/lo/dataset/sub-decree-no-168-on-the-management-of-plastic-bags/resource/ea9eeff5-78db-4ba9-b9c3-ad5135ba1d47?type=laws_record

Ministry of Environment Cambodia. (2019a). *Draft National Waste Management Strategy and Action Plan*. Phnom Penh: Ministry of Environment Cambodia.

Ministry of Environment Cambodia. (2019b). *Waste Management Data*. doi:10.1007/978-1-62703-176-9_8

Mongtoeun, Y., Fujiwara, T., & Spoann, V. (2019). Household Solid Waste Generation and Socioeconomic Factors in the Capital City of Cambodia. *International Journal of Environmental Sciences & Natural Resources*, *20*(1), 4. doi:10.19080/IJESNR.2019.20.556029

Mun, V. (2016). *Progress and Challenges of Deconcentration in Cambodia: The Case of Urban Solid Waste Management* (CDRI Working Paper Series No. 110). Retrieved from www.cdri.org.kh

Muny, M. (2016). *Survey Report on Waste Management Practices At Municipality/District Level*. Retrieved from https://www.kas.de/c/document_library/get_file?uuid=9e9be00d-4355-3a39-a006-9e856bb5eb58&groupId=252038

One Earth One Ocean. (2018). *PROJECTS - One Earth One Ocean*. Retrieved April 5, 2019, from https://oneearth-oneocean.com/en/the-association/projects/

Perrin, D. (2018). *Key results from a baseline study on waste generation of all the provinces*. Phnom Penh: UNDP.

Phnom Penh Capital Administration. (2017). *Phnom Penh Green City Strategic Plan 2017-2026*. Author.

Plastic Free Cambodia. (2018). *Plastic Free Cambodia – Reducing plastic pollution in SEA*. Retrieved April 13, 2019, from https://plasticfreecambodia.com/

PPCA. (2015). *Sixth Regional 3R Forum in Asia and the Pacific - City Report - Phnom Penh*. Retrieved from http://www.uncrd.or.jp/content/documents/25936-3R_City-Report_PhnomPenh.pdf

Sahmakum Teang Tnaut. (2019). *Urban Governance Waste Management in Phnom Penh A review of waste management from the streets*. Retrieved from http://teangtnaut.org/wp-content/uploads/2019/01/20190129_STT_Final_Annual_Report_English_version-compressed.pdf

Sang-Arun, J., Chau, K. H., Uch, R., & Sam, P. (2012). *A guide for technology selection and implementation of urban organic waste utilisation projects in Cambodia* (Vol. 2011–06). Institute for Global Environmental Strategies. Retrieved from https://pub.iges.or.jp/pub/guide-technology-selection-and-implementation

Sassoon, A. M., & Chheng, N. (2018). *Lack of control at landfill fire*. Retrieved April 12, 2019, from https://www.phnompenhpost.com/national/lack-control-landfill-fire

Seingheng, H., Chiemchaisri, C., Tanchuling, M., Xaypanya, P., & Takemura, J. (2018). Characterization of Landfill Leachates and Sediments in Major Cities of Indochina Peninsular Countries—Heavy Metal Partitioning in Municipal Solid Waste Leachate. *Environments*, *5*(6), 65. doi:10.3390/environments5060065

Seng, B., Hirayama, K., Katayama-Hirayama, K., Ochiai, S., & Kaneko, H. (2013). Scenario analysis of the benefit of municipal organic-waste composting over landfill, Cambodia. *Journal of Environmental Management*, *114*, 216–224. doi:10.1016/j.jenvman.2012.10.002 PMID:23168253

Seng, K. (2015). *L'analyse de la composition des dechets solide et la prevision de la production de dechets en Phnom Penh avec la production du methane des sites d'enfouissement Dangkao au Cambodge*. Institut de Technologie du Cambodge.

Seng, K. (2016). *L'analyse de la composition des dechets solide et la prevision de la production de dechets en Phnom Penh avec la prodution du methane des sites d'enfouissement Dangkao au Cambodge*. Institute of Technology of Cambodia.

Seng, K., Kouk, F., Hul, S., & Penh, P. (2015). *Solid Waste Composition, Forecasting of Waste Generation and Methane generation in Phnom*. Academic Press.

Sethy, S. (2017). *State of the 3Rs in Asiat and the Pacific Country Chapter - The Kingdom of Cambodia*. Retrieved from http://www.uncrd.or.jp/content/documents/5686[Nov 2017] Cambodia.pdf

Sethy, S., Sothun, C., & Wildblood, R. (2014). Municipal Solid Waste Management in Cambodia. In A. Pariatamby & M. Tanaka (Eds.), *Municipal Solid Waste Management in Asia and the Pacific Islands* (pp. 77–94). Singapore: Springer Singapore. doi:10.1007/978-981-4451-73-4_5

Singh, R. K., Gamaralalage, P. J. D., Yagasa, R., & Onogawa, K. (2018). *State of Waste Management in Phnom Penh, Cambodia*. Retrieved from https://pub.iges.or.jp/pub/state-waste-management-phnom-penh-cambodia

Spiess, R., & Sokhorng, C. (2017). *Cintri chided by former exec*. Retrieved April 12, 2019, from https://www.phnompenhpost.com/business/cintri-chided-former-exec

Suy, P. (2018). *Svay Rieng to build recycling plant - Khmer Times*. Retrieved April 13, 2019, from https://www.khmertimeskh.com/536529/svay-rieng-to-build-recycling-plant/

The Royal Government of Cambodia. Sub-decree on Management of Urban Garbage and Solid Waste (English Translation), Pub. L. No. No. 113 NKR.PR, 23 (2015). Cambodia.

The World Bank. (2019). *World Bank Open Data*. Retrieved March 11, 2019, from https://data.worldbank.org/

UNEP. (2011). *Near-term Climate Protection and Clean Air Benefits: Actions for Controlling Short-Lived Climate Forcers*. Retrieved from http://www.unep.org/publications/ebooks/SLCF/

UNEP. (2017). *The Emissions Gap Report 2017 A UN Environment Synthesis Report*. Retrieved from www.unenvironment.org/resources/emissions-gap-report

UNEP & WMO. (2011). *Integrated Assessment of Black Carbon and Tropospheric Ozone | Climate & Clean Air Coalition*. Retrieved from http://www.ccacoalition.org/ar/resources/integrated-assessment-black-carbon-and-tropospheric-ozone

UNEP-IETC. (2010). *Waste and Climate Change - Global Trends and Strategy Framework*. Retrieved from http://www.unep.or.jp/ietc/Publications/spc/Waste&ClimateChange/Waste&ClimateChange.pdf

Urban Voice Cambodia. (2018). *Flooding and Drainage Infrastructure Development in Phnom Penh - Urban Voice Cambodia*. Retrieved March 28, 2019, from https://urbanvoicecambodia.net/flooding-and-drainage-infrastructure-development-in-phnom-penh/?lang=en

World Bank. (2019). *Cambodia Overview*. Retrieved April 8, 2019, from https://www.worldbank.org/en/country/cambodia/overview

Yagasa, R., Sang-Arun, J., Uch, R., Sam, P., & Gamaralalage, P. J. D. (2018). *Participatory Waste Management Approach for Climate Change Mitigation: the Case of Battambang City*. Hayama, Kanagawa: Institute for Global Environmental Strategies. Retrieved from https://pub.iges.or.jp/pub/participatory-waste-management-approach

Yim, M., Fujiwara, T., & Sethy, S. (2013). *A Study on Household Solid Waste Characteristics in Phnom Penh City*. Academic Press.

Yim, M., Takeshi, F., & Sour, S. (2014). Current status of commercial solid waste generation, composition and management in Phnom Penh city, Cambodia. *Environment and Waste Management*, *1*(3), 31–38.

ENDNOTES

1 Dangkor Final Disposal Site in Phnom Penh was originally designed as Sanitary Landfill site, but is currently operated as a controlled dump site.

2 Source: (Plastic Free Cambodia, 2018).

3 Source: (One Earth One Ocean, 2018)

4 Source: (Suy, 2018)

5 Discussion at the Internal Review Committee Meeting on Waste Management Strategy and Action Plan for Phnom Penh at PPCH, 22 August 2018.

Chapter 4
Municipal Solid Waste Management and Treatment in China

Mi Yan
Zhejiang University of Technology, China

Haryo Wibowo
Zhejiang University of Technology, China

Qike Liu
Zhejiang University of Technology, China

Yi Cai
Zhejiang Energy Group R&D, China

Dicka Ar Rahim
Zhejiang University of Technology, China

Hu Yanjun
Zhejiang University of Technology, China

ABSTRACT

This chapter focuses on the generation, treatment, and management policy, but excluding reduction and recycling information, of municipal solid waste (MSW) in China. Cities in China generated 215 million tons of MSW in 2017. It presents a big challenge for sustainable development of cities. Currently, sanitary landfill is the dominant method for MSW treatment, treating 57.2% of total waste. MSW composting has seen a decrease in application, only used to treat less than 2.5% of total MSW. Meanwhile, waste to energy in the form of incineration for energy production has developed significantly in the last decade. The percentage of MSW amount treated by incineration is 40.2% from the total amount of treated MSW in 2017, a sharp increase from only 4.9% in 2003, due to China's commitment to WtE. The content of this chapter shall include comprehensive data from government and industry, including plant capacity, quantity, technology, emission, policy and regulation promoting, and guiding WtE in China.

DOI: 10.4018/978-1-7998-0198-6.ch004

Figure 1. MSW generation and GDP growth in China (1979 - 2017)

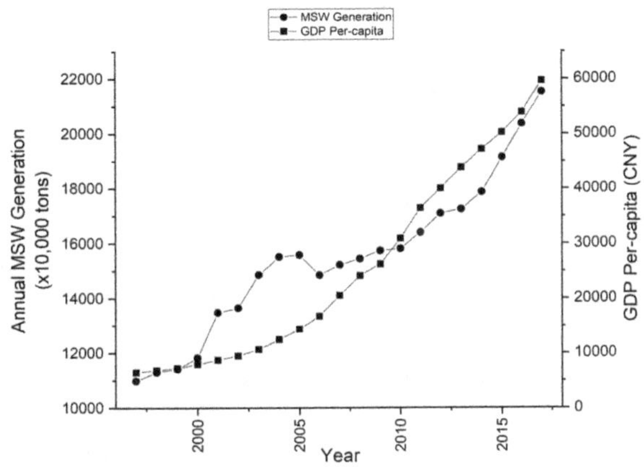

MSW GENERATION AND COMPOSITION

Municipal solid waste (MSW) in China is generated from household and commercial activities for providing goods and services, including offices and shopping malls; but excluding the waste from hospital, industry and construction. MSW sorting is still not currently well operated in China, and the Chinese government is recently struggling to promote waste sorting. This result in MSW that consist of various components mixed together, making the composition complex and the amount large. These create a huge challenge on efforts on recycling or reutilization of MSW in China.

MSW Generation in China

According to China Statistical Yearbook (NBSC, 2018), the total population of China in 2017 is 1.39 billion. 813.47 million of it is citizen, accounting for 58.52% of total population. Due to urbanization development in China, the population in urban area has quickly increased, creating population concentrations in major cities. The economic development has also been significant in the last 40 years, with gross domestic product (GDP) increasing from CNY 41.01 billion (6.12 billion USD) in 1979 to CNY 8271.22 billion (1233.33 billion USD) in 2017 (NBSC, 2018). Both of increasing citizen population and GDP are expected to cause more MSW to be produced in cities. Reviewing the MSW annual generation data as shown in Figure 1, the amount increased from 25.08 million ton in 1979 to 215.2 million ton in 2017 (NBSC, 2018). This is an average increase of 6.06% per year.

Analyzing the correlation between citizen population and GDP to MSW generation as presented in figure 1 (NBSC, 2018), it was calculated that China's MSW generation per-capita was 0.725 kg/person/day and 26.02 kg/CNY 10,000 GDP nationally in 2017. China's MSW generation per-capita is lower compared to other developed countries such as USA (2.01 kg) and Japan (0.96 kg). MSW generations per 10,000 RMB GDP for the two countries are 20.6 kg (USA) and 14.9 kg (Japan), respectively. MSW generation per 10,000 GDP is an index used by the Chinese government to measure and promote sustainable development. MSW management system tied to economic model has been well applied in Japan, and China is developing plans of the same model.

Figure 2. Daily MSW generation per-capita and annual MSW generated / CNY 10,000 GDP for 31 provinces in China in 2017 (NBSC, 2018)

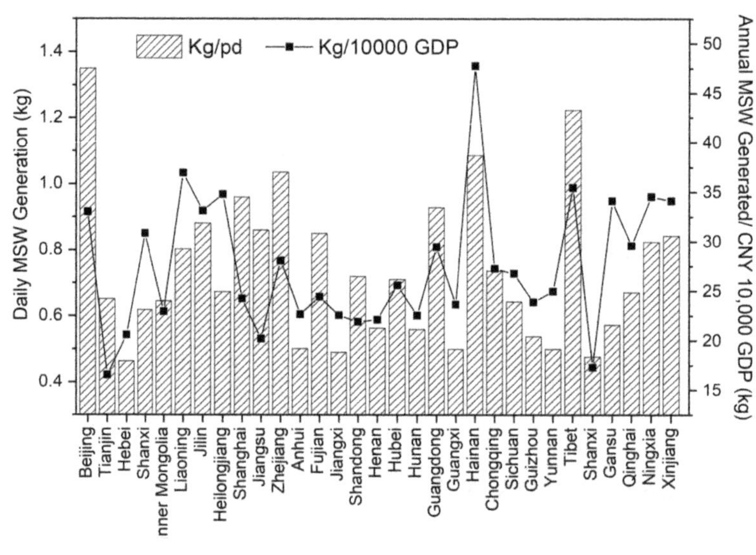

On the provincial level, the numbers related to MSW generation are quite varied. Mainland China comprise of 31 provincial-level administrative areas, with different economic level and population counts. The comparison of daily MSW generation per-capita and annual MSW generation per CNY 10,000 GDP in provinces in China is provided in figure 2. Beijing has the highest value of per-capita each day with 1.35 kg/day, and Tianjin has the lowest value of per 10,000 RMB GDP with 16.55 kg/10,000GDP (NBSC, 2018).

While presented as provincial data, the statistical data of MSW generation presented in Figure 2 only cover the waste produced in cities. It most likely did not account for over 576.61 million people living in rural areas or villages, whom also produce huge amount of solid waste. According to a report by Ministry of Housing and Urban-Rural Development, over 150 million tons of waste was produced in rural areas in 2017 (Ministry of Housing and Urban-Rural Development, 2017). With the program of integrated treatment of city and rural waste, the total annual amount of waste generated could be up to 365 million tons. It presents a big challenge for waste disposal in China, necessitating the development of advanced management and treatment system.

Prior to 2018, China imported a huge amount of waste from overseas market to be used as source of raw materials. It was reported that the annual MSW import increased from 4.5 million ton in 1995 to 45 million ton in 2016 (Seldman & Morris, 2018) Due to the fact that not all parts of the imported waste were able to be fully utilized, the imported waste inevitably caused extra waste to be generated in China. In order to protect the environment and promote domestic MSW recycling, Chinese government prohibited the import of overseas MSW since January 1st, 2018 (Office of the State Council, 2017).

MSW Composition and Potential Value

In order to formulate the best solution for MSW management, it is important to understand the characteristics of MSW. For that purpose, numerous studies have been conducted in determining the composition of MSW in China. The compiled results from studies of MSW composition in various cities in China are listed in Table 1.

From Table 1, it is clear that while the exact composition of MSW varies from city to city, the main components are mostly the same. The main constituent of MSW in all cities is organic waste (42.2% - 67.5%wt.), in which food waste is a major component. This presents a challenge for MSW management, as food waste usually consist of organic materials with low heating value and has high moisture content. China's food waste is also considerably more complex compared to other developed countries, with high usage of spices and cooking oil.

The presence of combustible components in MSW raises the possibility of its utilization as combustion fuel. In order to learn more about this possibility, a lot of research was done to analyze the elemental content of MSW in China. The resulting data from proximate and ultimate analysis of MSW in as-received basis can be found in table 2.

Table 1. Components of MSW from major cities in China

City	Composition (%wt.)							
	Organic	Paper	Fabric	Wood	Plastic	Glass	Metal	Inorganic
Shenyang	67.5	7.6	1.7	1.5	11.0	2.8	0.5	2.3
Qingdao	42.2	4.0	3.2	-	11.2	2.2	1.1	36.1
Nanjing	66.0	9.0	1.0	-	12.0	9.0	1.0	2.0
Wuhan	52.0	7.1	1.4	1.7	9.9	7.1	1.0	19.8
Shanghai	67.3	8.8	1.9	1.3	13.5	5.2	0.7	1.4
Hangzhou	58.2	3.7	2.2	1.2	7.6	2.1	1.0	24.0
Guangzhou	60.2	5.4	3.4	1.1	9.0	3.4	0.5	17.1
Average	**59.1**	**6.5**	**2.1**	**1.4**	**10.6**	**4.5**	**0.8**	**14.7**

Table 2. Heating value and ultimate analysis of China's MSW

Component	Proximate analysis, %				Heat Value	Ultimate Analysis, %				
	M_{ar}	A_{ar}	V_{ar}	Fcar	$Q_{net,ar}$(cal/g)	C_{ar}	H_{ar}	N_{ar}	St,ar	O_{ar}
Food waste	75.0	6.3	15.3	3.4	396.0	9.0	1.1	0.4	0.1	8.1
Others	42.6	28.1	24.6	4.7	1155.6	14.4	1.7	1.0	0.3	11.9
Plastic	46.9	22.7	27.9	2.5	1857.8	19.3	2.7	1.0	0.3	7.2
Paper	54.0	7.9	32.6	5.5	1251.4	17.8	2.3	0.4	0.1	17.5
Wood	57.9	3.2	31.5	7.4	1269.2	19.3	2.3	0.7	0.2	16.4
Textile	54.4	11.9	30.8	2.9	1460.0	17.1	1.8	0.7	0.5	13.6
Mixture	52.6	22.8	21.2	3.4	998.9	13.0	1.6	0.7	0.2	9.1

Table 2 shows that food waste has the lowest heating value in as-received basis compared to other MSW components due to its high moisture content (75%) and low carbon content (9%). As was pointed out earlier, food waste makes up a major part of MSW in China, and these food wastes should be prioritized in order to effectively manage MSW. In average, the lower heating value (LHV) of MSW in China is in the range of 1000 to 1500 kcal/kg, which is much lower than the value of MSW LHV in USA and Japan (>2000 kcal/kg). This heat value is significantly affected by economic well-being and lifestyle of the population, and is one of the determining factors in the development of waste to fuel industry.

In China, the MSW recycling is mostly conducted by pickers who mostly consist of low-income people in an unofficial capacity. These pickers often gather resalable parts from waste bins before collection by city administration. The recyclable components, which mainly include paper, metal and beverage plastic bottle, would be collected by these by pickers, sold to bigger collectors, and finally sent to factories for reuse or recycle. The implication is that the MSW generation data collected through government and city services exclude the amount of waste collected by these pickers. There are no official data tracking the amount of MSW recycling in China, although certain reports suggested that up to 48 million tons of MSW is recycled nationally in China every year.

NATIONAL PLAN FOR MSW MANAGEMENT AND TREATMENT

Chinese government is committed towards environmental protection to improve the quality of environment, conversion to environmentally-friendly economic model, and improvement of public life quality. This commitment is formalized in national plans, laws, regulations, and notices issued by the government. The environmental protection law of China is one of the principal laws designed to meet those objectives. First issued in 1989, it has been revised and updated often to keep up with the latest in environmental study findings and technological advancements. The latest revision of the law has been implemented since January 1st 2015, introducing more comprehensive and much stricter regulations compared to the previous revisions.

China also has a set of solid waste pollution prevention and control law, which determines the government's duty and responsibility in MSW collection, transportation and disposal. MSW management and treatment requires the contribution of both public and government, with the government's role being to set guidelines and provide support for clean treatment and utilization of MSW. The key national plans related to MSW management and treatment are laid out in China's 13th five-year plan, which contains the Chinese government's blueprint in various fields from 2016 to 2020. A few key plans related to MSW management shall be briefly introduced in this section.

The Chinese Government's 13th Five-year Plan (2016 -2020)

The "13th Five-year Plan for China's National Economic and Social Development" (State Council, 2016a) provides the groundwork for a comprehensive environmental governance between 2016 and 2020. The main focus of the plan includes acceleration of the development of MSW treatment facilities, improvement of MSW collection and transportation systems, the increase in waste incineration ratio, and the application of proper treatment for MSW leachate.

In the "Announcement on Accelerating the Construction of Ecological Civilization" by the Central Committee of the Communist Party of China and State Council (Central Committee of the Communist Party of China and State Council, 2015) a concept for the development of a circular economy model is introduced in accordance with the principles of reduction, reuse, and resource reutilization. For 2016 – 2020, the focus is on accelerating the establishment of a recycling-oriented industrial, agricultural, and service industry system and to increase the resource output rate of the whole society. It also mandated that the existing recycling system of renewable resources be improved, better implementation of waste separation and recycling, as well as the development and utilization of "urban minerals".

"13th Five-Year Plan for Ecological Environmental Protection Plan" (State Council, 2016b) instructed the government to build an operational network of MSW disposal and disposal facilities that would provide full coverage all over the country. This objective calls for the government of all levels and appointed private companies to accelerate the construction of MSW disposal facilities in the county and create an interconnecting network of MSW disposal facilities. The plan also aims to improve the level of urban domestic waste treatment through reduction, resource utilization, and harmlessness. Currently, the harmless treatment rate of MSW has nationally reach more than 95% of the MSW collected.

In the "13th Five-year Plan for Energy Saving and Emission Reduction" (State Council, 2016c), the Chinese government has declared some specific focus related to waste-to-energy's (WtE) twofold potential as a waste-reducing and energy-generating industry. For the year 2016 to 2020, the focus is on accelerating the construction of MSW recycling facilities. The facilities shall also be supported by an easily applicable management system for waste separation on the household level along with its supervision, MSW collection and transportation improvement, and the establishment of a more reliable system of MSW recycling.

In the rural areas, which often has less advanced MSW management facilities compared to urban areas, the government plans to comprehensively promote specific rural solid waste control. The goal is to establish a village cleaning system and promote waste sorting as well as resource recycling and utilization. By 2020, it is expected that more than 90% villages in China would be equipped with operational solid waste treatment facilities.

The "13th Five-year Plan for Housing and Urban-rural Construction" (Ministry of Urban-Rural Development, 2016a) determines the direction of MSW management system application in housing and construction. This plan allows some flexibility for provincial governments to provide and execute regional plans according to local conditions. The regional plans however, must comply to the objectives determined by the government through the five-year plan. For 2016 to 2020, the objectives related to MSW management in housing and construction are to ensure every city has the ability to carry out comprehensive waste sorting; improve the statistical index system of MSW generation for every province; creation of an efficient network of MSW sorting, recycling, and renewable resource recovery system; and to improve the recycling rate.

Cities also work in concert with the Chinese government to create an efficient network of MSW incineration treatment technology in cities with scarce free area and high population density. The goal is to increase the capacity of MSW incineration nationally by 220,000 tons/day from 2016 to 2020 (Ministry of Urban-Rural Development, 2016a). The network is expected to provide all cities with the ability to efficiently treat MSW and all counties to have the access to harmless MSW treatments.

In the "13th Five-year Plan for Construction of MSW Treatment Facilities" (National Development and Reform Commission, 2016), the Chinese government has set up a target 100% MSW harmless treatment rate in first-rank cities and provincial capitals by 2020. The harmless treatment rate of MSW is targeted to reach over 95% in cities, 80% in counties, and 70% in towns (National Development and Reform Commission, 2016). The capacity of MSW treatment facilities is going to increase to 340,000 tons/day in the 13th Five-year. By 2020, no MSW shall be directly dumped into landfills after collection in first-class cities and provincial capitals. Instead, it shall first go through separation, recycling, and incineration process; with the only the unusable solid incineration byproducts going to landfills. To support this program, the Chinese government has mandated that the capacity of MSW incineration be increased to at least be 50% of total treatment capacity nationally and over 60% in the east coast area. While MSW recycling rate must go over 35% of the total MSW generated (National Development and Reform Commission, 2016).

No-Waste Demo City

State Council implements a plan for "No-waste Demo-City" to be realized in 2020 (Office of the State Council, 2019), which is a concept city with comprehensive and efficient waste management and re-utilization system. It is based on the construction of an integrated management and technical system for MSW management. The demo-cities are expected to make significant progress in key areas such as: prevention of any increase in total storage of solid waste, full reuse of agricultural wastes, reduction of generated MSW, comprehensive safety control of hazardous wastes, zero-occurrence of illegal solid waste dumping, and the cultivation of solid waste resources for reutilization.

Through these pilot experience and practices, a comprehensive management reform of solid waste can be formulized in the form of a "no-waste city" model, which can be replicated and promoted. These cities and their societies are expected to practice a green lifestyle, promote the reduction of MSW generation, and increase resource reutilization. The government would do its part in supporting these no-waste societies by applying full implementation of MSW charging system and waste metering charges; building resource recycling bases; provide guidelines for classification of domestic waste; and promote resource reutilization methods such as recycling of recyclable materials, incineration power generation, and biological treatment.

From the policies and regulations discussed above, it is clear that China is vigorously promoting waste reduction and sorting, improving resource recovery ratio and reuse level, developing waste treatment technologies according to local conditions, and encouraging the development of waste incineration. It is expected that by 2020, the MSW recycling rate will exceed 35%. National waste disposal rate should exceed 95% nationally and 90% in the rural areas. Reduction of direct MSW dumping to landfill would be accompanied by the increase in waste incineration treatment capacity, anticipated to account for more than 50% of the total amount of MSW treatment in 2020. Special attention is also given for MSW disposal byproducts such as leachate and fly ash.

MSW SORTING REGULATIONS & EXPERIENCES

Sorting Regulations in State and Provinces

The national, provincial, and municipal level governments in China regularly issue updated targets, requirements, and measures for waste sorting in order to reduce the amount of disposed waste and encourage waste recycling. In 2017, the National Development and Reform Commission and the Ministry of Housing and Urban-Rural Development jointly issued the “Implementation Plan for Municipal Solid Waste Classification System” (National Development and Reform Commission (NDRC), 2017a).

The plan stipulates that cities must maintain municipal waste recycling rate over 35%. The program provides guidelines in the form of procedures for waste-classification related works, mandatory classification requirements, direction for residents to properly carry out separation of domestic waste, and the blueprint for the construction of domestic waste separation support system. 46 key cities were selected as demonstration cities for this waste sorting system.

As of March 2019, 29 provinces and municipalities across the country have created work plans and regulations for the classification of municipal solid waste and recycling targets based on the instructions issued by the ministries and commissions of the state. Some of those guidelines are the “Notice on Promoting the Classification of Wastes in Public Institutions such as Party and Government Organs” (Guoguan Energy Conservation [2017] No. 180), “Opinions on the Implementation of Waste Classification System by Military Units” (Army Construction) [2017] No. 485), “Notice on Promoting the Classification Management of Waste in Medical Institutions” (National Health Office Medical Development [2017] No. 30), and “Notice on Accelerating the Classification of Waste in Some Key Cities” (Jiancheng [2017] No. 253).

Sorting Regulation Cases for Cities and Villages

In 2014, Shanghai has issued the “Shanghai Municipal Solid Waste Classification and Reduction Measures”. This measure was put into application in 2018 with the “Implementation Plan on Establishing and Improving the Whole-process Classification System of Municipal Solid Waste” and the “Action Plan for the Construction of the Whole-process Classification System of Municipal Solid Waste (2018-2020)”.

According to the “Implementation Plan on Establishing and Improving the Whole-process Classification System of Municipal Solid Waste”, by the end of 2020 Shanghai would mandate compulsory waste classification on all levels, starting from household waste classification that would be implemented in residential areas. The legal basis of the program is formalized in “The municipal solid waste management regulation of Shanghai municipality”, which will be implemented in 2019 (Shanghai City Government, 2019). Now, the waste separation is strictly forced to practice in Shanghai.

It is expected that the quality of household waste classification will be significantly improved by the creation of a front-to-end separation system of household waste, segregated collection, segregated transportation, and separate treatments. The target is for harmless disposal capacity of household waste to reach over 32,800 tons/day, including 7,000 tons/day for wet waste. Currently, the integration of waste collection and transportation system has been successfully applied and the reutilization rate of domestic waste resources in Shanghai has reached 35%. The full brunt of municipal solid waste management regulation of Shanghai will come into effect on July 1, 2019.

Waste classification has also been effectively implemented in rural areas. The villages in Jinhua, Zhejiang province, established the rural household waste management model of “household collection, village collection, township transfer and county treatment” in 2014. This model was aimed at solving the problem of rural household waste outlet. Based on local needs and situation, Jinhua implemented the “four points method”, “two classification methods”, and “waste not fall to the ground” transfer method. The village also built “sunshine composting houses” to ferment organic waste aerobically and use the product as fertilizers.

Jinhua has formulated "Standard for the Management of Rural Household Waste Classification" and "Regulations on the Management of Rural Household Waste Classification in Jinhua City". In 2016, the ministry of housing and urban-rural development issued a notice on promoting the experience of waste classification and resource utilization in rural areas of Jinhua city as an example for other villages in China (Ministry of Housing and Urban-Rural Development, 2016b).

Sorting Case for Recycling and Incineration by Company

Waste classification is currently done by the residents and the government but the Chinese government is now looking to get enterprises involved. This move is done with the assumption that the government has laid out the groundwork to make waste classification profitable, and the involvement of enterprises is expected to ensure that waste classification is done efficiently and professionally. Opportunities are now provided for enterprises' to be involved in waste processing through either resource recycling or incineration. One example of a recycling company is the Zhejiang Jiucang Renewable Resources Development Co., ltd. which carries out resource-recycling waste classification in Hangzhou. While Hangzhou Jinjiang Environment Group which carries out waste upgrading for use as incineration fuel.

Zhejiang Jiucang Renewable Resources Development Co., ltd. operates a network of dry/ wet waste separation facilities in Hangzhou city. The company pays for dry waste and sends it to their facilities for sorting and recycling of useful resources, while wet waste is separated and enters the municipal waste treatment and disposal system.

Hangzhou Jinjiang Environment Group is a well-known waste incineration power plant company in China. MSW in China is not immediately conditioned for incineration after collection due to the complex composition of waste, the high moisture content, and the low calorific value. Therefore, Hangzhou Jinjiang Environment Group explored the practice of crushing and sorting waste to turn it into high-value fuel for waste incineration power plant. The sorting process flowchart is provided in figure 3. After going through this series of processes, the lower calorific value of waste could increase from 1000 kcal/kg to 2500 kcal/kg.

Figure 3. Waste separation process applied in Hangzhou Jinjiang Environmental Group

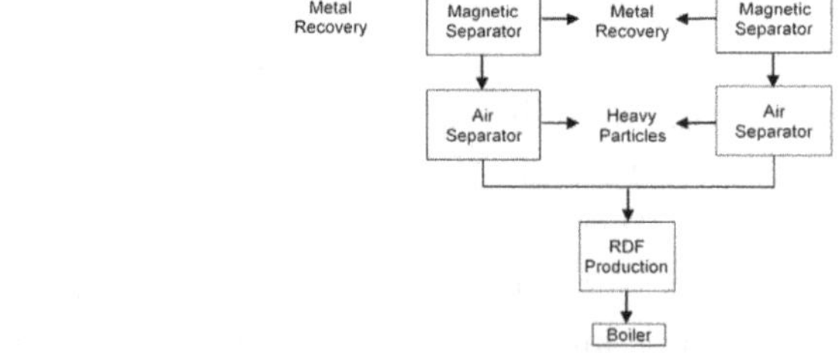

MSW TREATMENT BY NON-THERMAL TECHNOLOGIES

MSW Treatments Overview

There are three main MSW disposal methods used in China: landfill, composting, and incineration. In the last decade, Chinese Waste Treatment has rapidly developed. The percentage of MSW treated increased from 50.4% in 2003 to 97.75% in 2017 (NBSC, 2018). These percentages translate to 120.38 million tons disposed by landfill, 84.63 million tons by incineration, and 5.33 million tons by composting and others. The total amounts to 210.34 million tons of treated MSW as of 2017, as shown in Figure 4 (NBSC, 2018).

Figure 4 shows the level of commitment by the Chinese government in addressing the issue of MSW management, with the current MSW treatment rate reaching around 97% of collected MSW. It is also apparent that due to area limitation and safety concerns, the rate of MSW treated by landfilling is relatively stagnant. In its place, the use of incineration has seen a growth from less than 5 million tons in 2003 to close to 80 million tons in 2017 (NBSC, 2018).

The growth of incineration uses to process MSW is also obvious when evaluating the relative usage data of each MSW disposal method in the last few years. This data is presented in figure 5. Figure 5 shows that while landfill is still the dominant method of MSW disposal at 57.2% in 2017, the percentage of MSW disposed by incineration continues to increase from 4.9% in 2003 to 40.2% in 2017. Since 2011, the ratio of MSW disposed of by incineration has increased by around 3% per year. The situation of waste treatment in China could be described as "mainly treated by landfill, with a quick development of incineration".

Figure 4. Amount of MSW processed by different methods and total treatment rate (2003 – 2017) (NBSC, 2018)

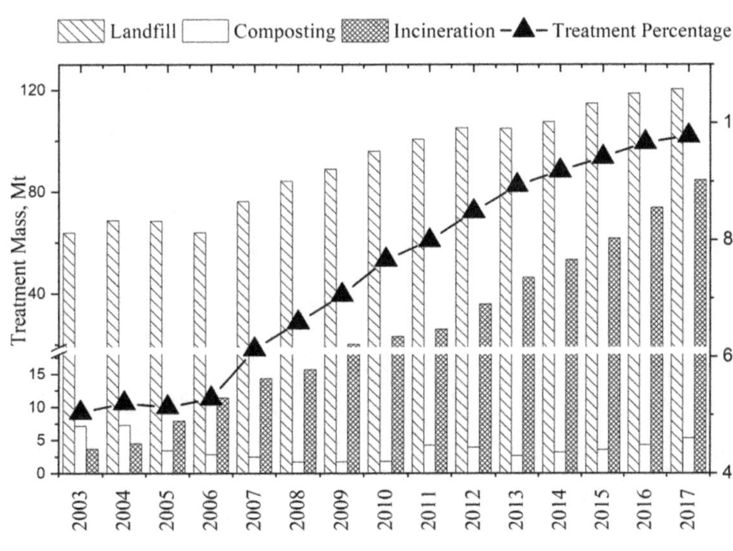

Figure 5. Usage percentage of each MSW disposal method in China (2003 - 2017) (NBSC, 2018)

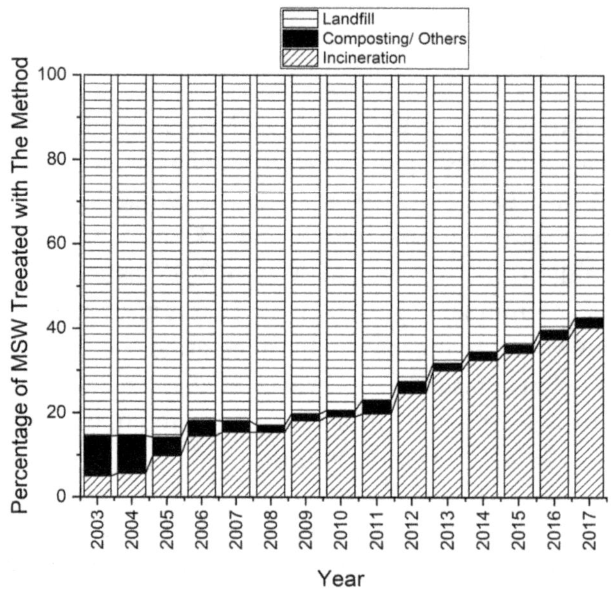

Sanitary Landfill

Landfill Situation and Biogas Collection

Landfill is the most commonly used method for MSW disposal in China and many other countries. As seen in Figure 6, landfill is still the most dominant method of MSW disposal in China. There are currently 654 active landfill sites in China, with the total MSW disposal capacity of 360,524 tons/day (NBSC,

Figure 6. Number and capacity of landfills in China (2003 - 2017) (NBSC, 2018)

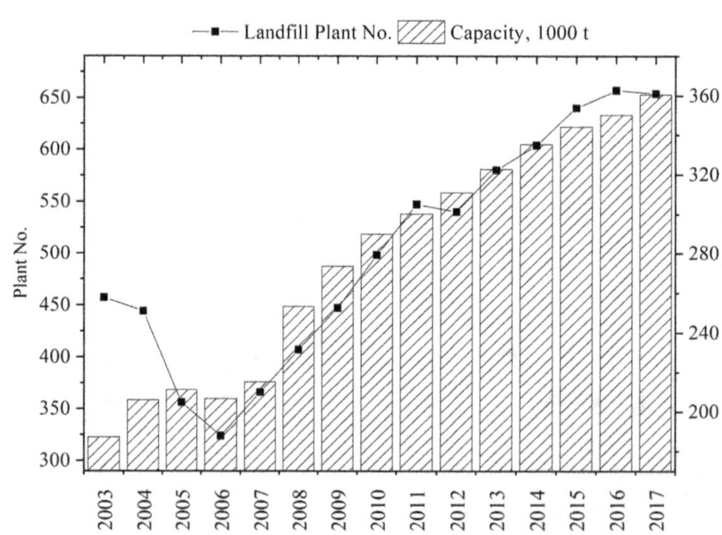

2018). Recently, more landfill plants are constructed and operational, as shown in Figure 6. But even that might not be enough. Calculations suggested that MSW amount to landfills are currently at 91.5% of their capacity. It shows that the development and construction of MSW management technology and facilities are urgently needed.

Outside of the capacity problem, landfill in China is being scrutinized due to its inefficient land use and potential environmental pollution including leachate and gas emission. The concern of methane emission, odor, health problem for workers and people living near the site, as well as the risk of landfill fires and explosion have pushed the government to mandate the reduction of landfill use. This mandate is formalized in the aforementioned 13th five-year plan that instructed that no new landfills can be opened in the provincial capitals.

In order to avoid the greenhouse gas of CH_4 emission, biogas collection and electricity production are now conducted in landfills. This effort was first started in Tianziling landfill, Hangzhou City, in 1998. Biogas collection is expected to bring benefits in form of usable biogas and the prevention of landfill gas accumulation that could result in fire and explosion hazard. With the current landfill situation in China, where more landfill is in full capacity and need to be closed, new applications are available for the development of landfill remediation technologies. A few landfill operators have also started to consider landfill mining for material recycling and fuel utilization in incineration. This method could make landfill act as a new source of material, while slowly reopening land previously used by landfill for other more valuable usages.

Standards & Guidelines of Landfill Construction & Operation

There are a series of standards and guidelines for landfill construction and operation that are continuously updated. The main standards include "Technical Code for Municipal Solid Waste Sanitary Landfill" (GB50869-2013) that governs the design process, site selection, and construction of landfill. 'Technical Specification for Operation and Maintenance of Municipal Solid Waste Sanitary Landfill' (CJJ93-2003) provides guidelines for landfill operation. "Standard for Pollution Control on the Landfill Site of Municipal Solid Waste" (GB16889-2008) is the standard for pollution control (leachate) in landfill and the specific requirement of feeding waste (such sludge, fly ash and other wastes). "Technical Code for Closure of Municipal Solid Waste Sanitary Landfill" (GB51220-2017) lists the requirements and procedure to ensure safe landfill closure at the end of its lifetime.

According to 'Standard for Pollution Control on the Landfill Site of Municipal Solid Waste' (GB 16889-2008), the leachate after treatment should meet the limits as shown in Table 3.

Landfill Closure and Land Reclamation

China is moving away from direct dumping of waste to the landfill. The guidelines to the closure of a landfill in provided in the Chinese standard GB 51220-2017, "Technical code for municipal solid waste sanitary landfill closure". Even after closure, landfills are usually still productive, producing landfill gas or being mined for resources.

One case study would be the Tianziling landfill in Hangzhou city. The first part of this landfill was opened in 1991 and closed in 2004. Since the beginning, the landfill was designed to be a sanitary landfill, with HDPE cover and lining creating airtight cells from which landfill gas is able to be collected. The surface of the first section of Tianziling landfill has now been converted to a park. However, the

Table 3. Standard limit for leachate from landfill according to GB 16889-2008

Item	Limit value	Item	Limit value
Chroma (dilution multiples)	40	Number of coliform Flora mg/L	10000
COD_{Cr},(mg/L)	100	Total Mercury,mg/L	0.001
BOD_5, mg/L	30	Total Cadmium,mg/L	0.01
Suspended Matter,mg/L	30	Total Chromium,mg/L	0.1
Total nitrogen,mg/L	40	Cr^{6+},mg/L	0.05
Ammonia nitrogen,mg/L	25	Total Arsenic,mg/L	0.1
Total phosphorus,mg/L	3	Total lead,mg/L	0.1

waste underneath the surface continues to produce landfill gas which is channeled to the on-site WtE plant (Zhou, et. al., 2018). The HDPE lining also works to prevent odor and gases from leaking out in an uncontrolled manner.

Currently, there is a push to promote landfill/ urban mining in China. This method offers the advantage of recovery of resources as well as the gradual recovery of space. A study has suggested that the main cost of landfill mining is in capital expenses for landfill mining equipment (Zhou, et. al., 2015). This seems to suggest that landfill mining in China is currently only feasible for reclamation of large sized sanitary landfills.

Composting/ Fermentation

Composting Situation for MSW

Composting is not a popular MSW disposal method in cities in China, as shown by the low percentage of MSW treated by composting in figure 5. In 2017, composting was used to treat only 2.5% of total MSW treated, for a total capacity of 21303 tons/day (NBSC, 2018). In the 2018 Chinese statistic yearbook, compositing is not even calculated as its own category, instead being lumped together in the "others" category outside of landfill and incineration. It is however, still practiced in the rural and village areas. This is due to the low volume of waste and the high percentage of organic waste suitable to be composted in rural waste. Anaerobic composting is done mainly to produce biogas, and aerobic composting is mainly for waste disposal and produce fertilizer. The application of the two types of composting are briefly explained below.

Anaerobic/Aerobic Composting in Rural Area

Anaerobic composting/fermentation of village waste is primarily done for biogas production. The Chinese government showed support for this program with CNY 14.2 billion allocated fund regulated in the 12th five year (2010-2015). Currently, the biogas production from village waste composting is over 15.8 billion m^3. It is mainly used to supply clean fuel for cooking in villages, with an estimated 41.93 million families reaping the benefits (NDRC, 2017b).

Anaerobic composting/ fermentation is now faced with new challenges, in the form of urbanization and higher labor cost. Small family-scale composting/ fermentation process that was popular a decade ago, is now almost entirely not practiced due to its low efficiency and high man-hour requirement. It prompted the Chinese government to adjust its policy in 2015, to focus more on supporting the larger scale composting/ fermentation facilities. According to the "13th Five-year Plan of Biogas Development in Village" (NDRC, 2017b), the target would be to achieve 22.77 million m^3 of annual anaerobic composting/ fermentation capacity with 4.9 billion m^3 biogas product in 2020.

In Jindong district of Jinhua City, aerobic composting with solar assistance for waste treatment is being used. The solid product can be used as garden fertilizer or sent to final disposal site. Depiction of the facility used to perform the process is depicted in Figure 7.

WASTE TO ENERGY APPLICATION

As shown in figure 5, incineration is the second most used technology for MSW treatment in China. It accounts for 40.2% of total MSW disposal (NBSC, 2018). Other than effective disposal of MSW, waste-to-energy (WtE) technologies such as incineration could also contribute to a nation's power generation. With its twofold benefit, WtE is expected to resolve the increasing generation of MSW and energy shortage in China. With the potential to generate 300 – 400 kWh/ ton of MSW, electricity generated from WtE plants is expected to have a bigger contribution towards China's energy production in the future.

Based on a simple equation of heat value, the current MSW yield (1000 Kcal/kg of low heat value) is equal to 30.74 million tons of standard coal (7000 Kcal/kg of low heat value) (Ansar, et. al., 2014). Energy generation from MSW instead of coal could also result in lower CO_2 emission. WtE also offers the benefit of more efficient land use. While 10 tons of MSW directly disposed to landfills requires 10 m^2 of dumping area, fly ash from the incineration of 10 tons of MSW only requires 0.3 m^2. This more efficient use of land is particularly important for China. At 138 capita/km^2, available land per capita in China is lower compared to countries such as USA and France. It forces China to develop WtE technology and application for MSW disposal, energy production, and emission control within the limited available land areas around its cities.

Figure 7. Facility for aerobic composting with solar assistance

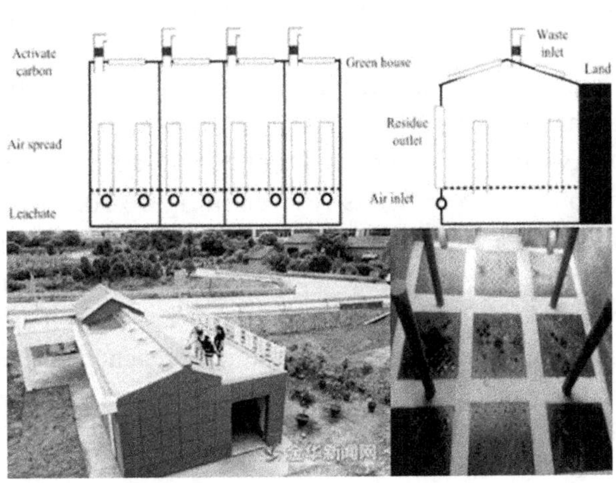

Figure 8. Number and total capacity of WtE plants in China (2003 -- 2017)

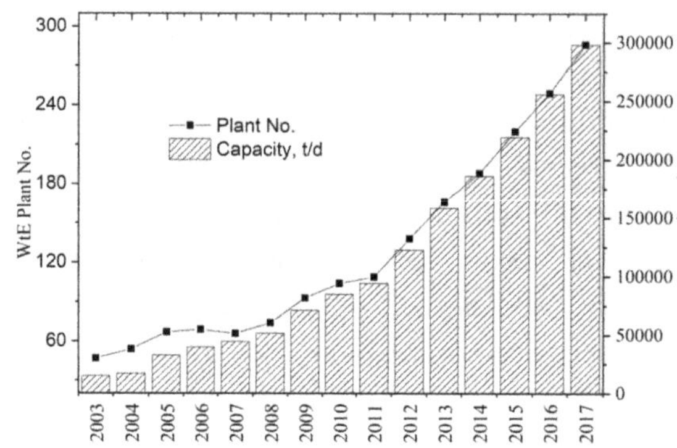

WtE Plants & Power Generation

WtE Plants & Location

WtE is developing rapidly in China, especially in the last decade. According to the 2018 China Statistical Yearbook, the number of WtE plants increased from 47 plants with the total capacity of 15000 t/d in 2003 to 286 plants with the total capacity of 298062 t/d in 2017. Growth of plant number and capacity in China from 2003 to 2017 is provided in figure 8 (NBSC, 2018).

From the data in figure 8, it could be calculated that the average capacity of WtE plants increased from 319 t/d in 2003 to 1042 t/d in 2017. According to an industry survey at the end of 2018, there are 356 WtE plants with 425,495 t/d capacity operational in China. The locations of these plants are shown in Figure 9 (Wuhu Ecology Center (WEC), 2019).

From Figure 9, it can be observed that more WtE plants are located in the eastern provinces. This is due to the higher economic level of those provinces. The provinces with the regional domestic product (RDP) levels in China are mostly located in the eastern part of the country. These provinces include Zhejiang, Shanghai, and Guangdong. As previously mentioned, there is proof that economic level has a strong influence on waste generation in the region. This data also shows that the Chinese government has been able to identify the specific regional needs in term of MSW management and has acted upon them properly.

According to the annual report by China Electricity Council (China Electricty Council, 2018), power generation from WtE plants has increased from 209.74×10^4 Kw (from a total capacity of 97.23×10^8 Kwh) in 2011 to 725.3×10^4 Kw (from a total capacity of 375.2×10^8 Kwh) in 2017. The data on yearly growth of electricity generation from WtE plants is given in figure 10.

One example of good WtE progress at the provincial level is the Zhejiang province. According to the report data from Zhejiang Provincial Development and Reform Commission, the generation of power from WtE is in the range of 200 -380 Kwh/t (on the basis of raw MSW entering plant), with the average of 287 Kwh/t. If this power generation efficiency could be reached nationally, the amount of power generated from WtE plants could exceed the 480,000 t/d processing capacity and 750×10^4 Kw generated that was set by the 13th Five-year Plan for Biomass Energy (NDRC, 2017c).

Figure 9. Number and locations of WtE plants in China

Figure 10. Level of generation and capacity of energy from WtE plants (2011 - 2017)

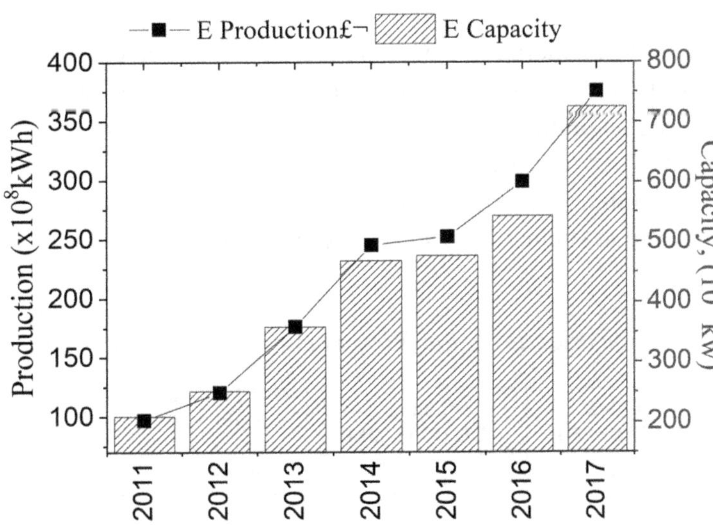

WtE Incineration Technology

There are two main types of incineration technologies applied in China, mechanical grate and circulating fluidized bed (CFB). At end of 2018, the capacities of grate and CFB are 344,145 t/d and 78,850 t/d, respectively. This number translates to 80.9% market share for grate and 18.5% for CFB. This number becomes interesting when considering that in 2010, the number of facilities utilizing grate and CFB technology were almost even.

The reason for the development that seems to favor grate technology could be linked to two factors: the reliability of grate technology, and the smaller amount of fly ash produced by grate technology compared to CFB. Data from the industry shows that grate facilities could operate up to 8000 hours/year, higher compared to 7000 hours/year for CFB facilities. CFB technology also requires MSW to go through a preliminary shredding process before incineration, adding cost and complexity to the process.

Even though CFB offers process-related benefits as it is more effective in handling the composition variation in MSW and its ability to process low heat value MSW better than grate, it also produces more flue gas and fly ash. This inherent disadvantage runs against emission control standard that is getting stricter by the year, forcing CFB facility operators to invest in additional utility system to handle the incineration byproducts. The technical regulations for the construction and operation are provided in "Technical Guideline on Municipal Solid Waste Incineration" (RISN-TG009-2010)for grate incinerator design and construction, and "Technical Guideline on Municipal Solid Waste Fluidized Bed Incineration" (RISN-TG016-2014)for CFB incinerator design and construction.

As mentioned before, the Chinese government is embracing private companies in running MSW management facilities. The biggest CFB MSW incinerator operator is China Jinjiang Environment Holding Company Limited, while China Everbright International Limited is the biggest MSW grate operator in China.

WtE Flue Gas Pollution Control

Flue gas emission from MSW incineration China is regulated by the Standard for Pollution Control GB 18485. This standard was issued in 2000 and revised for the first time in 2001. The second revision was done in 2014, lowering the maximum emission levels compared to the 2001 standard and adding restrictions for heavy metals. The 2014 revision also included site selection guidelines, technical requirements, operating procedures, pollutant emissions, and pollutant monitoring. A comparison of the old and new Chinese standards compared to standards used by other developed countries is presented in table 4.

Table 4. Comparison of various standards for flue gas emission (Zhang, 2016)

Pollutants	China (GB 18485-2001)	China (GB 18485-2014)	USA	EU (2010/75/EU)	Germany (17th BlmSchV)
PM (mg/m^3)	80	20	20-30	10	10
HCl (mg/m^3)	75	50	30	10	10
HF (mg/m^3)	-	-	-	1	1
SO_X (mg/m^3)	260	80	60	50	50
NO_X (mg/m^3)	400	250	215-500	200	200
CO (mg/m^3)	150	80	-	50	50
TOC (mg/m^3)	-	-	-	10	10
Hg (mg/m^3)	0.2	0.05	0.03	0.05	0.03
Cd + Tl (mg/m^3)	0.1	0.1	0.07	0.05	0.05
Pb (mg/m^3)	1.6	< 1.0	0.14	≤ 0.5	≤ 0.5
Dioxins (ng TEQ/m^3)	1.0	0.1	0.2	0.1	0.1

Table 4 shows how the Chinese standard has changed over the years. The 2014 standard is much stricter than the 2001 version, with standard limit values approaching the standard values applied by other developed countries.

In 2013, the main pollution control system used in China consists of semi dry scrubber + active carbon spray + bag house. While in USA, 57.5% of WtE plants utilized selective non-catalytic reduction (SNCR) + semi dry scrubber + active carbon spray + bag house (Yan, Xiong & Li, 2015). With the revision of "Standard for Pollution Control on the Municipal Solid Waste Incineration", more pollution control technology and equipment are installed in WtE plants in China. The additional utilities are mostly installed in the post combustion zone.

According to a survey of 283 plants with 371,858 t/d capacity in 2018, most of WtE plants use semi dry scrubber + activated carbon spray (AC Spray) + bag house. SNCR use also rose significantly. Other technologies such as electrostatic precipitator (SEP) is still rarely used in WtE plants in China. The full data is depicted in Figure 11.

In order to increase public acceptance of WtE incinerators, a few local governments have required WtE plants to work on modifying their white smoke emission and make it colorless. This is done with the use of gas-gas heater (GGH) to increase the flue gas temperature before emission. The heater prevents partial condensation of water vapor in the gas emission that is responsible for the white color in the smoke, making the smoke completely colorless.

Fly Ash Characteristics and Treatment

Fly ash is one of the solid byproducts of the incineration process. It is finer compared to bottom ash and could only be collected by bag house filtering of flue gas. The production of fly would be different based on the type of incinerator used. Grate incinerators producing 30-50 kg fly ash per-ton of MSW incinerated and CFB incinerators producing 50-150 kg fly ash per-ton MSW incinerated (Nie, 2008).

Fly ash consist of solid particles with the size in the range of 40 – 200 μm. It is also known to adsorb some pollutants, such as heavy metal and dioxins, during the post combustion. These properties make it unsafe for people and the environment. Fly ash is defined as hazardous waste in China and has to be

Figure 11. Available flue gas cleaning technologies used in WtE plants in China

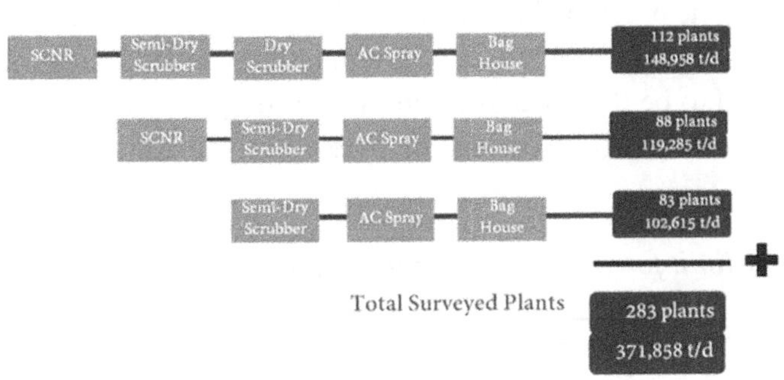

properly treated avoid secondary environmental pollution. This treatment of fly ash has become the biggest challenge for sustainable development of WtE in China, as the cleaning technology for flue gas and leachate are much more mature and well applied.

According to "Standard for Pollution Control on the Landfill Site of Municipal Solid Waste (GB16889-2008)", the requirements for fly ash from WtE facilities to be treated in landfill include: less than 30% moisture content, less than 3 μg I-TEQ/kg dioxin concentration, 3) leaching concentration of heavy metal less the values shown in table 5 (Ministry of Environmental Protection (MEP), 2018).

In 2018, Ministry of Ecology and Environment conducted a trial for the implementation of "Environmental Access Conditions for Waste to Energy Construction Project" (Ministry of Ecology and Environment, 2018). One of the results was a recommendation that treatment of fly ash as hazardous waste be carried out in strict accordance with the relevant national hazardous waste management regulations for transportation and harmless safety disposal. After treatment according to the requirements of "Standard for pollution control on the landfill site of municipal solid waste (GB16889-2008)", fly ash can be disposed at landfills. It could also be used for co-processing in cement kiln after being treated to meet the requirements of "Standard for pollution control on co-processing of solid wastes in cement kiln" (GB30485-2013).

Leachate Treatment

According to "Environmental access conditions for waste to energy construction project (trial)", leachate and vehicle-washing wastewater should be collected and disposed in plant or sent to an external treatment plant. The leachate release has to meet the requirement set by "Standard for pollution control on the landfill site of municipal solid waste" (GB16889-2008) and "Standard for pollution control on the municipal solid waste incineration" (GB18485-2014).

New Technology Research & Development

While incineration is efficient in its function of converting MSW to electricity, new attention is given towards the possibility of using MSW as a potential source of material to produce high value products. To achieve this goal, there is a focus on the research of MSW advanced conversion technologies including gasification and pyrolysis.

Pyrolysis is expected to be the process to produce high quality liquid fuel from MSW, commonly referred to as bio-oil. While the potential is there, bio-oil is not readily available yet to be used as gasoline or diesel engine fuel. Further study into bio-oil refinement and de-sulfurization is necessary.

Gasification is projected to be the process to produce high quality syngas usable for a variety of applications. Syngas from MSW gasification has the potential to be used as a substitute for syngas from

Table 5. Leaching toxicity limit for fly ash treated in landfill (MEP, 2018)

Heavy metals	**Hg**	**Cu**	**Zn**	**Pb**	**Cd**	**Be**
Concentration(mg/L)	0.05	40	100	0.25	0.15	0.02
Heavy metals	Ba	Ni	As	Total Cr	Cr^{6+}	Se
Concentration(mg/L)	25	0.5	0.3	4.5	1.5	0.1

coal or synthesis gas from natural gas for a wide range of applications such as methanol and ammonia synthesis. Syngas, with its high content of hydrogen (up to 30%), is also being researched as a potential source of hydrogen for fuel cell applications.

The application of MSW gasification is still met with a few hurdles currently being worked on by researchers. Chief among them is the presence of tar byproduct in syngas, the problem of high moisture content in MSW making gasification process inefficient, and the upgrading of syngas to make it suitable for more high value applications. The problem of tar in syngas has been addressed in various methods. The most recent development involves tar cracking to break tar and form more combustible gases. Tar cracking could be achieved either by thermal cracking or catalytic cracking with rare earth catalysts.

Recently, supercritical water gasification (SCWG) has been suggested as an alternative process to gasify MSW. The inherent advantage of SCWG compared to thermal gasification is that, since it is done in supercritical water environment, it is not necessary for the MSW to be dried before gasification. Considering the high volume and moisture content of MSW, this advantage would save significant time and cost. However, research into SCWG is still in its infancy. With major obstacles to be overcome such as the energy needed to make supercritical water, the complex equipment needed to make this high temperature and high-pressure process work in a continuous flow, and the problem of wastewater it generates.

Syngas cleaning or upgrading is defined as the process of removing unwanted gases from syngas to make it suitable for high value applications. For syngas from gasification process, these gases are usually CO_2 and H_2S. The removal of acid gases from syngas differs from flue gas especially in their concentration and composition. The current focus of syngas upgrading is on making syngas available for hydrogen fuel, be it for use in fuel cells or as material for hydrogen carrier synthesis.

Figure 12 describes the trend in publications regarding pyrolysis and gasification in China from 2013 to 2018. While the amount of publications regarding pyrolysis is growing at an increasing rate since 2016, publications on gasification only seem to grow by a small rate. Considering the potential of gasification process, the lack of growth in publications on the subject seems to indicate problems relating to the process that still needs more work in future.

Figure 12. Number of scientific publications on gasification and pyrolysis by Chinese affiliated authors (Scopus, 2019)

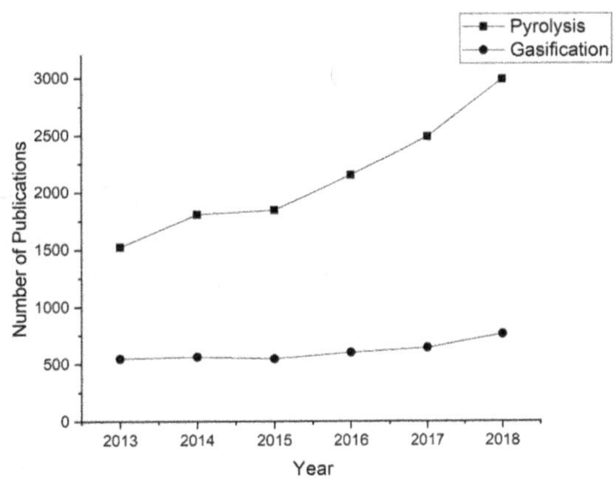

WTE-PROMOTING POLICY & REGULATIONS

The Chinese government has set and implemented a series of policies for promoting the development of WtE technology and applications, encouraging private capital investment in WtE industry, optimizing approval process, along with reducing taxes and fees. In China, WtE is a classified as renewable energy technology and source, making the electricity from WtE eligible to receive subsidy from The State Grid Corporation of China.

Promoting Investment by Tax Reduction

According to the Law of the People's Republic of China on Circular Economy Promotion, WtE meets the requirements of circular economy development. In the law, governments (city level and above) are responsible in building MSW collection and utilization facilities in addition to establishing and improving sorting collection and recycling system.

According to the Law of the People's Republic of China on Renewable Energy, electricity and energy generated from WtE are considered renewable energy. It means that the scientific and technological research of WtE must be given priority by science and technology institutions, as well as high-tech industries. Special funding is provided to support and promote technological advancements in the development and utilization of WtE, and to decrease its cost.

The integration of renewable energy enterprises in electricity generation requires grid enterprises to carry out effective and active grid-connected absorption, and support the sale subsidy of WtE generated electricity together with the government. The Renewable Energy Industry Development Guidance Catalogue also lists 88 renewable energy industries, which includes WtE and landfill biogas power generation.

According to the "Notice on the Comprehensive Utilization of Resources and Other Products VAT Policies", another policy in promoting the development of WtE is implemented in the form of tax break. The government guaranteed that VAT from the sale of "electricity or heat generated by MSW as fuel", where MSW makes up over 80% of the total fuel, will be refunded immediately.

Tipping Fee and Electricity Sale Subsidy

The tipping fee of MSW incineration is calculated based on regional characteristics. It needs to be determined by the government according to investment and operation cost assessment, which are significantly affected by geographical differences. The tipping fee of MSW also varies according to the quality of MSW transported to the WtE plant. With the maturity of technology and full market competition, tipping fee of MSW has shown a downward trend. It is now around CNY 100/tons, down from the initial CNY 200/ton. In large cities with relatively developed economies, tipping fee for MSW can be as high as CNY 150/ ton, with the average fee around CNY 100/ton. In other cities, tipping fee of MSW is generally around CNY 50-100/ ton as our survey of the business news. There are also very few cities with overly competitive market, where tipping fee of MSW is as low as CNY 30/ton. Tipping fee of MSW is supplied by the local government from the financial budget.

Through a regulation dubbed "Measures for the Full Purchase of Renewable Energy Electricity by Power Grid Enterprises", The Chinese government mandated that power grid enterprises purchase the electricity from renewable energy power generation projects within their grid coverage. The measure has solved the problem of power generation grid-connected obstacles and sales difficulties for WtE.

Notice on Improving the Price Policy for WTE specifically quantifies the on-grid price of waste incineration power generation. The notice stipulates that power generation from WtE is first converted into electricity based on the amount of waste disposal in the plant. For each ton of MSW processed, 280 kWh is given the national unified electricity price of CNY 0.65/ kWh. The rest of the production is priced the same as electricity from coal power plant according to the local provincial power grid charge, which is CNY 0.1/ kWh. The government also ensure that power generation projects will receive electricity subsidy for 15 years since the date of first production. A co-firing power generation project with more than 20% conventional fossil energy consumption in power generation is treated as a conventional energy power generation project, and the benchmark electricity price of a local coal-fired power plant is implemented, and the subsidized electricity price is not provided.

Others

In 2017, the National Development and Reform Commission issued the "Notice on Further Planning and Site Selection of MSW Power Plants" (NDRC, 2017d). It laid out the detailed plan for WTE plant construction nationally until 2020. The notice clearly defines the construction scale and site, completion time, and disposal capacity. It also contains a list of WTE plant construction target by 2030, including the construction scale, site, etc., while incorporating the latest version of city development plan. This effectively guarantees the availability of land required for WTE plant construction in the future.

The National Climate Change Plan (2014-2020), the 13th Five-Year Plan for Renewable Energy Development and the 13th Five-Year Plan for Biomass Energy Development, etc., propose to vigorously develop WtE, to effectively treat MSW for environmental protection, and to expand power supply for fossil use reduction. By 2020, the capacity of WTE will be 480,000 tons/day, and the installed capacity of electricity production will be 7.5 million kilowatts (National Development and Reform Commission (NDRC), 2016).

OTHER WASTES

Other than MSW which is the main concern in waste management in China due to its volume, other types of wastes are also being managed by the government in China. According to China statistical yearbook (2018), industrial solid wastes are divided into common industrial solid waste (CISW) and hazardous waste (HW). In 2017, the generation of CISW was up to 3315.9 million tons. Of that amount, 1811.9 million tons was comprehensively utilized, 798.0 million tons was disposed of, and over 784.7 million tons was stored or discharged (NBSC, 2018).

69.4 million tons of HW was generated in 2017. 40.4 million tons of it was utilized, 25.5 million tons was disposed of, and 8.71 million tons temporarily stored (NBSC, 2018). In additional, both of CISW and HW significantly increased, as 992.6 million ton of CISW and 11.7 million ton of HW was generated, respectively.

The Chinese government is also very concerned about the utilization and treatment of industrial waste. During the 12th five year (2010-2015), "12th five-year plan on environmental protection of hazardous waste pollution" and "National Plan on construction of hazardous waste and medical waste disposal facilities" and other specific regulation and plan were implemented to prevent the risk of industrial solid waste. In the year 2015, 50 hazardous waste central disposal center and 273 large scale medical waste disposal facilities were installed and they are currently in operation (State Council, 2016b).

13th Five-Year Plan for Ecological Environmental Protection Plan also calls for an increase in the disposal capacity of industrial waste and hazardous waste. This includes the production monitoring, pollution prevention, disposal facility and smart management system. It is projected that in 2020, the precentage of comprehensive utilization of industrial solid waste and agricultural waste will be over 73% and 85%, respectively, based on the 13th Five-year Plan for Energy Saving and Emission Reduction. The main groups of other specific wastes, based on commonly used classification for waste streams, and their management shall be discussed in this section.

Hazardous Waste

According to the 2018 Chinese Statistical Handbook, 69.36 million ton of hazardous waste was produced in China in 2017 (NBSC, 2018). The preferred method of disposal for hazardous waste in China is by thermal method including incineration or co-disposal in cement kiln.

The Chinese government paid special attention to the disposal of hazardous waste, especially hazardous chemical waste, by issuing two mandates in the country's 13th five-year plan: "Plan for Hazardous Chemical Safety" and "Comprehensive Plan for Hazardous Chemical Safety Management" (Ministry of Industry and Information Technology, 2016). These plans have spawn 92 regulations related to hazardous chemical management in China, with its implementation mainly done by 8 governmental departments (Wang et. al., 2018).

While incineration process eliminates most hazardous properties from hazardous waste and reduces their volume by up to 90%, there is the problem of secondary pollution from incineration. In the incineration of hazardous waste, this shows up in the form of heavy metal released from the incineration in the form of ash (Wi et. al., 2018). Therefore, further research on the distribution and characteristics of heavy metal from hazardous waste incineration is required.

In China, medical waste is also considered hazardous waste and cannot be mixed with municipal solid waste. Medical waste is usually treated by independent incinerator or disposal facility. According the research by Mi Yan (2011), the annual generation of MW reached 0.97 million tons in China in 2008. The major components of MW are fibrous materials, plastic and rubber (which have a high caloric value,10,000–16,000 kJ/kg) and chlorine content (4.5–7.5%) compared to municipal solid waste. This results in MW giving a very good energy yield as incineration fuel, but on the other hand producing higher amount of pollutants compared to MSW (Yan, 2011).

Agricultural Waste

According to a study, 60% of agricultural waste in China is disposed without treatment (Zhang & Sun, 2018). The other 40% is mostly treated with composting, which has been extensively covered in section 4.3.

Based on the assessment of crop straw utilization in 12th five year (2010- 2015), there was 1.04 billion tons of crop straw waste in China in 2015, with only 0.9 billon tons of it collected properly. Chinese government urges for the utilization of agricultural waste as fertilizer, feed, fuel, and raw material. The government also strictly prohibits open burning in most areas in China. According to Notice on Further Accelerating the Comprehensive Utilization of Crop Straw and Prohibition of Burning, the current ratio of crop straw utilization is up to 85%, and the burning area reduced 6% compared to the value in 2016.

The Chinese government has targeted that 75% of livestock and poultry waste would be processed before disposal in 2020. To meet that end, 95% of small and medium-sized breeding farms must be equipped with a facility capable of processing its own waste by 2020, while large scale farms are required to do so earlier, in 2019 (OSC, 2019).

The responsibility of creating a plan for agricultural waste recycling falls to the city-level government. For the current period (2016-2020), city governments have worked to execute their plans with the Ministry of Agriculture since 2017 (OSC, 2019).

E-Waste

E-waste is defined as waste from electronic devices and their components. It was reported that in 2016, China produced over 6 million tons of e-waste. On top of it, China imported nearly 70% (which amounted to nearly 30 million tons) of the world's e-waste under the guise of second-hand electronics (Wang, Zhang, and Guan, 2016). It is also predicted that in 2030, China's e-waste production would surpass 28 million tons (Gu et al., 2016). E-waste is a complex problem since it is wanted for its valuable metal components. At the same time, the low-technology process used to extract these metals often leaves behind secondary wastes that are dangerous to the environment and human health.

To counter this problem, the Chinese government issued a series of regulation since 2007. It includes "Administrative measures on the collection of renewable resources" (2007), "Administrative measures on the prevention and control of environmental pollution by WEEE" (2007), "Administration regulation for the collection and treatment of waste electronic and electrical equipment" (2009), "Administrative measures on the qualification of WEEE treatment" (2010), "Administrative measures on the distribution of used electrical and electronic products" (2013), and "The restriction of hazardous substances in electrical and electronic equipment" (2016) (Song et al., 2019). These regulations govern clean production methods, usage, proper disposal, and treatment of e-waste.

As a result of the aforementioned measures, China's e-waste recycling industry is on the rise. It has grown from only a handful to over 100 official companies in 2015 (Li et al., 2015). However, reliable data regarding their capacity is difficult to obtain due to their different procedures which resulted in different level of resource recoveries (Zeng et. al., 2017).

Marine Waste

Marine waste/ debris has received relatively little attention in China. In was reported that in 2014, the weight densities of submerged marine debris in China was 22.6 g/ 100 m^2. This value was considered low level compared to other countries in the world. However, the beached marine debris density was high, at 133.80 g/ 100 m^2 (Zhou et. al, 2016). This was thought to be caused by the southeast wind during the wet seasons carrying litter to accumulate mainly on the northern South China Sea region (Zhou et. al, 2016). The Management of marine debris in China has not been regulated by any specific law.

CONCLUSION

MSW disposal and utilization in China has greatly progressed in the last decade in China, and 97.75% of total MSW was treated in 2017. City environment quality is significantly improved with the technical development of sanitary landfill and incineration, and thus all citizen stands to gain benefit from MSW treatment system and construction of disposal facilities.

Landfill is the treatment most used for MSW treatment in China, with 57.2% of the total treated MSW (NBSC, 2018). The current model of MSW treatment is "mainly treated by landfill, with a quick development of incineration", for resolving the challenge for lack of areas around the cities. With 91% of landfill capacity filled up, new and innovative processes are required to handle the MSW generation, which is expected to continuously increase due to population and economic growth. The Chinese government has embraced the industry and people to overcome this problem. With strong stimulus and incentives provided by the government, the private enterprises are investing in the field of waste-to-energy (WtE) in droves. This is evident from the percentage of MSW disposed by incineration, which continue to increase from 4.9% in 2003 to 40.2% in 2017 (NBSC, 2018). The involvement of private companies with the support of the government is expected to provide an effective and professional approach to waste management through WtE. Waste management is also being reinforced from the point of disposal with the participation of the people. With the involvement of the population, it is expected that the waste sorting and recycling process would be much more efficient.

The currently designed plan for MSW treatment in China would first be to reduce the flow of waste to landfills while gradually reducing the number of active landfills as formulated in the 13th five-year plan. In this waste management plan, there would be no direct dumping of MSW to the landfills. Instead, MSW would first be directed towards recycling centers where the recyclable parts such as metals and plastics are separated for reutilization. MSW components that could not be recycled would be taken to MSWI, where it would be used as fuel to generate electricity. The fly ash byproduct would be sent to landfill. With this scheme, it is expected that the volume of the fly ash dumped into landfills would only be around 10% of that from the same amount of MSW if it was directly deposited in the landfill.

The National Climate Change Plan (2014-2020), the 13th Five-Year Plan for Renewable Energy Development, and the 13th Five-Year Plan for Biomass Energy Development proposed to vigorously develop WtE. This would be achieved by effectively treating MSW for environmental protection and to expand power supply for reduction of fossil fuel use. By 2020, the capacity of WTE is projected to be 480,000 tons/day, and the installed capacity of electricity production to be 7.5 million kilowatts (NDRC, 2016). WtE will play an ever-increasing role in waste management in China, and the percentage of disposal MSW amount in incineration is set to be over 50% in the end of 2020. China would also actively promote the export of Chinese WtE technology and equipment to overseas market, as it is now able to provide high quality product and service with rich experience.

The emission limit will be continuously revised for stricter control, as Chinese government completely focuses on the environmental protection and sustainable development. The present technology for flue gas emission is "SNCR + Semi-dry Scrubber + Dry Scrubber + Activated Carbon Sprayer + Bag House". With the stricter limit, more SCR and gas heating are predicted to be installed in the new plants.

WtE plants main avenues for profit are from MSW tipping fee and electricity subsidy. Due to the market competition and technology development, tipping fee has seen a downward trend recently. On the electricity subsidy side, subsidy per Kwh is fixed with no excess of 280 Kwh/ton MSW. In the opinion of the authors, the tipping fee will soon become stable. While a reduction in subsidy would be happen, as was the case with solar energy in China. Further economic gains would have to rely more on scientific progress to promote high efficiency energy conversion.

REFERENCES

Ansar, M., Ani, F., & Kader, A. (2014). The Energy Potential of Municipal Solid Waste for Power Generation in Indonesia. *Jurnal Mekanikal*, *37*, 42–54.

Central Committee of the Communist Party of China and State Council. (2019). *Announcement on Accelerating the Construction of Ecological Civilization*. Retrieved July 1, 2019, from: http://www.xinhuanet.com/politics/2015-05/05/c_1115187518_3.htm

China Electricity Council. (2018). *Annual Report of Electricity 2018*. Retrieved July 1, 2019, from: http://www.cec.org.cn/guihuayutongji/tongjxinxi/niandushuju/

Gu, Y., Wu, Y., Xu, M., Mu, X., & Zuo, T. (2016). Waste Electrical and Electronic Equipment (WEEE) Recycling for a Sustainable Resource Supply in The Electronics Industry in China. *Journal of Cleaner Production*, *27*, 331–338. doi:10.1016/j.jclepro.2016.04.041

Li, J., Yang, L., & Liu, L. (2015). Development Potential of E-waste Recycling Industry in China. *Waste Management & Research*, *33*(6), 533–542. doi:10.1177/0734242X15584839 PMID:25990983

Li, W., Ma, Z., Huang, Q., & Jiang, X. (2018). Distribution and Leaching Characteristics of Heavy Metals in a Hazardous Waste Incinerator. *Fuel*. doi:10.1016/j.fuel.2018.06.041

Ministry of Ecology and Environment. (2018). *Environmental Access Conditions for Waste to Energy Construction Project*. Retrieved July 1, 2019, from: http://www.sdfgw.gov.cn/art/2018/3/20/art_4996_817510.html

Ministry of Environmental Protection. (2018). *Standard for Pollution Control on the Landfill Site of Municipal Solid Waste*. Retrieved July 1, 2019, from: http://codeofchina.com/standard/GB16889-2008.html

Ministry of Housing and Urban-Rural Development. (2016). *13th Five-year Plan for Housing and Urban-rural Construction*. Retrieved July 1, 2019, from: https://max.book118.com/html/2017/1219/145056289.shtm

Ministry of Housing and Urban-Rural Development. (2018). *Notice on Promoting the Experience of Waste Classification and Resource Utilization in Rural Areas of Jinhua City*. Retrieved July 1, 2019, from: http://www.mohurd.gov.cn/wjfb/201612/t20161228_230118.html

National Bureau of Statistics of China. (2018). *China Statistical Yearbook 2018*. Beijing: China Statistics Press.

National Bureau of Statistics of China. (2018). *National Data*. Retrieved may 28, 2019, from: http://data.stats.gov.cn/

National Development and Reform Commission. (2016). *Notice on The Publication of the 13th Five Year Plan for The Development of Renewable Energy by The National Development and Reform*. Retrieved June 18, 2019, from: https://chinaenergyportal.org/en/13th-fyp-development-plan-renewable-energy/

National Development and Reform Commission. (2017). *13th Five-year Plan of Biogas Development in Villages*. Retrieved June 20, 2019, from: http://www.ndrc.gov.cn/fzgggz/fzgh/ghwb/gjjgh/201706/t20170607_850195.html

National Development and Reform Commission. (2017). *13th Five-year Plan for Biomass Energy*. Retrieved June 20, 2019, from: http://www.ndrc.gov.cn/fzgggz/fzgh/ghwb/gjjgh/201708/20170809_857320.html

National Development and Reform Commission. (2017). *Notice on Further Planning and Site Selection of MSW Power Plants*. Retrieved June 19, 2019, from: http://www.ndrc.gov.cn/ zcfb/gfxwj/201712/t20171220_ 871129.html

National Development and Reform Commission, Ministry of Housing and Urban-Rural Development. (2016). *13th Five-year Plan for Construction of MSW Treatment Facilities*. Retrieved April 20, 2019, from: http://www.ndrc.gov.cn/zcfb/zcfbtz/201701/t20170122_836016.html

National Development and Reform Commission, Ministry of Housing and Urban-Rural Development. (2017). *Implementation Plan for Municipal Solid Waste Classification System*. Retrieved June 20, 2019, from: http://www.gov.cn/zhengce/content/2017-03/30/content_5182124.htm

National Development and Reform Commssion. (2015). *Notice on Further Accelerating the Comprehensive Utilization of Crop Straw and Prohibition of Burning*. Retrieved June 25, 2019, from: http://www.ndrc.gov.cn/zcfb/zcfbtz/201511/t20151125_759523.html

Nie, Y. (2008). Development and Prospect of Municipal Solid Waste (MSW) Incineration in China. *Frontiers of Environmental Science & Engineering*, *2*(1), 1–7. doi:10.100711783-008-0028-6

Office of the State Council. (2016). *Pilot Work Plan for None Waste City Construction*. Retrieved June 18, 2019, from: http://www.gov.cn/xinwen/2019-01/21/content_5359705.htm

Office of the State Council. (2017). *Prohibition of Foreign Garbage Entry and Promotion of Solid Waste Import Management System Reform Implementation Plan*. Retrieved June 20, 2019, from: https://www.isri.org/docs/default-source/int'l-trade/170727chinaannouncement-en.pdf

Office of the State Council. (2017). *China to Boost Recycling of Agriculture Waste*. Retrieved June 18, 2019, from: http://english.gov.cn/policies/latest_releases/2017/06/12/content_281475684141592.htm

Scopus. (2019). *Advanced Search*. Retrieved July 20, 2019, from: http://www.scopus.com/

Seldman, N., & Morris, D. (2019). *China's New Recycling import Policy and the Future of US Municipal Recycling*. Retrieved July 2, 2019, from: https://ilsr.org/china-recycling-policy-opportunity/

Shanghai City Government. (2019). *The Municipal Solid Waste Management Regulation of Shanghai Municipality*. Retrieved June 20, 2019, from: http://www.shanghai.gov.cn/nw2/nw2314/nw2319/nw12344/u26aw58275. html

Song, X., Lu, B., & Wu, W. (2019). Environmental Management of E-waste in China. *Electronic Waste Management and Treatment Technology*, 285–310.

State Council. (2016a). *13th Five-year Plan for China's National Economic and Social Development*. Retrieved June 20, 2019, from: https://wenku.baidu.com/view/aafe5d3549d7c1c708a1284ac850ad02de8007bf.html

State Council. (2016b). *13th Five-Year Plan for Ecological Environmental Protection Plan*. Retrieved June 20, 2019, from: http://www.gov.cn/xinwen/2016-12/05/content_5143464.htm

State Council. (2016c). *13th Five-year Plan for Energy Saving and Emission Reduction*. Retrieved June 21, 2019, from: http://www.gov.cn/zhengce/content/2017-01/05/content_5156789.htm

Wang, C., Wu, G., Reniers, P., Huang, L., Kang, L., & Zhang, L. (2018). The Future of Hazardous Chemical Safety in China: Opportunities, Problems, Challenges and Tasks. *The Science of the Total Environment*, *643*(12), 1–11. doi:10.1016/j.scitotenv.2018.06.174 PMID:29935358

Wang, Z., Zhang, B., & Guan, D. (2016). Take Responsibility for Electronic-Waste Disposal. *Nature*, *536*(7614), 23–25. doi:10.1038/536023a PMID:27488785

Wuhu Ecology Center. (2018). *WtE Location Map*. Retrieved April 1, 2019, from: http://www.waste-cwin.org/

Yan, M., Li, X., Lu, S., Chen, C., Chi, Y., & Yan, J. (2011). Persistent Organic Pollutant Emissions from Medical Waste Incinerators in China. *Journal of Material Cycles and Waste Management*, *13*(3), 213–218. doi:10.1007/10163-011-0020-2

Yan, M., Xiong, Z., Li, X., & Hu, Y. (2015). Development Trend and Present Status of Municipal Solid Waste Incineration in China and USA. *Environmental Engineering*, *3*, 87–91.

Zeng, X., Duan, H., Wang, F., & Li, F. (2017). Examining Environmental Management of E-Waste: China's Experience and Lessons. *Renewable and Sustainable Energy Review, 72*, 1076-1082.

Zhang, D., Huang, G., Zu, G., & Gong, Q. (2015). Waste to Energy in China: Key Challenges and Opportunities. *Energies*, *8*(12), 14182–14196. doi:10.3390/en81212422

Zhang, L., & Sun, X. (2018). Influence of Sugar Beet Pulp and Paper Waste as Bulking Agents on Physical, Chemical, and Microbial Properties During Green Waste Composting. *Bioresource Technology*, *267*, 182–191. doi:10.1016/j.biortech.2018.07.040 PMID:30021150

Zhang, X. (2016). *Emission Standards and Control of PM2.5 From Coal-Fired Power Plants*. London: IEA Clean Coal Center.

Zhou, C., Gong, Z., Hu, J., Cao, A., & Liang, H. (2015). A Cost-Benefit Analysis of Landfill Mining and Material Recycling in China. *Waste Management (New York, N.Y.)*, *35*, 191–198. doi:10.1016/j.wasman.2014.09.029 PMID:25453315

Zhou, C., Liu, X., Wang, Z., Yang, T., Shi, L., Wang, L., ... Zhang, C. (2016). Assessment of Marine Debris in Beaches or Seawaters Around the China Seas and Coastal Provinces. *Waste Management (New York, N.Y.)*, *48*, 625–660. doi:10.1016/j.wasman.2015.11.010 PMID:26589868

Zhou, Z., Tang, Y., Dong, J., Chi, Y., Ni, M., Li, N., & Zhang, Y. (2018). Environmental Performance Evolution of Municipal Solid Waste Management by Life Cycle Assessment in Hangzhou, China. *Journal of Environmental Management*, *227*, 23–33. doi:10.1016/j.jenvman.2018.08.083 PMID:30172156

Chapter 5
Existing Realities and Sustainable Pathways for Solid Waste Management in Ghana

Kwaku Oduro-Appiah
https://orcid.org/0000-0001-8710-6470
University of Cape Coast, Ghana

Anne Scheinberg
Springloop Cooperative U. A., The Netherlands

Kodwo Miezah
Kwame Nkrumah University of Science and Technology, Ghana

Anthony Mensah
Ministry of Sanitation and Water Resources, Ghana

Nanne K. de Vries
https://orcid.org/0000-0002-4348-707X
Maastricht University, The Netherlands

ABSTRACT

The demands of the circular economy and the sustainable development goals call for a critical appraisal of the solid waste management systems of developing economies to divert decision-making towards the development of sustainable strategies and support systems for purposes of modernization. This chapter uses a combination of literature review, network queries to key experts within municipalities, and the authors' experiences and outputs of an ongoing collaborative research to evaluate and present the realities of solid waste management and recycling in Ghana. The goal is to chart a locally appropriate and sustainable pathway to drive system improvement. The analysis and evidence suggest that structural and coordinated cooperation between researchers of higher educational and research institutions and municipal authorities creates an enabling platform to build human resource capacities, to bridge data gaps, to identify what works, and to drive decision-makers' commitment towards the development of realistic action plans and continuous policy strategies.

DOI: 10.4018/978-1-7998-0198-6.ch005

INTRODUCTION

This chapter presents the existing realities of solid waste management (SWM) and recycling in Ghana and proposes an evidence-based and locally appropriate pathway to inform sustainable system modernisation. The authors have elaborated an assessment of the physical components and governance aspects of the solid waste management (SWM) system seen through the frame of Integrated Sustainable Waste Management (ISWM). Moreover, the authors have relied on data and research outputs of an ongoing collaborative and participatory research project with selected municipalities within the country that seeks to develop action plans and interventions to modernise the SWM system (Miezah, Obiri-Danso, Kádár, Fei-Baffoe, & Mensah, 2015; Oduro-Appiah, Afful, Kotey, & de Vries, 2019; Oduro-Appiah, Scheinberg, Mensah, Afful, et al., 2017).

Consistent with an ISWM methodology, the authors have analysed the physical system using materials flow diagramming based in part on the Wasteaware benchmark indicator approach (Wilson et al., 2015) for the six metropolitan cities of the country. Wasteaware is increasingly seen as a tool both to assess the physical systems and the governance aspects of the SWM systems of cities and to allow comparability of urban waste systems across countries, regions, and similar income levels. The goal is to aid readers and stakeholders to comprehensively visualise the realities and performance of the SWM system on the ground, and open up the opportunity for stakeholders to contribute to planning.

Although there is a robust literature on the SWM system in cities of Ghana (Boadi & Kuitunen, 2003; Oduro-Kwarteng & van Dijk, 2013; Oteng-Ababio, Owusu-Sekyere, & Amoah, 2017), this chapter seeks, for the first time, to integrate many aspects, and address municipal solid waste (MSW), agricultural and biomass utilisation, biomedical solid waste, waste of electrical and electronic equipment (WEEE), industrial solid waste, and construction and demolition waste. Secondly, the authors of this chapter have relied on the ideas of a working group on system modernisation, that is, involving practitioners in the discussion and the choices made. The objective is twofold:

- To provide a quick and easy reference resource for academia, research institutions, policy makers, solid waste managers, non-governmental organisations (NGO's) and the international community; to support them to identify points of entry for the development of sustainable interventions and support systems for the country's SWM system.
- To create a participatory basis for advocacy and policy development purposes amongst policy-makers.

BACKGROUND

Increasing population growth, rapid urbanisation, technological advancement and the apparent rise in living standards in the 21st century have led to an unprecedented consumption of natural resources and a subsequent increase in the quantities and diversity of solid waste (SW) generated across the globe (Kaza, Yao, Bhada-Tata, & Van Woerden, 2018). In developed and highly industrialised countries, this increase in waste generation has been associated with a range of political and environmental crises, which in turn stimulated the development of a set of integrated waste management and recycling policies and practices to promote public health, environmental control and efficient resource management which have become recognised as good practice in high-income countries in Europe, Oceania, Asia and North

America (Marshall & Farahbakhsh, 2013). Ideally, the result of applying governance and financial tools will work to create virtuous circles of self-correction and continuous improvement (Scheinberg, 2011).

Importing these Integrated Waste Management (IWM) tools and approaches to upgrade waste systems in developing and transition economies has generally failed. The boundary conditions that have made sub-systems of IWM (such as internalising the environmental costs by charging fees for disposal of mixed waste; and diverting materials from land disposal through municipally sponsored waste prevention, recycling, composting, or energy recovery) thrive in developed countries are not present in the current SWM systems in (African) emerging economies. More often than not, the SWM systems of developing countries are characterised by a variable political commitment and a lack (or absence) of clearly defined strategies to deliver an efficient service and charge its real cost price (Bundhoo, 2018; Ezeah & Roberts, 2012; Guerrero et al., 2013). As a result, despite multiple attempts to transfer the wealth of information, approaches, and best practices developed and institutionalised in the global North, cities in Ghana (and other cities in lower middle-income countries) are caught in a downward spiral of overcapitalisation of technology without the fiscal systems to raise revenues and pay operating costs. Transfer of technology has thus partly contributed to a persistent inability to create sustainable systems or achieve balanced national policies that protect environmental and public health (Diaz, 2017).

The point of departure for this chapter is to acknowledge that Ghana, like other developing countries, is faced with numerous challenges in solid waste management (SWM) and recycling systems (Guerrero et al., 2013; Oteng-Ababio et al., 2013). More than 60 years after independence from colonial rule, the country continues to suffer from the absence of clear policies and an uneven political commitment towards SWM (Asase et al., 2009; Oduro-Appiah et al., 2017). Despite the continual investment in personal and institutional capacity development, those charged with the management of solid waste have neither been able to generate the necessary baseline data for planning purposes; nor to create the enabling conditions for support from academia and research institutions towards the development and implementation of locally appropriate strategies. Technology transfer and murky Private Public Partnerships (PPP) have been accepted as the path of least resistance to address challenges. The international loans and Extended Producer Responsibility (EPR) funds have been grabbed and used for modernising landfills and constructing state-of-the-art composting and recycling plants, even when the solid waste characteristics, handling conditions and market forces do not support the application of such technological approaches. Analysis of the solid waste landscape in Ghanaian cities continues to document a basic lack of appropriate strategies or support systems that could stimulate sustainable system modernisation (Government of Ghana, 2017).

The SWM system of the country has thus evolved through a series of external pressures, developments, policies, and *ad hoc* interventions. The result for Ghanaian cities is a mix of weak policies and regulatory frameworks, partial collection coverage, low recovery and recycling rates dependent on informal recycling enterprises, under-regulated private sector participation, and the total unavailability of environmentally appropriate disposal facilities. The results can be widely perceived in the prevalence open-burning of MSW, industrial solid waste, and WEEE with its attendant health- and environmental-related burdens (Feldt et al., 2014; Oduro-Appiah et al., 2017; Tue et al., 2016).

Moreover, the six Ghanaian cities profiled are carrying heavy financial burdens related to overambitious technology choices, and paying (or defaulting on payment) for systems that do not support them to meet their own performance standards. In this situation, there is little convincing reason for residents to comply with rules or change their environmental behaviours. In short the prevailing situation is chaotic, inadequate, and a health hazard.

One measurable consequence of the existing situation has been poor public and individual health (Boadi & Kuitunen, 2005). Sanitation related diseases constitute more than 70% of the health problems reported at outpatient health facilities in the country with persistent seasonal epidemic outbreaks of cholera (Government of Ghana, 2010). Solid waste in drains and gutters and along beaches has been a major cause of floods with its attendant annual loss of lives and property (Asumadu-Sarkodie et al., 2015; Gaisie et al., 2019). The visibility of solid waste all over the place further interferes with the tourism potential of the country.

In the quest to modernise the SWM system, municipal authorities in Ghana have tended to place excessive confidence in the ideas and commitment of international financial institutions and development partners. Rather than taking control of the situation, authorities allow the panic associated with a series of (real or media) crises to guide their choice for quick fixes, which further contribute to the vicious circle of capitalising unsustainable technologies and infrastructure whose operating costs are not affordable. The resulting models and master plans fail to address local waste conditions and realities (Oteng-Ababio et al., 2013).

Ghana continues, therefore, to witness a confusing proliferation and overlap of SWM intervention projects leaving in its path a cycle of perpetual dependency on short-term projects from development partners and international organisations. Unfortunately, there is no evidence that this has translated to improved system performance, but it does appear to have diminished the institutional space for public and private-sector specialists to chart a locally responsive, holistic and long-term pathway towards system modernisation. If circular economy policies and the Sustainable Development Goals (SDGs) goals are to be achieved, sustainable responses to the aforementioned challenges have to be built in this regard.

OVERVIEW OF GHANA

Ghana is a lower-middle income republic along the West African Coast with a gross national income (GNI) per capita of US$1880.00 (World Bank, 2019). It is bounded by Ivory Coast to the West, Togo to the East, Burkina Faso to the North and the Gulf of Guinea to the South (Figure 1). It comprises 16 administrative regions, 260 Metropolitan, Municipal and District Assemblies (MMDAs), and covers an area of 238,539 km^2. Ghana is home to an estimated 30,280,811 inhabitants (Ghana Statistical Services, 2019). The daily average per capita municipal solid waste (MSW) and health care waste generation rates are 0.47 kg and 1.2 kg respectively (Asante et al., 2014; Miezah et al., 2015). The four largest household

Figure 1. Map of Ghana showing the regions and boundaries to neighbouring countries

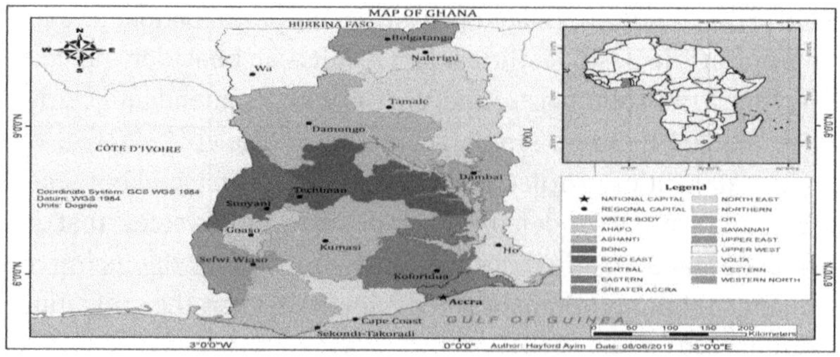

waste composition categories in Ghana are biodegradable waste (61%), plastic (14%), paper (5%) and metals (2%) (Miezah et al., 2015). Ghana was the first country in Sub-Saharan Africa to gain independence from British colonial rule.

STATUS OF SOLID WASTE MANAGEMENT AND RECYCLING IN GHANA

Solid waste management in Ghana is presented here in relation to the three key physical components (collection, treatment and controlled disposal, and the 3Rs (Reduce, Reuse and Recycle) and the three governance aspects (stakeholder inclusivity, financial sustainability, and sound institutions with proactive policies) of the ISWM concept (Wilson et al., 2013). The ISWM framework allows for a comprehensive assessment of the system elements, the stakeholders, and the governance aspects required to deliver a well-functioning solid waste and recycling system in a city (Schübeler et al., 1996). For this chapter, the authors are extrapolating from six cities (Accra, Cape Coast, Kumasi, Sekondi-Takoradi, Tamale and Tema) to benchmark the entire country's MSW system performance; using the Wasteaware benchmark indicators.

First elaborated in the 2010 publication "Solid Waste Management in the World's Cities", global ideals of good practice such as universal collection coverage, controlled disposal, resource efficiency, activity-based budgeting, service provider and user inclusivity, and sound institutions and proactive policies constitute the basis for the indicators (Scheinberg et al., 2010). The indicators (which come with automated excel sheets to compute scores) have proved useful and responded to the robust demand for a unified, freely available and real-time diagnostic tool. They are increasingly utilised by consultants and public authorities to assess SWM systems, identify strengths and weaknesses, and to make informed decisions for improvement. Since their introduction, more than 45 cities have been assessed using the indicators (Kabera et al., 2019; Wilson et al., 2015).

The performance of the six metropolitan cities in the physical components and the governance aspects of MSW including detailed background and waste related data is shown in *Table 1*. The quantitative physical indicators of collection coverage and waste capture, controlled treatment and disposal, and resource recovery are represented by the numbers (1.1 and 1.2), 2 and 3 respectively. The corresponding qualitative physical indicators are represented by the number and letter combinations of 1C, 2E, and 3R respectively. The combinations (4U and 4P), 5F, and (6N and 6L) represent the 'qualitative governance indicators of user and provider inclusivity, financial sustainability, and sound institutions and proactive policies' respectively. The colours are reflections of the performance of the cities in the indicators, where red, orange and green denote low (L), medium (M) and high (H) performances respectively. The letters B1, B2, B3 and W1 and W2 represent relevant background and waste-related information of the cities such as income level, population, total annual MSW generation rate, per capita annual generation rate and MSW composition respectively. In the following sections, the status of MSW management systems in the six cities, in relation to the indicators are discussed. The institutional and policy indicator is however discussed later in addition to the other forms of solid waste for the sake of brevity and organisation.

Collection of Municipal Solid Waste

In spite of 20 years of official franchising of MSW collection services to formal service providers, average collection coverage in the country stands at 60%. Informal service providers (ISPs) have partially filled the gap to push collection coverage to an average of 75% within metropolitan and urban centers (Government of Ghana, 2017); and in Accra, the capital city, locally responsive interventions targeted at providing a response to the informal service sector since 2016 has led to a coverage of 90% (Oduro-Appiah, Afful, et al., 2019). At a national per capita generation rate of 0.47 kg/day (Miezah et al., 2015), the total daily average MSW generation rate stands at 14,232 tons (Figure 2). An estimated 75% of all generated MSW is collected with the rest ending up on beaches, in drains, on vacant plots, or in unofficial dumpsites; and where there are no other options, it may be burned in households. The respective population, per capita and annual MSW generation rates, and the four major MSW compositions of the six metropolitan cities are provided in *Table 1*.

Figure 2. Sankey-type material flow diagram for solid waste collection, disposal and recycling in Ghana. The thickness of the flow corresponds to the weight of the waste (tons per day)

Source: The authors, 2019. Developed using STAN 2.6© (Institute for Water Quality Resources and Waste Management, TU-Wien, Austria).

Legend: C&D: Construction and demolition waste; E-Waste: Waste of electrical and electronic equipment (WEEE); MSW: Municipal solid waste.

Colour Legend, Flowlines: Black: MSW; Blue: WEEE; Magenta: Industrial solid waste; Orange: Construction and demolition waste; Red: Healthcare-related waste

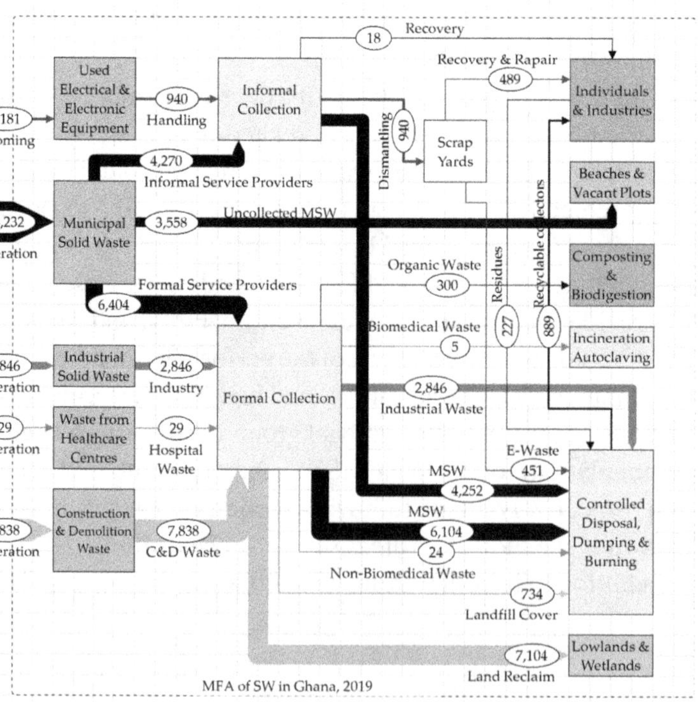

Two main MSW collection systems exist in Ghana: house-to-house collection by formal companies on the one hand, and combined house to house collection, street sweeping, and related services (including some separate collection) provided by the ISPs. However, in most rural and urban areas (outside of central business districts and high-income neighborhoods), skip containers (12 m^3) are provided at key intersections and other convenient points within communities to allow for secondary collection by formal service providers. Informal actors alone contribute to an average MSW collection rate of 40% in the six metropolitan cities of the country (Oduro-Appiah, Afful, et al., 2019).

Disposal of Municipal Solid Waste

Landfilling and open dumping remains the largest and inevitably preferred MSW disposal methods in Ghana. There are four engineered landfills with leachate collection systems and 172 official dumpsites in the country (Environmental Protection Agency, 2017). With the exception of 300 tons of organic materials valourised in two state-of-the-art composting and recycling plants, one windrow composting and two fecal sludge digestion plants in the Greater Accra Region, all collected MSW is disposed of in landfills and open dumpsites. Routine landfill operations such as daily covering, compaction and leachate recycle are seldomly adhered to (see Figure 3). This in addition to poor management has contributed to the deterioration of all four engineered landfill sites within the metropolitan cities to controlled landfill sites; a reason for the low to moderate environmental control score for the six cities (*Table 1*).

Although all the engineered landfills have gas collection systems in place, there are no records of landfill gas emission measurements to support the development of a climate change action plan for the cities; nor are there any future post-closure energy recovery plans. Currently, most landfills and disposal sites have exceeded their design capacities, leaving the cities and towns in a precarious situation with regards to disposal of MSW (Government of Ghana, 2017). Years of poor governance and a lack of system user inclusivity in planning of MSWM facilities have denied the capital city, Accra, of a planned engineered landfill under the erstwhile Urban Environmental Sanitation Project (Oteng-Ababio, 2011, 2014); compelling city authorities to rely on landfills within neighbouring municipalities.

Recovery and Recycling of Municipal Solid Waste

Apart from the publicly funded but privately managed Accra Composting and Recycling Plant (ACARP), the recently commissioned Integrated Recycling and Composting Plant (IRECOP), the Ga Mashie Aerobic Composting Plant, the JVL Fortifer Compost Plant, and the Safi Sana biogas digester, which together process only 18% of the daily input of 300 tons of organic waste (Figure 2), all recovery and

Figure 3. Disposal patterns and informal waste recovery and recycling activities within Ghana

Table 1. Wasteaware benchmark indicators for MSWM in the six metropolitan cities of Ghana

No.	Category	Indicator	Results					
City/Country			Accra	Cape Coast	Kumasi	Sekondi Takoradi	Tamale	Tema
Year of assessment			2018					
Background information								
B1	Income level	WB income category	Lower middle-income					
		GNI/capita	US$1880.00					
B2	Population	City population	2,087,668	186,159	2,105,382	726,905	275,547	353,086
B3	Waste generation	MSW generation (tons/year)	536,670	45,609	576,875	188,995	33,065	92,862
Key MSW-related data								
W1	Waste/capita	MSW/capita Kg/year	270	245	274	260	120	263
		MSW/capita Kg/day	0.74	0.67	0.75	0.70	0.33	0.72
W2	SW composition	Composition (%)						
W2.1	Organic	Food and green waste	65.8	63.2	48.4	60.0	58.6	55.0
W2.2	Paper	Paper and Cardboard	5.3	4.1	6.5	7.3	3.2	4.6
W2.3	Plastics	Plastics	10.4	10.6	17.6	11.5	10.9	14.0
W2.4	Metals	Metals	3.1	2.1	4.5	2.4	2.8	2.5
Physical components								
1.1	Public health –waste collection	Waste collection coverage	90% (M/H)	65% (L/M)	80% (M)	75% (M)	60% (L/M)	85% (M)
1.2		Waste captured by the system	90% (M/H)	65% (L/M)	75% (M)	80% (M)	80% (M)	85% (M)
1C		Quality of waste collection	M/H (63%)	M (50%)	M/H (67%)	M/H (63%)	M (58%)	M/H (71%)
2	Environmental control –waste treatment and disposal	Controlled treatment and disposal	87% (M/H)	72% (L/M)	80% (M)	77% (M)	75% (M)	87% (M/H)
2E		Degree of environmental protection in waste treatment and disposal	M/H (65%)	M (60%)	M/H (70%)	M (60%)	M (60%)	M/H (65%)
3	Resource management	Recycling rate	18% (L/M)	3% (L)	10% (L/M)	5% (L)	5% (L)	15% (L/M)
3R		Quality of the 3Rs- Reduce, reuse, recycle –provision	M (46%)	L/M (25%)	L/M (38%)	L/M (33%)	L/M (29%)	M (46%)
Governance aspects								
4U	Inclusivity	User inclusivity	M (46%)	L/M (33%)	M (42%)	M (42%)	L/M (38%)	M (46%)
4P		Provider inclusivity	M/H (75%)	M (60%)	M (60%)	M (60%)	M (60%)	M (60%)
5F	Financial sustainability	Financial sustainability	M (50%)	L/M (25%)	M (42%)	L/M (33%)	L/M (29%)	M (42%)
6N	Sound institutions and proactive policies	Adequacy of national SWM framework	M (54%)	M (50%)	M (50%)	M (50%)	M (50%)	M 50%
6L		Local institutional coherence	M (42%)	L/M (38%)	M (42%)	M (42%)	L/M (33%)	M 42%

Legend: Performance levels in physical and governance indicators and corresponding colour codes: Low performance (L)-red; Low/ Medium performance (L/M)-red/orange; Medium performance (M)-orange, Medium/High performance (M/H)-orange/green; and High performance (H)-green

WB: World Bank; GNI: Gross national income; MSW: Municipal solid waste

Source: The authors, 2019.

recycling activities within the country take place in the private informal service and value chains. The combined MSW recycling rate in Ghana is estimated at 8.6% with the informal sector contribution at 6.2% (Oduro-Appiah et al., 2019), which is classified as low on the Wasteaware indicators but is at least based on some real evidence-based information (*Table 1*). However, the real solid waste recycling rate is probably higher, since this rate calculation does not include the recovery of metals and e-waste taking place at the Old Fadama Scrap Yard.

The 3Rs (reduce, reuse and recycle) are not supported by laws or policies, although there is some activity at the Government level to change this. There is no formal MSW segregation at source at the household and institutional levels in Ghana (Oduro-Appiah & Aggrey, 2013); the exception is a private formal collection company, Jekora Ventures, which encourages its clients within a municipality to separate biodegradable solid waste from others before collection. This means that a significant amount of recyclable material travels as mixed MSW to formal and informal disposal sites (see Figure 2). In larger cities, much of this material is recovered by waste pickers on the disposal sites and sold to recycling companies via junk dealers and processors. These together with the absence of policies to integrate and support the waste pickers explains the relatively low to moderate scores in the quality of recycling (*Table 1*).

The market value of recyclables `inspires some householders to separate plastics for sale to itinerant buyers. The Greater Accra Region, which harbours both the capital city, Accra, and the industrial hub of the country, Tema, accounts for 90% (Government of Ghana, 2017) of all the recycling industries and activities in Ghana. High production costs and erratic energy supplies have drastically reduced investments into the recycling sector since 2012. The recycling efficiency of industries in Ghana stands at 75%; meaning that 25% (227 tons) (see Figure 2) of MSW recovered are returned as residues to landfills and dumpsites.

Level of User and Provider Inclusivity in Municipal Solid Waste Management

Service users – the households and businesses whose waste is removed for disposal -- are seldom involved in the planning or implementation of MSWM decisions and programs in the six cities. Four of the six municipalities lack a stakeholder's platform, further denying system users of transparency and improved MSWM governance (Alamgir, Bidlingmaier, & Cossu, 2012). There is little equity in service provision —poorer clients have a lower level of service and are seldom listened to (even if they take the time to complain). Most users within the urban and peri-urban centres pay a fee (directly to the service provider) to pay for solid waste collection services, but they are not normally consulted, nor invited to participate in determining, either the level of service or the price for the level in their area. Although sanctioned by law, public review of MSWM budgets is seldom implemented in Ghana cities, earning them a low score on user inclusivity. What is missing is the type of participatory research, appraisal, monitoring and feedback that serve to stimulate the feeling of ownership that leads to compliance with norms and other forms of good solid waste behaviour. It is therefore not surprising that littering, illegal dumping, resistance to pay user charges and the prevalence of the Not In My Backyard (NIMBY) syndrome towards the construction of waste treatment and disposal facilities within the country are recurrent issues. Formal service providers are known to be moderately involved in MSW decision-making (*Table 1*), but this mild commitment to participation (provider inclusivity) does not apply to any of the informal waste and recycling service providers. Still, formal service providers do have a platform that relates to all MSW tendering processes, with clear guidelines and franchise agreements provided to support them in their service delivery.

Municipal Solid Waste Management Funding and Financing

Solid waste financing remains a major challenge to national and local authorities. Loans and grants from development partners and donor agencies constitute the major source of capital funding, and user pay systems are underdeveloped, and, where they exist, the revenues seldom cover the costs. Tax revenues from municipalities, supplemented by user fees, theoretically cover budgeted operation and maintenance costs for collection services, especially in urban centres of the country, but this does not "reach" incidental or irregular expenses such as clearing away illegal dumpsites, even when such activities are technically part of the collection service. Subsidy schemes for landfill operations are politically popular, but they hide the true cost of the service from users, so that there is no incentive to divert recyclables from disposal, nor any perceived financial advantage for waste prevention. The result is that waste disposal is chronically underpriced, leaving the local and national government in a vicious cycle of debt to service providers and landfill operators (Oduro-Appiah, Afful, et al., 2019). The absence of a MSWM strategy, political hesitation to charge the full cost for disposal, and the lack of activity-based budgeting means that the total cost of MSWM within the country has not been known since independence.

AGRICULTURAL BIOMASS AND UTILISATION, WASTE-TO-ENERGY

Residual biomass from economic activities is an important source of fuel for renewable energy and sustainable development and its use is expected to increase in the future. At the present time, it is estimated that biomass residues contributes as much as 64% of primary source of energy supply in Ghana (Kemausuor, Kamp, Thomsen, Bensah, & Østergård, 2014). Several forms of biomass can be identified in Ghana and they include: forest residues; wood processing residues; agricultural residues deriving from the cultivation from sugar cane, maize, oil palm, coconut, rice, cocoa, sorghum, millet, grasses; manures and slaughterhouse wastes from domestic livestock; urban waste including municipal solid waste, food waste, and sewage sludge (Duku, Gu, & Hagan, 2011; Kemausuor et al., 2014; Präger, Paczkowski, Sailer, Derkyi, & Pelz, 2019). Kemausour et al. (2014) determined the technical potential of biomass residues in Ghana as 96 PJ in 2700 Mm^3 of biogas or 52 PJ in 2300 ML of cellulosic ethanol and this can replace more than a quarter of wood fuel use, a major source of deforestation in Ghana.

The challenge to sustainable use of biomass residues in Ghana lies in the mapping of these resources, and the organisation of logistics, transport, and financing systems. This is the case despite successful examples of conversion of biomass from pit latrine emptying into biogas. There is no specific policy nor incentive defining biomass; the rules for classifying and distinguishing between harvesting biomass or collecting residual biomass are not clear; and there are few guidelines for its utilization in Ghana. Even though the Ghana Energy Commission strives to gain maximal use of biomass as the main renewable source of energy (10% renewable energy use by 2020), there is much to be done if these goals will be met, and there are basic challenges still to be addressed.

SOLID WASTE FROM HEALTHCARE CENTRES

In Ghana biomedical waste is generated by over 8,318 healthcare centres (Government of Ghana, 2018). The type of wastes generated depends on the status and activities of the healthcare centres, but legally, all waste from the healthcare system is classified as either biomedical or non-biomedical. The biomedical wastes from laboratories, pharmacies, surgical theatres, delivery rooms, etc. are defined to include pathogenic wastes, infectious waste, pharmaceutical wastes, laboratory waste, and other materials which may have health consequences. The daily waste generation from health care activities stands at 29 tons (Figure 2), and this has been estimated from the per capita biomedical waste generation rate of 1.2 kg per bed per day (Asante et al., 2014) and the 23,829 beds occupied per day within the country (Government of Ghana, 2018).

Storage and Collection of Biomedical Waste

Per regulation (based on accepted good practice,) biomedical waste should be comprehensively separated at the point of generation into appropriately coded containers and directed to proper treatment and disposal centres. This currently occurs only in a few metropolitan and larger health care institutions. Notwithstanding, sharps and pathogenic wastes are generally separated in most hospitals. Other biomedical wastes are normally not separated, but stored and collected together with MSW. With the exception of hospitals within some tertiary institutions which collect their wastes, all other healthcare centres within urban areas have agreements with formal waste collection companies for collection and disposal. More often than not, biomedical waste is transported in ordinary MSW waste skip containers and compaction trucks for treatment and disposal at landfills.

Treatment and Disposal of Biomedical Waste

An estimated five tons of biomedical waste is treated daily via incineration and autoclaving (Figure 2). The rest, estimated on average to be 24 tons of both pathogenic and infectious wastes, is mixed with non-biomedical waste from hospitals and disposed of on municipal landfills and dumpsites. Larger and older healthcare centres in the regions make use of brick-lined incinerators for the burning of infectious and pathological waste, but this is done without temperature monitoring and without treatment of flue gases. In the capital city, Accra, there is a privately managed autoclave facility situated on the premises of one of the MSW bulk transport stations for the treatment of biomedical waste. Designed for a capacity of 1.2 tons per hour (230 tons per month), the facility has been increasing its client base and treatment capacity, and had increased from 15 healthcare centres and 300 kg per month to 110 centres and 10.2 tons per month by mid-2019. Biomedical waste received at the facility is autoclaved at temperatures ranging between 121°C and 131°C. This renders it inert and enables shredding to reduce the volume by 70%, as well as making it safe for land disposal. The cost of treatment for a kilogram of biomedical waste is US$1.2, which is deemed unsustainable considering the fact that most public and private healthcare centres in the country are only able to pay for services through a national health insurance policy which is constrained by financial difficulties and delayed payments. The autoclave facility provides training to clients to ensure effective segregation of waste streams and also makes use of specialized vans with freezing compartments for the collection and transportation of biomedical waste to the treatment centre. The lack of enforcement of the regulation on biomedical waste disposal further affects the rate at which medical centres subscribe to appropriate facilities. In place of facing sanctions for improper management of their wastes, managers of healthcare facilities remain free to choose less expensive, non-specialised management of their wastes -- at poorly functioning landfills -- in the normal municipal waste system.

MANAGEMENT OF WASTE FROM ELECTRIC AND ELECTRONIC EQUIPMENT (WEEE)

Ghana and especially the Old Fadama dumpsite in Agbogbloshie (a slum area within the capital city, Accra) has achieved worldwide notoriety (Oteng-Ababio, 2018) and a great deal of "disaster tourism" for claimed abuses in labour, human rights, and environmental aspects of the dismantling, recycling, and disposal of end-of-life electronics. There is a great deal of written information and many unverified claims about what is happening at the site, and these create a climate of scandal and confuse real information on the economic activities in the technology and recycling sectors.

The scholarly literature alone since 2010 is replete with more than 50 articles on e-waste in Ghana with major themes ranging from assessment (Amoyaw-Osei et al., 2011; Atiemo, Ofosu, Aboh, & Kuranchie-Mensah, 2012; Caravanos, Clarke, Osei, & Amoyaw-Osei, 2013), public health impacts and environmental burdens (Amuzu, 2018; Feldt et al., 2014; Huang, Nkrumah, Anim, & Mensah, 2014; Tue et al., 2019), occupational health and safety (Burns, Sayler, & Neitzel, 2019; Caravanos, Clark, Fuller, & Lambertson, 2011), socio-economic contributions, (Daum, Stoler, & Grant, 2017; Grant & Oteng-Ababio, 2012) to policy initiatives (Atiemo, Faabeluon, Manhart, Nyaaba, & Schleicher, 2018; Manhart, Meinel, & Walgenbach, 2014). The gloomy e-waste picture in the literature on Ghana and especially on Agbogbloshie as a dumping ground for the e-waste from the global North has however been recently contested (Akese & Little, 2018; Davis, Akese, & Garb, 2019). The handlers and recyclers of e-waste in Ghana are predominantly value chain actors, many of whom are Muslim young men from Northern tribes around Tamale, who come to the cities and begin with recycling activities as a form of livelihood.

Because of the large amount of negative publicity and contested numbers, this Chapter will not go into much detail on the management of e-waste in Ghana, but will attempt to orient the reader to some basics about the recycling activities at Agbogbloshie/Old Fadama.

Old Fadama is the name given to several areas which are behind the Agbogbloshie onion market, which is the site of a great deal of economic activity based on the value of metal scrap, especially ferrous scrap from automobiles, and non-ferrous scrap derived from waste from electrical and electronic equipment (WEEE), otherwise known as e-waste. Old Fadama and two adjacent areas in Agbogbloshie, one called Galloway and the other called Mortuary Road, is an informal area where a great many *valorisation* activities take place. Most fall under the labels of recycling, repairing and disassembly. The goal of most of the activities is to recover materials for sale to the recycling value chains, ranging from components from electrical and electronic equipment to automobiles on the one side of the road, to the pre-processing of plastic and waste oils on the other (Lepawsky, Akese, Billah, Conolly, & McNabb, 2015), personal observation of the first two authors.

What is not contested is that the original source of much of this material is used vehicles and devices (and some parts and components) which have been imported to Ghana by Ghanaian entrepreneurs who are in the business of wholesaling and retailing second-hand (working) equipment. These entrepreneurs choose to import used high-quality devices from Europe, North America and Asia because they have solid state components, are repairable, and have a longer useful life than new devices made for African markets. Export of devices from Europe is strictly controlled and all devices have to be complete and working. Import of all devices is controlled by the Ghana customs service on arrival, and non-working items are not allowed.

The electronic equipment (laptops, mobile phones, cameras, television sets, central processing units, monitors, game consoles, DVD players, MP3 players, washing machines and microwaves) and other metal devices and vehicles at Agbogbloshie have been purchased by the scrap dealers and metal recyclers with premises there: materials do not come to the scrap area unless they have value. There is one exception, there is a small former legal dumpsite in the same area, and Accra authorities have not been successful in closing it, so a small amount of municipal solid waste is discharged at that specific dumpsite by small waste collection companies and especially by some informal service providers.

After a period of 10-30 years of use, at the end of useful life, these items or their parts reach Old Fadama and similar recycling areas in Kumasi in the Ashanti region (Atiemo et al., 2018). Devices and components are either repaired for sale and/or for reuse; dismantled, processed (often by hand) and sold to metal recycling firms, which may be local intermediaries or global firms via a chain of scrap dealers, middlemen and international firms (Daum et al., 2017). Wire which is PVC coated is sometimes burned with attendant public- and environmental health risks to adjacent settlements (Caravanos et al., 2011). Projections from reported estimates indicate that the e-waste reuse, recovery and recycling sector currently employ more than 16,000 informal sector actors and indirectly sustains the lives of more than 200,000 people; contributing a projected US$500 million annually to Ghana's economy. (Amoyaw-Osei et al., 2011; Grant & Oteng-Ababio, 2012; Oteng-Ababio, Amankwaa, & Chama, 2014).

Collection, Recycling and Disposal of E-Waste

The Ghana technology sector is very active, and many second-hand devices are in use. Two systems of e-waste and valuable metal collection exist in Ghana: collection predominantly from electrical and electronic equipment retail and repair shops by scrap collectors, and limited collection from households, cooperative organizations and institutions by itinerant waste buyers (IWBs). Whilst the scrap collectors, who work for the informal recyclers, often know their clients and move straight to their shops, the IWBs are known to travel for relatively longer distances within cities and peri-urban areas, using push carts to collect all sorts of metal and electronic equipment (Grace Abena Akese, 2014). Informal dump pickers (recyclers) also collect unknown amounts of WEEE and metals from the country's landfills and sell to junk dealers for processing at the scrap yards.

There is a lack (or absence) of public sector and institutional oversight of the activities of the informal e-waste sector in Ghana. Estimating the flow of Electrical and Electronic Equipment (EEE) in such circumstances becomes an arduous venture to undertake. Nonetheless, the authors have relied on the recommendations of the most extensive and participatory work so far on e-waste in Ghana (Amoyaw-Osei et al., 2011) to conservatively estimate the current flows. The total tonnage of used EEE reaching the various scrapyards, the fractions dismantled, repaired, recovered, assembled into "new" products for reuse and for sale to individuals and the industrial value chain; in addition to the e-waste disposed and burnt daily in Ghana are shown in Figure 2. An estimated total of 1181 tons of used electrical and electronic equipment per day reach the scrap yards and processing sites at Agbogbloshie, Pantang and Ashaiman; all within the greater Accra region, and Dagomba Line and Suame Magazine within the Ashanti Region. At an estimated repair and recovery rate of 52%, the rest (451 tons) are disposed of at nearby dumpsites with some plastics being recovered and some burned (see Figure 2).

The advent of e-waste regulation policy has seen few formal companies showing interest in recycling. Both government agencies, especially the Environmental Protection Agency (EPA) and selected private companies are in the process of establishing collection and recycling systems to handle e-waste in Ghana. The government of Ghana in 2019 launched a "national integrated e-waste management plan" which is seeking to establish a facility at Agbogbloshie to create 22,000 self-sustaining jobs.

INDUSTRIAL SOLID WASTE AND EXTENDED PRODUCER RESPONSIBILITIES

There are no special collection, transportation, treatment and disposal systems for industrial solid waste in Ghana. Industrial solid waste is collected mostly by formal waste collection service providers and sent to MSW landfills and dumpsites, with some being disposed informally in nature or at unauthorised dumps and heaps. The authors' estimate of the daily generation rates of 2,846 tons (Figure 2) originates from publically available records documenting the operations of the country's landfills and dumpsites. There is, however, no specific nor separate regulation of documentation of disposal of the hazardous components of industrial waste, nor does this occur in special facilities or segregated cells.

Plastic importers and manufacturers under the umbrella of a plastics, beverage, and chemical industry plastic waste management program (that seeks to push back the central government's threat of banning plastics) contribute towards a SWM fund as part of an Extended Producer Responsibility system overseen by the Ministry of Environment, Science, Technology and Innovation (MESTI). The application of the fund however, for research, establishment of recycling plants and support of private recycling efforts to address the country's ever-growing plastic waste challenges has been associated with managerial challenges and murky value-for-money Public Private Participation (PPP) recycling and composting arrangements.

CONSTRUCTION AND DEMOLITION WASTE

Construction and demolition (C&D) debris is generated when new structures are built and existing structures are renovated or demolished. Sources of construction and demolition waste are homebuilders, commercial developers, highway and street contractors, site grading activities, demolishing, roofing, excavating companies among others (Prairie Village, 1998). In Ghana, major activities that generate C&D waste include construction, demolition and renovation of buildings. Debris from C&D wastes may include hazardous and non-hazardous materials. Non-hazardous substances may be wood, dry wall, plastics, roofing, bricks, glass, ceramic tiles, asphalt and metals. Hazardous materials from C&D however include batteries, fluorescence light bulbs, paints, thermostat, lead coated surfaces, cleaning agents, among others. Ghana chronically lacks data on C&D waste. The estimate of C&D waste generation rate on the mass flow diagram in Figure 2 is generated based on recommendations of the scholarly literature on developing countries (Al Masha'an & Mahrous, 1999; Kartam, Al-Mutairi, Al-Ghusain, & Al-Humoud, 2004; Lu et al., 2011). The daily C&D generation rate in Ghana is estimated at 7838 tons which amounts to 30% of all SW generated in the country (Al Masha'an & Mahrous, 1999; Kartam et al., 2004). The predominant fractions of C&D waste after demolitions of buildings are: 70% concrete; 15% wood and 4% iron. However, during construction of buildings, there are wastages leading to losses such as 36% concrete, 10% reinforcement materials, 14% bricks and blocks, 18% form works and 22% tiles (Awuah, 2017).

Storage, Collection and Management of C&D Waste

Waste from construction activities is generally used as filling material, often on site and without regulation or permitting. In contrast, demolition wastes are normally gathered in heaps for onward collection. The normal practice has been collection for disposal to reclaim excavated lands and to fill lowlands. Most developers in urban areas where land for construction is scarce illegally use C&D waste and soil to fill wetlands; a practice which contributes to the incidence of flooding in Ghanaian cities. The author's assessment of landfill records shows that 9.3% (734 tons) of C&D waste ends up daily on municipal landfills and dumpsites (Figure 2). In case of natural or man-made disasters, heavy earth-moving equipment is employed to clear debris into heaps for transport to landfill sites. There is no known C&D waste recycling company in Ghana. Metal components of demolition waste are sold to informal scrap dealers for recycling. Another popular option for managing C&D waste is reuse of the waste in new buildings or construction works. Metal parts, floor tiles, roofing materials, block or bricks are often reused. Alternative practices are the use of inert aggregate wastes for gravelling road surfaces, as base materials for building foundations, for filling of trenches, and for pothole patches.

POLICY, INSTITUTIONAL AND LEGAL FRAMEWORK FOR SOLID WASTE MANAGEMENT IN GHANA

Sound policies translated into strong legislative and regulatory frameworks coupled with pro-active institutions form the backbone of effective SWM systems (Rodić & Wilson, 2017). The MSWM system of Ghana is not driven by an integrated and sustainable strategy (Asase et al., 2009; Government of Ghana, 2017). In Ghana, as in most places in the world, significant progress in legislative reform and/or system upgrading and modernization are triggered by crisis management of disasters and disease outbreaks (Asumadu-Sarkodie et al., 2015).

Ghana is also not alone in that its solid waste operations are the responsibility of a fragmented institutional and bureaucratic landscape. Three separate ministries are responsible for solid waste management (SWM) in Ghana. The Ministry of Environment, Science, Technology and Innovation (MESTI) formulates policies and develops plans and strategies for management of industrial waste; WEEE and all hazardous waste. The Ministry of Health (MoH) develops policy and guidelines for the regulation of biomedical waste and a newly created Ministry of Sanitation and Water Resources (MSWR) is in charge of policy formulation and strategy development in MSW and C&D waste. Amongst the three, the MESTI is relatively well resourced in terms of sufficiency of human resource.

As part of efforts towards decentralization, waste management and environmental health departments within the Metropolitan, Municipal and District Assemblies (MMDAs) are the implementing agencies for MSW policies and strategies. At the national level, there is no national strategic municipal solid waste management plan. There is a policy on environmental sanitation with supporting strategic and investment plans, but the absence of a comprehensive legislative framework and a sustained funding source has hampered its operational usefulness to the waste system. The municipalities, the governance institutions tasked with the implementation of the policies for waste management, are under-resourced and over-committed. Cities and towns have limited budgetary authority and too little trained staff, but the core problem is lack of political commitment and space to develop long-term strategic plans. Nor are the waste management departments granted the requisite autonomy to seek research support nor supported to take politically difficult but economically and environmentally sustainable decisions for system modernization. The new ministry for Sanitation and Water Resources shows some promise, having recently bucked this trend by showing an appreciable level of commitment to develop new waste policies and master plans, and a willingness and capacity to open decision-making to involve relevant stakeholders., It has commenced the strengthening of capacities of waste management personnel as initial steps towards a continuous and sustained MSWM strategy.

The MSW system is managed with a limited number of acts and guidelines, which are also scattered in different documents within different departments. There is no specific regulation on C&D waste but the Ghana Building Code and the Local Government Authority Act have some statements that seek to provide a guide for C&D regulation.

Ghana is a signatory to two international regulatory frameworks on the control of transboundary movement of hazardous waste and disposal, namely: the Basel and the Bamako Conventions (Amoyaw-Osei et al., 2011). The country has since 2016 ratified a hazardous and electronic waste control and management act, Act 917 (2016), that seeks to regulate the control, management and disposal of hazardous waste, in addition to the establishment of an EPR system to provide the necessary finance for the sustainable management of WEEE accompanied by public education (Daum et al., 2017; Manhart et al., 2014). Act 917 (2016) classifies hazardous waste to include an ambitious list of various waste materials namely: biomedical wastes, industrial wastes, agrochemical wastes, laboratory wastes from research and academic institutions, photographic chemical wastes, flammable solids, oxidizing wastes, poisonous wastes, corrosive wastes, and toxic wastes among others.

Act 917 (2016) also provides the necessary directives for the handling of hazardous waste in addition to prohibiting importation, exportation, transportation, selling, purchasing, and trading of hazardous waste and WEEE in the country. Those contravening the directives are required to pay fines and/or even serve jail terms but enforcement has little relationship to what happens on the ground (Daum et al., 2017). The permitting system does not work, with the result that the informal recycling of non-ferrous metals coming from e-waste is largely an unpermitted activity. Although Act 917 (2016) stipulates that manufacturers, distributors and wholesalers of electrical and electronic equipment should take back used or discarded devices from their customers, this process is far from universal (Daum et al., 2017; Manhart et al., 2014).

Biomedical waste management in Ghana is regulated by the Ghana Health Service with the Ministry of Health Policy and Guidelines for Healthcare Institutions (2006). The guidelines require hospitals to use appropriate treatment options to handle biomedical waste. The generators are required to separate waste at source for collection to treatment centres, and use the correctly coded bins for storage of waste. However, there is no compliance to the regulation. The Healthcare Facility Regulation Authority (HEFRA) does not have the required capacity to enforce the regulation. They are however, faced by numerous challenges including human capacity and reliable funding sources.

REALITIES AND SUSTAINABLE PATHWAYS FOR SYSTEM MODERNISATION

The scholarly literature is replete with various recommendations of potential strategies to modernise the SWM of developing countries (Guerrero et al., 2013; Marshall & Farahbakhsh, 2013). Notably, most of these recommendations partly or wholly fall under the Integrated Waste Management (IWM) and the Integrated Sustainable Waste Management (ISWM) concepts which, for more than two decades, have gained prominence in the solid waste management systems of low and middle-income countries

(Daskalopoulos, Badr, & Probert, 1998; McDougall, White, Franke, & Hindle, 2008; Scheinberg et al., 2010; Schübeler et al., 1996) .

The IWM concept evolved in the 1980s in early adopter high-income countries in Western Europe, North America, Oceania, and high-income Asia. It advances a systems approach to solid waste management in which the system elements (storage, collection, transportation, recovery, recycling, treatment and disposal) are considered in a holistic manner with the ultimate goal of achieving environmental effectiveness, economic sustainability and social acceptability (White, Dranke, & Hindle, 2012). In a developing country context, an IWM strategy systematically recommends the:

- Separation at source of biodegradable kitchen and garden waste from all other waste to promote composting and anaerobic digestion for potential treatment and energy recovery options.
- Further separation of the non-biodegradable streams at designated central sorting points to support recycling and alternative treatment options.
- Collection and transportation of the separated waste for treatment and disposal.
- Gradual shift from use of dumpsites to controlled and sanitary landfill sites for solid waste disposal purposes.

The concept further advocates for all beneficiaries (local government, businesses, individuals, etc.) to equitably contribute towards the payment and financial sustainability of the system.

The ISWM framework (developed in relation to low- and middle-income countries and emerging economies), however envisions a city's SWM system as the intersection of two overlapping triangles (Wilson et al., 2015): the physical components and the governance aspects. The edges of the physical triangle comprise of collection, controlled treatment and disposal, and resource recovery as drivers of public health, environmental control, and resource management respectively. The extent to which users and providers of the physical services are involved in decision-making, the budgeting processes adopted to achieve financial sustainability, and the proactive policies and sound institutions put in place to support the delivery of a well-functioning SWM system form the edges of the governance triangle. Attention to these approaches is not well-established in the cities in lower middle-income Ghana and other developing countries of Sub-Saharan Africa (Asase et al., 2009; Henry, Yongsheng, & Jun, 2006; Oteng-Ababio et al., 2013). Whilst the literature based especially on the ISWM concept suggests that there are structural improvements in collection (Wilson et al., 2013), Ghana and other developing countries are experiencing recurring challenges in all the physical indicators and the governance aspects of SWM, and especially in relation to modernising disposal and making a policy commitment to diversion of recyclables (Grant & Oteng-Ababio, 2019; Oduro-Appiah, Scheinberg, Mensah, Afful, et al., 2017).

A review of the scholarship on SWM in Ghana and similar developing economies, enriched by the authors' experience with working with the selected municipalities in Ghana, suggests that the appearance of recurrent themes in relation to the seemingly endless challenges of SWM is a structural characteristic of these systems. At the top of the hierarchy of these challenges are the:

- Political unwillingness of decision-makers to take structural measures for improvement, complicated by the tendency to wait for crises and then to look for overnight solutions to fix the system (Ali, 2010; Oteng-Ababio et al., 2013).
- Absence of a consistent and realistic strategic policy and regulatory framework for implementation of waste management system improvements (Guerrero et al., 2013).

- Vulnerability of political authorities to accept "donations" of unsustainable technologies from developed countries without knowing what operating costs they are committing to, and in general without analysing their usefulness given the characteristics and composition of local solid waste streams (Ali, 2010; Oteng-Ababio et al., 2013).
- Lack (or absence) of commitment to charge for the real cost of SWM (even when there is evidence of users' ability to pay) to effectively fund the system, and the related overdependence on donor financial support for both capital and recurrent investments (Ali, 2010).
- Financial constraints and underfunded infrastructure (Ezeah & Roberts, 2012)
- Chronic lack of data to plan and develop locally appropriate and long-term interventions (Oduro-Appiah, Scheinberg, Mensah, Afful, et al., 2017; UNEP, 2018).
- Lack of sustained research into locally responsive treatment options (Guerrero et al., 2013; Marshall & Farahbakhsh, 2013).
- Weak and limited human resource capacity, especially in the SWM and sanitation sectors (Bundhoo, 2018; Wilson et al., 2013).

It is clear, that these observations and the evidence from the Wasteaware benchmark indicators (*Table 1*) show that governance, administrative and institutional constraints prevail as recurring nemesis to sustainable SWM systems delivery in Ghana and other developing countries (Tot, Vujić, Srđević, Ubavin, & Russo, 2017; Wilson, 2007; Wilson et al., 2013). Finding an evidence-based path to addressing these governance-related challenges is an urgent priority, and one which is highly relevant to Ghana (Wilson et al., 2017).

The authors' experience with working with the largest private sector actor in solid waste management in Ghana, along with supporting the waste management departments (WMDs) of selected municipalities within the country, suggests that coordinated and participatory cooperation and collaboration of research institutions and Higher Educational Institutions (HEIs) with municipalities of Ghana has a high potential to stimulate improvements in both solid waste governance and service delivery (Oduro-Appiah, Afful, et al., 2019).

The authors, as researchers, since 2016, have (through participatory and action research methods) been working with the Accra Metropolitan Assembly (AMA) and recently with the Greater Accra Metropolitan Area (GAMA) - an agglomeration of 25 municipalities - to systematically:

- Close reliable SWM data gaps.
- Assess the SWM systems.
- Identify the mix of operator models that suits the system.
- Prioritise and develop participatory and locally responsive interventions.
- Create ownership of interventions and empower staff for effective implementation.
- Provide the platform for successful execution and scale-up of donor supported SWM projects.
- Win the trust and commitment of decision makers for improved governance and locally appropriate solutions.
- Make available through peer-reviewed publications sustainable entry points for support of national and international stakeholders.
- Draw relevant stakeholders together to support the development of a locally appropriate long-term strategy for system modernisation and sustainability.

In the process, staff and professional capacities are being strengthened and systematic paths for informal waste sector recognition have been agreed upon (Oduro-Appiah, Afful, et al., 2019). This has opened the opportunity for support from a not-for-profit international organisation for further research towards informal waste sector integration and the development of a climate change action plan for the GAMA.

Along such a path of cooperation and sustained research, Ghana would not be unique: there is a great deal of scholarship and evidence that shows that higher educational and research institutions can be and have been key and central contributors to the continuous improvement of strategies for delivering sustainable, equitable and resource-efficient SWM systems in developed economies (Asase et al., 2009; Leal Filho, Vargas, et al., 2019; Marshall & Farahbakhsh, 2013). The authors have deliberately refrained from reiterating the sequence of conventional recommendations that seek to provide direct answers to the physical and governance aspects of the numerous SWM challenges in Ghana. Along that path, there is no paucity in the scholarly literature (Oteng-Ababio et al., 2017; Wilson, Araba, Chinwah, & Cheeseman, 2009); the evidence however, is that such recommendations have neither been adhered to by decision-makers nor have they led to sustained modernisation of the country's SWM system.

The dynamics of the SWM system of Ghana and other developing countries, and the evidence available show that a coordinated pathway of participatory cooperation from universities and research institutions along the lines of sustained research, data management, implementation support and monitoring and evaluation, especially in the waste- and sanitation-related SDGs, will bridge the existing gap in which academia, industry and local municipalities operate in isolation to a sustainable pathway in which the scholarship and know-how of academia and research institutions would become more relevant and practicable to compliment the efforts of development partners and NGO's towards SWM system modernisation (Leal Filho, Vargas, et al., 2019).

- The right mix of public and private (formal and informal) actor involvement in service delivery (Cossu, Grossule, & Lavagnolo, 2019; Wilson et al., 2017).
- The right mix of locally appropriate and sustainable treatment and recycling technologies (Grossule & Lavagnolo, 2018).
- The charging of realistic cost for service provision, treatment and disposal.
- The establishment of a capable, strong and transparent client base working within realistic and flexible policy and regulatory frameworks (Soós et al., 2017).
- The integration of all relevant stakeholders' intervention projects and support.

A sustainable future for ISWM in Ghana will include, at a minimum, the following features provides a huge potential to modernise and deliver a sustainable SWM system (Soós et al., 2017; Wilson et al., 2017), rather than the prevailing situation where foreign strategies are adopted for use without reference to the goals, conditions and characteristics of the solid waste for which such strategies were developed. The existing dynamics and the evidence available in Ghana and similar developing economies in Sub-Saharan African show that higher educational institutions (HEIs) and research institutions would have to provide leadership to initiate the cooperation and collaboration link; partly because of their position as the bastion of knowledge and their inherent capability to undertake applied research, and partly because of their social responsibility to help society to modernise and sustain the SWM systems and the related SDGs (Leal Filho et al., 2019; Leal Filho et al., 2019).

Higher educational and research institutions also may be able to host and keep the institutional memory of municipalities to forestall the too frequent shift and abandoning of policies, often associated with changes in government in Ghana and in most developing countries.

CASE STUDIES

Integrated and Improved Municipal Solid Waste Collection within the Greater Accra Metropolitan Area

Waste collection coverage has improved in the Greater Accra Metropolitan Area (GAMA), consisting of 25 municipalities, including the capital city, Accra, and the industrial hub of the country, Tema, (Oduro-Appiah, Afful, et al., 2019). The successful modernisation is partly due to the participatory and research-driven development of a locally appropriate strategy (Oduro-Appiah, Scheinberg, Mensah, Kotey, et al., 2017) that recognises and further unleashes the contribution of informal service providers (ISPs) and legalises their important role in solid waste collection service delivery. Substantial credit is also due to the willingness of a key private waste management contractor (Zoom Pak) to construct two state-of-the-art bulk transport stations with a combined capacity of 2400 tons to receive the collected waste of the ISPs. This strategy (Oduro-Appiah, Scheinberg, Mensah, Kotey, et al., 2017), which sought to improve MSW collection by including providers of informal services, was itself informed by an earlier research project (Oduro-Appiah, Scheinberg, Mensah, Afful, et al., 2017) that brought researchers to collaborate with municipalities, government agencies and other relevant stakeholders to assess the MSWM system, identify strengths and system failures and together develop and implement locally appropriate interventions to drive system modernisation.

Throughout a four-year period of such cooperation and strategy development, the working group of researchers, municipal officials and informal sector representatives has witnessed a positive change both in the intended modernisation of collection coverage, and also in improved recyclables capture, improved disposal, improved service provider inclusivity and improved tolerance between both the formal and informal service providers. The ISPs now work alongside the formal waste collection companies and sometimes serve as backup collectors, especially in poor communities. The commitment and support from municipal decision-makers has also improved and the narrative of the ISPs has moved from neglect to recognition with an ongoing development of action plans towards integration (Oduro-Appiah, Scheinberg, et al., 2019). The Accra Metropolitan Assembly in 2018 registered all 720 ISPs within its boundaries as part of efforts towards their integration and also as a best practice to the other municipalities. There are about 1618 ISPs within the GAMA contributing to the collection of 1320 tons (45%) of all MSW collected within the area; and their activities alone saves the 25 municipalities of US$20 million in annual MSW operations cost (Oduro-Appiah, Scheinberg, et al., 2019). The ISPs now have access to improved and affordable tricycles for MSW collection and travel shorter distances to disposal sites.

Prior to this strategy, MSW collection coverage along the formal channels was 60%. The ISPs had contributed to an overall coverage of 75% but were in the habit of illegally dumping collected waste along roadsides, bushes, vacant plots and on unofficial dumpsites. Whilst the strategy has been successful, there still remain challenges with respect to service user fee harmonisation and training of the ISPs to promote further quality of the collection service. Moreover, much more advocacy needs to be undertaken to solicit the support of all the formal service providers. Issues of apparent delays and random increase in disposal cost at the bulk transport stations would need further attention to support the ISPs to increase their turnover and also prevent them from abandoning the facilities for environmentally unsustainable disposal methods.

The Accra Composting and Recycling Plant (ACARP)

The Accra Composting and Recycling Plant (ACARP) is one of the only two state-of-the art Mechanical and Biological Treatment (MBT) facilities in the country. The ACARP has been in operation since 2012, with the goal and promise of producing high value compost for application by peri-urban farmers within the Akuapim South Municipality and its environs. Although the plant recovers plastics, metals, textiles and contributes towards youth employment in the country, it can be considered as one of the MSWM overcapitalised technologies within a developing country context. First and foremost, the plant was designed for a shift capacity of 600 tons but has since been operating at only half its capacity despite the fact that the GAMA generates 3293 tons of MSW. The concept of MSW segregation to readily make available biodegradable waste to the plant and for improved compost has not been implemented yet. Perhaps the market for the compost was not found to be viable even before project commencement since the government of Ghana subsidises 67% of the cost of US$7.00 for a 50 kg finished compost as an incentive to attract local farmers. Subsidiary formal service providers whose services are paid for by the clients and service users dispose collected MSW at no cost at the plant. As of 2016, the operations of the plant were reported to cost the central government US$320,000.00 per month under a PPP arrangement. Perhaps the PPP arrangement might be the platform that sustains the plant instead of a resource recovery drive. Paradoxically, a similar plant with almost similar goals and arrangement has been commissioned in 2019.

CONCLUSION

Modernising the solid waste system in Ghana has been a major challenge, and will remain so for many years to come. This chapter has shared with the reader, the realities of SWM in Ghana and further discussed the nexus between system failures and success. The authors have used participatory and action research methods to assess the SWM system and have recommended locally appropriate and evidence-based paths that have the potential to stimulate modernisation. The underlying hypothesis is that participation and consultation between researchers and public officials in Ghanaian cities creates a higher probability towards improvement in both the physical aspects and the governance components of solid waste. When researchers in both higher education and research institutions collaborate with local authorities to support policies driven by the concerns of public health, environmental control, and resource recovery; they are able to introduce improvements in existing management approaches in a systematic way, rather than waiting for modernisation by crisis management, and triggered by disasters. Acting upon the hypothesis has proven that such a path opens up for trust, transparency, and commitment; three key ethical and governance factors virtually absent in the SWM system of most developing countries. The existing cooperation has been supporting capacity building of municipal SWM staff to close data gaps and develop locally responsive strategies for implementation; two factors that are chronically lacking in the SWM system of Ghana. The Wasteaware indicators and the mass flow diagram are intended to support stakeholders to comprehensively understand the performance and the various waste flows of the system; to enable them to determine points of entry for support. Whilst the path recommended may not be able to solve all the complex challenges of SWM in Ghana overnight, it has the potential to systematically improve and sustain the system for future generations.

REFERENCES

Akese, G. A. (2014). *Price realization for electronic waste (e-waste) in Accra*. Memorial University of Newfoundland.

Akese, G. A., & Little, P. C. (2018). Electronic Waste and the Environmental Justice Challenge in Agbogbloshie. *Environmental Justice*, *11*(2), 77–83. doi:10.1089/env.2017.0039

Al Masha'an, M., & Mahrous, F. (1999). Environmental Strategy for Managing the Kuwait's Solid Wastes until 2020. *Proceedings of Arab and Middle East International Conference and Fair on Solid Waste Management.*

Alamgir, M., Bidlingmaier, W., & Cossu, R. (2012). Successful waste management strategies in developing countries require meaningful involvement of the concerned stakeholders. *Waste Management (New York, N.Y.)*, *32*(11), 2007–2008. doi:10.1016/j.wasman.2012.07.028 PMID:22959155

Ali, A. (2010). Wasting time on solid waste in developing countries. *Waste Management (New York, N.Y.)*, *30*(8-9), 1437–1438. doi:10.1016/j.wasman.2010.04.025 PMID:20493678

Amoyaw-Osei, Y., Agyekum, O. O., Pwamang, J. A., Mueller, E., Fasko, R., & Schluep, M. (2011). Ghana e-waste country assessment. *SBC e-waste Africa Project.* Available online: http://www.basel.int/portals/4/basel%20convention/docs/ewaste/e-wasteassessmentghana.pdf

Amuzu, D. (2018). Environmental injustice of informal e-waste recycling in Agbogbloshie-Accra: Urban political ecology perspective. *Local Environment*, *23*(6), 603–618. doi:10.1080/13549839.2018.1456515

Asante, B., Yanful, E., & Yaokumah, B. (2014). Healthcare Waste Management; Its Impact: A Case Study Of The Greater Accra Region, Ghana. *International Journal of Scientific & Technology Research, 3*(3).

Asase, M., Yanful, E. K., Mensah, M., Stanford, J., & Amponsah, S. (2009). Comparison of municipal solid waste management systems in Canada and Ghana: A case study of the cities of London, Ontario, and Kumasi, Ghana. *Waste Management (New York, N.Y.)*, *29*(10), 2779–2786. doi:10.1016/j.wasman.2009.06.019 PMID:19615883

Asumadu-Sarkodie, S., Owusu, P. A., & Rufangura, P. (2015). Impact analysis of flood in Accra, Ghana. *Advances in Applied Science Research*, *6*(9), 53–78.

Atiemo, S., Faabeluon, L., Manhart, A., Nyaaba, L., & Schleicher, T. (2018). *Baseline Assessment of E-Waste Management in Ghana*. Academic Press.

Atiemo, S. M., Ofosu, F. G., Aboh, I., & Kuranchie-Mensah, H. (2012). Assessing the heavy metals contamination of surface dust from waste electrical and electronic equipment (e-waste) recycling site in Accra, Ghana. *Res J Environ Earth Sci*, *4*(5), 605–611.

Awuah, H. K. (2017). *Waste reduction strategies on construction projects: the case of Accra metropolis.* Academic Press.

Boadi, K. O., & Kuitunen, M. (2003). Municipal solid waste management in the Accra Metropolitan Area, Ghana. *The Environmentalist*, *23*(3), 211–218. doi:10.1023/B:ENVR.0000017283.09117.20

Boadi, K. O., & Kuitunen, M. (2005). Environmental and health impacts of household solid waste handling and disposal practices in third world cities: The case of the Accra Metropolitan Area, Ghana. *Journal of Environmental Health*, *68*(4). PMID:16334095

Bundhoo, Z. M. (2018). Solid waste management in least developed countries: Current status and challenges faced. *Journal of Material Cycles and Waste Management*, 1–11.

Burns, K. N., Sayler, S. K., & Neitzel, R. L. (2019). Stress, health, noise exposures, and injuries among electronic waste recycling workers in Ghana. *Journal of Occupational Medicine and Toxicology (London, England)*, *14*(1), 1. doi:10.118612995-018-0222-9 PMID:30647766

Caravanos, J., Clark, E., Fuller, R., & Lambertson, C. (2011). Assessing worker and environmental chemical exposure risks at an e-waste recycling and disposal site in Accra, Ghana. *Journal of Health and Pollution, 1*(1), 16-25.

Caravanos, J., Clarke, E. E., Osei, C. S., & Amoyaw-Osei, Y. (2013). Exploratory health assessment of chemical exposures at e-waste recycling and scrapyard facility in Ghana. *Journal of Health and Pollution, 3*(4), 11-22.

Cossu, R., Grossule, V., & Lavagnolo, M. C. (2019). Sustainable low cost waste management: Learning from airlines. *Detritus*, *6*, 1–3.

Daskalopoulos, E., Badr, O., & Probert, S. (1998). An integrated approach to municipal solid waste management. *Resources, Conservation and Recycling*, *24*(1), 33–50. doi:10.1016/S0921-3449(98)00031-7

Daum, K., Stoler, J., & Grant, R. (2017). Toward a more sustainable trajectory for E-waste Policy: A review of a decade of E-waste research in Accra, Ghana. *International Journal of Environmental Research and Public Health*, *14*(2), 135. doi:10.3390/ijerph14020135 PMID:28146075

Davis, J.-M., Akese, G., & Garb, Y. (2019). Beyond the pollution haven hypothesis: Where and why do e-waste hubs emerge and what does this mean for policies and interventions? *Geoforum*, *98*, 36–45. doi:10.1016/j.geoforum.2018.09.020

Diaz, L. F. (2017). Waste management in developing countries and the circular economy. SAGE Publications.

Duku, M. H., Gu, S., & Hagan, E. B. (2011). A comprehensive review of biomass resources and biofuels potential in Ghana. *Renewable & Sustainable Energy Reviews*, *15*(1), 404–415. doi:10.1016/j.rser.2010.09.033

Environmental Protection Agency. (2017). *Ghana State of the Environment (2016) Report, Environmental Protection Agency, Ministry of Environment, Science, Technology and Innovation, Accra.* Retrieved from http://www.epa.gov.gh/epa/publications/state-environment-report

Ezeah, C., & Roberts, C. L. (2012). Analysis of barriers and success factors affecting the adoption of sustainable management of municipal solid waste in Nigeria. *Journal of Environmental Management, 103*, 9–14.

Feldt, T., Fobil, J. N., Wittsiepe, J., Wilhelm, M., Till, H., Zoufaly, A., ... Göen, T. (2014). High levels of PAH-metabolites in urine of e-waste recycling workers from Agbogbloshie, Ghana. *The Science of the Total Environment, 466*, 369–376. doi:10.1016/j.scitotenv.2013.06.097 PMID:23921367

Gaisie, E., Kim, H. M., & Han, S. S. (2019). Accra towards a city-region: Devolution, spatial development and urban challenges. *Cities (London, England), 95*, 102398. doi:10.1016/j.cities.2019.102398

Ghana Statistical Services. (2019). *Ghana Population and Housing Census: Projection_Districts_2015_2020*. Retrieved from http://www.statsghana.gov.gh/

Government of Ghana. (2010). *National Environmental Sanitation Strategic Action Plan*. Retrieved from https://www.ircwash.org/sites/default/files/MLGRD-2010-National.pdf

Government of Ghana. (2017). *Enhancing urban resilience in the Greater Accra Metropolitan Area. Global Practice on Social, Urban, Rural and Resilience*. The World Bank Group. Retrieved from http://documents.worldbank.org/curated/en/949241495793834492/Enhancing-urban-resilience-in-the-Greater-Accra-Metropolitan-Area

Government of Ghana. (2018). *The Health Sector in Ghana: Facts and Figures*. Retrieved from https://www.ghanahealthservice.org/

Grant, R., & Oteng-Ababio, M. (2012). Mapping the invisible and real" African" economy: Urban e-waste circuitry. *Urban Geography, 33*(1), 1–21.

Grant, R., & Oteng-Ababio, M. (2019). Electronic-Waste Circuitry and Value Creation in Accra, Ghana. In Value Chains in Sub-Saharan Africa (pp. 115-131). Springer.

Grossule, V., & Lavagnolo, M. C. (2018). From 3R to 3S: An appropriate strategy for developing countries. *Detritus, 4*, 1–3. doi:10.31025/2611-4135/2018.13703

Guerrero, L. A., Maas, G., & Hogland, W. (2013). Solid waste management challenges for cities in developing countries. *Waste Management (New York, N.Y.), 33*(1), 220–232. doi:10.1016/j.wasman.2012.09.008 PMID:23098815

Henry, R. K., Yongsheng, Z., & Jun, D. (2006). Municipal solid waste management challenges in developing countries–Kenyan case study. *Waste Management (New York, N.Y.), 26*(1), 92–100. doi:10.1016/j.wasman.2005.03.007 PMID:16006111

Huang, J., Nkrumah, P. N., Anim, D. O., & Mensah, E. (2014). E-waste disposal effects on the aquatic environment: Accra, Ghana. *Reviews of Environmental Contamination and Toxicology, 229*, 19–34. doi:10.1007/978-3-319-03777-6_2 PMID:24515808

Kabera, T., Wilson, D. C., & Nishimwe, H. (2019). Benchmarking performance of solid waste management and recycling systems in East Africa: Comparing Kigali Rwanda with other major cities. *Waste Management & Research, 37*(1_suppl), 58-72.

Kartam, N., Al-Mutairi, N., Al-Ghusain, I., & Al-Humoud, J. (2004). Environmental management of construction and demolition waste in Kuwait. *Waste Management (New York, N.Y.)*, *24*(10), 1049–1059. doi:10.1016/j.wasman.2004.06.003 PMID:15567670

Kaza, S., Yao, L. C., Bhada-Tata, P., & Van Woerden, F. (2018). *What a Waste 2.0: A Global Snapshot of Solid Waste Management to 2050. Urban Development Series*. Washington, DC: World Bank. The World Bank. doi:10.1596/978-1-4648-1329-0

Kemausuor, F., Kamp, A., Thomsen, S. T., Bensah, E. C., & Østergård, H. (2014). Assessment of biomass residue availability and bioenergy yields in Ghana. *Resources, Conservation and Recycling*, *86*, 28–37. doi:10.1016/j.resconrec.2014.01.007

Leal Filho, W., Doni, F., Vargas, V. R., Wall, T., Hindley, A., Rayman-Bacchus, L., ... Avila, L. V. (2019). The integration of social responsibility and sustainability in practice: Exploring attitudes and practices in Higher Education Institutions. *Journal of Cleaner Production*, *220*, 152–166. doi:10.1016/j.jclepro.2019.02.139

Leal Filho, W., Shiel, C., Paço, A., Mifsud, M., Ávila, L. V., Brandli, L. L., ... Vargas, V. R. (2019). Sustainable Development Goals and sustainability teaching at universities: Falling behind or getting ahead of the pack? *Journal of Cleaner Production*, *232*, 285–294. doi:10.1016/j.jclepro.2019.05.309

Leal Filho, W., Vargas, V. R., Salvia, A. L., Brandli, L. L., Pallant, E., Klavins, M., ... Conticelli, E. (2019). The role of higher education institutions in sustainability initiatives at the local level. *Journal of Cleaner Production*, *233*, 1004–1015. doi:10.1016/j.jclepro.2019.06.059

Lepawsky, J., Akese, G., Billah, M., Conolly, C., & McNabb, C. (2015). Composing urban orders from rubbish electronics: Cityness and the site multiple. *International Journal of Urban and Regional Research*, *39*(2), 185–199. doi:10.1111/1468-2427.12142

Lu, W., Yuan, H., Li, J., Hao, J. J., Mi, X., & Ding, Z. (2011). An empirical investigation of construction and demolition waste generation rates in Shenzhen city, South China. *Waste Management (New York, N.Y.)*, *31*(4), 680–687. doi:10.1016/j.wasman.2010.12.004 PMID:21208794

Manhart, A., Meinel, J., & Walgenbach, S. (2014). *Legal and institutional requirements for e-waste recycling in Ghana: Work package 3.5of the Bo2W project*. Retrieved from http://www.oeko/de/oekodoc/2142/2014-695-en.pdf

Marshall, R. E., & Farahbakhsh, K. (2013). Systems approaches to integrated solid waste management in developing countries. *Waste Management (New York, N.Y.)*, *33*(4), 988–1003. doi:10.1016/j.wasman.2012.12.023 PMID:23360772

McDougall, F. R., White, P. R., Franke, M., & Hindle, P. (2008). *Integrated solid waste management: a life cycle inventory*. John Wiley & Sons.

Miezah, K., Obiri-Danso, K., Kádár, Z., Fei-Baffoe, B., & Mensah, M. Y. (2015). Municipal solid waste characterization and quantification as a measure towards effective waste management in Ghana. *Waste Management (New York, N.Y.)*, *46*, 15–27. doi:10.1016/j.wasman.2015.09.009 PMID:26421480

Oduro-Appiah, K., Afful, A., Kotey, V., & de Vries, N. (2019). Working with the Informal Service Chain as a Locally Appropriate Strategy for Sustainable Modernization of Municipal Solid Waste Management Systems in Lower-Middle Income Cities: Lessons from Accra, Ghana. *Resources, 8*(1), 12. doi:10.3390/resources8010012

Oduro-Appiah, K., & Aggrey, B. E. (2013). Determinants of source separation of municipal solid waste in developing countries: The case of Ghana. *Journal of Sustainable Development in Africa, 15*, 47-60.

Oduro-Appiah, K., Scheinberg, A., Afful, A., & de Vries, N. (2019). *The contribution of participatory planning processes to reliable data gathering and inclusive policy development in municipal solid waste management in the greater Accra metropolitan area of Ghana.* Manuscript submitted for publication.

Oduro-Appiah, K., Scheinberg, A., Mensah, A., Afful, A., Boadu, H. K., & de Vries, N. (2017). Assessment of the municipal solid waste management system in Accra, Ghana: A" Wasteaware' benchmark indicator approach. *Waste Management & Research, 35*(11), 1149–1158. doi:10.1177/0734242X17727066 PMID:28931348

Oduro-Appiah, K., Scheinberg, A., Mensah, A., Kotey, V., Afful, A., & de Vries, N. (2017). *Locally responsive intervention to improve municipal solid waste collection coverage in Accra, Ghana.* Paper presented at the 40th WEDC International Conference, Loughborough, UK.

Oduro-Kwarteng, S., & van Dijk, M. P. (2013). The effect of increased private sector involvement in solid waste collection in five cities in Ghana. *Waste Management & Research, 31*(10_suppl), 81-92.

Oteng-Ababio, M. (2011). Beyond technical details: The stalled Kwabenya engineered sanitary landfill project in Accra, Ghana. *Geografisk Tidsskrift-Danish Journal of Geography, 111*(2), 169–179. doi:10.1080/00167223.2011.10669531

Oteng-Ababio, M. (2014). "Guilty with explanation": Rethinking the destiny of landfills in a Millennium City in Ghana. *Management of Environmental Quality, 25*(2), 200–215. doi:10.1108/MEQ-11-2012-0074

Oteng-Ababio, M. (2018). Crossing conceptual boundaries: Re-envisioning coordination and collaboration among women for sustainable livelihoods in Ghana. *Local Environment, 23*(3), 316–334. doi:10.1080/13549839.2017.1418847

Oteng-Ababio, M., Amankwaa, E. F., & Chama, M. A. (2014). The local contours of scavenging for e-waste and higher-valued constituent parts in Accra, Ghana. *Habitat International, 43*, 163–171. doi:10.1016/j.habitatint.2014.03.003

Oteng-Ababio, M., Arguello, J. E. M., & Gabbay, O. (2013). Solid waste management in African cities: Sorting the facts from the fads in Accra, Ghana. *Habitat International, 39*, 96–104. doi:10.1016/j.habitatint.2012.10.010

Oteng-Ababio, M., Owusu-Sekyere, E., & Amoah, S. T. (2017). Thinking globally, acting locally: Formalizing informal solid waste management practices in Ghana. *Journal of Developing Societies, 33*(1), 75–98. doi:10.1177/0169796X17694517

Präger, F., Paczkowski, S., Sailer, G., Derkyi, N. S. A., & Pelz, S. (2019). Biomass sources for a sustainable energy supply in Ghana–A case study for Sunyani. *Renewable & Sustainable Energy Reviews*, *107*, 413–424. doi:10.1016/j.rser.2019.03.016

Prairie Village, K. (1998). *Characterization of building-related construction and demolition debris in the United States*. The US Environmental Protection Agency Municipal and Industrial Solid Waste Division Office of Solid Waste.

Rodić, L., & Wilson, D. C. (2017). Resolving governance issues to achieve priority sustainable development goals related to solid waste management in developing countries. *Sustainability*, *9*(3), 404. doi:10.3390u9030404

Scheinberg, A. (2011). *Value added: modes of sustainable recycling in the modernisation of waste management systems* (PhD thesis). Wageningen University.

Scheinberg, A., Wilson, D. C., & Rodic-Wiersma, L. (2010). *Solid waste management in the world's cities. In UN-Habitat's State of Water and Sanitation in the World's Cities Series*. Earthscan.

Schübeler, P., Wehrle, K., & Christen, J. (1996). *Conceptual framework for municipal solid waste management in low-income countries*. Academic Press.

Soós, R., Whiteman, A. D., Wilson, D. C., Briciu, C., Nürnberger, S., Oelz, B., ... Schwehn, E. (2017). Operator models for delivering municipal solid waste management services in developing countries: Part B: Decision support. *Waste Management & Research*, *35*(8), 842–862.

Tot, B., Vujić, G., Srđević, Z., Ubavin, D., & Russo, M. A. T. (2017). Group assessment of key indicators of sustainable waste management in developing countries. *Waste Management & Research*, *35*(9), 913–922. doi:10.1177/0734242X17709911 PMID:28606005

Tue, N. M., Goto, A., Takahashi, S., Itai, T., Asante, K. A., Kunisue, T., & Tanabe, S. (2016). Release of chlorinated, brominated and mixed halogenated dioxin-related compounds to soils from open burning of e-waste in Agbogbloshie (Accra, Ghana). *Journal of Hazardous Materials*, *302*, 151–157. doi:10.1016/j.jhazmat.2015.09.062 PMID:26474377

Tue, N. M., Matsushita, T., Goto, A., Itai, T., Asante, K. A., Obiri, S., ... Kunisue, T. (2019). Complex Mixtures of Brominated/Chlorinated Diphenyl Ethers and Dibenzofurans in Soils from the Agbogbloshie e-Waste Site (Ghana): Occurrence, Formation, and Exposure Implications. *Environmental Science & Technology*, *53*(6), 3010–3017. doi:10.1021/acs.est.8b06929 PMID:30793891

UNEP. (2018). *Africa Waste Management Outlook*. United Nations Environment Programme. Retrieved from http://hdl.handle.net/20.500.11822/25514

White, P., Dranke, M., & Hindle, P. (2012). *Integrated solid waste management: a lifecycle inventory*. Springer Science & Business Media.

Wilson, D. C. (2007). Development drivers for waste management. *Waste Management & Research*, *25*(3), 198–207. doi:10.1177/0734242X07079149 PMID:17612318

Wilson, D. C., Araba, A. O., Chinwah, K., & Cheeseman, C. R. (2009). Building recycling rates through the informal sector. *Waste Management (New York, N.Y.)*, *29*(2), 629–635. doi:10.1016/j.wasman.2008.06.016 PMID:18701272

Wilson, D. C., Kanjogera, J. B., Soós, R., Briciu, C., Smith, S. R., Whiteman, A. D., ... Oelz, B. (2017). Operator models for delivering municipal solid waste management services in developing countries. Part A: The evidence base. *Waste Management & Research*, *35*(8), 820–841. doi:10.1177/0734242X17705723 PMID:28617175

Wilson, D. C., Rodic, L., Cowing, M. J., Velis, C. A., Whiteman, A. D., Scheinberg, A., ... Oelz, B. (2015). 'Wasteaware' benchmark indicators for integrated sustainable waste management in cities. *Waste Management (New York, N.Y.)*, *35*, 329–342. doi:10.1016/j.wasman.2014.10.006 PMID:25458855

Wilson, D. C., Velis, C. A., & Rodic, L. (2013). Integrated sustainable waste management in developing countries. *Proceedings of the Institution of Civil Engineers-Waste and Resource Management*, *166*(2), 52–68. doi:10.1680/warm.12.00005

World Bank. (2019). *GNI per capita, Atlas methods (current US$)*. Retrieved from https://data.worldbank.org/indicator/ny.gnp.pcap.cd

ADDITIONAL READING

Akese, G. A., & Little, P. C. (2018). Electronic Waste and the Environmental Justice Challenge in Agbogbloshie. *Environmental Justice*, *11*(2), 77–83. doi:10.1089/env.2017.0039

Cossu, R., Grossule, V., & Lavagnolo, M. C. (2019). Sustainable low cost waste management: Learning from airlines. *Detritus*, *6*, 1–3.

Daum, K., Stoler, J., & Grant, R. (2017). Toward a more sustainable trajectory for E-waste Policy: A review of a decade of E-waste research in Accra, Ghana. *International Journal of Environmental Research and Public Health*, *14*(2), 135. doi:10.3390/ijerph14020135 PMID:28146075

Davis, J.-M., Akese, G., & Garb, Y. (2019). Beyond the pollution haven hypothesis: Where and why do e-waste hubs emerge and what does this mean for policies and interventions? *Geoforum*, *98*, 36–45. doi:10.1016/j.geoforum.2018.09.020

Kaza, S., Yao, L. C., Bhada-Tata, P., & Van Woerden, F. (2018). *What a Waste 2.0: A Global Snapshot of Solid Waste Management to 2050. Urban Development Series*. Washington, DC: World Bank. The World Bank. doi:10.1596/978-1-4648-1329-0

Leal Filho, W., Vargas, V. R., Salvia, A. L., Brandli, L. L., Pallant, E., Klavins, M., ... Conticelli, E. (2019). The role of higher education institutions in sustainability initiatives at the local level. *Journal of Cleaner Production*, *233*, 1004–1015. doi:10.1016/j.jclepro.2019.06.059

Oteng-Ababio, M., Arguello, J. E. M., & Gabbay, O. (2013). Solid waste management in African cities: Sorting the facts from the fads in Accra, Ghana. *Habitat International*, *39*, 96–104. doi:10.1016/j.habitatint.2012.10.010

Wilson, D. C., Rodic, L., Cowing, M. J., Velis, C. A., Whiteman, A. D., Scheinberg, A., ... Oelz, B. (2015). 'Wasteaware' benchmark indicators for integrated sustainable waste management in cities. *Waste Management (New York, N.Y.), 35*, 329–342. doi:10.1016/j.wasman.2014.10.006 PMID:25458855

KEY TERMS AND DEFINITIONS

Collection Coverage: The percentage of total households or waste generating units in a city that has access to a regular service of waste collection and street sweeping, irrespective of whether it is provided by formal or informal service providers.

Controlled Landfill Site: A landfill site which enjoys some appreciable level of waste compaction, waste covering if even irregularly and also some level of leachate containment and treatment.

Formal Service Providers: Private companies who have official franchise agreements with municipalities and use different categories of solid related vehicles (skips, compactor trucks, etc.) to collect MSW from household, institutions and light industries to disposal and recycling centres.

Informal Service Provider (ISP): An individual entrepreneur or enterprise which provides a service -- in this case of waste collection or street sweeping -- which is paid by the clients but which is not planned, authorized, supervised, paid, or acknowledged by the formal authorities to be part of the official waste system. Specifically, the term ISP is designed to distinguish the waste collection informal sector, which is providing the service of removal of waste, from the recycling informal sector, which is extracting materials that have value and valorizing them by trading them to the recycling industry.

Municipal Solid Waste (MSW): MSW in the Ghanaian context of SWM is defined to include household solid waste, institutional waste, commercial waste, waste from light industries, and construction and demolition (C&D) waste.

Waste Captured: The percentage of waste generated that is actually handled completely by the waste management and recycling system. It includes all waste material collected and delivered to official treatment, recycling and disposal facilities. Any waste that is lost through illegal burning, burying or dumping in unofficial locations within a city or country is excluded in the estimation or computation of waste capture.

Chapter 6
Waste Management Under the Legal Framework in India

Sadhan Kumar Ghosh
Jadavpur University, India

ABSTRACT

India, the second biggest country in the world, has nearly 1.25 million people living in 29 states and seven union territories covering an area of 3,287,000 sq. km. India's economy grew at an impressive 8.2% in the first quarter of 2018-19. Traditionally, India has the habit of reuse and recycling the materials wherever possible. As the city agglomeration is increasing the waste generation is increasing. The number of towns/cities have increased from 5,161 in 2001 to 7,935 in 2011, whereas the number of metropolitan cities having million plus population has increased from 35 to 53 number as per 2011 census. It is projected that half of India's population will live in cities by 2050. Waste management in India has been experiencing a paradigm shift through the establishment of Swachh Bharat Mission in urban and rural India in 2014 and the revision and establishment of waste management rules in six types of wastes including transboundary movement in 2016. This study presents the overall waste management scenario and the legal framework in India.

INTRODUCTION

India, the second biggest populous country in the world has nearly 1.25 million people living in twenty-nine states and seven union territories covering an area of 3,287,000 sq. km. India's economy had an impressive 8.2% in the first quarter of financial year 2018-19 ending June 30, which was on the foundation of a strong core performance and a healthy base. Traditionally, Indians have the habit of reusing and recycling the materials wherever possible. This leads to the generation of lower amount of wastes per capita. The average solid waste generation in rural areas varies from 100 g to 300 g and in urban areas varies between 400 g and 900 g due to the urbanization and industrialization. The rate of waste generation is increasing at moderate rates in last few years because of higher GDP and the increase in purchasing capacity. The total amount of solid waste generated in India is nearly 62 million a year with an increase rate of 1.33% annually. The number of towns/cities have increased from 5161 in 2001 to

DOI: 10.4018/978-1-7998-0198-6.ch006

7935 in 2011 whereas the number of metropolitan cities having million plus population has increased from 35 to 53 number as per 2011 census. Projection forecasts that half of India's population will live in cities by 2050. With the enhanced city agglomeration, the waste generation is expected to increase. This study will present the overall Waste Management scenario and the Legal Framework in India.

WASTE DEFINITIONS, POPULATION, GDP, ANNUAL WASTE GENERATION, DAILY/ANNUAL PER CAPITA WASTE GENERATION, COMPOSITION OF SOLID WASTE/MSW, WASTE GENERATION NATIONWIDE (URBAN VS RURAL)

Solid wastes have been classified in different regulated categories in India, namely, a) solid waste, b) hazardous waste, c) electronic waste, d) biomedical waste, e) construction and demolition waste (C&D), f) plastic wastes and g) lead acid batteries. The recently revised Solid Waste Management (SWM) Rules 2016 defines "solid waste" as the solid waste, which includes solid or semi-solid domestic waste, sanitary waste, institutional waste, commercial waste, catering and market waste and other non-residential wastes, silt collected from the surface drains, street sweepings, agriculture and dairy waste, horticulture waste, treated bio-medical waste excluding industrial waste, bio-medical waste, e-waste, battery waste and radio-active waste generated in the area under the local authorities and other entities. The general waste flow methods in India starts from segregated storage at source and/or storage mixed at source – collection - transportation followed by disposal to recycling & treatment facilities and/or dumping to specified open ground or to sanitary landfill sites. Government of India has a regulation for the underground isolation lining in the landfill for avoiding leachate emissions and regular application of soil coverage over disposed waste.

SWM rules 2016 have given directions for storing, collection, transportation, technologies to be adopted and also the final disposal. Three bin system has been introduced for the storing of wastes at sources in segregated manner for biodegradable, dry, non-biodegradable and domestic hazardous wastes. In the same way, the collection and transportation are carried out for separate types of wastes. Although the SWM Rules 2016 requires the separation of household wastes in three bins, there are many Urban Local Bodies (ULBs) in the country where mixed wastes are collected, the intervention of informal sectors of waste pickers are observed to separate and collect the portion of valuable dry wastes for selling it to the recyclers for their livelihood earning. The biodegradable wastes are treated in facilities of composting, bio-methanation, and any other energy recovery processes. In composting the informal sector has a significant contribution as they remove the dry portion of the waste in case of mixed wastes. Dry wastes go for different types of recycling facilities including energy recovery processes, namely, waste to energy plants of incineration types, small pyrolysis plants, RDF (Refuse Derived Fuel) plants, metal processing plants, plastic recycling plants, paper plants, cement plants and many others. In India nearly 57% of solid waste is still destined to landfill and open dumping. Initiatives have been taken to reduce the disposal to landfill. The SWM Rules 2016 encourages the implementation of different treatment technologies of waste and disposal for only less than 15% of wastes to sanitary landfill. The implementation of this policy is in the process. At present more than 4,237 cities are trying to practice the same as per the survekshan report of 2019. The mixing of C&D wastes in the waste streams was problem for the treatment processes but introduction of C&D Waste Management Rules 2016 has enhanced the efficiency of treatment process. The co-processing in cement plants have emerged as the implementable sustainable process of energy recovery with the support of newly introduced rules. Significant interest in biological treatment of wastes is prevalent in the country which is being supported by the government at present. There are several provisions by the government to encourage the waste treatment. INR 1500 is given per ton of compost production as a support mechanism. There are several other support systems including the system of introducing tipping fees have been incorporated in the rules to enhance the recovery efficiency and reduce the wastes going to landfill.

Solid waste generation status as per the report of CPCB 2016 has been presented in the Table 1. Between 2016 and 2019 March, there is a significant change in the waste management in Indian cities because of the incorporation of yearly assessment survey, called Swachh Survekshan. At present the collection efficiency is increased to nearly 85% in 4237 cities covering nearly 4.3 billion citizens. Similarly, waste processing has also been improved substantially.

To fit in the globalization of the economy, culture, and technology, the concept of Megacities which is termed as Major cities in India have been evolved. All the Major cities have faster and dynamic economic growth, faster agglomeration of population, shifting from rural India, and high waste generation rates together with higher per capita waste generation (Table. 2).

Twenty-eight states and nine union territories of India have been taking up significant number of initiatives in the waste management arena to reduce the dumping and enhance the resource recovery. Figure 1 gives an idea of processing of SWM in the country. According to the data of National Sample Survey Office, NSSO, Government of India, a few states and union territories (UTs), namely, Delhi, Telangana, Goa, Manipur, West Bengal, Karnataka and Kerala have more percentage of processing

Table 1. Solid waste generation status in India as per CPCB report, 2016

Waste Generation	Waste Collected	Waste Processed	Landfilled Waste:	House–to-house collection	Segregation
1,41,064 tons per day	1,27,531 tons per day (90% of generated wastes),	34,752 tons per day (27% of generated wastes),	4,515 tons per day	Done partially varies 40-90%) in 18 States	Done partially; varies from 20 - 80%) in 5 States

Table 2. Waste generation data in major cities in India (2010–2011) (CPCB Report 2011) (Census done at each 10 years interval)

Megacity	Population (2011)×10^6 **	Waste Generation in tons per day*	*Waste Generation (kg* **per capita** *per day)*
Greater Mumbai	18.4	6500	*0.35*
Delhi	16.3	5800	0.41
Kolkata	14.1	3670	0.26
Chennai	8.6	4500	0.52
Bangalore	8.4	3700	0.44
Hyderabad	7.7	4200	0.54
Ahmedabad	6.3	2300	0.36

Figure 1. Percentage of processing of SWM in different states & Territories (Data Source: NSSO)

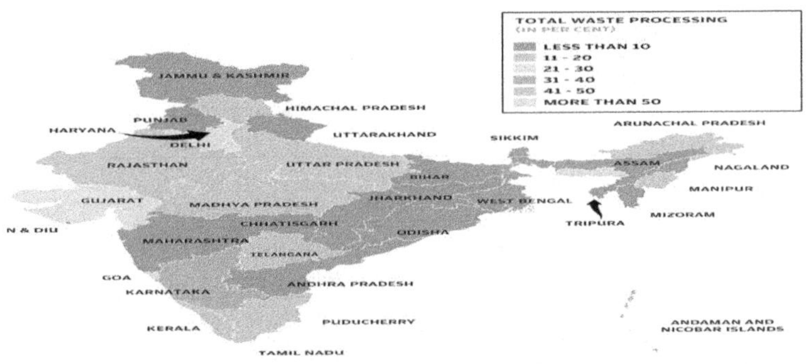

of wastes. However, with the intervention of Swach Survekshan, most of the states and UTs are now involved in improving the efficiency of collection, segregation, processing and recycling of municipal solid wastes (MSW), leading to a minimum percentage of final sink to the landfills. Demographic data of India is presented in Table 3.

Composition and Characteristics of Municipal Solid Waste

The MSW in India varies from location to location with its wide spread of culture, weathers, geographic condition and many other factors. The major categories of wastes, that are generally found in Indian cities, comprise of biodegradable wastes, non-biodegradable or recyclable wastes, domestic hazardous wastes, composite wastes, garden wastes and inert waste matter. Table 3 shows further breakup of all the types of wastes found within municipal solid waste and applicable legislation. However, on an average, MSW consists of 40–60% compostable, 30–50% inert waste and 10% to 30% recyclable waste. The MSW consists of Nitrogen (0.64% ± 0.8), Phosphorus (0.67% ± 0.15), Potassium (0.68% ± 0.15), and C/N ration (26% ± 5). The characteristics in the composition of wastes have been changed with time due to many factors (Joshi & Ahmed, 2016). As per the report of the Planning Commission, government of India (Planning Commission Report, 2014), major categories of waste in India and the change in the physical and chemical composition of MSW in Indian with time is shown in Table 4 and 5 respectively.

Table 3. India - demographic data [Sources: (S. K. Ghosh, 2016) & (WWE, 2019)]

Description	Demographic Data
Population (Census 2011)	1.25 billion
Population Increase	1.24%
Population density	386 people/ km^2
Geographical area	3,287,240 km^2
Rate of increase of MSW generation	1% - 1.33%
MSW generation, Per capita	0.2 to 1.0 kg
GDP grows at in 2018-19	8.2%
GDP [Source: World Economic Outlook Database, April 2019". IMF. Retrieved 9/04/2019.	$2.716 trillion (nominal, 2018) $10.505 trillion (, 2018)

Table 4. Major categories of wastes in Indian cities and applicable legislation

Major Types	Different Composition	Applicable Legislation
Biodegradable Wastes	Food waste, kitchen waste, green waste, (e.g., vegetables, flowers, fruits, leaves), paper waste etc.	Solid Waste Management (SWM) Rules 2016;
Non-Biodegradable or Recyclable wastes	Cans, metals, paper, glass, bottles, certain plastics, diapers, sanitary napkins, etc.	SWM Rules 2016; Plastics Waste Management (PWM) Rules 2016;
Composite wastes	Tetra packs, Waste clothing, waste plastics such as toys etc.	SWM Rules 2016; PWM Rules 2016
Domestic Hazardous Wastes	Waste medicine, syringes, Saline bottles, cans, paints, chemicals, spray, fertilizer & pesticide containers, E-waste, light bulbs, fluorescent tubes, shoe polish, batteries, etc.	SWM Rules 2016; PWM Rules 2016; e-waste Management (EWM) Rules 2016; Hazardous Waste Management & Trans boundary Movement Rules 2016; Battery Rules 2001 as amended;
Garden Wastes	Dry and green leaves, cut or broken limbs of trees,	SWM Rules 2016;
Inert Waste Matter	Construction and Demolition Waste (CDW), dirt, debris, etc.	CDW Management Rules 2016

Table 5. Percentage change in municipal solid waste composition with time

	Percentage							
Year	**Biodegradable**	**Paper**	**Plastic/ rubber**	**Metal**	**Glass**	**Rags**	**Others**	**Inert**
1996	42.21	3.63	0.60	0.49	0.60	–	–	45.13
2005	47.43	8.13	9.22	0.50	1.01	4.49	4.02	25.16
2011	42.51	9.63	10.11	0.63	0.96	-	-	17.00

WASTE GENERATION / 3R PRACTICES, WASTE MANAGEMENT TECHNOLOGIES

In India the total solid waste generation is 960 million tons whereas nearly 62 million tons are generated in Urban India. As per the twentieth report of standing committee on energy (2015-16) (Sixteenth Lok Sabha February 2015), waste generation in the country was estimated at nearly 0.143 million metric tons per day and 1.18 lakh metric tons (82%) is being collected and the remaining 18% is littered. Only 33,000 metric ton (28%) was being treated and disposed.

Waste Flow Path and the Supply Chain

The common waste treatment facility (ENVIS Newsletter, December 2010) is widely promoted and accepted. This facility utilizes waste as a resource by using it as a co-fuel or co-raw material in different manufacturing processes. Public Private Partnership (PPP) models in waste management have become a popular concept of doing business in waste management.

The solid waste generated in the households and municipal areas flow in the path described in Figure 2. The wastes are stored in bins. All the three types of wastes, namely, dry, wet and domestic hazardous wastes are collected by the municipalities or authorized service providers. The dry recyclables are taken to the material recovery facilities and the wet wastes are taken to the waste processing units where composting and/or bio methanation plants are operated. The dry recyclables are separated to as many compositions as possible and sent to the recyclers, namely, metals, plastics, papers, clothes, glass, wood, PVC, and others. The combustible wastes having more than 1500 calorific value per kg of wastes are sent to RDF units, cement plants and waste to energy plants. For composting plants, the government of India pays a subsidy amount of INR 1500/- per ton of compost produces. The least amount or inert materials ranging from 20% to 5% is sent to the landfill. India has a very good market for recycled products. India has a potential to generate about 583 MW of green energy daily from the bio-degradable portion of MSW, and produce about 5.1 million tons of bio fertilizer per annum, which may go up to 2,273 MW and 19.5 million tons, respectively, in 2030. Clean energy generated from OFMSW may save coal consumption of about 3.04 million tons and mitigate CO_2 emissions by 35.42 million tons per annum in India. Utilization of the organic fraction of MSW will be able to save about 550 acres (2.23 sq. km) of landfill area per year and around 8,182 hectares (82.5 sq. km) of landfill area during the period of next 15 years (Aich, 2016). A few waste-to-energy plants through incineration path generating nearly 84 MW of electricity are working in India with 700 tpd to 2000 tpd capacity generating 8 MW to 18 MW of power in Jabbalpur and Delhi.

POLICIES AND REGULATIONS (INCLUDING EXISTING SOLID WASTE NATIONAL PLANS, NATIONWIDE TARGETS ETC.)

With the latest development and changes in lifestyle, the existing Environment Protection Act 1986, Municipal Solid Waste (Management and Handling) Rules 2000, Hazardous Waste Management Rules 1998 and Bio Medical Waste Management Rules 1998 were needed to be revised to cope up with the present situation. Based on 5R and circular economy concepts a set of six rules were introduced in 2016 for the control of Solid Wastes, namely, the municipal wastes, construction & demolition wastes, biomedical wastes, plastics wastes, electronic wastes and hazardous and other types of wastes, including transboundary movement of wastes in effective manner. These Rules define the roles and responsibilities of different stakeholders including waste generators, operators of waste management facilities, and

Figure 2. Supply chain of municipal waste in municipalities as per existing rules (S. K. Ghosh, 2017)

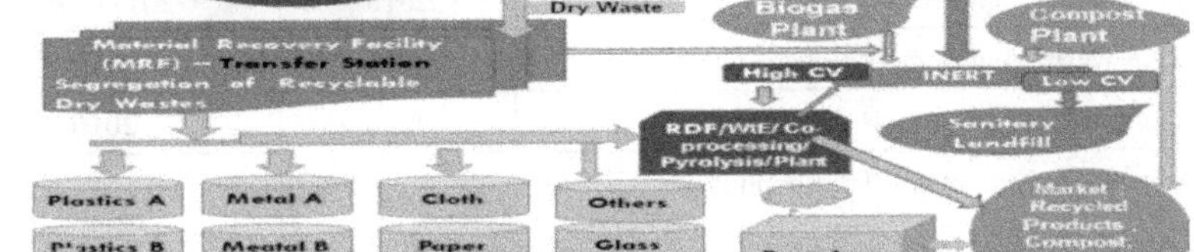

central and state implementing and regulatory agencies. Figure 3 gives the glimpses of the basis of the Indian policies and rules related to wastes management per se environmental protection and resource circulation. Very recently, the national resource efficiency policy has been released as draft for the feedback and finalization.

Rules related to waste management have been amended several times to cope up with the present day need from both national and international perspective. The changes in waste management rules in India have been shown in Table 6.

Environmental protection as well as waste management was covered in the factories Act 1881 that was finally enacted in Indian parliament in 1948 and amended in 1958. A major change in Indian environmental legislation took place after the Stockholm conference in 1972 and after the 42nd amendment, article 48 part IV and article 51 A(g) in the Indian Constitution. Common Biomedical Waste Treatment Facility (CBWTF) at every 150 kms is required as per Bio-medical waste (management and handling) rules, 1998, which was revised in 2016. CBWTFs have been set up and are functioning in cities and towns for quite a long time. However, new initiatives of setting up functional CBWTF, integrated common facilities for hazardous waste management, TSDF (Transfer, Storage and Disposal Facility), combine secured landfill facility, solidification/stabilization and incineration for treating the hazardous wastes from industrial units are in the process. The major change in waste management legislation in India was

Figure 3. Base and recent revision of Indian policies and rules related to wastes management

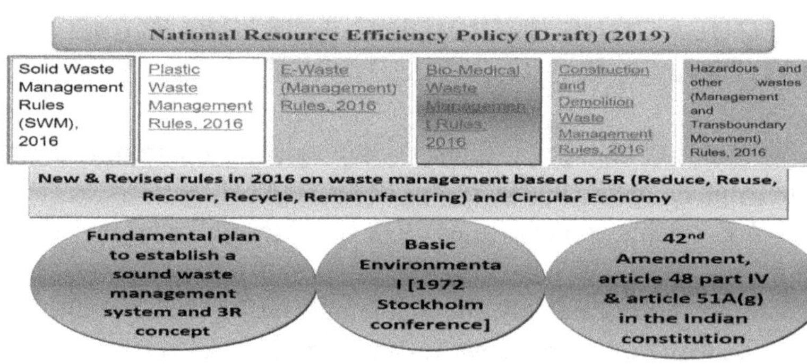

Table 6. Changes in waste management rules in India

Rules	Latest Release	Year of Amendments	Year of Original Release
Solid Waste Management Rules,	2016		2000
E-waste Management (Amendment) Rules,	2018	2016, 2011	2011
Plastics Waste Management (Amendment) Rules,	2018	2011, 2016	1987
Construction and Demolition Waste Management Rules,	2016	2016,	2016
Bio Medical Waste Management (Amendment) Rules,	2018	2016,	1998
Hazardous and Other Wastes (Management and Transboundary Movement) Rules	2016	2016,	1998
Battery Waste Management Rules,	2001	1991	1979

observed as a paradigm shift in 2016 based on 5R (Reduce, Reuse, Recover, Recycle, and Remanufacturing) and circular Economy and follow up actions of Swachh Bharat Mission 2014.

Swachh Bharat Mission

To enhance the cleanliness and improvement in sanitation of both rural and urban areas, Swachh Bharat Mission (Urban and Rural) {SBM-U&R} was launched in 2014 with a target time of 5 years ending in October 2019, and has performance targets to improve main objectives of the mission. The main objectives are as followed:

- Elimination of open defecation.
- Eradicating manual scavenging.
- Modern scientific Management of Solid Wastes in Municipalities.
- Initiating and establishing behavioral change towards healthy sanitation practices.
- Creating awareness on sanitation and its linkage with public health.
- Capacity building and enhancement for ULB.
- Creation of enabling environment for private sector participation in Opex (operation and maintenance) and Capex (capital expenditure).

The Swachh Bharat Mission (SBM) focused on MSW management and bringing in a behavioral change of people regarding healthy sanitation practices, training & awareness generation, ensuring door-to-door garbage collection and proper disposal of MSW in all urban areas by 2019. Swachh Bharat Mission has gained a big momentum in last four years. The flagship campaign of the present government aims at covering 10.4 million households, providing 260,000 public toilets, 250,000 community toilets, and a solid waste management facility in each town. Covering 4,401 towns over a period of five years. To strengthen the implementation of the SBM-U&R, and as a prelude to encouraging cities to improve urban sanitation, government introduced an assessment based survey, called, 'Swachh Survekshan' for the rating of 4237 cities in 2019 (Swachh Survekshan Toolkit, 2019). The swach survekshan started in 2016 and it increases the depth of survey with respect to its spearhead, assessment criteria and coverage described in Table 7.

Swachh Bharat Mission encourages large scale citizen participation and also creates awareness amongst different sections of society, making towns and cities a better place to live in by working together. The surveys on sanitation and cleanliness intend to foster a spirit of healthy competition among towns and cities to improve their services towards citizens, and thus creating cleaner cities. Swachh Bharat Mission (Urban) was launched in January 2016. Table 7 shows the details of the 'Swachh Survekshan' conducted from 2016 to 2019 revealing the number of cities surveyed, number of indicators on which survey was made, number of citizens impacting in the survey and the results of the top five cities.

Swachh Survekshan Grameen (Rural) - SBM-G: The Ministry of Drinking Water and Sanitation (MoDW&S), in July 2018, has commissioned "Swachh Survekshan Grameen-2018" (SSG 2018) through an independent survey agency to develop ranking of all districts of India on the basis of quantitative and qualitative sanitation (Swachhata) parameters to rank all 698 districts in the country on prevalent levels of cleanliness in the rural areas. The ranking was done based on a comprehensive set of parameters. The surveys took place at public places in Haat/Bazaars, schools, Anganwadis, PHCs and Panchayats. The perception of Swachhata and their recommendations for improvement of the program were gauged

Table 7. Swachh Survekshan 2016 to 2019 survey details at a glance

Swachh survekshan year	No. of cities included	Assessment Weightage of Ranking Criteria	Percentage of Service Level Progress and number of Indicators for rank assessment	Impacting number of citizens	Top five cleanest cities
2019	4237	• Service Level Progress 25% • Direct Observation 20% • Citizen Feedback 25% • Certification 25%	Collection & Transportation 27%: 7 indicators Processing & Disposal 30%: 8 indicators Sustainable Sanitation 25%: 7 indicators:3 indicators IEC & Behavioral Change 5%: 3 indicators Capacity Building 3%: 2 indicators Bye-Laws 5%: 4 indicators Innovation 5%: 2 indicators	0.653 million survey locations; Impacting 4.3 billion of people	Indore, Ambikapur, Mysore, Ujjain, New Delhi (NDMC)
2018	4,203	• Service Level Progress 35% • Direct Observation 30% • Citizen Feedback 35%	Collection & Transportation 30%: 13 indicators Processing & Disposal 25%: 8 indicators Sanitation & ODF 30%: 11 indicators:3 indicators IEC & Behavioral Change 5%: 7 indicators Capacity Building 3%: 4 indicators Innovation & best practices 5%: 1 indicators	0.22 million Survey Locations; Impacting 4.0 billion of people	Indore, Bhopal, Chandigarh,New Delhi, Vijayawada
2017	434	• Municipal Documentation:45% • Citizen Feedback:30% • Independent Observation:25%	Information, Education and Behavior Change Communication (IEBC) activity: 5% Collection - Door to door, Sweeping and Transportation: 40% MSW Processing and Disposal: 20% ODF & Toilet: 30% Capacity Building- ELearning Portal Provision: 5%	Cities with population of 0.1 million and above	Indore, Bhopal, Visakhapatnam, Surat, Mysuru
2016	73	• Service Level Status Data, 50%; • Independent Observation Data, 25%; • Data of Citizen Feedback, 25%;	Collection - Door-to-door,, Sweeping, Collection & Transportation: 40% Solid waste Processing and Disposal: 20% Provision of Public & Community Toilet: 15% Strategy for Open Defecation Free town and Integrated SWM: 5% Information, Education and Behavior Change Communication (IEBC) activity: 5% Individual Toilet:15%	73 million plus cities and state/UT capitals	Mysore, Chandigarh, Tiruchirappalli, Delhi (NMDC), Visakhapatnam

from citizen's. Similar to the model along the lines of ones undertaken in urban areas every year since 2016. 600-700 surveyors would fan out randomly to selected rural habitats from among the 6,900 odd villages across the country. Every district and state in the country has been included for ranking on the performance of SBM criteria to measure the on-ground effectiveness of the Swachh Bharat Mission. It has been conducted by the ministry of drinking water and sanitation (MoDWS). A combination of citizen feedback and direct observation by surveyors are used in ranking process. The data on the number of toilet available and number of toilets in use were also considered in order to arrive at the rankings.

In Swachh Survekshan Gramin (SSG) 2018, 27,963 public places in these 6786 villages in 685 Districts across India were visited by an independent agency for survey. Interview were taken for feedback from 182,531 citizens either personally or online in villages on SBM-G related issues. Data from Direct Observation based on the sanitation status in public places were also taken. Feedbacks were taken using community wide open meetings, personal interviews and focused group discussions (FGD) in each village involving more than 15 million people. Haryana ranked the best State while Satara District of Maharashtra ranked the best district in 2018. Table 8 shows the zonal ranking of top 3 states and UT level. Table 8 shows the top 3 districts – zonal and UT level. SSG 2018 Overall Ranking are as below (Swachh Survekshan Grameen Awards, 2018).

- Haryana, Gujarat and Maharashtra are the three top states.
- Satara in Maharashtra, Rewari in Haryana and Pedapalli in Telangana are the three top districts.
- Uttar Pradesh, Gujarat and Maharashtra are the States with maximum citizen participation.
- Nashik in Maharashtra, Solapur in Maharashtra and Chittorgarh in Rajasthan are the districts with maximum citizens' participation.

Integrated management information system (IMIS) of the MoDW&S in SSG 2018 will supply the data with 65% weightage to the findings and outcome and 35% to service level parameters. The weights to different elements of SSG have been defined as 30% from the direct observation of sanitation in public places, 35% from the citizen's feedback on sanitation parameters, and 35% from service level progress on sanitation as per SBMG-MIS.

Table 8. Top 3 states and union territory (UT) level in zonal ranking

Zones/UT	State Rank-1	State Rank-2	State Rank-3
Northern	Haryana	Rajasthan	Himachal Pradesh
Southern	Andhra Pradesh	Telangana	Karnataka
Eastern	Chhattisgarh	West Bengal	Jharkhand
Western	Gujarat	Maharashtra	Madhya Pradesh
North-East	Sikkim	Mizoram	Meghalaya
Union Territories	D & N Haveli	Daman & Diu	Chandigarh

Table 9. Zonal and union territory (UT) level - top 3 districts

Zones/UT	District Rank-1	District Rank-2	District Rank-3
Northern	Rewari, Haryana	Gurgaon, Haryana	Karnal, Haryana
Southern	Peddapalli, Telangana	Thoothukudi, TamilNadu	Warangal, Telangana
Eastern	Surajpur, Chhattisgarh	Surguja, Chhattisgarh	Hazaribagh, Jharkhand
Western	Satara, Maharashtra	Patan, Gujarat	Nashik, Maharashtra
North-East	Tawang, Arunachal	Aizawl, Mizoram	East Sikkim,
UTs	D & N Haveli	Daman	Diu

Smart City Mission

For an integrated, automated and controlled solid waste management system, smart city development (SCD) involves the applications of information communication technology (ICT) and Internet of Things (IoT). United Nations Development Programme (UNDP) projected 17 sustainable development goals (SDGs) 2030 with 169 indicators focusing on sustainable development, democratic governance, peace building, and climate & disaster resilience for sustainable smart city development in developing nations (Figure 4).

The Ministry of Urban Development (MoUD), and government of India (GOI) has broadcasted the formation of 100 Smart Cities through Smart City Mission Programme 2015, where the inhabitants are likely to practice information and communication technology (ICT) with the gain of internet. With growing inhabitants, varying policy and stagey requirements, innovative sustainability, reprocessing goals, resource utilization and improved technologies, municipalities throughout the world are assembling towards the "smart cities" movement to become more effective in handling and management of solid waste.

Agricultural wastes with wide variety are the most widely available and the most wasted energy source. Large amount of agricultural waste is used for cattle fodder while huge amounts of paddy straw, cane trash and other farm wastes are burnt in the fields, unlike wheat straw that is completely utilized as fodder. China and India are among the world's largest producer of paddy (Table 10). Table gives the idea of the production of straw and cane in India.

The huge quantity of straw generated is used as fodder in India and similarly the cane produced in India will yield huge quantity of cane trash which is also an excellent biomass fuel. With high silica content it has no commercial use and is therefore almost entirely burned. The agro wastes like, maize, cotton, millets, pulse, sunflower and other stalks, bull rushes (sirkanda), groundnut shells, coconut trash,

Figure 4. Location in Indian map showing location of first 20 smart cities in India agricultural biomass and utilization, waste-to-energy

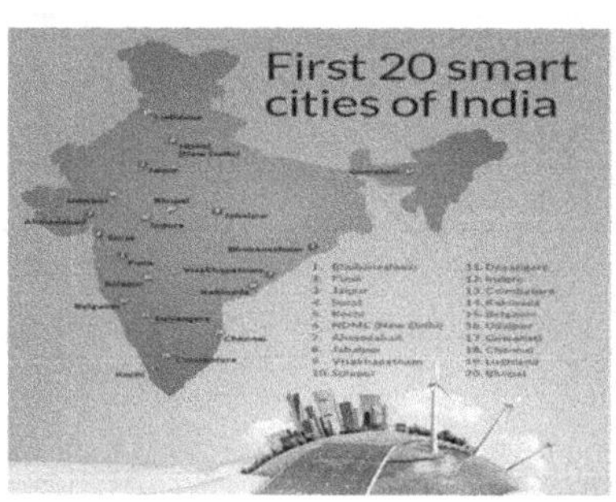

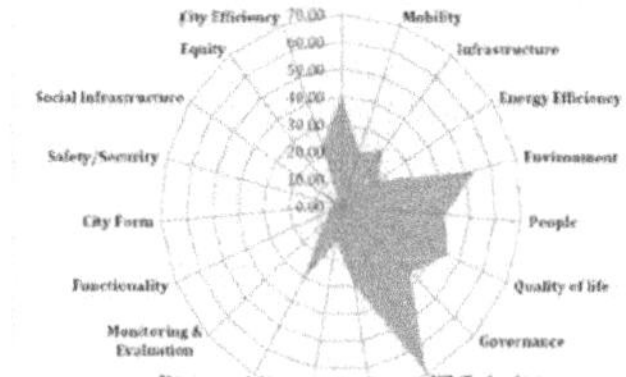

Table 10. Production of straw and cane in India

Country	*Paddy Production*	*Straw Generation*	*Straw used for Fodder*	*Cane Production*
India	98 million tons	130 million tons	60 million tons	350,000 tons/year

etc., act as good feedstock for biofuel production. Utilization of agricultural waste is a major concern especially when there is a scenario of energy demand gap. The resolution to mitigate this gap is utilization of biomass as an alternative source of energy (Harshwardhan, 2017).

It is estimated that about 500 million tons of agricultural and agro-industrial residues are generated annually in India. About 70% of these residues are used as fodder, as fuel for domestic and industrial sectors and for other economic purposes. The amount of surplus agro industrial and agriculture residues is nearly 120-150 million tons per year that could be utilized for power generation. Efficient utilization of biomass is being promoted in the country for power generation. Projects based on biomass combustion and biomass co-generation with total capacity of over 3,700 MW was set up on 31st January 2014 (PIB, 2014). In April, the government of India called for a doubling of India's nonconventional energy supply, including biomass, from 25,000 megawatts in 2012 to 55,000 megawatts by 2017.

Plastics Waste Management Status in India

The plastics waste management policies in India has taken an effective shape by introduction of respective rules in 2016. The Plastic Waste Management Rules 2016 replaced the Plastic Waste (Management and Handling) Rules 2011 as notified by the Ministry of Environment and Forest & Climate Change (MoEF&CC) that provide a regulatory framework for management of plastic waste by urban local bodies. India has taken a special focus to curb the use of single-use-plastic items to reduce the effect of plastics pollution. India hosted the World Environment Day, 2018 with the theme, "Beat Plastic Pollution". Very recently on 15th August 2019, the hon'ble prime minister of India announced that the initiative to stop the single use plastics will be started from 2nd October 2019 and very soon the use of single use plastics will come to an end. Having the recycling economy predominated by informal waste pickers, India has imposed a complete ban on single plastic use in several states, taking the commitment of World Environment Day 2018 theme forward. Several uses of plastics wastes are being implemented with the support of the government. However, the awareness of the ill effect of plastics littering is growing faster. On the other hand, India is the third largest manufacturer of plastics and plastics items in the world. E-waste rules have fixed some targets for collection and treatment. As per the CPCB Report, LATS/21/2015-16, the estimated plastic waste generation in India is 25,940 tpd (based on per capita) and plastics waste generation in 60 major cities is 4,059 tpd. On an average, the plastic waste contents in MSW is nearly 6.92%. The composition of recyclable and non-recyclable is 94%, and 6%, respectively. The Ministry of Environment, Forest and Climate Change has notified the Plastic Waste Management (Amendment) Rules 2018. The amended Rules lay down that the phasing out of multilayered plastic (MLP) is now applicable to MLP, which are "non-recyclable, or non-energy recoverable, or with no alternate use".

The Prime Minister of India vowed to ban single-use plastics in the last Independence Day on 15th August 2019 in Delhi. He called on Indians for an effort with a strong first steps towards bidding adieu to single-use plastic on October 2, 2019. India's target is to eliminate single use plastic by 2022. Eighteen states have already imposed a complete ban on single use plastic (in green) and other states has made partial ban on single use plastic (in red) (as shown in Figure.5). The ban has been imposed for manufacturing, using, transportation, distribution, wholesaling & retail sailing and storage. Importing of plastic bags with handle and without handle and disposable products manufactured from plastic has also been banned.

Figure 5. Status of plastic ban across various states in India

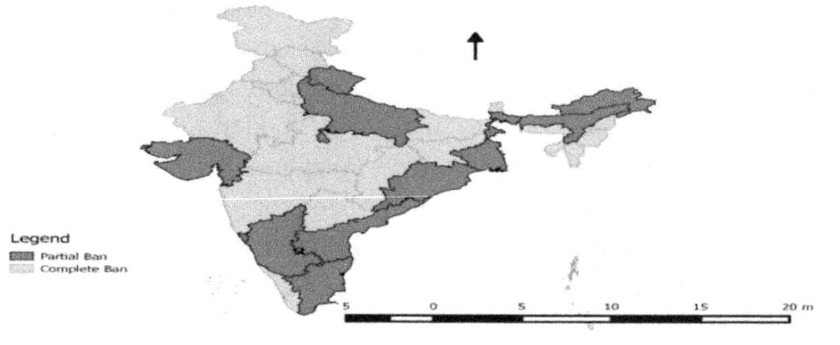

The new rules do not allow manufacture, use and storage of plastics sheet greater than 50 microns in thickness with a specified size. Recyclable plastic waste should be channelized to registered plastic waste recyclers. ULBs shall encourage the use of plastic for different validated recycling processes. Thermoset plastic waste are processed and disposed of as per the guidelines issued by the central pollution control board. The inert from recycling facilities should be disposed of in accordance to solid waste management rule, 2016. Every ULB is responsible to set up plastics recycling infrastructure. The rules have been amended in the year 2018, incorporating the phasing out of manufacture and use of MLP which are non-recyclable or non-energy recoverable or with no alternative use. The status of plastics waste generation and other aspects in megacities in India has been collated in Table 11 from different sources of CPCB.

HAZARDOUS AND INDUSTRIAL WASTE GENERATION AND MANAGEMENT, POLICY AND REGULATIONS, TRANSBOUNDARY MOVEMENT OF HAZARDOUS WASTE

The amount of hazardous waste (HW) generation is 7.90 million tons per annum. The HW generation is increasing at faster rate from various industrial and anthropogenic activities, e.g., from mining, tailings from pesticide based agricultural practices, industrial processes of textile, pesticides, tannery, petrochemicals, pharmaceuticals, paints, oil refineries and petroleum processing, fertilizers, asbestos, caustic soda and in production of many chemicals. The Hazardous and Other Wastes (Management and Transboundary Movement) Rules were revised in 2016 amended in 2018, and imposes restrictions in import and export of wastes, specifically hazardous wastes in line with the requirements of the Basel Convention. There are several TSDF (Transfer, Storage and Disposal Facilities) of HW in many of the states. Resource recovery out of waste is the focused policy in the Indian industries which is practiced in many units leading to employment generation. Around 3.98 million tons is recyclable and can be used for resource or energy recovery (Ghosh, 2017). Major states generating HW are Gujarat 23.47%, Maharashtra 22.72%, Rajasthan 12.08%, Jharkhand 5.44%, Tamil Nadu 5.0%, Andhra Pradesh 4.58% and Telangana 4.06%. Nearly 0.0074 million tons of HW are used as alternative fuel and raw material (AFR) in cement plants. Huge potential has been explored for utilization of HW in cement kilns as AFR. This may be noted that in Netherlands nearly 83% of total HW, is used as alternate fuel in cement kilns, that is the highest amount in the World.

Table 11. Status of PW generation and other aspects in megacities in India

Cities	PW Generated kg/MT/day	MSW Generated MT/day	PW % in MSW	Other Information/ Remarks
Delhi	102.98	6800	10.29%	Rag pickers voluntarily collecting the valuable approximately 200-250
Chennai	95.42	4500	9.44%	
Remarks	84% of HDPE/LDPE waste (carry bags, milk pouches, packing films); Tamil Nadu generated nearly 1,50,323 tons of PW in 2015-16; 301 registered & 112 unregistered plastic manufacturing/ recycling units and five compostable units in the state. In 2015-16, nearly 26.8 tons of PW were used for road construction in the state.			
Bangalore	84.83	3700		
Remarks	Nearly 93% of total PW (HDPE/ LDPE/ LLDPE). Total PW generated in Karnataka in 2015-16: 1,29,600 tons. Nearly 205 ULBs perform source segregation of waste. The state imposed complete ban on use of plastic carry bags and 167 ULBs have fixed prices for carry bags to be provided by shopkeepers and vendors.			
Hyderabad	47.46	4200		
Remarks	About 63.21% of PW (HDPE/ LDPE) was generated. 50 families of ragpickers segregate PW and sell it to re-processors at Jawaharnagar landfill site. Total PW generated in Telangana in 2015-16: 1,20,961 tons. The state has 195 registered and 15 unregistered plastic manufacturing/ recycling units.			
Ahmedabad	105.05	2300		
Remarks	Of total PW, 88% comes from HDPE/LDPE. The estimated plastic waste generation in Gujarat: 2,69,294 tons/annum. There are 689 registered plastic manufacturing/recycling units and nine multilayer plastic units. Approximately 52,746 MT of plastic waste is co-processed in cement plants. The use of plastic carry bags is banned in the city of Gandhinagar. Pricing of carry bags has been implemented by a few Urban Local Bodies (ULBs). Marking and labelling provisions, mentioned under PWM Rules, 2016 are followed by most of the manufacturers present in the state			
Mumbai	62.81	6500		
Remarks	Dumpsites have huge number of rag pickers, collecting polybags, PET bottles, sacks, milk pouches, rubber & foam slippers etc.			
Pune	77.96			
Other Information/ Remarks	About 95% of total PW comes from HDPE/LDPE. Total MSW of the city is 1300 MT/day. MSW collected from 85 wards of Pune city are transferred to four transfer locations, before being transported to the Fursungi Processing Plant. All PW in MSW are treated to produce plastic recycled granules, refused derived fuel and compost fertilizer from MSW. In 2015-16, Maharashtra generated approximately 4,69,098 tons/annum of PW. Maharashtra recently imposed a ban on use of plastic carry bags and disposable products manufactured from plastic and thermocol.			
Kolkata	116.09	3670		
Remarks	About 88.7% is from HDPE/LDPE material. There are about 471 plastic manufacturing/recycling units in West Bengal polymer bitumen roads have been constructed by a few ULBs. There is a ban on paper on the use of plastic carry bags in ecologically fragile areas and certain heritage and tourist places.			

Nearly 350 million tons of organic waste is produced from agricultural sources whereas, the total of 960 million tons of solid waste is generated annually as by-products during industrial, mining, municipal, agricultural and other processes including the agricultural wastes. A significant portion of agricultural wastes in rural India is wasted by open burning producing GHG emission. There are efforts for recycling different wastes and utilize them in value added applications. A portion of biomass and agricultural waste are naturally converted into organic fertilizer under favorable condition and a portion is being used effectively in energy conversion. Cattle feeding are also an area of concern for agricultural wastes.

E-Waste Generation, 3r, Disposal Practices, Policies And Regulations

Waste electrical and electronic equipment (WEEE) or electronic waste (E-waste) refers to end-of-life electronic products (Ghosh, 2016). Out dated electronic goods are becoming obsolete and are discarded in significant amounts (Wong et al., 2007). The global quantity of WEEE generation in 2014 was around 41.8 million tons. Only 6.5 million tons are collected by official take-back systems and treated (Baldé et al., 2015). WEEE contains different critical, valuable and hazardous substances, that need sustainable and environmentally sound recycling processes (Menikpura et al., 2014). E-waste is one of the fastest growing waste streams in emerging as well as developed regions. It is a complex problem to manage because of its duality as both a valuable commodity and source of HW. This dichotomy is the driving force behind the constant challenge to the most effective utilization of the potential economic resource while sufficiently mitigating environmental and social damage. There are a variety of economic, environmental and social factors that influence E-waste management and recycling practices.

India has a lucrative market for reusable products through repair, reconditioning and component reuse, remanufacturing and refurbishing shops. Nearly 200 enterprises are authorized by the respective governments to recycle E-waste in India. However, the PRO (Producer Responsible Organization) in ERP system has an important role to play in India and per the new rules on E-waste management. The situation is changing because of the implementation of E-waste management rules 2016 amended in 2018 with targets of implementation in phased manner. The existence of the unorganized collection and recycling are one way recovering a major portion of E-waste and on the other hand following crude processes of recycling and recovery create adverse impact on environment and health hazards in most of the large cities, namely, Delhi, Kolkata, Mumbai, Chennai, Coimbatore, Bhopal, and Bangalore. Through these hazardous processes, valuable materials namely, copper, silver, a small portion of gold, plastics, and others are recovered and a group of poor people earn their livelihood. Electronic waste (Management and Handling) rules 2011 was drafted by the government of India and implemented from 1 May 2012, emphasizing extended producer responsibility (EPR) and restriction on E-waste import, where producers are directed to finance the collection, treatment, recovery and environmentally sound disposal of electronic waste (E-Waste Handling and Disposal Rules, 2011); the rules have been revised in 2016.

The supply chain of an E-waste management system includes source identification, collection, inward trans-boundary movement, if exists, transportation, segregation for appropriate reuse, dismantling, recycling, and final sink. E-waste generated from televisions by 2020 will be 1.5 - 2 times higher in both China and India together. In India, very soon E-waste out of discarded refrigerators will be raised to two to three times (United Nations Environment Program, 2010). Highly toxic substances like cadmium, mercury and lead (EU Directive, 2002) are available in E-waste. Valuable substances such as gold and copper are also available in E-waste recovering which has become a profitable business, resulting in worldwide trans-boundary trade on e-waste (Widmer et al., 2005).

E-waste rules have fixed some targets for the collection and treatment. There are more than 200 registered E-waste recyclers in India but the management of E-waste has not reached to its significant stage of improvement. EPR has been introduced for the E-waste sectors but yet to take the momentum. The phase-wise collection targets for E-waste have been set based on the percentage of the quantity of waste generation as indicated in the annual EPR Plan until 2023 and beyond 2023. A number of E-waste recycling plants are operating in several states in the country.

Policies of Extended Producer Responsibility (EPR) for E-waste in India

India has introduced the EPR for E-wastes in 2018. In case, the producer has started sale recently, the EPR target shall be applicable as per E-waste management (amended) rules 2018 Schedule- III (A) and these targets are applicable from financial year 2018-2019 and once the product achieves its average life as fixed by the Central Pollution Control Board, the collection target shall be revised as per Schedule III (Table 12). EPR targets has been revised that is applicable from the 1st October 2017 as per Schedule III.

CASE STUDIES OF SUCCESS/FAILURE

Initiative in Housing Society in Kolkata

The bulk waste generating consumer need to have its own waste management system as per the SWM rules 2016. There are initiatives taken to have its own waste management system by the bulk waste generators, namely, housing complex and market complex in some parts of India. Bangalore is the pioneer in this aspect. The residents' society of Diamond City South housing complex in the south Kolkata has taken up one initiative "ISWMAW-DCSRA Waste Management Project" to segregate the wastes in wet and dry wastes at the source by incorporating bin culture among the dwellers. Customized composter has been developed and installed for composting (WtC Facility), using nearly 75 kg per day of waste flower, garden wastes, green vegetable & fruit wastes, egg shells etc., within the premise. The produced compost substituted all the chemical fertilizer previously used for gardening. The collected dry wastes segregated in twelve streams are sold weekly to the recyclers, reducing the burden of waste disposal to landfill site, generating revenues by recycling and effectively utilizing the resources, an example of

Table 12. EPR target schedule as per E-waste management (amended) rules 2018

Sl.	Year	E-Waste Collection Target (by Weight)	E-Waste Collection Target (by Weight)
(i)	2017-2018	10 percent of quantity of e-waste generation as per EPR Plan of organization	
(ii)	2018-2019	20 percent of quantity of e-waste generation as per EPR Plan of organization	5 percent of sales figure of FY 2016-17
(iii)	2019-2020	30 percent of quantity of e-waste generation as per EPR Plan of organization	5 percent of sales figure of FY 2017-18
(iv)	2020-2021	40 percent of quantity of e-waste generation as per EPR Plan of organization	10 percent of sales figure of FY 2018-19.
(v)	2021-2022	50 per cent of quantity of e-waste generation as per EPR Plan of organization	10 percent of sales figure of FY 2019-20.
(vi)	2022-2023	60 per cent of quantity of e-waste generation as per EPR Plan of organization	15 percent of sales figure of FY 2020-21.
(vii)	2023- 2014 & onwards	70 per cent of quantity of e-waste generation as per EPR Plan of organization	15 percent of sales figure of FY 2021-22.
(viii)	2024-2025		20 percent of sales figure of FY 2022-23.
(ix)	2025 onwards		20 percent of sales figure of the year preceding the previous year.";

recirculation of resources and the 3Rs (see Figure 6 and 7). This was possible with the support of all the dwellers in the complex and their higher level of awareness towards waste management. The project was supported by International Society of Waste Management, Air and Water (ISWMAW) and at present it runs sustainably.

E-waste Recycling Plant in Bangalore

This unit is one of the most important and oldest facilities in India. The company provides the necessary logistics for the collection of E-waste (Table 13) through e-auction or direct purchase depending on the sources, such as government offices, IT majors, private & public sectors, and others. It also has written agreements with its customers and has been approved by the Environmental Health and Safety audits conducted by its customers and their counter parts.

Table 13. Type of E-waste processed by the e-waste plant

Computers & Peripherals	Telecom and Lighting Equipment	Household Appliances	Other WEEE
Monitors, CPUs, CRT's, Lap tops, Printers, Servers, Printers, Plotters, Scanners, Floppies Printed Circuit Boards, Copiers, Cartridges, Mother Boards and Components, Dry cells and batteries	Fax machines, Fluorescent & CFL lamps Telecom equipt, Telephones, Cell phones, Tube lights	Audio & Video Devices, Air Conditioners, Tapes, CDs, Card Readers, Refrigerators, household Microwave, fans, Washing Machines, LCDs, Vacuum Cleaners, Swipe machine.	Regulators, Meters, Switches, Starters, Chokes, converter, Wires, Cables, Transformers and Condensers containing cooling oil, Industrial, Medical, Military & Space Electronics

Figure 6. Capacity building workshop for the implementers at DCS housing complex, Kolkata

Figure 7. 3R initiative to implement SWM Rules 2016 at DCS housing complex, Kolkata

Process/Methodology: Collected E-waste is stored at the recycling facility spread over an area of 1.5 acres of land with 25,000 sq. ft. closed area and 60, 000 sq. ft. of open area. It has 10 ton/day recycling capacity and is currently handing 6 ton /day. E-waste is first checked for any radioactive materials before acceptance in to the dismantling line. Segregation of different types of E-wastes are carried out at the first level before actual pre-processing. Most of the E-waste received at the facility undergoes mechanical treatment, which includes manual dismantling, sorting and segregation, recycling, inspection and storage of products and disposal of residue. Figure 8 demonstrates the average value share percentage of valuable and precious elements in EEE/E-waste per unit of e-waste in the company. Table 14 shows specific challenges for e-waste system

Policy Implementation in Bangalore

In April 2019 a fine of INR 45,000/- was charged for deviation in waste management practice as per SWM Rules 2016 as below.

First Concrete Action

- 6 housing societies penalized for not taking measures to dispose bulk garbage.
- 45 collected from the societies in Hadapsar, Mundhwa and Kondhwa in Bangalore city, India.
- Each society has at least 100 flats.

Figure 8. Per unit average value share % of valuable and precious elements in EEE/E-waste

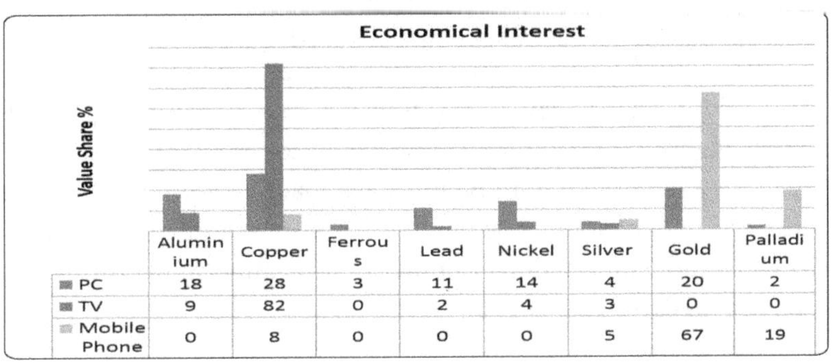

Table 14. Specific challenges for E-waste system in India

Sl.	Scientific Challenges	Engineering Challenges
1	Eco-friendly recovery solutions o Precious metals o Base metals	Scientific collection, transport, handling, segregation and disposal of E-waste
2	Value addition to recyclables for reuse o Plastics o Glass o other recyclables	Integrated/distributed processing facilities o Unorganized to organized o Involvement of SMEs o Involvement of NGOs
3	Disposal of process waste & residues o Size reduction o Toxic reduction	Sustainable Techno-economical solutions o Processing o Recycling o Recovery

Bulk Procedures

- Establishments producing 100kg garbage/day are considered as bulk garbage generators.
- They should make sure that they dispose of their garbage within their premises.
- Such buildings include housing, societies, hospitals, nursing homes, government offices, sport complexes, social clubs and religious establishments. Table 15 gives an idea of waste generation in the city.

Municipal Waste Management at Indore: The Cleanest City in India

The city generates nearly 1,100 tons of MSW every day. The city administration took three-pronged approaches: bin free, litter free and dust free city. The city installed nearly 3,053 litter bins in all commercial areas, parks and gardens and in all high footfall areas. For improving the standards of public health and environmental quality, efficient mechanism for collection and transportation of MSW have been established. Domestic hazardous waste is stored in separate bin and taken weekly to TSDFs. Treatment facilities like compost, bio-methanation and bioremediation have been established and operated regularly (Figure 9). Composts are sold to the farmers and other users, whereas the gas generated from bio-methanation plants are used for running public vehicles such as state operated buses. The dry waste collected is separately in three binned vehicle (Figure 10). Plastics are recycled to make granules and in one small plant to convert into liquid fuel. Dry wastes are separated in several streams and each of those including textiles, metals, papers, plastics, glass and others go to respective recycling plant.

Indore municipal corporation encourages home composting providing 50% subsidy. In Indore nearly 28,000 households are doing home composting and pot composting of household waste. Whereas, home composting has been carried out in all 85 wards with the help of volunteers. The city has a target of reduction of 150 ton solid waste per day. A 100 tons per day capacity C&D waste plant runs in the city. All the rag pickers have been rehabilitated and brought to the main stream by employing them in the formal sectors. Grievance redressing, complaint & feedback system have been established through "Mayor Helpline Indore 311" application. The application also monitors the collection and transportation system of waste in the whole city. Indore won the first position in the all India competition called Swachh Survekshan in 2017, 2018 and 2019.

Table 15. Amount of trash disposal in Bangalore

Daily Waste	Generation (Tons)	Trash Disposal Methods
Source		
House hold	950	Mechanical composting
Hotel and restaurants	150	Vermicomposting
Street sweeping and drainage cleaning	140	Any other method that allows disposal of garbage in scientific manner without hampering the environment
Construction / demolition waste	75	
Market / commercial areas	50	
Fruit, vegetable, fish, meat market	7.5	
Biomedical waste	4.5	

Figure 9. Compost plant at Indore

Figure 10. Author with the waste handlers at dry waste collection center (left). Two compartmented dry and wet waste collection vehicle (right).

CONCLUSION

Waste management is becoming a new business venture in India. Huge investment potential through composting, waste-to-energy, co-processing and recycling are a few areas of business potential. Being the biggest democracy, there are several significant cases of effective and successful waste management including, Indore, Mysore, Bhopal, Vishakhapatnam, Tiruchirapalli, Surat, New Delhi, Navi Mumbai, Vadodara and many others. The waste management in India can only be sustainable if the waste management can be based on sustainable business model. Endeavour has been initiated to implement in the same way.

There is an increasing demand of establishing waste-to-energy and waste-to-compost plants across the country. Concerted efforts have been taken recently for using incinerable wastes including a part of hazardous waste in cement plants. It has been studied that co-processing of wastes using refuse derived fuel (RDF) in cement plants can utilize nearly 100% of the wastes, using the generated ash for production of cement. Government of India, through an expert committee and the Cement Manufacturers Association, are working for a feasible business solution for introducing the co-processing in bigger way.

India is moving faster towards resource efficient country through the implementation of country-wide effective waste management systems. Considering its vast size and number of municipalities and panchayats, it will take some time. In next decade, India will be in an improved state of the art of waste management and resource circulation.

REFERENCES

Aich, A., & Ghosh, S. K. (2016). Organic fraction of municipal solid waste – a valuable source of green energy in India. *International Journal of Energy Sector Management, 10*(4), 526–545. doi:10.1108/IJESM-06-2015-0001

Batteries Rules. (2001). *India.* Retrieved from http://www.pccdaman.info/pdf/Batteries_Rules.pdf

BMWM Rules. (2016). *India.* Retrieved from http://www.indiaenvironmentportal.org.in/files/file/BMW%20Rules,%202016.pdf

Construction and Demolition Waste Management Rules. (2016). *India.* Retrieved from http://sscgj.in/wp-content/uploads/2016/07/Waste-Management-Rules-2016.pdf

CPCB Report. (n.d.). *LATS/21/2015-16, Plastic Waste Generation in India.* Retrieved from http://www.cpcbenvis.nic.in/publication_list.html

E-waste Management (Amended) Rules. (2018). *India.* Retrieved from http://sscgj.in/wp-content/uploads/2016/07/Waste-Management-Rules-2016.pdf

ENVIS Newsletter. (2010). Retrieved from http://mahenvis.nic.in/Pdf/Environemnt%20Literacy_Eng2011.PDF

Ghosh, S. K. (2016). Sustainable SWM in developing countries focusing on faster growing economies, India and China. *Procedia Environmental Sciences, 35*, 176–184. doi:10.1016/j.proenv.2016.07.073

Ghosh, S. K., Debnath, B., Baidya, R., De, D., Li, J., Ghosh, S. K., ... Tavares, A. N. (2016). Waste electrical and electronic equipment management and Basel Convention compliance in Brazil, Russia, India, China and South Africa (BRICS) nations. *Waste Management & Research, 34*(8), 693–707. doi:10.1177/0734242X16652956 PMID:27422616

Greenpeace Report. (2008). *Hazardous other wastes management and Transboundary movement rules 2016, India.* Retrieved from https://www.greenpeace.org/international/publication/19714/annual-report-2008/

Harshwardhan & Upadhyay. (2017). Effective Utilization of Agricultural Waste [Review]. *J Fundam Renewable Energy Appl, 7*, 5. doi:10.4172/2090-4541.1000237

Joshi & Ahmed. (2016). Status and challenges of municipal solid waste management in India: A review. *Cogent Environmental Science, 2*, 1139434. doi:10.1080/23311843.2016.1139434

Ministry of New and Renewable Energy Sadhan Kumar Ghosh. (2017). *India Report on 3R Status.* UNCRD. Retrieved from http://www.uncrd.or.jp/content/documents/5688[1121]%20India.pdf

Planning Commission Report. (2014). *Reports of the task force on waste to energy (Vol-I) (in the context of Integrated MSW management).* Retrieved from http:// planningcommission.nic.in/reports/genrep/rep_wte1205.Pdf

Plastics Waste Management Rules. (2016). *India.* Retrieved from http://sscgj.in/wp-content/uploads/2016/07/Waste-Management-Rules-2016.pdf

Press Information Bureau, Ministry of New and Renewable Energy, Government of India. (2014). *Production of Energy from Agricultural Waste*. Retrieved from https://www.progressivegardening.com/agricultural-wastes/energy-production-from-agricultural-waste.html

Sixteenth, L. S. F. (2015). *Standing committee on energy (2015-16)*. Government of India. Retrieved from http://www.wwfenvis.nic.in/Database/ConsulativeCommittee_5204.aspx

SolidW. M. R. (2016). *India*. Retrieved from http://sscgj.in/wp-content/uploads/2016/07/Waste-Management-Rules-2016.pdf

Swachh, S. G. A. (2018). *Press Information Bureau*. Retrieved from http://pib.nic.in/newsite/PrintRelease.aspx?relid=183949

Swachh, S. G. A. (2018). *Press Information Bureau*. Retrieved from http://pib.nic.in/newsite/PrintRelease.aspx?relid=183949

Toolkit, S. S. (2019). *Ministry of Housing and Urban Affairs, Government of India*. Retrieved from http://164.100.228.143:8080/sbm/content/writereaddata/Survekshan%20Survey%202019%20Toolkit%20 13.09.2018.pdf (Accessed on 4 October 2018)

Chapter 7
Challenges of Waste Management in a Developing Context:
Lessons From Lebanon

Mutasem El-Fadel
https://orcid.org/0000-0003-2379-9006
American University of Beirut, Lebanon

Amani Maalouf
https://orcid.org/0000-0002-4949-7746
American University of Beirut, Lebanon

ABSTRACT

Over the past years, several efforts were undertaken in an attempt to improve solid waste management in Lebanon. However, with the lack of a well-established national plan and defined policies and legislation, all attempts achieved little or no progress. Several challenges continue to hinder the sector's successful development at various levels: politically, legally and institutionally, socially, land use, and last but not least, technically. This chapter provides an overview of waste management practices in Lebanon such as waste generation, waste composition, current status of waste management practices, policy, and legislation, while highlighting the main challenges for implementing an integrated solid waste management towards a sustainable system.

INTRODUCTION

Worldwide, over 2 billion tons (BT) of solid waste are generated annually in large cities with projections to reach 3.4 BT by 2050, particularly in developing economies where waste generation rate is expected to increase by three folds by 3050 (Kaza et al., 2018). This can be mainly attributed to the rapid population growth, increasing urbanization, and socio–economic development. Along with inefficient waste

DOI: 10.4018/978-1-7998-0198-6.ch007

management systems, these factors are increasingly raising management challenges and environmental concerns.

In Lebanon, municipal solid waste (MSW) disposal has been a persistent and a chronic challenge for a long time particularly in urban areas with high population density, high production of refuse, and low availability of land adequate for landfills. In the context of the capital Beirut and surrounding areas in Mount Lebanon, an Emergency Plan was developed in 1997, which remained effective until July 2015. This culminated a national waste crisis, which was mainly triggered by the premature closure of the main landfill (Naameh) that served as the primary largest sanitary landfill in the country serving more than half the total population. The site was closed in July 2015 without a designated alternative due to social opposition leaving piles of uncollected waste along the streets. This crisis has reportedly led to serious health, social, economic and environmental impacts.

In other parts of the country, open burning and uncontrolled dumping on hillsides and on seashores are common practice equally resulting in serious land, sea, and air pollution problems. Sea pollution along the Mediterranean shores is extensive and the focus of regular complaints by local fishermen and the general public. The unsanitary disposal of municipal wastes constitutes a pressing environmental stress at the national scale. The waste crisis has evolved with the expansion of uncontrolled dumpsites that increased from 670 in 2011 to 941 in 2016 (MOE/UNDP, 2017; MOE/UNDP/ELARD, 2011). The main existing landfills in Lebanon since 1998, namely the Naameh landfill, the Zahle landfill, and the Tripoli controlled dumpsite, have only served around 55% of the total MSW generated. The remainder is partially recycled/composted and partially disposed of in open dumpsites, by local authorities, such as municipalities and/or unions of municipalities (MOE/UNDP, 2017; MOE/GEF/UNDP, 2015).

Several factors were identified as key shortcomings in the waste sector in Lebanon, particularly the lack of an integrated waste strategy, the provision of comprehensive waste data, and the failure to allocate roles and responsibilities. In addition, social and political interventions prevent regulatory compliance, monitoring and enforcement exacerbating waste management challenges. This chapter provides a historical overview including current waste management practices in Lebanon while highlighting the main challenges for implementing an integrated solid waste management (ISWM) towards a sustainable system.

BACKGROUND

Country Profile

Lebanon is located in Western Asia along the shores of the eastern Mediterranean (Figure 1). It is as the smallest country in the Asian Continent with a surface area of 10,452 km^2. The country is bordered by Syria to the north and east, Palestine to the south, and the Mediterranean Sea to the west. A long narrow coastal strip characterizes its Mediterranean shores with a complex rich mountain area ranging from 400 to 3088 m in altitude. The country is divided into 8 Governorates (or Mohafazat) that is headed by a Governor (Table 1). The Governorates/administrative areas are further subdivided into 25 Districts (or Caza) with Figure 2 highlighting the boundaries of both Governorates and Districts and the location of main waste management facilities and disposal sites under different plans.

While the population of Lebanon was estimated at nearly 5 million including Palestinian refugees and foreign workers (World Bank, 2019), the influx of refugees as a result of the Syrian conflict coupled with emigration make it harder to accurately estimate the population with values ranging between 6 to

Figure 1. The location of Lebanon

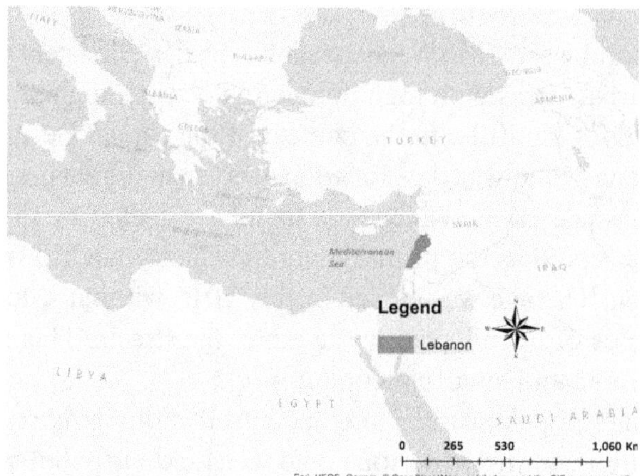

Table 1. Governorates and Districts in Lebanon

Governorate (Mohafazat)	District (Caza)	Municipalities	Surface Area (km²)
Beirut	Beirut	1	19.3
Mount Lebanon	Baabda, Meten, Chouf, Aley, Kesrouane, Jbeil	341	1,973.3
North	Tripoli, Koura, Zgharta, Batroun, Bcharre, Dannieh-Mnieh	261	1,551.6
Akkar			786
Bekaa	Zahle, West Bekaa, Rachaya	165	1,419.9
Baalbak-Hermel	Baalbeck, Hermel		2,735.2
South	Saiada, Sour, Jezzine	192	923.7
Nabatiyeh	Nabatiyeh, Bint Jbeil, Marjayoun, Hasbaya	148	1,095.3

Source: Laceco/Ramboll, 2012

Figure 2. Lebanon administrative boundaries and location of main waste management facilities and disposal sites under different plans

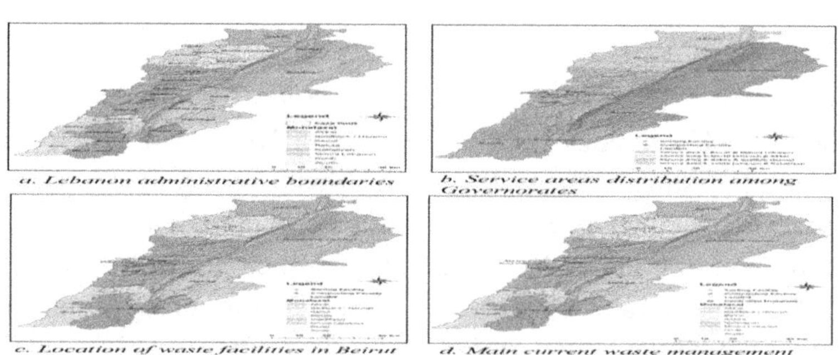

7 million. The population density stands at ~580 persons/km^2, which is generally high by international standards compared to France 94 and Germany 248 (CEDARE/Blue Plan/ UNEP, 2000). Furthermore, the population is unevenly distributed whereby about 50% of the general population resides in Beirut and Mount Lebanon governorates that account for only 20% of the territory (MOE/GEF/UNDP, 2011). The large majority (about 90%) of the population is centered in urban areas, which are mainly concentrated in cities along the coast (MOE/UNDP/GEF, 2016).

Economically, Lebanon is considered as an upper middle-income country according to the World Bank classification (World Bank, 2019). It has a free market economy whereby the supply and demand determine the costs of most goods and services. Moreover, the country imports more than it exports and is principally reliant on imports for food and fuel. The economic productivity is greatly subject to the regional and international shocks due to the dependence on imports and services (including banking and tourism) (MOE/UNDP/GEF, 2016). The current Gross Domestic Product (GDP) in Lebanon increased from US$ 38 billion in 2010 to an estimated US$ 53.6 billion in 2017 (World Bank, 2018). The service sector is the largest contributing share to the GDP (~31%), followed by the commercial trade sector that accounts for 28% of the GDP, the real estate at 14%; with the lowest contribution from the agriculture, forestry & fishing sector at 5% of the GDP (MOF, 2011).

Waste Characterization

The MSW generated are measured by the contractor carrying out collection, treatment and disposal services mostly in compliance with contractual obligations that levy charges on the basis of per ton of waste managed. It is worth highlighting that generated MSW is not just municipal but it is mixed with other sources (such as waste from public institutions, commercial and industrial enterprises, bulky waste, street sweeping waste etc.). The MSW compositions adopted are based on analysis after collection as opposed to the point of waste generation at source. It is equally important to highlight that to date, waste separation at source is practically absent in the country except for few voluntary small-scale pilot projects. Accordingly, the waste collected is composed of mostly commingled/unsorted waste.

Waste Generation

The waste generation rates are influenced by population size, seasonal variation, degree of urbanization, public habits, economic development, and affluence (Bogner et al., 2007). In turn, different indicators of affluence such as the gross domestic product (GDP), income per capita, energy and food consumption, and trends in lifestyle can affect waste generation rates (Bogner et al., 2007). Generally, the higher the rate of urbanization and economic status, the greater the volume of MSW generated. The income level and urbanization are closely interrelated whereby the increase in living standards and consumption of goods and services result in an increase in waste generated. In this context, the average amount of waste generated has ranged from 93 million tons per year (MT/yr) in low-income countries (equivalent to 5% of global waste) to 586-655 MT/yr in middle-income countries and more than 683 MT/yr (equivalent to 34% of the world's waste) in high-income countries (Kaza et al., 2018).

The current waste generation rate in Lebanon was calculated by extrapolation and interpolation based on MSW management data and information collected from several studies since 1994 (Ayoub et al., 1994; Ayoub et al., Ayoub et al., 2006; El Fadel & Sbayti, 2000; CDR, 2006; SWEEP-Net, 2010). Lebanon has a current estimated daily waste generation that varies from 0.85 kg/capita in rural areas to

0.95-1.2 kg/capita in urban areas, with a national weighted average of 1.05 kg/capita/day (GIZ/SWEEP-Net, 2014). Furthermore, the daily solid waste generation rates and trends showed a 26% increase in the overall MSW generation rate in Lebanon between 1994 and 2011 (MOE/GEF/UNDP, 2015). Beirut and Mount Lebanon governorates constitute about 50% of the total MSW generated in Lebanon (Table 2). The quantities of MSW generated within Beirut and Mount Lebanon governorates in 2013 was estimated to be 2,850 tons per day, compared to 2000 tons per day in 1999 (GIZ/SWEEP-Net, 2014). This can be mainly attributed to an increase in migration to major cities, from rural to urban areas, during the past 20 years. Currently, the country produces ~7,500 tons per day of various types of solid waste (MOE, 2018) and distributed as shown in Table 2.

Since the Syrian conflict, Lebanon has witnessed an influx of displaced Syrians that led to serious implications on several vital sectors. For instance, the incremental daily MSW reached nearly 900 tons per day in 2014 equivalent to 15.7% of the total MSW stream prior to the crisis (MOE/EU/UNDP, 2014).

Waste Composition

The starting point of variation between developed and developing countries is the difference in composition of the waste generated and subsequently in waste density, size distribution, moisture content, biodegradable fraction, and energy content. Waste composition is dynamic and varies according to many elements such as income level, culture, season, residence type, affluence, and location. Moreover, waste composition, type and distribution greatly influence the potential for greenhouse gas (GHG) emissions reduction in any country and waste collection frequency as well as treatment and disposal methods (Chen and Lin, 2008). Generally, the organic waste constitutes the highest proportion of MSW in developing economies (Kaza et al., 2018). In contrast, developed economies have the highest proportion of paper, plastics, and other inorganic materials (Kaza et al., 2018).

The average waste composition in Lebanon is expressed in terms of the percentage of the wet weight (w/w) as presented in Table 3. The waste composition consists mainly of about 53.4% organic fraction; 34.8% plastic, paper, cardboard, glass and metal; and 11.8% inert and other materials. The organic fraction is relatively high exceeding 50% of the total waste, particularly in rural areas where it reaches 60%. As such, the prevalent moisture content of the waste is naturally high (> 60%) (MOE/UNDP, 2017). The

Table 2. MSW generation

Governorate (Mohafazat)	% of Country
Beirut	10
Mount Lebanon	40
North	14
Akkar	6
Bekaa	8
Baalbak-Hermel	7
South	9
Nabatiyeh	6

Source: MOE, 2018

Table 3. Average MSW composition (% w/w)

Waste category	(% w/w)
Organics	53.4
Glass	3.4
Metals	2
Nappies	3.6
Papers	15.6
Plastics	13.8
Textiles	2.8
Wood	0.8
Others	4.6
Total	*100*

Source: Laceco/Ramboll, 2012; Maalouf and El-Fadel, 2019

waste composition in Lebanon varies slightly depending on geographical areas (e.g. paper and plastics are more produced in cities and commercial centers than rural areas) and seasons (e.g. during summer season the organic fraction and moisture content is the highest).

Existing Legal and Institutional Framework

National Legislative Framework

There are significant gaps on the solid waste management (SWM) legislative framework in Lebanon. The major existing legislation related to SWM are the Circular N° 8/1 (dated 16/11/2015) of the Ministry of Environment (MOE) related to guidance on ISWM as published in issue N° 47 of the Official Gazette (dated 19/11/2015) and the integrated SWM Law N⁰ 80/2018 that was prepared in 2005, approved by the Council of Ministers (COM) in 2012 (under decree N° 8003 dated 23/4/2012), and later ratified by the Parliament in 24 September 2018 (GFA, 2017). The Law defines solid waste management responsibilities and institutional framework, planning, permitting, information management, environmental insurance, private sector participation as well as financing and cost recovery mechanisms. It addresses several waste management principles that are widely accepted internationally such as: Principles of ISWM, Proximity Principle, Polluter Pays Principles, Producers' Responsibility, and Precautionary Principle. However, the Law itself is not sufficient for the implementation of an ISWM system because it lacks the essential application decrees permitting its effective enactment. In this context, a successful implementation of laws requires the setting and issuing of the needed applicable decrees, strategies and plans in different fields to ensure the sustainability of the system. Specifically, in terms of well-defined institutional framework, financing and cost-recovery system, and setting a SWM strategy with consensus at both national and local levels. Other legislation comprises fragmented regulations not specifically dealing with solid waste despite a number of laws, decrees, and ministerial decisions that govern environmental and solid waste management. Table 4 summarizes key legislation that have gradually developed over the years in Lebanon reaching the integrated solid waste Law that was ratified by the Lebanese Parliament in 2018.

Table 4. Key legislations with relevance to the waste sector

Legislation Number	Description
Decree 8735/1974	Assigns to Municipalities the roles, of public cleanliness and of collection and disposal of waste. It also recognizes punishments on entities that dispose waste in areas not intended for that purpose.
Law 64/1988	Preservation against harmful and hazardous waste pollution
Decree 9093/2002	Provides municipalities with incentives to host solid waste facilities
Law 444/2002	Defines the basis and norms for Environmental Protection
Decree 8006/2002 Decree 13389/2004	Identifying the categories of health care waste and means their disposal
Decree 1117/2008	Provides municipalities with incentives to host solid waste facilities
Decree 8471/2012	Environmental compliance for establishments
Decree 8633/2012	Fundamentals for Environmental Impact Assessment.
Law 280/2014	Exempts Ain Drafil Municipality from its solid waste debt, and provides incentives for municipalities hosting solid waste facilities
COM Decision 45/2018	Integrated solid waste management policy, provides provisions related to solid waste management
Law 80/2018	Integrated solid waste law provides frameworks and provisions for solid waste management

Source: GIZ/SWEEP-Net, 2014; GFA, 2017

Despite that Lebanon is signatory to several international conventions related to waste and environmental protection (Table 5), such as the Barcelona (1976), Basel (1994) or Stockholm (2001) conventions, it does not follow a strict regulatory framework for environmental assessment and protection. While recent Legislation provide environmental policies to be enforced through administrative and legal procedures, implementation is lacking and weak.

Table 5. Relevant international treaties

Treaty	Description
Marpol (1973/1978)	Prohibits waste dumping in the sea and prevents marine pollution
Barcelona Convention (1976)	Prevent and abate pollution from ships, aircraft and land-based sources in the Mediterranean Sea.
LBS protocol (1980)	Protects the Mediterranean Sea against Land Based Pollution
Basel Convention (1994)	Regulates transboundary movement of hazardous waste and their disposal
Mediterranean special protected area protocol (1994)	Prohibits dumping of waste into the sea and combats pollution within the sea area
Rotterdam Convention (1998)	Promote shared responsibilities in relation to importation of hazardous chemicals
Stockholm Conventions (2001)	An Environmental treaty that aims to eliminate or restrict the production and use of persistent organic pollutants (POPs)
Minamata Convention (2013)	Protects human health and the environment from anthropogenic emissions and releases of mercury and mercury compounds

Source: GIZ/SWEEP-Net, 2014; GFA, 2017

Nonetheless, enforcement of these laws and regulations is relatively weak, practically non–existent, mainly due to political interferences, non-defined responsibilities, and minimal coordination between authorities. Equally important is the lack of proper training, low level of fines, and the lack of awareness of regulations amongst personnel who are supposed to enforce them (i.e. health inspectors, police officers, as well as the general public that is supposed to abide by them). In short, existing regulations lack clarity and precision. Consequently, the country suffers a lack of national leadership in the solid waste sector.

Institutional Framework

Several governmental institutions are involved in SWM and planning with overlapping responsibilities (Table 6). Two ministries have authority on the policy, legislation, strategy and planning of solid waste. The MOE, responsible for setting SWM strategies and standards, and the Ministry of Interior and Municipalities (MOIM) representing the municipalities who are legally responsible for waste collection, disposal and management operations at their level. On the other hand, the Council for Development and Reconstruction (CDR) is the governmental entity responsible for reconstruction and development projects in the country, acting under the authority of the Prime Minister. The CDR was responsible for carrying out the Emergency Waste Management Plan in 1997 and has been responsible since then for upgrading and implementing SWM projects throughout the country. Basically, the CDR was primarily responsible for SWM in Beirut, Tripoli and most Mount Lebanon, followed by the MOE and the

Table 6. Relevant Stakeholders involved in the SWM sector in Lebanon

Entity	Main Responsibilities, Roles, and Achievement
MOE	- Establishment of waste management standards and guidelines - Approval of Environmental Impact Assessment projects - Review of all studies and tender documents related to waste management facilities and authorization of these facilities - Define environmental limit value - Establishment of waste management master plans - Supervision and inspection of compliance
MOIM	- Manages the allocation and distribution of funds from the Independent Municipal Fund to the municipalities - Coordinates and assists in the implementation of local waste management plans - Establishment and implementation of waste management programs
Local Authorities (municipalities, union of municipalities)	- Implementation of local waste management plans for non-hazardous municipal waste - Management of waste collection
CDR	- Assistance in procurement of waste management projects - Assistance in the development of waste management plans - Facilitate construction, planning, and execution of plans in line with ministries in Lebanon
Private Sector	- Companies such as Averda Group (incl. Sukleen and Sukomi) in Beirut and Mount Lebanon, IBC company in Saida, Lavajet in North Lebanon, Sanitek in Zahle, and various recycling industries has played a significant role in the design and provision of solid waste management services in Lebanon (such as waste treatment/disposal and collection)
Donors, civil society & NGOs	- Donors such as the European Union, World Bank, Italian Development Cooperation, and USAID have provided much needed funding and challenged the existing conditions by recommending alternative approaches to national SWM strategies

MOE: Ministry of Environment; MOIM: Ministry of Interior and Municipalities; CDR: Council for Development and Reconstruction
Source: GIZ/SWEEP-Net, 2014; MOE/UNDP, 2017

MOIM. Whereas in the rest of Lebanon, the municipalities are solely responsible for SWM starting from waste collection, disposal and management operations (SWEEP-Net, 2010; GIZ/SWEEP-Net, 2014). In addition, the Office of the Minister of State for Administrative Reform (OMSAR) is a governmental organization that aims to develop the institutional and technical capacities of ministries, has also been involved in the SWM projects under EU financing.

The private sector has also played an important role in the provision and design and of SWM services in the country. The most prominent illustration of the participation of private sector is associated to the Emergency Plan in 1997 whereby the local contractors (Sukleen and Sukomi) were awarded large contracts for waste collection and treatment. Elsewhere, the unions of municipalities (such as Tripoli, Byblos and Zahle), have also provided SWM services (including street sweeping, collection and landfilling) to private sector companies. Beyond these large municipalities, the informal sector and waste entrepreneurs are representing the private sector with informal track record (MOE/UNDP, 2017).

In summary, at the institutional level, overlaps in responsibilities are common occurrence. For instance, the legal framework assigns the MOE with the responsibility of regulating the sector and the municipalities with collection and disposal responsibilities, while the CDR was appointed by the COM to develop a country wide plan and to launch international tenders for its implementation, thus presenting a clear overlap in institutional responsibilities. The involvement of several institutions in SWM has resulted in uncertainties in the levels of authority and the relative responsibility that affects any SWM plan or system negatively. This problem has been tackled by the introduction of the new Law N^0 80/2018 on ISWM that unites the different stakeholders into a single joint system represented by the SWM Board, headed by the MOE which holds exclusive responsibility for SWM at the national level, and should include members from relevant public authorities as well as the private sector (academics, consultants) and Non-Government Organizations (SWEEP-Net, 2010; GIZ/SWEEP-Net, 2014).

Historical Overview on Relevant Solid Waste National Plans & Policies

Historically, prior to 1975, the refuse collection and disposal has been the responsibility of municipalities (Decision N^0 188, April 1920). However, the decreasing financial and human resources of municipalities coupled with an increasing population and waste stream limited their capacity to effectively manage solid waste at the local level. Consequently, many municipalities became unable to provide a much-needed service and have in most cases resorted to open burning and uncontrolled dumping on hillsides and seashores, resulting in land, sea, and air pollution problems.

In response to public concerns, the Government adopted various plans with implementation starting from Beirut and Mount Lebanon governorates due to its high population and where the SWM system was facing the most serious problems. A historical overview of the various attempts to develop a SWM plans are summarized in Table 7. Despite various efforts and attempts in this regard, it is generally accepted that to date, there is lack of a well-defined national SWM strategy in the country. All invested attempts were hindered by the lack of agreement between the different key players of the SWM system in the country.

The 1997 Emergency plan that covers Beirut and Mount Lebanon governorates (Figure 2c) was effectively the only implemented plan until the closure of Naameh landfill in July 2015 after it exceeded its design capacity. All other plans were recognized by the relevant authorities but not implemented for different reasons including financial, technical, and political. Other regions in Lebanon have experienced limited progress, with the exception of few secondary cities. The 1997 Emergency plan that

Table 7. Historical overview on policy initiatives for the SWM sector in Lebanon

Policy/ Plan	Status	Legal Basis	Source
The 1997 Emergency Plan	Implemented in Beirut and Mount Lebanon governorates in 1997 until July 2015	COM Decision N°58 (dated 2/1/1997)	Government of Lebanon/ Council of Ministers (1997)
The 2006 Master Plan for MSWM	Very limited implementation	COM Decision N° 1/4952 (dated 18/8/2005)	Government of Lebanon/ Council of Ministers (2005)
The 2010 Waste to Energy Plan	No implementation	COM Decision N°55 (dated 1/9/2010)	Government of Lebanon/ Council of Ministers (2010)
The 2013 Draft National Master Plan	No implementation	COM Decision N° 55 of (dated 1/9/2010) as amended by COM N° 52 (dated 9/1/2013)	Government of Lebanon/ Council of Ministers (2010)
The 2014 MSWM Roadmap	No implementation (CDR prepared Tender Documents for the provision of SWM services)	COM Decision N° 46 (dated 30/10/2014) as amended by COM Decision N° 1 (dated 12/01/2015)	Government of Lebanon/ Council of Ministers (2014)
The 2015 MSWM Plan	CDR prepared Tender(s) covering solutions for MSWM (Six Service Areas) were aborted	COM Decision N° (dated 12/01/2015)	Government of Lebanon/ Council of Ministers (2015)
The 2016 Emergency Plan	Implemented in Beirut and Mount Lebanon governorates	COM Decision N° 1 dated 12/03/2016, updated 17/03/2016	MOE/UNDP (2017)

COM: Council of Ministers; CDR: Council for Development and Reconstruction; MSWM: Municipal Solid Waste management

indeed spread to cover Beirut and most of Mount Lebanon areas (except Jbeil Caza) consisted of closing existing dumps (i.e. Bourj Hammoud dump with the Normandy dump having closed well before). The plan implemented what can be referred to as an ISWM involving: (1) Collection and transport of raw commingled municipal waste (by Sukleen contractor); (2) Sorting and processing of raw municipal waste at two transfer facilities (Karantina and Amrousieh with design capacity of design capacity of 1,700 tons/day and 1,250 tons/day, respectively); (3) Recycling of the waste fraction composed of glass, metals, papers, and plastics (4) Composting (open windrow) of the waste rich in highly biodegradable organic content (Coral facility with design capacity of 750 tons per day); (5) Transport and disposal of sorted-baled and bulky waste at two old nearby quarries converted to landfills (Naameh and Bsalim). The Emergency Plan was partly implemented despite controversies associated mainly with system costs, and the effectiveness of sorting and composting plants. The COM extended the management contracts for Sukleen several times after 1997. To date, this plan represents the most significant infrastructure investment in the solid waste management sector in Lebanon.

In an effort to organize the SWM system following a request from the COM (Decision 1/4952 (CDR, 2006)), the Ministry of Environment (MOE) and the CDR prepared an IMSWM 10-year plan (proposed for the period 2006-2016) that covers all Lebanon involving collection and sorting, recycling, composting, and landfilling. The plan recognized four service areas (Figure 2b), divided according to governorates: (1) Beirut & Mount Lebanon, (2) North & Akkar, (3) South & Nabatiyeh, and (4) Bekaa & Baalbak-Hermel. The Master Plan proposed that every service area would have a sanitary landfill (about 6 to 7 landfills in total) and that every Caza would be equipped by its own waste treatment plant for sorting and composting (around 12-14 facilities in total). The proposed Master Plan was approved by the COM in June 2006, however the consequent war in July 2006 drained the resources of the government and weakened the political decision for its implementation. The GOL did not succeed in building any of the proposed plants and landfills. In addition, all prospects for its implementation were further eroded by the lack of agreement on the foreseen site locations of proposed facilities and the available public funding.

For the reasons mentioned above, the 2006 Master Plan had a limited implementation during the period of 2006 and 2010. Instead, the GOL built several small-scale facilities with grant funding. Recognizing this impasse, the COM issued Decision N° 55 (Government of Lebanon/Council of Ministers, 2010) to amend and complement the 2006 Master Plan. The 2010 Waste-to-Energy Plan, based on ministerial committee, advocated the use of incineration and waste-to-energy (WtE) technologies in large cities, and resorted to the 2006 Master Plan in the rest of the country, while also studying the feasibility and possibility of WtE systems in these areas. This Plan was accepted by the COM in its updated form given the difficulty for siting landfills in large cities.

In February 2013, a proposal was issued to prepare a national SWM strategy for Lebanon based on item 4 of the COM Decision N° 55 in 2010 (Government of Lebanon/Council of Ministers, 2010), which was further mandated by the ministerial committee by Decision N° 52 in 2013. The proposal addressed the urgent need to develop a SWM Plan in Lebanon while tackling various aspects affecting its execution, mainly technical, financial, institutional, awareness, and legal. In addition, the proposed Draft Plan of 2013 presented the proposed locations and type of waste management facilities in Lebanon. Despite that this proposal was formally submitted in March 2013, the Plan could not be formally endorsed due to the subsequent resignation of the Prime Minister.

In February 2014, a new Government was formed considering SWM as its priority following more than 11 months of resignation. The COM issued Decision No. 46 (dated 30/6/2014), later amended by COM Decision N° 1 in 2015 (Government of Lebanon/Council of Ministers, 2014; 2015) to cover all Lebanon. The ministerial committee commissioned the CDR to prepare national tenders following stringent requirements for the procurement of SWM services in the six service areas. Specifically, the contractor is required to locate sites for landfills and treatment facilities, provide solutions starting from waste separation to landfilling, and achieve ambitious waste diversion targets corresponding to 60% recuperation of total MSW generated (first three years of operation) and 75% in the following years, until Lebanon integrates the WtE technologies (after 7 years of the contract service period). However, the tenders were aborted by the COM in Decision N° 1 dated 25 August 2015 and waste was piling in the streets (mid-July 2015) of Beirut and Mount Lebanon governorates after the closure of Naameh Sanitary landfill. Following that in September 2015, the Minister of Agriculture was mandated to propose a solution to the waste crisis and to develop a new Emergency National MSWM plan that consists of adopting the principles of decentralization and recognizing the role of municipalities in upholding responsibility of SWM. The COM then issued Decision N⁰ 1 on 12 March 2016 for adopting the Emergency Plan that consists of establishing one temporary sanitary landfill in Bourj Hammoud, Jdeideh – Bouchrieh – El Sed, and another temporary sanitary landfill in Ghadir River estuary, which will be accompanied by handling the removal of the Open Dump in Bourj Hammoud and embankments near Ghadir River. Unlike former plans, this plan was immediately executed with a lifetime of four years, after which the WtE facilities foreseen under the COM Decision N⁰ 55 in 2010 would become operational.

CURRENT SITUATION AND PLANS

Currently, about 7,500 tons of MSW per day is generated in Lebanon (including refugees' generation), of which about 50% is disposed in uncontrolled dumpsites (about 940 dumpsites), about 35% in sanitary landfills (Bourj Hammoud & Jdaideh, Ghadir River estuary, and Zahle) (MOE, 2018). The remaining ~15% of MSW is recovered for recycling (e.g. paper and cardboard, plastic, metal, glass, etc.) or converted into organic soil enhancer/fertilizers in around 55 treatment facilities across the country (Table 8). In addition, another 12 facilities (under construction or already established) are scheduled to be operational soon (Table 8). Table 9 shows existing landfills, which are currently operational or under construction.

Table 8. Overview of current and proposed waste treatment facilities in Lebanon

Governorate (Mohafaza)	Current Waste Treatment Facilities (Operational)	Waste Treatment Facilities (Under Construction)
Beirut & Mount Lebanon	19	2
Kesrouane-Biblos	4	2
North	6	0
Akkar	2	2
Bekaa	7	3
Baalbak-Hermel	1	0
South	4	1
Nabatiyeh	12	2
Total	*55*	*12*

Source: MOE, 2018

Table 9. Overview of current waste sanitary landfills in Lebanon

Governorate (Mohafaza)	Name/Location	Capacity	Status
Beirut & Mount Lebanon	Bourj Hammoud	1,400 tons/day	Operational
	Costa Brava	1,500 tons/day	Operational
Kesrouane-Biblos	Ghosta	-	Operational
Akkar	Srar	-	Under construction
Bekaa	Zahle	300 tons/day	Operational
	Bar Elias	150,000 m^3	Operational
	Jeb Jannin	200,000 m^3	Operational
Baalbak-Hermel	Baalbak	1,00,000,000 m^3	Operational

Source: MOE, 2018

MSWM Plan in Beirut and Mount Lebanon

In March 2016, the COM approved a four-year SWM Plan, following the closure of the Naameh landfill that served Beirut and Mount Lebanon Governorates. Accordingly, the CDR on behalf of GOL and in accordance to the CDR's Board of Directors Decision No 589/2016A dated 08/09/2016 signed a contract for the treatment of MSW in Beirut City and Mount Lebanon governorate. The plan consisted of the development of an ISWM system that consists of collecting raw MSW and transfer to two sorting facilities. Figure 3 depicts the routing of MSW through various process activities at various sites. About 2,600 tons per day of MSW (1,350 tons/day at Amrousieh and 1,250 tons/day at Karantina facility) are sorted manually and mechanically into reject, recyclable, and compostable material. Reject materials (~1,560 tons/day) are transported for ultimate disposal in Bourj Hammoud, Jdeideh – Bouchrieh – El Sed landfill with a design capacity of 1,250,000 tons (Figures 4-5) and Costa Brava landfill in Ghadir River estuary with a design capacity of 1,000,000 tons (Figure 6). These shall not exceed 60% according to the CDR's decision (Laceco, 2018).

Recyclables are extracted for onward sale to appropriate consumers whereby organic materials are transported to the Coral composting plant for further processing. The amount of recovered recyclables is around 290 tons /day (11.15% of the incoming waste) whereby the amount of organic materials is about 750 tons/ day (equivalent to 28.85% of the incoming solid waste). The produced compost is distributed to farmers while the discarded rejects are sent to landfills. Bulky items collected under a separate system are picked for recyclables and the discard is sent to Bsalim landfill together with the products of shredded tires and wood. Accordingly, 40% landfill diversion should be achieved through composting and retrieving of recyclable material (Laceco, 2018).

Figure 3. Routing of the MSW through various sites (solid waste processing, composting, and disposal) under the current management plan in Beirut and Mount Lebanon

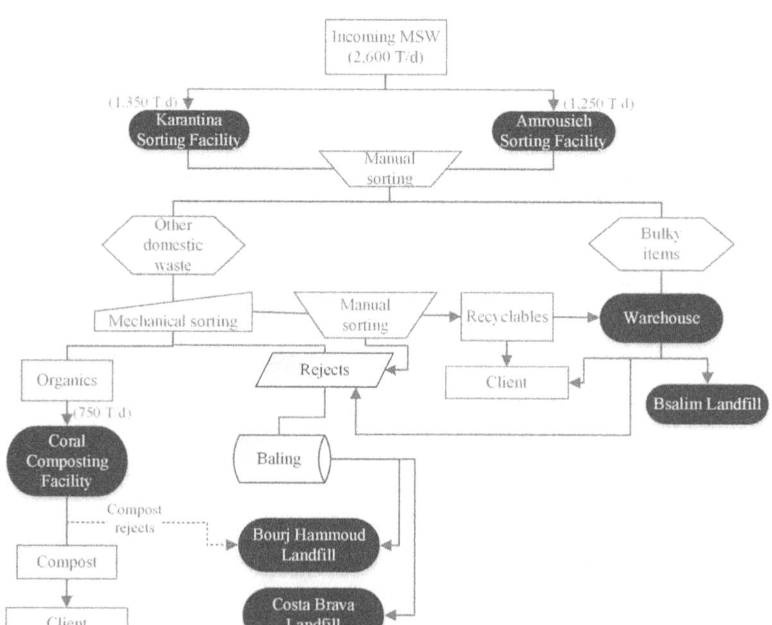

Figure 4. Bourj Hammoud-Jdeideh land reclamation in process
(Source: Google Earth)

Figure 5. Plan of the Bourj Hammoud-Jdeideh sanitary landfill and land reclamation project
(Source: Lebanese Republic Council for Development and Reconstruction)

Figure 6. Site location of Costa Brava Landfill
(Source: Laceco, 2018)

MSWM Outside of Beirut and Mount Lebanon

Outside Municipal Beirut and Mount Lebanon, no comprehensive SWM plan exists and the areas are characterized by rudimentary collect and dump practices with few exceptions. Full or partial SWM systems exist in the main cities (Saida, Tripoli, and Zahle) (Figure 2.d).

The main program (funded by the GOL through a loan from the World Bank) that is worth mentioning consists of the Zahle experience where a controlled landfill was built and the old one was closed with its waste transferred to one cell in the new landfill. This program included collection-transport-processing-recycling and landfilling. Similar to the Beirut-Mount Lebanon experience, this program faced various problems at the startup phase (particularly leachate management and an improper-low capacity processing facility). With time, the program was improved and recently further developed with USAID funding to include a new processing and composting facilities as well as leachate treatment and is targeted for additional funding from the EU for medical waste treatment.

In Saida, the facility (endorsed and financed by the municipality) consists of a sorting plant with a design capacity of 300 tons/day and an anaerobic digestion unit mechanical biological treatment (MBT), both of which are operated by a private company-IBC. More recently, it introduced a refuse derived fuel (RDF) component.

As for Tripoli, a private company- Lavajet also manages collection of MSW, and unsanitary waste landfill is managed by the private company-Batco. The latter has reached its design capacity but continues to receive waste waiting for an alternative treatment and disposal site.

Worth noting as well are various small waste management facilities funded by international donors that remain largely non-sustainable and plagued with technical, financial, and / or operational difficulties. An ongoing and promising EU funded program (through OMSAR) is another recent initiative offering potential partial solutions at selected locations but is equally facing various institutional and socio-political constraints-delays and may become a success story to emulate if these constraints are alleviated and the initiative is integrated within the GOL plans. The previous experience however, raises much skepticism about the long-term fate of this initiative as well.

In the rest of Lebanon, burning and open dumping prevails. According to the Updated Masterplan for the Closure and Rehabilitation of Uncontrolled Dumpsites (MOE/UNDP, 2017), Lebanon has 617 Municipal solid waste dumpsites latest distributed as shown in Table 10. The geographical distribution, status and volumes of these dumpsites are shown in Figures 7. About 55% (341) of the MSW dumpsites were identified as operational and 43% (263) as non-operational MSW dumpsites. Of the surveyed MSW dumpsites, 2% (13) were inaccessible (Table 10).

Table 10. Total current number of non-sanitary waste dumps and dumping sites operating in Lebanon, which have been discontinued and have not yet been rehabilitated

Governorate (Mohafaza)	Number of Dumpsites	Operational	Non-Operational	Inaccessible
Beirut & Mount Lebanon	132	80	50	2
North	51	19	31	1
Akkar	36	19	15	2
Bekaa	89	54	29	6
Baalbak-Hermel	71	42	28	1
South	207	119	87	1
Nabatiyeh	31	8	23	0
Total	*617*	*341*	*263*	*13*

Source: MOE/UNDP, 2017

Figure 7. Geographical locations, status, and volume of MSW dumpsites throughout Lebanon in 2016 *(Source: MOE/UNDP, 2017)*

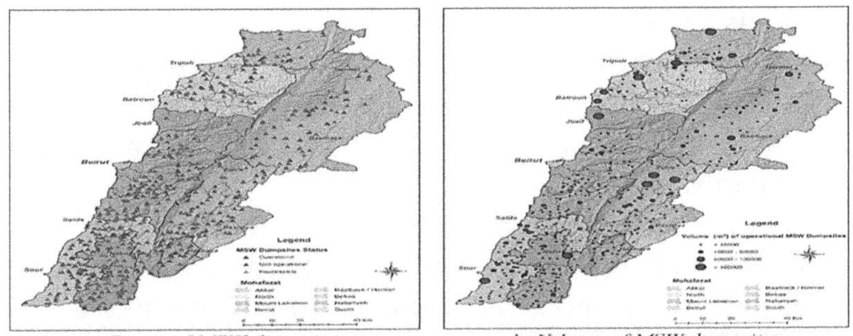

a. Status of MSW dumpsites *b. Volume of MSW dumpsites*

IMPACT OF WASTE MANAGEMENT ON GLOBAL WARMING /GHG

Lebanon is one of 195 countries that have ratified the United Nation Framework Convention on Climate Change (UNFCCC) and are known by the Parties to the Convention (UNFCCC, 2012). The convention is an international environmental treaty between countries with the aim to limit the average global temperature increases and thus climate change, and to cope with its impacts. Moreover, Lebanon has signed the Paris Agreement at the UN Ceremony in April 2016 awaiting final parliamentary ratification. At the very heart of this agreement, countries will reduce emissions in the form of mitigation targets and/or actions, which are referred to as intended nationally determined contributions (UNFCCC, 2015).

According to the latest National Greenhouse Gas Inventory Report, greenhouse gas (GHG) emissions from the waste sector increased by 47% from 1994 to 2011 (MOE/GEF/UNDP, 2015). The waste sector was the second largest emitting sector (11% of the overall national GHG emissions) and the highest source of methane emissions (with ~90% of total national methane emissions) in 2011 (MOE/GEF/UNDP, 2015). This number is mainly attributed to landfills (such as Naameh and Zahle) and unmanaged dump sites (open dumpsites with a depth of more than five meters such as Tyre, Saida, and Tripoli dumpsites) and other land disposal methods. Accordingly, solid waste disposal (landfilling and open dumping) is the largest emitting category accounting for 80% of total GHG emissions from the waste sector (MOE/GEF/UNDP, 2015).

MAIN ISSUES AND CHALLENGES IN WASTE MANAGEMENT

Evidently the challenges towards the development of a sustainable solid waste management plan in Lebanon are multi-faceted with political interference and limited resources being the most significant to overcome. The absence of coherent legal and institutional frameworks and the limited public awareness, trust and participation exacerbate these challenges particularly with the prevalence of the emerging "not-in-my-backyard" (NIMBY) syndrome throughout the country.

The current SWM consists mainly of waste collection, some recovery of recyclables and waste treatment and disposal of the largest fraction in landfills and dumpsites posing high risk on public health and the environment. The operation of existing facilities is substandard with low efficiencies and non-compliant with existing state of practice.

To date, and despite several attempts, there is still no single overarching legislative framework for SWM in the country, and no national strategy albeit many legal instruments that indirectly govern the SWM. Consequently, efforts by various authorities to improve SWM services have been ineffective at best and exacerbated with operational challenges. Generally, law enforcement remains weak and institutional responsibilities not well defined. Many governmental institutions are involved in the planning and management of the solid waste sector with overlapping mandates and responsibilities that have led to ambiguous lines of authority.

Historically, the absence of a comprehensive strategy constitutes the main reason that prevented sustainable solutions to waste management in Lebanon while considering various environmental, scientific, economic, social and ethical aspects. In addition, the weak implementation of proper methods of waste treatment and hierarchy curtailed achieving targeted goals. These methods are based on a general philosophy, leading to a comprehensive environmental strategy that includes waste management within specific principles that evolve into policies and draft special laws developed by the Ministry of Environment and adopted by the Councils of Ministers and Representatives) and translated into decrees, ministerial decisions, tenders, applications, and monitoring mechanisms, which are subject to periodic review, amendment and development.

Knowledge and technology transfer from developed economies is another reason that prevented sustainable solutions in the context of Lebanon, although the challenge is not as much technical as it is mismanagement and planning of public resources. The latter transformed the country and society into a vast market/consumer that is increasingly producing waste without controls.

The strategy focuses on the importance of sustainability in solid waste management and adopts a logical and practical methodology covering the economic, social and environmental sectors. The most important strengths of sustainability are to balance the economic, social and environmental situation of any society through the optimal use of raw materials and other available resources.

Proper waste management goes beyond the disposal or recovery of waste generated and needs to address the problem at its root. Efforts at the national level can be evaluated objectively to develop strategies and plans with clear objectives and standards, and to examine their applicability, dimensions and implications in the near and long term. In this context, an integrated waste management system must focus on waste reduction from the stage of receiving waste at source to final disposal, while ensuring community and stakeholder's participation in the planning and implementation processes, and considering updated data on various elements.

CONCLUSION

Based on the Beirut-Mount Lebanon Emergency Plan as well as the latest approved plans and on-going initiatives funded by international donors, it is evident that the GOL and the general public are emphasizing the role of ISWM in managing the solid waste sector. While it is commonly accepted that this approach constitutes a significant step in the right direction, several challenges continue to hinder the sector's successful development at various levels: politically, legally and institutionally, socially, land use, and last but not least technically. In addition, there is a trend of multiple and at times incompatible solutions without providing preferred alternatives highlighting the importance of public awareness on the advantages and disadvantages of solid waste management alternatives. The poor process of monitoring and environmental control has also eroded public trust in solutions proposed by the government.

The experience to date raises concerns leaving much to be decided to overcome these challenges. Current waste management activities, whether viewed as a hierarchy or as complementary components, particularly source separation and recycling have not measured up favorably with the steps outlined in an ISWM system. The difficulty associated with locating and approving suitable sites for landfilling has been continuously increasing, which dictates the adoption of policies that will minimize the amount of waste to be disposed of in a landfill. Despite the legacy of uncontrolled incineration practices (or better described as near open burning at old facilities) that has apparently resulted in the elimination of this option in the first place, today, the technology is available (perhaps at a premium cost but seemingly within the range of what is currently being paid or will be paid in the future) to build thermal systems with high efficiency and adequate air quality control devices. While this option does not completely eliminate the need for a landfill because of ash generation and other residues, it minimizes the amount of end waste that should be disposed of in a landfill and hence the need for a land is kept at a minimal level that can indeed be sustainable in the long term. Such a plan has been adopted in many developed countries particularly where land is scarce or prohibitively expensive. The limited suitable land available for landfilling, particularly along the coastal region, creates a necessity for considering thermal systems, at least in regions such as the Beirut and Mount Lebanon and Tripoli. It is non-sustainable to continue with the same policy of locating new landfills in the future.

Last but not least, for a waste management alternative to be successful (recycling, biological/thermal processes, or landfilling), adequate efforts and resources must be allocated for the site selection process. The ISWM plan for the Beirut and Mount Lebanon and the rest of Lebanon has failed most in this regard. It is equally important to consider waste minimization/ reduction principle as a basis to develop any waste management plan. In light of the arguments described above, it is envisaged that all possible alternatives be re-considered on the basis of past experience gained as well as the best available technology that would be most suitable for the case at hand.

REFERENCES

Ayoub, G., Acra, A., Abdallah, R., & Merhebi, F. (1994). *Fundamental Aspects of Municipal Refuse Generated in Beirut and Tripoli. Technical Report*. Beirut: Department of Civil and Environmental Engineering, American University of Beirut.

Ayoub, G., Acra, A., Abdallah, R., & Merhebi, F. (1996). *Fundamental Aspects of Municipal Refuse Generated in Beirut-Phase II. Technical Report*. Beirut: Department of Civil and Environmental Engineering, American University of Beirut.

Bogner, J., Abdelrafie Ahmed, M., Diaz, C., Faaij, A., Gao, Q., Hashimoto, S., Pipatti, R., … Zhang, T. (2007). Waste Management, In Climate Change 2007: Mitigation. Contribution of Working Group III to the Fourth Assessment Report of the Intergovernmental Panel on Climate Change. Cambridge University Press.

CDR. (2006). *Municipal Solid Waste Management in Lebanon*. Beirut: Council for Development and Reconstruction.

CEDARE/ Blue Plan/ UNEP. (2000). *Policy and institutional assessment of solid waste management in five countries Cyprus, Egypt, Lebanon, Syria, Tunisia*. Blue Plan- Regional Activity Centre.

Chen, T., & Lin, C. (2008). Greenhouse gases emissions from waste management practices using Life Cycle Inventory Model. *Journal of Hazardous Materials*, *155*(1-2), 23–31. doi:10.1016/j.jhazmat.2007.11.050 PMID:18164811

El Fadel, M., & Sbayti, H. (2000). Economics of mitigation greenhouse gas emissions from solid waste in Lebanon. *Waste Management & Research*, *18*(4), 329–340. doi:10.1034/j.1399-3070.2000.00140.x

GFA (Consulting Group GmbH). (2017). *Support to Reforms – Environmental Governance, Beirut, Lebanon. Assessment of Solid Waste Management Practices in Lebanon in 2015*. Hamburg, Germany: European Union.

GIZ/SWEEP-Net. (2014). *Country profile on the solid waste management situation in Lebanon*. Retrieved from http://www.sweep-net.org/sites/default/files/LEBANON%20PROFIL.pdf

Government of Lebanon/ Council of Ministers. (2015). *Amendment of the Council of Ministers' decision No. 1 dated 12/01/2015*. Retrieved from http://www.pcm.gov.lb/arabic/landing.aspx?pageid=9

Government of Lebanon/Council of Ministers. (1997). *COM Decision No.58 (dated 1/2/1997)*. Retrieved from http://www.pcm.gov.lb/arabic/landing.aspx?pageid=9

Government of Lebanon/Council of Ministers. (2005). *COM Decision No.1/4952 (dated 8/18/2005)*. Retrieved from http://www.pcm.gov.lb/arabic/landing.aspx?pageid=9

Government of Lebanon/Council of Ministers. (2010). *COM Decision No.55 (dated 9/1/2010)*. Retrieved from http://www.pcm.gov.lb/arabic/landing.aspx?pageid=9

Government of Lebanon/Council of Ministers. (2014). *COM Decision No.46 (dated 10/30/2014)*. Retrieved from http://www.pcm.gov.lb/arabic/landing.aspx?pageid=9

Kaza, S., Yao, L. C., & Bhada-Tata, P. Van, Woerden, F. (2018). What a Waste 2.0: A Global Snapshot of Solid Waste Management to 2050. World Bank, Washington, DC.

Laceco. (2018). *Supervision of Greater Beirut Solid Waste Treatment Plants Contract N 19942*. Beirut, Lebanon: CDR.

Laceco/Ramboll. (2012). *Preparation of Pre-qualification documents and Tender Documents for Solid Waste Management in Lebanon, Sub Report 1, Baseline Study*. Beirut, Lebanon: CDR.

MOE. (2018). *Policy Summary on Integrated Solid Waste Management*. Beirut: Republic of Lebanon – Ministry of Environment.

MOE/EU/UNDP. (2014). *Lebanon Environmental Assessment of the Syrian Conflict & Priority Interventions*. Retrieved from http://www.undp.org/content/dam/lebanon/docs/Energy%20and%20Environment/Publications/EASC-WEB.pdf

MOE/GEF/UNDP. (2011). *Lebanon's Second National Communication to the UNFCCC*. Retrieved from http://www.lb.undp.org/content/lebanon/en/home/library/environment_energy/lebanon_s-second-national-communication-to-the-unfccc.html

MOE/GEF/UNDP. (2015). *National Greenhouse Gas Inventory Report and Mitigation Analysis for the Waste Sector in Lebanon*. Retrieved from http://climatechange.moe.gov.lb/viewfile.aspx?id=222

MOE/UNDP. (2017). *Nationally Appropriate Mitigation Action in Lebanon's Municipal Solid Waste Sector NAMA Proposal and Design Document*. Beirut: Republic of Lebanon – Ministry of Environment.

MOE/UNDP. (2017). *Updated master plan for the closure and rehabilitation of uncontrolled dumpsites throughout the country of Lebanon*. Beirut: Republic of Lebanon – Ministry of Environment.

MOE/UNDP/ELARD. (2011). *Provision of consultancy services for the Preparation of a Master plan for the Closure and Rehabilitation of Uncontrolled Dumps throughout the country of Lebanon*. Retrieved from http://www.moe.gov.lb/The-Ministry/Reports/Master-plan-for-the-closure-and-rehabilitation-of.aspx

MOE/UNDP/GEF. (2016). Lebanon's third national communication to the UNFCCC. Beirut, Lebanon: Author.

MOF. (2011). *Lebanon Country Profile 2011*. Beirut: Ministry of Finance. Retrieved from http://www.finance.gov.lb/en-us/Finance/Rep-Pub/DRI-MOF/Sovereign%20and%20Investment%20Reports/Lebanon%20Country%20Profile%202011.pdf

SWEEP-Net. (2010). *Country Report on the Solid Waste Management in Lebanon*. Retrieved from http://www.sweep-net.org/ckfinder/userfiles/files/country-profiles/CountryreportLebanon-En-mai2011,pdf

UNFCCC (United Nations Framework Convention on Climate Change). (2012). *Background on the UNFCCC: The international response to climate change*. Retrieved from https://unfccc.int/essential_background/items/6031.php

UNFCCC (United Nations Framework Convention on Climate Change). (2015). *Adoption of the Paris Agreement*. Retrieved from https://unfccc.int/resource/docs/2015/cop21/eng/l09r01.pdf

World Bank. (2018). Lebanon's *Economic Outlook*. Retrieved from https://www.worldbank.org/en/country/lebanon/publication/economic-outlook-october-2018

World Bank. (2019). *The World Bank In Lebanon*. Retrieved from https://data.worldbank.org/country/lebanon

Chapter 8
Waste Management Challenges in Malaysia

Agamuthu Pariatamby
https://orcid.org/0000-0001-9346-6843
Sunway University, Malaysia

Mehran Sanam Bhatti
https://orcid.org/0000-0001-9165-7195
University of Malaya, Malaysia

ABSTRACT

Malaysia, an upper-middle class country, populated with approximately 31 million people generated nearly 13.9 million tonnes of municipal solid waste (MSW) at per capita of 1.23 kilogram per day in 2016. Landfills and open dumps, being the absolute opposite of sustainable waste management, received about 80% of this generated MSW. Waste collection is on a par with developed nations, and almost all urban MSW is being collected for disposal. However, illegal dumping still occurs sporadically, and it can account for 10% of the total MSW generated. Hence, Malaysia is facing a stiff challenge in reducing the amount of waste sent to landfills and adopting sustainable waste management. National Solid Waste Management Department of Malaysia targets to divert 40% of MSW from landfills by 2020. There are total 296 landfills in Malaysia, and only 165 landfills are currently functional whereas the number of sanitary landfills is 8 out of 165 operating landfills. The national recycling rate of MSW was 17% in 2016 and the national recycling target in 11th Malaysian Plan is 22% by 2020.

COUNTRY SITUATION

The federation of Malaysia practices a form of constitutional monarchy, and is situated in South East Asia (Figure 1). The number of states and federal territories that constitute Malaysia are 13 and 3, respectively (Figure 2). The total landmass of Malaysia is 329,847 square kilometres where Peninsular Malaysia and East Malaysia or Malaysian Borneo (Sabah and Sarawak states) are separated by South China Sea. Malaysian population in 2017 was 32 million (Department of Statistics, 2018) and the official

DOI: 10.4018/978-1-7998-0198-6.ch008

Figure 1. Map of south east Asia region (United Nations, 2012)

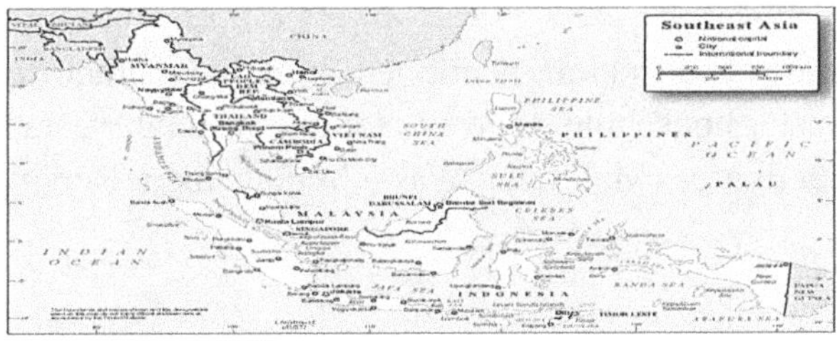

Figure 2. States of Malaysia (UNCRD, 2017)

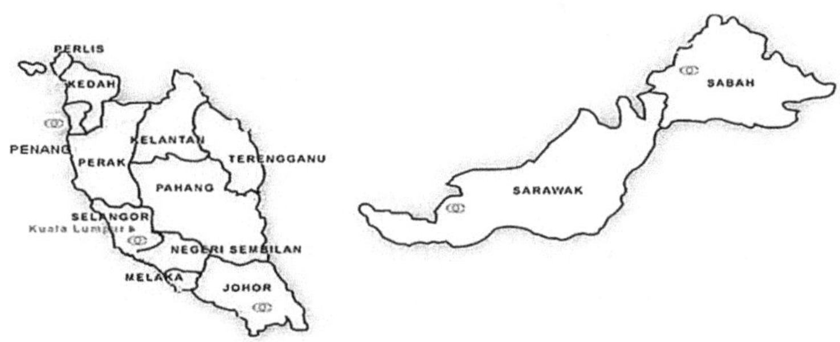

religion of the country is Islam. Kuala Lumpur is the capital city of Malaysia. The government system in Malaysia is highly based on Westminster parliamentary system, since Malaysia was ruled by British empire before independence. While Malaysia still has a King or commonly known as Yang di-Pertuan Agong to head the country, its executive powers lie with Prime Minister of the cabinet. The economy of Malaysia was predominantly based on agriculture before 1970s, followed by a transition to multi-sector economy until 1980s and industrial sector leading the national economy since then.

WASTE DEFINITION

According to Malaysian regulation, Environmental Quality Act 1974 (Act 627), and Solid Waste and Public Cleansing Act 2007, solid waste is defined as:

- Any by-product i.e. scrap material, rejected products or unwanted surplus material that results from a process.
- Any material that is broken, spoiled, contaminated or worn out and thus needs to be disposed of.
- Any other substance that according to abovementioned law or any other written law demands the authority to dispose of, but it does not come under Act 127 as scheduled waste, or under Water Services Industry Act 2006 (Act 655) as sewage or under Atomic Energy Licensing Act 1984 (Act 304) as radioactive waste.

Solid waste can be classified into municipal solid waste (MSW), industrial waste, hazardous waste, agricultural waste and electronic waste (E-waste).

Municipal Solid Waste

MSW is also defined as solid waste as aforementioned. It includes imported solid waste, public solid waste, institutional solid waste, household solid waste, and special waste arising from construction, commercial and controlled activities (Malaysian Waste Management Report, 2010).

Biomass

The definition of biomass is variable but some of the widely agreed definitions at International and European level are given below:

a. According to United Nations Framework Convention on Climate Change (UNFCCC), non-fossil component of municipal or industrial waste is called Biomass.
b. Whereas, Biomass is characterized as any plant matter that can be utilized as fuel or transformed into other products before consumption, by Organization for Economic Co-operation and Development (OECD).
c. Waste Framework Directive of European Union defines Bio-waste as biodegradable park and garden waste, kitchen and food waste from restaurants, caterers, households, comparable waste from food processing plants, and retail premises.

Industrial Waste

Under Solid Waste and Public Cleansing Management Act 672, 2007, Industrial waste is referred to solid waste that is generated from all industrial activities. Thus, industrial waste may include ashes, rubbish, demolition and construction waste, as well as, hazardous and special waste.

Hazardous Waste

Environment Quality (Scheduled Wastes) Regulations 2005 defines Hazardous waste as any waste that comes under the waste categories listed in first schedule of regulation. Schedule waste IS deemed hazardous due to the presence of substances that could harm or pose danger to the life form i.e. human beings, plants, animal, or to environment. The properties of schedule waste are corrosivity, ignitability, reactivity, toxicity or infectivity. Generally, clinical waste is characterized separately due to the potential infectivity. However, clinical waste might also be classified as radioactive waste, biological waste, chemical waste, explosive waste or flammable waste.

Electronic Waste

Environmental Quality Act 1974 defines Electronic waste (E-waste) as waste generated from electrical and electronic assemblies that contains certain components. These components may be mercury-switches, accumulators, glass from cathode-ray tubes, and other activated glass or polychlorinated biphenyl capacitors, or contaminated with mercury, lead, cadmium, nickel, copper, chromium, silver, lithium, polychlorinated biphenyl or manganese.

WASTE MANAGEMENT IN MALAYSIA

Due to the onset of industrialization and development, generation of waste is increasing more than urbanization. Nevertheless, there has been a dramatic evolution in solid waste management of Malaysia since 1970s, despite having low population at that time and manageable amount of waste generated.

Municipal solid waste generated in Malaysia used to be managed by state and local governments before the enforcement of Solid Waste and Public Cleansing Management Act (Act 672) in 2011. Local authorities, by hiring small contractors, used to provide solid waste management services. Then Solid Waste and Public Cleansing Management Act was enacted by the parliament in 2007, followed by enforcement of Act on 1st September 2011 which granted executive authority to the federal government on solid waste management of eight states and federal territories.

The municipal solid waste collection was privatized in Peninsular Malaysia. Three companies were awarded the concessions to provide waste collection in three different regions. Southern Waste Management Pvt Ltd (SWM) Enviro was assigned the southern region including Johor, Malacca, and Negeri Sembilan. On the other hand, E-Idaman Pvt Ltd was given Perlis and Kedah and Alam Flora Pvt Ltd was assigned in charge of Kuala Lumpur, Putrajaya, and Pahang. The timeline in Figure 3 summarizes the solid waste management history of Malaysia. The National Solid Waste Management Department has set a goal of reducing 40% of landfilling and 38% of greenhouse gas emissions from landfills by 2020. Consequently, it would lessen the burden on landfills and reduce generation of greenhouse gas emissions from landfills in Malaysia.

The department of National Solid Waste Management is going to collaborate with Solid Waste and Public Cleansing Cooperation as required by Solid Waste and Public Cleaning Management Act 2007. The responsibilities of department of National Solid Waste Management is to propose policy, plan and strategies for the nationwide management of solid waste. Additionally, their responsibilities also include setting standards, specifications, and codes of practices, along with plan formulations for solid waste management facilities. Concurrently, the responsibility of Solid Waste and Public Cleansing Corporation (SWPCC) is to implement department's policy and plans. In addition to monitoring, the compliance of codes of practice and standards, promotion of public participation and improvement of public awareness of solid waste management is carried out by SWPCC. From 1st September 2012, National Solid Waste Management department has been implementing a "2+1 collection system" in eight states of Malaysia,

Figure 3 The timeline of solid waste management history

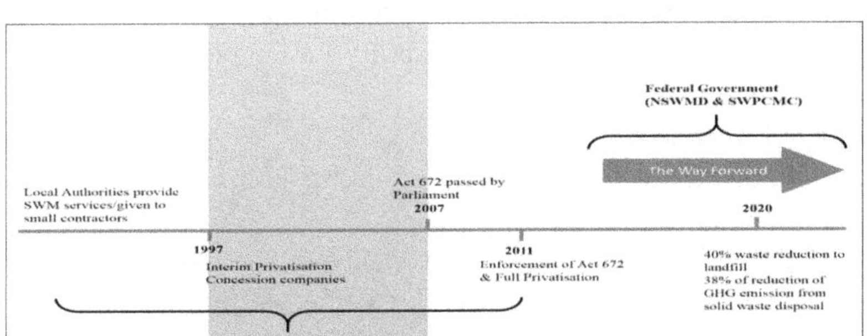

where residual waste is collected two times a week and recyclable, bulky and green wastes are collected one a week. Additionally, new standards were also implemented on waste bins and garbage collection trucks. Moreover, Key Performance Index (KPI) was also prescribed to garbage collection schedule.

The waste recovery and treatment facilities are inadequate in Malaysia as only one waste to energy (WtE) plant is established in Kajang, whereas mere five incinerators are constructed in Langkawi Island, Labuan, Pangkor Island, Tioman Island and Cameron highlights, respectively. Malaysia's first and only WtE plant, that was commissioned in 2009, converts MSW into refuse derived fuel (RDF) to provide fuel to the integrated steam power plant. However, the latest government in Malaysia aims to establish a total of 8 incinerators in coming two years (Malaymail, 2018). There will be one incineration plant in Kuala Lumpur with a capacity of 1000 tons of MSW per day. The other planned incineration plants are for the cities including Selangor, Malacca and Negri Sembilan. In the long run, the government wants to install one incineration plant in every state of Malaysia with a maximum capacity of 1500 tons of MSW per day.

Presently, the most deployed technology for the waste disposal in Malaysia is landfilling. There is a total of 296 landfills, where only 165 are currently operating and eight of those are sanitary landfills. It is estimated that about 95% of MSW is disposed in landfills and only 5% is recycled. On the contrary, national recycling rate is reported to be higher than 15%. Hence, it must be highlighted that the regulation of recycling activities in Malaysia is absent. Therefore, proper data collection may not occur, thus leading to inaccurate national recycling rates. Additionally, market in Malaysia determines the prices of recyclables that makes it seasonal in nature. So, the prices of recyclables shift according to the demands, as collectors and traders hoard the recyclables until their preferred selling price is attained. Glass bottles, old newspapers, aluminium tin cans, vehicle tires, and plastic PET bottles are few recyclables that are collected and recycled in Malaysia, in addition to food waste. For example, a program named "Local Agenda 21 Kuala Lumpur (LA21KL)" was launched by Kuala Lumpur City Council (DBKL) that organized campaigns with the purpose of cleaning and beautifying the Malaysian capital by converting food waste and organic waste into fertilizers. The aim of the program was promotion of cleaner and greener city through effective waste management.

The National Solid Waste Management Department apprehended the unsustainability of continued utilization of landfills. A few factors also directed the department away from choosing landfilling such as unavailability of land for establishment of new landfills in major urban cities, negative perception of public for landfills and subsequent emission of greenhouse gases from landfills. Hence, a framework has been formulated for recovery and treatment facilities with the following objectives:

1. Minimum impact to the environment and natural resources.
2. Commercially proven technology and high reliability for long term solution.
3. Minimum risk exposure to the government.
4. Sustainable model of implementation.
5. Maximum reduction of waste.

Similarly, to improve environmental situation in Malaysia, the Government has been introducing various policies and plans. One such example of policy is the Master Plan on National Waste Minimization, devised by Ministry of Housing and Local Government, to provide visions, strategies and roles for stakeholders, as well as to promote awareness of waste reduction among the public. The Solid Waste Management Corporation Strategic plan was also initiated to increase public awareness for establishment of environmentally sustainable solid waste management. Moreover, Malaysian government also instituted the National Policy on climate change in 2009 to promote climate resilient development in Malaysia with the aim of reducing 2005 greenhouse gas emissions by 50%.

Furthermore, various campaigns and programs were also introduced by the government, aiming at creating awareness of 3R (reduce, reuse and recycle). The reception of these campaigns had been inadequate by the public as 3R campaigns were launched in 1996 and 2000. Alam Flora Pvt Ltd, being one of the main players in the solid waste management scene in Malaysia, has recently worked together with the Solid Waste Management and Public Cleansing Corporation to promote awareness program in a bid to minimize waste generation and indiscriminate waste disposal. Community programs involving recycling centre and mobile recycling centres exist in the cities, along with the recycling banks that were promoted in schools in Putrajaya. Congruently, Alam Flora Pvt Ltd also set successful 3R programs from neighbouring countries as a benchmark of its performance. For instance, Thailand's Wongpanit recycling system and Indonesia's 3R implementation in Surabaya were used.

Malaysian government also introduced a green technology soft loan scheme of MYR 1.5 billion (2.8 million USD) in the 10th Malaysian Plan (2012 – 2016), aiming to supply and employ green technology. According to this program, 2% of the total interest rate and a guarantee of 60% on the financing amount is borne by the government, whereas 40% is borne by banking institutions.

Additionally, Ministry of Domestic Trade, Cooperative and Consumerism introduced the No Plastic Bag Day Campaign in 2011, to reduce the excessive consumption of plastic bags in Malaysia. MYR 0.20 (0.05 USD) was charged on the purchase of plastic bags on every Saturdays, which are No Plastic bag day. Through this concept of taxation on plastic bags, revenues from plastic bag levy were intended for funding nationwide environmental programs. This campaign was further incorporated by Penang (Malaysia state) by implementing No Plastic Bags campaign throughout the year in major shopping malls.

The difference between capital cost and operating cost of various types of disposal methods that were considered by the National Solid Waste Management Department of Malaysia are given in Table 1. As it can be seen, landfilling is the most economical mode of waste disposal in Malaysia, costing less than MYR 50 (12 USD) in comparison to other methods.

Table 1. Various types of disposal methods used in Malaysia

Disposal Method	Capital Expenditure		Operating Expenditure	
	MYR million	USD million	MYR per ton	USD per ton
Landfill	30	7.2	28.80 – 49.00	6.9 – 11.7
Waste to Energy	250	60	101	24.2
Plasma gasification	650	156	202	48.5
Mass burn stoker	550	132	124	29.7

One USD = 4.21 MYR (as at 01/9/2019)
Source: (NSWMD, 2012)

MUNICIPAL SOLID WASTE

National Solid Waste Department of Malaysia reported that each Malaysian generated 1.2 kg of MSW every day. In 2014, an approximate amount of 12,982,685 tons of MSW was generated in Malaysia. It must be noted that this value is estimated (an increase of 3.59%) from original MSW data available for 2012 where per capita MSW generation was 1.17 kg. The availability of reliable data on MSW in Malaysia is a matter of concern (Agamuthu, 2014) and therefore latest data is not accessible every year in Malaysia unlike that in member states of EU, Japan, USA and others. According to an estimate (Table 2), Malaysia generated approximately 13 million tons (12,982,685 million tons) of MSW in 2014. However, according to Solid Waste Management and Public Cleansing Corporation (SWCorp), Malaysian generated approximately 38,200 tons per day which would amount to approximately 14 million tons in 2016 (Pauze, 2016).

State wise projected generation of MSW for the year 2014 is shown in Figure 4. Whereas, the amount of solid waste generated from 1970 to 2012 by major cities of Malaysia is demonstrated in Figure 5. As it can be seen that highest amount of solid waste is generated by Kuala Lumpur every day. Urbanization and sharp increment in population are the two factors that have increased waste generation in Kuala Lumpur. The Malaysian urban population of 14,429,641 people generates 21,918 tons of MSW every day where each city dweller is responsible for producing 1.52 kg of MSW per day (World Bank, 2014). Table 3 enlists the daily average of MSW disposed in Malaysia.

The composition of the MSW generated in Malaysia varies with the economic status of the household as seen in Table 4. The percentage of dry recyclables such as glass, plastic and paper increase with the increase in socio-economic status. On the other hand, food waste decreases with the increase in socio-economic status because affluent households purchase processed products in greater quantity as compared to less affluent households.

WASTE FLOW IN MALAYSIA

MSW generated in Malaysia is managed by using three technologies mainly as they are landfilling, recycling and composting. Figure 6 demonstrates the flow of all streams of MSW from generation to final disposal. An overall management scenario of MSW management in Malaysia is depicted in Figure 7. About 81.5% of MSW is landfilled (including unspecified and sanitary landfills), whereas 17.5% of MSW is recycled and only 1% is composted.

Table 2. Summary of MSW generation and disposal

Indicators	Data	Unit	Year	Reference
Total MSW Generation	12,982,685 (estimation)	Ton/year	2014	UNCRD, 2017
MSW Generation per Capita	1.17	kg/capita/day	2012	
Total MSW Disposal	12,982,685 (estimation)	Tons/year	2014	Agamuthu and Tanaka (2014)
	35,569	Tons/day		

Figure 4. Percentage of solid waste generated by states in Malaysia in 2014

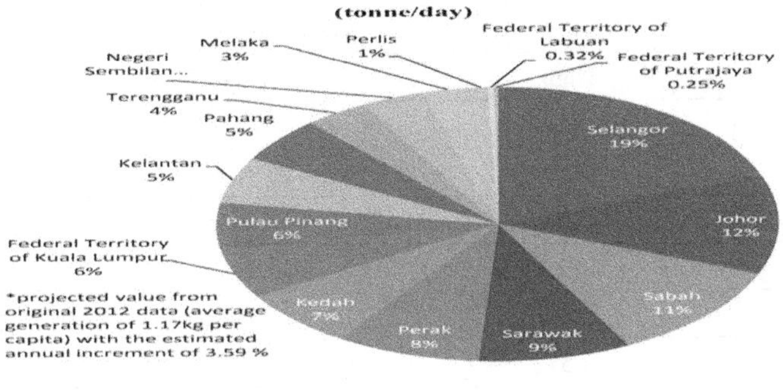

Figure 5. Generation of MSW in major urban areas in Peninsular Malaysia (1970 - 2012) (Source: Agamuthu & Tanaka, 2014)

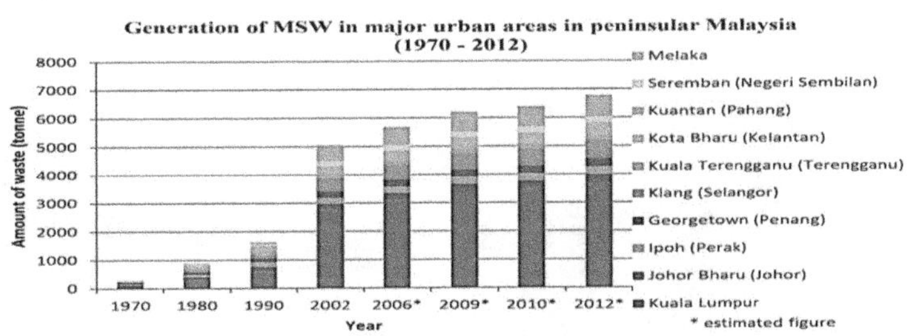

Table 3. Average municipal solid waste disposal (tons/day) in Malaysia 2013

State	Average Disposal (Tons/day)
Selangor	6,855.11
Johor	4,205.97
Sabah	4,030.20
Sarawak	3,101.12
Perak	2,950.46
Kedah	2,448.25
Federal Territory of Kuala Lumpur	2,096.71
Pulau Pinang	1,958.60
Kelantan	1,933.49
Pahang	1,883.27
Terengganu	1,305.73
Negeri Sembilan	1,280.62
Melaka	1,092.52
Perlis	288.77
Federal Territory of Labuan	113.00
Federal Territory of Putrajaya	87.89
Total	35,568.71

Table 4. Composition of the waste generated in Malaysia based on the economic status of the household

Composition (%)	Socio-Economic Status		
	High	**Middle**	**Low**
Paper products	19.79	15.73	13.04
Plastic and rubber	21.05	18.61	13.01
Glass and ceramics	14.99	9.42	7.57
Food waste	24.13	29.77	31.86
Metals	8.80	12.75	9.15
Textiles	1.57	3.87	3.08
Garden waste	5.50	6.95	15.56
Wood	3.45	2.90	6.72
Total	**100.00**	**100.00**	**100.00**

Source: (Agamuthu & Tanaka, 2014)

Figure 6. Municipal solid waste flow in Malaysia

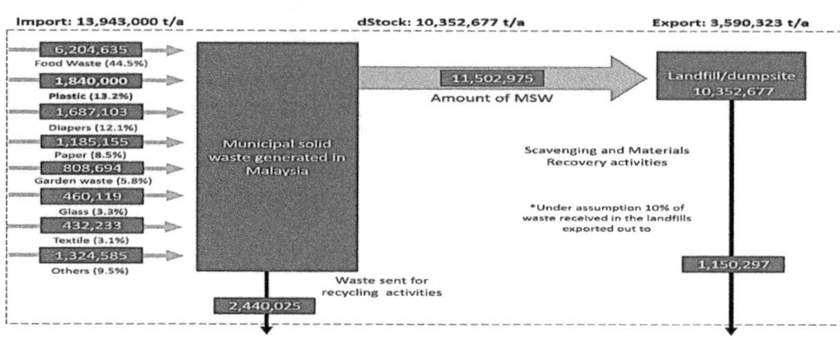

Figure 7. Municipal solid waste flow

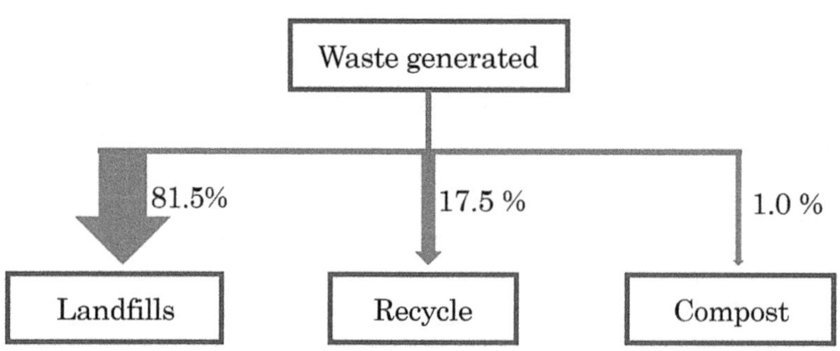

RECYCLING OF MUNICIPAL SOLID WASTE

Act 672 defines recycling as collection and separation of solid waste with the intention of creating new products, whereas it defines recycling centres as the venues for dropping or selling recyclables. The estimated value of Malaysian recycling industry in 2005 was MYR 476 million (114.24 million USD) from where in 2011 it increased to MYR 600 million (144 million USD). High variability in recyclables' prices due to the dependency on market and prices of commodities, that are seasonal, result in absence of incentives for the end customers. Consequently, recycling establishments face undersupply of recyclables, thus hampering recycling efforts. Table 5 and 6 enlist recycling rates in Malaysia, however it must be noted that these figures are underestimated as recycling activities are still not regulated. Lastly, Table 7 enlists projected values of recycling for year 2020.

Table 5. Summary of the recycling rate

Indicators	Data	UNIT	Year	Reference
Overall Recycling Rate	5	%	2005	Table 6
	10.5	%	2012	
	15	%	2014	Agamuthu & Tanaka 2014
	17.5	%	2016	UNCRD, 2017
Target of Recycling Rate	22 by 2020	%	2002	UNCRD, 2017

Table 6. Waste composition and recycling rate (Agamuthu and Tanaka, 2014)

Year	2005	2012
Population	26,600,000	28,300,300
Waste Generation (ton/day)	19,000	33,000
Generation Rate	0.8 kg/day/person	1.1 kg/day/person
Waste Composition (%)		
Food Waste	45	44.5
Plastic	24	13.2
Diapers	-	12.1
Paper	7	8.5
Garden Waste	-	5.8
Glass	-	3.3
Textile	24	3.1
Others	-	9.5
Recycling Rate (%)	5	10.5

Table 7. Waste generation and recycling rate trend towards 2020 (Agamuthu and Tanaka, 2014)

Year	2005	2012	2020 (Projection)
Total Waste Generation (ton/day)	19,000	33,000	30,000 (projected in 2005)
Recycling Rate (%)	5	10.5	22 (targeted)

HAZARDOUS WASTE

The terminology used in Malaysia for hazardous waste is Scheduled waste. The legislation, Environmental Quality (Scheduled Wastes) Regulations 2005 defines any waste that may belong to any of the 107 categories (grouped into specific and non-specific sources) of hazardous waste. These categories are listed in the First Schedule of EQA, 1974 and by law, scheduled waste ought to be rendered as inert as possible before final disposal. Table 8 enlists the amount of scheduled waste generated and managed by prescribed facility, Kualiti Alam Pvt Ltd.

Table 9 registers the amount of hazardous waste generated in Malaysia from year 2007 to 2017. As it can be seen in Figure 8, the generation of hazardous has been roughly increasing since 2007 until 2017. Due to growth in industrializations, the environment has been potentially at increasing risk from generation of hazardous waste. The quantity of scheduled wastes generated in 2017 was 2,017,280.76 metric tons, which is approximately 27.08% lesser than 2016, 2,766,613,45 metric tons.

The regulations that govern the management of scheduled waste are listed below:

1. Environmental Quality Act 1974.

Table 8. Summary of hazardous waste generation

Indicators	Quantity (Tons)	Year
Amount of Hazardous Waste Generation	2,918,478.34	2015
	2,766,613.45	2016
	2,017,280.76	2017
Amount of Hazard Waste Disposal (to Kualiti Alam Pvt Ltd)	120,221.72	2015
	143,063.4	2016
	133,507.97	2017

Table 9. Scheduled wastes generated in Malaysia, 2005 – 2012 (Environmental Quality Report, n.d.)

Year	Waste generated (tons)
2007	1,138,839.49
2008	1,304,898.77
2009	1,705,308.14
2010	1,880,928.53
2011	1,659,537.00
2012	1,708,708.00
2013	2,965,611.65
2014	2,541,762.34
2015	2,918,478.34
2016	2,766,613.45
2017	2,017,280.76

Figure 8. Hazardous waste generation in Malaysia from 2000 to 2012 (Environmental Quality Report, n.d.)

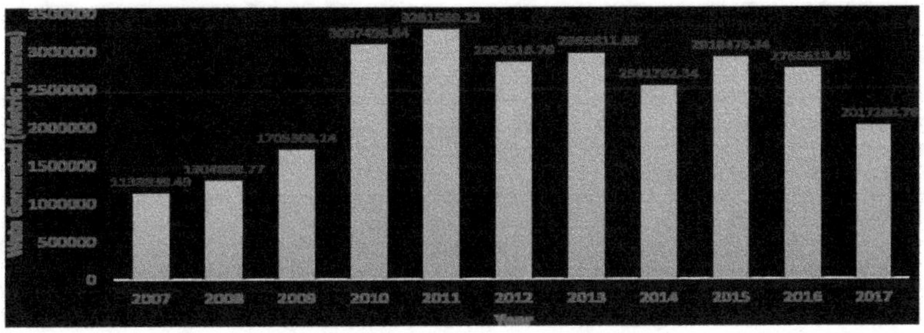

2. Environmental Quality (Prescribed Premises) (Scheduled Wastes Treatment & Disposal Facilities) Regulations 1989.
3. Environmental Quality (Prescribed Premises) (Scheduled Wastes Treatment & Disposal Facilities) (Amendment) Regulations 2006.
4. Environmental Quality (Scheduled Wastes) Regulations 2005.
5. Environmental Quality (Scheduled Wastes) (Amendment) Regulations 2007.
6. Environmental Quality (Dioxin and Furan) Regulations 2004.
7. Environmental Quality (Prescribed Conveyance) (Scheduled Wastes) Order 2005.
8. Environmental Quality (Prescribed Premises) (Scheduled Wastes Treatment and Disposal Facilities Order) 1989.

As of 2015, a total of 606 off-site recovery facilities have been awarded license by Department of Environment (DOE), Malaysia. The most license issued to recovery facility according to waste category is for e-waste, 140 facilities. Table 10 gives the number of recovery facilities for each category of scheduled waste. All prescribed premises must conform to the legislations, Environmental Quality (prescribed activities) (Environmental Impact Assessment) Order 1987. If required Environmental Impact Assessment (EIA) ought to be submitted to the DOE. Moreover, the process of licensing is administered by Environmental Quality (Prescribed Premises) (Scheduled Waste Treatment and Disposal Facilities) Regulations, 1989. Table 11 describes the amount of each type of scheduled waste generated in Malaysia for the year 2017.

DOE Malaysia is encouraging the execution of regulation 7 of Environmental Quality (Scheduled Waste) Regulation 2005, which concerns with 'Special Management' of scheduled waste. It is mainly referred for diverting the scheduled waste towards unlicensed facilities with the aim to recycle and reuse, as well as, to render scheduled waste unhazardous in order to dispose at sanitary landfill. Table 12 reports the type of scheduled waste treated under 'Special Management' and subsequent mode of disposal. Approximately 9,353.67 metric tons of schedule waste were exported the same year. The exportation and recovery of scheduled waste require written permission from the authorities in Malaysia including Ministry of Natural Resources and Environment, Department of Environment and Hazardous Substances Division. Lastly, an overall management of scheduled waste for the year 2017 is enlisted in Table 13.

Table 10. List of off-site recovery facilities according to waste category

Waste Category	Number of Recovery Facilities
e-waste	140
Dross / Slag / Clinker	64
Oil / Mineral Sludge / Spent Coolant	51
Heavy Metal Sludge	49
Used Container	48
Paper & Plastic	41
Spent Solvent	39
Acid / Alkali	34
Phenol / Adhesive / Resin	32
Mixed Wastes	27
Ink / Paint	23
Catalyst	19
Photographic	12
Batteries	11
Rubber Sludge	10
Gypsum	6

CLINICAL WASTE

The waste generated from hospital is heterogenous in composition as clinical waste, hazardous chemicals, pharmaceutical waste, radioactive waste and general waste constitute the hospital waste. Often clinical waste and pharmaceutical waste are reported together. General waste comprises 75 to 90% of total hospital waste, which is non-hazardous waste, whereas remaining 10 to 25% of hospital waste may consist of hazardous or clinical waste. Thus, it may pose a variety of health risks. Table 14 summarizes the WHO categories (Health-care waste, 2018) of hospital waste.

The diverse composition of hospital waste requires proper management of clinical waste in order to avoid health risks and potential threats to flora, fauna and environment. Moreover, pathogenic microorganisms are abundant in hospital waste, thereby necessitating appropriate, safe and reliable management. The threats, such as disease or injury, from the clinical waste have been established due to the characteristics such as genotoxicity, pathogenicity, and hazardousness from chemicals. Additionally, pharmaceutical or radioactive residues, as well as, sharp items pose even greater threat. The transportation and disposal of clinical waste in Malaysia is administered under the Environmental Quality (Scheduled Waste) Regulations 2005.

According to Environmental Quality (Scheduled Waste) Regulations 2005, the definition of clinical waste is waste comprised completely or partly of human or animal tissue, of blood or other body fluid, excretions, drugs or other pharmaceutical products, syringes, needles or other sharp instruments, swabs or dressings, or any waste that may prove hazardous to any person upon contact without rendering it safe. Table 15 describes the clinical waste categorized according to four groups. Moreover, clinical waste also refers to any other waste originating from nursing, dental, medical, veterinary, pharmaceutical or similar

Table 11. Quantity of scheduled wastes generated by category, 2017

No.	Waste Categories	Quantity of Waste	
		Metric Tons	Percentage
1	*Dross / Slag / Clinker/ Ash*	706,750.00	35.03
2	*Gypsum*	510,724.45	25.32
3	*Heavy Metal Sludges*	226,747.90	11.24
4	*Spent Lubricating oil*	94,582.12	4.69
5	*Spent Acids*	86,664.98	4.30
6	*Used Container*	47,509.28	2.36
7	*Mixture of Scheduled Waste & Non-Scheduled Waste*	31,962.60	1.58
8	*E-Waste*	28,604.15	1.42
9	*Pathogenics Clinical Waste*	28,375.24	1.41
10	*Rubber / Latex Waste Containing Heavy Metal*	27,243.89	1.35
11	*Spent mineral oil-water emulsion*	26,113.03	1.29
12	*Rags / Plastics / Papers contaminated with Scheduled Waste*	23,262.15	1.15
13	*Waste of Non-Halogenated Solvent*	20,815.18	1.03
14	*Residue from Recovery*	18,809.16	0.93
15	*Mineral Sludges*	16,443.87	0.82
16	*Waste oil / Oily Sludges*	16,421.67	0.81
17	*Lab Waste*	11,546.37	0.57
18	*Waste of Inks & Paints*	8,370.65	0.41
19	*Ink & Paints Sludges*	8,060.58	0.40
20	*Adhesive / Glue Containing Organic Solvent*	6,598.62	0.33
21	*Sludge Containing Flouride*	6,153.01	0.31
22	*Waste Catalyst*	5,921.14	0.29
23	*Spent Alkalis*	5,880.38	0.29
24	*Discarded of Ink / Paint / Pigment / Lacquer Containing Organic Solvent*	4,455.15	0.22
25	*Oil -Water mixture*	4,437.57	0.22
26	*Contaminated Active Carbon*	4,361.88	0.22
27	*Spent Hydraulic oil*	3,934.79	0.20
28	*Waste of Halogenated Solvents*	3,576.95	0.18
29	*Mixture of Scheduled Waste*	3,389.63	0.17
30	*Spent Alkalis with PH > 11.5*	3,139.91	0.16
31	*Waste of acid plumbum batteries*	2,954.29	0.15
32	*Oil / Sludges from Oil Refinery*	2,390.59	0.12
33	*Contaminated Land / Oil*	2,170.63	0.11
34	*Waste of Thermal Fluids*	1,943.32	0.10
35	*Waste of Batteries Containing Cadmium / Hg / Lithium*	1,917.30	0.10
36	*Waste of Resin Containing Organic*	1,864.74	0.09
37	*Spent Organic Acids*	1,743.44	0.09
38	*Sludges from mineral oil storage tank*	1,486.01	0.07
39	*Waste containing Mercury/Compound*	1,478.05	0.07
40	*Photographic Waste*	1,299.52	0.06
41	*Waste of halogenated or unhalogenated non aqueous distillation residues arising from organic solvents recovery process*	1,070.27	0.05

continued on the following page

Table 11. Continued

No.	Waste Categories	Quantity of Waste	
		Metric Tons	Percentage
42	*Oily Residue from Workshop*	1,047.95	0.05
43	*Clinker / Slag / Ashes from Incinerator*	976.34	0.05
44	*Spent Aqueous alkaline Containing Cyanide*	653.46	0.03
45	*Asbestos*	642.72	0.03
46	*Tar Residue from Oil Refinery/Petrochemical Plant*	529.13	0.03
47	*Expired Drug*	458.97	0.02
48	*Discarded Drug*	298.53	0.01
49	*Chemical Waste*	291.76	0.01
50	*Waste Containing Formaldehyde*	207.86	0.01
51	*Contaminated Oil from re-refining / used lubricating Oil*	196.28	0.01
52	*Waste Containing Peroxides*	168.13	0.01
53	*Stabilized Sludges*	168.11	0.01
54	*Residue from Recovery of Acid Pickling Liquor*	99.00	0.00
55	*Oil tankers sludges*	97.86	0.00
56	*Flux Waste*	88.53	0.00
57	*Pesticide*	78.40	0.00
58	*Waste containing arsenic*	45.72	0.00
59	*Spent di-Isocyanates*	13.37	0.00
60	*Spent Oxidizing Agent*	12.23	0.00
61	*Waste of Phenols/Its Compound*	9.67	0.00
62	*Waste Containing Dioxins or Furans*	7.49	0.00
63	Acid Sludges	7.05	0.00
64	*Waste of Organic phosphorus compound*	2.74	0.00
66	*Slag of Copper*	1.24	0.00
67	*Sludges Containing Cyanide*	0.80	0.00
68	*Waste containing PCB or PCT*	0.75	0.00
69	*Zink Residue*	0.35	0.00
70	*Spent salt containing Cyanide*	0.29	0.00
71	*Galvanic sludges*	0.20	0.00
72	*Spent quenching Oil Containing Cyanide*	0.08	0.00
Total		2,017,280.76	100

practices, care, treatment, investigation, teaching or research, or from blood collection for transfusion, that may infect any person upon contact. Table 16 demonstrates the major classification of the waste and its recommended management procedure in Malaysia.

Clinical waste which comes under scheduled waste according to the Environmental Quality (Scheduled Wastes) Regulations, 2005 has been awarded codes by DOE depending on the composition. The codes are as followed:

- SW403: Discarded drugs containing psychotropic substances or containing substances that are toxic, harmful, carcinogenic, mutagenic or tetrogenic.

Table 12. Generated scheduled wastes managed under special management, 2017

Waste Category	Source	Metric Tons	Percentage	Method of Disposal
Fly Ash & Bottom Ash	Coal-Fired Power Plant	363,087.02	71.16	Reuse as raw material for product
	Industry	18,515.50		
Heavy Metal Sludge	Drinking Water Treatment Plant	26,032.33	5.10	Sanitary Landfill
	Industry	63,822.59	12.51	
		30,515.88	5.98	Reuse as raw material for product
Ash of Paper Sludge	Industry	4,824.78	0.95	Sanitary Landfill
Mineral Sludge	Industry	1,954.05	0.38	Reuse as neutralizing agent
Waste Containing Formaldehyde, resin, discarded epoxy powder	Industry	885.78	0.17	Sanitary Landfill
Mixture Of scheduled waste	Industry	501.35	0.10	Reuse as raw material for product
Equipment contaminated with scheduled waste	Industry	23.40	0.00	Reuse as raw material for product
Waste from manufacturing of explosive	Industry	10.39	0.00	Slow Burning method
Discarded Pharmaceutical Product, Discarded Product	Industry	10.00	0.00	Sanitary Landfill
Total		510,206.65	100	

Table 13. Facilities handling scheduled waste, 2017

Type of Industries	Quantity of Waste	
	Metric tons	Percentage
Local Off-site Recovery Facilities	661,557.87	32.79
On-site Treatment	510,724.45	25.32
Special Waste Management	510,206.65	25.29
On Site Storage	152,161.45	7.54
Kualiti Alam Sdn Bhd	133,507.97	6.62
Off-site Clinical Waste Facilities (Incinerator, Microwave, Sanitary Landfill)	28,375.24	1.41
Trinekens (Sarawak) Sdn Bhd	11,393.47	0.56
Foreign Facilities	9,353.67	0.46
Total	**2,017,280.76**	**100**

Table 14. The classification of hospital waste by WHO

No.	Waste Category	Description	Examples
1	General Waste	Domestic or general waste. Potentially no risk to human health	office paper, wrappers, kitchen wastes, general sweeping
2	Pathological Waste	Human tissues or fluids	body parts, blood and other body fluids, fetuses
3	Sharps Waste	Waste with sharp edges or parts	needles, infusion sets, scalpels, knives, blades, broken glass
4	Infectious Waste	Potentially pathogens containing waste	laboratory cultures, waste from isolation wards, tissues (swabs), materials or equipment that have been in contact with infected patients, excreta
5	Chemical Waste	Chemicals or chemical substances containing waste	film developer, laboratory reagents, expired or unrequired disinfectants, solvents
6	Radioactive Waste	Waste containing radioactive substances	unused liquids from radiotherapy or laboratory research, contaminated glassware, packages, or absorbent paper, urine and excreta from patients treated or tested with unsealed radionuclide, sealed sources
7	Pharmaceutical Waste	Pharmaceuticals containing waste	Expired or unrequired pharmaceuticals, items contaminated with pharmaceuticals or pharmaceuticals containers
8	Cyctotoxic Waste	Waste containing material with genotoxic properties	waste containing cytostatic drugs, genotoxic chemicals

Table 15. Categories of clinical waste in Malaysia

Categories	Type of Waste
A	Blood, body fluids, excretions, soiled surgical, dressings, swabs, material from infectious diseases
B	Discarded syringes, needles, broken glass, and other contaminated disposable sharp instruments
C	Waste from lab or post-mortem room
D	Pharmaceutical and cytotoxic waste
E	Used disposal bed pan liners, urine containers, incontinence and stoma pads

- SW404: Pathogenic and clinical wastes and quarantined materials.
- SW421: A mixture of scheduled wastes.
- SW424: A mixture of scheduled and non-scheduled wastes.

The grouping of healthcare establishments in Malaysia is based on the source and hence are labelled as large source, medium source, and small source. Table 17 reveals the sources.

The amount of clinical waste generated is determined by the development of healthcare services. Hence, the establishment of healthcare institutions in Malaysia is increasing at a fast rate due to refocus and upgradation of operating activities by the hospitals. Additionally, introduction of new services and changes in procedures every year also contribute to the establishment. 28,375.24 tons of clinical waste were generated in 2017 in Malaysia. The quantity of clinical waste disposed at incinerators in 2013 increase by 17.5% as compared to 2009 (Figure 9).

Table 16. The classification and recommended management of clinical waste in Malaysia

Waste Category	Waste Type	Examples	Management
1. Blood and body fluid waste	Soiled surgical dressings	Cotton wool, gloves, swabs, plasters, bloodied bandages or bandages removed from wounds, cloths or wiping materials cleaning up body fluids and spills of blood, all types of contaminated waste generated from treatment area	The main objective of management is to prevent infection. This type of waste must be incinerated completely.
	Material from cases involving infectious diseases, except reusable linen	Human biopsy materials, blood, urine, stools	
	Pathological waste	All types of human tissues (uninfected or infected), organs, limbs, body parts, placenta and human fetuses, animal carcasses and tissues from laboratories, All types of related swabs and dressings	
2. Sharps	All objects and materials generated from any kind of healthcare activities that have potential risk of injury and/or infection	Needles, scalpel blades, blades and saw, any other instrument that may cause a cut or puncture	The collection and management of this type of waste must be performed separately from other waste. The collection container must be puncture resistant and leakproof. This category of waste has to be disposed / destroyed completely as to prevent potential risk of injury / infection.
3. Infectious waste		Waste generated from haematology, pathology, microbiology, blood transfusion or histology laboratories) or from post mortem rooms, and clinical waste other than waste included in Category 1 waste.	The main objective of management is to prevent infection. This type of waste must be incinerated completely.
4. Pharmaceutical and Cytotoxic Pharmaceutical Waste	Unopened or opened expired pharmaceuticals or user prepared ready-to-use pharmaceutical or use is not possible for other reasons such as call–back campaign	Class I - pharmaceuticals such as cough syrup, chamomile tea etc. or unhazardous pharmaceuticals	This type of waste may be managed together with municipal waste
		Class II - pharmaceuticals which may pose a potential hazard when used improperly by unauthorized persons	This type of waste must be managed in an appropriate waste disposal facility
	Wastes generated from utilization, manufacturing and preparation of pharmaceuticals with cytotoxic, and/or from oncological treatment of patients with pharmaceuticals with a cytotoxic effect i.e. carcinogenic, mutagenic, and teratogenic properties)	Class III - Heavy metal – containing unidentifiable pharmaceuticals	This type of waste must be managed in an appropriate waste disposal facility. Intermediate storage of this type of waste must be controlled and in locked conditions. Due to occupational safety, cytotoxic pharmaceutical wastes must be collected separately from pharmaceutical waste and incinerated.
6. Other infectious waste	All healthcare waste known or clinically assessed by a medical practitioner or veterinary/ surgeon to be potentially infectious to humans or animals.	Used disposable bed-pan inners, incontinence pads, urine containers, and stoma bags.	This type of waste must be incinerated in a scheduled waste incinerator licensed by the Department of Environment.

Table 17. Source wise classification of healthcare establishments

Source	Example
Large Source	University hospitals and clinics, Maternity hospitals and clinics, General hospitals
Medium Source	Medical centres, out-patient clinics, mortuary/autopsy facilities, farm and equine centers, hospices, medical laboratories, medical research facilities, animal hospitals, blood banks and transfusion centres, emergency services
Small Source	General medical practitioners, convalescent homes, nursing and remedial homes, medical consulting rooms, dental practitioners, animal boarding and hunt kennels, acupuncturist, veterinary practitioners, pharmacies, cosmetic piercers

Figure 9. Quantity of clinical waste handled for destruction at incinerator, Malaysia

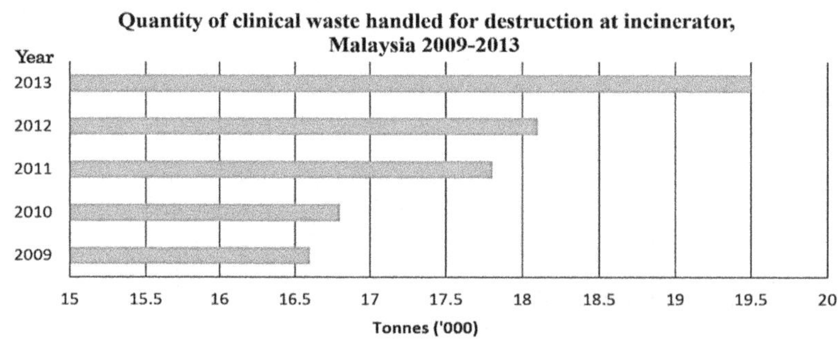

Serious apprehensions concerning dispersal of pathogens along with propagation of environmental contamination from improper handling of biomedical and clinical waste were brought up more than ten years ago. While developed countries have implemented guidelines and regulations for the management of clinical and biomedical waste, Malaysia had only prepared preliminary guidelines for the management of hospital waste in 1998. Currently, Ministry of Health and concession companies are responsible for clinical waste management and for providing all the equipment, as well as, appropriate containers to healthcare facilities for clinical waste, respectively. The ministry of Health and three concession companies had signed concessionaire agreements for clinical waste management in health care facilities in October 1996. These concession companies are:

1. Faber Medi-Serve Sdn Bhd for states of Perlis, Kedah, Penang, Perak, Sabah and Sarawak.
2. Radicare (M) Sdn Bhd for Wilayah Persekutuan Kuala Lumpur and Putrajaya, Selangor, Pahang, Kelantan and Terengganu.
3. Pantai Medivest Sdn Bhd for Negeri Sembilan, Melaka and Johor.

Environmental Quality (Scheduled Waste) Regulations 1989 administers the handling of clinical waste in Malaysia from labelling, identification, on-site storage and management, transportation, treatment and disposal. The concessionaires bear most of the responsibilities for every step in the life cycle of medical waste, which includes supply of the consumables, collection of waste at the generators' sites, storage, transportation to treatment sites, and disposal in treatment plants. However, healthcare providers perform the initial stage of waste management, which is waste segregation.

Agreement between concessionaires and government signed in 1992 stated that the service charge for handling and managing these clinical wastes is RM 5.20 per kilograms. These charges are including services from collection of waste in hospitals until the waste is incinerated.

BIOMASS

Agricultural sector contributes 11% to the Gross National Income (GNI) of Malaysia, highlighting tremendous amount of generation of biomass. The major sources of biomass generation are plantation and agricultural residue i.e. palm oil, rubber and rice (Shafie et al. 2012). The major agriculture biomass in Malaysia and their quantities are listed in Table 18.

The biggest contribution to GNI comes from Palm oil sector, hence indicating potentially the major biomass generator. Total annual estimated biomass generation in Malaysia is shown in Figure 10. The schematic flow demonstrating a general concept of major agriculture biomass waste flow in Malaysia is depicted in Figure 11.

CURRENT MALAYSIAN SCENARIO OF BIOMASS

The oil palm is the largest crop cultivation that generates massive amounts of biomass. Therefore, biomass generated from oil palm has been recognised as potentially productive for reutilization and have

Table 18. The generation and source of biomass in Malaysia

Biomass Source	Quantity (Tons)	Year	Reference
Oil palm waste (empty fruit bunch (EFB), fibre, shell, palm oil mill effluent (POME), trunk and fronds)	129,498,829	2013	MPOB
Rice (husk, straw)	1,669,000	2014	Abdul, 2014
Coconut trunk fibers	273,000	2012	UNCRD, 2017
Sugar cane waste (bagasse, molasses)	14,500	2013	
Livestock	21,739,603	2013	

Figure 10. Total projected annual biomass availability in Malaysia

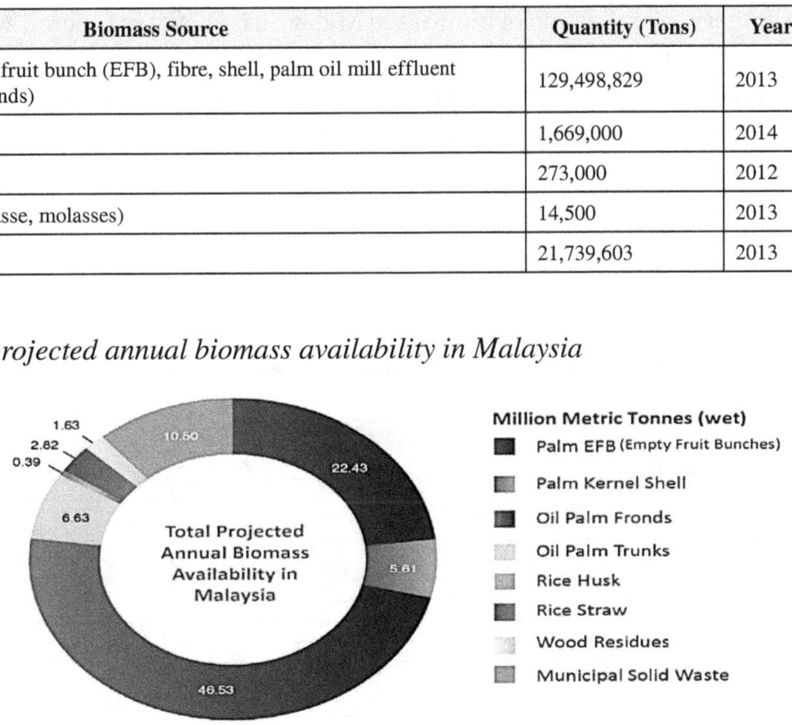

Figure 11. Agriculture biomass flow in Malaysia

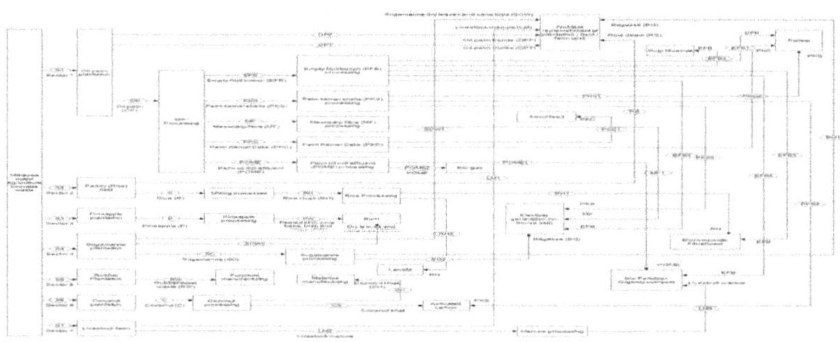

Figure 12. Current scenario of biomass utilization in Malaysia

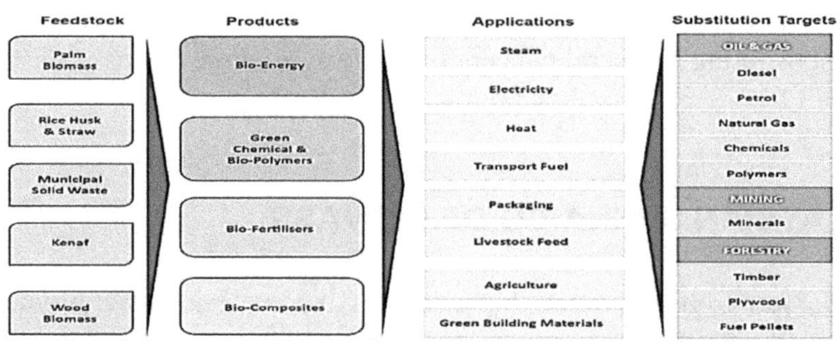

received higher attention in Malaysia. Moreover, lack of landfill space, ban on agricultural open burning and large number of palm oil mills also influence usage of biomass as resource or energy (Shafie et al., 2012). The present scenario of utilization of biomass in Malaysia is shown in Figure 12. The end products yielded from biomass exploitation are bio-energy, green chemical and bio-polymers, bio-fertilizer and bio-composites products. Whereas, the developing stages of each types of biomass in different sectors is portrayed in Figure 13. Lastly, the financial gains from each category of commercialized biomass are presented in Figure 14.

Figure 13. Developing stages of each types of biomass in different sector (Tang, 2014)

	PELLETS	BIOFUELS	BIOGAS	GREEN CHEMICAL	BIOFERTILISERS	BIOCHAR	BIOCOMPOSITES	OTHERS
EFB	EFB Pellets	Bioalcohol	Syngas	Industrial Sugars/ Chemical	Organic Compost	Carbon Fibers	Fibreboard	Pulp Fibremat
PKS	Coal substitute					Activated Carbon		
OPT	OPT Pellets	Bioalcohols	Syngas	Industrial Sugars/ Chemical	Organic Compost	Biochar	Engineered lumber	
OPF	OPF Pellets	Bioalcohols	Syngas	Industrial Sugars/ Chemical	Organic Compost	Biochar		Phytochemicals
PKC	PKC Pellets			Biopolymers				Animal Feed
POME		Bioalcohols	Methane	Biopolymers	Organic Compost			
SAW DUST	Wood Pellets	Bioalcohols	Syngas	Industrial Sugars/ Chemical	Mushroom Cultivation	Biochar	Fiberboard	
RICE HUSK							Biocomposite	Silica Aerogel
PADDY STRAW	Straw Pellets							
KENAF							Biocomposite	
SAGO WASTE	Fuel Pellets	Bioalcohols	Methane	Biopolymers				Animal Feed
MSW	RDF Pellets		Methane		Organic Compost			
SEWAGE								Phosphate Recovery

LEGEND ■ Commercialised ■ Development Stage ■ Potential

Figure 14. Monetary value generated from biomass utilization (Tang, 2014)

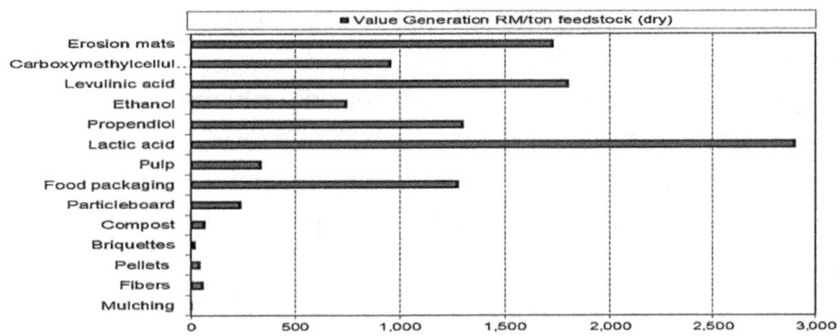

Recent Policy Trends

The main focus of the government has been on renewable energy as depicted by the recent government strategies, policies and actions concerning utilization of biomass (Tang, 2014). These are as followed:

1. Renewable energy policy & act
   - Feed-in-tariff for generation of renewable energy.
2. National biotechnology policy
   - Bio-conversion of biomass into high value chemicals and liquid fuels.
3. Green technology policy
   - Availability of investment tax incentives for green projects in Malaysia.
   - Financing scheme of USD 1 billion as subsidized loans for green technology.
4. Palm oil industry biogas power generation
   - Utilization of biogas from effluent waste to produce energy for national electricity grid.
5. Biomass industry strategic action plan
   - Encourage small and medium enterprises to utilize high value products of biomass.
6. National biomass strategy
   - Development of market players and technologies for biomass pallets and biochemical.
7. National Biomass Strategy 2020: New wealth creation for Malaysia's palm oil industry
   - Make an assessment of revenue generation from utilization of biomass from palm oil industry.

Challenges

A total of six major challenges of utilizing agricultural biomass, highlighted by Tang (2014) are summarized below:

1. Availability of biomass feedstock - volume & pricing
   - Existing agricultural practice – returning biomass to the field.
   - Absence of commodity market for trading of biomass.
   - High cost of transportation due to remote locations of biomass sources.
   - Unfavorability of biomass owners towards long-term supply contract due to rising demand and pricing of biomass.

2. Availability of finances for biomass endeavours
    - Risks in supply of biomass feedstock supply and pricing.
    - Unfamiliarity of local financiers with biomass industry.
    - Relative high investment and long payback period for biomass endeavours.
3. Absence of support from domestic market
    - Non-concurring approach to sustainable production and consumption locally.
    - Allocation of high percentage of biomass products for export in order to supply the premium 'green' products.
    - Lack of commitment from government and businesses lead to poor execution of sustainable market initiatives.
4. Commercialisation of local technologies
    - Several factors such as research and development programs without consideration of market applicability and lack of multi-disciplinary approach to commercialisation lead to disparity between research and commercialisation.
    - Reluctance of local market and financiers to deploy local technologies and products.
5. Sustainability requirements and certifications
    - Availability of required sources for current and future usage for biomass such as pugnacious issues on exploitation of primary forests and peatland.
    - Absence of local standards and certifications concerning biomass feedstock and products.
    - Inadequate awareness and sluggish adoption of carbon foot print to encourage business decisions on sustainability.
6. Low-value utilization of biomass by local corporations
    - Current biomass endeavours are mainly focused on manufacturing of commodity-based products such as compost, fibre, and fuel pellets.
    - Inadequate access to technologies for creating higher value products from biomass feedstock including biochemical, eco-products, and polymers.
    - Absence of technological capacities in local companies, especially in small and medium sized enterprises, to incorporate and execute high technologies in their schemes.

MARINE PLASTIC WASTE

A study on plastic debris reported that out of 37 researches, 33 had stated percentage of plastic to be more than 50% (Derraik, 2002). Moreover, the percentage of plastic waste recorded from coastal areas has been increasing over the years (Figure 15). Whereas, the quantity of non-plastic waste had reduced in the same period. The turnover of plastic industry in Malaysia was RM27.32 billion in 2016, recording an increment of 10.3% from last year, where turnover was RM24.77 billion 2015 (Malaysia Petrochemical Country Report, 2017). The plastic production in Malaysia has been growing steadfastly, achieving production rate of 8.2%, partly due to increase in in demand from Japan and Europe (Department of Statistics, 2014). Ironically, the increase in plastic production, indirectly, leads to increase in the amount of plastic waste accumulation in marine and coastal areas. An estimate of 60 to 80% of marine litter originates from land (ICC, 2010). At land, lakes, streams, rivers, storm drains as well as wind, participate in transportation of litter to oceans, resulting in generation of marine litter. Subsequently, ocean currents and winds haul the marine litter all around the world. The significance of marine litter data is invaluable as it could assist in

Figure 15. Percentage of marine and coastal plastic waste in Malaysia from 2009 to 2013 (ICC, 2010; ICC, 2012; ICC, 2014)

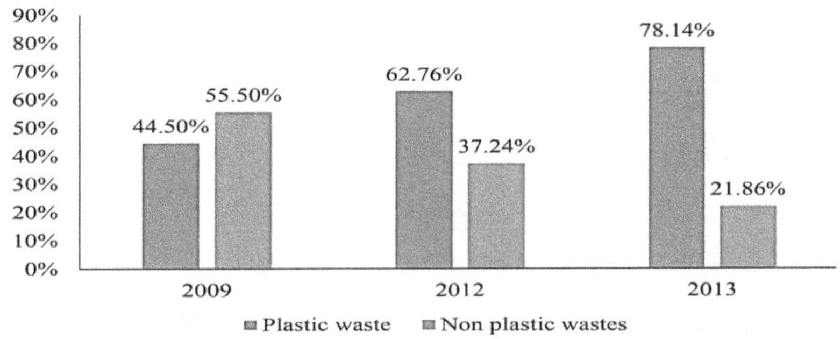

Table 19. Summary of marine and coastal waste

Indicators	Data	UNIT	Year	Reference
Marine/Coastal Waste Generation	1.52	kg/person/day	2010	Jambeck et al., 2015
Collected marine/coastal waste	0.14 to 0.37	million metric tons /year	2010	

formulating a framework to eliminate marine litter by reducing the plastic waste at source, and altering the behaviours that cause it. The generation of marine plastic litter is given in Table 19.

ICC (2010) reported plastic to be comprising 44.50% of the total collection marine litter in Malaysia in 2009. Unfortunately, an increase of 62.76% of plastic waste was recorded in marine litter in 2012 (ICC, 2012). The composition of marine plastic litter sampled was food wrappers, plastic bags, plastic tarps, bait packaging, beverage bottles, cleaner bottles, oil bottles, straws, tobacco packaging, toys and caps or lids. Figure 16 depicts the composition of marine plastic litter found in Malaysia in 2009 where three major constituents were plastic bags (26.65%), food wrappers (24.18%) and beverage bottles (18.02%). As established before, marine litter comes mainly from land-based activities i.e. discarding fast food wrappers after eating, littering during picnics, beach trips, sports and festivals. In comparison to 2009, three major constituents of marine plastic litter in 2012 were beverage bottles, plastic bags and food wrappers with percentages of 42.23%, 33.74% and 7.77% respectively. The composition of marine plastic litter in 2012 in Malaysia is shown in Figure 17. An estimate of total retail sales of food and beverages

Figure 16. Percentage of marine and coastal plastic waste in Malaysia in 2009 (ICC, 2010)

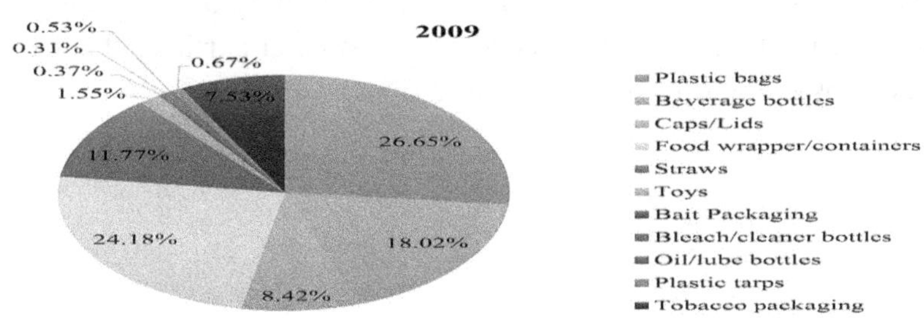

Figure 17. Percentage of marine and coastal plastic waste in Malaysia in 2012 (ICC, 2012)

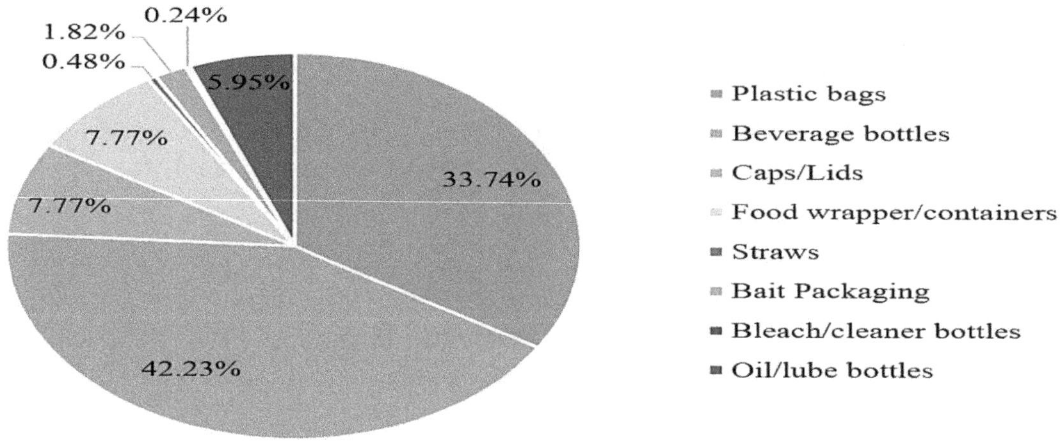

was stated to be USD 11 billion (Market Watch, 2012) which is expected to grow 10% every year for next three to five years. It may be the probable explanation of increase in amount of beverage bottles recorded in marine and coastal areas. Additionally, increase in import of beverages to Malaysia has also possibly resulted in more beverage bottles to be littered in Malaysia (Market Watch, 2012). The increase is import has been from USD 295,955,939 to USD 373,696,894 for the year 2009 and 2010, respectively.

In 2010, approximately 8 million tons of plastic waste were found in the world oceans, that is equivalent to five grocery bags filled with plastic for every foot of the coastline in the world (Fauziah et al., 2015; Jambeck et al., 2015). Malaysia had been listed among the twenty countries that dumped the most plastics in the ocean. Malaysia was ranked at eight in terms of the amount of plastic waste discarded into the oceans (Jambeck et al., 2015). About 22.9 million people live on the coast in Malaysia with a per capita MSW generation of 1.52 kg every day and approximately 13% of it is composed of plastic. Hence, the total quantity of marine plastic litter generated in Malaysia is 0.14 to 0.37 million tons annually, that is released into the Malaysian seas.

Since marine litter comes from land, one mitigation measure proposed is cleaning of trash from beaches and waterways within the cities. Thus, the yearly cost of cleaning beach and waterways is listed in Table 20 based on Stickel et al. (2012). Based on these calculations, the clean-up in coastal areas of Malaysia, populated with 22.9 million inhabitants, would cost from USD 0 to USD 1,837,398. Where the average per capita cost of beach and waterway clean-ups in Malaysia would be USD 0.83.

Marine plastic litter floats on the surface of oceans, or gets washed up on the beaches and amasses on sea floors. There have been cases of entanglement of marine fauna such as dolphins, turtles, as well as whales by the floating large items of marine plastic litter including bags, wrapping and fishing gears. With passage of time, marine plastic litter breaks down into smaller particles that can be unintentionally or intentionally ingested by the marine animals (Ian, 2015).

Table 20. Annual cost for beach and waterway clean-ups

City Size	Population Range	Range of Annual Costs Reported	Average Annual Cost	Average per Capita Cost
Largest	Over 250,000	$0 – $1,837,398	$422,185	$0.83
Larger	75,000 to 249,999	$0 – $17,500	$3,329	$0.03
Mid-sized	15,000 to 74,999	$0 – $112,549	$12,746	$0.28
Smaller	Under 15,000	$0 – $114,005	$6,418	$1.28

ELECTRONIC WASTE

Due to the increasing usage of electrical and electronic gadgets, electrical and electronic waste (E-Waste) generation has been increasing rapidly. Consequently, in 2006 DOE incorporated E-waste in the Environmental Quality (Scheduled Waste) Regulations 2005 as a new category of waste (Environmental Quality Report, n.d.). The increment in E-waste generation over the years had been the main reason of development of waste and environmental management policies (Agamuthu & Victor, 2011). Table 21 summarizes the framework related to E-waste.

The responsibility of planning and enforcement of regulatory requirements concerning E-waste is on DOE along with Ministry of Natural Resources and the Environment. While no direct regulations exist for handling E-waste in Malaysia, the management of E-waste is administered under the Environmental Quality (Scheduled Waste) Regulations, 2005, as well as Environmental Quality (Prescribed Premises) (Treatment, Disposal Facilities for Scheduled Waste) Regulations, 1989, for collection, treatment, recycling and disposal.

The generation of E-waste from 2006 to 2017 is shown in Figure 18. Malaysia generated approximately 40,275 tons of E-waste in 2006 (Environmental Quality Report, n.d.), whereas in 2017 E-waste generation has been 31,999.5. However, it is projected that 1.2 million tons of E-waste would be generated in 2020 (Agamuthu & Victor, 2011). Thus, it is suspected that this drastic increase in E-waste generated may result in major issues in waste management.

There are three types of E-waste that are categorized by Environmental Quality (Scheduled Wastes) Regulations 2005. These are as followed:

- SW103: Waste of batteries containing cadmium and nickel or mercury or lithium.

Table 21. Summary of electronic waste regulations

Available Framework	Title of Regulations/Guideline
Policies for e-waste management	Environmental Quality (Scheduled Wastes) Regulations 2005 has categorized e-waste into i. SW103: Waste of batteries containing cadmium and nickel or mercury or lithium ii. SW109: Waste containing mercury and its compound iii. SW110: Waste from electrical and electronic assemblies containing components such as accumulators, mercury-switches, glass from cathode-ray tubes and other activated glass or polychlorinated biphenyl-capacitors, or contaminated with cadmium, mercury, lead, nickel, chromium, copper, lithium, silver, manganese or polychlorinated biphenyls.
Guideline for e-waste management	Following the regulations stipulated under Environmental Quality (Scheduled Wastes) Regulations 2005 particularly of SW110

Figure 18. Electronic waste generation in Malaysia from 2006 to 2017

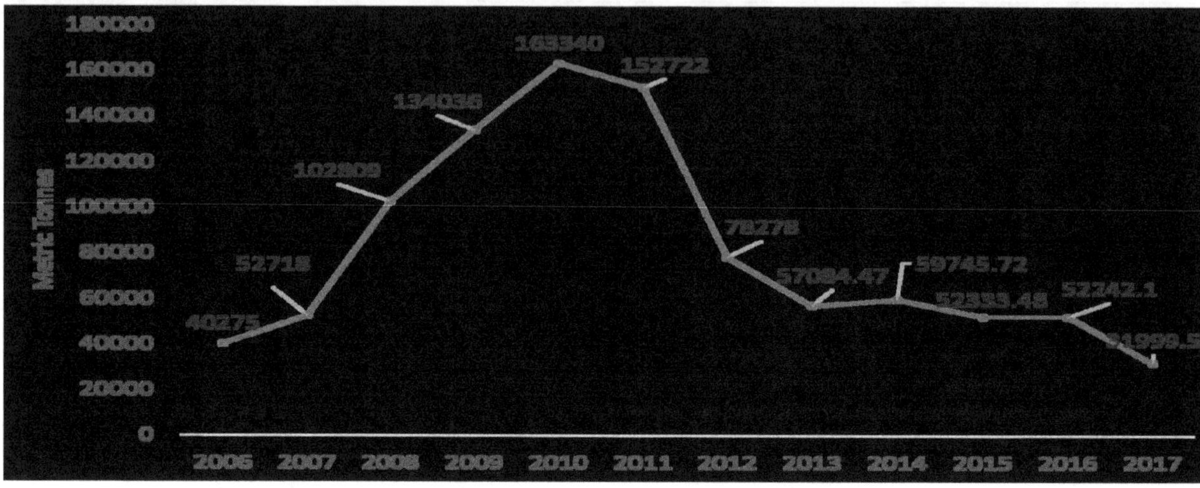

- SW109: Waste containing mercury and its compound.
- SW110: Waste from electrical and electronic assemblies containing components such as accumulators, mercury-switches, glass from cathode-ray tubes and other activated glass or polychlorinated biphenyl-capacitors, or contaminated with cadmium, mercury, lead, nickel, chromium, copper, lithium, silver, manganese or polychlorinated biphenyls.

The classification of used electrical and electronic equipment in Malaysia provide guidance for determining the used electrical and electronic equipment as e-waste or second-hand goods. Such a criteria is important for determining the import of second-hand electrical and electronic equipment for direct use.

As mentioned earlier, electrical and electronic assemblies are assigned a code SW 110, where such a waste contains mercury-switches, accumulators, glass from cathode-ray tubes and other activated glass or polychlorinated biphenyl-capacitors, or contaminated with cadmium, mercury, lead, nickel, chromium, copper, lithium, silver, manganese or polychlorinated biphenyls.

Different codes are also assigned to E-waste as A1180 and A2010 according to Basel Convention on the Control of Transboundary Movements of Hazardous wastes and their Disposal and can be referred in Annex VIII, List A. Since Malaysia has ratified to this convention, the export and import of E-waste, as well as hazardous waste must follow the protocol of the convention. Section 34B (1)(b)&(c) of Environmental Quality Act 1974, mandates written approval of DOE for import and export of E-waste and/or other hazardous waste. Any person who disobeys this section would be held guilty for the offence and be convicted for punishment of imprisonment for a term not exceeding five years and would be liable to a fine not exceeding RM5,000 (approximately USD 1,200).

These guidelines are intended to assist all concerning parties to identify and classify a used electrical and electronic equipment or component as E-waste. If it is classified as E-waste, then it should be handled according to the first schedule of the Environmental Quality (Scheduled Waste) Regulations 2005. The concerning parties are:

a. Waste generators.

b. Waste transporters.
c. Importers or exporters of waste.
d. Relevant authorities involved in the management of e-wastes.

The management of E-waste generated in Malaysia was reported by Japan International Cooperation Agency (JICA) in 2005, which is shown in Figure 19. The junkshops, recycling centres and scrap collectors perform an invaluable task of collecting E-waste generated from multiple sources and providing it to E-waste recyclers, thereby closely associating waste generators and recyclers. Therefore, E-waste generators must not segregate E-waste with other streams of MSW, and instead sell or give it to related collectors or middlemen for recycling who will then pass E-waste to E-waste recovery facilities. In addition to recovery of recyclable materials i.e. plastics and common metals, several precious metals can also be extracted from the circuit boards of E-waste. The precious metals are mainly gold, platinum, silver, and lead.

Since early 1980s, Malaysian policy and regulations have incorporated extended producer responsibility (EPR). The regulation, Environmental Quality (Recycling and Disposal of End-of-Life Electrical Electronic Equipment) Regulations, introduced by DOE encourages electronic manufacturers to design electrical and electronic equipment such that they have least number of hazardous components so that at the end of life, treatment and recycling of E-waste is efficient (Agamuthu & Victor, 2011).

The essentials of EPR were incorporated in the earlier Malaysian policies such as Action Plan for a Beautiful and Clean Malaysia (ABC) in 1988 and Environmental Quality Act (EQA) 1974 in 1996. However, succeeding policies failed to assimilate it, thereby missing the opportunity of establishing sustainable waste management system in Malaysia. These policies were National Strategic Plan for Solid Waste Management (NSP) in 2005, the Master Plan on National Waste Minimization (MWM) in 2006 and the National Solid Waste Management Policy (NSWMP) in 2006. Nonetheless, following legislation and policy such as Solid Waste and Public Cleansing Management Act (SWMA) in 2007 and Tenth Malaysian Plan in 2010, clearly assimilated the EPR. EQA and SWMA, both have detailed requirements for execution of EPR such as tack-back systems, deposit refund system and minimum

Figure 19. Material flow of E-waste in Malaysia (Theng, 2008)

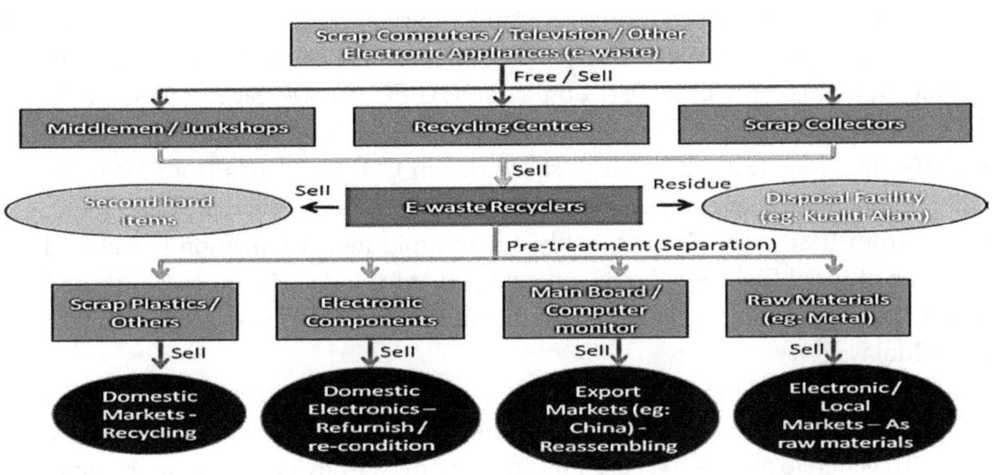

recycled content. Whereas, NPE, NSP and MWM only mentioned EPR as a general concept that could be executed on a voluntary basis, even though EPR is key component in the policy documents for waste management in Malaysia.

Ironically, while two core legislations of solid waste and scheduled waste management award power to Department of Solid Waste Management (DWSM) in 2007 and DOE for the implementation of EPR in Malaysia as early as in 1990, but both departments have not ratified related legislations to implement EPR in Malaysia. It indicates that the policymakers in Malaysia are still focused on the basic protocols of post-use collection, recovery and disposal of both solid waste and scheduled wastes. As stated by DOE, illegal import of E-waste in Malaysia still occurs despite the strict legislation of transboundary movement of E-waste.

A total of 28 containers containing E-waste have been intercepted from 2008 to 2011 only and were returned back to exporting countries (Environmental Quality Report, n.d.). Whereas in 2009, a local company manager in Malaysia was sentenced to one-day jail sentence and was fined RM 180,000.00 (US60, 000) for illegally importing e-waste (Environmental Quality Report, n.d.).

The challenges of E-waste management in Malaysia are given below (Ibrahim, 2013):

- Capacity building to manage household e-waste in an environmentally sound manner.
- Collection, segregation and transportation of household e-waste.
- Disposal/ collection fee for household e-waste.
- Legislation and policy.
- Transboundary movement of e-waste.
- Managing the informal sectors.

EXTENDED PRODUCER RESPONSIBILITY IN MALAYSIA

A summary of regulations and policies on EPR are given in Table 22.

GREENHOUSE GAS EMISSIONS FROM WASTE SECTOR

Three major greenhouse gases (GHGs) are covered in Malaysian GHG inventory including Carbon dioxide (CO_2), Methane (CH_4) and Nitrous oxide (N_2O). Whereas, the other GHG are Sulphur hexafluoride (SF_6), Hydrofluorocarbons (HFC), and Perfluorocarbons (PFC). A 40-year inventory of GHG emissions in Malaysia is shown in Figure 20. Growth in urban population is the credited to be the main cause of increase in GHG emissions (Figure 21). The sharp rise in CO_2 emissions from 1990 onwards are associated with high increase in urban population in the same period. Whereas, rapid increase in overall GHG emissions from 1980 onwards is probably due to rapid industrialization in Malaysia. Moreover, due to combustion of fossil fuels in order to generate power for industrial processes is credited to be the main cause of increment in emissions of carbon dioxide. Table 23 enlists the emissions of GHG from waste sector in Malaysia.

The emissions of carbon dioxide from multiple sectors in Malaysia is depicted in Figure 22. Heat production and electricity have been contributing mainly to the emissions of carbon dioxide for the 20-year period. To generate electricity, coal-powered and diesel-powered generators are predominantly

Table 22. Summary of policies, guidelines, and regulations based on the principle of EPR

Status of Implementation	Name of the Policies (Year)	Product Items Covered by the Policy
Fully Implemented	Environmental Quality Act (1974)	□ Section 30A: gives authority to DOE to advise for any substances to be reduced, recycled, recovered or regulated and to propose producers on the minimum recycled content in their products as well as matters relating to environmental labelling. □ Section 30B: grants power to DOE for specifying rules on deposit and rebate schemes especially for the disposal of products that are considered environmentally unfriendly or disposal of such products may harm the environment.
	Solid Waste and Public Cleansing Management Act (2007)	□ Section 101: grants authority to the Department of Solid Waste Management (DSWM) to impose the usage of environmentally friendly material, to stipulate the amount of recycled materials for specified products and limit the use of specified products or materials. □ Section 102: gives power to DWSM to necessitate taking back of the products or goods for recycling or disposal by the manufacturers and producers at their own cost and establish a deposit refund system including the deposit refund amount.
Postponement period before full implementation	-	-
Under preparation of specific legislations	-	-
Existence of provisions supporting EPR principle	Action Plan for A Beautiful and Clean Malaysia (ABC) (1998)	□ To reduce the generation of municipal solid waste, especially packaging waste and household chemical waste by bringing together manufacturers, distributors and the consumers.
	Master Plan on National Waste Minimization (2006)	□ Proposes introduction of regulations on green purchasing and EPR as regulatory options are to be explored □ Recommends the examination of voluntary take-back of containers and packages based on EPR for manufacturers and retailers, but fails to deliver a detailed initiative on EPR
	The Tenth Malaysian Plan (10MP) (2011)	□ Explicitly mentions provisions of EPR as part of the Malaysian government's strategy for sustainable waste management and includes the initiative to obtain producer's commitment to implement a take-back system for producers along with a deposit refund system
Based on voluntary approach/ agreement	National Strategic Plan for Solid Waste Management in Malaysia (2002)	□ Suggests regulations to place responsibility on the manufacturer of packaged goods for packaging waste but fails to devise a detail strategy on EPR.

deployed in Malaysia, thus leading to increment in generation of carbon dioxide as by-product. The steadfast growth of fossil fuel-based vehicles in Malaysia have also contributed to GHG emissions. Figure 23 shows the increase in vehicles in Malaysia This increase in number of vehicles unavoidable due to fast paced urbanization in Malaysia, for vehicles are the main source of intracity transportation. Moreover, unreliability and inadequate coverage of public transport also plays a role in the increment of vehicles in Malaysia.

When it comes to disposal of waste, dumpsites are the most common in Malaysia. There are about 212 dumpsites and disposal sites are the major contributors of methane emissions in waste sector, except sanitary landfills with gas capture. Methane emissions from landfills and dumpsites in Malaysia is demonstrated in Figure 24. Higher emission of methane for southern and eastern regions is probably due to absence of sanitary landfills for final waste disposal. Open dumpsites do not have proper linings and

Figure 20. Greenhouse emissions in Malaysia (World Bank, 2014)

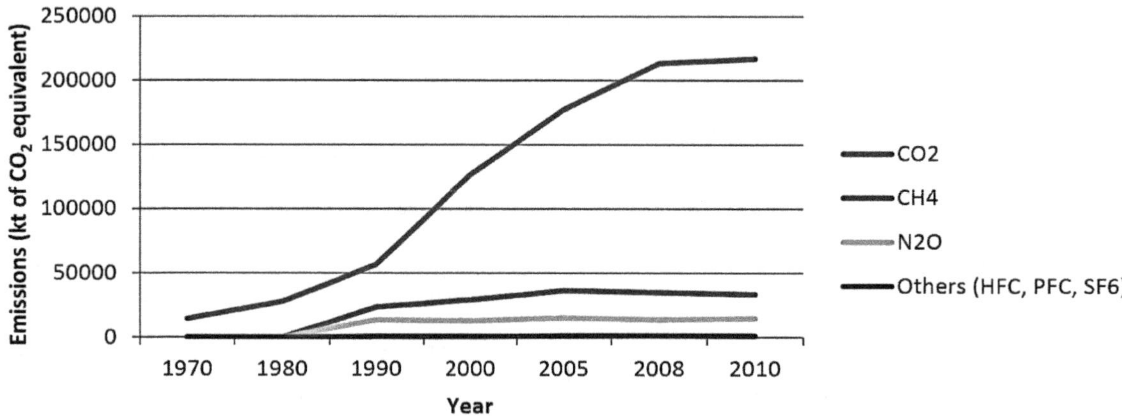

Figure 21. Urban population in Malaysia (World Bank, 2014)

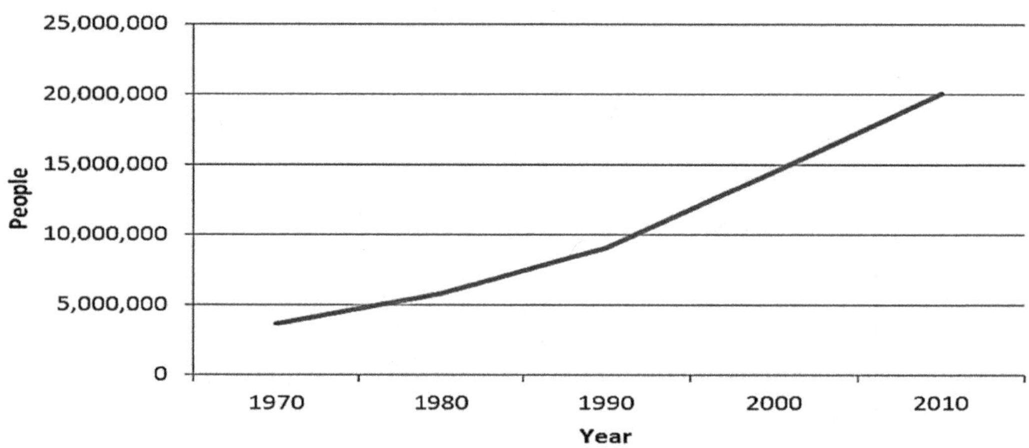

Figure 22. CO_2 gas emissions from various sectors (World Bank, 2014)

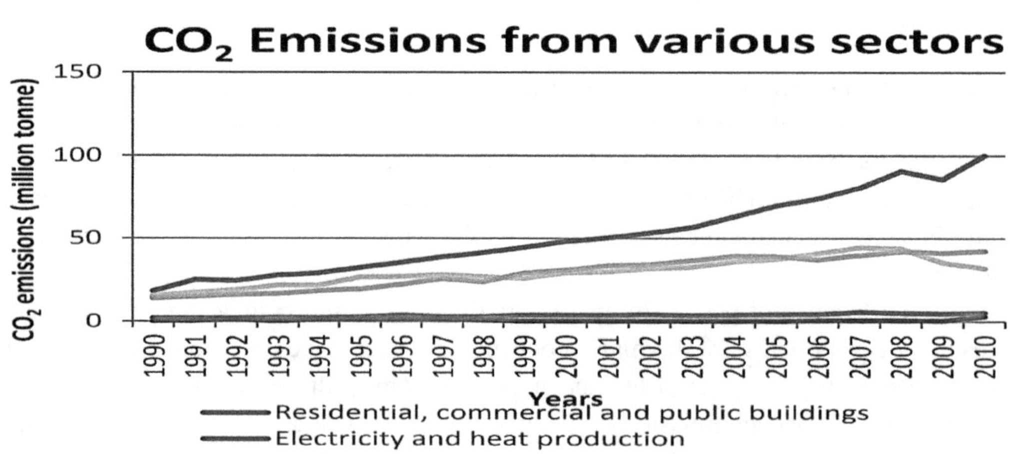

Table 23. Summary of the greenhouse gas emissions from waste sector

Indicators	Data	UNIT	Year	Reference
Total GHG Emissions from Malaysia	2.17×105	Gg of CO2-eq	2010	World Bank, 2014
Methane Emissions from landfill	318.0	Gg of CO2-eq	2011	Ministry of Housing and Local Government, 2009

Figure 23. Number of motor vehicles from 2000 to 2011 (World Bank, 2014)

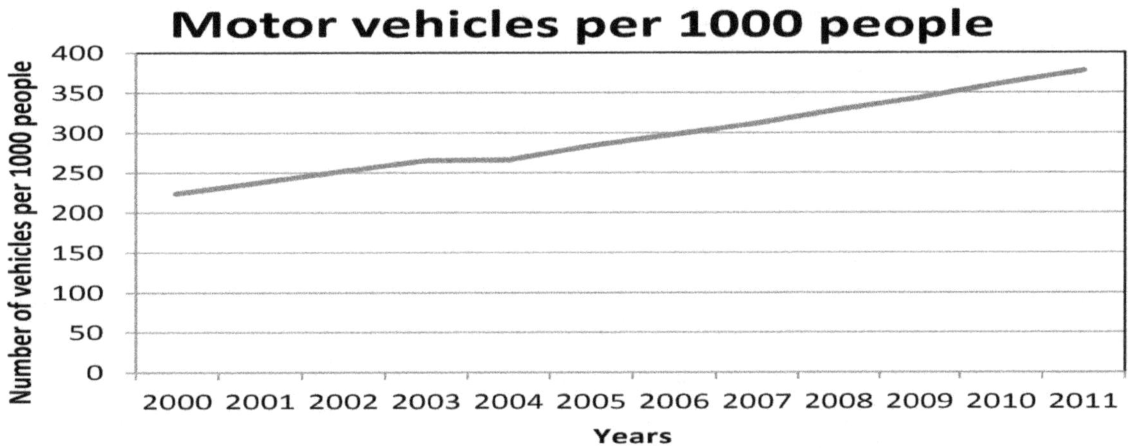

gas capture system to avoid methane emission into the atmosphere. While, sanitary landfills have proper linings and gas capture system, the containment and collection of landfill gas is inefficient in Malaysia. On another hand, greater number of sanitary landfills in northern and central regions (four in each region) result in collective higher amount of methane emission. As categorial representation of methane emission are shown in Figure 24, the five regions and their respective states are enlisted in Table 24.

According to 2011 emissions (as shown in Figure 25), developed states including Selangor, Penang and Johor, are the major contributors of methane emissions. The cause of high emissions from these states is higher rate of urbanization thereby leading to increased rate of waste generation. As reported by Ministry of Housing and Local Government, Labuan state in Malaysia does not have data on methane emission as no landfill is established in this state.

EXPERT SUMMARY / RECOMMENDATIONS

The management of municipal solid waste in Malaysia is at crossroads due to several changes implemented or planned for improving overall waste management services available to the public. While, collection of waste in Malaysia is at the same level of developed countries and approximately entire generated MSW from cities is collected for disposal, illegal dumping still occurs intermittently, accounting for 10% of the total MSW generated.

Figure 24. Methane emissions from landfill and dumpsites (UNCRD, 2017)

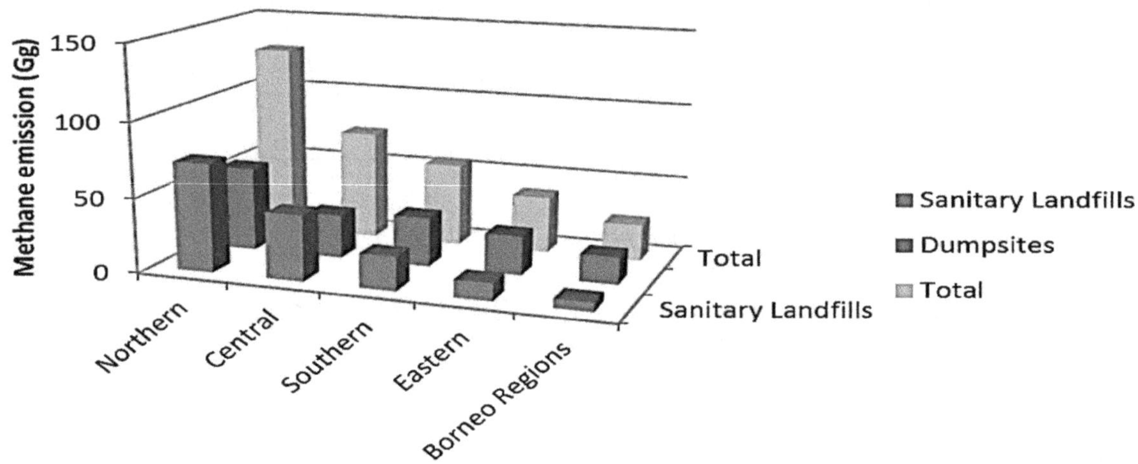

Table 24. State categorization according to region

Regions				
Northern	**Central**	**Southern**	**Eastern**	**Borneo**
Kedah	Selangor	Malacca	Pahang	Sabah
Penang	Negeri Sembilan	Johor	Terengganu	Sarawak
Perak	Kuala Lumpur	Kelantan	Labuan	
Perlis				

Figure 25. Methane emissions in Malaysia for year 2011

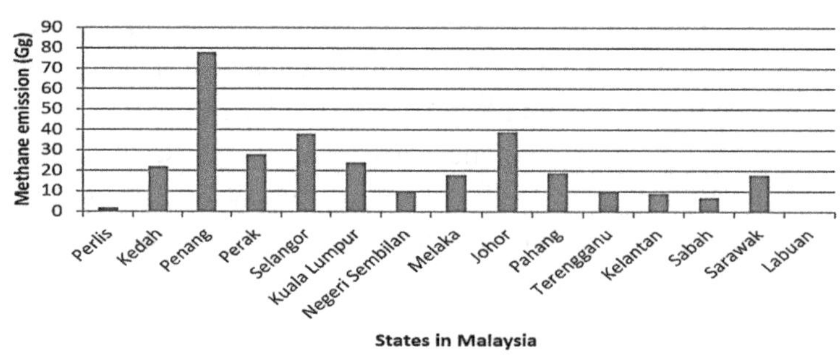

Several recommendations are proposed for further improving the quality of waste management in Malaysia with a special focus to 3R activities.

a. An urgent need for implementation of an Integrated Waste Management policy especially 3R inclusive, at an early stage.

b. Execution of waste segregation must be enforced throughout the country uniformly. As it is only imposed on some states, as well as is not for all types of households. Therefore, 3R goals have been achieved only partially.
c. A clear transparent strategy on incineration or waste to energy should be established along with relevant public education regarding the concerned matter to ensure public support.
d. The utilization of biomass should be increased since there is great potential in biomass utilization for bioenergy, bio-chemicals, and others.
e. A review of policy on plastic bag usage ought to be carried out. A complete ban on plastic bags should be considered and gradual replacement of plastic bags with biodegradable starch-based plastics ought to be achieved.
f. Formalization and subsequent coordination of role of informal recyclers should be carried out. Their contributions ought to be recognized and recycling data should be included in the national reports.
g. Above mentioned recommendations will also contribute to achieving goals related to reduction in GHG emissions. Thereby, climate change or global warming could be reduced significantly.
h. While the management of scheduled waste is on the same level with developed countries, it should be further improved with mainly focusing on 3R within Malaysia. It would also reduce transboundary movement of hazardous waste. Explicit policies on E-waste can result in execution of 3R for E-waste and reduce the wastage of resources.

REFERENCES

Abdul, G. W. (2014). *USDA Global Agricultural Information Network (GAIN) Report: Malaysia Grain and Feed Annual 2014. Foreign Agricultural Service, MY400*. Retrieved from http://www.ers.usda.gov/media/978994/rcs-121-01.pdf

Abor, P. A. (2007). *Medical Waste Management at Tygerberg Hospital in the Western Cape, South Africa* (Unpublished doctoral dissertation). Cape Peninsular University of Technology.

Abor, P. A., & Anton, B. (2008). Medical Waste Management Practices in a Southern African Hospital. *International Journal of Health Care Quality Assurance*, *4*(4), 356–364. doi:10.1108/09526860810880153 PMID:18785461

Abushammala, M. F. M., Basri, N. E. A., Basri, H., El-Shafie, A. H., & Kadhum, A. A. H. (2011). Regional landfills methane emission inventory in Malaysia. *Waste Management & Research: The Journal of the International Solid Wastes and Public Cleansing Association, ISWA*, *29*(8), 863–873. doi:10.1177/0734242X10382064 PMID:20858637

Agamuthu, P. (2001). *Solid Waste: Principles and Management: With Malaysian Case Studies*. Kuala Lumpur: Institute of Biological Sciences, University of Malaya Press.

Agamuthu, P., & Tanaka, M. (Eds.). (2014). *Municipal Solid Waste Management in Asia and the Pacific Islands*. Singapore: Springer.

Agamuthu, P., & Victor, D. (2011). Policy trends of extended producer responsibility in Malaysia. *Waste Management & Research*, *29*(9), 945–953. doi:10.1177/0734242X11413332 PMID:21771873

Azizin, S. M. (2010). *Isolation of hemicellulose from sugarcane bagasse by alkaline extraction*. Faculty of Applied Sciences, Universiti Teknologi MARA. Retrieved from http://eprints.uitm.edu.my/2366/

Bahafun, K. A. (2007). *Analysis of renewable energy potential in Malaysia*. Postgraduate Coursework Student Faculty of Electrical Engineering and Telecommunications (Eet), University of New South Wales.

Barker, J.C., & Walls, F.R. (2011). *North Carolina agricultural chemicals manual: livestock manure production rates and nutrient content December 2001*. Academic Press.

Biomass-SP. (2014). *Overview of biomass industry in Malaysia. EU-Malaysia Biomass Sustainable Production Initiative*. Retrieved from http://biomass-sp.net/about/overview/

Birpinar, M. E., Bilgili, M. S., & Erdorgan, T. (2008). Medical waste management in Turkey: A case study of Istanbul. *Waste Management (New York, N.Y.)*, *29*(1), 445–448. doi:10.1016/j.wasman.2008.03.015 PMID:18550354

Chaerul, M., Tanaka, M., & Shekdar, A. V. (2008). A System dynamic approach for hospital waste management. *Waste Management (New York, N.Y.)*, *28*(2), 442–449. doi:10.1016/j.wasman.2007.01.007 PMID:17368013

Crouse, D.A., Smyth, T.J., & Crozier, C.R. (2014). North Carolina Agricultural Chemicals Manual 39: Livestock & Poultry Manure Production Rates and Nutrient Content. *Soil Science, Biological & Agricultural Engineering*.

Department of Statistics Malaysia Official Portal. (2018). Retrieved from https://www.dosm.gov.my/v1/index.php?r=column/cthemeByCat&cat=155&bul_id=c1pqTnFjb29HSnNYNUpiTmNWZHArdz09&menu_id=L0pheU43NWJwRWVSZklWdzQ4TlhUUT09

Derraik, J. G. (2002). The pollution of the marine environment by plastic debris: A review. *Marine Pollution Bulletin*, *44*(9), 842–852. doi:10.1016/S0025-326X(02)00220-5 PMID:12405208

Fauziah, S. H., Liyana, I. A., & Agamuthu, P. (2015). Plastic debris in the coastal environment: The invincible threat? Abundance of buried plastic debris on Malaysian beaches. *Waste Management & Research*, *33*(9), 812–821. doi:10.1177/0734242X15588587 PMID:26092255

Guilberto, B., & Okumura, S. (2010). *Municipal Waste Management Report: Status-quo and Issues in Southeast and East Asian Countries*. AIT/UNEP Regional Resource Center for Asia and the Pacific (RRC.AP). Thematic Working Group on Solid and Hazardous Waste. Retrieved from http://www.environment-health.asia/twg.cfm?themeid=3

Health-care waste. (2018). Retrieved from https://www.who.int/news-room/fact-sheets/detail/health-care-waste

Hoornweg, D., & Bhada-Tata, P. (2012). *What a Waste: A Global Review of Solid Waste Management*. Washington, DC: World Bank.

Hosseini, S. E., & Abdul Wahid, M. (2014). The role of renewable and sustainable energy in the energy mix of Malaysia: A review. *International Journal of Energy Research*, *38*(14), 1769–1792. doi:10.1002/er.3190

Ian, S. (2015). *Coastal communities dumping 8m tons of plastic in oceans every year*. Academic Press.

Ibrahim, C. A. (2013). *Policy Update from Malaysia*. Department of Environment. Retrieved from https://19january2017snapshot.epa.gov/sites/production/files/2014-05/documents/malaysia.pdf

ICC. (2010). *Trash travels. International Coastal Cleanup, Ocean Conservancy*. Retrieved from https://oceanconservancy.org/wp-content/uploads/2017/04/2010-Ocean-Conservancy-ICC-Report.pdf

ICC. (2012). *The ocean trash index*. International Coastal Cleanup, Ocean Conservancy. Retrieved from https://oceanconservancy.org/wp-content/uploads/2017/04/2012-Ocean-Conservancy-ICC-Report.pdf

ICC. (2014). *Turning the tide on trash*. International Coastal Cleanup, Ocean Conservancy. Retrieved from https://oceanconservancy.org/wp-content/uploads/2017/04/2014-Ocean-Conservancy-ICC-Report.pdf

Ishak, M. B., & Samah, M. A. A. (2010). Strict liability versus policy and regulation for environmental protection and agricultural waste management in Malaysia. *Environment Asia*, *3*, 11–19.

Jambeck, J. R., Geyer, R., Wilcox, C., Siegler, T. R., Perryman, M., Andrady, A., ... Law, K. L. (2015). Plastic waste inputs from land into the ocean. *Science*, *347*(6223), 768–771. doi:10.1126cience.1260352 PMID:25678662

MalaysiaP. C. R. 2016. (2017). Retrieved from http://cpmaindia.com/pdf/apic-country-2017/apic2017-malaysia-report.pdf

National Biomass Strategy. (2020). *New wealth creation for Malaysia's palm oil industry*. Agensi Inovasi Malaysia.

NSWMD. (2012). *Solid Waste Management in Malaysia: The Way Forward*. Retrieved from http://ensearch.org/wp-content/uploads/2012/07/Paper-13.pdf

Omar, D., Razali, S. N., & Karunpannan, S. (2012). Clinical Waste Management in District Hospital of Tumpat, Batu Pahat and Taiping. Asia Pacific International Conference on Environment-Behavior Studies. *Procedia: Social and Behavioral Sciences*, *68*, 134–135. doi:10.1016/j.sbspro.2012.12.213

Pauze. (2016). *Integrated Solid Waste Management: Challenge and Future*. NEHAP Conference 2016. 26 September 2016. Putrajaya International Convention Centre. Retrieved from http://nehapmalaysia.moh.gov.my/wp-content/uploads/2016/03/Paper-2-Solid-Waste.pdf

Shafie, S. M., Mahlia, T. M. I., Masjuki, H. H., & Ahmad-Yazid, A. (2012). A review on electricity generation based on biomass residue in Malaysia. *Renewable & Sustainable Energy Reviews*, *16*(8), 5879–5889. doi:10.1016/j.rser.2012.06.031

Stickel, B. H., Jahn, A., & Kier, W. (2012). *The Cost to West Coast Communities of Dealing with trash, reducing marine debris*. Retrieved from https://www.coastal.ca.gov/publiced/coordinators/WestCoast-CommsCost-MngMarineDebris.pdf

Tang, K. M. (2014). *Towards Environmental & Economic Sustainability in Malaysia via Biomass Industry*. Retrieved from https://www.oeaw.ac.at/forebiom/WS2lectures/02-01-TKMUN.pdf

Tiong, C. S., Latiff, P. A., & Karupannan, S. (2012). Medical waste management in private medical clinics Taiping. *Perak. International Conference on Ecological, Environmental and Bio-Sciences (ICEEBS'2012)*, Pattaya.

UNCRD (United Nations Centre for Regional Development). (2017). *State of 3Rs in Asia and the Pacific Country Report: Malaysia*. Secretariat of the Regional 3R Forum in Asia and the Pacific, UNCRD, Nagono, Japan, and Institute for Global Environmental Strategies, Kamiyamaguchi, Japan.

UNFCCC. (2005). *Sixth compilation and synthesis of initial national communications from Parties not included in Annex I to the Convention*. Retrieved from http://unfccc.int/resource/docs/2005/sbi/eng/18a02.pdf

United Nations. (2012). *Department of Field Support, Cartographic Section, Map No. 4365 Rev. 1*. Author.

World Bank. (1999). *What a Waste: Solid Waste Management in Asia*. Washington, DC: The World Bank.

World Bank. (2014). *World Bank Development Indicators*. Retrieved from http://data.worldbank.org/country/malaysia

Yusop, M. H. (2006). *Production of bio-ethanol from biomass waste(sugarcane)*. Faculty of Chemical and Natural Resources Engineering, University College of Engineering & Technology Malaysia. Retrieved from http://umpir.ump.edu.my/611/

Chapter 9
Status of Solid Waste Management in Myanmar: Key Challenges and Opportunities

Premakumara Jagath Dickella Gamaralalage
Institute for Global Environmental Strategies, Japan

Ohnmar May Tin Hlaing
Environmental Quality Management Co., Ltd, Myanmar

Aung Myint Maw
Yangon City Development Committee (YCDC), Myanmar

Matthew Hengesbaugh
Institute for Global Environmental Strategies, Japan

ABSTRACT

Myanmar, the largest country in mainland Southeast Asia, has been facing considerable challenges with the management of solid waste in the recent past because of increasing income and consumption patterns, urban growth, and lack of effective waste management policies, treatment, and disposal methods. Waste management is also a crosscutting issue that touches on many aspects of social and economic development, and as such is widely associated with a range of global challenges including public health, climate change, poverty reduction, food security, resource efficiency, and sustainable production and consumption. This country chapter therefore presents an overview of the current waste management in Myanmar, discusses key challenges and opportunities, and identifies some policy recommendations towards its improvement.

DOI: 10.4018/978-1-7998-0198-6.ch009

INTRODUCTION

Extending 677,000 sq. km (km^2), Myanmar is the largest country in mainland Southeast Asia based on land area. Strategically located between China and India (See Figure 1), Myanmar's wealth of natural resources, abundant labour force and fast-growing consumer market signal that the country is positioned to become the region's next *"Asian giant"*. Myanmar's estimated population was 51.48 million in 2014 (Ministry of Labour Immigration and Population, 2015). More recent figures indicate that the population grew to 54 million in 2018, having increased by 0.71 percent annually, and may reach as high as 65 million by 2050 based on current birth rate trends (UNFPA Myanmar, 2017). A heterogeneous population made up of diverse ethnic groups speaking over 100 languages and dialects, one-third of Myanmar's inhabitants reside in cities; the remaining 70 percent are located in rural areas and remain largely dependent on subsistence farming (UNDP, 2019).

Myanmar is subdivided into one union territory (Nay Pyi Taw), seven states (Chin, Kachin, Kayah, Kayin, Mon, Rakhine, and Shan) and seven regions (Ayeyarwady, Bago, Magway, Mandalay, Sagaing, Tanintharyi, and Yangon) for administrative purposes. Myanmar's smallest administrative unit is designated as a village. Several villages grouped together are classified as a township, of which the lowest division of government is generally stationed; as such, townships carry out many key public functions including birth and land registration, tax collection as well as the provision of some public services such as waste management. Clusters of several townships are organised into districts, which in turn form the

Figure 1. A location of Myanmar in the context of southeast Asia

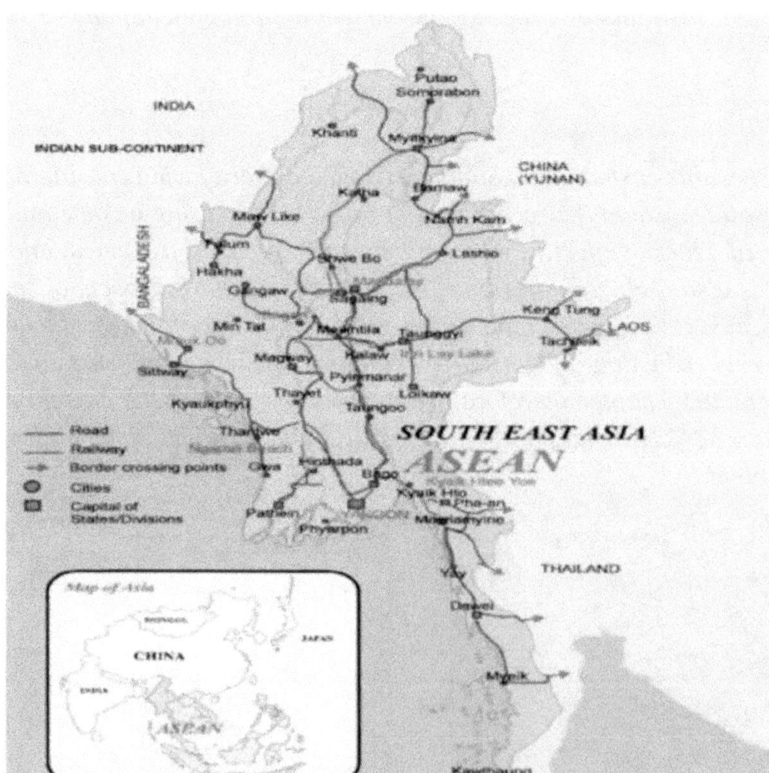

states or regions. Districts thus comprise a middle tier of government administration connecting state/regional governments to townships (Nixon, Joelene, Chit, Aung & Arnold, 2013; Bingham Centre for the Rule of Law, 2013).

A lower-middle income economy, Myanmar's gross national income (GNI) per capita was 1,455 USD in 2017. However, the country is also one of the fastest expanding economies in the world, averaging about 6.4 percent gross domestic product (GDP) growth in the period of 2016-2017, with expectations upwards of 7 percent for 2019-2020 (World Bank, 2018). Myanmar's agriculture sector remains the backbone of its economy, contributing to 37.8 percent of its GDP and accounting for 25 to 30 percent of total export earnings; the sector also employs roughly 70 percent of the labour force (FAO, 2018). Nevertheless, the share of GDP accounted for by the country's industrial sector has more than doubled over the past few years, reaching 25 percent in part because of recent market-oriented reforms including Myanmar's opening up to foreign direct investment (FDI). This has precipitated rapid growth of the country's industrial base, most notably concentrated around the continued expansion of natural gas exports. Myanmar's market transformation has also attracted attention to the enormous potential of further developing its tourism and hospitality-related sectors as well.

However, underinvestment in urban infrastructure and associated services including access roads, energy, water supply, sanitation, drainage, wastewater, and solid waste management has resulted in significantly deficient public utilities throughout Myanmar especially in the larger cities of Yangon and Mandalay. Currently, Myanmar generates an estimated 20,000 tons of solid waste per day and this volume is expected to increase rapidly in the coming years due to urban growth. In many townships, typically one to two-thirds of generated waste remains uncollected. Residents frequently dump uncollected waste on the streets and in water bodies or resort to openly burning it, subsequently polluting air, soil and ground water. On the other hand, waste that is collected by respective township development committees (TDCs) and city development committees (CDCs) is regularly transported to final disposal sites that are open dumpsites lacking environmentally appropriate controls. In addition, all major cities experience challenges with regard to managing industrial, medical, and other emerging waste streams such as e-waste and plastic waste, all of which are often disposed without proper treatment.

Inadequate waste, water, drainage and sanitation services, coupled with insufficient investment in preventive health care, have resulted in significant health challenges at many levels. The incidence of diarrhoea among children under five years of age is considerably higher than elsewhere in Southeast Asia, contributing to a high child mortality rate. Out of every 100 children, 6.2 die before their first birthday and 7.2 before their fifth (Government of Myanmar & UNFPA, 2014). The high prevalence of debilitating water-related vector borne diseases, such as malaria, dengue, and Chikungunya fever are also directly related to the poor state of critical urban services (ADB, 2017). Myanmar is one of the world's most disaster-prone countries and stands vulnerable to multiple hazards, including floods, cyclones, earthquakes, landslides, and droughts, ranking 2nd out of 187 countries in the 2016 Global Climate Risk Index and 9th out of 191 countries in the INFORM Index for Risk Management (2019).

The Intergovernmental Panel on Climate Change (IPCC) estimates that solid waste management (SWM) accounted for around three percent of greenhouse gas (GHG) emissions in 2010, with most of this percentage attributed to methane emissions associated with landfill disposal. According to the World Resources Institute Climate Analysis Indicators Tool (WRI-CAIT), total GHG emissions in Myanmar were calculated at 201.5 million metric tons of carbon dioxide equivalent ($MtCO_2e$) in 2013 (WRI-CAIT 2.0, 2017), making up 0.48% of total global GHG emissions. Out of this figure, 11.7 (5.8%) was from the waste management sector. Myanmar's Initial National Communication (INC) under the United Nations Framework Convention on Climate Change (UNFCCC) also identified that open dumping of solid waste contributes to emitting roughly 2,799.51 gigagram of carbon dioxide equivalent (Gg of CO_2e) into the atmosphere (MOECAF, 2012).

Against this background, both national and subnational governments in Myanmar have undertaken efforts to improve existing waste management policy, regulations and infrastructure. For example, the Environmental Conservation Department (ECD) of the Ministry of Natural Resources and Environmental Conservation (MONREC), in partnership with the International Environmental Technology Centre of UN Environment (UNEP-IETC) and the IGES Centre Collaborating with UNEP on Environmental Technologies (CCET) has developed the county's first National Waste Management Strategy and Master Plan (2018-2030). In addition, ECD has been working with the Norwegian research organization SINTEF as well as Norway's Environment Agency on the development of a Master Plan for Hazardous Waste Management starting in 2019. Both strategic documents are awaiting official approval from Myanmar's Cabinet Office in advance of their implementation. Together they provide a supportive policy framework for sustainable waste management based on waste hierarchy and 3R (reduce, reuse and recycle) principles, aimed at directing city-level waste management strategies that focus on piloting and execution of concrete actions led by responsible TDCs and CDCs.

Accordingly, this chapter seeks to review the current solid waste management situation in Myanmar, discussing key challenges and opportunities as well providing several policy recommendations towards its improvement. Relevant information was accumulated through the use of both primary and secondary data collection methods. A number of informal discussions and interviews with relevant staff of ECD, Mandalay City Development Committee (MCDC), Yangon City Development Committee (YCDC) and other concerned institutions were conducted along with site visits to respective waste management facilities. The authors' experience, knowledge and related support for ongoing activities in Myanmar also contributed to the identification of practical recommendations.

This chapter is structured into five sections. Following this brief introduction, Section 2 presents an overview of the existing waste management situation in Myanmar. The current status of Myanmar's waste management system is contextualized through a description of relevant data on waste generation and composition, transportation, treatment, disposal and 3R activities across all waste streams including municipal, industrial, medical and hazardous waste. Section 3 discusses the country's existing policy, legal, and institutional framework. Key challenges and gaps to be addressed are thereafter highlighted in Section 4. Finally, section 5 concludes with a discussion on how to ensure the aforementioned strategies can effectively address the waste management challenges facing Myanmar by offering specific policy recommendations.

CURRENT STATUS OF WASTE MANAGEMENT

Municipal Solid Waste Management (MSWM)

MSW Generation

There is no official definition of MSW in Myanmar. MSW is generally defined as the non-gaseous and non-liquid fractions of waste that is generated because of daily household activities (60%), as well as public operations, such as public markets, street sweeping etc. (15%), commercial producers (10%),

Table 1. Project population and daily waste generation in Yangon and Mandalay cities (2015-2025)

Year	Yangon			Mandalay		
	Population (millions)	Total waste generation/ day (tons)	Per capita waste generation	Population (millions)	Total waste generation/ day (tons)	Per capita waste generation
2015	7.4	5,920	0.80	1.5	1,200	0.80
2016	7.6	6,220	0.82	1.5	1,261	0.84
2017	7.8	6,531	0.84	1.6	1,324	0.83
2018	8.0	6,853	0.86	1.6	1,389	0.86
2019	8.2	7,188	0.88	1.6	1,457	0.91
2020	8.4	7,535	0.89	1.7	1,527	0.90
2021	8.6	7,895	0.91	1.7	1,600	0.94
2022	8.8	8,268	0.94	1.7	1,676	0.98
2023	9.0	8,656	0.96	1.8	1,755	0.98
2024	9.2	9,057	0.98	1.8	1,836	1.02
2025	9.4	9,473	1.01	1.8	1,920	1.06

hotels (2%), gardens (5%) and other institutions (8%) (Premakumara et al., 2016a; Premakumara et al., 2017). Moreover, there also exists no accurate data on total waste generation in Myanmar (Premakumara & Hengesbaugh, 2016b). According to preliminary estimations of the World Bank (2012), total waste generation in Myanmar was 5,616 tons/day in 2012 with projections that this may reach 21,012 tons/day by 2025 (World Bank, 2012). A recent assessment conducted by the World Bank (2018) identified that total waste generation in Myanmar was about 20,415 tons/day in 2016. In addition, another study led by the Netherland's Enterprise Agency (2017) estimated that Myanmar produced approximately 20,000 tons of solid waste per day in 2017. Based on the above studies, the authors have calculated that Myanmar's population of 54 million produces around 20,000 tons of MSW per day with a per capita waste generation of 0.38 kg/person/day.

When it comes to a closer examination of city-level per capita waste generation, YCDC and MCDC—the agencies responsible for carrying out waste management in Myanmar's two largest urban centres—report a per capita waste generation of 0.64 and 0.41 kg/person/day, respectively. However, it is generally understood that these figures represent the total volume of waste transported to final disposal sites and do not include waste volumes that remain uncollected, burned or self-managed by waste generators, including waste illegal dumped, recycled or reused (Netherland Enterprise Agency, 2017). Based on the GDP of these two cities and existing international data, the Netherland Enterprise Agency study also identified that average per capita waste generation should approximate 0.8 kg/person/day in both cities and may potentially increase to one kg/person/day by 2025. Respective waste generation rates of both cities are illustrated in Table 1. The data indicates that in view of Myanmar's total waste generation, approximately 42 percent of waste was produced in both cities including Yangon (6,853 tons/day) and Mandalay (1,389 tons/day) in 2018. However, as current waste collection rates in Yangon and Mandalay cities stand at about 2,500 and 940 tons per day respectively, it can be assumed that roughly 3,500 tons and 300 tons of waste in Yangon and Mandalay do not reach final disposal sites and remain unaccounted (Netherland Enterprise Agency, 2017; Premakumara et. al., 2017).

Figure 2. Average per capita waste generation in townships in Mandalay region

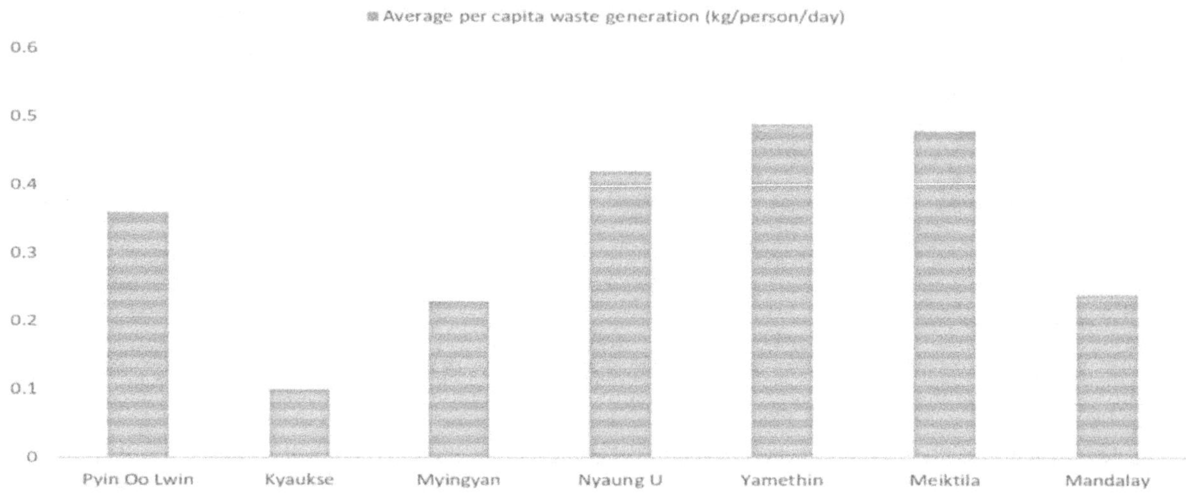

A recent sample survey conducted by the authors in Mandalay Region identified that average per capita waste generation in affiliated townships (with the exception of MCDC) approximated roughly 0.36 kg/person/day (Premakumara & May Tin, 2019). Figure 2 provides the average per capita waste generation across different townships in the Mandalay Region.

MSW Composition

MSW in Myanmar is mainly comprised of organic materials (77%). The balance is made up of plastic (13%), paper (7%), and other residuals (3%). As shown in Figure 3, these values have been confirmed by observation as well as evaluation of the existing waste composition of Mandalay and Yangon cities (Premakumara, et. al., 2017; Premakumara & Hengesbaugh, 2016b; Netherland Enterprise Agency, 2017).

Figure 3. Waste composition in Mandalay (left) and Yangon (right) cities

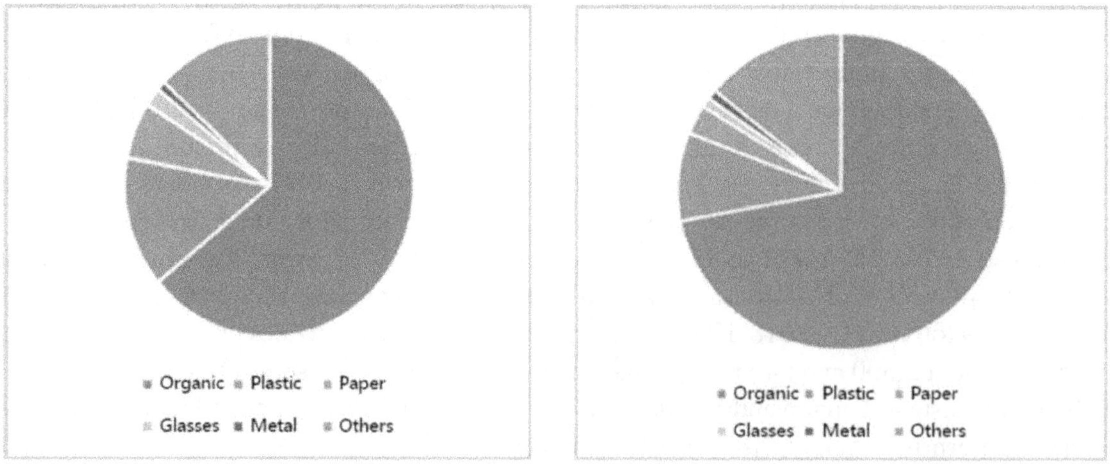

City Cleaning and Waste Collection

As highlighted previously, MSW management is the responsibility of respective CDCs and TDCs. Involvement of private sector remains virtually absent from waste collection and treatment activities with the exception of a few informal or community-based initiatives on *"cash for trash"* (Netherland Enterprise Agency, 2017). Current methods of waste collection led by the cities include primary collection, transport to transfer stations, secondary collection as well as final disposal.

Primary waste collection is conducted via door-to-door (bell collection), kerb side, bin container collection, as well as by way of open or communal waste collection points. Door-to-door collection is particularly labour intensive, making use of various methods such as bell ringing and loudspeaker announcements. Such collection is carried out making use of a combination of pushcarts, tri-bikes and tipper trucks depending on the layout and conditions of respective areas (See Figure 4).

According to YCDC, current waste collection rates in Yangon stand at about 90 percent. The city is divided into four districts, including north, east, south and west, covering all 33 townships, all of which have access to city cleansing and collection services. Nevertheless, it has been estimated that only 45 percent of collected waste is transported to final disposal sites, following initial waste separation and recycling activities conducted by the informal sector for materials not exceeding 5% of the city's total waste volume (Netherland Enterprise Agency, 2017). These figures indicate that roughly 50 percent of MSW in Yangon City remains outside of formal collection routs. Consequently, uncollected waste is often observed accumulating in residential areas, market places and drainage systems. Although daily

Figure 4. Solid waste collection system in Mandalay City

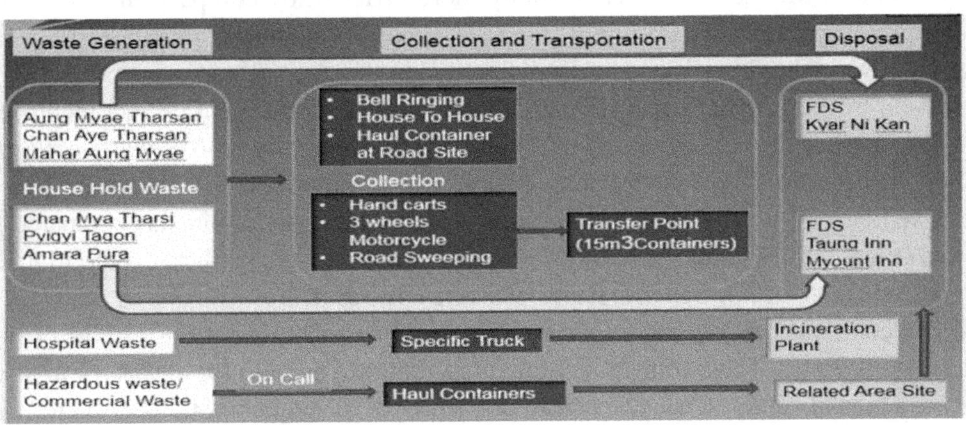

Table 2. Number of employees, vehicles and pushcarts deployed for waste management in Yangon and Mandalay

City	Employees		Vehicles		Push Carts	
	Total	Inhabitants per employee	Total	Inhabitants per vehicle	Total	Inhabitants per push cart
Yangon	4,900	1,429	355	19,718	1,086	6,446
Mandalay	2,200	682	211	7,109	179	8,380

Figure 5. Average waste collection rate in townships in Mandalay region

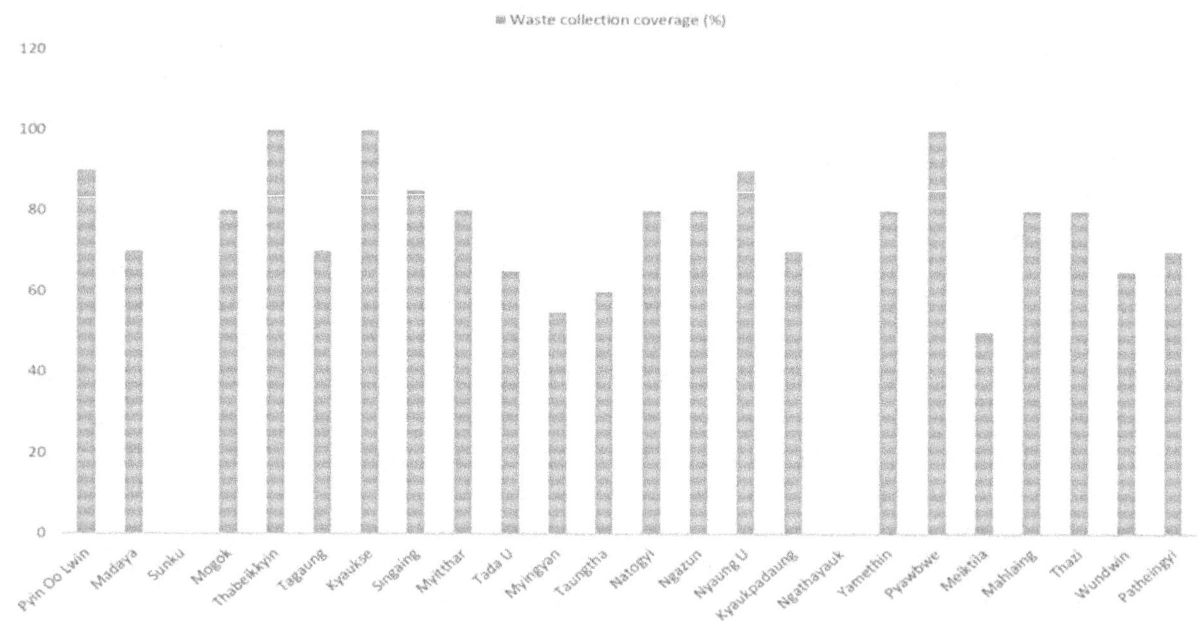

sweeping and cleaning is conducted in the city's central business area, outside of this zone, such services are not provided; this is especially the case in less affluent parts of the city, as well as in industrial areas.

Waste collection in Mandalay City is generally more efficient by comparison. The average waste collection ratio in Mandalay City currently stands at about 80 percent (Premakumara et. al., 2017), thus 20 percent of waste has been found to remain behind in the city without proper disposal. A sample survey carried out by the FASEP/French Grant Urban Service Improvement Project identified that this figure also varies between urban centre and peri-urban areas, where the ratio of waste collection ranges between 50-70 percent. Furthermore, Mandalay City benefits from a high staff rate of waste employees as well as vehicles for waste collection services in view of its population size compared to Yangon City, as shown in Table 2 (Netherland Enterprise Agency, 2017). These indictors help to illustrate why Mandalay performs comparably better in terms of waste collection.

As large city centres, both Yangon and Mandalay maintain transfer stations as part of their overall waste management systems. These transfer stations allow for the consolidation of waste, initially collected through the use of small vehicles, tricycles and carts, which is subsequently transferred to large volume trucks that transport the waste to final disposal sites. In view of existing traffic conditions, as well as other geographical factors, such as location of final disposal sites far away from city centres, these transfer stations represent an important strategy for both cities moving forward. Beyond designated city transfer stations, there also exist a large number of roadside transfer stations (i.e., secondary collection points) where waste from tricycles and other small capacity collection vehicles is bulked and moved into larger containers and collection trucks. Managing these transfer stations is a largely labour-intensive activity given their function as *in situ* waste separation centres for informal waste workers. Field observations have noted the importance of considering the occupational safety and health of workers at these facilities.

A sample survey carried out in the townships in Mandalay Region (excluding MCDC) identified that the average waste collection rate in urban areas stands at about 77 percent. As shown in Figure 5, existing waste collection and coverage rates in these townships can be viewed as adequate when considering the figures provided by respective TDCs. Although field observations have documented a large degree of litter in different parts of the city, waste collection services appear to be effective in business centres.

Practice of 3Rs (Reduce, Reuse and Recycle)

Most cities in Myanmar conduct information sharing, public awareness raising, and environmental education for local residents aimed at mobilising support for 3R (reduce, reuse and recycling) activities. As shown in Figure 6, the most popular information and education campaign methods employed by CDCs and TDCs include advertising/sign boards, public meetings, an announcement by ward leaders, social media, engaging Hluttaw (community representatives), newspapers and working with charity groups. However, most residents appear to be reluctant (or unaware) about cooperating in city clean-up efforts and waste is often seen scattered in public places. Unmanaged waste can lead to contamination of soil and drinking water, contribute to waterborne diseases and, when supportive infrastructure is not in place, they blocking of sewerage systems, resulting in overflow flooding during the monsoon season. Apart from impacts on health and safety, poor waste management reduces the attractiveness of the city and results in environmental degradation from discharge of waste (especially plastics) into the country's rivers, deltas and sea. Myanmar's largest water body— the Irrawaddy— has been found to be one of the 20 most polluting rivers in the world in terms of ocean plastic leakage (Lebreton et al., 2017).

In addition to the efforts of respective CDCs and TDCs, some charity groups and local non-governmental organizations (NGOs) are involved in educating residents and supporting community-led activities around cleaning and greening the city. For example, YCDC's Alley Garden Project is initiative focused on converting back alleys previously used as rubbish dumps into gardens aimed at promoting clean and healthy recreational spaces in Yangon. It was first started in 2016 on Yangon's 27th street when residents transformed a 20-meter alley into a vegetable plot, a playground and place for community gathering.

Figure 6. Popular methods used for information sharing and public awareness at townships

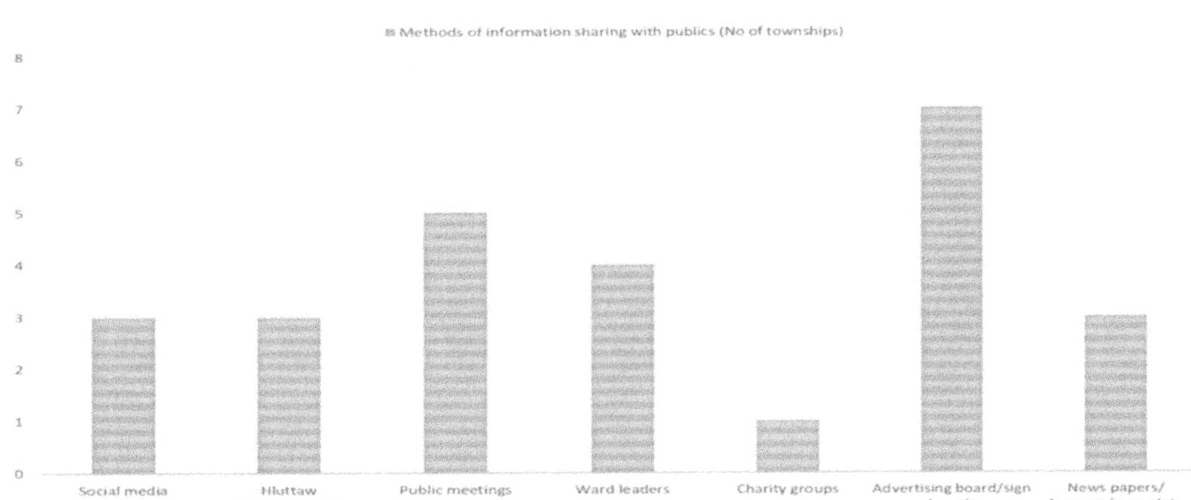

Since then, the initiative has been receiving wide exposure from communities, YCDC representatives and local media; community organizers have announced their intention to scale up the initiative with a view towards covering another one km of city alleyways (More information on this program can be followed through, www.doheain.com).

Similarly, small-scale recycling activities are carried out in many townships across Myanmar, particularly by the informal sector: this includes scavengers, waste collectors, and waste dealers. Scavengers and waste collectors recover some valuable recyclable materials such as newspapers, metal, plastic bottles, tin and glass from households, communal depots, streets, commercial areas and final disposal sites and in turn sell these items to waste dealers who subsequently clean, sort, store and sell them in bulk to the recycling industry both locally as well as for export.

At present, there is a lack of accurate and reliable data on recycling volumes, ratios and number of recycling facilities in Myanmar cities. However, a sample survey carried out in Yangon City identified that 86 tons of recyclable materials are directed to the waste dealers per day by the informal sector. Out of this total, 57% was glass; cardboard and paper accounted for 15% and 13%, respectively. Plastic (7%), tin cans (7%) and other materials such as metal, iron and steel, (1%) were also observed to be collected (Premakumara & Hengesbaugh, 2016). In this context, the importance of this informal sector should not be underestimated. Informal recycling activities provide a primary source of income to thousands of families in Myanmar. According to the Netherland Enterprise Agency (2017), the waste pickers operate at the small transfer station in Mandalay city can earn about 10,000 MMK per person per day by sorting out the recyclables from the waste transported to the location.

Myanmar also hosts a sizable reuse and recycling supply chain. Some sources attribute these operations to Myanmar's decades-long isolation from the international community (Netherland Enterprise Agency, 2017). According to this idea, because the country's lack of trade often resulted in many materials such as iron, paper and plastics being in short supply, Myanmar aspired towards self-sufficiency: this in turn opened a market for both reuse and recycling products and services. However, new pressures brought on by the country's exposure to the global recycling industry suggest that the sector may have to restructure and modernize in order to stay competitive. For instance, plastic recycling is quickly be-

Figure 7. Waste treatment and disposal facilities in townships in Mandalay region

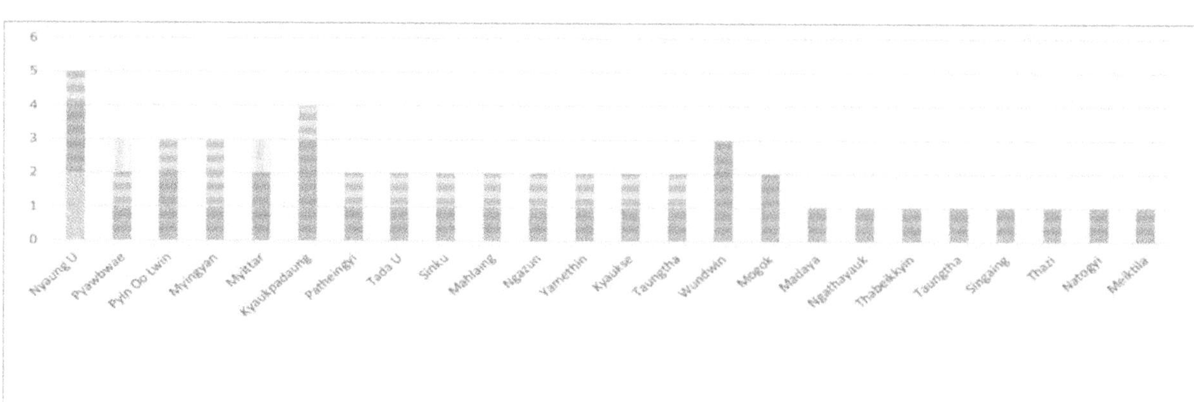

coming an emerging business opportunity within Myanmar. Yangon currently hosts three large plastic factories as well as a number of smaller preprocessing facilities. The estimated production output of shredded plastic bottles from these operations approximates 1,000 tons/month. A similar situation can be found in Mandalay where the estimated production has been calculated at 700 tons/month (Netherland Enterprise Agency, 2017).

Waste Treatment and Disposal

Waste is generally collected by respective townships and transported to designated final disposal sites located within their city limits. These final disposal sites are largely open dumpsites with almost no environmental measures or controls; as such, there is little in the way to prevent trespassing, avert landfill fires, collect leachate, reduce dust or capture gas emissions. A sample survey carried out in Mandalay Region identified that all townships have at least one active disposal site, in addition to some closed dumpsites that have been decommissioned without any effective closure plan (Figure 7).

The situation is the same in Myanmar's larger cities. Both Yangon and Mandalay host a number of active and closed dump sites. These sites are often located about 10 to 25 km away from the cities' central business districts (CBD) and encompass an area ranging from approximately less than one hectare to over 60 hectares. The typical duration of time waste dumping can be conducted in these sites in line with average capacity has been observed to range from one to two years to a maximum of five to ten years. Most disposal sites face challenges in terms of operations and management with work predominantly limited to the registration of incoming waste. Transported waste to disposal sites is first unloaded on the ground or onto existing waste piles. Waste pickers subsequently sort this waste by hand or use manual tools, such as rakes. Following manual sorting, unsorted waste is moved further inside the dumpsite through the use mechanical equipment, such as bulldozers. Sorted recyclable waste is thereafter packed, stored and transported back to the city for resale.

With a view towards improving the environmental and sanitary conditions associated with MSW treatment and disposal, the Governments of Myanmar and Japan jointly financed and constructed a waste to energy (incinerator) model plant in Yangon city at the cost of 16 million USD. Total funding for the facility comprised 8 million USD provided by YCDC and 8 million USD earmarked by Japan under its financial support scheme, the Joint Crediting Mechanism (JCM), which seeks to support developing countries with the transfer of Japanese low-carbon technologies. The model plant was planned and constructed by the JFE Engineering Corporation (JFE), and has a capacity of 60 tons/day (1 unit). The facility began operations in April 2017 on a trial basis, with YCDC and JFE accepting waste from the nearby townships of Insein, Shwe Pyi Thar, North Okkalar and Mingalardon. In addition to contributing to the reduction of an estimated 4,800 tons of CO_2/year, the plant has been designed to generate 700 kilowatts (KW) of electricity of which 300 KW are used for its own operations with the remainder (400 KW) fed into the national grid. As the incinerator is still in its initial stage of operations, lessons learned, including those relevant to its technical applicability, environmental considerations, social acceptability and financial sustainability should be studied and documented before commissioning other plants in Yangon City and elsewhere.

It has also been identified that some cities in Myanmar have working towards upgrading their final disposal sites, including by constructing semi-aerobic landfills based on the Japanese Fukuoka method. This approach seeks to accelerate the decomposition of organic waste materials, which in turn can improve site drainage and suppress the release of methane gas. For instance, an overseas development assistance agreement was signed between Japan's Ambassador to Myanmar and the Country Programme Manager for UN-Habitat's office in Myanmar amounting to 5.5 million USD for introducing the Fukuoka method to the Htein Bin dumpsite in Yangon, where a large fire took place in April 2018 (Embassy of Japan in Myanmar, 2019). In addition, both Mandalay and Bagan Cities have undertaken independent initiatives; both have utilised local budget and expertise towards constructing semi-aerobic landfills following a waste management-training workshops jointly organised by CCET in partnership with ECD, Mandalay Regional Committee, and MCDC.

Other Emerging Wastes

Industrial Waste

Although Myanmar remains largely a primary based economy at present, the government is placing increasing emphasis on shifting towards industrialisation. In 2018, the national government approved the Myanmar Sustainable Development Plan (MSDP), 2018-2030, which provides directions for guiding sustainable and equitable development (Ministry of Planning and Finance, 2018). The MSDP encourages the formulation of trade policies based on liberal economic principles and seeks to support an investment climate that fosters the implementation of Public-Private-Partnerships. In addition, the government has also made efforts to enhance the country's basic infrastructure including by constructing bridges and communication systems to strengthen industrial development.

Myanmar's Ministry of Industry is responsible for managing state-owned enterprises (SOEs), including the country's 18 industrial zones and three special economic zones (SEZs). The Ministry also engages the industrial sector on issues of waste management. As a number of companies generate waste from their operations consisting both of recyclable materials and hazardous components, it has become incumbent on Myanmar to develop a treatment and recovery system as well as associated facilities for managing industrial waste.

According to SINTEF, there are approximately 50,000 private industries in Myanmar classified under 13 broad categories. The total volume of hazardous waste produced by these private industries approximates 2,000 tons/year (including Yangon and Mandalay). About 85 percent of total hazardous waste generated in Yangon City is made up of five waste streams: lead-slag from lead-acid battery recycling; sludge from wastewater treatment associated with textile dyeing processes, lead-acid battery manufacturing and paints production; and oil-based solvents from paints production. In Mandalay, the largest industrial hazard associated with these waste streams is sludge from textile dyeing industries, representing 74 percent of total industrial hazardous waste generated in the city. In addition, Thilawa SEZ currently generates 160 tons/year of hazardous waste. Moreover, several industries are expected to be operational in Dawei and Kyauk Phyu SEZs from the year 2025. Waste generation is projected to peak ten years from start of operations in all SEZs (ECD, MONREC, 2019).

Myanmar has limited infrastructure for treatment and environmentally friendly disposal of industrial waste at present. For instance, CDCs and TDCs are responsible for collecting and managing industrial waste generated by factories in their respective areas but only on an on-call basis (i.e., whenever requested to do so). Consequently, collected waste is often transported to landfill sites without any preliminary treatment. In this context, all cities in Myanmar face tremendous challenges with regard to managing industrial waste. For instance, levied disposal fees often only cover associated transportation costs; this issue, together with limited budgets and lack of technical expertise are reported as major obstacles to implementing environmentally sound industrial waste management. Forecasted growth of Myanmar's industrial sector suggests that hazardous waste issues will become increasingly more difficult to manage if actions are not carried out in the immediate term.

In Yangon, Golden Dowa Eco-System Co., Ltd., a subsidiary of Dowa Holdings Co., Ltd. in Japan has established and initiated operations of the country's first controlled landfill facility. Located in the Thilawa SEZ and spread over 42 ha, the facility was constructed at a total cost of 36 million USD. The landfill has been designed to address both hazardous and non-hazardous wastes in line with the United States Environmental Protection Agency (USEPA) standards, and includes a sorting/stabilization shed, wastewater and leachate treatment facility, waste laboratory and oil waste handling system. An area has also been demarcated for the future installation of an incinerator and material recovery facility. In this context, the landfill is currently receiving wastes from industries located within the SEZ as well as from several multinational oil and gas industries. Most of the received wastes are non-hazardous in nature; hazardous wastes include heavy metals and oil containing wastes. The company is further discussing with YCDC and MCDC about the potentially extending services to additional industrial zones located in Yangon and elsewhere.

Medical Waste

Myanmar generates an estimated 10 tons/day or 3,653 tons/year of hazardous, infectious and sharps waste (ECD, MONREC, 2019). Of this total, 50 percent of total hazardous healthcare wastes are produced by the cities of Yangon and Mandalay. The total generation of healthcare waste including non-hazardous general waste is estimated to be as high 67 tons/day. In line with these figures, 15 percent of total healthcare waste in Myanmar is hazardous whist 85 percent has been identified as general or non-hazardous waste. On average, YCDC and MCDC dispose of 15 tons/year of expired medicines, respectively, in cooperation with Myanmar's Food and Drug Administration.

Myanmar cities are responsible for collecting medical waste. Both YCDC and MCDC collect medical waste from large hospitals and special clinics on a daily basis while collection service is provided to smaller facilities once a week or on an on-call basis. Three different coloured bags are used for separating the waste; this includes (i) blue or green (YCDC)/ black (MCDC) for non-hazardous health care waste or domestic waste uncontaminated with infectious or pathogenic agents (food residues, paper, cardboard and plastic wrapping); (ii) yellow for pathological waste, infectious waste as well as items that have been used for medical care; and (iii) red for sharps, and mainly, but not exclusively, auto-disable or disposal syringes with needles and pharmaceutical waste that consists of outdated drugs or expired unfinished medical solvents. Infectious waste is incinerated at cemeteries whereas sharp wastes are buried underground in landfills: residual wastes are treated as domestic waste.

Plastic Waste

Plastic waste issues have been gaining increasing attention due the negative impacts such materials pose to the environment, public health and human wellbeing. According to Earth Day Network (2018), roughly 300 million tons of plastic is produced annually, half of which is used for the manufacture of single-use products. The World Economic Forum (2016) has reported that only 14 percent of plastic packaging is collected for recycling whereas 10 percent is actually recycled. The remainder, also referred to as end-of-life plastic, is landfilled (40 percent), incinerated (14 percent), and alarmingly, leaked into the environment (32 percent), where it may take up to 1,000 years to decompose, leaching potentially toxic substances into soil and water systems. Terrestrial micro plastic pollution contaminating soils, sediments and freshwater may have potentially long-term negative effects on ecosystems. The World Economic Forum report also highlights that more than 8 million tons of plastic enters the oceans each year. The total volume of mismanaged plastic has been estimated at 0.46 million tons per year, of which about 0.1 million tons is plastic marine debris. Myanmar has been ranked among the top 20 countries contributing to the mismanagement of plastic waste (Jambeck et al., 2015).

Against this background, key stakeholders in Myanmar have begun to pilot various initiatives to address the issue of plastic waste. YCDC officially announced a ban on business activities associated with the manufacturing, importing, trading and distribution of high-density polyethylene (HDPE) plastic bags in 2009, which was followed, two years later, by a ban on polyethylene bags by the MCDC. Since 2011, Yangon authorities have also worked to stem the production, storage and sales of non-biodegradable waste, such as polyethylene bags. A number of other important activities have also been launched with a view to raise public awareness about the problems surrounding plastic waste priorities. For instance, one food and beverage business association in Yangon initiated a campaign in 2017 entitled *"Straws Suck"*, aimed at curbing the practice of serving customers single-use plastic straws by introducing reusable, biodegradable alternatives. Another example can be seen through the efforts of the Yangon art studio and gallery, *"Wired 39"*, which is currently collaborating with a group called *"Beat Plastic Pollution in Myanmar"* on an art installation fashioned from plastic bags and beverage containers. Signs have also been situated on some roads announcing *'Plastic Bag Free Zone'* in certain areas (MYANMORE 2018, Mar, W. W, 2018).

Despite these positive developments, Myanmar continues to face significant challenges with addressing plastic waste. Unlike other neighboring countries in the region, Myanmar has yet to pass legislation prohibiting the import of waste plastics following China's recent import ban on such materials. Moreover, Myanmar's waste management policies and regulations are still in their infancy; the country's lack of supportive financial schemes and appropriate technologies for waste prevention indicate that it will be some time before collection and recycling activities effectively resolve issues of plastic pollution (MYANMORE 2018, Mar, W. W, 2018, Netherland Enterprise Agency, 2017).

Other Special Waste

In addition to the above waste categories, there also exist a number of emerging and special waste streams that require specific attention in Myanmar. This includes, electronic waste (E-waste or WEEE), as well as chemical and mercury containing wastes (ranging from items such as fluorescent lamps, thermometers, and dental amalgam). Due to their hazardous components, these types of waste cannot be sold or processed without proper disassembly and handling. Likewise, certain types of materials cannot be destroyed, and therefore must be immobilised, permanently encapsulated, or managed in perpetuity — all costly processes that illustrate the importance of changing product composition, modifying industrial processes or finding alternative sources of raw materials.

E-waste is one of the fastest-growing waste streams in Myanmar due to changing lifestyles, increased consumer demand, rapid changes in technology, inventions of new electronic devices and availability of inexpensive product imports, including computers, televisions, and video games, as well as all kinds of electrical equipment, often divided into large (washing machines, air-conditioners, freezers etc.) and small equipment (hairdryers, electric toothbrushes, vacuum cleaners etc.).

According to United Nations University (UNU) estimations, approximately 29,000 tons of e-waste were generated in 2014, averaging 0.4 kg per capita (Balde et al., 2015). However, SINTEF's Master Plan on Hazardous Waste Management identifies that e-waste generation in Myanmar currently stands at about 5,920 tons (ECD, MONREC, 2019). The main reason for this discrepancy may be Myanmar's general cultural tendency to repair rather than replace goods, as highlighted above. However, as no national data or information currently exists about these waste streams, it can be inferred that waste volumes will progressively increase into the future. In this regard, existing treatment and recycling practices in Myanmar are woefully inadequate for tackling special waste streams, posing risks to the environment and human health. Examples of operations include haphazard dismantling of items (metal frames, power supplies, circuit boards, plastics), often manually, but increasingly by way of automated shredding equipment; open burning for purposes of segregating organic and inorganic compounds (for example, torching cables to recover copper); and using acids to extract gold, silver, palladium and other metals, largely from printed circuit boards. Although Myanmar has yet to establish a legal framework for e-waste management, e-waste is currently controlled under Basel Convention, of which the country is signatory.

In addition, waste management systems in Myanmar will need to address the growing volume of hazardous waste generated by oil and gas facilities, mining as well as stockpiles of Persistent Organic Pollutants (POPs). It has been estimated that major exploration and production (E&P) companies operating in Myanmar including PTTEP, Total, Petronas, MPRL, Gold Petrol and Petro Brunei, generate approximately 43,000 tons/year of hazardous waste. The majority of these wastes include drilling mud (65 percent), as well as drill cuttings (19 percent), drilling fluids (15 percent) and others (1 percent). Hazardous waste attributed to mining activities in Myanmar include (i) waste solvents, waste acids and contaminated soil, mainly from copper mining (230 tons/year); (ii) waste oil from copper mining (450 tons/year) and (iii) lead mine tailings (98,000 tons/year). An ongoing United Nations Industrial Development Organisation (UNIDO) project collecting data on POPs in Myanmar has identified stockpiles totalling 23 tons of dichlorodiphenyltrichloroethane (DDT). Many older electrical transformers used in Myanmar also contain transformer oil with polychlorinated biphenyl (PCB) content. The total amount of PCB found in this transformer oil is estimated at 2,380 tons/year (ECD, MONREC, 2019).

SOLID WASTE MANAGEMENT POLICIES, REGULATIONS AND INSTITUTIONAL FRAMEWORK

Myanmar emphasises the importance of environmental protection in policies at both the national and local levels. Myanmar is a signatory of various multilateral environmental treaties and agreements such as the Convention on the Control of Transboundary Movements of Hazardous Wastes and their Disposal (Basel Convention) 1989, the Agenda for Sustainable Development (SDGs) 2015 and the Paris Agreement on Climate Change 2016. Myanmar is also a party to the Stockholm Convention on POPs and to the Vienna Convention and the Montreal Protocol. Myanmar is not yet a signatory of the Minamata Convention but has indicated that it is considering becoming a party. Table 3 provides an overview of key national policies and strategies relevant to the waste management in this context.

In addition to national policies and regulations, respective CDCs and TDCs have promulgated a number of local policies and bylaws aimed at establishing a legal basis for local level waste management actions (see Table 4). However, in practice, effective implementation and enforcement of these environmental

Table 3. Major waste management policies, regulations and strategies at national level

National Policies, Strategies and International Commitments	Brief Description
National Environmental Policy (NEP) (2018)	• NEP (2018) was developed by the ECD, MONREC with support of UNDP and the Ministry of Foreign Affairs of Finland aiming to place environmental considerations at the centre of efforts to promote economic and social development, reduce poverty, and mitigate and adapt to climate change and natural disasters. • It reaffirms and builds on the 1994 National Environmental Policy, the environmental responsibilities in the 2008 Constitution of the Republic of the Union of Myanmar, and the obligations contained in the 2012 Environmental Conservation Law. The Policy also recognises Myanmar's commitment to the UN Sustainable Development Goals (SDGs) and Paris Agreement
Myanmar Sustainable Development Plan (MSDP) (2018-2030)	• MSDP, 2018-2030 was prepared by the Ministry of Planning and Finance and ensures the importance of managing the environment and natural resources in a sustainable manner, promoting sustainable and transparent investments in ways that sustain the resource base and benefit the local and national population as a whole; reducing environmental health risks from air and water pollution with improved access to those services, and reducing vulnerability to climate change related disasters and impacts. • This MSDP is structured around 3 Pillars, 5 Goals, 28 Strategies and 251 Action Plans. It is also aligned with the SDGs, the 12-Point Economic Policy of the Union of Myanmar, and various regional commitments, which Myanmar has made as part of the Greater Mekong Sub region (GMS) Strategic Framework, the ASEAN Economic Community (AEC) and many others. It also considered the National Sustainable Development Strategy (NSDS) in 2009 and the Agenda 21 (1997).
National Waste Management Strategy and Master Plan (NWMSMP) (2018-2030)	• NSWMSMP (2018-2030) is developed by ECD, MONREC in cooperation with the UN Environment and IGES/CCET and other relevant ministries, departments, institutions, state/ regional governments and townships aiming to institutionalise the waste management and offers a visionary document and strategic guide to address key issues, needs and challenges whilst raising awareness amongst key stakeholders towards achieving a resource efficient and zero waste society. • It is respected a holistic nature of waste management, which need to address waste in all its forms (solid waste, liquid waste/wastewater, and gaseous emissions) and identify strategic programs and actions to improve solid waste collection, reduction through 3Rs (reduce, reuse and recycling), intermediate treatment and disposal. • In addition, it identifies actions to maximize proper collection and disposal of industrial, medical and other hazardous waste, proper disposal and treatment of liquid waste (waste water from domestic and industries) whilst ensuring waste management services are made sustainable over the long term through supportive financial mechanisms, sound policies, institutional and monitoring frameworks.
Myanmar Climate Change Strategy and Action Plan (MCCSAP), 2017-2030	• Myanmar has ratified the global Paris Agreement under the United Nations Framework Convention on Climate Change (UNFCCC) and complemented by the Myanmar Climate Change Strategy and Action Plan 2016-2030 (MCCSAP 2016-2030), put the commitments of the Agreement into practice. It sets the stage for a strategic answer to the severe impacts of climate change in Myanmar. It aims to ensure that Myanmar's socio-economic development goes together with an efficient use of its resources • The Strategy has identified six key sector entry points to deliver inclusive climate resilient and low carbon development outcomes. The entry points include primary, secondary and service sectors, which contribute significantly to current and future economic and social development
Master Plan for Hazardous Waste Management in Myanmar, 2019	• The Master Plan for Hazardous Waste Management in Myanmar is developed by the ECD, MONREC in partnership with SINTEF and Norwegian Environment Agency aiming to achieve a sustainable hazardous waste system, Myanmar must manage its hazardous waste in a manner that will not compromise future generations' ability to meet their own needs. The development of a proper hazardous waste management infrastructure is not only required to protect human health and the environment, but it is also necessary to sustain further development of the economy.

Table 4. Related policies and regulations for solid waste management at townships

	Mandalay City Development Committee (MCDC)	Yangon City Development Committee (YCDC)	Nay Pyi Taw City Development Committee (NDC)
Related Laws/ Acts	- MCDC Law 2015, - MCDC Environmental Conservation and Cleansing bylaws, 2015 - The City of Mandalay Development Law, 2002	- The Yangon Water -work Act, 1885 - The City of Yangon Municipal Act, 1922 - The Water Power Act, 1927 - The Underground Water Act, 1930 - The City of Yangon Development Law, 1990 - The Yangon Civil Development Law, 2013	- NDC Pollution Control and Cleansing Department bylaws - NDC Water and Sanitation Department bylaws - The Nay Pyi Taw Development Law, 2009

regulations has lagged behind general policy proclamations. It has also been observed that Yangon and Mandalay Cities have made efforts to develop waste management strategies and associated master plans with the technical and financial assistance from international partners. For example, with the support of the Japan International Cooperation Agency (JICA), YCDC is currently working on a draft Master Plan for Waste Management through to 2040.

In Mandalay, MCDC has developed its City Waste Management Strategy and Action Plan (2017-2030) with the technical assistance of UN Environment and CCET. The strategy provides a long-term vision for guiding MCDC efforts shifting from conventional waste management practices (waste collection and disposal) towards a more sustainable waste management system, including through the implementation of the 3Rs (reduce, reuse and recycle) with the greater aim of achieving a resource efficient and zero waste society. Specific targets and actions have also been identified via a consultative process with key stakeholders to ensure efficient operation of waste management and recycling activities over the long term (See Figure 8).

Figure 8. Overview of city waste management strategy of Mandalay City, 2017-2030

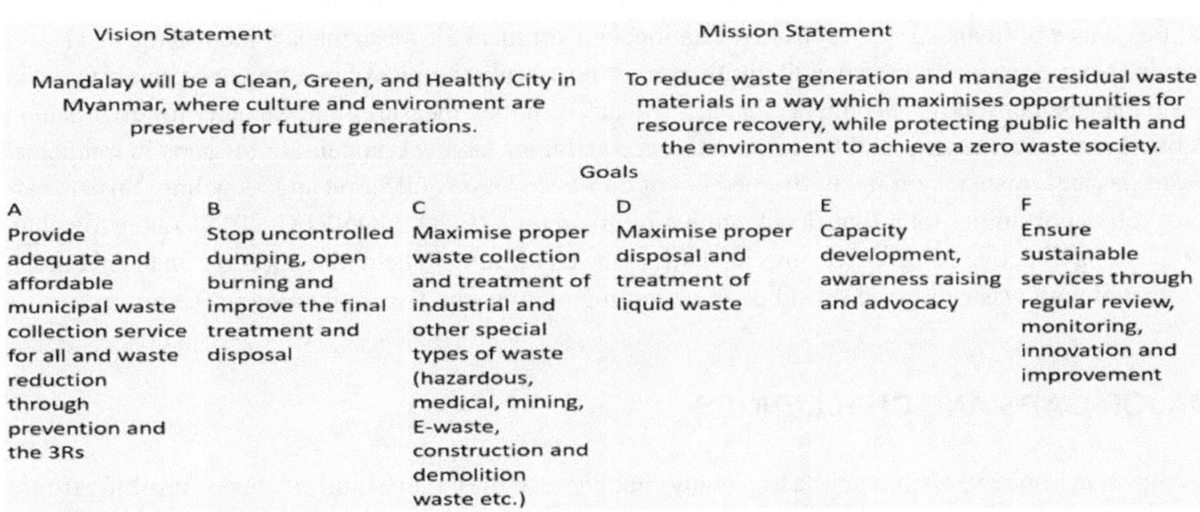

WASTE COLLECTION FEES, COSTS AND AFFORDABILITY

Although there is no current data on collection rate efficiency and corresponding fees, nearly all the cities in Myanmar have set user charges for the provision of waste collection services. Collection charges for household waste are based on a monthly fixed rate. Currently, MCDC and YCDC charge about 300-900 MMK/month (0.2-0.6 USD/month) and 300-600 MMK/month (0.2-0.4 USD/month) respectively for household waste collection. Field observations however have identified that average waste collection charges in Yangon fluctuated based on location, ranging from 600 MMK/month (0.4 USD/month) in CBD areas, 450 MMK/month (0.3 USD/month) in peri-urban areas and 300 MMK/month (0.2 USD/moth) in small townships (Netherland Enterprise Agency, 2017). Cities also provide waste collection services for other types of waste, of which customers such as commercial businesses, industries, medical facilities and others use on an on-call basis, with fees determined based on waste volumes or the fixed price for one truck (on average, three tons capacity). In Mandalay, collection charges for other waste range from about 11,000 MMK per ton (7.12 USD/ton) to about 35,000 MMK per trip (22.64 USD/trip) (MCDC and ECD, 2017).

Nevertheless, an analysis of the data reveals that the total revenue generated from the collection of waste remains quite low and thus is insufficient to cover total waste management expenditures. This is especially the case with regard to capital costs required for establishing new waste management infrastructure in respective cities. For example, annual operation and management costs of MCDC's cleansing department were identified to be 4,946 million MMK (3.2 million USD) in the year of 2015/2016. In contrast, total revenue collected from the provision waste collection services during the same period was only 651 million MMK (0.5 million USD). As such, municipal revenue covers only 16 percent of total costs of waste collection, despite including the public cleansing tax (320 million MMK), tax from public, commercial, and institutional building cleansing (32 million MMK), waste collection fees from medical waste collection (29 million MMK), hotels (120 million MMK), and commercial wastes (150 million MMK). The balance of approximately 2.7 million USD (84 percent of total costs) was subsidized by the MCDC from its annual operation budget (MCDC and ECD, 2017).

Given that waste service provision has conventionally been considered the responsibility of municipalities, a lack of financial support from the national government for waste management at the local level has also been identified. Correspondingly, there exist no useful examples of private sector involvement in waste management activities both due in part to a lack of policy measures and capacity for establishing Public-Private Partnerships (PPP). Informal sector activities largely compensate for gaps in municipal waste services, managing roughly five percent of city-level waste collection and recycling; this also affords job opportunities for a significant number of urban poor (ECD, MONREC, 2018) Taken together, establishing effective cost recovery mechanisms at the city level will help to reduce the financial burden associated with waste disposal subsidies, encouraging more sound fiscal planning in the process.

MAJOR GAPS AND CHALLENGES

As shown in Figure 9, Myanmar is facing many challenges at the national and city levels resulting from a range of social, economic, institutional and technical constraints (Premakumara & Hengesbaugh, 2016b; Premakumara et. al., 2017; ECD, MONREC, 2018).

Figure 9. List of key challenges and gaps

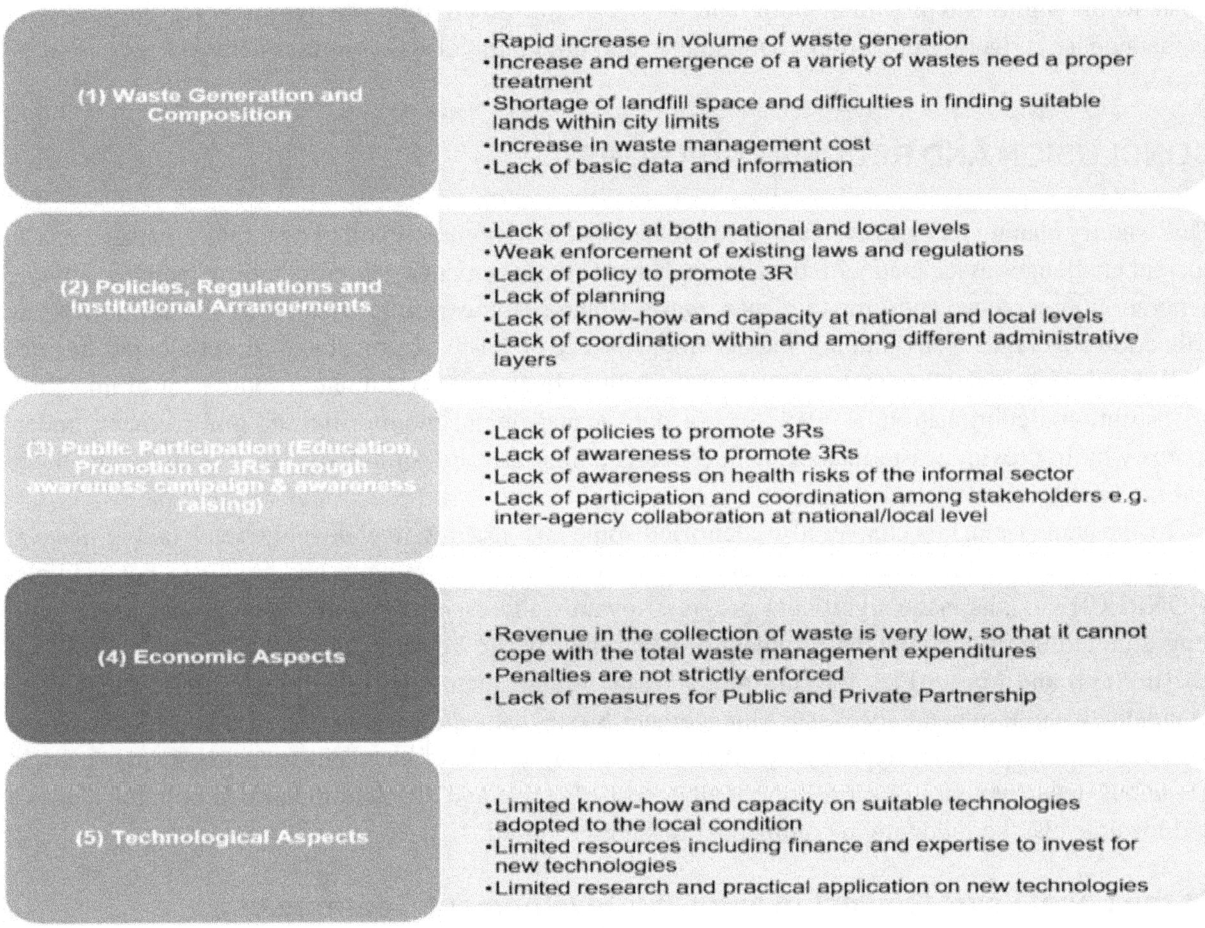

At the national level, ECD, MONREC reports that increasing consumption of resources is leading to growing waste generation and pollution, particularly of new and emerging types of waste, such as industrial, medical, plastic, e-waste and other hazardous waste. An attendant issue is the lack of budget for introducing proper treatment technologies and infrastructure, and the low capacity and knowledge to mobilise resources towards this end. Similarly, the Myanmar government is experiencing challenges with promoting the institutional coordination at horizontal (among relevant ministries and departments) and vertical (among central, regional and townships) necessary for managing waste related issues, as well as monitoring, evaluation, and enforcement of existing laws and regulations governing the waste sector.

Accordingly, townships have also identified a number of issues related to waste management and are making efforts to tackle them at the local level. In addition to increasing waste generation, these cities describe data gaps, low levels of public awareness about waste management and willingness to pay for waste services, a general lack of compliance with existing waste rules/ regulations, and poor cooperation with local authorities. At the same time, Myanmar cities also recognise that waste segregation practices are not being effectively implemented, nor is waste collection and transportation being carried out efficiently. This is resulting in the poor treatment of domestic, industrial and hazardous waste, as well as creating health and safety risks for waste collectors and operators. Following similar observations made by their national counterparts, another concern highlighted by cities is a deficiency of resources for waste technologies and procurement of land for the establishment of waste treatment facilities, as well as inadequate budget for operation, maintenance and upgrading of waste infrastructure.

CONCLUSION AND RECOMMENDATIONS

This country chapter recognizes that waste management in Myanmar is still in its infancy. In this context, current challenges associated with the country's waste management system, including increasing waste generation, emergence and growth of new waste streams (industrial, medical, and plastic), low waste collection coverage, open dumping and burning, absence of waste treatment and recycling options prior to disposal, lack of public awareness and participation, financing and human resource constraints, lack of institutional coordination, as well as poor data management, monitoring and enforcement, underscores why improving waste management remains a high priority for authorities at both the township and national levels.

At the same time, this chapter also identified some encouraging new developments taking place at the national and local levels aimed at improving waste management systems and practices. For example, MONREC-ECD has made significant progress towards addressing existing waste management challenges including through the development of a National Waste Management Strategy and Master Plan (2018-2030) and Master Plan for Hazardous Waste Management (2019). Similarly, both Yangon and Mandalay have designed City Waste Management Strategies and Master Plans in a consultative manner and taken some initial steps towards their implementation. In order to ensure waste management in Myanmar in carried out in a more coordinated, cost-effective and efficient manner at both the national and local levels, the following recommendations are proposed:

At the Local Level (Township and City Development Committees)

- As an initial step, TDCs and CDCs should prioritize extending waste collection services across all areas of towns and cities, particularly informal communities and peri-urban zones, where such collection is currently unavailable. TDCs and CDCs can work on identifying ways for improving the affordability and frequency of waste collection based on consultation with local residents and integrating the involvement of informal sector/ community groups. In addition, secondary waste collection and transportation to final disposal sites can be managed and operated by TDCs or CDCs with potential involvement of private waste collection companies. Establishing a proper waste collection planning including collection routing, frequency, transfer stations/ secondary collection points and waste collection fee system is also necessary to ensure effective services.
- Secondly, TDCs and CDCs should implement measures to eliminate illegal disposal and the open burning of waste as well as improve the operation of existing disposal sites. Illegal dumping and burning of waste in residential areas can be controlled by providing effective and affordable waste collection services, conducting educational and awareness on the negative impacts of such practices and issuing regulations supported by proper monitoring and enforcement mechanisms. Based on the current situation, all TDCs and CDCs should take immediate measures to stop open dumping and convert current dumpsites into controlled or engineered operations with proper access roads and fences as a first step.

- In addition, sanitary landfill sites should be equipped with impermeable liners to prevent liquid discharges from polluting ground and surface water, gas management systems to reduce risks of fire or explosion, sufficient soil cover to minimize odor, as well other environmental protection features. Considering the investment costs, lack of public lands and technical skills, regional or cluster landfill can be introduced with potential business models (public-private partnership). In the medium or long-term, other technical options such as Waste-to-Energy Technology (incinerator and landfill gasification) can be also be considered particularly in Yangon and Mandalay with requisite operational and financial recovery plans.
- Further, sustainable waste management does not imply an exclusive focus on waste collection and end-of-pipe treatment and disposal. It is necessary to integrate upstream pollution prevention actions based on the waste hierarchy, including waste prevention, minimization, reuse, recycling, and recovery including energy recovery, prior to the final disposal. This will involve setting waste targets and formulating appropriate strategies for encouraging waste segregation at source, promoting traditional repair and reuse practices, preventing the landfilling of food waste and maximizing the involvement of informal and small-scale entrepreneurial recycling within the conventional waste management sector. Considering that organic waste comprises a large volume (about 60%-70%), actions should be taken to reduce levels of organic waste (food waste) from households, markets, shopping malls and other enterprises.
- In addition, partnership building and awareness raising to encourage behavior change is crucial to moving from the waste disposal practices to more resource circular approaches. This requires implementing innovative and well-designed environmental education and training programs, with a view to both generate awareness and knowledge on new waste management practices, as well as motivate commitments and actions for sustainable lifestyles in line with the Sustainable Development Goals (SDGs).

At the National Level (ECD, MONREC and Other Relevant Ministries)

- Establishing an enabling environment including effective policies, legal systems, institutional framework, financing mechanisms and an enforcement/monitoring system for enforcing the successful implementation of National Waste Management Strategy and Master Plan (208-2019) and other related strategies, including the country's Master Plan for Hazardous Waste Management (2019). These actions should be aligned with related sustainable development and climate targets set out in the country's Intended Nationally Determined Contribution (INDC). The additional policies and strategies should also develop to address the emerging waste such as plastic waste, e-waste and industrial waste etc., in coordination with other relevant ministries/ departments and concerned stakeholders.
- ECD, MONREC should develop waste classification and clear performance indicators as well the requisite methodology to track progress against national and city waste management strategies and targets, such as a national waste database and monitoring mechanisms. This involves concentrating the majority of waste management responsibilities within a single government body or establishing an appropriate committee, independent department or agency and clearly defining the waste roles and responsibilities of other concerned institutions.
- Actions should also be taken to increase financial support as well as encouraging additional stakeholders, including the domestic and international financial community and/or other private sec-

tor actors to invest in the improvement of local waste management based on local strategies and action plans. Moreover, where appropriate, national authorities can work towards enabling the establishment of innovative and transparent funding approaches, including independent, blended or pooled funding entities, as well as pay for performance delivery mechanisms. Myanmar's Corporate Social Responsibility (CSR) regulation targeting private enterprises may also be considered for this purpose.

- Establishing end-of-life incentive policies aimed at stimulating market demand for recycling, increasing product recyclability and creating the appropriate conditions for encouraging investments in collection, sorting and environmentally-friendly waste treatment should be implemented within the country particularly considering the plastic waste import ban by China and some other Association of Southeast Asian Nations (ASEAN) countries. This includes offering financial incentives and rewards to industries aimed at nurturing climate-smart and resource-efficient initiatives, including the promotion of cleaner production practices and technologies. Other actions include supporting the informal and small and medium-scale recycling sectors to enter new service agreements for carrying out collection, recycling, composting and waste treatment in partnership with townships and civil society organizations, which can assist in improving health and safety practices while at the same time supporting economic livelihoods.
- Training and capacity building programs should be developed for township staff specifically with regard to institution strengthening aimed at promoting proactive policies and technologies that are acceptable, affordable, scalable, and appropriate to local circumstances. This should also include efforts aimed at incentivizing city- actions to improve waste management through national awards and certification programs.

International Cooperation

- While waste management in Myanmar is at an early stage of development, international cooperation has enormous potential for improving the country's governance system, including enabling policies, regulations, institution and capacity building as well as investments for infrastructure development necessary for effective waste management. Improving access to capital finance is essential for assisting the development of Myanmar's critical infrastructure aimed at managing increasing levels of waste generation, as well as ensuring proper collection, transport, treatment and disposal. Doing so will include improving access to loan funding at affordable interest rates from international development institutions (World Bank, ADB and JICA) as well as leveraging investments from private investors, philanthropic sources and other bodies supporting climate finance (UN Environment, GEF and concerned donor countries).
- However, international investments to improve waste management in Myanmar should be appropriate and affordable in line with local circumstances as opposed to being strictly determined by donor countries that have modernized their waste management systems over a longer period of time. In this regard, strengthening knowledge and performance of Myanmar's public sector, including by introducing more inclusive approaches through the endorsement of appropriate policies, regulations, and institutions will help to ensure that the country is equipped with the competencies and skills to deliver sustainable, locally-supported waste management systems.
- The experience in Yangon and Mandalay in collaboration with international partners illustrates the potential of intercity cooperation for mainstreaming and disseminating environmentally sound

waste management systems with Myanmar's TDCs and CDCs. Promoting city-to-city cooperation between Myanmar cities and other mentor cities--not only North-South but also South-South development assistance, especially from countries that possess longstanding-experience in terms of upgrading their waste management systems, offers a useful strategy for encouraging the sharing of experiences including best practices, and other technical support.

REFERENCES

ADB. (2017). *Urban Development and Water Sector Assessment, Strategy and Road Map*. Manila, Philippines: ADB.

ADB. (2018). *Myanmar Economy*. Retrieved from https://www.adb.org/countries/myanmar/economy

Baker, N. (2016). *Myanmar's air pollution among the worst in the world*. WHO. Retrieved from http://www.mmtimes.com/index.php/national-news/22840-m

Balde, C. P., Wang, F., Kuehr, R., & Huisman, J. (2015). *The global e-waste monitor – 2014, United Nations University*. IAS – SCYCLE.

Bingham Centre for the Rule of Law. (2013). *Constitutional Awareness Myanmar*. Retrieved from http://binghamcentre.biicl.org/binghamcentre/projects/burmamyanmar

Earth Day Network. (2018). *Fact Sheet – Single Use Plastic*. Retrieved from https://www.earthday.org/2018/03/29/fact-sheet-single-use-plastics/

ECD, MONREC. (2018). *National Waste Management Strategy and Master Plan for Myanmar, Nay Pyi Taw* (Waiting for official approvals). Author.

ECD, MONREC. (2019). *Master Plan for Hazardous Waste Management in Myanmar, 2019, Nay Pyi Taw* (Waiting for official approvals). Author.

Embassy of Japan in Myanmar. (2019). *Exchange of Notes on Grant Aid to Myanmar "the Project for the Urgent Improvement of Solid Waste Management in Yangon City"*. Retrieved from https://www.mm.emb-japan.go.jp/profile/english/press/20190218.html

FAO. (2018). *Myanmar Economy and the Role of Agriculture*. Retrieved from http://www.fao.org/myanmar/fao-in-myanmar/myanmar/en/

Government of Myanmar and UNFPA. (2014). *Population and Housing Census of Myanmar – Data Sheet*. Retrieved from http://myanmar.unfpa.org/sites/asiapacific/files/pub-pdf/Census%20Data%20Sheet%20-%20ENGLISH_0.pdf

INFORM. (2019). *Information Report (Canadian Wood Fibre Centre)*. Retrieved from http://www.inform-index.org/Portals/0/InfoRM/2019/Inform%202019%20WEB%20spreads%20(3).pdf?ver=2019-02-07-113610-123

Jambeck, J. R., Geyer, R., Wilcox, C., Siegler, T. R., Perryman, M., Andrady, A., . . . Lavender Law, K. (2015). *Plastic waste inputs from land into the ocean*. Retrieved from https://www.iswa.org/fileadmin/user_upload/Calendar_2011_03_AMERICANA/Science-2015-Jambeck-768-71__2_.pdf

Lebreton, L. C. M., van der Zwet, J., Damsteeg, J.-W., Slat, B., Andrady, A., & Reisser, J. (2017). River Plastic Emissions to the World's Oceans. *Nature Communications*, *8*, 15611. doi:10.1038/ncomms15611 PMID:28589961

Mar, W. W. (2018). *The 2018 World Environmental Day – Steps to beat plastic pollution*, Retrieved from http://www.globalnewlightofmyanmar.com/2018-world-environment-day-steps-beat-plastic-pollution/

MCD and ECD. (2017). *City Waste Management Strategy and Action Plan for Mandalay, the Republic of the Union of Myanmar*. Author.

Ministry of Planning and Finance. (2018). *Myanmar Sustainable Development Plan (2018-2030), the Government of the Republic of the Union of Myanmar*. Nay Pyi Taw.

MOECAF. (2012). *Myanmar's Initial National Communication under the United Nations Framework Convention on Climate Change*. Retrieved from https://unfccc.int/resource/docs/natc/mmrnc1.pdf

MYANMORE. (2018). *Myanmar targets plastic waste in growing green movement*. Retrieved from https://www.myanmore.com/yangon/2018/06/myanmar-targets-plastic-waste-in-growing-green-movement/

Netherland Enterprise Agency (2017). *Scoping Mission on Integrated Solid Waste Management (ISWM) to Myanmar.* Author.

Nixon, H., Joelene, K. C., Chit Saw, P., Aung Lynn, T., & Arnold, M. (2013). *State and Region Governments in Myanmar*. Yangon: MDRI-CESD and the Asia Foundation.

Premakumara, D. G. J., & May Tin, H. O. (2019). Results of Quick Survey on Solid Waste Management in Mandalay Region. A Presentation make at the Regional Workshop for Improving Municipal Solid Waste Management in Mandalay, Myanmar.

Premakumara, D. G. J., & Hengesbaugh, M. (2016). *Quick Study on Waste Management in Myanmar: Current Situation and Key Challenges*. IGES.

Premakumara, D.G.J., Hengseshaugh, M., Onogawa, K., & May Tin, H. O. (2017). *Waste Management in Myanmar, Current status, key challenges and recommendations for National and City Waste Management Strategies*. IGES and UN Environment.

Premakumara, D. G. J., Kataoka, Y., & Chowdhury, M. (2016a). *Development of Environmental Learning Programme for Establishing a Sustainable Solid Waste Management System in Mandalay City, Myanmar*. The HDCA 2016 Conference: Capability and Diversity in a Global Society, 1-3 September 2016, IGES, Japan.

UNDP. (2017). *UNDP – Myanmar Annual Report 2017*. Retrieved from http://www.mm.undp.org/content/myanmar/en/home/library/poverty/2017-annual-report.html

UNFPA Myanmar. (2017). *Census report projects population growth over 35 years, and supports long-term planning*. Retrieved from https://myanmar.unfpa.org/en/news/census-report-projects-population-growth-over-35-years-and-supports-long-term-planning

WHO. (2016). *Ambient air pollution: A global assessment of exposure and burden of disease*. Geneva: World Health Organisation.

World Bank. (2012). *What a Waste*. Retrieved from http://siteresources.worldbank.org/INTURBANDEVELOPMENT/Resources/336387-1334852610766/What_a_Waste2012_Final.pdf

World Bank. (2018). *Overview of Myanmar*. Retrieved from http://www.worldbank.org/en/country/myanmar/overview

World Bank. (2018). *What a Waste 2.0*. Washington, DC: World Bank.

World Economic Forum. (2016). *The New Plastic Economy – Rethinking the Future of Plastic*. Retrieved from http://www3.weforum.org/docs/WEF_The_New_Plastics_Economy.pdf

World Resource Institute. (2011). *Climate Analysis Indicators Tool (CAIT) Version 9.0*. Washington, DC: WRI.

WRI-CAIT. (2017). *Myanmar*. Retrieved from https://www.climatewatchdata.org/ndcs/country/MMR/full

Chapter 10
Solid Waste Management Practices in Pakistan

Maria Manzoor
National University of Sciences and Technology, Pakistan

Iram Gul
National University of Sciences and Technology, Pakistan

Irum Iqrar
https://orcid.org/0000-0001-8083-1813
Quaid-i-Azam University, Pakistan

Muhammad Arshad
National University of Sciences and Technology, Pakistan

ABSTRACT

This chapter covers different aspects of solid waste management (SWM) in Pakistan. Being a developing country, Pakistan is facing serious environmental issue as a result of improper and limited SWM practices. The current status of waste generation is far high than waste management operation working in the country. Detailed information has been provided on account of waste generation per capita, waste composition and types, current status of waste management practices, policy and legislations on solid waste management. Furthermore, particular challenges in local setups for achieving waste management goals have been highlighted. There is urgent need to look into the various options to deal with proper management of solid waste in the country. In this regard, various projects by governmental and private sector have been initiated. Particularly, projects related to waste to energy (WTE) have been planned and operational in certain parts of the country.

DOI: 10.4018/978-1-7998-0198-6.ch010

INTRODUCTION

Solid waste management is a key component in growth and development of countries. It not only address various environmental issue related to pollution control and management, but also contribute in social development, tourism, re-sale value of property and disease control thereby have direct and indirect impact on country's economy. However, solid waste management is a serious issue in developing countries as well as in Pakistan. Country planning directives have never considered poverty and environment as an important nexus to address. As a result, enormous increase in population, unplanned urbanization, agriculture and industry proliferation without necessary checks and regulation led to generation large amount of solid waste contaminating natural environment. Solid waste generation in Pakistan is 50,000 tonnes per day, out of which only 60% is collected and disposed off by local municipalities (IUCN, 2004). Like other developing countries, Pakistan lacks waste management infrastructure, creating serious environmental problems. Poor management of solid waste is threatening environment quality and health safety. Due to environmental issues regarding poor sanitation and pollution caused by improper waste management, more than 5 million people die out each year. The issues regarding environment protection and management are not adequately addressed by government during national decision making. The enormous increase in industrialization, miss managed urbanization; poor agricultural practices, improper construction and demolition are the major sources of producing solid waste in the country.

Unfortunately, there is no proper solid waste management system exist in any city of Pakistan. All procedure, from collection points to transportation and final disposal of solid waste are poorly managed. The collected waste is dumped on open ground which inturns contaminate environment and generated different environmental problems. The most important issues includes site quality degradation, hazardous waste pollution, seepage of leachate and ground water pollution, bad smell and breeding ground for rodents, flies and mosquitoes, that carry different diseases. The main problems in solid waste management in Pakistan are listed below:

- Lack of awareness in citizens about importance of proper waste disposal without compromising environment.
- Lack of awareness that environmental and public health problems are associated with improper waste management practices.
- Absence of proper waste collection, transportation and disposal system.
- Unplanned and open dumping of domestic waste on nearby streets or free land designated by community.
- Absence of waste separation facilities during waste dumping and collection.
- Absence of controlled (lined) waste disposal landfill sites.
- Lack of technical experience, human resources and equipment lower the capacity of municipalities.
- Absence of local taxation, employment and regulations in solid waste management (SWM).
- Less priority is given from political leaders and public.
- Lack of investment and review generation in waste disposal.
- Absence of relevant policies and regulations followed by ineffective implementation.
- Lack of public interest and incentives to adopt hygienic methods of waste disposal.

Figure 1. Population increase and solid waste generation (SWG) in major cities of Pakistan
(Source: USCS Pakistan contact from Project Procurement International, Pakistan, http://projectpi.pk/)

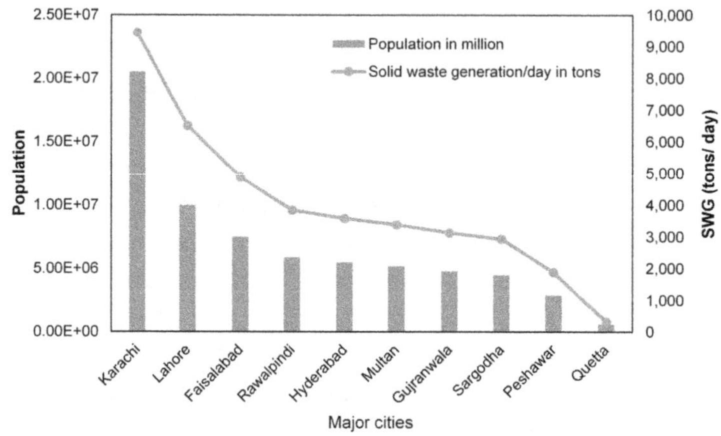

SOLID WASTE GENERATION

Pakistan is a developing country and facing serious environmental issues as a result of population increase. Currently, Pakistan is 6th most populous country in the world with estimated population of 204.60 million, approximately (Pakistan Bureau of Statistics, 2019) and contributes to 2.65% of the total world population. The total land area of Pakistan is 770,880 Km^2 (297,638 sq. miles) with population density of 265 per Km^2 (687 people per mi^2). Of the total population, 39.8% is urban (81,499,983 people in 2019). The Annual increase in population (population growth rate 4.1%) and GDP (GDP rate 6%) have increased the dependency on natural resources. Solid waste generation statistics in major cities of Pakistan in relation to population is given in Figure 1 and Table 1 respectively.

The exploitation of natural resource has resulted in increased environmental pollution and associated health issues. Among which hazardous and non-hazardous solid waste is the most prominent and widely encountered environmental issue. Owing to the significance of problem, it has been identified in core priority list by national environmental action plan (NEAP). The solid waste generation (SWG) in major cities of Pakistan is given in Figure 2. It is observed that major solid waste in country is contributed by food waste (30%) followed by brick and dirt (18%) and yard waste (14%). According to a statistic data, urban areas are major contributor of solid waste production and generate 54,850 thousand tonnes/day of solid waste with much lower collection efficiency (60%) by municipal solid waste management committee. It is estimated that in urban areas, the waste is generated at 0.283 – 0.612 kg/capita/day with a growth rate of 2.4% (UNEP, 2019).. Only 60% of solid waste is collected by town municipalities, while rest of the waste remains unattended which becomes a source of various environmental problems (USCS, 2019).

Solid waste collection in Pakistan is mainly done by local municipalities. The waste collection facilities are limited and do not exist in poor/ low income rural areas. In better off areas the facilities of waste collection are available but the frequency of collection is not adequate and the proportion of waste collected is much lower than the amount of waste generated. In poor areas, the only way of waste minimizing is the informal scavenging practices done by poor and needy people. Other activities include animal eating, biodegradation, mineralization, dispersion by wind and burning heaps of solid waste at disposal site.

Table 1. The population census along with amount of MSW generated in different cities of Pakistan in 2016 (Korai et al., 2017)

Cities in Pakistan	Population in Millions (2016)	Rate of Solid Waste Generation (Kg/Capita/Day)	Total Waste Generated (Million kg/Year)
Karachi	22.83	0.57	190.60
Lahore	10.36	0.75	113.40
Faisalabad	3.68	0.45	24.16
Hyderabad	2.99	0.80	34.92
Peshawar	1.79	0.38	9.92
Gujranwala	2.20	1.08	34.60
Quetta	1.14	0.38	6.28
Multan	1.95	0.53	15.08
Sialkot	0.58	0.31	2.68
Islamabad	0.74	0.53	5.72
Rawalpindi	1.77	0.21	5.44
Sukkur	0.59	0.45	3.84
Total	50.59	6.44	446.64
Other urban area	23.11	0.84	283.44
Total Rural area	114.32	0.30	500.72
Hazardous wastes	188.02	7.58	1230.56
Grass Total	-	-	61.52

Figure 2. Amount of MSW generated in different cities of Pakistan (Korai et al., 2017)

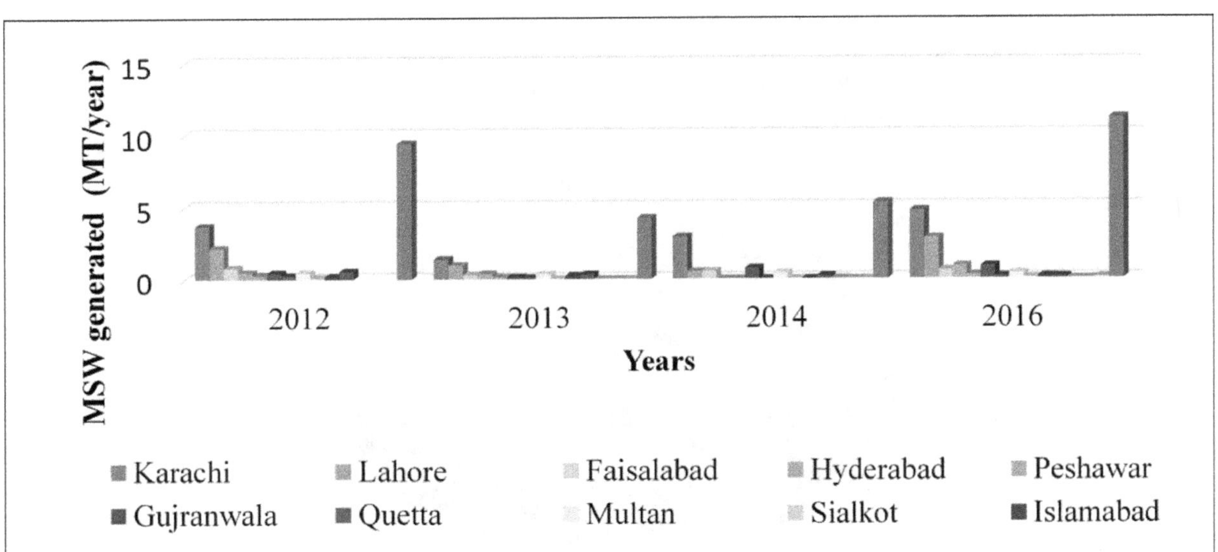

This informal waste collection system makes a large proportion of waste recycling sector in Pakistan. This is a common practice both in rural and urban areas throughout Pakistan. The scavengers collect these items for the sake of reuse or sell them in recycling market for the sake of income. The composition of waste collected varied with area, time and advancement in technology. Now–a-days the principal components of waste collected includes plastics, glass, aluminum cans, iron and paper. With the advancement in technology and living standards particularly in urban areas, traditional practice of solid waste disposal on open land poses serious health and environmental risks. Therefore, public sector is mainly and directly involved in solid waste management throughout Pakistan. With the growth in urbanization and industrialization the problems associated with urban and industrial waste management are also increasing particularly in major cities of Pakistan such as, Lahore, Karachi, Faisalabad and Rawalpindi. These problems include enormous waste generation, complex waste composition, hazardous waste, minimum waste collection and disposal. The public sector itself cannot manage solid waste alone.

Under this scenario, there is dire need of involving private sectors and non-governmental organizations (NGOs) to handle wide spread waste management problem by offering them incentives and projects. Up till now there is no organized private sector working with government to collect, reuse or recycle the waste generated. There are particular success stories like Waste Buster, but this problem is not one time, it need to be regularized and operational for timely solid waste management. The municipalities are now realizing the benefits of privatizing solid waste management including, collection, transportation, segregation and composting of organic waste. However, this requires financial support and transparency in project allocation based on build, operate and transfer (BOT) of solid waste management facility (GoP, 2019).

WASTE CLASSIFICATION AND FLOW

There is no proper solid waste classification system in Pakistan. All waste are collected at transported in similar fachion. Only hospital and hazardeous waste are given significant importance and are incinerated

Figure 3. Classification of MSW generation (%) in Pakistan
(Source: United Nations Environment Program, report on waste management in Pakistan. UNEP, 2019)

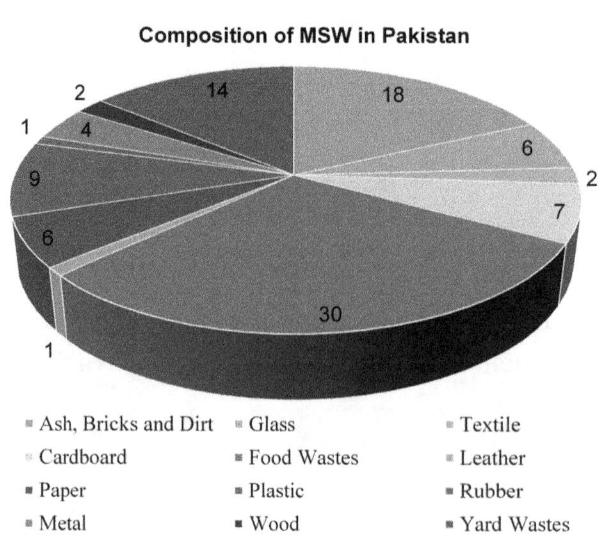

Figure 4. Steps in waste management in Pakistan
(Source: Pakistan Environmental Protection Agency, 2005)

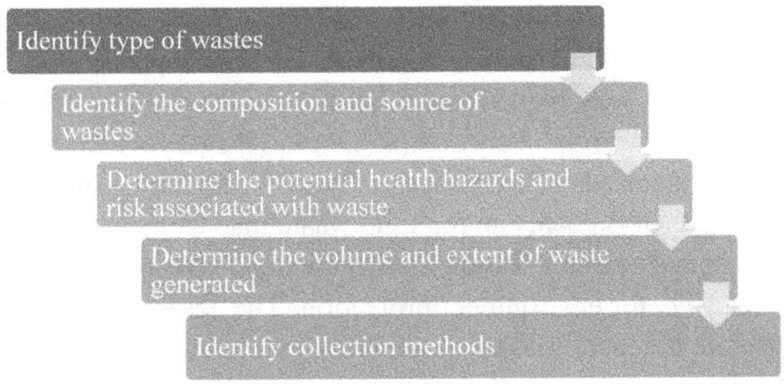

at the point or generation or carried and then incinerated at waste management facility. For all other types of waste, which are collected local municipalities and treated are given in Figure 3.

The Pakistan Environmental Protection Agency (2005) has identified waste classification and flow that must be considered in solid waste management process. The steps are given in Figure 4.

WASTE GENERATION / 3R PRACTICES, WASTE MANAGEMENT TECHNOLOGIES, POLICIES AND REGULATIONS

Solid waste generation in Pakistan is about 48.5 million tonnes/year with an annual increase of 2% (Pakistan, 2019). It is estimated that 87,000 tonnes of solid waste is generated per day in Pakistan (Pakistan, 2019). And most of this solid waste is contributed by major metropolitan cities. Among which Karachi and Lahore are big contributors. However only 60-70% of soilid waste is collected from cities and managed by the local mucicipalities and governmental agencies. The 3R practices are not properly followed by the concerned departments. Usually the scavangers and poor people collect and recycle few items from the dumping sites. The major achievement in 3R practices in Pakistan is the establishment of Solid Waste Management Strategy for Tehsil Municipal Administration (TMA) Barikot, Khyber Pakhtunkhwa (KPK). The project was implemented in 2014 by GIZ (German Cooperation) at Barikot, KPK. The purpose of this initiative was to transform solid waste into a viable and competitive industry relying upon a locally generated resource base. The strategy was developed to be implemented for five years and targeted 50% waste efficiency.

The waste management technology includes the collection of solid waste at open dumping site, from where the waste is transported by open trucks, tractors/trolleys system to the disposal site. Some municipalities also hire sweepers and workers to collect solid waste by wheelbarros, handcarts and brooms. Finally the collected waste is disposed off on open landfills. Close/lined landfill do not exist in Pakistan. The number of open landfills are also inadequate to mange the amount of qwaste generated in country. The largest metropolitan city of Pakistan is Karachi, which generated 13,000 tonnes of municipal waste daily but only have three sanitary landfills. Lahore, the 2nd largest city of Pakistan, has two landfills which are inadequate to manage the huge amount of waste generated. Generally, in most of locations, solid waste is simply dumped at open place just outside the city. Solid waste management facilities varies province wise. In Punjab, only Lahore have proper solid waste management plan which is achived through collaboration with Turkish comparies. The similar solid waste management plans have been planned for other cities of Pakistan. In Sindh, Asian Development Bank's (ADB) is sponsoring 400 million dollars to Sindh Cities Improvement Investment Program (SCIP) (Source: Government of Sindh, http://scip.gos.pk/website; http://scip.gos.pk/eoi.php). In Khyber Pakhtunkhwa, attempts have been done by Water and Sanitation Services Peshawar (WSSP) to build a sanitary landfill. Where asno significant infrastructure exists in Balochistan for soild waste management.

To address solid waste management, Pakistan Environmental Protection Act (PEPA) was enacted by government of Pakistan in 1997 for environment protection. The act provided guidelines for establishment and working framework of federal and provincial Environmental Protection Agencies (EPAs). The act addressed various environmental issues including biodiversity conservation and waste management and set nationwide targets for protection of environmental and health.

The government of Pakistan also address to multilateral environmental agreements as outlined in policy document (National environmental policy 2005)

- "Effectively participate in regional and international forum to foster cooperation for protection of environment and natural resources.
- Ensure effective implementation of all bilateral, regional and international multilateral environmental agreements, protocols and conventions to which Pakistan is a party, in line with national policies and priorities.
- Develop and implement national strategies and action plans for all multilateral environmental agreements, Johannesburg Plan of Implementation and Millennium Development Goals."

Case Study: Waste Disposal Strategy of Lahore Waste Management Company (LWMC) (GoP, 2019)

In Pakistan, SWM has long been a neglected sector. The reason is, lack of strong commitment on the part of government to introduce management reforms in institutions for managing waste on the basis of internationally accepted standards. In this context, the government has established Lahore Waste Management Company (LWMC) on 19th March 2011 with a vision to improve and modernize the SWM services in Lahore. This model of effective waste management has been replicated in other cities of Pakistan too and now the efficient collection and disposal of waste is trending in this country and efforts are being made to exploit the energy potential of MSW. MSW Generation in Lahore is:

- 2090620.633 tonnes/ year.
- 2090620633 kg/ year.
- 612 kg/capita/day.

Waste management practice: The Lahore Waste Management Company (LWMC) in consultation with M/s ISTAC, designed the new SWM system for out sourcing. The project was made after fair international competitive bidding. Briefly, two Turkish companies were selected and the total cost of the project was US$ 320 million for seven years. During project Lahore was divided in to teo zones and each zone was given to different companies selected. The M/s Albayrak has worked in zone 1 and M/s Ozpak in zone 2.

The scope of outsourcing of SWM services includes:

- Sweeping manually.
- Sweeping by mechanical operations.
- Washing by mechanical sources.
- Collection of solid waste.
- Door to door collection.
- Container based collection.
- Waste transportation to the disposal site selected.

The most preferred strategy to dispose of waste is waste reduction and recycling followed by biological and thermal treatment of waste, landfill gas (methane) capturing and then landfilling of remaining waste (Figure 5). Henceforth, making use of the 3R waste minimization strategy. The disposal of waste in landfill is the least preferred option in the context of sustainable waste management. The management hierarchy also explains the type of technologies to be used for treatment of waste. LWMC anticipates that its waste disposal strategy will lead it to efficient waste management operations, which are economically affordable, technically feasible and above all environmentally sound.

Figure 5. Waste DISPOSAL strategy of Lahore waste management company

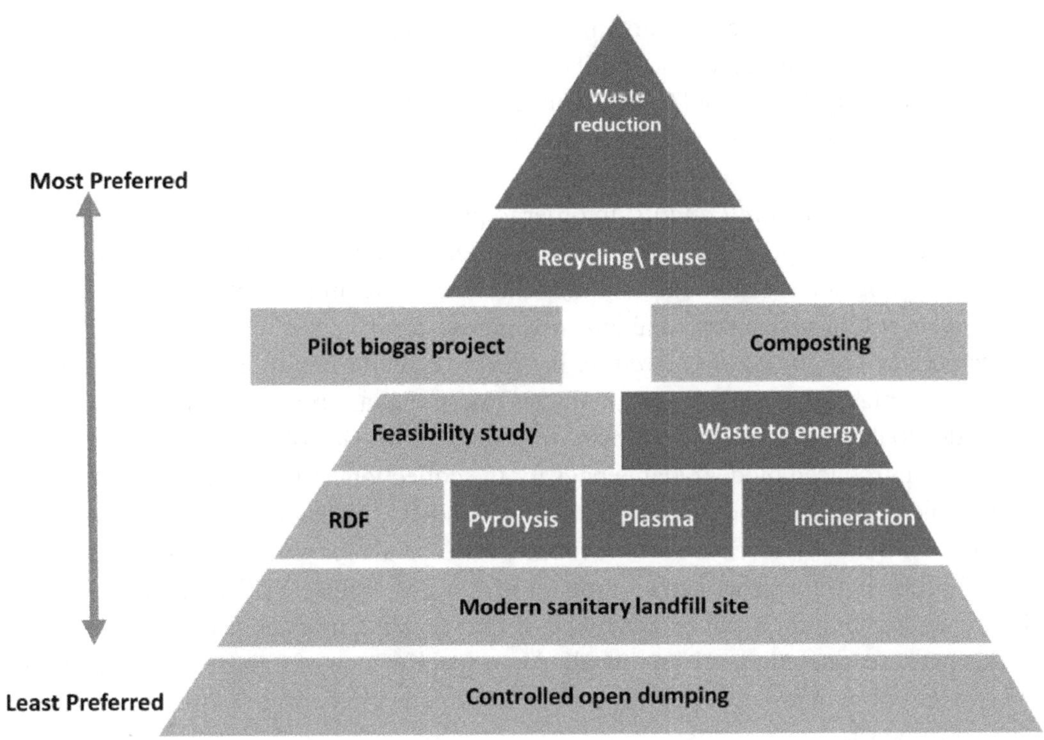

Formal Waste Sector: Lahore Waste Management Company (LWMC) is a PPP company operating in Lahore. In addition to handling the waste in Lahore, it is associated with the different projects. A brief discussion is given as follows:

a) *Generation of biogas from solid waste:* Currently LWMC is further exploring into the potential of generation of biogas from cattle's dung, for this plant has been set up at Dunya Pur (Chak 342 & Chak 327) and Ichogil. The aim is to generate biogas and provide it to the residents of the respective villages.
b) *Compost Formation:* LWMC after entrusting Lahore Compost with the composting project has established the first composting facility at Mahmood Booti under an agreement with the City District Government Lahore (CDGL).
The purpose of the project was to convert municipal solid waste to organic fertilizer of commercial value. For this purpose, aerobic windrow type composting technology was utilized for production organic fertilizer. This method is instrumental in reducing the risk of environmental degradation of the soil by conversion of organic waste to compost that can be utilized safely. Under natural conditions, the decomposition of municipal organic waste results in generation of toxic gases including methane. By composting, the methane emissions can be minimized and controlled.
Another objective of the project was to reduce greenhouse gas emission and thereby help in achieving the objectives of combating climate change under UNFCCC. The project also aimed at regional and national sustainable development.
c) *Refuse Derived Fuel:* LWMC signed an agreement with the M/s DG Khan cement in August 2011 for sale of waste. M/s D G Khan cement has built a Refuse Derived Fuel (RDF) plant for processing of 1000 tonnes of municipal waste. RDF is largely combustible components of municipal waste such as plastics and biodegradable waste. DG Khan Cement uses RDF in its cement plants as an additive fuel with coal.
d) *Waste Treatment and disposal site/Landfill:* In order to prevent the disposal of waste haphazardly in and around Lahore, now proper sites have been identified for waste treatment and disposal. One dumping and landfilling site is Lakhodair and the other one is Mehmood Boti. Arrangements are being made to recover biogas from the available organic waste.
e) *Using Bio-degradable bags for waste collection:* In order to fight the emerging issue of plastic pollution caused by the plastic bags which eventually end up in the landfills or are a source of havoc for aquatic life due to its flow to the main streams of water; LWMC in collaboration with Business Dynamics Pvt. has introduced biodegradable bags for the collection and transport of waste. These bags are readily bio degradable and thus do not lay unattended at the landfill sites for years.
f) *Separate Collection system for recyclables:* About 21.2% of all recyclable waste in Lahore is recycled. The informal sector generates an amount of Rs. 271 million (US dollars 4.5 million) per year. However, Rs.65 million can be generated, if the recycling practice is done by the formal sector. For this purpose, dual bin has been installed in a specific area of Lahore i.e. Shadman and Upper Mall as a part of pilot project. The major objective of this project is to reuse and recycle the waste produced at the site.
g) *Using C&D waste for construction of pathways:* A dedicated service for the collection of construction and demolition (C&D) waste has been introduced. The collected waste may be sold for the filling / leveling of low lying areas and /or for the construction of roads etc. Undergoing the present practices of construction, it is estimated that C & D waste generated from these activities would be

approximately 350 – 400 tonnes / day. Further option to recycle C&D will also be explored. C&D waste is collected and transported on full cost recovery basis by keeping in mind the following objectives:

- To avoid intermixing of C&D waste into household/street waste.
- To ensure in time collection of C&D waste.
- To improve the aesthetics of the city.
- Reusing the C&D waste.

HAZARDOUS WASTE GENERATION AND MANAGEMENT, POLICY AND REGULATIONS, TRANSBOUNDARY MOVEMENT OF HAZARDOUS WASTE

Hazardous wastes are generally defined as the waste or combination of wastes which adversely affect environment and human health (Misra and Pandet, 2005). Further these waste may contain one or more hazardous or radioactive characteristics such as oxidizing, carcinogen, irritant, flammable or explosive (Li et al. 2015). The materials from laboratories, clinics, hospitals, e-product, fuels, vehicles, industrial sludge, waste oils, chemical waste, health care units, batteries, florescent lamps and solvent based paints contain the hazardous waste (Gu et al., 2014). These waste are persistent, non-biodegradable, highly toxic at low concentrations and are biologically magnified (Saleh, 2016; Pichtel, 2014). These waste affect the soil, water, air furthermore presence of hazardous waste negatively affect the health of microorganisms, plants, animals and human as these water transfer through food chain. Therefore, proper treatment is required before releasing these wastes into the environment. Like other countries, hazardous waste is also generated in Pakistan from different sectors including: small and large industrial sectors, hospitals, clinics and laboratories, agriculture, household and commerce (EPA, 2005).

According to Environmental Protection Act (1997), hazardous waste is defined as "waste which is or which contains a hazardous substance, and includes hospital waste and nuclear waste. Hospital waste includes waste medical supplies and materials of all kinds, and waste blood, tissue, organs and other parts of the human and animal bodies, from hospitals, clinics and laboratories. Further nuclear waste, means waste from any nuclear reactor or nuclear plant or other nuclear energy system, whether or not such waste is radioactive". In Pakistan, the common hazardous wastes are radioactive wastes (hospitals, clinics, laboratories, hospitals, and nuclear power plants), toxic chemicals (chemicals shops, car wash, agriculture, nuclear power plants, pest control agencies, and battery shops), biological wastes (medical research units, hospitals, laboratories and clinics), flammable wastes and explosive. The statistics related to the hazardous waste generation by the industrial sector is not defined however, 250,000 tonns of waste is generated from all kind of medical facilities. The hazardous substance rules of Pakistan (2014) enlist a large number of substances (243 substances) under the category of hazardous waste. Some of the substances are enlisted in Table 2.

In Pakistan, there is no specific mechanism for the collection, transportation and disposal of hazardous waste. However, for the management of hospital waste guidelines and policies are avialble. But, unfortunately the guidelines are not imppemented and there is a lack for the proper disposal of hospital waste. Genreally, collection of wastes from hospitals, clinics and laboratories is the responsibility of Municipal Corporation. Practically, the wastes are disposed of without any treatment and without any knowledge of its hazardous effects on environment and humans. However, efforts have been made for the handling and treatment of hazardous waste generated from the industrial sectors.

Table 2. List of prescribed hazardous substances (Hazardous Substance Rules of Pakistan 2014)

S. No.	Chemical Name	S. No.	Chemical Name
1.	Acetaldehyde	21.	Dichlorvos
2.	Acetone cyanohydrins	22.	Dimethyl hydrazine
3.	Acrylamide	23.	Dimethylaniline
4.	3-Amino-1, 2,4 triazole	24.	Endrin
5.	Ammonium sulphamate	25.	Ethyl benzene
6.	Antimony & Compounds	26.	Ethyl ether
7.	Arsenic & Compounds	27.	Fluorine
8.	Asbestos	28.	Hexachlorobenzene
9.	Barium and Compounds	29.	Hydrogen sulphide
10.	Benzenamine 3-Trifluoromethyl	30.	Hydrogen cyanide
11.	Benzene	31.	Lead (Inorganic)
12.	Benzidine and Salt	32.	Lead arsenate
13.	Beryllium and Compounds	33.	Methyl Parathion
14.	Calcium Cyanamide	34.	Butyl alcohol
15.	Cadmium and Compounds	35.	Nitro benzene
16.	Carbachol chloride	36.	Nitrocyclohexane
17.	Chlorobenzene	37.	Nitroaniline
18.	Cyanides and Compounds	38.	Parathion
19.	DDT	39.	Styrene
20.	Dichlorobenzene	40	Trinitrotoluene (TNT)

Policies have been made for the control and management of hazardous waste in country. National environmental policy 2005 also gave brief policy guidelines on controlling and managing hazardous waste and trans-boundary movement of hazardous waste. Key policy items are given below:

- "Develop and enforce rules and regulations for proper management of municipal, industrial, hazardous and hospital wastes.
- Develop and implement strategies for integrated management of municipal, industrial, hazardous and hospital waste at national, provincial and local levels.
- Develop and enforce regulations to reduce the risk of contamination from underground storage tanks."

AGRICULTURAL BIOMASS AND UTILIZATION, WASTE-TO-ENERGY

Agricultural biomass is derived from living organisms such as plants, corn, and straw, animal waste, and grasses. The agricultural waste is rich in organic content and the biomass can be used for the production of energy or other useful product. This is the environment friendly method for reducing the pollution by utilizing waste produced from the biological organisms.

In Pakistan more than 70% of population live in rural areas and are farmers. Pakistan is an agricultural country, producing large amount of crops annually. Pakistan is generating millions of tonnes of biomass annually in the form of rice husk, crop residue, wood mills waste, corn pith and cotton processing waste which can be easily used in the production of energy (Saeed et al., 2015). According to the United Nation Industrial Development Organization (2016), the main constituent of biomass and their availability in Pakistan are: agricultural waste (20,494 tonnes), agro industrial wastes (25,271 tonnes), and wood based residues (1,121 tonnes). Cotton, rice, maize, wheat and sugarcane are the five major crops grown in Pakistan and produced large amount of agricultural waste. The annual production, types of resides and potential of energy produced from these five crops are mentioned in Table 3 (FAOSTAT, 2013). The production of energy from the biomass of crops is a renewable source, the production and consumption of biomass is carried out in cyclic manner therefore overall production of carbon dioxide in environment is negligible. It is environment friendly and economically feasible method. But, every thechnique has negative impacts, if the incineration of biomass isnot carried out in controlled conditions it may affect the quality of air.

Waste to Energy (WTE) is now considered as a reliable and renewable form of energy that has become the basis for many of the most successful Solid Waste Management Systems (SWM) in Europe and the United States. In the light of the existing national energy vacuum, the Government of Punjab is vigorously exploring viable, holistic energy generation. Converting waste to energy has always been of prime importance for LWMC. For this, LWMC has concentrated its efforts to make use of waste in order to create some RER. A Letter of Interests (LOI) was issued to Punjab Power Development Board (PPDB) for renewable energy project to a consortium of local and two Chinese Firms.

Table 3. Annual waste production from major crops of Pakistan (FAO 2013)

Crops	Annual Production 1000 MT	Types of Residues	Energy Potential MTOE = 11.6TWh
Cotton	6605	Stalks, Boll, Husk, Shell	17.039
Rice	6160	Stalks, Straw, Husk	8.475
Wheat	25214	Stalks, Pods	19.512
Maize	55309	Stalks, Cobs	4.223
Sugarcane	4270	Leaves, Bogasse	9.036

Table 4. Biomass power plants in Pakistan (Saeed et al., 2015)

Location	Province	Power Generation	Crop Residues
Jhang	Punjab	12 MW	Cotton stalk, rice husk, bagasse, sugarcane trash
Faisalabad		12 MW	Rice husk, cotton stick
Mirpurkhas	Sindh	12 MW	Wood chips, cotton stalk
Matli		9 MW	Sugarcane
Mardan	Khyber Pakhtunkhwa	200 MW	Sugarcane

Cultivation of banana trees is another important way for the energy production. The major province of banana production is Sindh and 90% of bananas is produced in this area. The waste of banana is thrown and after drying they are burnt. But it has a great potential for the production of energy from this waste. Approximately, 296 Km^2 area is cultivated with banana, 173000 trees are grown in 1 Km^2 area and total of 3 cultivars are grown in a year. After drying, the total waste produced from the banana is 1075 x 10^6 Kg/year and the energy content from this waste is 1908 x10^6 MJ/year (Saeed at al. 2015). Few examples of biomass power plants installed in Pakistan are summarized in Table 4.

Other plants established for management of agriculture waste are explained below:

- "Establishment of Waste Busters: Waste Busters is an independent private sector waste Management Company located at Islamabad, Lahore, Karachi, Gujarat, Peshawar and Quetta. The objective of this organization are to separate and recycle all the collected waste into various product streams such as organic fertilizers, refuse derived fuel as well as recycled plastic and tetrapak products.
- Refuse Derived Fuel (RDF) plant: has built for processing of 1000 tonnes of municipal waste. RDF is largely combustible components of municipal waste such as plastics and biodegradable waste. DG Khan Cement uses RDF in its cement plants as an additive fuel with coal".

E-WASTE GENERATION, 3R, DISPOSAL PRACTICES, POLICIES AND REGULATIONS

The production and generation of electronic waste is a worldwide issue. Organization for Economic Co-operation and Development (OECD) define e-waste as "any appliance using an electric power supply that has reached its end-of-life". The e-waste includes: mobile, computers, ovens, refrigerators and other appliances using electrical power. With increasing population and industrialization in Pakistan, the demand for the electronic appliance is increasing and ultimately e-waste generation is also increasing which is an emerging environmental issue (Khan, 2012). Pakistan is producing 315 metric kilo tonnes of e-waste. In the world, Pakistan is the largest dumping site of e-waste and importer of the used computers (Iqbal et al. 2015). As developing country, there are no proper practices for the management of e-waste. The recycling of e-waste is carried out in informal facilities for the recovery of valuable materials form the e-waste. The recovery of valuable metals such as copper is the main intersest and the remaining scraps are burned into open air (Sajid et al., 2018). The open burning or melting of e-waste is responsible for the generation of toxic into environment and also affect the labors dealing with these activities (Iqbal et al., 2015). Sites for the recycling of e-waste located at major cities of Pakistan including Karachi, Islamabad, Peshawar, Rawalpindi, Faislabad, Gujranwala and Lahore (Saeed et al. 2013). Being the developing country, the generation of e-waste in increasing day by day. But still there is a strong need for the development of regulation and their implementation for the management of e-waste (Kandil, 2013). There is need to improve the strategy for the management and recycling facilities of e-waste, as it will generate revenue for the Governmnet and can benefit the people involve in this business. For the proper management and recycling of e-waste different policies that directly and indirectly address 3R are as follows:

- The Pakistan Environmental Protection Act 1997.
- National Environment Policy 2005.
- National Sanitation Policy 2006.
- National Climate Change Policy 2012.
- National Development Strategy 2012.
- Punjab Environmental Policy 2015.
- Draft Hazardous Waste and Hazardous Substances Rules, 2016.
- Draft Pakistan Environmental Protection Motor Vehicle Regulations 2016.

INDUSTRIAL WASTE, POLICIES AND REGULATIONS, CASE STUDIES

Industrial pollution is localized due to increase in industries and discharge of untreated waste. The most polluting industries in Pakistan include: chemicals, pharmaceuticals, pesticides, electrical electronic equipment, glass, leather tanning, pulp and paper board, ceramics, pesticides, cements etc. In developing countries like Pakistan mostly two options i.e. discharge into open dumps or landfill and incineration are widely used for the disposal of solid waste. The solid wastes from industries are dumped openly or in landfills. The effluents from the industries are discharged into nearby ponds, rivers or streams without any treatment which ultimately affects the water, soil, air, plants and animals.

Table 5. City waste management in Pakistan (Source: Javied et al. 2014)

City	Organization Name	Waste Management Details
Rawalpindi	RSWMC	Collection, transportation and disposal by Rawalpindi solid waste management company 4000 manpower in use Waste is generally dumped into landfill (non-engineer sites) Improper treatment of waste
Lahore	LSWMD	Lahore solid waste management department Collection, transportation and disposal 630 waste collection sites 5500 manpower use Insufficient waste treatment 60-70% waste collecting 1 engineer landfill site
Gujranwala	TMA	Collection, transportation and disposal by TMA 1130 manpower 550 machinery use No treatment facility available
Faisalabad	TMA	Waste collection manually and through machines 3200 manpower 900 waste lifting by machinery Open dumping Improper treatment
Karachi	TMA & SWD	Collection, transportation and disposal by TMA 11200 manpower 4085 community bins used for waste collection 567 vehicles for waste transportation 2 incinerator plants 3 landfill sites operating by district Government

The collection, segregation, transport and disposal of solid waste vary from area to area as there is no standard policy or regulation for the management of solid waste. In most areas of Pakistan, the domestic waste is thrown at designated site and is collected by the municipality, and finally disposed into the landfill mostly open dumping sites. Municipal wastes management of some cities of Pakistan is given in Table 5.

Case study: Soild Waste Generation at Hayatabad Industrial Estate, Peshawar, Pakistan (Khan 2015)

Hayatabad Industrial Estate (33.9741°N and 71.4341 °E) is located in Peshawar, Pakistan and was established in 1963. Hayatabad Industrial Estate is spread on an area of 868 acres having total 550 industries including metals industries, pharmaceuticals, chemicals, fertilizers, and leather, textiles, and tobacco, match, ceramics, wood and paper industries. The area was analyzed for the solid waste generation from different industries.

Figure 6. Solid waste production in Hayatabad Industrial Estate, Peshawar, Pakistan (Khan 2015)

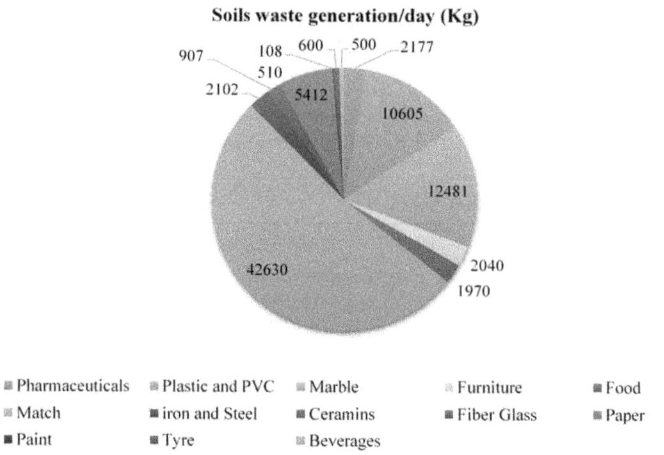

Figure 7. Composition of solid waste at dumping site of Hayatabad Peshawar, Pakistan (Khan 2015)

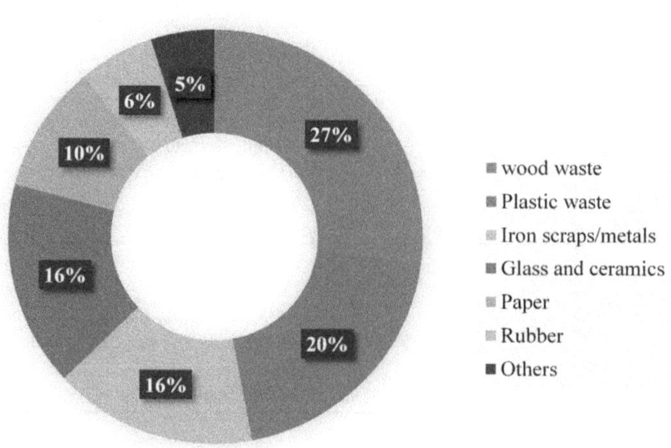

Solid Waste Generation: The total solid waste generated by different industries are presented in Figure 6. Match (42630 Kg) marble (12481 Kg) and plastic (10605 Kg) industry showed highest production of solid waste among all studied industries.

Solid waste disposal: Total 26,039 Kg/day of solid waste is collected from the industrial area of Peshawar and is transferred to the landfill sites. The collection of solid waste is carried out by the truck, trolleys and tractors. Peshawar Development Authority (PDA) is responsible for the collection and transfer of waste. Dumping site is used for the disposal of all waste form the industries as well as households.

Composition of Solid waste: composition of solid waste at dumping site is presented in Figure 7. The waste constitute of wood, plastic, metals, paper and rubber. Other include the non-recyclable waste which was 5% of total waste dumped at site.

CONSTRUCTION AND DEMOLITION WASTE

In Pakistan's construction industry, construction waste is of great concern as many other countries. This sector focuses on waste management efforts such as recycling and reusing programs. Environmental degradation is the final effect of these residues. Maintaining the environment by removing solid waste from construction from demolition waste from your work area is a major waste in some construction industries (Akhund et al., 2018). CDW often contain bulky, heavy materials includes concerts, bricks, wood, boards, plaster, asphalt and metals as shown in Figure 8.

Figure 8. Construction and demolition waste

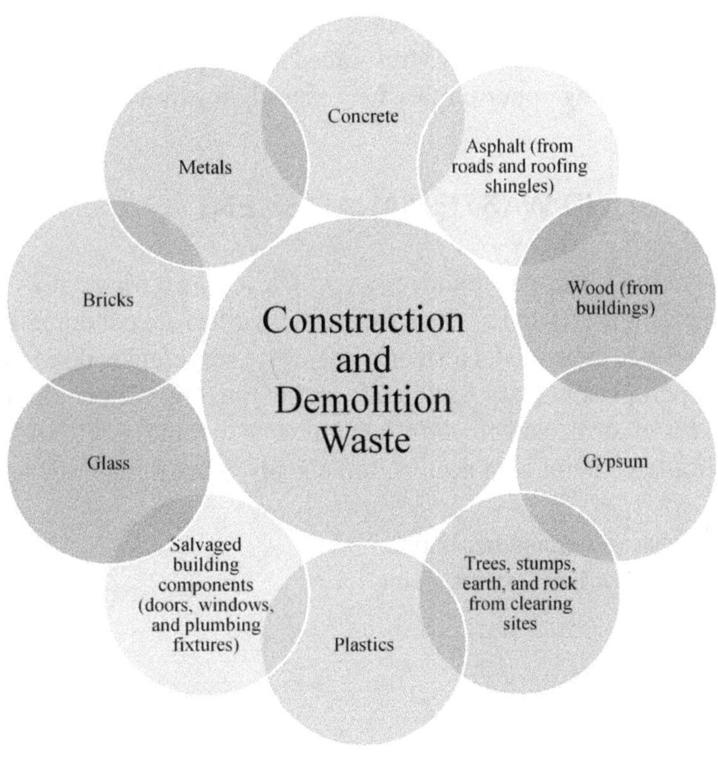

In Pakistan, the construction industry is the largest sector and shares 2.3% of the country's GDP (Arshad et al., 2017). But due to poor management, a large amount of CDW was produced. Millions of tons of construction and demolition waste (CDW) are produced every day, which represent a threat to the socio-economic growth of the country. Much has been spent on the management and recycling of this waste. There is a great need to identify significant attributes, what causes this threat. According to the recent study, seventy-eight attributes contribute mainly to the generation of CDW (Akhund et al., 2018). Some of the attributes are:

- Complicated design.
- Damages caused by the worker and during transportation, poor design quality.
- Poor quality and material management.
- Insufficient training of workers.
- Lack of project information and environmental awareness.
- Lack of waste management plans and projects without integrating the waste minimization strategy.

Despite regulatory guidelines and pollution control measures, industrial wastes are discharged to the ground and discharged into bodies of water without proper treatment. This serious misconduct creates serious dangers to public and environmental health. Construction and demolition waste (CDW), coming from the construction, maintenance, renovation and demolition of buildings and structures, represent a large percentage of waste in industrialized societies (Lauritzen, 2018).

Compared to other forms, such as household garbage, more than 90% of CDW can be used as a resource and as a substitute for building materials, particularly primary natural raw materials. The reuse, recovery and recycling depend on the quality and the market of the materials and the environmental impact of the processes to convert CDW from old structures to their use in new structures. However, the current use of CDW products as secondary resources is marginal. Most CDW is deposited or used as filler and high-quality recycling opportunities are generally overlooked (Lauritzen, 2018).

3R INDICATORS OF SOLID WASTE MANAGEMENT

Waste reduction and recycling performance and resource efficacy are the indicators that measure progress of the waste management (Figure 9). In Pakistan, the municipalities are not doing any practice to reduce, recycle or reuse of solid waste generated. Normally, the waste recycling is done by public and the main recyclable items like paper, plastic, glass and metals are retained by the people themselves, which are later sold to street hawkers/waste dealers for the sake of money/ income (GoP, 2019). Waste management indicators should cover a large number of sectors, but usually cover three basic areas:

- Amount of urban waste generated per capita.
- Quantity sent to landfill or incineration plant.
- Recycling rate

Figure 9. 3R indicators of solid waste management

ISSUES AND CHALLENGES IN WASTE MANAGEMENT

The circumstance of solid waste management in Pakistan is a reason of serious concern, as more than 5 million individuals die every year because of waste-related ailments (Jaffar, 2019). All major cities of Pakistan i.e. Islamabad, Lahore and Peshawar face tremendous challenges in dealing with the issues of solid waste management. The root causes of the worsening waste problem in Pakistan are the poor urban planning, outdated infrastructure, lack of public awareness, lack of knowledge, poor economy, malnutrition and endemic corruption.

Being the sixth (6th) most populous country in the world; there is a lot of industrialism and a lot of waste is generated. Like the other developing countries, the waste management sector in Pakistan is plagued with a wide range of social, cultural, legislative (authoritative and administrative) and financial issues. In the country, more waste is generated than the number of amenities available to manage them. All cities of Pakistan lack solid waste management and disposal facilities. There is no operational landfill for proper waste disposal. At present solid waste is dumped openly on any barren land, without consideration of its environmental and health impacts. Considering the situation, the main issues and challenges in waste management in Pakistan includes:

- There is no adequate waste collection system.
- The basket is poured into the streets.
- The different types of waste are not collected separately.
- There are no controlled dumps. The combustion opening is common.
- Citizens are not aware of the relationship between the reckless disposal of waste and the resulting public and environmental health problems.
- Lack of technical expertise and criteria for the implementation.
- Waste Agriculture Biomass (WAB) management capacities need to be built primarily within the agricultural sector.
- Lack of awareness and understanding of the use of WAB is very meager, because farmers are illiterate and environmental education is non-existent.
- The development of a functional solid waste management (SWM) system requires a major investment and it may be difficult to give it priority over other resource demands such as energy.

- Development of Strategy/Policy and Facilitation for establishing recycling plants including financial and institutional mechanisms needs to be focused.
- Promotion of Public Private Partnerships (PPP) and various incentives (providing land at low cost, Tax rebate, easy delivery of waste to recycling facility, etc.) needs to be realized to attract private investments.
- Source separation, Extended Producers Responsibility (EPR), awareness of people and capacity of government and private sector need to be improved.
- Already developed huge informal recycling sector should be brought under formal setting with necessary support of the government.
- Lack of modern Waste Electrical and Electronic Equipment (WEEE) management facilities and mechanism.

As a result of these issues and challenges, waste accumulates in roads, canals and other common areas, and the burning of garbage is common, resulting in exposure to dangerous toxins, which threaten human health and the environment. Among the few landfills present, they are even less in operation. Even within Pakistan's capital, Islamabad, no permanent landfills are found. The waste on the roads provides an ideal environment for the growth of numerous flies, which affect human health and environmental health for other species (Rachael, 2018).

ECONOMICS OF WASTE MANAGEMENT

In Pakistan, the solid waste management system is worse and, therefore, becomes an environmental hazard. In urban areas of Pakistan, only 57% of families have a formal SW elimination system, while in small cities only 40% of families have a formal waste collection system (SPDC, 2015). In Lahore (Pakistan), 1.97 million tons of wastes are generated each year, of which between 30% and 35% are not collected and 1.28 million tons are available. 21.2% of the available waste is recyclable, of which 15% is sold directly to industries to generate profits. The general calculation revealed that 21.2% of the total recyclable waste in Lahore is used for recycling and that it generates a quantity of Rs. 271 million (US $ 4.5 million) per year through the informal private sector. If recycling is adopted as an industry in Lahore, it can generate income of Rs. 530 million (US $ 8.8 million) per year. Value of income of Rs. 65 million (US $ 1.08 million) can be generated by reducing the cost of collecting solid waste generated in Lahore (Batool et al., 2008).

THE WAY FORWARD

The plans for solid waste management according to United Nations Centre for Regional Development (UNCRD) are listed below:

- Development of projects for the generation of energy from waste.
- Establishment of provincial centers for cleaner production for the minimization and recycling of waste.
- Installation of composting plants and scientific dumps.

- Remediation measures will be adopted for the treatment of wastewater.
- The government aims to develop more waste to energy projects (WTE), such as Khairpur WTE, and Lahore Waste Management Company (LWMC) WTE projects.
- The government aims to replicate the LWMC model in other parts of the country.

While there are SWM strategies, the dimensions at which they are applied and implemented are not adequate because of government organizations that don't have the assets and gears. These institutions/ organizations are mainly run by public sector policy makers and workers who are not necessarily the most informed about the management of waste. To improve the management of solid urban waste, it is necessary to involve experts and contribute to environmental governance (Rachael, 2018). It is imperative that public opinion acquires environmental education, change attitude and act.

REFERENCES

Akhund, M. A., Memon, N. A., Ali, T. H., Memon, A. H., & Imad, H. U. (2018). A Comprehensive Review on Waste Generating Attributes: Way Forward for Pakistan's Construction Industry. *Engineering Science & Technology International Research Journal, 2*(3), 1–7.

Arshad, H., Qasim, M., Thaheem, M. J., & Gabriel, H. F. (2017). Quantification of Material Wastage in Construction Industry of Pakistan: An Analytical Relationship between Building Types and Waste Generation. *Journal of Construction in Developing Countries, 22*(2), 19–34. doi:10.21315/jcdc2017.22.2.2

Batool, S. A., Chaudhry, N., & Majeed, K. (2008). Economic potential of recycling business in Lahore, Pakistan. *Waste Management, 28*(2), 294-298.

Ejaz, N., Hashmi, H.N., & Ghumman, A.R. (2011). Water quality assessment of effluent receiving streams in Pakistan: A case study of Ravi River. *Mehran University Research Journal of Engineering & Technology, 30*(3), 383-396.

Environmental Protection Agency. (2005). *Guidelines for solid waste management; present status of SWM in Pakistan and strategy for its improvement*. Author.

Environmental Protection agency. (2014). *The Hazardous Substances Rules. Sindh environment protection agency*. Government of Sindh.

FAO, Food and Agricultural Organisation of United States,FAO Statistics Division. (2013). Available: http://faostat.fao.org/site/567/DesktopDefault.aspx?PageI D=567#ancor

Government of the Punjab (GoP). (2019). *Climate and Clean Air Coalition Municipal Solid Waste Initiative*. Retrieved from http://www.waste.ccacoalition.org/

Gu, B., Zhu, W., Wang, H., Zhang, R., Liu, M., Chen, Y., ... Bi, J. (2014). Household hazardous waste quantification, characterization and management in China's cities: A case study of Suzhou. *Waste Management (New York, N.Y.), 34*(11), 2414–2423. doi:10.1016/j.wasman.2014.06.002 PMID:25022547

International Union for Conservation of Nature (IUCN). (2004). *AbbottabadfState of the Environment and Development*. Karachi: IUCN Pakistan. Available at http://cmsdata.iucn.org/downloads/abbottabad_soed.pdf

Iqbal, M., Breivik, K., Syed, J. H., Malik, R. N., Li, J., Zhang, G., & Jones, K. C. (2015). Emerging issue of e-waste in Pakistan: A review of status, research needs and data gaps. *Environmental Pollution*, *207*, 308–318. doi:10.1016/j.envpol.2015.09.002 PMID:26433180

Jaffar, I. (2019). *Current Scenario of Solid Waste in Pakistan*. Retrieved from https://www.technologytimes.pk/current-solid-waste-pakistan/

Javied, S., Hanan, F., Munawar, S., Qasim, M., Anees, M. M., Ghani, M. U., ... Ansar, A. (2014). Management of municipal solid waste generated in eight cities of Pakistan. *International Journal of Scientific & Engineering Research*, *5*(12), 1186–1192.

Kandil, S. (2013). *World e-waste map reveals national volumes. Int. Flows*. Retrieved from http://i. unu.edu/media/unu.edu/news/41225/World-E-Waste-Map-Reveals-National-Volumes-International-Flows.pdf

Khan, A. A., Ahmed, Z., & Siddiqui, M. A. (2012). Issues with solid waste management in South Asian countries: A situational analysis of Pakistan. *J. Environ. Occup. Sci.*, *1*(2), 129–131. doi:10.5455/jeos.20120731084356

Korai, M. S., Mahar, R. B., & Uqaili, M. A. (2017). The feasibility of municipal solid waste for energy generation and its existing management practices in Pakistan. *Renewable & Sustainable Energy Reviews*, *72*, 338–353. doi:10.1016/j.rser.2017.01.051

Lauritzen, E. K. (2018). *Construction, Demolition and Disaster Waste Management: An Integrated and Sustainable Approach*. CRC Press. doi:10.1201/b20145

Misra, V., & Pandey, S. D. (2005). Hazardous waste, impact on health and environment for development of better waste management strategies in future in India. *Environment International*, *31*(3), 417–431. doi:10.1016/j.envint.2004.08.005 PMID:15734194

Pak EPA Pakistan. (2019). *Waste Management*. Available at file:///G:/Sir%20arshad/book%202/Pakistan%20-%20Waste%20Management%20_%20export.gov.html

Pakistan Map and Satellite Image. (2019). Assessed on 29 March 2019. https://geology.com/world/pakistan-satellite-image.shtml

Pakistan National Report submitted to The United Nations Conference on Environment and Development. (1991). Environmental and Urban Division, Government of Pakistan (GOP).

Pakistan - Waste Management. (n.d.). Retrieved from https://www.export.gov/article?id=Pakistan-Waste-Management

Pichtel, J. (2014). *Waste management practices: municipal, hazardous, and industrial* (2nd ed.). Taylor & Francis Group. doi:10.1201/b16576

Rachael, L. (2018). *Solid Waste Management in Pakistan*. Retrieved from https://www.bioenergyconsult.com/solid-waste-management-in-pakistan/

Saeed, A. (2013). *Dangerous labour: the little known e-waste phenomenon.* Retrieved from http://www.dawn.com/news/1032967

Saeed, M. A., Irshad, A., & Sattar, H. (2015). Agricultural Waste Biomass Energy Potential In Pakistan. *Proceedings of the International Conference.*

Sajid, M., Syed, J. H., Iqbal, M., Abbas, Z., Hussain, I., & Baig, M. A. (2018). *Assessing the generation, recycling and disposal practices of electronic/electrical-waste (E-waste) from major cities in Pakistan.* Waste Mangement. doi:10.1016/j.wasman.2018.11.026

Saleh, H. M. (2016). *Introduction to hazardous waste management.* INTECH. doi:10.5772/61668

UNEP. (2019). Retrieved from http://www.unep.org/

United Nation Industrial Development Organization. (2016). *Policy on biomass energy technology.* Assessed 5 May 2019. https://www.aedb.org/images/UNIDOBiomassPolicyDocumentSecondDraftV1.pdf

Chapter 11
Waste Management in the Philippines

Vella Atienza
College of Public Affairs and Development (CPAf), University of the Philippines Los Banos (UPLB), Philippines

ABSTRACT

In response to the growing problems on waste management in the country, the Philippines' Republic Act 9003 (RA 9003), also known as the Ecological Solid Waste Management Act of 2000, was enacted in January 26, 2001. This chapter hopes to provide the brief background of the country particularly on the issues related to waste management such as population, economic situation, urbanization, and modernization, among others. It will also discuss the definition, classification, and generation of waste both in urban and rural areas. In addition, it will also review the existing policies and the current waste management practices, and the lessons learned based on case studies of both successful and failed experiences. Furthermore, this will also provide the latest updates on the compliance of the local government units to RA 9003, identify the challenges, opportunities, and the proposed recommendations on how the waste management in the country can possibly be improved and become more sustainable.

INTRODUCTION

The problems on waste management remain a challenge for many developing countries including the Philippines. Although various policies have been formulated to address these concerns, there are many underlying issues that affect their effective implementation such as the political will and capability of the leaders, the availability of appropriate waste management technologies, and the understanding and willingness of the members of the society to act cooperatively towards achieving more sustainable waste management.

The increasing population, urbanization, and changing lifestyle have contributed to the continued increase of waste generation in the Philippines. Based on the 2015 census of population (POPCEN), the country has a total population of 100,987,437 persons (Philippine Statistics Authority [PSA], 2017). The waste generation per person is 0.70kg/day in highly urbanized city areas, 0.60kg/day in urban city

DOI: 10.4018/978-1-7998-0198-6.ch011

areas, and 0.30 kg/day in rural areas (Aguinaldo 2009, as cited in Atienza, 2017). The National Capital Region (NCR) has the highest waste generation rate of 0.71 kg per capita per day while the Autonomous Region of Muslim Mindanao (ARMM) has the lowest with 0.30 kg per capita per day. It was estimated that the total waste generation was 35 tons per day or 13.1 million tons per year (National Solid Waste Management Commission [NSWMC], 2016).

With the above-mentioned data, there is indeed an urgent need to find ways on how to effectively manage waste in the country. As in many developing countries, a higher volume of waste generated was not properly collected and thus, ended up in open dumpsites or waterways which caused severe flooding especially during typhoons. The country experienced a waste management tragedy in 2000 when the former open dumpsite in Payatas collapsed due to heavy rains and thus, killed about 200 people (Atienza, 2013).

In response to the growing problems on waste management in the country, the Philippines' Republic Act 9003 (RA 9003), also known as the Ecological Solid Waste Management Act of 2000, was enacted in January 26, 2001. Unlike previous waste management policies which seemed to be a piecemeal approach, RA 9003 is considered as the most comprehensive act which declares the "policy of the state to adopt a systematic, comprehensive, and ecological solid waste management program which shall ensure the protection of public health and environment" (Republic of the Philippines, RA 9003, Article 1, Section 2). It has been eighteen years since it came into force, however, the rate of compliance is still low.

As mandated by the RA 9003, all local government units (LGUs) through its solid waste management (SWM) boards, shall prepare a 10-year solid waste management plan which shall ensure the efficient management of solid waste within their jurisdiction. But as of October 2018, the National Solid Waste Management Commission (NSWMC) reported that only 30.4% (497 LGUs) have approved SWM plans; 62.4% (1,080 LGUs) are for evaluation and pending approval; and 7.2% (117 LGUs) have not submitted their plans (NSWMC, 2018).

In terms of waste diversion, it is cited from RA 9003 that each LGU shall divert at least 25% of all waste from waste disposal facilities within five (5) years after the effectivity of the law in 2001 through re-use, recycling, and composting and other resource recovery activities; and that this 25% diversion goals should be increased every three (3) thereafter (Section 20) or after 2006. However, it was reported by the Metro Manila Development Authority (MMDA) that as of May 31, 2011, the average waste diversion rate in Metro Manila was only 33.92% (MMDA, 2011). It is assumed that there is even lower diversion rate in other parts of the country. Based on the report of the Philippines' Senate Economic Planning Office (SEPO), there was a 48% solid waste diversion rate in Metro Manila while 46% outside of Metro Manila as of 2015 (SEPO, 2017).

Based on the National Solid Waste Management Strategy 2012 - 2016 of the Philippines, it was reported that more than 50% of municipalities in the country are classified as low income communities and do not have sufficient funds for waste management activities. Others may have financial resources but lack the technical capacity to effectively implement these activities (NSWMC, 2016). This implies that the low compliance rate of LGUs may not necessarily be due to their unwillingness to comply with the law, but is dependent on their level of capacity. Hence, it is necessary to also review other underlying causes of the low compliance rate of the LGUs to RA 9003 and the possible strategies to effectively implement solid waste management policies and programs in a more sustainable manner.

OBJECTIVES

This chapter hopes to provide the brief background of the country particularly on the issues related to waste management such as population, economic situation, urbanization, and modernization, among others. It will also discuss the definition, classification, and generation of waste both in urban and rural areas. In addition, it will also review the existing policies and the current waste management practices, and the lessons learned based on case studies of both successful and failed experiences. Furthermore, this will also provide the latest updates on the compliance of the local government units to RA 9003, identify the challenges, opportunities, and the proposed recommendations on how the waste management in the country can possibly be improved and become more sustainable.

BRIEF BACKGROUND OF THE PHILIPPINES

Occupying more than 300,000 square kilometers territory, the Philippines is an archipelago with 7,641 islands and is located in Southeast Asia. It is divided into three island regions, namely Luzon, Visayas, and Mindanao (Republic of the Philippines, 2019). This archipelagic characteristic of the country affects the differing views of the citizens on various concerns including that of waste management. The buying and selling prices of recyclables also vary depending on the location. In addition, since most of the big recycling industries are located in Metro Manila and nearby provinces, the lack of recycling facility interferes with the recycling activities in other regions (Atienza, 2011). The limited opportunities in rural areas and the desire for other forms of livelihood are one of the contributing factors for the movement to urban cities. The poverty incidence among Filipino families, or the proportion of families whose income is below the poverty line to the total number of families was estimated at 16.1% in the first semester of 2018 while the unemployment rate was estimated at 5.2% in January 2019 (PSA, 2019).

Based on the 2015 Census of population (POPCEN) by the Philippine Statistics Authority (PSA), the Philippines had a total population of 100,981,437 persons as of August 01, 2015. More than half

Figure 1. Philippine population from 1990 to 2015
(Source: PSA, 2019)

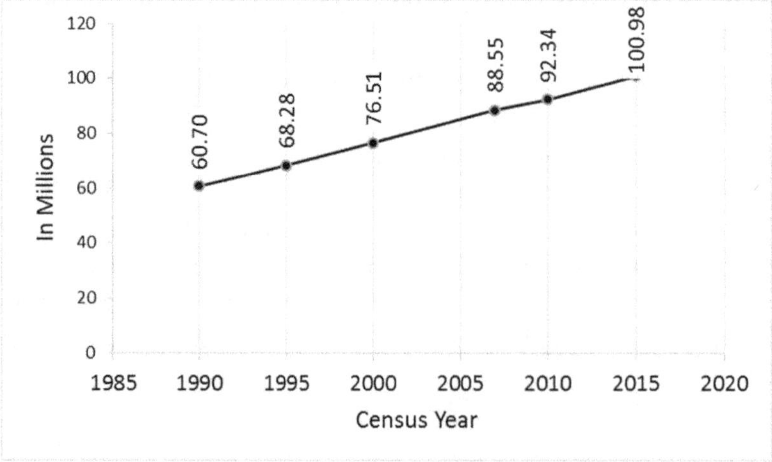

Figure 2. Philippine rural-urban population (2010 - 2015)
(Source: PSA, 2019)

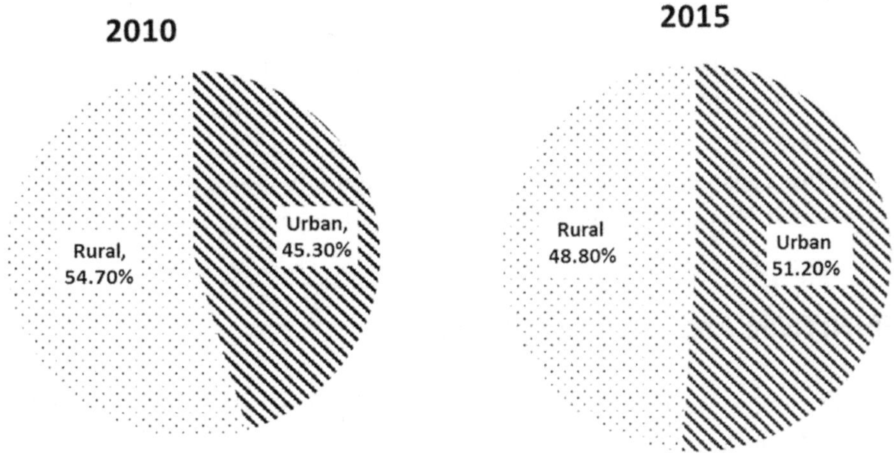

(56.9%) of the country's total population reside in Luzon, which comprise of eight regions; followed by Mindanao (23.9%) with six regions, and Visayas (19.2%) with four regions. As shown in Figures 1 and 2, the country's population and the level of urbanization continue to increase from 1990 - 2015. The country has 33 highly urbanized cities (HUCs); 1,489 municipalities; and 42,036 barangays (smallest political unit in the Philippines). Four HUCs surpassed the one million population mark, namely Quezon City (2.94 million), City of Manila (1.78 million), Davao City (1.63 million), and Caloocan City (1.58 million) (PSA, 2019).

Figure 3. Projected waste generation (tons per day)
(Source: NSWMC, 2016)

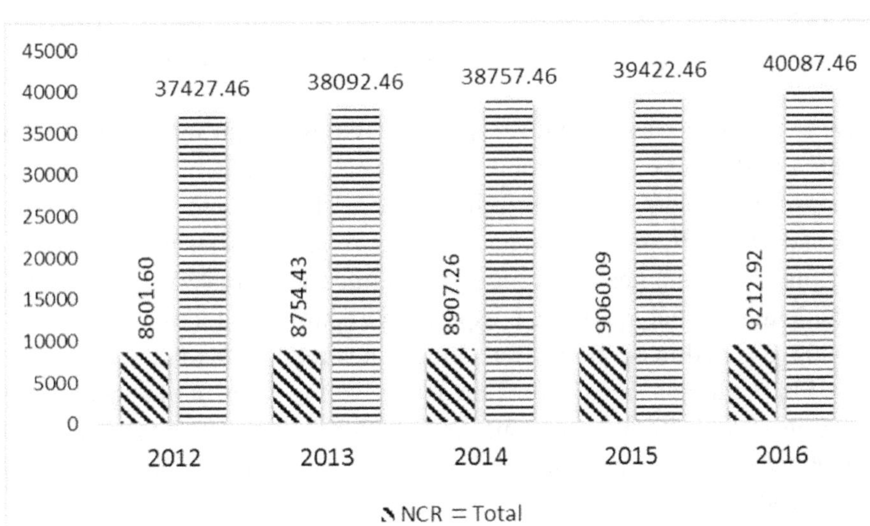

Based on the data from World Bank, the population of the Philippines has reached 106,651,922 in 2018 (World Bank, 2019). This increasing population and urbanization have contributed to the continued increase of waste generation in the Philippines. The waste generation per person is 0.70 kg/day in highly urbanized city areas, 0.60 kg/day in urban city areas and 0.30 kg/day in rural areas (Aguinaldo 2010, as cited in Atienza, 2017). The National Capital Region (NCR) had the highest waste generation rate of 0.71 kg per capita per day, while the Autonomous Region of Muslim Mindanao (ARMM) had the lowest with 0.30 kg per capita per day. It was estimated that the total waste generation was 35 thousand tons per day or 13.1 million tons per year (NSWMC, 2016). Figure 3 shows the projected waste generation from 2012-2016 with NCR contributing almost a quarter of the country's solid waste generation (Ibid.).

WASTE DEFINITION AND CLASSIFICATION

As directly quoted from the Article 2, Section 3 of the Philippines' Republic Act 9003 (RA 9003), the following terms are defined as the following:

Solid waste - refers to all discarded household, commercial waste, non-hazardous institutional and industrial waste, street sweepings, construction debris, agricultural waste, and other non-hazardous/non-toxic solid waste. Unless specifically noted otherwise, the term "solid waste" as used in this Act shall not include:

1) Waste identified or listed as hazardous waste of a solid, liquid, contained gaseous or semisolid form which may cause or contribute to an increase in mortality or in serious or incapacitating reversible illness, or acute/chronic effect on the health of persons and other organisms.
2) Infectious waste from hospitals such as equipment, instruments, utensils, and fomites of a disposable nature from patients who are suspected to have or have been diagnosed as having communicable diseases and must therefore be isolated as required by public health agencies, laboratory wastes such as pathological specimens (i.e. all tissues, specimens of blood elements, excreta, and secretions obtained from patients or laboratory animals) and disposable fomites that may harbor or transmit pathogenic organisms, and surgical operating room pathologic materials from outpatient areas and emergency rooms.
3) Waste resulting from mining activities, including contaminated soil and debris.

Municipal waste - refers to waste produced from activities within local government units which include a combination of domestic, commercial, institutional, and industrial wastes and street litters:

Solid waste management refers to the discipline associated with the control of generation, storage, collection, transfer and transport, processing, and disposal of solid wastes in a manner that is in accord with the best principles of public health, economics, engineering, conservation, aesthetics, and other environmental considerations, and that is also responsive to public attitudes.

Republic Act 9003, as cited in Section 17, also mandates the LGUs to conduct waste characterization of the said Act. For the initial source reduction and recycling element of a local waste management plan, the LGU waste characterization component shall identify the constituent materials which comprise the solid waste generated within the jurisdiction of the LGU. The information shall be representative of the solid waste generated and disposed within the area. The constituent materials shall be identified by

Figure 4. Sources of municipal solid waste in the Philippines, 2008-2013
(Source: NSWMC/EMB/DENR. 2015)

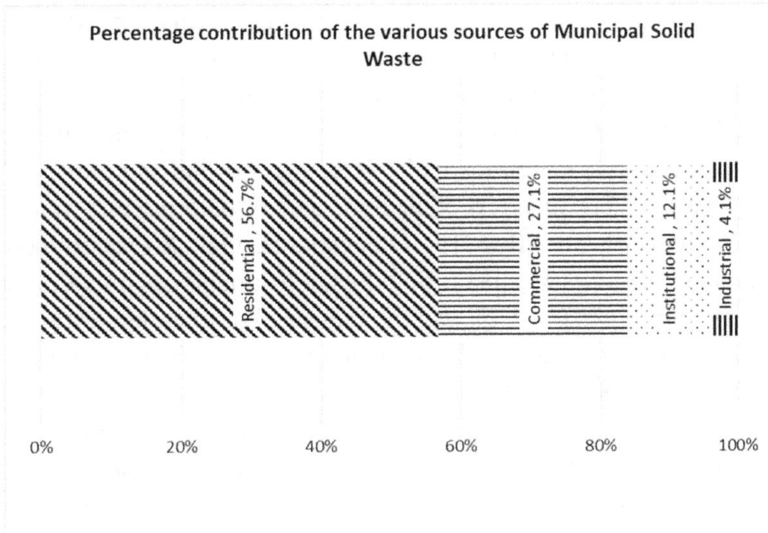

Figure 5. Composition of municipal solid waste in the Philippines, 2008-2013
(Source: NSWMC/EMB/DENR. 2015)

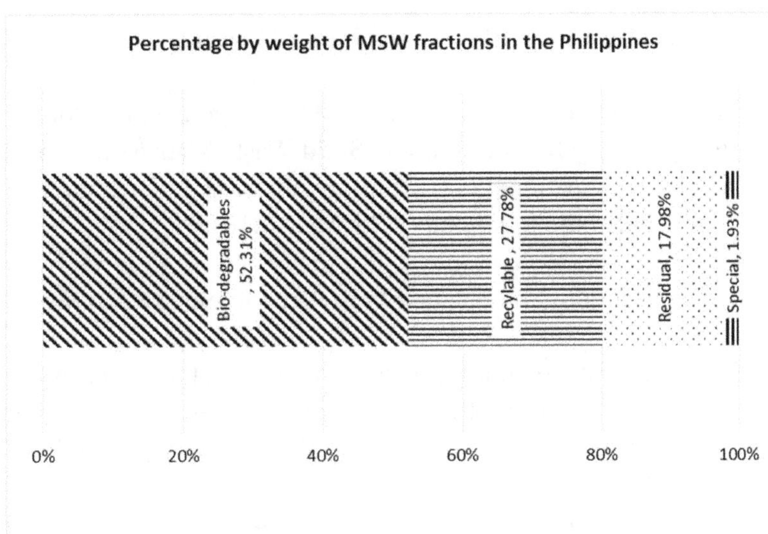

volume, percentage in weight or its volumetric equivalent, material type, and source of generation which includes residential, commercial, industrial, governmental, or other materials.

Based on the report of the National Solid Waste Management Commission (NSWMC), Environmental Management Bureau (EMB) of the Department of Environment and Natural Resources (DENR), the municipal solid waste in the Philippines (2008-2013) come from residential, commercial, institutional, and industrial sources (Figure 4). Residential waste accounts for 56.7% which includes kitchen scraps, yard waste, paper and cardboards, glass bottles, plastic containers and *sando* bags, foils, soiled tissues and diapers, and special wastes (i.e. containers of cleaning agents, batteries and waste electrical and electronic equipment) (NSWMC/EMB/DENR, 2015). Other sources include the commercial establishments and public and private markets (27.1%); institutional establishments such as the government offices, educational and medical institutions (12.1%); and the industrial or manufacturing sector (4.1%) (Ibid.).

In terms of classification, municipal wastes are categorized according to types: a) biodegradable/compostable; b) recyclables c) special waste and d) residuals (RA 9003). In the same report of NSWMC/EMB/DENR, it was cited that 52.31% of waste are biodegradables which include kitchen or food waste and yard or garden waste; 27.78% are recyclables which comprise of plastic packaging materials, paper and cardboard, metals, glass, textile, leather, and rubber; 1.93% are special wastes which include household healthcare waste, waste from electrical and electronic equipment (WEEE), bulky waste and other hazardous materials; and 17.98% are residuals (Figure 5) (NSWMC/Environmental Management Bureau/Department of Environment and Natural Resources [NSWMC/EMB/DENR], 2015).

Given the information on the composition and sources of wastes as shown in Figures 4 and 5, segregation of wastes from establishments and households has the potential of becoming composting and recycling industries (Atienza, 2017).

Existing Waste Management Policies and Regulations

This section reviews some of the relevant policies on waste management to address the growing problems of waste in the country. As cited in the National Solid Waste Management Strategy (2012-2016), the Philippines has enacted several national laws, rules, regulations, orders, and memoranda related to waste management which included the following (NSWMC, 2016):

a. Toxic Substances and Hazardous and Nuclear Waste Act of 1990 (RA 6969)

It declares the "policy of the State to regulate, restrict or prohibit the importation, manufacture, processing, sale, distribution, use and disposal of chemical substances and mixtures that present unreasonable risk and/or injury to health or the environment; to prohibit the entry, even in transit, of hazardous and nuclear wastes and their disposal into the Philippine territorial limits for whatever purpose; and to provide advancement and facilitate research and studies on toxic chemicals" (Republic of the Philippines, RA 6969, Section 2).

b. Local Government Code of 1991 (RA 7160)

Based on RA 7160, solid waste disposal system or environmental management system and services or facilities related to general hygiene and sanitation are included in the devolved functions of the local government units.

c. Philippine Clean Air Act of 1999 (RA 8749)

The Act stated that the "state shall protect and advance the right of the people to a balanced and healthful ecology in accord with the rhythm and harmony of nature". It also recognized the primary role of LGUs to deal with environmental problems and the "polluters must pay" principle (Republic of the Philippines, RA 8749, Section 2). It also prohibits the use of incineration which is defined as the burning of municipal, biomedical and hazardous waste or the process, which emit poisonous and toxic fumes. The Act also mandates LGUs to promote, encourage, and implement segregation, recycling and composting within their jurisdiction and require the phasing out of incinerators by July 2003 (NSWMC, 2016).

d. Philippine Clean Water Act of 2004 (RA 9275)

In pursuit of economic growth, this Act mandates the protection, preservation, revival of quality of fresh, brackish, and marine waters of the country to pursue economic growth (NSWMC, 2016).

e. National Environmental Awareness and Education Act of 2008 (RA 9512)

The act promotes environmental awareness through environmental education. It cited that educational and social welfare and development institutions in coordination with the DENR and the Department of Science and Technology (DOST) and other relevant agencies should "integrate environmental education in school curricula at all levels, whether public or private, including barangay daycare, preschool, non-formal, technical vocational, professional level, indigenous learning and out-of-school youth courses or programs" (Republic of the Philippines, RA 9512, Section 3).

f. Renewable Energy Act of 2008 (RA 9513)

The Act promotes the development, utilization, and commercialization of renewable energy and for other purposes.

g. The Ecological Solid Waste Management Act of 2000 (RA 9003)

The Ecological Solid Waste Management Act of 2000, also known as the Philippines' Republic Act 9003 (RA 9003) which came into force on January 26, 2001 is considered as the most comprehensive act on solid waste management. It declares the "policy of the state to adopt a systematic, comprehensive, and ecological solid waste management program which shall ensure the protection of public health and environment" (Republic of the Philippines, RA 9003, Article 1, Section 2).

However, in spite of these policies, the problems on solid waste management remain as one of the pressing issues in the Philippines. Atienza (2011) cited that one of the probable factors could be the archipelagic nature of the Philippines which made the implementation of policies more difficult. The lack of infrastructure for efficient transportation especially in the provinces and inner areas of cities and small villages could have contributed to the inefficient provision of waste management services and monitoring of waste management programs.

Another initiative of the Department of Interior and Local Government (DILG) is the Local Governance Performance Management System (LGPMS). It is an online national information system on local governments. This tool generates information which are useful for the LGUs to monitor their performance and assess the state of their development which will be useful in making decisions and actions in providing better and quality services (Department of the Interior and Local Government [DILG] Calabarzon, 2019). The LGPMS identifies the areas wherein LGUs have excellent performance and the areas for improvement. Solid waste management of LGUs is one of the factors under the environmental governance which are being looked into by this tool.

It is important to identify the possible reasons which cause delay or hinder the implementation of these policies. It is illogical to design a so-called "perfect technical system or set of policies if they cannot be implemented." By careful consideration of the available resources and the constraints, we can avoid the common mistake of "determining what should be, and instead concentrate on what is possible" (United Nations Environment Programme-International Environmental Technology Centre [UNEP-IETC] 1996: 16 as cited in Atienza, 2013).

The Informal Waste Sector and Its Role In Waste Management

The informal waste sector includes "individuals, families, groups, or small enterprises engaged in the recovery of waste materials with revenue generation as the motivation, either on a full-time or part-time basis. Members of this sector are classified as itinerant waste buyers, jumpers at collection trucks, garbage crew, waste reclaimers, and small and illegal junkshops" (NSWMC, 2016, p. 76).

Previously, the significant roles of the informal sector to the waste management were not recognized. However, the government has started to realize the contribution of this sector to the reduction of solid waste especially in urban cities. In 2009, a collaborative effort of the National Solid Waste Management Commission (NSWMC), the Solid Waste Management Association of the Philippines (SWAPP), and other international organizations helped in the formulation of the "National Framework Plan for the Informal Sector in Solid Waste Management". It envisages the informal waste sector as an empowered and recognized partner in the implementation of 3R (reduce, reuse, recycle) and it hopes to integrate this sector in the solid waste management system by "providing them with a favorable policy environment, avenues for skills development, and access to a secured livelihood, employment and social services (NSWMC, 2009: 34)."

One of the successful cases to illustrate how forming the informal sector into an organization can be more effective and sustainable is the experience of the KILUS (*Kababaihang Iisa ang Layuning Umunlad ang Sambayanan*) Foundation Environmental Multi-purpose Cooperative. Their experience will be discussed in the latter part of this chapter under the section on "Waste management practices and available technologies: Lessons learned from the experiences of selected cases".

Compliance Updates of the Local Government Units to Philippines' RA 9003

Republic Act 9003 is considered as the most comprehensive policies on solid waste management in the Philippines. Since its enactment in 2001, the implementation of the solid waste management programs is mainly based on the mandates cited in this Act. Pursuant to the Philippine Local Government Code of 1991, the LGUs are mandated in RA 9003 as the primary units responsible for the implementation of the Act (RA 9003, Section 10). They are tasked to establish provincial and city/municipal solid waste management boards and to prepare the 10-year solid waste management plan to ensure the efficient management of solid waste generated within their respective jurisdictions (Sections 11, 12, and 16). As of October 2018, there are 497 LGU plans approved (30.4%); 1,080 for evaluation and pending approval (62.4%); and 117 LGUs have not submitted their plans (7.2%) (NSWMC, 2018).

Section 20 of the Act cited that LGUs are mandated to divert 25% of their waste generated within five years after the implementation of the Act through composting, re-use, recycling, and other resource recovery activities. It further stated that the waste diversion goals should be increased every three years. As cited in the Philippine Development Plan (2011-2016), 50% of the total 19.7 million tons waste generated nationwide should have been diverted by 2016. Based on the 10-year SWM Plans submitted by the LGUs, the diversion target for 2017 was 52% of 20 million tons waste generated nationwide, and 55% of 3 million tons in Metro Manila. (NSWMC, 2018).

Upon the coming into force of RA 9003, it also prohibits the operation and establishment of open dumpsites. It further states that after three years, all open dumpsites shall be converted into controlled dumpsites, and that within five years of the implementation of the Act (Section 37), all controlled dumpsites should be closed. Open dumpsites refer to a disposal area wherein wastes are indiscriminately thrown or disposed of without due planning and consideration for environmental and health hazards while controlled dumpsites refer to disposal areas wherein solid wastes are deposited in accordance to the minimum prescribed standards of site operation (RA 9003 as cited in NSWMC, 2016).

As an alternative, the construction of sanitary landfill (SLF) is allowed as a final disposal site for residual wastes but this should be in accordance to the criteria provided by the Act (Sections 40, 41, and 42). The Act also mandates a segregation of solid waste at source (Section 21), and the creation of the material recovery facility (MRF) in every barangay or cluster of barangays (Section 32). The barangay is responsible for the collection of the biodegradable and recyclable wastes while the city or municipality for the collection of non-recyclable and special wastes (Section 10). As of October 2018, there are only 345 (21.11%) LGUs with access to SLFs. As of 2017, only 24% of 42,036 barangays have established MRF, servicing to 32% of 42,036 barangays (NSWMC, 2018).

The above-mentioned information reveals that although policies on solid waste management are available, the approaches on how these can be effectively implemented remain a challenge. In the next section, some of the issues and challenges towards achieving effective and sustainable waste management will be discussed.

Issues and Challenges in Waste Management

The policies on waste management are in place, however, the great challenge lies on their implementation. Atienza (2011) cited the following factors that have contributed to the weak compliance of solid waste management policies and program: the technical and policy issues, financial constraints and the not-in-my-backyard (NIMBY) syndrome, setting of unrealistic deadlines, and other institutional issues such as the political will and terms of office of the local officials, the non-mandatory or non-permanent position of the Municipal Environment and Natural Resources Officers (MENRO), among others.

In terms of technical and policy issues, the LGUs are the main implementers of RA 9003 but there is limited support provided to enable them to effectively comply with the law. Financial constraints on the construction of sanitary landfills (SLFs) as alternative disposal for residuals, establishment of material recovery facilities (MRFs) per barangays, and conducting various SWM activities are hurdles to proper implementation especially for low income municipalities (Atienza, 2011).

Considering these constraints, it seems that the Act provided unrealistic deadlines for LGUs to comply with these requirements. There are many technical, financial, environmental, and social issues that need to be considered for a safe and proper closure of dumpsites and for constructing SLFs (Ibid.). This was confirmed with the data mentioned earlier which showed that even about 17 years after the coming into force of the RA 9003, the number of LGUs with access to SLFs and MRFs are still low (NSWMC, 2018).

Other challenges in managing solid wastes are the proliferation of plastics and other waste such as the electrical waste. In the study "Plastic Waste Inputs from Land into the Ocean" (Jambeck, et al. 2015 as cited in Atienza, 2017), the Philippines next to China and Indonesia, ranked third among countries with the highest mass of mismanaged plastic wastes. Hence, there are many LGUs which have started banning the use of plastics in their localities. Furthermore, several public and non-government organizations are also promoting non-use of plastics.

E-waste is classified under special waste and is usually handled separately from other residential and commercial wastes as cited in Section 3 of the RA 9003. Special wastes consist healthcare wastes, waste electrical and electronic equipment (WEEE) and other hazardous materials account for about 1.93% of municipal solid waste (NSWMC 2016). There is no existing national policy on e-waste and on Extended Producer Responsibility (EPR) in the country. However, the DENR has developed Guidelines on the Environmentally Sound Management (ESM) of WEEE to provide framework for the appropriate mechanism of WEEE. The Lighting Industry Waste Management Guidelines were issued through the Joint DENR-DOE Administrative Order No. JAO 2013-09-0001 in 2013 as an initiative on EPR (Atienza, 2017). It provides a great challenge on how to reduce the production of these wastes from the sources or the producers of these wastes. This also entails for the strong information, education, and awareness (IEC) campaigns to influence positive change to the behavior of the users especially in this era of modernization.

Waste Management Practices and Available Technologies: Lessons Learned From the Experiences of Selected Cases

In spite of the issues and challenges presented in the previous section, it is important to note that there are also numerous innovative solid waste management approaches which are proven effective through the experiences of certain LGUs. This section provides some of the existing waste management practices and available technologies in the Philippines; and the lessons learned from the successful and failed experiences of selected cases.

Technology Interventions and 3R Strategies in Quezon City

Quezon City is the biggest and most populated city in Metro Manila occupying a total land area of 160,112 sq.km. or about one fourth of the total land area of Metro with 2.94 million population in 2015 (PSA, 2017). It is home to several national government agencies, various specialized hospitals and educational institutions. In terms of waste generation, more than half (53.95%) of the waste generated are biodegradables, and the remaining are recyclables (20.30%) and residuals (25.76%) (Po, n.d.). The average per capita waste generation is 0.88/person/day. Waste generation is 2,639 tons/day while the average volume of waste disposed is 1,609.79 tons/day (Ibid.).

The City implements the package clean-up system wherein the SWM services are contracted by private service provided through competitive bidding. The winning contractor takes care of the solid waste collection, cleaning and disposal; street sweeping; and the conduct of the information, education, and communication (IEC) campaigns. Segregation-at-source is strictly implemented following collection schedules for biodegradable and non-biodegradable wastes (Quezon City's Environmental Management Program, n.d.).

There are several technology interventions in the City which includes the community-based material recovery facility (MRF); the biogas emission reduction project; and the refuse derived fuel (RDF) project. As mandated by RA 9003, all barangays shall establish MRFs in their respective jurisdiction for the biodegradables and recyclables. In Quezon City, they implement the Material Recovery System (MRS) in the absence of MRF, wherein a separate collection of kitchen or food wastes and recyclables are applied at the community level. The Biogas Emission Reduction Project, the 1st CDM Project in Southeast Asia under the Kyoto Protocol was established in 2007, which captured 466,740 tons of CO_2eq and converted 28,645 KwH of electricity in 2013 for the offices at the disposal facility (Po, n.d). Through this project, more than PhP1 million since March 2013 has been remitted to the City Government out of the sales from Meralco, a power distributor in the Philippines. In 2013, an RDF project, a highly mechanized solid waste processing facility, was also established. This has produced 15,000 tons of RDF on its pilot stage and these were sold to La Farge Cement Factories (Ibid.)

Other waste reduction initiatives include community-based (i.e. kitchen waste collection), school-based (i.e. junior environment police), and institution-based activities (i.e. recyclables trading at Quezon City malls); and social marketing and advocacy campaigns. They have also deputized 216 barangay officials and personnel and 35 police officers. They also have the landmark legislation on plastic bag reduction ordinance which provides a mechanism to recover plastic waste. Through this, the plastic recovery system fee and green fund for environmental projects are collected. With its implementation, the City has recovered 2 million pieces of plastics bags and collected PhP60 million worth of green fund. With all these technology interventions and 3R strategies, the City achieved more than 99% garbage collection efficiency and reached the 42% waste diversion (Quezon City's Environmental Management Program, n.d.).

Breakthroughs in the Solid Waste Management in Los Baños, Laguna

The Municipality of Los Baños, a "Science and Nature City", is a first-class municipality in the province of Laguna which is situated 63 kilometers south of Metro Manila. It has a total land area of 5,650 hectares. It lies between the shore of Laguna de Bay and Mt. Makiling. The municipality is composed of 14 barangays where two barangays (Bagong Silang and Tadlac) are considered rural and the remaining barangays – Anos, Bambang, Bayog, Batong Malake, Baybayin, Lalakay, Maahas, Malinta, Mayondon, San Antonio, Timugan, and Tuntungin-Putho are considered urban. As of 2016, the municipality has a total population of 112,008 based on 2015 census (PSA, 2016 as cited in Dela Rueda, 2017). Los Baños is a home to various research institutions such as the University of the Philippines Los Baños, the International Rice Research Institute (IRRI), and the Philippine Council for Agriculture, Aquatic, and Natural Resources Research and Development of the Department of Science and Technology (PCAARRD-DOST), among others. In addition, it is also known for its hot spring resorts which attract many tourists especially during weekends and summer season which contributes to the increased generation of waste in the municipality. According to Perez (2006), there are about 2,500 commercial establishments, two public markets, and 17,030 households that contribute to the generation of approximately 33-35 tons of solid waste per day.

The influx of people and commercial establishments in the municipality for the past decades contributed to the increased generation of solid waste. Until the early 2000, there was no clear system on managing solid waste in the municipality and the people had no discipline in managing their own waste. Open dumping was the common practice by the community and even Mount Makiling, a forest reserve area had become the location of the municipality's dumping site. Thus, it became an eyesore to the municipality and posed threats on the health of the nearby community (Atienza, 2008).

To address this solid waste management concern, the then Mayor Caesar Perez initiated his program with the implementation of the Municipal Ordinance 2001-08 as of 17 August 2001, the "Anti-Littering and Waste Segregation Program". This mandated the strict implementation of waste segregation, the collection schedule for the biodegradable and non-biodegradable wastes, and the imposition of penalties for non-compliance (Municipality of Los Baños, Laguna, 2001). In addition, the project entitled "Malinis at Maayos na Kapaligiran, Malusog at Matinong Pamayanan" (*Clean and orderly surroundings, healthy and sensible community*) was also launched led by the Local Chief Executive (LCE) in partnership with the Los Baños Science Community Foundation Incorporated (LBSCFI), a multi-agency organization composed of 22 scientific, technological, academic, and developmental institutions in Los Baños (Perez, et al., 2002).

The two major breakthroughs in solid waste management in Los Baños, as cited by Atienza (2008), included the conversion of the open dumpsite into an ecological waste processing center (EWPC) which was also considered as an MRF by the DENR; and the formation of the informal waste sector into a people's organization, known as the Los Banos Solid Waste Organization (LB-SWO). The EWPC/MRF was inaugurated on June 14, 2014 and in the same year, the LB-SWO was formally formed and recognized as the official collectors of the recyclables in the municipality. These breakthroughs became possible through participatory decision-making, multi-sector dialogues, community mobilization, and strong IEC campaigns. The municipality also strictly implemented the "no segregation, no collection" policy; composting of biodegradable wastes as fertilizers for fruits, vegetables, and flowers; and the proper management of recyclables. Los Baños is also one of the first LGUs who started banning the use of plastics in commercial establishments. With its effective implementation of solid waste management programs, Los Baños was awarded as one of the model municipalities in the Philippines in 2005 for its Trailblazing Program on solid waste management.

However, these solid waste management programs were not sustained when the Local Chief Executive (LCE) or the Mayor finished serving the maximum three terms (three years per term) and thus, a new LCE was elected. According to study of Dela Rueda (2017), there were no policies on solid waste management introduced during the three-year term of the new LCE. According to the respondents of the study, it seemed that at that time, the solid waste management programs were not among the top priorities of the new leadership. The priorities of the newly elected LCE were geared towards the promotion of tourism and provision of livelihood for the residents of Los Baños, thus, infrastructure developments were given priority at that time. But when it came to waste management, it mainly involved the maintenance of existing programs. In addition, most tasks related to SWM were designated to the concerned departments with minimal intervention of the Mayor. Unlike the previous Mayor who was very hands-on when it came to solid waste management programs of the municipality. It was also observed that there was lower compliance in terms of waste segregation among households probably due to weak IEC campaigns and law enforcement in terms of waste management (Dela Rueda, 2017).

The experience of Los Baños, Laguna on its implementation of the solid waste management program is an example of both a successful and failed or non-sustainability or the lack of continuity of the program due to different priorities of the local chief executives (LCEs).

Waste Recovery Systems in San Carlos City, Negros Occidental

The San Carlos City, Negros Occidental is located at the Western Visayas (Region VI). It a second income class city occupying a total land area of 45,150 hectares and a population of 132,536 based on the 2015 POPCEN (PSA, 2017). It has 18 barangays with six (6) urban, six (6) rural, six (6) coastal and island barangays. Sixty nine percent of its total population is classified as agricultural land planted with sugarcane as the main crop and other high value crops (Atienza, Sison, & Cardenas, 2011). It has numerous agro-industrial and socio-economic development projects and initiatives supporting environmental sustainability (Acosta, Paul, & Hanuschke, 2013).

Based on the San Carlos City Ten Year Solid Waste Management Plan 2000-2010, the estimated total waste generation was 17,671kg in 2011 with a .44kg/day per capita waste generation. The solid waste management board of the City was created in 2002. Its creation has contributed to the successful implementation of the City's waste management programs. The City bagged the Hall of Fame Award for winning the Model Solid Waste Board award for the three (3) consecutive years 2008, 2009 and 2010 (Atienza, et al., 2011). Another Hall of Fame award for the City is the Seal of Good Local Governance (SGLG) award for 2015, 2016 and 2017 wherein solid waste management is one of essential areas under Environmental Management (San Carlos City, 2017).

The City has implemented various waste management recovery systems managed by LGU which includes the following: the barangay-based, school-based, and the waste recovery system at the Eco-Center or Centralized MRF. Aside from the barangay-based MRFs, many barangays in the City has established *purok* "zone" level MRFs on a voluntary basis. The biodegradable wastes are not collected in rural barangays because they are managed by the waste generators themselves while there are scheduled collections for urban barangays. For recyclables wastes, residents have option either to drop them off to the MRFs or sell them to the itinerant waste buyers locally called "canvassers" by the junkshops which provide the latter with buying capital (Atienza et al., 2011).

For school-based recovery system, the Solid Waste Management Office (SWMO) is collaborating with schools in promoting waste management activities such as the waste segregation, recycling and composting. They have also launched school competition and provided incentives or gift packages for schools with highest performing schools. The accredited junkshops were chosen by the City for the recovery of recyclables from schools. These junkshops are required by the SWMO to record the volume of recyclables recovered from their assigned schools (Ibid.). There were 29 schools in the City which were cited as Sustainable and Eco-Friendly schools for 2016 (Bulletin, 2017).

The eco-center facility in the City was inaugurated in September 2007 which introduced the integrated approach. The city introduced the novel technology called as the "centralized gravity-driven" MRF in addition to clay-lined landfill cells. This significantly increased the efficiency of material sorting without requiring electricity in its operation (Acosta et al., 2013). Another unique strategy in the management of the City's eco-center is the recognition of the role and skills of the informal waste sector. The City government hired some of the former waste pickers who used to work from the closed dumpsite to formally work at the eco-center (Ibid.).

Forming the Informal Sector into an Organization: The Experience of the KILUS Foundation Environmental Multi-Purpose Cooperative

The KILUS Foundation Environmental Multi-purpose Cooperative is an organization with 250 members (244 women and 6 men) in Ugong, Pasig City, a highly urbanized city in Metro Manila. Although the members of this cooperative were not formerly waste pickers, their experiences provide good examples of the potentials of waste not just as a source for alternative livelihood but also the benefits of formalizing waste recycling activities. Most of these women were previously unemployed. In the continuous search for providing alternative livelihood opportunities, then Barangay Chairman Alejandro Santiago initiated the development of various products from doy packs. The doy pack bags used for artificial juices are made of triple-laminated cello foil. They are non-biodegradable and are not bought by junkshops for recycling thus, these commonly end up in dumpsites or sanitary landfills. With doy packs and colored magazines as raw materials, the workers of the cooperative turned them into fashionable products such as bags, shoes, office and school supplies, necklaces, and other accessories (Atienza, 2010). A livelihood seminar was conducted to the members of the Foundation. For the year 2002, KILUS bought 3,100,000 pieces of doy packs. At 200 pieces of doy packs per kilo, this total to about 15.4 tons of garbage recycled. KILUS workers earn P350/day (about 7 USD/ day; or 154 USD monthly) for in-house workers while those working at home with other members of the family can earn about P3,000 - 7,000/week; or about 240 - 560 USD monthly). This income is comparable or even higher than those of office or factory workers since they do not pay for transportation and buy expensive clothes for work. As a cooperative, they can also apply for grants and/or loans. Their recycled products are being exported in various countries like Japan, US, Australia, and England, among others.

From the KILUS experience, there are a number of important lessons that other LGUs can learn from. First, the innovation was initiated by the local government unit. This contributed to its success because it has the authority and support of the local government leader. Second, the significant roles of other sectors such as the private or financial sectors and training institutions, among others, contributed to the success, continued operation, and improvement of the organization. Third, it has become sustainable because their work does not only provide livelihood to its members but it also contributes to the cleanliness of the community. Thus, it exhibits environmental, economic, and social benefits. Last but not the least, recognition and inclusion of the informal sector including women in the waste management system are promoted. Their experience also illustrates that waste is not really a problem if this would be managed properly. This also provides a transformative impact of organizing the informal sector into an organization and/or cooperative and the difference it can bring to the community.

THE WAY FORWARD

In the Philippines, the policies on solid waste management are already in place but the challenge is on how to strengthen its implementation. Furthermore, the problem on solid waste is a complex concern which entails a more integrative approach involving the environmental, political, economic, and social perspectives. Given this huge challenge, this calls for a more participative approach of governance. As shown in the earlier discussion, the LGUs cannot do it alone. Hence, the active participation of various stakeholders such as the private sector and NGOs is necessary to achieve a more effective and sustainable waste management in the country. In the case of Quezon City, some of the technology options cited are the plastic recycling facility which can process, laminate, and convert soft plastic waste into usable products such as chairs, pavements, etc.; and the Waste-to-energy (WTE) facility, which was believed to be a long term solution to the waste disposal needs of the City. Currently, a feasibility study is being explored and a public-private-partnership project is being considered (Po, n.d.).

Since the term of the Mayor is only three years with a maximum of three terms, the institutionalization of the waste management program should be promoted to ensure its continuity despite the change in leadership. Given the sources and types of waste generated and the capability of the LGUs, alternatives that are simple, cost effective, and sustainable, such as segregation-at-source and fervent application of 3R should be considered and aided with appropriate technologies, machineries, and/or equipment.

REFERENCES

Acosta, V., Paul, J., & Hanuschke, K. (2013). Eco-center: Integrated solid waste management facilities with sanitary landfill and resource recovery technologies. Version 1. Manila: Deutsche Gesellchaft fur Internationale Zusammenarbeit (GIZ) GmbH.

Aguinaldo, E. (2009). *Philippines solid waste management.* In Regional 3R Forum in Asia, Tokyo City.

Atienza, V. (2008). Breakthroughs in solid waste management: Lessons from selected municipality and barangay in the Philippines. *Asian Review of Public Administration*, *XX*(1-2), 82–98.

Atienza, V., Sison, E., & Cardenas, L. (2011). *Review of the Philippines' waste segregation and collection system and the trading of recyclables. Joint Research Program Series 158.* Institute of Developing Economies, Japan External Trade Organization.

Dela Rueda, J. B. (2017). *The effects of change in political leadership on service delivery of a program: the case of municipality of los baños' solid waste management program (SWM)* (Unpublished masteral field study). College of Public Affairs and Development, University of the Philippines Los Baños. Los Banos, Laguna, Philippines.

Department of the Interior and Local Government (DILG) Calabarzon. (2019). *Local government performance management system.* Retrieved August2, 2019, from http://calabarzon.dilg.gov.ph/programsprojects/265-lgpms

Jambeck, J. R. (2015). *Plastic waste inputs from land into the ocean.* Retrieved, June 2016 http://www.iswa.org/fileadmin/user_upload/Calendar_2011_03_AMERICANA/Science-2015-Jambeck-768-71__2_.pdf

Manila Bulletin. (2017). *San Carlos City honors eco-friendly schools.* Retrieved August 22, 2019, from https://news.mb.com.ph/2017/02/01/san-carlos-city-honors-eco-friendly-schools/

Municipality of Los Banos, Laguna. (2001). Los Banos Municipal Ordinance No. 2001-08.

National Solid Management Commission. (2016). *National Solid Waste Management Strategy 2012-2016.* Retrieved from nswmc.emb.gov.ph/wp-content/uploads/2016/07/NSWM-Strategy-2012-2016.pdf

NSWMC/Environment Management Bureau/ Department of Environment and Natural Resources. (2015). *National solid waste management status report (2008-2014).* Author.

Perez, C. (2006). *Local ecological governance: Viable solutions to the solid waste management problem of Los Banos*. Academic Press.

Perez, C., Faylon, P., Pantua, L., & Valdez. (2002). *Managing town-community solid waste: Malinis at maayos na kapaligiran malusog at matinong pamayanan.* Paper presented at the 2002 PACS Seminar Series of the Philippine Association of Career Scientists, Philippines.

Philippine Statistics Authority. (2017). Retrieved June 2019, from https://psa.gov.ph/population-and-housing/node/120080

Po, A. V. A. (n.d.). *Technology interventions and 3R strategies to address Quezon City's solid waste generation*. Academic Press.

Republic of the Philippines. RA 6969. Toxic Substances and Hazardous and Nuclear Waste Act of 1990.

Republic of the Philippines. (2019). *Official Website of the Republic of the Philippines*. Retrieved August 2019, from https://www.gov.ph/about-the-philippines

San Carlos City. (2017). *San Carlos City wins anew in SGLG. Tribute: Official publication of San Carlos City, Negros Occidental*, 11, 4.

Senate Economic Planning Office (SEPO). (2017). *Philippine solid waste at glance*. Retrieved August 20, 2019, from https://www.senate.gov.ph

United Nations Environmental Programme- International Environmental Technology Centre (UNEP-IETC). (1996). *International Source Book on Environmentally Sound Technologies for Municipal Solid Waste Management*. Technical Publication Series. Available at UNEP website: http://www.unep.or.jp

World Bank. (2019). Retrieved August 21, 2019, from https://data.worldbank.org/country/philippines

Chapter 12
Solid Waste Management in Rwandа:
Status and Challenges

Telesphore Kabera
College of Science and Technology, Rwanda

ABSTRACT

This chapter aims to describe the status and challenges of waste management (WM) in Rwanda. Currently, waste is managed by the Ministry of Local Government, with the participation of private companies which are only in charge of waste collection. In the city of Kigali, waste is managed by the city of Kigali whereas in other four remaining provinces waste is managed by the districts. Implementation of waste management policy is carried out by a government-owned company called Water and Sanitation Corporation (WASAC) Ltd. The per capita solid waste generation rate in the city of Kigali is equal to 0.57 kg/person/day. A lot of legislations and regulations on WM are in place but their enforcement is weak. The Government of Rwanda should do more in terms of enforcing WM legislations and regulations.

INTRODUCTION

General Introduction

Globally, large quantity of Municipal Solid Waste (MSW) is generated every day (Gupta et al., 2015). Domestic waste (DW) is the waste generated and thrown as useless or unwanted from activities in households, however DW is also a resource for reuse, recycle, and recovery of materials (Tchobanoglous et al., 1993). The problem associated with solid, liquid, and toxic-wastes management in Africa has mainly arisen from urbanization in the developing countries of Africa. The rapid growth of cities is one of the characteristics of urbanization in developing countries (Yoada et al., 2014). The rapid African urbanization implies a rapid accumulation of unwanted waste material (Ahmed & Ali, 2006).

The world is now experiencing private and public sector partnership to provide the services for domestic waste collection, disposal and management. But for this partnership to succeed and to show an

DOI: 10.4018/978-1-7998-0198-6.ch012

impact on total service coverage and environmental cleanliness, it will have to be supported by formal rules, policies, legislations and regulations that created, enforced and maintained by the governments. The later initiative will provide the required impetus to encourage people for more investment, improved service quality, and regular price adjustment (Fobil *et al.*, 2008). Recently, the world has experienced a high incidences of sanitation related illness such as cholera, intestinal worms and typhoid due to poor waste management. Cholera, intestinal worms and typhoid are among top ten diseases recorded, which raises the alarm of public health crisis (Achankeng, 2003).

Most of developing countries are still experiencing poor waste collection and management (Tanskanen, 2000). This is due to the failure of the waste sector in developing countries which has not been able to provide adequate and sustainable solid waste management services to the citizens (Laurent et al., 2014). Rwanda is no exception specifically in Kigali city where urban population growth and economic development lead to the increase of MSW generation. This growth has also increased the use of products that generate hazardous waste which contribute to the pollution and public health hazards in the localities. Kigali city has not yet established a proper and comprehensive waste management practice (Isugi & Niu, 2016).

In some countries due to poor state of solid waste management, waste collection is infrequent, has inadequate coverage, leads to indiscriminate dumping, and lacks septic tanks locations (some of the septic tanks in residential homes are not accessible by the desludging vehicles), thus manual methods are employed which is a health risk. Sludge treatment in urban areas in developing countries is often not appropriately carried out and it is often disposed off in an unhygienic way in the dumpsite. Rivers (some rivers pass through urban areas) and wetlands are often polluted by some chemical wastes disposed improperly. Rwanda lacks proper facilities and treatment skills to deal with waste management (REMA, 2009). The improper waste management can result in bad odor, methane gas explosions, risks of garbage landslides and groundwater pollution.

The rapid increase of population has exerted pressure on infrastructure and this results in many complex problems regarding settlement, solid and liquid waste management, therefore solid waste management has become major concern for the City of Kigali (CoK). The Government of Rwanda has planned to have sufficient sewerage and disposal systems by 2020. Each city in Rwanda is planned to be endowed with an adequate solid waste treatment system. All the households in Rwandan cities will get proper services from waste collecting companies.

When dealing with Municipal Solid Waste Management (MSWM), economically developing countries encounter the following typical problems: (i) inadequate service coverage and operational inefficiencies of services; (ii) limited utilization of recycling activities; (iii) inadequate landfill disposal and (iv) inadequate management of hazardous and healthcare waste (UNEP, 2005).

In general, waste in Rwanda is classified in four categories as solid, medical, hazardous and liquid wastes.

For Rwanda, some solid waste management (SWM) studies have been conducted, thus only one peer reviewed paper was available, which focused on SWM in Kigali city (Kabera et al., 2019), along with limited 'grey' literature and a few student projects including Isugi and Niu (2016), Nshimiyimana (2015), Bazimenyera et al, (2012).

In a proper solid waste management system, the ultimate step is the disposal of solid waste generated in a community. In the advanced technologies for solid waste management, the waste disposal is preceded by engineering activities such as sorting, volume reduction and/or receding (Getahun et al., 2012).

All governments are willing to provide proper waste management services to their people. Looking at different municipal services, solid waste can possible be ranked as the most important service within the municipality. Due to fast development of developing countries and migration from rural areas to urban areas, the quantity of MSW is growing faster than the urbanization rate. The rapid growth of urban population leads to high socio-economic and environmental impact in solid waste management practices. Because of enhanced economic activities and rapid urbanization MSWM has become an acute problem. MSW is a source of renewable energy in the form of biogas; high moisture and organic nutrient rich solid waste provide renewable biogas (Zaman, 2014). The purpose of this chapter is to describe the status and challenges of waste management (WM) in Rwanda

Background Information

Rwanda is situated in the central/eastern of Africa with an area of 26,338 km^2. The capital city of Rwanda is known as the City of Kigali. Rwanda is a landlocked country and is surrounded by four countries (Figure 1). To the east of Rwanda is the United Republic of Tanzania, to the west is the Democratic Republic of Congo (DRC), to the north is Uganda and to the south is Burundi. Rwanda's latitude and longitude are 2° 00' S and 30° 00' E respectively.

Figure 1. Map of the Republic of Rwanda

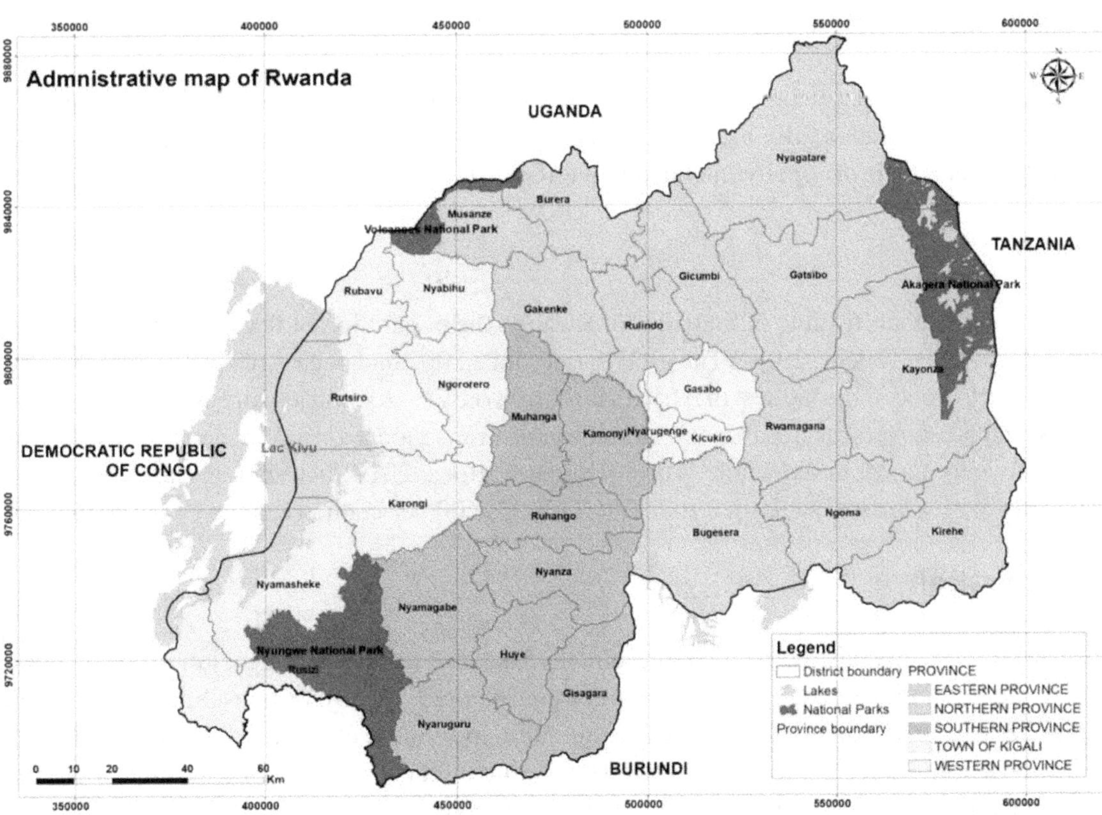

The temperatures in Rwanda are the product of a tropical highland climate and high-altitude temperatures in Rwanda that are lower than the typical temperatures of equatorial countries. The variation in temperature depends on the altitude. The larger part of the country is situated on a plateau of about 1,500 m above sea level and the weather is pleasantly warm all year round. The lowest point in Rwanda is located near Lake Kivu along the Rusizi river in Rusizi District. Altitude decreases to 915 m and the temperature can sometimes reach 35°C. The average temperature in the capital City (Kigali City) is 21°C.

There are four climatic seasons in Rwanda, a long dry season which ranges from June to August, a short rainy season which is from September to November, a short dry season which is from December up to February and a long rainy season which ranges from March to May.

SOLID WASTE

Understanding Domestic Waste

Domestic waste is the waste from households and it normally includes biodegradable, electrical and electronic, and composite wastes such as clothing, hazardous waste (paints, spray, and chemicals), and medical waste. These types of wastes consist of everyday items that are discarded by the public.(Pirsaheb et al., 2013)

Waste Definition

Waste in Rwanda is defined as a product or a substance that is no longer suited for its intended use. The legal definition often includes fully usable substances but defines them as waste if they are to be used in contexts other than their originally intended one (AOG, 2016).

Population

According to the National Institute of Statistics of Rwanda the population of Rwanda this year (2019) is estimated at 12,374,397 (NISR, 2019). The growth rate since 2002 had been 2.6% per annum. Rwanda's population density is one of the highest in Africa (475/km^2), only two countries surpass Rwanda, Mayotte (694/km^2) which is an overseas department and region of France located in Africa and Mauritius (622/km^2) and it is twenty fourth in the whole World (United Nations, 2019). Rwanda is classified as a low-income country with a Gross Domestic Product (GDP) of $700 per capita (Kabera et al., 2019).

Waste Generation

For a successful waste management planning, reliable data such as (i) waste generation, (ii) factors influencing waste generation and (iii) waste quantities are required. In case of household related studies, relationship between individual habits and waste quantity, individual characteristics of the representatives of the household itself have to be analyzed. Depending on the type of house like single household, dwelling or an entire community, predication is mainly based on income level of the households (Lebersorger & Beigl, 2011).

In Rwanda, the government is obliged to consider stakeholders' opinions regarding solid waste management (SWM) systems. Public is well involved in SWM in Rwanda because of environmental committees present at each administrative level from the village up to the province or Kigali city level. This is easy because from each village level up to the level of the City of Kigali (the lowest level of administrative entities in Rwanda are the village, followed by cell, sector, district, province and then the country) environmental committees are present. These committees formed by local people address the environmental problems including SWM. They raise their concerns when they meet authorities in every last Saturday of each month during a community service or once a month (usually Wednesday) when locals meet the executive secretary of the sector.

The quantity of SW generated is the fundamental information in Solid Waste Management (Tchobanoglous et al., 1993). Annual waste generation of the whole country is not known and the only known amount is for the City of Kigali. The annual waste generation of Kigali was obtained by weighing vehicles bringing SW to the only dumpsite for which data are obtained. It was estimated at 638 tons per day (tpd), or 233,000 tons per year. This corresponds to a per capita MSW generation of 0.57 kg/day, or 205 kg/year (Kabera et al., 2019). For the other remaining places (other cities together with rural areas), data on waste generation are yet to be obtained.

Waste Collection

There are various collection schemes, with differences in the types of targeted recyclable materials and this comes from the different characteristics of the households within the targeted area and the inter-relationships between the various collection systems, sorting methods, the composition of MSW and the frequency of collections. Waste collection frequencies of recyclable materials in the developed countries such as United Kingdom can be weekly, fortnightly or monthly. Increasingly common practice in those developed countries (like United Kingdom) is alternate weekly collection (AWC), where each household has its recyclable waste and residual waste collected fortnightly basis on alternating weeks (McLeod & Cherrett, 2013). According to Takele (2004) in solid waste management report, with the Ethiopia Public Health Training Initiative, Solid waste collection systems may be classified from several points of view. These are:

a) Mode of operation.
b) Equipment uses.
c) Type of waste collected.

Mode of Operation: Solid waste collection systems are classified according to their operation into two categories. These are:

- **Hauled Container System:** Which is a system where the containers used for storage of wastes are hauled to the disposal site, emptied and returned. This system is ideally suited for the removal of wastes from sources where generation rate is high.
- **Stationary Container System:** Where the containers used for the storage of waste remain at the point of generation, except for occasional short trip to the collection vehicle.

Equipment Used: Some of the equipment used are containers, machineries, hand tools and others. Containers for the hauled systems or operation have various sizes and shapes. Basically, since the system is used to haul wastes from sources where the route of generation is high, large containers are provided. Hauled containers system requires only one track, and one driver, to accomplish the collection cycle.

Composition of Solid Waste

In 2012, a detailed design study was conducted for a new sanitary landfill site and recycling center (WAT, 2012) yet to be constructed in Kigali. During that study the composition of SW was determined (for the city of Kigali only). The study found that waste composition in the city is as followed: (i) organics 70% (by weight), (ii) paper and cardboard 5%, (iii) plastics 5% and metals 1% (WAT, 2012). Solid waste in the City of Kigali contains high concentrations of organic waste and this indicates that moisture content would also be high (unfortunately data for moisture content and waste density are not available). This might be generalized because the lifestyle of Rwandans is more or less the same.

Waste Classification

There is no administrative order classifying waste in Rwanda. In Rwanda there are mainly four categories of waste: (a) Solid, (b) Medical, (c) Hazardous waste and (d) Liquid wastes.

(a) Solid wastes are composed of (1) waste generated in households which results from household activities, (2) waste from streets sweeping and street litter bins, (3) waste from commercial areas and markets.

(b) Medical waste (*Waste from health centers and hospitals):*
Medical waste is defined as all wastes produced in health-care establishments, research facilities, diagnostic activities and laboratories. These wastes also include waste from sources like those produced during health-care activities undertaken from homes i.e insulin injection, dialysis (ICRC, 2011).
75% to 90% of medical wastes are similar to household waste (they can be handled in a similar way) whereas the remaining 10% to 25% are hazardous waste or special waste and need a special handling and treatment (WHO, 2019).

(c) The Rwanda Health Sector Strategic Plan 2012–2018 planned to increase healthcare facilities together with effective medical waste management systems from 55% in 2012 to 88% in 2018. In 2016, 35 health facilities (3 referral hospitals, 2 provincial hospitals and 31 district hospitals) were assessed for waste quantities and types of waste generated. The assessment found: (i) an average of 5.168 kg/day of medical waste generated by inpatients and outpatients, (ii) that from the total volume of medical waste produced in one year by health facilities in the country, 74% is non-infectious waste, 24% is liquid waste and 1.3% is infectious and/or hazardous waste (Ministry of Health, 2017).

(d) Hazardous Waste:
In Rwanda these wastes are composed of used oil such as engine lubrication oil, and gear oils used in cars and bikes. If these used oils are not properly handled, they can pollute the environment. They are mainly used as a termite-resistant building material in construction. Wood used for roofs are coated with the used engine oil in order to protect them against termites.

(e) Liquid waste:

Figure 2. Sankey type material flow diagram (MFD) for MSW in Kigali

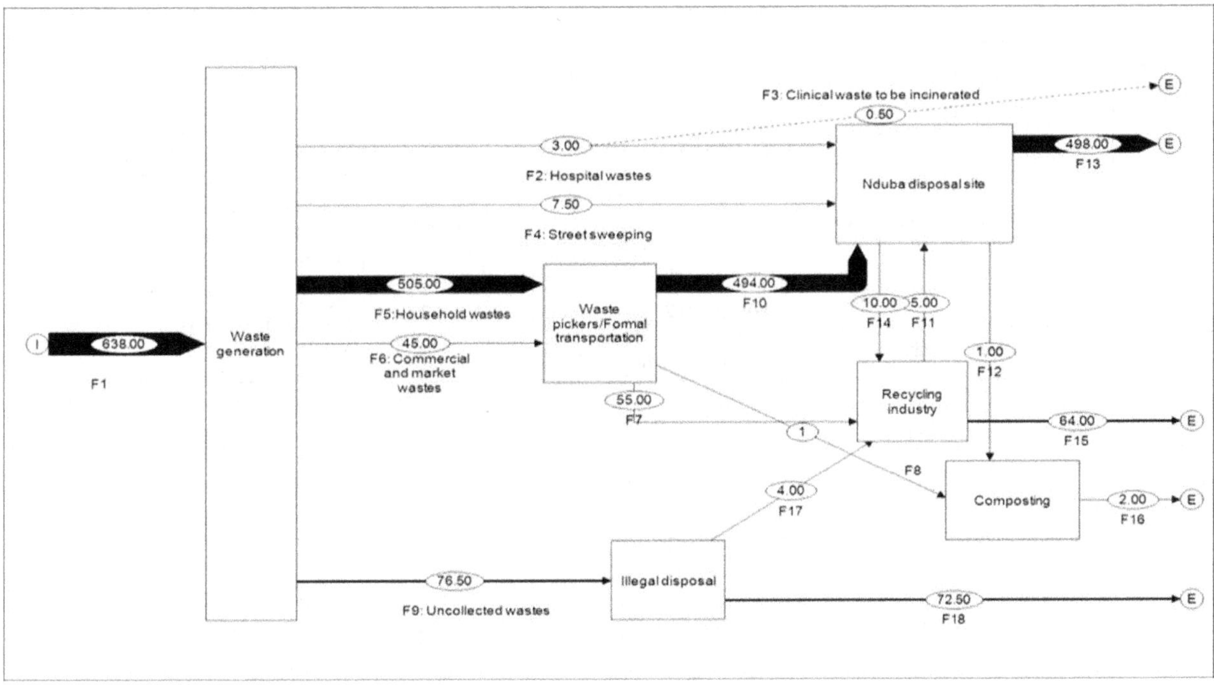

This also known as sewage and it is a viscous mixture of by-products of human digestion (faeces & urine) deposited in toilets (AOG, 2016). Data on liquid waste are not yet available for the country.

Waste Flow

There is no data on MSW generation for the whole country, but and the only waste generation data available are the ones for Kigali City. In order to measure the MSW generated, researchers used the number of vehicles transporting waste to the dumpsite. Many of the flows in the MDF diagram were measures based on the observations and responses from interviews with stakeholders and others were calculated from the flow diagram (Figure 2) (Kabera et al., 2019).

The thickness of the flow corresponds to the weight of the waste (tons per day). The sources and methods used estimate each flow in the MFD are summarized in Table 1. (Source: Kabera et al., 2019)

All procedures covered to calculate all flows present in the flow diagram are explained in Table 1.

3R PRACTICES WASTE MANAGEMENT TECHNOLOGIES

Recycle: The recycling rate calculated from the MFD diagram (Figure 2) is equal to 12% (just for the City of Kigali). In rural areas people choose to use their lands as waste disposal sites, because a high percentage of their waste constituting of organic matter will be used as fertilizer.

Items such as plastic and paper are very easy to recycle because they are made of a single material whereas other items, such as electronic materials, are very difficult to recycle because there will be a

Table 1. Basis for calculating the waste flows for Kigali in the MFD (Figure 2) (Source: Kabera et al., 2019)

Waste Flows	Description	Calculation	Basis for Data
F1	Total Quantity of waste generated in Kigali	Official data give collection coverage as 88%. So F1=(F2+F4+F5+F6)/0.88	Back-calculated from collection coverage and other flows in MFD
F2	Waste from official Health Centers and Hospitals	Calculated based on the average load delivered by the vehicle. 3 vehicles a week each carrying an average load of 5 tons. 15 tons over 6 days, 2.5 tpd collected.	Vehicle numbers and load size provided by the person in charge of Nduba disposal site
F3	Hazardous clinical waste treated onsite by simple combustion	Assumes 0.25 kg per bed per day (cf. WHO guideline 0.2 (Haifete et al., 2016)). There are 2000 beds in the 5 main hospitals.	Estimate by authors, informed by WHO data
F4	Waste from street sweeping/ street litter bins	9 vehicles per week each carries 5 tons. 45 tons over 6 days = 7.5 tpd	Vehicle numbers arriving at Nduba disposal site
F5	Waste collected from door to door by the contracted private companies.	30 vehicles per day carrying 10 tons + 41 vehicles per day carrying each 5 tons.	Vehicle numbers and sizes arriving at Nduba disposal site
F6	Waste collected from commercial areas/markets	9 vehicles per day carrying each 5 tons	Vehicle numbers arriving at Nduba disposal site
F7	Wastes collected by the private companies and separated by their employees for recycling	Estimated at 10% of flows F4 and F5 F7 = 0.1*(F5+F6) = 0.1*550 = 55 tpd	Estimate by authors based on observation and interviews
F8	Organic wastes collected directly from the source and transported to compost plant	Very small flow. Estimated at 1 tpd	Estimate by authors based on observation and interviews
F9	Uncollected wastes	Waste collection coverage equals 88%. So, 12% is not collected. Then F9 = 12% of F1	Collection coverage calculated from official data)
F10	Waste delivered to the Nduba disposal site from households, commercial areas and markets	Waste collected less the quantity separated for recycling F10= F5 + F6 – F7	Calculated from the MFD
F11	Reject materials from recycling, sent to Nduba disposal site	Estimate	Estimate by authors based on observation and interviews
F12	Organic waste sorted by waste pickers from Nduba disposal site, taken to a composting plant	Estimate	Estimate by authors based on observation and interviews
F13	Total waste deposited at Nduba disposal site	F13= (F2-F3 + F4 +F10 + F11) – (F14 + F12)	Calculated from the MFD
F14	Waste sorted for recycling by pickers at Nduba disposal site	Estimate. Mainly plastic bottles, exported to Uganda and China	Estimate by authors based on observation and interviews
F15	Recycled materials	F15 = (F7 + F17+ F14) – F11	Calculated from the MFD
F16	Fertilizer from Compost plant.	F16 = F8 + F12. Quantities very small, so no attempt to correct for process losses.	Calculated from the MFD.
F17	Waste sorted for recycling by pickers on illegal dumpsites	Estimate	Estimate by authors based on observation and interviews
F18	Waste left in illegal dumpsites after waste picking	F18 = F9 – F17	Calculated from the MFD

need to dismantle and separate them. In Rwanda recycling takes place both formally and informally. For formal recycling, there are two companies which recycle small quantities of waste. Whereas, for informal recycling: there are people who pick valuable wastes from mixed waste at dumpsites and sell them at their collection points, which are near the dumpsites. From there, they are transported to recycling shops.

In Rwanda some private companies such as COPED, AGRUNI and COCEN do some recycling through making briquettes and composting. Metals are also recycled by STEELRWA Company, metal waste are recycled into steel reinforcing bars. Waste from demolished houses or waste from excavation activities are used to fill open pits resulting from sand/stone removals. Although there are two companies which recycle, recycling is still considered as informal and no source separation occurs except those bought from sources by private companies for waste recycling e.g: papers, bottles, and others. Although there is awareness support for waste separation at the source (householders and offices).

In the city of Kigali, awareness is carried out through Media (Radio, TV and Newspapers), the mandatory community clean-up campaign *Umuganda* taking place at every last Saturday of each month, community gathering, parents' evenings and sometimes direct meetings with the person in charge of environment protection in the City of Kigali are organized. Every citizen from both rural and urban Rwanda areas is concerned about the environment. Government has succeeded in educating people about the protection of environment.

Reuse: An example of waste which is mainly reused in Rwanda is construction waste. In Rwanda, the demolished road waste is mainly used to pave mud roads.

Reduce: Plastic bottles pollute oceans and experts warn that these bottles have found their way into the human food chain. Researchers in Belgium (at Ghent University) showed that those who eat seafood ingest around 11,000 plastic tiny pieces every year (The Telegraph, 2017). In another study in Plymouth University reported the presence of plastic in a third of fish caught in United Kingdom (The Guardian, 2017). In order to tackle the problem posed by plastic bottles, the Government of Rwanda has started to implement some strategies. For example, during meetings the Government has decided to stop using small plastic bottles. Water will be in a big bottle (as shown in Figure 3) and it will be served in glasses.

Figure 3. Big plastic bottle that is used in Government meetings

Regarding plastic bags, they are known to be difficult and costly to recycle and they take up to 300 years to degrade, they contaminate both waterways and soil and they enter the food chain once ingested by animals (Jacobsen, 2005). In order to tackle the problem Rwanda has banned plastic bag use since 2008. There is no formal composting in Rwanda. Some private companies attempt to compost but the quality is unknown since compost has not been tested.

Waste Management Technologies

- **Uncontrolled Dumpsite**: The Government of Rwanda has committed itself to stop using dumpsites and replace them with landfills (MININFRA, 2018).
  In Rwanda, a landfill is defined as any location within a SW disposal site that is used to permanently dispose waste and in which the organic portion of the disposed waste is subject to natural processes of aerobic and anaerobic decomposition (RURA, 2009). In all dumpsites there is no thermal control, no gas or liquid/ leachate control. Some compaction and covering are done occasionally.
- **Incineration**: The only use of thermal treatment (incineration) in Rwanda is for the incineration of some hospital wastes on site (without any emissions control). There are no incinerators in all dumpsites in Rwanda.
- **Semi-controlled Sites**: There are many semi-controlled dumping sites in Rwanda, where one example is Nduba semi-controlled dumpsite which is the only facility in Kigali city dealing with collected solid waste.

POLICY AND REGULATIONS

In order to have proper waste management system, existing laws, legislations and regulations have to be updated and at least minimum levels of service and environmental protection should also be established. These levels can be improved over time. Enforcement of legislations, which is a serious problem in developing countries, should be accompanied by user-friendly advice and guidance that address the aspects such as operating licenses, producer responsibilities, landfill regulations, tariffs, and disposal of hazardous, industrial and agricultural waste, illegal dumping, prosecution and recovery of cleanup costs (OAG, 2016). Moreover, public awareness and attitudes towards waste management can affect the whole solid waste management system of an area.

Public awareness and public participation are very important in solid waste management because all steps in solid waste management services, except waste treatment and disposal facilities, all depend on them. Thus, it is also a crucial issue, which determines the failure or success of a solid waste management system. It is also necessary to extend active support to the existing consumer awareness and anti-litter campaigns, which are mainly voluntary. Additionally, few co-coordinated anti-litter campaigns and consumer awareness campaigns should also be developed in the country. It must be noted that Rwanda is doing well regarding consumer awareness. While consumer awareness and anti-litter campaigns are often the interventions most favored by most industries and institutions, it still has its shortcomings especially in the developing countries, where a segment of the population will always litter and it cannot prevent inadequate littering. Education includes workshops and trainings, and is nevertheless an important support for other initiative aimed to reduce waste and its impact (Nilsson et al., 2009).

In Rwanda, the organization in charge of managing water and sanitation, in addition to being responsible for implementing SWM policy, is the Water and Sanitation Corporation (WASAC) Ltd, a government-owned company. On the other hand, for environmental regulating activities, Rwanda Environmental Management Authority (REMA) is responsible for standards and guidelines development for SWM, whereas Rwanda Utilities Regulatory Agency (RURA) is in charge of balancing the profit of private companies (in this case the waste collectors) and the quality to the consumers.

The Government of Rwanda has developed a number of legislative regulations, policies, and strategies on waste management, which not only protect both public health and the environment but which also provide guiding principles for all sanitation aspects. Unfortunately, their implementation and enforcement are sometimes weak. Table 2 summarizes the Rwandan Framework for Waste Management.

The national sanitation policy has already addressed solid waste management, delegating the task to households, community, non-government organizations (NGOs), the private sector, community as-

Table 2. National framework for waste management

Short Name	Observations
Legislation and regulations	Rwanda has got important SWM-related legislation/regulations, including: • Law N° 04/2005 of 08/04/2005 on the protection, conservation and promotion of the environment in Rwanda • Law N° 39/2001 of 13 September 2001 establishing as utility regulator the Rwanda Utilities Regulatory Authority (RURA) • LAW N° 57/2008 OF 10/09/2008 prohibiting the manufacturing, importation, use and sale of polythene bags in Rwanda • Regulations Governing of Solid Wastes Recycling in Rwanda, April 2015 (RURA, 2015) • Governing Liquid Waste Collection and Transportation, November 10, 2016. No: 005/R/SANEWS/RURA/2016 (GoR, 2016) • Law Relating to the Prohibition of Manufacturing, Importation, Use and Sale of Polythene Bags in Rwanda, October 9, 2008. No: 57/2008 (Ministry of Natural Resources, 2008) • Ministerial Order Establishing Special Regulations Relating to Burying Toxic Wastes, July 1 2010 No: 006/16.01 • Draft Regulations Governing E-waste Management in Rwanda, Republic of Rwanda, Rwanda Utilities Regulatory Authority, 2016 (RURA, 2016)
Strategy/Policy	Relevant strategies and policies include: • Economic Development and Poverty Reduction Strategy – The Republic of Rwanda (MINECOFIN, 2007). • National Strategic Plan, and five-year operational plan, for the Management of Healthcare Waste (Ministry of Health, 2012) • National Industrial Policy (Ministry of Trade and Industry,2011) • National Sanitation Policy (MININFRA, 2016). • National Policy and Strategy for Water Supply and Sanitation, February 2010 (MININFRA, 2010) • National e-Waste Management Policy for Rwanda, September 2015 (Ministry of Natural Resources, 2015) However, none is fully implemented.
Guidelines and implementation procedures	Clear SWM guidelines and implementation procedures include: (i) Practical Tools on Solid Waste Management of Imidugudu, Small Towns and Cities: Landfill and Composting Facilities (REMA, 2010). (ii) Standards on the Management of Waste Disposal Sites (RURA, 2009).
National institution responsible for implementing SWM policy	Water and Sanitation Corporation (WASAC) Ltd is a government owned company set up to manage services for water and sanitation in Rwanda. Also responsible for implementing SWM policy.
Regulatory control	Environmental regulator: Rwanda Environmental Management Authority (REMA). Responsible for standards and guidelines development for SWM. Utility regulator: Rwanda Utilities Regulatory Agency (RURA) - balances the profit of private companies (in this case the waste collectors) and the quality to the consumers.

sociations and district authorities, which are supervised by the City of Kigali. The environmental law of Rwanda approved in 2005 defines all the rules to be applied in order to preserve the living environment for the whole population of Rwanda, especially in urban areas.

As for SWM, it especially gives preventive provisions like the prohibition of any dumping or disposal of any solid, liquid or gaseous substances in a stream, river, lake and their surroundings, and also in wetlands. Another article prohibits the piling, importing, immersing, burning or eliminating of waste by any process which does not respect the rules applied in Rwanda (Article 89 of N° 04/2005 of 08/04/2005**:** Organic Law determining the modalities of protection, conservation and promotion of environment in Rwanda).

In the towns of Rwanda, private companies are responsible to collect waste. For example, in the City of Kigali, there are now 12 private solid waste collecting companies and 7 private liquid waste collecting companies. The 2010 National Sanitation Policy addressed SWM, delegating the task to households, communities, NGOs, the private sector, community associations and district authorities. There is an integrated solid waste management approach in the Cities Master Plan but it is not yet implemented. (Kigali Master Plan Report, 2013). Regulations mainly concern private waste collecting, but not consumers. For instance, there is no regulation on how to sort waste at source, no obligation to comply with or to pay for selective collection (National Sanitation Policy, 2016). In the cities where private companies are operating, regulations and guidance for service contracts and their duration are clear and when the contact is not respected it is annulled.

Integrated solid waste management approach has been developed in Kigali master plan report 2013: Environment Treatment Zone. Decision on what solid waste facility, if any, to be included in each ETZ will depend on the actual type of waste generated by area contributing to it.

HAZARDOUS WASTE GENERATION

The manufacturing sector in Rwanda is not well-developed. Therefore, the amount of hazardous waste from the Rwandan industrial sector is very low. No study has been conducted to determine the amount of hazardous waste generated. A lot of so-called industrial waste comes from food processing industries and these industries produce mainly waste liquid waste.

AGRICULTURAL BIOMASS AND UTILIZATION, WASTE-TO-ENERGY

The only use of biomass is through converting solid waste into briquettes and pellets. This is only done in the city of Kigali by two companies, namely Compagnie pour la Concervation de l'Environment (COCEN) which makes briquettes and the Company for Environment Protection and Development (COPED) which makes pallets. From the discussion with COCEN's Chief Executive Officer it was found that five tons of solid waste are converted into briquettes daily, on average.

E-Waste

In 2015, the Government of Rwanda developed a National E-Waste Management Policy:

(http://www.minirena.gov.rw/fileadmin/Environment_Subsector/Laws__Policies_and_Programmes/Laws/E-waste_policy_-_FINAL.pdf)

In Rwanda e-waste management is in the hands of a private company known as Enviroserve Rwanda Green Park, which is a subsidiary of the Emirati company Enviroserve Services LLC Dubai. From the discussion with the Enviroserve country general manager (CGM) it was found that the total e-waste generated in Rwanda is 10,000 to 15,000 tons per year and all of this is either reused or recycled. The e-waste mainly reused in schools (secondary schools) comprises of computers and printers. For recycling, TVs are the main recycled e-waste and the remaining hazardous e-wastes are treated through incineration.

ISSUES AND CHALLENGES IN WASTE MANAGEMENT IN RWANDA

- **Valorization**: Valorization of waste in Rwanda is not yet developed. Only a small quantity of waste is composted or reused or recycled.
- **Disposal**: Waste disposal in Rwanda is a challenge because waste is mainly discarded in dumpsites.
- **Separation**: Separation at source is not yet implemented.
- Waste collection fees are not enough, private collection companies are always complaining that waste collection fees are very low.

ECONOMICS OF WASTE MANAGEMENT IN RWANDA

In towns, citizen pay for the collection services where people pay according to their Ubudehe categorization (which are similar to social classes). There are four categories. Category 1 is the poorest whereas category 4 is the richest. People who belong to category 1 do not pay anything, whereas the other three categories pay differentially. The amount to pay differs by sectors but the payment is in the range of Frw 1000 (USD 1.2) for category 2 to Frws 5000 (USD 6) per month. Even if private collection companies complain about collection fees, they still make profits.

CASE STUDY OF SUCCESS: PUBLIC-PRIVATE SECTOR PARTNERSHIP

The City of Kigali is chosen in order to show the success in Public-Private sector partnership.

Private Sector Representation

The city of Kigali is the one which sets the rules of collection, treatment and disposal of solid wastes as well as liquid wastes. Additionally, in accordance to the decentralization law, the three districts of Kigali (Gasabo, Kicukiro and Nyarugenge) participate in in making choices and are responsible for the implementation of the rules defined. The city also plays the role of collecting taxes.

The operational level of SW collection and communication towards the population is on sector level, and in coordination with the districts. The sectors coordinate the SW collection (SW is collected by the cooperatives or companies), try to impose some price framework, control the effectiveness and even impose fines on reluctant households (SWM strategic plan, 2012).

Private sector is well represented in the whole process. Private companies participate in each and every step of development of community associations, private sector providers are given the zone in the different part of 3 districts of Kigali city to pick and collect waste.

The Balance of Public vs Private Sector Interests in Delivering Services

Rwanda Utilities Regulatory Authority (RURA) is in charge of balancing the profit of private companies and the quality of the services given to the consumers. Citizens always get waste collection services at fair prices.

Legal Framework

In the City of Kigali, there are now 12 private solid waste collecting companies and 7 private liquid waste collecting companies. Every household is supposed to have a private company to collect and/or pick waste to legal dumpsite. The 2010 National Sanitation Policy addressed solid waste management, delegating the task to households, communities, NGOs, the private sector, community associations and district authorities where some of them operate with limited technical and financial means. There is an integrated solid waste management approach in Kigali City Master Plan but it has not yet been implemented (Kigali Master Plan report, 2013). Regulations are primarily targeted to private waste collecting and recycling companies, who are obligated to deliver sorted waste, and does not involve consumers since there are no guidelines on sorting waste at source, no obligations to comply with regulations or a scheme to pay for selective collection (National Sanitation Policy, 2016). There are very clear regulations and guidance for service contracts and their duration is three years. When the contact is not respected it is annulled. The contract does not specify the money to be paid.

Rwanda has not implemented systematically the integrated solid waste management approach but it has seen different interventions carried out by districts and the private sector, the latter mostly in Kigali, in line with the (global) principles set out in the policy 2010. Kigali city is undertaking considerable efforts to maintain the urban environment clean, but financial treatment and disposal is lagging behind schedule.

The Bidding Process

Contacts for MSW collection lasts for three years and the bidding process is very clear and done properly. The only important aspect is the fact that the contract does not mention the money to be paid for collection service. For payment, there are fixed prices according to the social class (Ubudehe category) of a household in each sector, as mentioned earlier.

Payment

All citizens receive a 'good' level of services, regardless of the social class. From households, the collection is carried out once a week, whereas in the restaurants, markets and hotels, waste collection is performed every day. Citizens pay for the collection service.

Right To Be Heard-Participation

The government of Rwanda is obliged to ask for stakeholders' opinion in the SWM systems. There are environmental committees from cell level up to the City of Kigali (i.e. Cell, Sector, District and City of Kigali). These committees formed by local people address environmental problems including SWM. They raise their concerns when they meet authorities in every last Saturday of each month during a community service or once a month (usually Wednesday) when citizens meet executive secretary of the Sector.

Capital For Service Delivery

Private companies rely on private investments or loans from the bank for their infrastructures whereas public sector relies on central government funds.

THE WAY FORWARD

Potential Opportunities: 70% of solid waste in Rwanda is organic waste, therefore processing organic waste via composting into a soil conditioner for crops, or via anaerobic digestion into both gas/electricity and a soil conditioner, would be an effective waste management strategy.

Government Initiatives for Achieving Sustainable Waste Management: The Government is determined to fund the design and the construction of landfills and recycling centers. As an example, the detailed design report of a landfill in the city of Kigali has been completed but so far there has been no auction held for any contract.

CONCLUSION

Improper management of solid, liquid and hazardous waste in Rwanda is due to urbanization and like in most other developing countries, solid waste management in Rwanda presents serious challenge.

The only solid waste disposal site for the capital of Rwanda, City of Kigali, is a dumpsite called Nduba dumpsite. In Rwanda, solid waste management data quality is limited. Data such as waste generation, waste composition are not measured. Disposal sites do not have weighing balance to use when weighing waste quantities entering the disposal sites.

Legal framework in Rwanda is excellent with different entities enforcing them such as the Rwanda Utilities Regulatory Authority (RURA), which is a governmental institution to regulate all types of services by balancing the profit of private companies and the quality to the consumers. Water and Sanitation Corporation (WASAC) LTD is in charge of implementing solid waste policies. In 2008, the Government of Rwanda put in place a law prohibiting manufacturing, importing, use and sale of polythene bags in Rwanda. Partnership between private and public sectors works very well and households pay an affordable fee and the fee to pay is determined by the social class (Ubudehe) classification. Fee ranges from 1000 (USD 1.2) for category 2 to Frws 5000 (USD 6) per month. Category 1 which represents poor people is exonerated. In Rwanda, communication channels between authorities, private collection companies and citizens are well established, of which is an example of umuganda, which is a mandatory community clean-up exercise which takes place every last Saturday of each month.

In the end, it is recommended that the separation of waste at source be made compulsory. It should mainly be implemented at the household level. It is also recommended to urgently design and construct proper waste disposal sites such as landfills, to provide suitable place for waste disposal.

REFERENCES

Achankeng, E. (2003). *Globalization, Urbanization and Municipal Solid Waste Management in Africa*. African Studies Association of Australasia and the Pacific. Retrieved from http://www.inclusivecities.org/pdfs/achankeng Globalization Urbanization and MSW Mgmt in Africa.pdf

Ahmed, S. A., & Ali, S. M. (2006). People as partners: Facilitating people's participation in public-private partnerships for solid waste management. *Habitat International*, *30*(4), 781–796. doi:10.1016/j.habitatint.2005.09.004

Bazimenyera, J. D., Qiang, F., & Karangwa, A. (2012). Solid waste management in Kigali City, Rwanda. *East African Journal of Science and Technology*, *1*, 46–58.

Fobil, J. N., Armah, N. A., Hogarh, J. N., & Carboo, D. (2008). The influence of institutions and organizations on urban waste collection systems: An analysis of waste collection system in Accra, Ghana (1985-2000)'. *Journal of Environmental Management*, *86*(1), 262–271. doi:10.1016/j.jenvman.2006.12.038 PMID:17291677

Getahun, T., Mengistie, E., Haddis, A., Wasie, F., Alemayehu, E., Dadi, D., ... Van der Bruggen, B. (2012). Municipal solid waste generation in growing urban areas in Africa: Current practices and relation to socioeconomic factors in Jimma, Ethiopia. *Environmental Monitoring and Assessment*, *184*(10), 6337–6345. doi:10.100710661-011-2423-x PMID:22042042

Government of Rwanda. (2016). *Governing Liquid Waste Collection and Transportation, November 10, 2016 No: 005/R/SANEWS/RURA/2016*. Retrieved from http://www.rura.rw/uploads/media/Regulations_Governing__liquid_waste_collection_and_transportation.pdf

Gupta, N., Yadav, K. K., & Kumar, V. (2015). A review on current status of municipal solid waste management in India. *Journal of Environmental Sciences (China). Elsevier B.*, *37*, 206–217. doi:10.1016/j.jes.2015.01.034

International Committee of the Red Cross. (2011). *Medical Waste Management*. International Committee of the Red Cross, Geneva, Switzerland. Retrieved from https://www.icrc.org/en/doc/assets/files/publications/icrc-002-4032.pdf

Isugi, J., & Niu, D. (2016). Research on Landfill and Composting Guidelines in Kigali City, Rwanda Based on China's Experience. *International Proceedings of Chemical, Biological & Environmental Engineering*, *94*, 62–68. doi:10.7763/IPCBEE

JacobsenS. (2005). Plastic Bag Pollution. Retrieved from http://dpw.lacounty.gov/epd/PlasticBags/Articles/Googobits_07-21-05.pdf

Kabera, T., Wilson, D. C., & Nishimwe, H. (2019). Benchmarking performance of solid waste management and recycling systems in East Africa: Comparing Kigali Rwanda with other major cities. *Waste Management & Research, 37*(1), 58–72. doi:10.1177/0734242X18819752 PMID:30761955

Laurent, A., Ioannis, B., & Julie, C. (2014). *Review of LCA studies of solid waste management systems - Part I: Lessons learned and perspectives*. Waste Management. doi:10.1016/j.wasman.2013.10.045

Lebersorger, S., & Beigl, P. (2011). Municipal solid waste generation in municipalities: Quantifying impacts of household structure, commercial waste and domestic fuel. *Waste Management (New York, N.Y.), 31*(9–10), 1907–1915. doi:10.1016/j.wasman.2011.05.016 PMID:21689921

McLeod, F., & Cherrett, T. (2013). *Reverse logistics for sustainable waste management processes, Sustainable Practices: Concepts*. Methodologies, Tools, and Applications. doi:10.4018/978-1-4666-4852-4.ch084

MININFRA. (2018). *Solid Waste Management*. Retrieved from http://www.mininfra.gov.rw/index.php?id=90

MININFRA – Ministry of Infrastructure. (2016). National Sanitation Policy. Kigali, Rwanda: Retrieved from http://www.mininfra.gov.rw/fileadmin/user_upload/new_upload/NATIONAL_SANITATION_POLICY__DECEMBER_2016.pdf

Ministry of Health. (2017). *Medical Waste Management Plan (MWMP)*. Retrieved from http://www.rbc.gov.rw/fileadmin/user_upload/SPRP_MWMP_November_21_2017.pdf

Ministry of Natural Resources. (2015). *National e-Waste Policy for Rwanda*. Retrieved from http://www.minirena.gov.rw/fileadmin/Environment_Subsector/Laws__Policies_and_Programmes/Laws/E-waste_policy_-_FINAL.pdf

Ministry of Trade and Industry. (2011). *National Industrial Policy*. Retrieved from http://www.minicom.gov.rw/fileadmin/minicom_publications/policies/Industrial_Policy-2.pdf

National Institute of Statistics of Rwanda. (2012a). *Size of the resident population*. Retrieved from http://www.statistics.gov.rw/publication/size-resident-population

National Institute of Statistics of Rwanda. (2012b). *The Fourth Population and Housing Census in Rwanda*. Retrieved from http://microdata.statistics.gov.rw/index.php/catalog/65

Nilsson, M., Eklund, M., & Tyskeng, S. (2009). Environmental integration and policy implementation: Competing governance modes in waste management decision making. *Environment and Planning. C, Government & Policy, 27*(1), 1–18. doi:10.1068/c0794j

Nshimiyimana, V. (2015). *Assessment of municipal solid waste management practices case study: Nyarugenge district-Rwanda (Master's Thesis)*. University of Makelele. Retrieved from http://makir.mak.ac.ug /handle/10570/5863

Office of the Auditor General of State Finances. (2016). *Performance audit report on management of solid and liquid (sewage) waste in city of Kigali, Kigali-Rwanda*. Retrieved from http://www.oag.gov.rw/fileadmin/user_upload/Performance_Reports/MANAGEMENT_OF_SOLID_AND_LIQUID

Office of the Auditor General of State Finances. (2016). Performance audit report on management of solid and liquid (sewage) waste in city of Kigali, Kigali-Rwanda. Retrieved from http://www.oag.gov.rw/fileadmin/user_upload/Performance_Reports/MANAGEMENT_OF_SOLID_AND_LIQUID __SEWAGE__IN_CITY_OF_KIGALI.pdf

Pirsaheb, M., Khosravi, T., & Sharafi, K. (2013). Domestic scale vermicomposting for solid waste management. *International Journal of Recycling of Organic Waste in Agriculture*, *2*(1), 4. doi:10.1186/2251-7715-2-4

REMA. (2009). *Rwanda State of Environment and Outlook*. Ministry of Natural Resources. doi:10.13140/RG.2.1.5148.6328

Republic of Rwanda Ministry of Infrastructure. (2010). *National Policy & Strategy for Water Supply and Sanitation Services*. Retrieved from http://www.rura.rw/fileadmin/docs/Board_Decisions/WATSAN_Policy_Strategy.pdf

RURA. (2009). *Guidelines on the management of waste disposal site (landfill)*. Retrieved from http://www.gicumbi.gov.rw/fileadmin/templates/document/GUIDELINES_Landfill_.pdf

RURA. (2015). *Regulations for the Governing of Solid Wastes Recycling in Rwanda, April 24, 2015. No: 002/ESWASTAN/SW/RURA/2015*. Retrieved from http://www.rura.rw/uploads/media/CLEANED_FINAL_SOLID_WASTES_RECYCLING_REGULATIONS.pdf

RURA. (2016). *Draft Regulations Governing E-waste Management in Rwanda*. Retrieved from http://www.rura.rw/fileadmin/laws/Ewaste_regullations_Rwanda_with_technical_requirements_.pdf

Takele. (2004). *Solid waste and liquid waste management*. Addis Ababa: Ethiopia Public Health Training Initiative.

Tanskanen, J.-H. (2000). Strategic planning of municipal solid waste management. *Resources, Conservation and Recycling*, *30*(2), 111–133. doi:10.1016/S0921-3449(00)00056-2

Tchobanoglous, G., Theisen, H., & Vigil, S. A. (1993). *Integrated Solid Waste Management: Engineering: Principles and Management Issues*. New York: McGraw Hill Inc.

The Guardian. (2017). *A million bottles a minute: world's plastic binge 'as dangerous as climate change*. Retrieved from https://www.theguardian.com/environment/2017/jun/28/a-million-a-minute-worlds-plastic-bottle-binge-as-dangerous-as-climate-change

The Telegraph. (2017). *Seafood eaters ingest up to 11,000 tiny pieces of plastic every year, study shows*. Retrieved from https://www.telegraph.co.uk/science/2017/01/24/seafood-eaters-ingest-11000-tiny-pieces-plastic-every-year-study/

United Nations. (2019). *World Countries by Population Density 2019*. Retrieved from http://worldpopulationreview.com/countries/countries-by-density/

WAT. (2012). Report on consulting services for detailed design and supervision for the construction works of a new landfill site and recycling centre. Kigali, Rwanda: Report to City of Kigali, funded by the United Nations Development Programme.

World Health Organization (WHO). (2019). *Definition and characterization of health-care waste*. Retrieved from https://www.who.int/water_sanitation_health/medicalwaste/002to019.pdf?ua=1

Yoada, R. M., Chirawurah, D., & Adongo, P. B. (2014). Domestic waste disposal practice and perceptions of private sector waste management in urban Accra'. *BMC Public Health*, *14*(1), 697. doi:10.1186/1471-2458-14-697 PMID:25005728

Zaman, A. U. (2014). Measuring waste management performance using the "Zero Waste Index": The case of Adelaide, Australia. *Journal of Cleaner Production*, *66*, 407–419. doi:10.1016/j.jclepro.2013.10.032

Chapter 13
Solid Waste Management in the Republic of Serbia

Goran Vujić
Faculty of Technical Sciences, Department of Environmental Engineering and Occupational Safety and Health, University of Novi Sad, Serbia

Bojana Tot
Faculty of Technical Sciences, Department of Environmental Engineering and Occupational Safety and Health, University of Novi Sad, Serbia

ABSTRACT

In most of the countries of South East Europe, the current situation in the field of municipal waste management is characterized by unreliable and incomplete data about the physical characteristics of generated municipal waste. The first and the second part of the chapter contain the description of specific project activities that are related to measuring quantity and composition of municipal waste in municipalities in the Republic of Serbia. The methodology description for the determination of the quantity and composition of waste is analyzed in this chapter. The potential of implementing the circular economy concept in developing countries is reflected primarily in the optimization of production and industrial activities; the example of Serbia in the third part of chapter shows the opposite. In the fourth part, results of application of the AHP method are presented. Taking the obtained results as a basis, it can be concluded that the applied model can be efficiently used to support decision-making in this kind of problem.

WASTE DEFINITIONS IN THE REPUBLIC OF SERBIA

The level of waste management system in Serbia is similar to those of almost all South-Eastern European (SEE) countries. Based on the results of measurements in reference municipalities, about 2,374,374 tons of waste were generated in 2008. The results of the 2014 modeling, obtained for the needs of innovating the waste management strategy, indicate that the amount of generated municipal waste has increased by about 0.5% on the annual basis compared to 2008. This means that 2,448,566 tons of municipal waste, i.e. 340.7 kg/inhabitant per year (0.93 kg/inhabitant/day) are currently generated in Serbia. Measurements and control measurements were done for the first time in the Republic of Serbia in 2008 and 2014,

DOI: 10.4018/978-1-7998-0198-6.ch013

respectively, according to the methodology developed at the Department of Environmental Engineering of Faculty of Technical Sciences, University of Novi Sad, with the help of German Corporation for International Cooperation.

The waste landfilling is the dominant waste treatment in Serbia and in all Southeastern European countries as well. In Serbia, about 20,000 tons of waste is burned in cement factories, which makes up about 1% of the total generated amount of waste.

The national target for 2015 was 31% of packaging waste recycling and the previous target of 25% for 2014 had been reached. In Serbia, a total of 5% of the total generated waste is recycled; if we add extraordinary activities of the informal sector, this percentage doubles and reaches 10% of recycling. The composition of waste recycled consists primarily of PET, and other plastics, paper, and paperboard. Table 1 gives the generation of municipal waste and reduction of waste in accordance with EU Landfill Directive and EU Packaging Directive. Whereas, Table 2 enlists the estimation of the amounts of biodegradable and packaging waste that are expected to meet the EU targets for the City of Novi Sad and Novi Sad region.

The average morphological composition of waste in Serbia is shown in Figure 1 (it should be noted that the said waste composition figure seemingly differs from more detailed figures in the following chapters, which is due to the client's request to classify a large percentage of individual waste streams as 'other', and in more detail, it is defined individually), which is dominated by biodegradable waste that accounts for more than 50% of the total generated waste, due to which Serbia can be classified as a developing country characterized by a high share of biodegradable waste in the waste morphological composition (Hoornweg and Bhada-Tata, 2012; IMG, 2014).

Table 1. The estimated amounts of waste in accordance with the goals of the EU landfill directive, the framework directive and the packaging and packaging waste directive

			Reducing the amount deposited in relation to the amount generated in the reference year, according to the EU Landfill Directive				Increasing the recycling of packaging waste according to the EU Packaging Directive	
		Total generated amount of municipal waste - 2014	**Total generated amount of biodegradable municipal waste - 2014**	**1 target 2022 - to 75%**	**2 target 2026 - to 50%**	**3 target 2030 - to 35%**	**Total amount of packaging waste - 2014**	**Target 2025 - Recycling of at least 55%**
Region	Municipality	t/yrs	t/yrs	2022	2026	2030	t/yrs	2025
Novi Sad	Novi Sad	133939	90409	33360	62836	84767	27190	17565
	Bačka Palanka	19108	12898	4807	9045	12213	3879	2518
	Bački Petrovac	4115	2778	939	1790	2396	835	519
	Beočin	5407	3650	1356	2553	3446	1098	711
	Žabalj	8071	5448	1950	3688	4963	1638	1044
	Vrbas	14034	9473	3336	6323	8498	2849	1802
	Srbobran	5625	3797	1289	2455	3289	1142	710
	Temerin	10062	6792	2238	4282	5722	2043	1255
Total		200361	135244	49276	92971	125592	40673	26124

Table 2. Estimation of the amounts that will meet the EU targets for the City of Novi Sad and the region

Waste Framework Directive	EU Targets	Proposed National Targets	Estimated Amounts (tons)
Biodegradable waste (garden waste and food waste)	By 2020, achieve a minimum recycling rate of 50% of the total generated amount of recyclable waste	By 2025, achieve a minimum recycling rate of 50% of the total generated amount of recyclable waste	92,971
Packaging waste			26,124
Total			124,466

In total, 19% of waste, or 420,000 tons, is currently deposited in 7 sanitary landfills that have been designed and operating in accordance with the provisions of the Landfill Directive, or in accordance with Article 14, while the rest is deposited on non-sanitary landfills. Non-sanitary landfills are not dumpsites: a large percentage of urban landfills have a fence, scales, some of them have passive drainage of landfill gas, so a large number of landfills do have certain elements of sanitary landfills. However, all those that are called non-sanitary do not have the prescribed built landfill bottom with foil and all the necessary layers required by the Landfill Directive regulations. Some of these landfills are built on impermeable soil and do not pose a threat to the environment. However, a greater number of these landfills are dumpsites that do not have the basic elements to be called sanitary landfills and they endanger the environment. Also, the completion of the construction of two more regional landfills is expected soon: a regional sanitary landfill in Inđija, and a regional waste management center in Subotica, which will increase the percentage of waste that is deposited on sanitary landfills to around 22% (Vujic et al., 2017).

Figure 1. The average composition of waste generated in Serbia

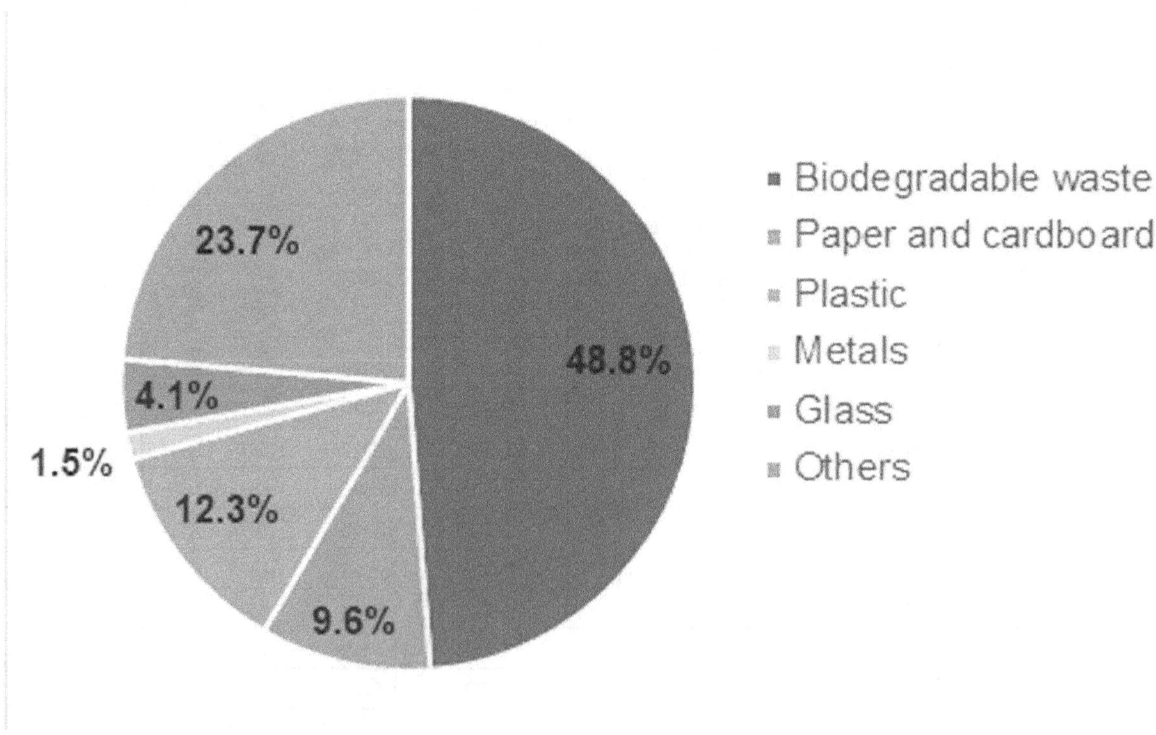

The remaining generated amounts of waste are deposited at municipal non-sanitary landfills. Of the 26 waste management regions, as envisaged in the National Waste Management Strategy, nine regions have not yet initiated activities on the preparation of planning documents and technical documents, and still do not have a Regional Waste Management Plan, while seven regions have planning documents that have not been adopted yet. Also, six regions still do not have a signed inter-municipal agreement, Figure 2.

Figure 2. The regions in Serbia with status of sanitary landfilling

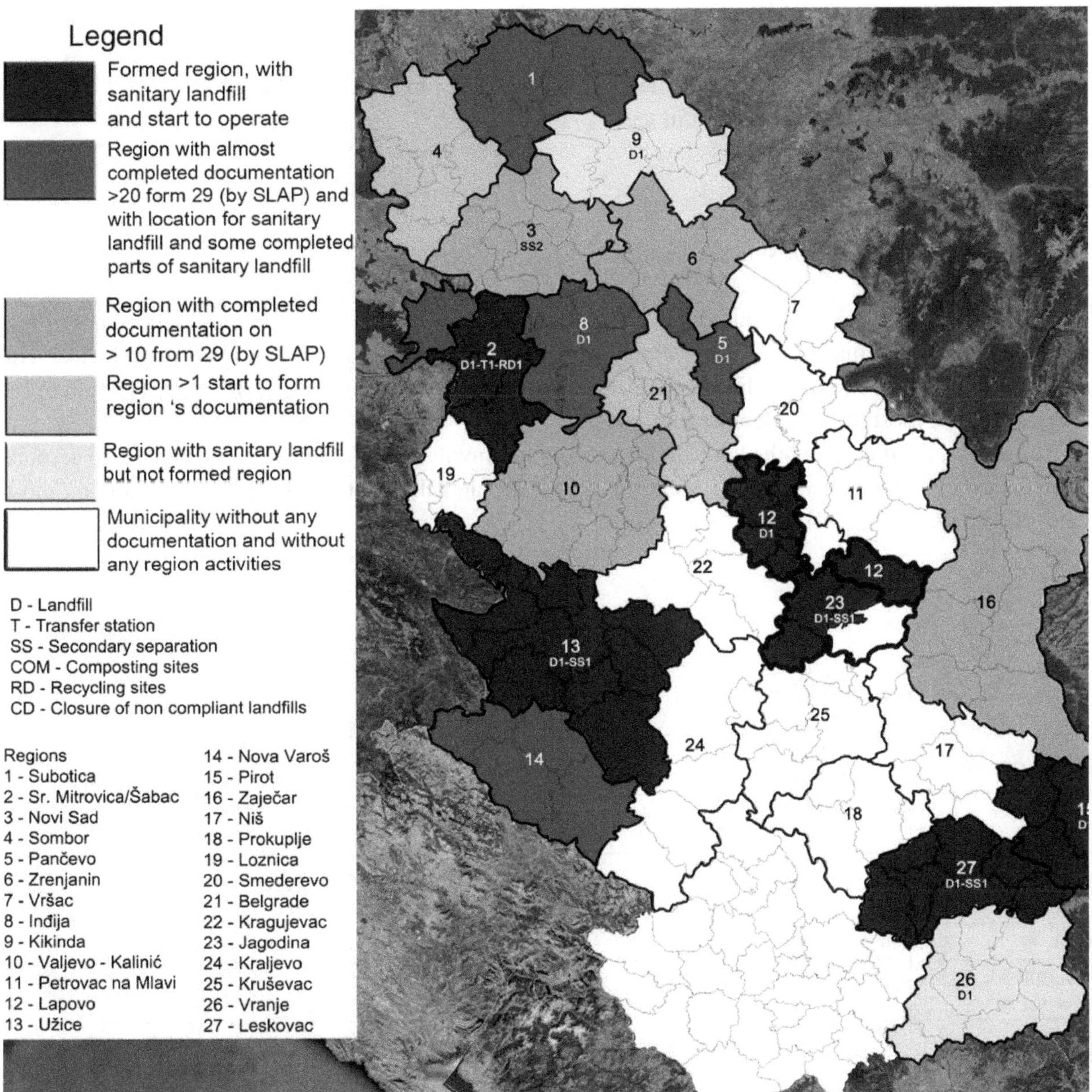

One of the major problems encumbering the waste management system is a large number of non-sanitary landfills and dumps. According to the data from 2009 shown in Table 3, there are 3,582 landfills and dumps. They vary considerably in size and quantity of deposited waste and affect the environment in different ways. Legislation in the Republic of Serbia and the European Union (EU) requires the reclamation of dumps, as well as landfills that do not meet the required conditions. Reclamation includes several elements (for example):

- Providing a location for newly-generated waste (prevent further disposal of waste, fence the damp/ landfill in and reduce the risk of unauthorized access to landfill and injuries).
- Designing of the dump body.
- Drainage (including drainage of collected leachate) and circumferential channels.
- Degassing and treatment of landfill gas.
- Laying of the upper impervious layer.
- Restoration of vegetation.
- Monitoring.
- Access and use after closing the landfill.

The choice of certain elements depends on the specific situation in a particular location and the available sources of funding. In some cases, the landfill reclamation requires significant investment costs while in others it requires high operational costs. Few landfills need landfill degasification to be introduced, but most of them do not require it.

The reason for this is that the landfill practice did not involve landfill covering, which led to aerobic degradation of biodegradable waste, so the potential for landfill gas development is very low on most landfills (Džolev et al., 2016).

Table 3. Categorization of dumps in Serbia (Stanisavljević et al, 2012)

Criterion	Number of Landfills/Dumps	% of the Total Number	Surface Area (ha)	Volume (m3)	% of the Total Volume
Waste layer thickness (m)					
< 1 m	3,302	92	819	3,640.193	8.3
< 1 < 3	210	6	200	4,344.805	9.9
> 3	70	2	321	36,050.680	81.8
Landfilled waste volume (m^3)					
<1,000	2,702	75	154	604,629	1.4
<1,001<10,000	698	19	480	2,251.995	5.1
<10,001<100,000	131	3.5	313	4,087.590	9.3
<100,000	51	1.5	393	37,091.464	84.2
Total	3,582	100	1,340	44,035.678	100

Table 4. The landfill closure method (Off. Gazette RS, 2010)

Landfill Class			
Measures applied regarding the formation of upper covering layer	For non-hazardous waste	For hazardous waste	For inert waste
Landfill gas drainage layer ≥ 0.3 m	Required	Not required	Not required
Artificial watertight coating - foil	Not required	Required	Not required
Impermeable mineral layer ≥ 0.5 m	Required	Required	Not required
Reclamation layer ≥ 0.5 m	Required	Required	Not required

The application of the Water Framework Directive can also create additional requirements. Waste landfilling is treated as part of the "pressure analysis" during the preparation of the river basin management plan, thus requiring a risk assessment for groundwater flows.

The scope of application of the landfill reclamation requirements directly affects the reclamation costs. Accordingly, lowering the required standards for landfill reclamation, where this risk analysis shows it possible, achieves cheaper but still sufficient environmental protection. This approach requires the establishment of a methodology for risk assessment and assessment of landfills against potential risks to environmental pollution. Based on such an assessment, various technical methods for landfill reclamation can be applied.

The Regulation on Waste Landfilling (No 36/2009 i 88/2010) describes the requirements for reclamation and closure of landfills. The recommendations are the same as the requirements of the EU Landfill Directive and are shown in Table 4.

Serbia has a large number of landfills and dumps which have to be considered within the implementation of the requirements laid down in the Directive. According to the data from 2009 presented in Table 3 (above), there are 3,582 landfills and dumps. They considerably vary in size and quantity of deposited waste and affect the environment in different ways.

The methodology (developed at the University of Novi Sad, Department of Environmental Engineering), within MOPRORK (Determination of contamination from landfills and monitoring models, risk assessment, determination of amounts of waste with modern satellite information technology to support the implementation of legislation) and UPIOV (Determination of data on wastewater infrastructure in the Republic of Serbia with online measurements of the composition and amount of wastewaters in representative municipalities) projects financed by the Ministry of Environment of the Republic of Serbia and the German Corporation for International Cooperation (GIZ)) for the assessment and quantification of potential environmental risks from landfills covers the assessment and ranking of 12 different factors, including:

- The volume or surface area of the landfill.
- The amount of landfilled waste.
- Distance from buildings/settlements.
- Distance from surface waters.
- Distance from arable land.
- Periodic landfill covering.
- Number of *biotrns (*biotrn - an individual technical facility, such as a degassing well).*

- Groundwater level.
- The existence of the landfill fence and guard.
- Landfill accessibility.
- Measured/theoretical potential of methane production.
- Quality of leachate.

In the first step, each of these factors is considered separately and divided into five ranges of possible values. Different ranges represent different levels of potential environmental impact, where value 1 signifies the smallest impact, while value 5 signifies the greatest impact on the environment. By observing the data of a specific landfill, the appropriate values from 1 to 5 are assigned to all 12 factors.

In the second step, an overall environmental risk assessment of the landfill can be obtained by adding up the assessments of all the above factors. In this respect, the overall risk is categorized in percentages as follows:

- **Low Risk**: From 0 to 40
- **Medium Risk**: From 41 to 60
- **High Risk**: From 61 to 100

It should be noted that the values for the "Leachate quality" parameter were obtained from the laboratory analysis of the leachate samples for ten landfills only (Subotica, Sombor, Novi Sad, Kikinda, Novi Bečej, Zrenjanin, Šabac, Pančevo and Inđija), while this parameter was not considered for any of other landfills due to the lack of laboratory analysis. In these cases, the overall risk assessment rate is defined based on the extrapolation of the remaining eleven parameters. In some cases, for landfills less than 10,000 m^3 in size, a simplified assessment was made based on six or three most important factors.

After the risk assessment, all landfills get classified into four groups by volume and three groups by ranking based on the risk levels.

The methodology processed 214 landfills and dumps. This number covers all municipal landfills, as well as dumps greater than 10,000 m^3. Out of this group, 10 are ranked as high-risk landfills and dumps, 66 as high-risk of second priority, 65 as high-risk of third priority, 71 as medium-risk and 2 as low-risk landfills and dumps. It was suggested that the same reclamation conditions should be applied to all high-risk landfills, and that the 60 - 100 risk level should be used to prioritize the activities in terms of the start of reclamation. The Table 5 gives an overview of the prices by regions, as well as a proposal as to the period in which it is possible to do the rehabilitation.

The calculation included costs for transportation of waste from small landfills amounting to 7 EUR/m^3, while the price of simple covering of landfills without a significant risk amounts to 9 EUR/m^3; the closure with a complex cover required by the EU High-Level Landfill Directive amounts to 21 EUR/m^3, or 30 if the landfill gas extraction system is established, which is envisaged only for extremely large landfills.

The calculation indicated that only the funds necessary to close the landfills amounted to EUR 248 million, but the calculation was based on the existing data. After carrying out exploratory works and chemical analyses, a potentially larger number of high-risk landfills were identified, which would significantly increase the funds necessary for closing the landfills. The landfill rehabilitation will be a major burden for the improvement of the waste management system, since the necessary funds need to

Table 5. The estimate of financial resources necessary for the rehabilitation of existing landfills and the realization schedule proposal

Region	Number of Controlled Municipal Landfills	Number of Illegal Landfills	Total to Close	Stage1 2014 - 2020	Stage2 2021 - 2028	Total Price (EUR)	23,000,740
2	S. Mitrovica – Šabac	2	54	56	14,899,358		14,899,358
3	Novi Sad	8	74	82		20,641,540	20,641,540
4	Sombor	5	50	55		16,192,710	16,192,710
5	Pančevo	2	13	15	6,026,400		6,026,400
6	Zrenjanin	4	58	62		6,146,800	6,146,800
7	Vršac	4	71	75		8,586,738	8,586,738
8	Inđija	6	108	114	18,642,960		18,642,960
9	Kikinda	6	53	59	13,107,990		13,107,990
10	Valjevo – Kalenić	9	208	217		7,542,134	7,542,134
11	Petrovac na Mlavi	4	138	142		2,023,414	2,023,414
12	Lapovo	4	237	241	3,890,983		3,890,983
13	Užice	8	161	169	7,578,137		7,578,137
14	Nova Varoš	3	57	60		1,124,129	1,124,129
15	Pirot	4	23	27	2,836,776		2,836,776
16	Zaječar	8	464	472		14,043,610	14,043,610
17	Niš	5	167	172		8,505,525	8,505,525
18	Prokuplje	3	47	50		4,724,429	4,724,429
19	Loznica	3	110	113		5,989,949	5,989,949
20	Smederevo	5	227	232		17,876,211	17,876,211
21	Beograd	3	150	153		17,706,705	17,706,705
22	Kragujevac	5	193	198		6,561,488	6,561,488
23	Jagodina	3	95	98	5,784,262		5,784,262
24	Kraljevo	4	114	118		4,843,743	4,843,743
25	Kruševac	7	201	208		3,579,185	3,579,185
26	Vranje	7	156	163	2,452,067		2,452,067
27	Leskovac	6	160	166	4,137,688		4,137,688
Total for all the regions		135	3,447	3,582	79,356,621	169,089,050	248,445,671

be allocated from the current financial revenues, which will further reduce the funds for improving the collection and treatment of waste that is carried out on the daily basis.

WASTE CLASSIFICATION

One of the basic prerequisites for the planning and implementation of an integrated waste management system at the local, regional and national level is the knowledge of reliable information on the amount and composition of generated waste (Beigl et al., 2008; Sharma and McBean, 2007; Chung S and Poon, 2001; Batinić et al., 2015). Unfortunately, for most developing countries, the current situation in the field of municipal waste management is characterized by unreliable and/or incomplete data on the physical characteristics of waste (Hristovski et al., 2007).

Reliable data on the amount and composition of generated waste are the basis for the development of appropriate planning documents, as well as for the determination of long-term objectives and rational and sustainable waste management at the national level. Knowledge of relevant indicators of physical characteristics of waste is of utmost importance for the successful functioning of all the waste management system elements which include collection, transport, treatment and final disposal (Beigl et al., 2008).

In addition to influencing the choice of equipment and process optimization in technical terms, the assessment of future investments is also closely related to information about the amount and composition of the municipal solid waste components. By defining a standardized method for monitoring the generated amounts and composition of waste, the conditions are created for comparing the available data between municipalities, i.e. at the national level, as well as for the fulfillment of the reporting obligation towards the European Environment Agency (EEA), or the comparison of data between different countries (FTN, 2009).

The application of a standardized methodology, that is, the knowledge of reliable information regarding the amount and composition of municipal waste, obtained from measurements, is necessary in the context of planning and optimizing virtually all elements of the waste management system.

However, despite the apparent need for a standardized methodology that should enable municipalities to obtain reliable data on the physical characteristics of municipal waste, the fact is that in most developing countries, the largest number of municipalities, or PUCs, still do not carry out specific measurements of the amount and composition of municipal waste generated on the territory of the local self-government unit.

The insufficiently developed awareness of the municipal and PUC representatives about the need to determine the amount and composition of municipal waste, but also the lack of financial resources and basic infrastructure as the necessary preconditions for the successful realization of measurements, are some of the basic reasons for this situation. The biggest problem in this regard, especially when it comes to measuring the amount of generated waste, is the fact that a very small number of PUCs have weighbridges on landfills so as to measure the weight of waste collection trucks. Also, in order to analyze the composition of waste, it is necessary to provide logistics and additional conditions (suitable location), machinery and equipment (sample collection truck, sorted waste bins, weighing grids and scales), as well as the workforce that will work on sorting waste samples, which are additional obstacles to analysis implementation (IMG, 2014).

Another problem is reflected in the fact that the laws and other regulations regulating waste management in most countries have not adequately defined and regulated the problems of determining the amount and composition of municipal waste. Within the SEE region, the Republic of Serbia belongs to rare countries that, based on the Law on Waste Management and the Rulebook on the methodology for collecting data on the composition and amounts of municipal waste in the territory of the local self-government unit (RS Official Gazette 61/2010), have introduced the obligation of local self-government units to keep records of collected municipal waste.

Description of The Methodology For Determining The Amount and Composition of Waste Implemented in Pilot Municipalities

Based on the Law on Waste Management ("RS Official Gazette", No. 36/09, 88/2010), the Rulebook on the methodology for collecting data on the composition and amounts of municipal waste in the territory of the local self-government unit ("RS Official Gazette", No. 61/10) has been developed.

The methodology essentially consists of two segments:

Determining the Amount of Generated Waste – by Measuring the Weight of Waste Collection Trucks

In the observed municipality, measuring of generated amounts of waste is done by weighbridge-measuring of the weight of all waste trucks that collect municipal waste in the territory of the local self-government unit. All the truck measurements are carried out within a period of 7 days, that is, in the period covering the provision of waste removal services for each household in the observed municipality. The first step in the measurement process is to measure the tare weight of all waste collection trucks, or their weight without loaded waste.

The tare weight of all trucks needs to be determined before going to the field and collecting waste, and it is important to emphasize that it is enough to measure the weight of empty trucks just once. After that, the same trucks are measured when taking their regular waste collection routes and when they use their full capacity (gross weight). The obtained net weight of the collected waste is of interest for further analysis.

The waste whose weight is measured is then landfilled (or taken to a specific waste treatment plant), and the procedure is repeated until each truck completes the planned collection, measurement and disposal of waste for that day, according to the regular collection schedule within the municipal services in a given municipality. By the same principle, the process is repeated for a period of 7 days, while the measurement of the waste weight in this period is done at least 4 times per year due to the influence of seasonal variations (Vujic et al., 2010)

Analysis of the Morphological Composition of Waste - by Classifying the Waste Sample to Fractions Defined by the Catalog

For the analysis of the municipal waste morphological composition, it is necessary to deliver waste samples weighing approximately 500 kg to the location for analysis (preferably covered). The procedure envisages the analysis of a total of 3 samples of this weight, depending on the housing sector in the observed municipality. Two samples are taken from two different housing sectors within the urban zone (collective housing and individual housing), while the third sample is taken from the predominantly rural part of the municipality, that is:

1. **Urban Zone**: Collective housing and commercial zone (settlements with residential building blocks).
2. **Urban Zone**: Individual housing (settlements with houses that have a yard/garden and are located in the wider urban zone).

3. **Rural zones** within the municipality (settlements with houses that have a yard/garden and are located in the rural zone of the municipality).

The waste samples sent to analysis need to be selected randomly from each sector by choosing different streets, which represent the chosen sector in the most representative way, and then within them, randomly select bins/containers the contents of which are analyzed. The desired sample waste of 500kg is reached by collecting about 50 bins volumes 80 l, 35 bins volumes 120 l, or 6-8 containers volumes 1.1 m^3. Each sample is collected and analyzed separately, by dividing the delivered waste sample to different fractions, defined by the sorting catalog, and the weight of each fraction is measured.

CIRCULAR ECONOMY AND EXTENDED PRODUCER RESPONSIBILITY

Circular Economy Change or Illusion

The concept, phenomenon or system of circular economy has become extremely popular and widespread throughout the world. In almost all countries, politicians and economists view it as a new opportunity for economic development and creation of new jobs.

It should be noted that the Circular Economy model is not the first model of that kind that appeared in the world, nor is it a completely new model. There were a number of similar ideas, and when it comes to those dealing with waste, it can be said that circular economy is based on two key systems governing the field of waste - the polluter pays and the extended producer responsibility principle.

The concept of circular economy has experienced great popularity following the publication of the famous study by the Ellen MacArthur Foundation “Towards the Circular Economy - Opportunities for the Consumer Sector”. It states that an inflow of 1.5 billion pounds could be expected in the UK from applying the circular economy system in the field of waste originating from food. The inflow is based primarily on the revenue that would be achieved by avoiding the payment of landfilling taxes and partially on the inflow generated by the production and sale of green energy at the feed-in tariff. This calls the possibility of implementing a circular economy in countries such as Serbia into question. In Serbia, the fee for organic waste landfilling is 0 dinars, while in the United Kingdom it amounts to 100 euro. A major difference in calculation is immediately seen and it is only a matter of time when this type of fee will be introduced in Serbia. In all the SEE countries, the landfilling fee is extremely low and goes up to a maximum of 15 euro in Romania.

The same study notes that the Organization for Economic Cooperation and Development (OECD) expects that the number of people belonging to the middle class will go from 1.9 billion in 2009 to 4.9 billion in 2030. The largest percentage, almost 90% of this increase, will be found in the countries belonging to the Asia-Pacific region. This information should be observed from the side, or what the expected result of an increase in the number of middle-class citizens is, which is precisely the increase in their purchase power. The introduction of a circular economy means advanced treatments, primarily of waste, and the introduction of advanced treatments requires additional funding that needs to be provided. The provision of financial resources will be enabled by the citizens, as citizens are generators of waste and they must be able to pay for the introduction of a circular economy.

When considering the reasons for the introduction of a circular economy, it is necessary to consider the fact that there is still room in the developing countries for the development of the linear economy, as in these countries neither labor nor capital has been organized on the basis of a biological or process cycle. Certainly, the linear economy leads to the destruction of natural resources and large environmental pollution, but it must be pointed out that the fight among the advocates of both economies, i.e. linear economy, which, if preservation of natural resources and pollution are left aside, brings faster economic growth and new jobs compared to the circular economy, leans on the side of the linear economy advocates. Politicians in developing countries need faster results in the fight against unemployment and gross domestic product (GDP) growth; the linear economy gives faster and better results in the short term, and it is, therefore, more acceptable for politicians who want to be favored by the public opinion to support the development of a linear economy.

The concept of the circular economy, is neither new nor the only one based on the principle of sustainability of the entire society. In the past, other concepts have been developed aimed at improving the sustainability of the society as a whole. However, no concept has been accepted by the public as a circular economy concept and none has gained its place in the legal regulations. The concept of the circular economy has been recognized as one of the main developments. The EU and the European Commission adopted the EU Action Plan for the Circular Economy on 2/12/2015. This is the first time such a concept has been adopted and regulated.

The state of the circular economy in developed and developing countries varies substantially. Considering the fact that production processes in developed countries are in most cases at the highest technological level, when it comes to the circular economy concept in these countries, the focus is on the further product management, after the end of its life cycle. Due to the poor technological development and the developing industry, the initiator of the circular economy in developing countries is still in the process of production.

The definition of the concept of biological and technical material circulation is based on the fact that waste does not exist and that the basic natural and anthropogenic components of a particular product are designed with the aim of aligning with the life cycle of the material. That is, all products must be designed in a way that will enable more efficient recycling or reuse (design for recycling/reuse). Biological materials are not toxic and can be composted in an easy way, while anthropogenic materials (various alloys, etc.) are designed so that they can be very easily reused with the minimal use of energy. In many countries around the world with prevailing existential problems, it has been very difficult to establish such priorities so far.

Extended Producer Responsibility (EPR)

The possibility of using a circular economy must be viewed only segmentally. Approaching the circular economy fictitiously or as a hurricane, on the wave of its successful application in developed countries, will lead to a complete failure in its application in developing countries.

Looking at some basic facts, that the linear economy still has its place in developing countries and that, for the introduction of a circular economy, it is necessary to have citizens who are able to pay for its introduction, the circular economy must be introduced gradually. If the management of individual waste streams by the extended producer responsibility principle could be considered a circular economy, which it essentially is in certain waste flows, the circular economy has been established in Serbia as well. Examples of the circular economy established in Serbia can be seen in the management of individual specific waste streams (electronic waste, tires, batteries and packaging waste).

Figure 3. The flowchart of management of packaging waste in Serbia

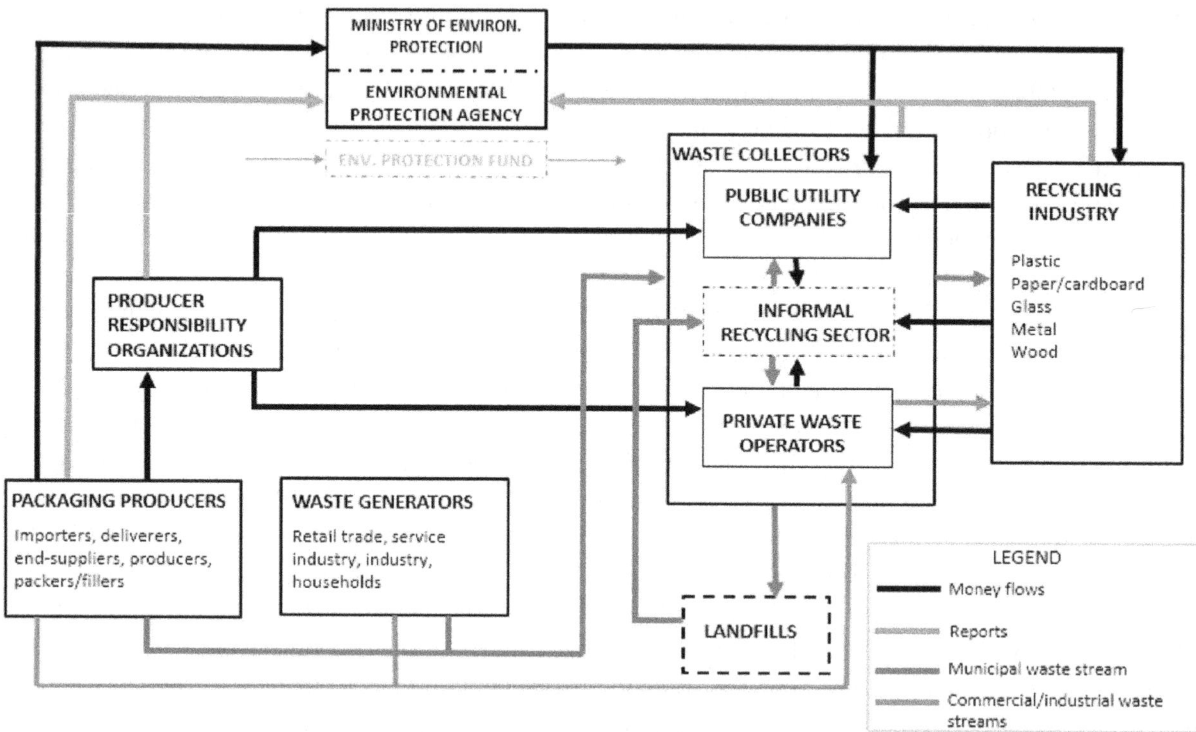

The management of electronic waste, tires and batteries is organized in such a way that when buying a new electronic device, the purchaser pays the recycling fee defined by the regulation of the Government of the Republic of Serbia. Funds from the collected recycling fees are transferred to recyclers based on the amount they collect and recycle. Another example of managing a special waste stream that can be classified as a circular economy phenomenon is the management of packaging waste as shown in Figure 3.

The system is set up on the principle of a collective operator, but this system is based on the polluter pays principle and the extended producer responsibility principle. The buyer does not pay a visible recycling fee, but from the total price of the product, one part is separated for recycling. The system is formed in a way that the funds do not flow into the budget, but the financing of recycling is done through a collective operator. Through the line ministry, the state has a role of controlling and setting the annual targets the producers combined in collective operators need to achieve.

The concept of a circular economy can be seen on individual waste streams, never in its entirety. A separate system needs to be developed for each waste stream. The circular economy concept has become popular and attractive and is very acceptable for economists and politicians around the world.

WASTE MANAGEMENT SYSTEM DEVELOPMENT INDICATORS

Serbia and the Southwestern European countries have been recently facing various problems in the waste management system: a large number of illegal landfills, a small percentage of collected and disposed waste, low level of recycling, unreliable and inaccurate data on amounts of generated waste, lack of legislation, lack of public participation in the decision-making process in the field of waste management, unclear delineation of responsibilities and obligations of various institutions, lack of monitoring and equipment, lack of cooperation among stakeholders (Government, non-governmental organizations, citizens, etc.), lack of economic initiative (fees, the polluter pays principle), insufficient training and lack of expert knowledge, poor management, lack of staff and financial resources, lack of adequate training, insufficient interest in waste management issues, and lack of general public awareness (Tot et al., 2017).

Defining and evaluating important indicators potentially having the greatest impact on the state and prospects of waste management in the region is the first step in developing a waste management strategy.

Although conditions for the waste management vary among countries, it is necessary to define some common indicators that are responsible for the current state and future directions that promote and limit effective waste management. A new approach to improving the waste management system is something that needs to be achieved. Since there is no ideal solution, but only an optimal one, it is clear that any selected model for evaluating possible solutions will have certain shortcomings, to a greater or lesser extent (Tot et al., 2016).

The basic indicators affecting the poor and slow improvement of waste management systems in developing countries include:

1. Institutional and administrative.
2. Technological.
3. Economic.
4. Socio-cultural.

Each of these four basic indicators has its own subindicators that further define the impact on the waste management system development (Tot et al., 2016).

Institutional and Administrative Indicator

The following subindicators have been defined within the institutional and administrative indicator: the degree of development of democracy, then the degree of development of legal regulation and its implementation, and the level of expertise and motivation of administration (Tot et al., 2016).

The Degree of Development of Democracy: In "young democratic" countries, including the countries of the Southeastern European region, a new electoral cycle is implemented every four years (sometimes even earlier), as in all democratic countries of the world, but elections entail a change of the management team (director, technical director, sector manager) in utility companies. These changes bring new ideas, and projects begin to evolve from the beginning. Unfortunately, this means that with each change, projects start from scratch. Therefore, it is impossible to implement projects in the field of waste management, as their realization takes six to ten years.

The degree of development of legislation and its implementation - The problem is reflected in the erratic work of state bodies and institutions in this field. Although the harmonization of national laws with the EU waste legislation is partial, there are still parts of the legislation that are lacking, in particular, the secondary regulations necessary for the full implementation of laws at the local level. In addition, the lack of strategic guidelines results in shared responsibilities and lack of communication and coordination between different institutions at the national level and between local government units. Inappropriate implementation of waste management regulations can be to a significant extent be attributed to poor planning, i.e. planning methodology and non-observance of the already planned.

The level of expertise and motivation - Strengthening of the administration capacity to establish an adequate waste management system is necessary. Stability of competent institutions, expertise, and credibility of public administration are necessary conditions for building partnerships as the highest form of cooperation between public authorities and citizens (Tot et al., 2017).

Technological Indicator

The technological indicator is determined by the subindicator of the possibility to apply modern technological solutions based on available financial resources, then on the possibility of applying technological solutions based on the amount and composition of waste, as well as on the subindicator of the lack of standardized solutions.

The indicator of the possibility of applying new technologies based on the available financial resources and the quality and quantity of waste relates to the following problem: the more efficient solutions in the field of waste management, the higher investment, as well as the operating assets they require, compared to the value of the investment. As the users of services are citizens, they are the ones who need to pay the costs of introducing new technologies. Poorly educated staff are not ready to analyze the companies from developed countries that advertise their technologies in the right way and to guarantee for them. The decision makers get lost in all the offered solutions, which seemingly appear to be acceptable. It often happens that the local administration and decision makers find a foreign company's solution ideal, after visiting a place in Europe where it has been successfully implemented. However, the operating conditions of the plant at the site visited and the conditions in the place in Serbia are totally different. In most cases, the amounts of waste our cities dispose of are small, and the composition of waste is not suitable for the treatments offered or used in the plant the members of the local authorities visited (Tot et al., 2016).

Economic Indicator

The impact shown by the economic indicator is determined by the level of gross domestic product as a subindicator, then by the insufficient funding sources, due to the poor situation in all segments of the utility sector (water, wastewater, green areas), as well as by other segments of the public importance such as roads, healthcare, schools.

The GDP level is a parameter that reflects the economic growth of a particular country. The economic crisis and unstable political situation condition the slow growth of GDP, which further conditions the lack of funds for investments. There are significant differences in average incomes both between different regions and within one region. Due to the low GDP, most households cannot afford to pay for waste management-related costs. The growth of GDP increases the standard of living of citizens, which leads to an increase in their purchasing power, and thus the ability to pay, but the increase in GDP also affects the increase in the amount of generated waste. In addition, the state budget increases, resulting in greater available resources for waste management.

Socio Indicator

The socio indicator includes aspects (sub indicators) of waste management such as the attitude of people to waste or public perception of the waste management system, then the subindicator which includes perseverance in project implementation, and this group also includes the level of corruption subindicator.

The problem of public perception can be considered in two aspects. One is that citizens do not consider waste as a resource but rather something to throw away from their field of vision, and this is the way they treat waste, so the waste separation at its source finds great resistance among citizens who are not ready to fully accept the new system of removal, i.e. separation of waste at the source. The second aspect of this problem is the perception of the public good. Citizens do not consider public as their own good. Citizens think that all which is their property, that is, "everything in their backyard" is what they are responsible for; everything else that is beyond their ownership, roads, parks, rivers, sewage, and waste, is not considered their good, "it's not mine, it will be solved by someone else."

There is no persistence in the project implementation. At the beginning of the project, all participants are highly motivated to successfully implement the project. However, as the project runs, many barriers appear in the form of corruption, complex legislation, choices and changes of people involved in the project, lack of funds, and many others. As a result, the enthusiasm and motivation of the participants decrease. Because of the frequency of elections, key people change, and ultimately, energy, desire, and enthusiasm disappear, and the project is not brought to its end.

Corruption is a problem of a society that mainly strikes developing countries. There are many reasons why it exists and, depending on many factors, its percentages can be high and low in different countries, but it is definitely present in all the segments of society.

Evaluation of Key Aspects

Socio-economic aspects that influence the project implementation are extremely complex, interconnected, and their solving requires a detailed analysis and, above all, their prioritization in solving. Inevitably, there is a need for a new approach to determining priorities in addressing the waste management issues. The needs of the research are reflected in the necessity of establishing priorities, which will be used as a support for decision-making when it comes to the directions of the strategy for the development of a waste management system. Preliminary research in this field indicates that the application of the multi-criteria method contributes to detecting inefficiency and potential improvements in order to develop sustainable waste management systems.

Based on the above, a research was conducted using the method for group assessment of indicators and subindicators, which the analytic hierarchy process (AHP), a multi-criteria method for determining the decision-maker weight (DM), is integral part of (Forman and Peniwati, 1998). In the case of the AHP method application, a research was conducted based on the experience of the author of the text, as well as the available data, and it was suggested that defining the waste management problem should be carried out on the basis of 14 subindicators (there are insignificant differences compared to the presented indicators in the previous part of the text) defining waste management problems in Serbia and countries in the region (Bosnia and Herzegovina, Macedonia and Montenegro). In accordance with the requirements of the applied AHP method, the sub-indicators are grouped into four indicators (institutional and administrative, economic, technological and social) and in this way, a hierarchy of the decision-making issues has been formed (Figure 4).

The research implementation and data collection process are based on filling out a questionnaire. The questionnaire consisted of several basic elements: introduction, questions, and instructions for respondents. Given that it was about evaluating predefined indicators and subindicators, respondents (decision-makers) should have only arranged items by importance. The number of respondents who took part in the questionnaire was 21 (Tot et al., 2016).

In the second phase, a comparison of elements in pairs at each level of the hierarchical structure was done, with selected aspects of the decision makers expressed according to the Saaty's scale of relative importance, consisting of 5 degrees and 4 interdegree intensities, which are verbally described and have corresponding numerical values ranging from 1-9 (Saaty, 1980). However, interdegree intensities were not taken into account in this study.

In the third phase, the global weights of the indicators were calculated with respect to the target, the global weights of the subindicators were calculated with respect to the target, and the local weights of the subindicators were calculated with respect to the indicator to which they belong. During the evaluation of the hierarchy elements, until the end of the procedure and the synthesis of the results, the consistency of the selected decision-maker indicators was checked and the correctness of the obtained ranges of indicators and subindicators, as well as their weight values, were determined. In order to reduce the risk posed by subjective thinking, it is desirable to make the decision in a group context (Ramanathan & Ganesh, 1994). In this way, the risk of a wrong estimation will be reduced, as the approach to certain problems is based on different knowledge and experiences of decision-makers. The group has a greater knowledge of the decision-making issue than the individual. One of the ways is to unify individual decisions in a group, by geometric averaging, which was done in this case as well.

Figure 4. Four indicators of analytic hierarchy process (AHP) method

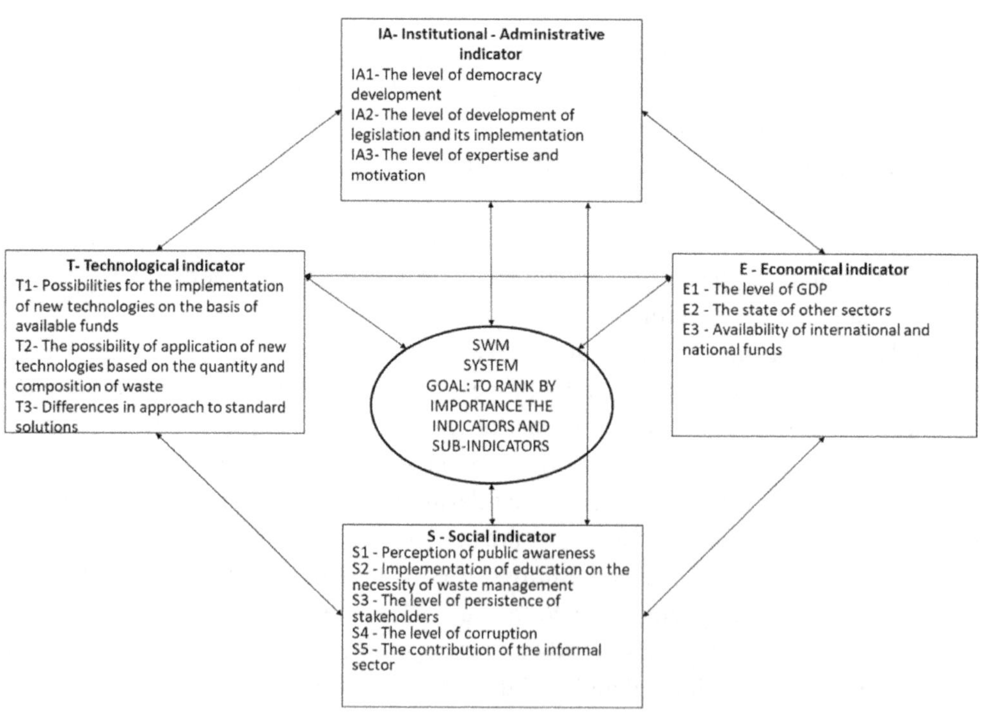

Figure 5. The values for indicators and sub-indicators of AHP method from the questionnaire

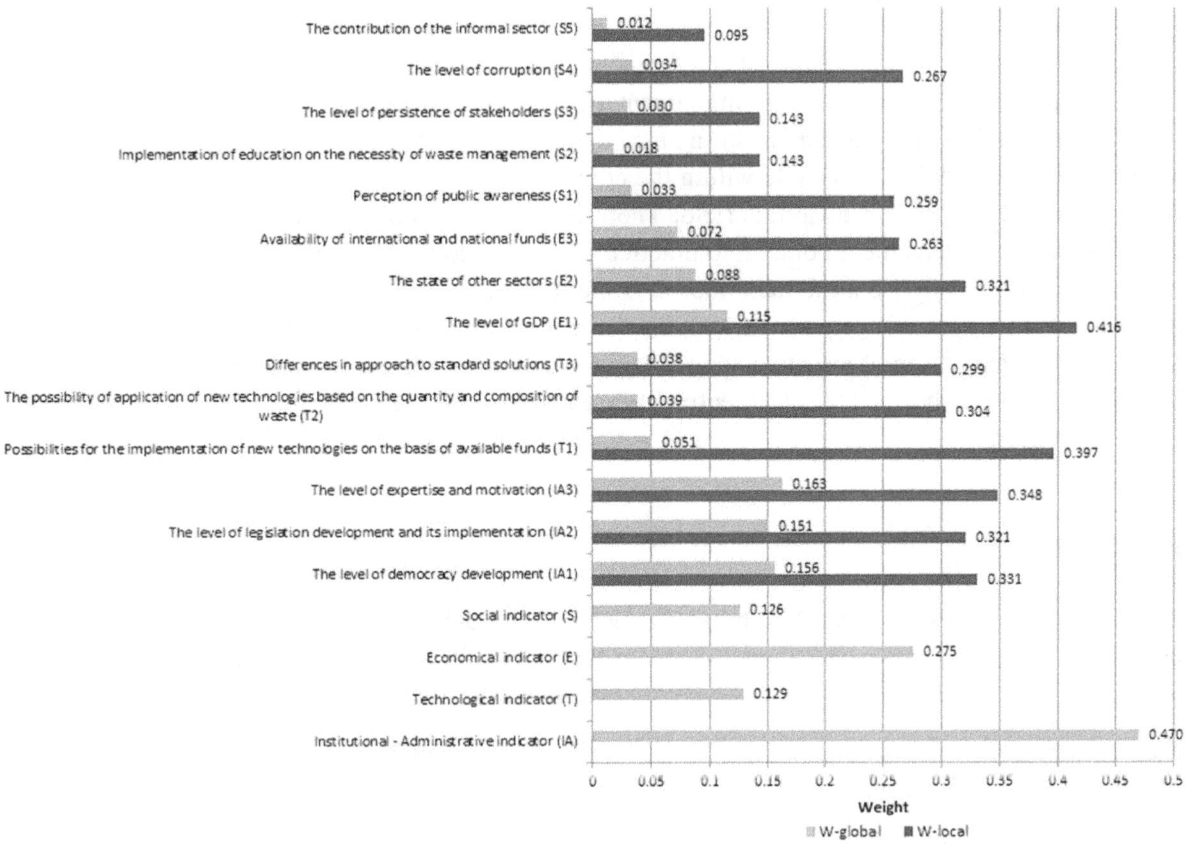

The problem that emerged was the inconsistency of the decision makers (DMs) who participated in the research. Inconsistency can occur when decision makers have varying degrees of expertise, when they are insufficiently concentrated in deciding or when they are not sufficiently motivated. For these reasons, a degree of consistency has been adopted in the range of 0.00-0.20. The decision makers with a degree of consistency greater than 0.20 were considered highly inconsistent and were not taken into account when determining the group weights of indicators or subindicators of their individual weights (Saaty and Kearns, 1985).

Considering the set goal, from the set of defined indicators, the decision makers found the "institutional and administrative" indicator the most important for the current problems of waste management, thus pointing what should be paid attention to first, in order to improve the waste management situation in the region.

The results can have practical implications in terms of supporting people in government administrations, decision makers, waste management experts, to better understand the environment and problems of the environment in which they operate, and to make better strategic decisions on the basis of formal decision-making models.

CONCLUSION

Improving the waste management system and achieving the EU targets in the Republic of Serbia requires exceptional reforms from both technical and financial point of view. However, the most challenging are social system reforms that will enable the implementation of technical improvements in waste management.

The transition from the current waste management practice to a modern integrated management system will be difficult but inevitable within the country's efforts to become a member of the EU and fully comply with EU standards in this field. There are many reasons for the slow implementation of the existing waste management policy into practice - in some cases, it is compliance (non-compliance) with the regulations only - but the main reason is the lack of control over the implementation of legal obligations and the lack of adequate waste management infrastructure, as well as the lack of financial resources. Implementation of a sustainable management system waste presented in the scenario that is in line with the targets defined by EU directives presented in Table 1 and Table 2 will require additional funds from citizens.

REFERENCES

Batinić, B., Vujić, G., Zivancev, M., Ubavin, D., & Stanisavljević, N. (2015). Knowledge of the MSW quantity and composition as a basic prerequisite for the successful implementation of the EU Directives. *Proceedings of the Symposium "The Role of Communication in Waste Management"*.

Beigl, P., Lebersorger, S., & Salhofer, S. (2008). Modelling municipal solid waste generation: A review. *Waste Management (New York, N.Y.)*, *28*(1), 200–214. doi:10.1016/j.wasman.2006.12.011 PMID:17336051

Chung, S. S., & Poon, C. S. (2001). Characterization of municipal solid waste and its recyclable contents of Guangzhou. *Waste Management & Research*, *19*(6), 473–485. doi:10.1177/0734242X0101900603 PMID:12201677

Džolev, N. M., & Vujić, G. V. (2016). Impact assessment of concentrate re-circulation on the landfill gas production. *Thermal Science*, *20*(4), 1283–1294. doi:10.2298/TSCI160401137D

Eurostat. (2016). *Municipal Waste Statistics*. Retrieved from http:// http://ec.europa.eu/eurostat/statistics-explained/index.php/Municipal_waste_statistics

Forman, E., & Peniwati, K. (1998). Aggregating individual judgments and priorities with the analytic hierarchy process. *European Journal of Operational Research*, *108*(1), 165–169. doi:10.1016/S0377-2217(97)00244-0

FTN – Faculty of Technical Sciences. (2009). *Determination of waste composition and quantity estimation in order to define a strategy for the management of secondary raw materials within the sustainable development of the Republic of Serbia*. FTN - Department of Environmental Engineering and Occupational Safety and Health, University of Novi Sad.

GIZ - Deutsche Gesellschaft für Internationale Zusammenarbeit - German Corporation for International Cooperation. (2016). *Solid Waste Data Collection in South East Europe*. Belgrade, Serbia: Author.

Hoornweg, D., & Bhada-Tata, P. (2012). *What a Waste: A Global Review of Solid Waste Management, Urban development series, knowledge papers no. 15*. Washington, DC: World Bank.

Hristovski, K., Olson, L., Hild, N., Peterson, D., & Burge, S. (2007). The municipal solid waste system and solid waste characterization at the municipality of Veles, Macedonia. *Waste Management (New York, N.Y.)*, *27*(11), 1680–1689. doi:10.1016/j.wasman.2006.09.003 PMID:17098409

IMG - International Management Group. (2014). *Municipal Solid Waste Information – Final Report*. Belgrade. Serbia: Author.

Mrkajić, V., Stanisavljevic, N., Wang, X., Tomas, L., & Haro, P. (2018). Efficiency of packaging waste management in a European Union candidate countries. *Resources, Conservation and Recycling, 1*(136), 130-141.

OECD. (2008). *Measuring Material Flows and Resource Productivity, Volume 1*. The OECD Guide. Retrieved from http://www.oecd.org/environment/indicators-modellingoutlooks/MFA - Guide.pdf#page=1&zoom=auto,-82

Off. Gazette RS, No 36/2009 i 88/2010. Law on Waste Management

Off. Gazette RS, No 61/2010. The Rulebook on the Methodology for Collection of Data on Composition and Quantities of Municipal Waste at the Territory of Local Self-Government Units

Ramanathan, R., & Ganesh, L. S. (1994). Group preference aggregation methods employed in AHP: An evaluation and an intrinsic process for deriving members' weightages. *European Journal of Operational Research*, *79*(2), 249–265. doi:10.1016/0377-2217(94)90356-5

Saaty, T. L. (1980). Multicriteria Decision Making: The Analytic Hierarchy Process. RWS Publications.

Saaty, T. L., & Kearns, K. P. (1985). *Analytical Planning: The organization of system* (1st ed.). Pergamon Press.

Sharma, M., & McBean, E. (2007). A methodology for solid waste characterization based on diminishing marginal returns. *Waste Management (New York, N.Y.)*, *27*(3), 337–344. doi:10.1016/j.wasman.2006.02.007 PMID:16600585

Stanisavljević, N., Ubavin, D., Batinić, B., Fellner, J., & Vujić, G. (2012). Methane emissions from landfills in Serbia and potential mitigation strategies: A case study. *Waste Management and Research*, *30*(10), 1095-1103. doi:10.1177/0734242X12451867

Tot, B., Srđević, B., Vujić, B., Tavares Russo, M. A., & Vujić, G. (2016). Evaluation of key driver categories influencing sustainable waste management development with the Analytic Hierarchy Process (AHP): Serbia example. *Waste Management & Research*, *34*(8), 740–377. doi:10.1177/0734242X16652963 PMID:27288689

Tot, B., Srđević, Z., Tavares Russo, M. A., Dejan, U., & Vujić, G. (2017). Group assessment of key indicators of sustainable waste management in developing countries. *Waste Management & Research*, *35*(9), 913–922. doi:10.1177/0734242X17709911 PMID:28606005

Vujić, G., Stanisavljević, N., Batinić, B., Jurakić, Ž., & Ubavin, D. (2017). Barriers for implementation of ''waste to energy'' in developing and transition countries: a case study of Serbia. *Journal of Material Cycles and Waste Management, 19*(1), 55-69. DOI doi:10.100710163-015-0377-8

Vujić, G.V., Jovičić, N.M., Babić, M.J., Stanisavljević, N.Z., Batinić, B.M., & Pavlović, A.R. (2010). Assessment of plastic flows and stocks in Serbia using material flow analysis. *Thermal Science, 14*, S89-S95. DOI: doi:10.2298/TSCI100621031V

Chapter 14
Waste Management in South Africa

Joan Mwihaki Nyika
https://orcid.org/0000-0001-8300-6990
University of South Africa, South Africa

Ednah Kwamboka Onyari
University of South Africa, South Africa

Shivani Mishra
https://orcid.org/0000-0001-7200-2679
University of South Africa, South Africa

Megersa Olumana Dinka
https://orcid.org/0000-0003-3032-7672
University of Johannesburg, South Africa

ABSTRACT

Solid waste management (SWM) is a challenge in developing countries such as the Republic of South Africa (RSA). This book chapter highlights the drivers and state of SWM in RSA and suggests alternatives to make solid waste a resource. The SWM strategy of the country has a role in pushing waste up its hierarchy towards minimal generation, reuse, and recycling through extended producer responsibility and economic instruments. However, the lack of an all-inclusive planning and management has challenged the success of these initiatives. In recognition of these flaws, the private sector is teaming up with the government and individuals to bridge service and value chains in sustainable SWM by formalising waste pickers, initiating waste-to-energy initiatives, promoting recycling at all stages of the waste cycle, and adopting practices that divert wastes from landfills. These initiatives if taken up will promote better economic turnover through the production of alternative energy, environmental conservation, and creation of employment opportunities in RSA.

DOI: 10.4018/978-1-7998-0198-6.ch014

INTRODUCTION

Urbanization, population increase and economic development are growing trends in contemporary society that have resulted to improvements in human well-being. Evenly, resource consumption has escalated and so are solid waste (SW) generation trends (Chen, 2018). This is a serious problem in developing countries whose capacity to manage such waste is limited. In these countries per capita waste generation is lower compared to developed countries; but the capacity to handle, landfill, recycle and reuse effectively and efficiently is challenging (Ahsan et al., 2014). Variations in SW handling capacity arise because developing countries focus on a linear resource consumption model involving processing, production, use and discard of products to nature (Garces-Ayerbe et al., 2019). The model has negative effects on the environment through greenhouse gas (GHG) emissions, land and water resource pollution and enhanced climate change effects. SW management (SWM) has technologically and operationally improved in the last decade to respond to resultant environmental issues as Bello, Ismail and Kabbashi observed (2016). Although this is the case, the focus has been on end-of-pipe solutions that focus on waste reduction rather than sustainability whose focus is to prevent waste (Singh et al., 2014). Additional initiatives such as site specific environmental designs, cleaner productions, industrial symbiosis and extended producer responsibility (EPR) are in place though they focus on isolated systems and individual products rather than integrated systems (Ahsan et al., 2014). In integrated SWM systems, the focus is accounting and controlling all kinds of wastes and their resultant environmental effects. In addition, process changes that control environmental waste flows and reflect on its upstream processes with precautionary actions are prioritized in an integrated system. In this chapter, the status, drivers and key challenges of Republic of South Africa (RSA) SWM system are discussed towards improvements and sustainability. Additionally, case studies of successes and failures in this sector are highlighted for in-depth insight on the issue.

BACKGROUND

Waste generation in RSA has been on a rising trend. The country produced 121 million tons of waste in 2017 (Department of Environmental Affairs, 2018). These estimates are higher compared to those documented in 2011 that showed total waste generation to be 108 million tons (DEA, 2011). The trend could arise due to increased waste production from the rising population and improved economic wellbeing, which has resulted to unaccounted for waste that is neither grouped as hazardous nor general (Fakoya, 2018). More than 60% of general waste and about 95% of hazardous is landfilled (DEA, 2018) despite the fact that most filling facilities are not managed in accordance to stipulated regulations. According to Mannie and Bowers (2014), 95% of generated waste is landfilled and 87% of municipalities do not have infrastructure and capacity to manage and pursue minimization strategies effectively. Additionally, SWM is poorly funded and uncoordinated, which makes the country 2 to 3 decades behind developed worlds such as Europe (Godfrey & Oelofse, 2017). Key issues include poor collection services, unlicensed SWM activities, illegal dumping, poor waste data management and non-enforcement of existing waste regulations (Abdel-Shafy & Mansour, 2018). Cognizant of these challenges, the national and municipal governments are advocating for a trend towards waste minimization, reuse and recycling under the National Waste Management Strategy (NMWS) (Dlamini et al., 2019). With increased awareness of these challenges, the focus will shift from landfill disposal to the view of waste as a resource.

DEFINITION OF WASTE AND PRESSURES OF GENERATION

South African Definition of Waste

South Africa has two legal definitions of waste. According to the NEMWA[1] amendment act (Government Gazette, 2014), waste is any superfluous, discarded, abandoned, rejected and unwanted substance. This definition perceives waste as useless and is incognizant of its pollution potential (Dlamini et al., 2019). With the broad nature of this definition, up-taking of re-using and recycling practices is discouraged.

The second definition describes waste as any solid substance transported, dissolved or suspended in water such as sediment and that is deposited or spilled into water or on land in such a manner, volume and composition that causes pollution. This definition by the National Water Act (DWA, 1998) limits the uptake of reuse as the burden lies in proving waste as pollutant cause. With this definition, all waste is considered harmful unless treated and re-classified. These processes are expensive and involve specialised testing of both waste and the treated product, which is still considered waste according to the waste classification and management regulations (Dlamini et al., 2019). The restrictive nature of defining waste in RSA could thus be an impediment to effective uptake of recycling and reuse initiatives.

Pressures of Waste Generation

The RSA is the most Southern country of Africa subdivided to nine provinces; Western Cape, North West, Northern Cape, Mpumalanga, Limpopo, KwaZulu-Natal, Gauteng, Free State and Eastern Cape (Figure 1). The country has several human induced agents and processes that promote waste generation including population growth, economic development and urbanization (DEA, 2018).

Figure 1. Map of South Africa's nine provinces (Singh, 2016)

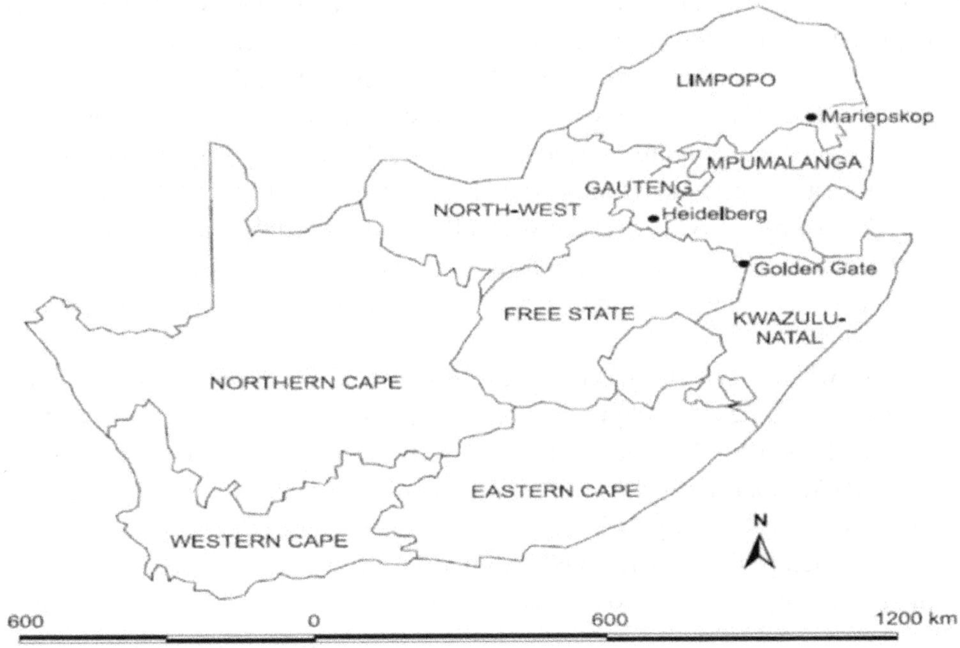

Figure 2. Changes in population and its growth between 2013 and 2017

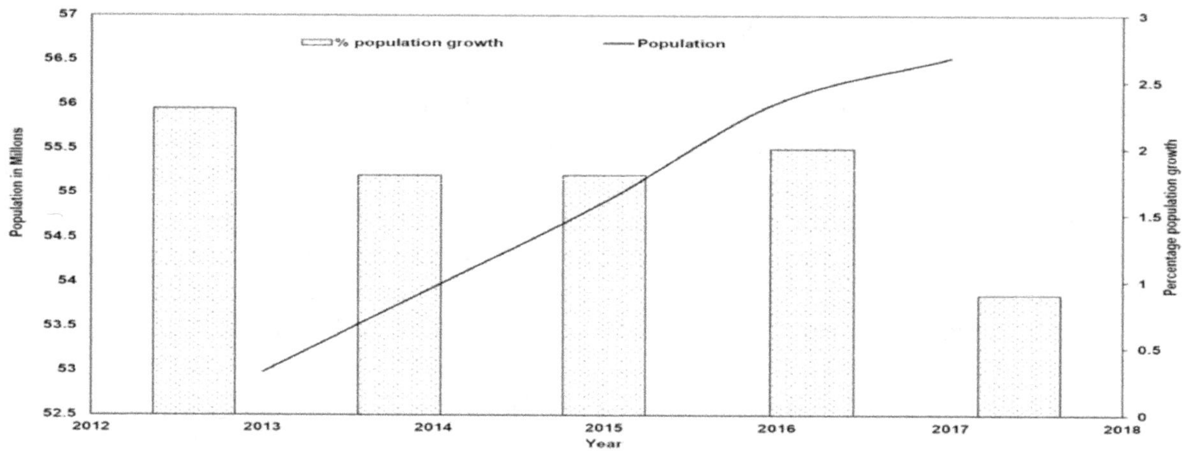

Population growth is directly proportional to SW since it promotes consumption of natural resources. South Africa has seen progressive increase in population and in a five-year period from 2013, population rose by more than three million (Figure 2). Although the country had notable population increase in the period, there was a decline in population growth, a trend that is replicated worldwide (StatsSA, 2017a). Gauteng and Western Cape provinces had the highest population growth at 3.1 and 2.2 percent, respectively, which corresponded to high waste generation potential compared to Northern Cape's 0.7% growth rate (StatsSA, 2017a). Furthermore, Gauteng had the highest population density at 785.5/km^2 and generated more waste despite the great pressure to provide waste services compared to other provinces (StatsSA, 2017a).

Economic development is another driver to waste generation and its impacts are affected through higher incomes and increased industrial and manufacturing activities. South Africa's economic performance is depicted by gross domestic product (GDP) in a given period. In the period between 2013 and 2017, the GDP of RSA increased from rand (R) 3.54 (USD 240 billion) to 4.65 trillion (USD 320 billion) (Figure 3) (DEA, 2018). Gauteng had the highest GDP contribution to the economy and could be the highest waste generator (Figure 4a) (DEA, 2018). Real estate business, general government services and trade, catering and accommodation sectors were high contributors to GDP at 20, 18 and 15 percent, respectively (Figure 4b) (DEA, 2018). The World Bank (2019) established a relationship between waste generation and income levels, consumption rate for goods, services, and living standards. The trend applies in South Africa where per capita generation of waste increases with improved living standards (Dlamini et al., 2019)

Urbanization, which results from economic and industrial development, is another pressure to waste generation (World Bank, 2019). With urbanization, rural-to-urban migration is on the rise, where migrants are in search for better livelihoods. Consequently, the population size per capita waste generation is on the rise. SWM is an 'urban' issue because wealthier residents with high purchasing power and less recycling and reuse capacity dwell in these areas (Tsheleza et al. 2019). Urbanization in RSA is on the rise and United Nations (2015) predicted that 71.3% and 80% of the country's population would reside in urban areas by 2030 and 2050, respectively. Some of established urban areas include Nelson

Figure 3. GDP of RSA over a five-year period from 2013 (StatsSA, 2017a)

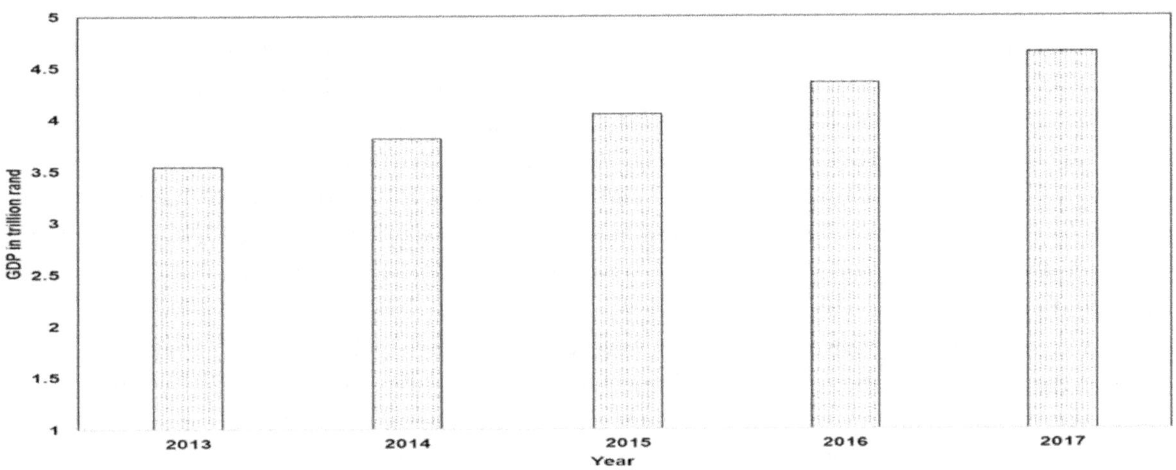

Figure 4. Percentage GDP contribution based on a) provinces and b) economic sectors (StatsSA, 2017b)

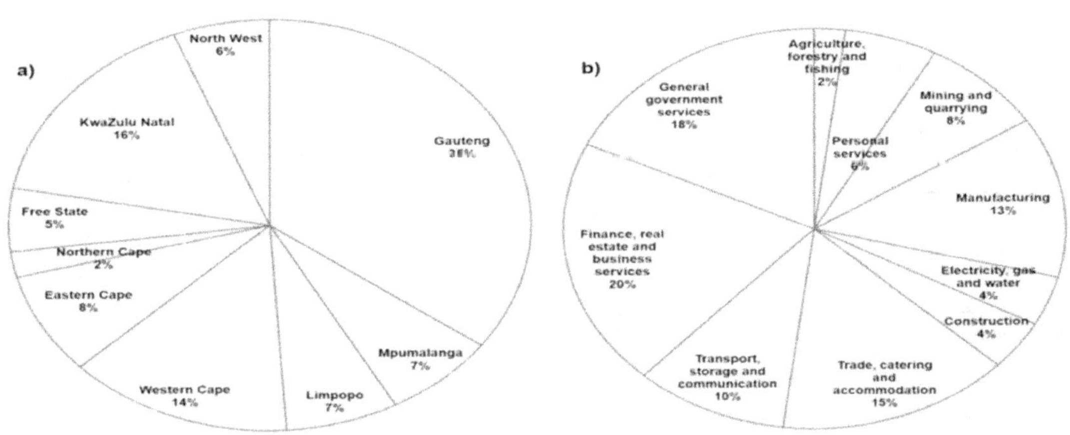

Mandela Bay, eThekwini, Cape Town, Ekurhuleni, City of Tshwane and Johannesburg in addition to growing regions such as Sol Plaatjie, Mbombela, Rustenburg, Polokwane, Msunduzi, Buffalo city and Manguang (StatsSA, 2017b).

The growing urban population is characterised by a younger population, whereby 64% of youth dwell in urban areas (Department of Cooperative Governance and Traditional Affairs, 2015). Educated youths are seeking employment in urban areas for higher incomes compared to the aged who are economically inactive. Urban areas of RSA contribute 80% of its gross value added and their growth is twice as that of rural areas (COGTA, 2015). The trend of urbanisation has led to increased poverty particularly in townships evident from informal settlements in major cities and towns. Consequently, waste generation is on a rising trend and so is the pressure to management it as DEA (2018) highlighted.

Waste Classification

Waste classification in RSA is outlined in the country's classification and management regulations, R634 (SAWIC, 2013) and based on risk since no material is considered entirely non-hazardous and even if the risk posed is very remote, it exists (Gutberlet, 2018). Classification thus assesses risks of wastes to distinguish extremely hazardous waste that requires precaution during disposal from less risky waste. Waste is divided into two classes: hazardous and general waste. Classification of these wastes is as shown in Table 1. Out of the estimated 126.9 million tons of waste generated in 2017, 42.6% is believed to be general while 57.4% was estimated to be hazardous. Distribution of generated wastes based on the classification are shown in Table 2

General wastes do not pose environmental and public health risks if properly managed and can be landfilled according to the classification and management regulations, (Gutberlet, 2017). However, the emphasis in contemporary SWM trends is not on disposal of general waste, but rather its reuse and recycling. Domestic waste though classified as general waste can contain hazardous components though their qualities and quantities have minor potential risk. Some hazardous wastes can be toxic even in low concentrations. Such wastes require strict control and management. Furthermore, they should be disposed in specified landfills and in adherence to the precautionary principle to minimize their associated environmental and public health risks (Tsheleza et al., 2019).

Table 1. Classification of wastes and their percentage contribution to total amount (DEA, 2018)

Municipal Waste		Hazardous Waste	
Type	**% Contribution**	**Type**	**% Contribution**
Municipal	8.9	Gaseous waste	8.2
Commercial and industrial	6.7	Mercury enhanced waste	0.05
Organic	56.5	Batteries	0.05
Construction and demolition	0	POP[2] waste	0.1
Paper	8.3	Inorganic waste	1.1
Plastic	4.1	Asbestos enhanced waste	0.1
Glass	2	Oils	0.1
Metals	4.6	Organics	0.7
Tyres	7.4	Bituminous and tarry waste	0.4
Other	1.5	Brine	8
		Fly dust and ash	60.4
		Bottom ash	8.2
		Slag	10.8
		Mineral waste	0.1
		WEEE[3]	0.4
		HCRW[4]	0.1
		Sewage sludge	0.8
		Others	0.4
Total	100	Total	100

Table 2. Amounts of wastes generated based on their type (DEA, 2018)

General Waste		Hazardous Waste	
Type	**Amount (Million Tons)**	**Type**	**Amount (Million Tons)**
Municipal	4.8	Gaseous waste	6
Commercial and industrial	3.6	Mercury enhanced waste	0.0001
Organic	30.5	Batteries	0.004
Construction and demolition	0	POP waste	0.0006
Paper	4.5	Inorganic waste	0.8
Plastic	2.2	Asbestos enhanced waste	0.006
Glass	1.1	Oils	0.1
Metals	2.5	Organics	0.5
Tyres	4	Bituminous and tarry waste	0.3
Other	0.8	Brine	5.8
		Fly dust and ash	44
		Bottom ash	6
		Slag	7.9
		Mineral waste	0.1
		WEEE	0.4
		HCRW	0.05
		Sewage sludge	0.6
		Others	0.3
Total	54		72.9

Waste Cycle

The flow of waste in RSA is divided into two components: the service and value chains (Godfrey & Oelofse, 2017). The latter involves the generation of waste, its collection and disposal at municipal level. The value chain processes involve adding value to generated waste by recovering it for reuse and recycling at buy-back centres, exporting waste for further use and reuse of recycled end-products. The informal sector is essential in bridging the service and value chains through its role in supplementing municipal waste services and promoting reuse and recycling through waste picking. According to Godfrey and Oelofse (2017), RSA has approximately 90,000 informal waste pickers who work on landfills and at kerbsides though they are not integrated to the mainstream municipal SWM systems. This number is set to rise in response to high unemployment levels (Schenck et al., 2016; Blaauw, 2017). Figure 5 represents the waste flow and the role of waste pickers in bridging its components.

Waste Management

Waste management system in RSA is structured around the waste hierarchy (Figure 6) that was a hallmark of the 1999 National Waste Management Strategy (NWMS) (DEA, 2018). The hierarchy orders management options of the waste flow based on their preference in a descending order. In this case, waste

Figure 5. A representation of waste flow in RSA and the role of waste pickers in bridging its service and value chains (Adopted from Godfrey and Oelofse, 2017)

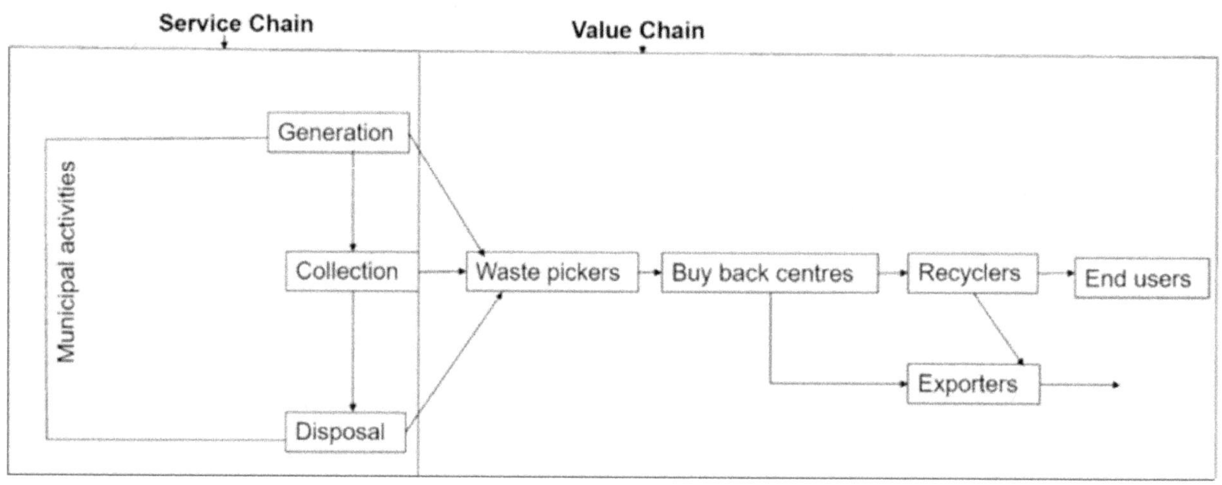

Figure 6. A representation of RSA waste management approach using the hierarchy (DST, 2014)

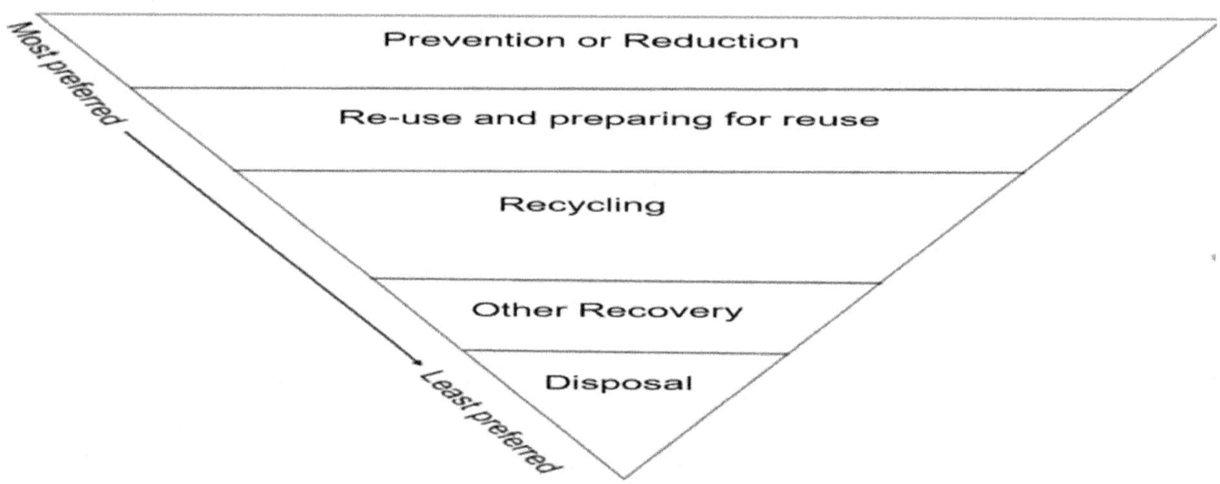

avoidance and reduction are precedence followed by reuse, recycling, recovery and treatment. Disposal is the last resort to waste management while reduction and prevention are the priority (Department of Science and Technology, 2014). The entire hierarchy is translated to the waste act. This implementation stimulates EPR during material designing, composition decisions, production and packaging to prioritize on waste reduction, reusability and recyclability of products. Additionally, the hierarchy supports the shift from the concept of 'cradle-to-grave' to the 'cradle-to-cradle (Zhakata et al., 2016). In the latter, products are recovered, re-used or recycled at the end of their life span to make new products while in the former, a product's cycle ends with its disposal. The next section discusses the components of the waste hierarchy.

Prevention and Reduction

Prevention and reduction form the waste hierarchy foundation, are the most preferred management options and aim at minimizing products entering the waste stream. If products cannot be prevented, they should be recovered, reused and recycled. Additionally, if recovery practices for particular wastes are expensive; their initial use should be reduced or even prevented in preference to alternatives. Although quantifying these components is difficult, statistics show a rising trend in waste generation for all provinces of RSA (DEA, 2018). This could be because waste prevention and reduction results from producer responsibility, economic incentives and competitive pressures that are implemented at industrial levels. Examples of practices in this level include composting organic waste at home instead of directing it to landfills (DST, 2014).

Table 3. Statistics of recycled waste in 2017(DEA, 2018)

General Waste			Hazardous Waste		
Type	**% Recycled/Recovered**	**% Landfilled**	**Type**	**% Recycled/Recovered)**	**% Landfilled**
Municipal	0	100	Gaseous waste	0	96
Commercial and industrial	10	90	Mercury enhanced waste	0	95.6
Organic	31.1	68.9	Batteries	0	26.9
Construction and demolition	90	10	POP waste	0	100
Paper	58	42	Inorganic waste	0.6	99.4
Plastic	43.7	56.3	Asbestos enhanced waste	0	100
Glass	78.4	21.6	Oils	80	20
Metals	75	25	Organics	14.2	85.8
Tyres	100	0	Bituminous and tarry waste	0	100
Other	9.1	90.8	Brine	0	100
			Fly dust and ash	5.8	93.4
			Bottom ash	8.3	91.7
			Slag	4.1	95.9
			Mineral waste	2.8	94.4
			WEEE	9.7	90.3
			HCRW	0	100
			Sewage sludge	15	85
			Others	0.9	97.6
Total	38.6	61.4		6.3	93.7

Re-Use, Preparing for Re-Use and Recycling

Although they are different, recycling, reuse and recovery are physical processes that reclaim materials perceived as waste to reduce the stream directed to landfills. Prior to recycling, waste is prepared through processes such as refurbishing, repairing and cleaning to be reusable and recyclable. Recycling is carried out by non-governmental organisations, industries and volunteers with demand for certain resources and social needs being key motives (Godfrey & Oelofse, 2017). Recycling companies include Collect-a-can and Steelrec that recycle beverage used cans. Collaborative efforts by RSA government and donor fund organisations such as the DANCED[5] are positive efforts to push waste up the hierarchy (Godfrey & Oelofse, 2017). The setting of recycling targets in the 2001 Polokwane Declaration also depicts government commitment to support recycling. According to Sentime (2014), the contents of the declaration are debatable and were never legalised. Table 3 shows that RSA was only able to divert 11% and 7% of general and hazardous waste, respectively to recycling in 2017 (DEA, 2018). More efforts towards strengthening recycling and integrating it in the local economy is needed to create green jobs and promote economic growth according to Godfrey and Oelofse (2017).

Other Recovery and Disposal

To ensure non-recyclable materials are the only ones disposed of, other recovery processes such as pyrolysis, gasification, incineration and anaerobic digestion are on the rise (DST, 2014). Adopting these practices are positive efforts towards the waste hierarchy's desired outcomes. The processes are alternatives to disposal (Simelane, 2016) and they are discussed later in this chapter. Disposal is the least preferred approach to waste management. However, landfilling remains the most practiced method of management according to the 2017 statistics that showed that 61.4% of general waste and 93.7% of hazardous waste was disposed (DEA, 2018).

Legislation, Policies and Regulations on Waste Management

Existing concerns regarding increased waste flow and the need to manage it effectively have led to the introduction of a range of legislative measures. These regulations aim at driving waste up its hierarchy towards reuse, recycling and reduction (Godfrey & Oelofse, 2017). Table 4 shows the chronological development of these regulations. These regulations are many and some even have duplicative roles, which has been counterproductive in the country's SWM sector because efforts towards minimization are difficult to implement without triggering extensive conditions (Zhakata et al., 2016). The private sector has also been burdened with requirements to evolve with these regulations to be compliant and concurrently remain competitive internationally and locally. In addition to the above regulations, specific provinces and municipalities manage waste under several by-laws. Re-evaluating these regulations towards and integrated system is essential for their effective enforcement in SWM activities (Godfrey & Oelofse, 2017).

Waste Management Strategy of RSA

The NWMS has its history from 1999 when it was developed to implement the white paper on integrated pollution control and waste management (IPC-WM) (DEA, 2019). Under the NEMWA provisions, it

Table 4. Chronological changes in waste management regulations of RSA (SAWIC, 2015)

Year	Waste Policy and Regulations
Legislations **1973** **1977** **1989** **1993** **1996** **1998** **2000** **2004** **2006** **2008** **2014**	Hazardous substance act no. 5 Health act no. 63 Environmental conservation act no. 73 Occupational health and safety act no. 85 South African constitution (act no. 108) National water act no. 1998 National environmental management act no.107 Municipal structures act no. 117 Municipal systems act no. 32 Air quality act no. 39 Amendment notices for notice no. R386-7 of NEMA act 1998 NEMWA act no. 59 NEMWA act no. 26 Amendments on the waste act implementation guidelines Amendments on NEMA act no. 25
Regulations **2003** **2008** **2009** **2012** **2014** **2015** **2016** **2017**	Plastic bag regulations Asbestos regulations Waste tyre regulations National waste information regulations Regulations to phase out PCB[6] materials Regulations to phase out substances that deplete the ozone Amended regulations on waste tyre management Regulations on management of residue deposits and stockpiles Waste tyre regulation Regulations to control export and import waste
Policies **2008** **2009** **2011**	Integrated waste management policy Thermal treatment of hazardous and general waste policy Policy on the provision of basic waste services to poor households

was improved to the 2012 NWMS. The strategy is part of a plan to absorb generated waste despite the growing challenges of population increase, urbanization and consumerist trends. The plan emphasizes the national plan towards sustainable development through environmental resource protection (Dlamini et al., 2019). Such an initiative demands effective use of raw materials, waste prevention, efficient use of resources and sustainable material designing (DEA, 2018). The country hopes to achieve these action plans through systematic use of the waste hierarchy. In the plan, eight goals and their associated targets that were to be realized by 2016 are outlined as summarized in Table 5.

Although the motive of the NWMS outlined by its goals was to improve waste services, most of these targets have not been met according to Dlamini et al. (2019). Only 61% of households had access to waste services and the number was skewed in favor of the affluent in urban areas (Zhakata et al., 2016; Gutberlet, 2018). Budgetary allocations for waste management at local levels were insufficient to meet the demand. The plan overburdens the local municipalities to steer up its goals despite the lack of corporative governance with provincial and national governments. As such, municipalities are overwhelmed as evident from the high quantity and diversity of wastes, and thus their inability to serve its residents leading to the rise of illegal dumping (Zhakata et al., 2016). These flaws call for re-planning that involves collaboration of all governing spheres, integrated legislation and collaborations with the private sector in SWM systems of the country. The government is currently reviewing this NWMS to formulate realistic goals and action plans towards sustainable SWM (SAWIC, 2018).

Table 5. Goals and targets of the NWMS of RSA as summarized by DEA (2011)

Goal	Description	Targets
1.	To promote waste recovery, reuse, recycling and minimization.	-divert 25% of disposable waste for reuse and recycling. -Introduce waste separation sources at all cities and metropolitan municipalities.
2.	To promote efficient and effective waste service delivery.	-ensure that 75% and 95% of rural and urban households respectively have adequate waste services. -ensure that 80% of landfill facilities have permits and adhere to set disposal standards.
3.	Enhance the contribution of SWM systems to green economy.	-Creation of 70,000 new jobs in the sector. -Incorporation of at least 2,500 new cooperatives and small and medium enterprises in the waste flow system.
4.	Enhance community awareness on the effects of waste to environment, human well-being and health.	-80% of municipalities and 80% of schools have active awareness campaigns.
5.	Promote integrated SWM and planning.	-all municipalities should develop their integrated waste management plans (IWMPs) and align them to integrated development plans (IDPs) that have specific targets. -All SWM facilities to provide SA waste information system (SAWIC) with their waste quantification reports.
6.	Promote realistic financial management and budgeting of waste services.	-Implement cost reflective tariffs and conduct cost-benefit analysis of waste services at municipality level.
7.	Implement measures to remediate and rehabilitate contaminated land.	-Assess at least 80% of sites documented in the contaminated land database. -Approval of 50% of rehabilitation plans of confirmed polluted sites.
8.	Promote effective adherence and compliance to the waste act.	-At least 50% rise in successful actions against non-compliance at all waste flow stages. -Appointment of 800 environmental management inspectors at national, provincial and local government levels to enforce the waste act.

Hazardous Waste Management

Hazardous waste management (HWM) is one of RSA's environmental concerns due its pollution potential to land and water resources. It is thus essential to dispose and handle this waste in a legitimate and responsible manner, which entails understanding its nature to make informed decisions on disposal options. The norms and standards of disposal of waste to landfill R636 (DEA, 2013) provide guidance on HWM. In the regulation, landfills are classified into four classes based on their waste containment barriers and engineering designs. Class A has the most advanced engineering with a geotextile barrier layer while class D is the least advanced. Before waste is accepted, it must be passing the acceptance criteria where waste is categorized to 5 groups: type 0, 1, 2, 3 and 4.

Type 0 is not acceptable for landfilling. Usually, it is treated and re-classified to a class that allows disposal, while type 1-4 should be disposed in classes A-D landfills, respectively. Types 1-4 wastes are disposed in either H:H[7] or H:h[8] landfills according to the minimum regulations of handling waste by the department of water affairs and forestry (DWAF) in 1998. The typing of wastes and classes of landfills to dispose them is on the basis of their quantified toxicity. Hazardous wastes fall under type 1 to be disposed in Class A landfill (DEA, 2013). According to the regulations, asbestos, expired, unusable and spoilt hazardous waste is disposed in Class A landfills. PCBs, hazardous chemicals and mixed hazardous wastes are landfilled in accordance to minimum regulations in the H:h/H:H landfill facilities.

Before disposing the hazardous waste, it is classified and assessed to dispose it of in the appropriate landfills. Wastes with hazard rating 1 or 2 can only be disposed in Class A landfills particularly the H:H category while those with of rating 3 or 4 are disposed in H:h class A landfills. Hazardous and general wastes, which are re-classified in less toxic types, are disposed in class B landfills. In all these cases, wastes must meet the norms, standards and regulations of landfilling according to DWAF 1998 prescriptions (DEA, 2013). HWM should adhere to some restrictions and prohibitions of compliance before landfilling. Immediate compliance is required for tyres, explosives, oxidizing waste, corrosives, wastes with pH of <6 to >12, flammables, reactive wastes, lead acid batteries, compressed gases untreated and HCRW. POPs and other batteries have 8 years to comply with landfilling regulations while other pesticides and recyclable waste oils have 4 years. WEEE wastes have 3-8 years while hazardous wastes of calorific value have 4 years. Generally, all waste under type 1 category have five-year disposal timeframe (DEA, 2013).

Waste-to-Energy

The estimated financial value of RSA waste sector was R15.2 billion (0.51% of GDP) in 2012 (DST, 2014). This value can be increased through waste-to-energy (WtE) initiatives that generate energy (gaseous/liquid fuels or electricity) using waste as feedstock. According to Godfrey and Oelofse (2017), WtE projects in RSA can be enhanced through the uptake of high-temperature waste treatment approaches such as gasification, pyrolysis and incineration in the mainstream waste cycle. These initiatives will unlock opportunities for RSA to adopt technologies that enhance waste recovery. Such initiatives include WtE-focused technologies. Based on processes involved, WtE forms are grouped to physical, thermal and biological (Tan et al. 2015). Thermal processes produce heat and electricity and include pyrolysis, gasification and incineration while biological involve production of biogas through anaerobic digestion of waste. Physical processes entail the recovery of methane gas from landfills to produce energy (Tan et al., 2015).

Composting and anaerobic digestion in RSA is practiced both in small and large scales according to Mutezo (2016). However, the latter is hardly practiced because of the poor conversion efficiencies of biogas-to-electricity (Mutezo, 2016). An example is a 4.4 MW[9] biogas plant in Tshwane city that supplies BMW Rosslyn factory with about 30% of its energy needs (Business Sweden, 2017). The plant generates power from industrial food wastes and cow manure. Another example is the Northern water works project of the city of Johannesburg that generates 4.5MW of electricity from wastewater and leachate biogas (Business Sweden, 2017). The RSA local governments are collaborating with department of energy (DoE) to discuss WtE interventions in most of its wastewater treatment plants that though installed, have stalled due to lack of human capacity and poor maintenance.

Initiatives focusing on the use of landfill gas (LFG) as energy are also underway in RSA. The gas can be compressed, purified and used to power motor vehicles. A collaboration between Ekurhuleni municipality and Global Infrastructure Basel company processed LFG and used it to fuel fleet vehicles in a project costing USD 5 million (Njoku et al., 2018). LFG can also be used to generate up-to 45 megawatts of electricity through a steam engine. In eThekwini municipality of RSA, an LFG plant has been set up to generate electricity using an economical and environmental friendly process incorporating steam turbines and reciprocal engine combustion among other conditions (Njoku et al. 2018). In collaboration with the EnerG systems company, the Johannesburg city converts landfill gas to electricity, an initiative that contributes to the clean development mechanism (CDM). Case examples are Robinson Deep, Marie Louise and Goudkoppies landfills that are expected to produce 3 MW of electricity each, which will ease pressure on existing power grid (Business Sweden, 2017; Dlamini et al., 2019).

Incineration of non-recyclable waste is practiced in RSA but on an ad hoc basis. The process facilitates waste pre-processing, separation and treatment. Dlamini et al. (2019) suggested that incineration will increase landfill space, help create employment and reduce fuel costs though the capital costs of such initiatives demand careful consideration. In Nelson Mandela Bay municipality, 2,200 tons of solid waste are incinerated daily to produce 50 MW of electricity (Sustainable Energy Africa, 2017). However, the uptake of the problem is challenged by air emission control regulations and high capital costs of the process.

Pyrolysis and gasification of organic materials including plastic waste, wood and biomass at temperatures >700°C without combustion is another sustainable SWM approach practiced in RSA. The result is production of Syngas that can power combustion engines or substitute hydrogen, transportation fuel, fertilizer, chemical and natural gas products (Sustainable Energy Africa, 2017). Although the technology to run these processes is available in RSA, ineffective waste pre-processing and separation lengthens the involved processes. Through the renewable energy independent power producer procurement program (REIPPPP), which is spearheaded by the DoE (Anton & Raine, 2016), gasification and pyrolysis projects are growing in RSA. Through the REIPPPP project, 145 MW is allotted for large-scale pyrolysis and gasification projects. Already, 59 MW of the 145 MW has been allocated and the remainder is potential. These opportunities will push waste up the hierarchy and create jobs towards a green economy in RSA. Since 2011, more than 90 WtE projects have been awarded 6,327MW and attracted R192.6 billion investments (Business Sweden, 2017). Additionally, larger and small-scale developers of renewable energy have collaborated to produce more than 200 MW. These WtE projects will diversify the energy generation mix from the state-owned company (Eskom[10]).

A notable initiative by the country towards sustainable SWM is the waste tyre (WT) management. The country is known to have more than 60 million WT that are illegally dumped and pose as environmental threats (Muzenda, 2014). To deter this problem, an initiative by the name, the recycling and economic development initiative of South Africa (REDISA), which aims at using WT as a supplementary energy source in WtE pyrolysis and incineration was developed. This project is in line with pushing this waste up the hierarchy. WT has high calorific value compared to coal and its use as supplementary feedstock is cost efficient. On pyrolysis, WT derived oil can substitute liquid fuels and its carbon black is a smokeless fuel (Muzenda & Popa, 2015). However, the uptake is not sustainable since the demand for WT will rise, producers will incur the cost, and ultimately the cost and benefit of its use will be inequitable (Maisiri, 2016). Initiatives towards WtE processes using plastics are also evident in the country but propagated by the private and informal sectors (DST, 2014).

Issues and Challenges of SWM

Although RSA has made great strides in developing legislative frameworks to support SWM, much more needs to be done to enforce these regulations, improve waste services and recovery considering that landfilling still remains the predominant waste solution (Godfrey, 2019). Some of the challenges the country faces in SWM are in planning, financial management, interpretation of existent SWM legislation and in provision of waste services (DEA, 2016a). According to Oelofse (2018), these problems are classified into four broad themes: financial, institutional, labor and equipment management.

SWM is a multi-dimensional issue incorporating technological, economic, institutional, legal, socio-cultural and environmental factors. Linking these aspects to a sound-functioning system and concurrently involving all relevant stakeholders in RSA is intricate and challenging since no optimal strategies are laid out to assess the current and future needs of the sector at all governing levels. According to Gutberlet (2018), municipalities charged with the role of providing waste services do not have baseline data and decision-making tools to assist them in making informed decisions on SWM. Additionally, they do not have formal IWMPs or strategies, which by law, should link existent structures to governance and address SWM issues effectively. Consequently, only 524 landfills of the known 1,203 facilities are registered and even those properly registered are not managed according to stipulated standards (Niekerk & Weghmann, 2019). These planning flaws are attributable to poor prioritizing of SWM and provision of waste services that are not responsive to the community needs according to Dlamini et al. (2019).

Misinterpretation and poor understanding of existent legislative frameworks, guidelines and policies is a setback in SWM. Poor decentralization and coordination national, provincial and local laws has hampered implementation of appropriate decisions and strategies in the sector. There is inadequate cooperative governance and existent policies remain as promises with little implementation and enforcement (Zhakata et al., 2016). Existent laws also require sufficient enforcement, which in most cases lacks. Although existent laws have made efforts to control RSA's waste sector, they have created an environment where waste recovery, reuse and recycling is difficult unless new legislative requirements are included (Godfrey & Oelofse, 2017). The associated flooding of legislation has further made it difficult for the private sector to participate in SWM due to the huge compliance burden amidst efforts to become competitive locally and worldwide (Niekerk & Weghmann, 2019).

Poor financial management characterized by inadequate funding, embezzlement of funds and poor waste service recovery deters effective SWM by interfering with institutional behavior (in planning and management), equipment, infrastructure and labor management (Mannie & Bowers, 2014). Poor waste tariff allocation and financial constraints in SWM sectors of many municipalities of the country confirm the challenge as serious (Niekerk & Weghmann, 2019). The challenges have made operational expenses such as capital expenditure, employee remuneration, maintenance and fuel unrealistic to meet. Poor financial accounting of allocated funds mulled by corruption leads to poor management of the waste cycle by non-investment based on service demand. At institutional level, SWM is challenged by inconsistent waste collection schedules, unreliable services, inadequate organizational capacity, ineffective sanitation laws and an ambiguous authority line (Gutberlet, 2018).

The national government affirms that provision of waste services is a right though its treatment to the local governments that are financially constrained by rising populations and ever-increasing waste quantities is insufficient (Tsheleza et al., 2019). Poor waste services are predominant in rural areas and residents have turned to alternatives such as burning and illegal dumping that have negative environmental effects. Additionally, these flaws have resulted to unpleasant and unhealthy environments. Even in urban areas such as Johannesburg, waste services are hampered by a complex waste flow due to increased middle class citizens and informal settlements that pressure local authorities with increased generation (Dlamini et al., 2019). This challenge depicts a failure by the government to plan effectively and apply cost effective approaches to bridge the SWM gaps (Simelane, 2016). To these waste service and cost recovery challenges, improved governance and planning incorporating all stakeholders including informal waste pickers is essential (Godfrey, 2019).

Waste transport/equipment are the main cost drivers as landfills are located distances away from waste generators. Vehicles and equipment used in waste collection processes are deemed inefficient. According to Fakoya (2014), most of these equipment and vehicles are imported or donated from international sources and their maintenance costs are very high. Complex and diverse nature of waste generated exacerbate the current situation and makes separation and sorting difficult to implement. The equipment do not consider the peculiar characteristics of various waste location and their proximity to disposal sites. These trends have disabled their effective use, which often have a short lifetime. According to Thornhill (2012), the purchase of large and high number of waste vehicles, their ineffective use and overuse has resulted to misuse of funds at municipality level, which depicts institutional inefficiency. Corruption and the assumption that equipment work well in all situations has undermined the effectiveness of SWM systems in RSA (Fakoya, 2014).

SWM in RSA faces a challenge in labor management due to limited or relevant skills to manage the involved activities (DST, 2016). Although the waste sector is known to have created more than 60,000 jobs by 2016 through informal waste picking, recovery and recycling, it is absorbing unskilled labor, which is not sustainable when compared to the rising SWM demands. Inadequate human capacity to handle technical issues such as maximum space use, environmental compliance and compaction ratios at municipality level ail the country's waste sector. Although majority of unskilled labors are required due to the labor-intensive nature of SWM activities, equipping them with necessary skills would improve the management of its value chain (Godfrey et al., 2016).

WtE initiatives towards sustainable SWM also face challenges that hinder the full output of their benefits. The limited data on waste generation and unsuitable waste feedstock deters constant supply of WtE feedstock (Dlamini, 2016). Bureaucratic processes involved in purchasing electricity from other sources other than Eskom discourages the uptake of such initiatives for financial benefit (Mutezo, 2016). This is because WtE projects are not integrated to mainstream energy, SWM and planning programs. Other challenges that hamper these initiatives include limited knowledge and technological capacity by implementers, lack of political will from responsible authorities, low landfill tariffs, poor public awareness on the initiatives and limited capital investments on such projects (Maisiri, 2016).

Although the waste classification and management (R634) regulations have streamlined the amounts of landfilled waste streams, they have led to prohibitions of others. Consequently, more streams are being prohibited at landfills though production is on the rise. These measures protect both the environment and those involved in physical disposal. However, they come with increased costs of waste treatment and the need to re-plan the management of such wastes sustainably. According to Stubbs (2019), South African industries must adapt and adopt to greener products and a circular economy that shifts the focus from landfilling.

Economics of Waste Management

The national pricing strategy for waste management (NPSWM) (No. 904) by DEA (2016a) directs the costing of waste in RSA. NPSWM aims at waste reduction through its diversion from landfilling to reuse, recovery and recycling, support development of waste to a secondary economic resource and mainstreams the polluter pay principle (DEA, 2016b). Additionally, NPSWM minimizes the environmental impact associated with waste. According to this legislation, waste pricing is based on sound evidence, a cost-benefit analysis and maximum returns that promote social-economic development. The pricing strategy in RSA is based on several economic instruments in the product-to-waste value chain.

Table 7. Economic instruments (EIs) adopted in RSA SWM sector (DEA, 2016b)

Category	Instruments
Downstream instruments	1. Incineration, landfill and other disposal taxes 2. Pay-as-you-throw/ volumetric tariffs
Upstream Instruments	1. EPR fees 2. Deposit-refund system 3. Advanced disposal/recycling fees 4. Product taxes 5. Virgin/hazardous material and input taxes
Subsidy-based instruments	1. Capital financing 2. Tax rebates and benefits 3. Recycling subsidies

Traditionally, SWM and pricing in RSA was regulated by command and control (CAC) measures that streamline waste handling behaviors through predetermined standards, laws and by imposing fines and penalties (Luken & Clarence-Smith, 2019). CAC approach is highly predictable and though it has met some environmental objectives, it has not addressed economic sustainability objectives of providing equitable waste services to the poor. Due to these weaknesses, SWM in the country is shifting to the use of economic instruments (EIs) or waste management charges that focus on indirect behavioral change in SWM (GreenCape, 2017). EIs are incentives for recyclers, consumers and manufacturers to seek novel alternatives to disposal. Categories and examples of EIs adopted in RSA's SWM sector are as shown in Table 7.

Downstream charges are incurred after production and consumption processes. They are the typical charges of waste collection services that are a monthly flat rate based on location, property size and not weight/volume related. Even if the flat rate is increased, waste generators do not incur additional cost for producing more. The solution to this is the introduction of volumetric tariffs where weight of waste produced is standardized and additional volume is charged (DEA, 2016b). Landfill tipping fees could be introduced, which will increase disposal fee and make waste reduction, recycling and recovery economically viable. Benefits reaped from these downstream EIs include funding to clean up engineered sites, landfill closure and maintenance.

Upstream EIs are incurred at during transformation of raw materials to products. These charges supplement disposal taxes and volumetric tariffs that are sometimes impractical to implement due to administrative and political factors (DEA, 2016b). Taxes in this case are accessed based on upstream activities such as products' purchase to incentivize waste generators to produce less and manufacturers to use inputs that generate less waste and concurrently, reuse other products. Input and material taxes are imposed on hazardous, packaging and virgin materials. In hazardous and virgin materials, the aim is to reduce their use and encourage recycling while in packaging material, it is to discourage over-packaging using non-recyclable materials.

Product taxes are applied on end-products based on their capacity to generate waste during manufacturing processes. For instance, products that generate high non-recyclable waste are highly priced to reduce their demand. Advanced recycling fees (ARF) are similar to product taxes and aim at raising funds for downstream waste collection services. Product, material and recycling fees encourage considerable waste reduction but not significant recycling unless supplemented with EPR schemes that support infrastructure to reprocess recyclables. Importers and producers pay EPR fees (DEA, 2013). A combination of product tax and recycling subsidy system known as deposit-refund scheme is used as an upstream EI. When purchasing a product, a deposit that can be returned on repackaging, recycling or reusing the product is paid. This charge system has been implemented in beverage bottles, batteries, cars and tyres in RSA (DEA, 2016a).

Subsidies supplement both up-and-down-stream waste charges. For instance, a recycling subsidy is given as a government payment based on amount recycled or as lump sum at community-based recycling centers. Additionally, the government provides rebates and tax credits for recycling investment. Subsidies are also provided as grants to promote research and development on SWM and provide capital to fund such projects (DEA, 2016b).

EIs in RSA have grown to push waste up the hierarchy and generate revenue to fund waste services and minimize distortionary levies in SWM sector (Hettiarachchi et al., 2018). Their effective implementation however requires numerous pre-conditions such as adequate institutional management, well-functioning markets, monitoring of illegal activities and stringent enforcement of waste regulations. In RSA, where CAC mechanisms dominate local SWM, these preconditions are unlikely to be met without compromising on equity and competitiveness. Hettiarachchi et al. (2018) who evaluated the opportunities and constraints of adopting EIs in RSA's waste sector advised on the need to consider the context of individual municipal governance carefully prior to their implementation.

Initiatives Towards Sustainable Waste Management

The apparent challenges in SWM in RSA are gaining public and research attention as several counter measures are in place. Recycling projects mainly driven by social needs and demand for some resources are on the rise. In South Africa's Bloemfontein and Pretoria cities, buyback centers (BBCs), that are privately owned, have been used to improve recycling of cans, glass, plastics and paper, which has created employment for more than 300 people (Hettiarachchi et al., 2018). According to Godfrey and Oelofse (2017), formation of cooperatives as a strategy to integrate the informal SWM sector has stimulated enterprise development and job creation, but these have proven to be mostly unsustainable. A joint collaboration by Mama-She recyclers, PIKITIUP, Mondi paper and Nampak packaging companies has spearheaded paper recycling at collection centers. Consequently, paper recycling has risen from 41% to 57% between 2007 and 2015 (Godfrey & Oelofse, 2017). Through the bottle-2-bottle recycling plant, PET recycling company (PETCO), city of Johannesburg and Coca Cola company have channeled 22,000 tons of plastic waste for recycling to produce bioplastics (Infrastructure News, 2015). Although the SWM sector has been largely CAC skewed, alternative policy instruments to incentivize secondary economic resources are emerging through EPR. In these trends, the responsibility of SWM is being diversified to producers through an EPR tax by the government to enhance responsibility in processes towards waste reduction (DST, 2014; Bell & Russell, 2018). Discussions are ongoing to extend EPR schemes to WEEE, lighting, paper and packaging industries.

SWM sector in RSA is turning to public private partnerships (PPP) commonly known as municipal service partnerships (MSPs). MSPs are collaborations between private and public institutions where the former assumes considerable operational, technical and financial risk in the operation, building, financing and designing a project. For instance, Buffalo city metropolitan municipality of RSA held a waste conference in partnership with Border-Kei business chamber to exchange ideas on implementing sustainable waste management practices (Buffalo City News, 2018). Issues discussed were on how to supplement waste funding through the partnership and an exhibition was held to promote recycling waste to new products. In the city of Johannesburg, an initiative dubbed Sisonke, which was a partnership between the city, department of science and technology and Pikitup was initiated to build underground waste bins and a recycling center at Slovo park to reduce littering and encourage reuse (Gumbi, 2016). South African government has partnered with international companies such as Coca-Cola and Danish International Development Agency (DANIDA) to fund and promote waste reduction and recycling projects (Godfrey & Oelofse, 2017). Countries such as Sweden have facilitated the country to develop green economy through the shared knowledge and funding of REIPPPP program aimed at leveraging WtE opportunities (Business Sweden, 2017). The effectiveness of these MSPs is pegged on their strategic structuring and implementation with holistic stakeholder incorporation and external auditing for accountability.

CASE STUDIES

E-Thekwini Municipality

The municipality is located in Kwa-Zulu Natal province and covers an area of 2297km^2. SWM largely relies on disposal at the Buffelsdraai and Bisasar road landfills (E-Thekwini Municipality, 2016). The municipality has made great strides in moving waste up the hierarchy. First, the municipality formulated an IWMP with defined goals running between 2016 and 2021 focusing on intensified waste reduction, reuse and recycling (E-Thekwini Municipality, 2016). Towards implementing the plan, the municipality collaborated with Mondi, a pulp and paper company to form the Orange bag recycling initiative where paper wastes are collected at generation points and recycled. The municipality rehabilitated its Mariannhill landfill site, which was initially closed to be a conservancy and educational center on sustainable SWM and a source of income (E-Thekwini Municipality, 2016). This project is unique worldwide. The municipality under Durban Cleaning and Solid Waste (DSW) agency is running a WtE project at Bisasar road landfill, which produces 7.5 MW of electricity from methane gas towards CDM and climate change mitigation (Simelane, 2016). The project is expected to reduce CO_2 emissions by more than 340,000 tons annually in its first 7 years of operation. These successes are a result of good political will, effective financial management and accounting and proper engineering of landfills. Expansion and sustainability of these projects is however challenged by capital expenditure and fiscal constraints as well as their institutional mismanagement (Simelane, 2016).

West Rand District Municipality

West Rand District Municipality (WRDM) is found in the southwest region of Gauteng province and covers an area of 4,095km^2. The area is subdivided into four municipalities namely Westonaria, Merafong city, Mogale city and Randfontein (Ginindza & Muzenda, 2016). About 98% of generated waste is disposed at 4 landfills: Lebanon, Uitvaalfontein, Raipoort and Luipaardsvlei facilities and only 2% is channelled to recycling (Ginindza & Muzenda, 2016). Informal waste picking propagates SWM in the municipality, though it is unregulated and characterized by informal settling at areas near landfills. Illegal dumping due to the small size and inadequate number of waste containers is a common phenomenon. About 20% of the population is not covered by municipal waste services and in areas where it is done, residents have to contend with odors and spillages due to poor infrastructure, an overwhelmed transport system, vehicle breakdowns and spare parts non-availability (Ginindza & Muzenda, 2016). In recognition of these failures, the municipality in collaboration with the West Rand Development Agency has plans to build two buybacks centers and a recycling plant in Westonaria to reclaim disposed waste. In addition, the municipality is conducting trials for methane monitoring at Luipaardsvlei landfill in a WtE pilot project and educating the residents using street clean-up campaigns to minimize generation, illegal dumping and promote alternative waste treatment techniques (Ginindza & Muzenda, 2016). These opportunities can be enhanced to successes if integrated waste management planning is incorporated.

CONCLUSION

This book chapter makes the following conclusions:

- SWM is influenced by technological, economic, institutional, legal, socio-cultural and environmental factors and due to this multifaceted nature; it requires proper planning and governance.
- In RSA, SWM is a mandate of municipal governments that face infrastructural, institutional, human capacity, labor and financial management challenges propagated by drivers to waste generation.
- Involvement of the private sector as well as international partners and integrating waste pickers could be a viable strategy towards sustainable SWM.
- In evaluated case studies, counter measures to SWM challenges are now opportunities to renewable power sources through WtE initiatives and sources of employment.
- As such, the country is moving from a linear economy based on production, consumption and disposal to a circular economic outlook that prioritizes on recovery, reuse and recycling.
- To reap the benefits of sustainable SWM, RSA must inculcate strategic planning, invest more in waste recovery/recycling projects and use an integrated management system.

REFERENCES

Abdel-Shafy, H., & Mansour, M. (2018). Solid waste issue: Sources, composition, disposal, recycling and valorisation. *Egyptian Journal of Petroleum*, *27*(4), 1275–1290. doi:10.1016/j.ejpe.2018.07.003

Ahsan, A., Alamgir, M., El-Sergany, M. M., Shams, S., Rowshon, M. K., & Daud, N. N. N. (2014). Assessment of Municipal Solid Waste Management System in a Developing Country. *Chinese Journal of Engineering*, *14*, 1–11. doi:10.1155/2014/561935

Anton, E., & Raine, N. (2016). The South African renewable energy independent power producer procurement programme: A review and lessons learned. *Journal of Energy in Southern Africa*, *27*(4), 1–14. doi:10.17159/2413-3051/2016/v27i4a1483

Bell, R. G., & Russell, C. (2018). Environmental Policy for Developing Countries. In G. E. Helfand & P. Berck (Eds.), *The Theory and Practice of Command and Control in Environmental Policy* (1st ed.; pp. 239–246). By P. Berck. doi:10.4324/9781315197296-10

Bello, I., Ismail, M., & Kabbashi, N. (2016). Solid waste management in Africa: A review. *International Journal of Waste Resources*, *2*, 216–221.

Blaauw, P. (2017). Informal employment in South Africa: Still missing pieces in the vulnerability puzzle. *Southern African Business Review*, *21*, 339–360.

Buffalo City News. (2018). *Partnering for sustainable waste management*. Retrieved 8 July 2019 from: https://infrastructurenews.co.za/2018/06/27/partnering-for-sustainable-waste-management/

Business Sweden. (2017). *High potential opportunity waste to energy South Africa: Energy for Africa*. Retrieved 5 May 2019 from https://www.business-sweden.se/globalassets/international-markets1/reports-and-facts---markets/waste-to-energy-in-south-africa.pdf

Chen, Y. (2018). Effects of urbanisation of municipal solid waste composition. *Waste Management (New York, N.Y.)*, *79*, 828–836. doi:10.1016/j.wasman.2018.04.017 PMID:30078496

DEA. (2011). *South Africa state of waste report, first draft. DEA*. Pretoria: RSA.

DEA. (2013). *National norms and standards for disposal of waste to landfill*. Retrieved from https://www.environment.gov.za/sites/default/files/gazetted_notices/nemwa59of2008_norms_standards_fordisposa_0.pdf

DEA. (2016a). *Policy brief 8: Transitioning South Africa to a green economy: opportunities for green jobs in the waste sector*. Retrieved 15 July 2019 from: https://www.sagreenfund.org.za/wordpress/wp-content/uploads/2016/11/Policy-Brief-No-8.pdf

DEA. (2016b). *National pricing strategy for waste management*. Retrieved 7 July 2019 from: https://cer.org.za/wp-content/uploads/2010/03/National-Pricing-Strategy-for-Waste-Management.pdf

DEA. (2018). *South Africa state of waste. A report on the state of the environment. First draft report*. Pretoria: RSA.

DEA. (2019). *National waste management strategy*. Retrieved on 6 July 2019 from: https://www.environment.gov.za/documents/strategicdocuments/wastemanagement

Department of Cooperative Governance and Traditional Affairs (COGTA). (2015). *Integrated urban development framework: Draft for discussion. COGTA*. Pretoria: RSA.

Department of Environmental Affairs. (2011). *National waste management strategy. DEA*. Pretoria: RSA.

Department of Science and Technology. (2012). *A National Waste R&D and Innovation Roadmap for South Africa: Phase 1 Status Quo Assessment. Current and required institutional mechanisms to support waste innovation*. Department of Science and Technology.

Department of Science and Technology (DST). (2014). *A National Waste R&D and Innovation Roadmap for South Africa: Phase 2 Waste RDI Roadmap. Trends in waste management and priority waste streams for the Waste RDI Roadmap*. Department of Science and Technology.

Department of Water Affairs (DWA). (1998). *National Water Act*. Retrieved 8 August 2019 from: http://www.dwa.gov.za/Documents/Legislature/nw_act/NWA.pdf

Dlamini, S. (2016). *Solid waste management in South Africa: Exploring the role of the informal sector in solid waste recycling in Johannesburg* (MSc. Thesis). University of Witwatersrand, RSA.

Dlamini, S., Simatele, M., & Kubanza, M. (2019). Municipal solid waste management in South Africa: From waste-to-energy recovery through waste-to-energy technologies in Johannesburg. *Local Environment, 24*(3), 249–257. doi:10.1080/13549839.2018.1561656

DST. (2016). *A waste RDI roadmap for South Africa*. Retrieved from https://www.wasteroadmap.co.za/download/presentation_06.pdf

E-Thekwini Municipality. (2016). *E-Thekwini municipality integrated waste management plan 2016-2021 (*Project no. 4024-022*)*. Durban, RSA: Author.

Fakoya, M. (2014). Institutional challenges to municipal waste management service delivery in South Africa. *Journal of Human Ecology (Delhi, India), 45*(2), 119–125. doi:10.1080/09709274.2014.11906685

Garces-Ayerbe, C., Rivera-Torres, P., Ines, S., & Dante, L. (2019). Is it possible to change from a linear to a circular economy? An overview of opportunities and barriers for European small and medium sized enterprise companies. *International Journal of Environmental Research and Public Health, 16*(5), 851. doi:10.3390/ijerph16050851 PMID:30857193

Ginindza, B., & Muzenda, E. (2016). *Waste management challenges to opportunities in the west rand district municipality, Gauteng, South Africa: initiatives*. Johannesburg: RSA.

Godfrey, L. (2019). Waste plastic, the challenge facing developing countries-ban it, change it, collect it? *Recycling, 4*(3), 1–6.

Godfrey, L., & Oelofse, S. (2017). Historical Review of Waste Management and Recycling in South Africa. *Resources, 6*(4), 57. doi:10.3390/resources6040057

Godfrey, L., Strydom, W., & Phukubye, R. (2016). *Integrating the informal sector into the South African waste and recycling economy in the context of extended producer responsibility*. Retrieved on 12 July 2019 from: https://www.wasteroadmap.co.za/download/informal_sector_2016.pdf

Government Gazette. (2014). *Act No. 26: National Environmental Management: Waste Act (2014)*. Retrieved 6 July 2017 from: https://cer.org.za/wp-content/uploads/2010/03/Waste-Amenment-Act-2014.pdf

GreenCape. (2017). *Waste economy-2017: A market Intelligence Report. GreenCape*. Cape Town: RSA.

Gumbi, S. (2016). *Current waste management and minimisation patterns and practices: an exploratory study on the Ekurhuleni metropolitan municipality in South Africa* (MSc Thesis). University of South Africa, RSA.

Gutberlet, J. (2018). *Waste in the city: Challenges and opportunities for urban agglomerations, urban agglomeration Mustafa Ergen*. IntechOpen. Retrieved 09 July 2019 from: https://www.intechopen.com/books/urban-agglomera

Hettiarachchi, H., Meegoda, J., & Ryu, S. (2018). Organic waste buyback as a viable method to enhance sustainable municipal solid waste management in developing countries. *International Journal of Environmental Research and Public Health, 15*(11), 1–15. doi:10.3390/ijerph15112483 PMID:30405058

Infrastructure News. (2015). *First bottle-2-bottle plant in Africa*. Retrieved 14 July 2019 from: https://infrastructurenews.co.za/2015/05/12/first-bottle-2-bottle-plant-in-africa/

Luken, R., & Clarence-Smith, E. (2019). *Green industrialisation in sub-Saharan Africa: A guide for policy makers. The Institute of African Leadership for Sustainable Development.* Tanzania: Uongozi Institute.

Maisiri, W. (2016). *A techno-economic evaluation of a waste-to-energy grate incineration power plant for a small South African city* (MSc. Thesis). North West University, RSA.

Mannie, N. M., & Bowers, A. (2014). Challenges in Determining the Correct Waste Disposal Solutions for Local Municipalities - A South African Overview. *Proceedings of the 20th wastecon conference.*

Mutezo, G. (2016). *Challenges impeding South African municipalities from adopting waste-to-energy schemes: an exploratory approach* (Mphil Thesis). University of Cape Town, RSA.

Muzenda, E. (2014). A Discussion on Waste Generation and Management Trends in South Africa. International Journal of Chemical. *Environmental and Biological Sciences*, *2*(2), 105–113.

Muzenda, E., & Popa, C. (2015). Waste tyre management in Gauteng, South Africa: Government, industry and community perceptions. *International Journal of Environmental Sciences and Development*, *6*(4), 311–315. doi:10.7763/IJESD.2015.V6.609

Niekerk, S., & Weghmann, V. (2019). *Municipal solid waste management services in Africa.* Working Paper, Public Service International, Johannesburg, RSA.

Njoku, P., Odiyo, J., Durowoju, O., & Edokpayi, J. (2018). A review of landfill gas generation and utilisation in Africa. *Open Environmental Sciences*, *10*(1), 1–15. doi:10.2174/1876325101810010001

Oelofse, S. (2018). Opportunities and challenges in waste management. *CSIR*. Retrieved 9 July 2019 from: http://sustainabilityweek.co.za/assets/files/Day%203%20-%20CSIR%20-%20Waste.pdf

SAWIC. (2015). *Waste policy and regulations*. Retrieved 12 July 2019 from: http://sawic.environment.gov.za/?menu=13

SAWIC. (2016). *Approach to waste in South Africa*. Retrieved 5 July 2019 from: http://sawic.environment.gov.za/?menu=60

SAWIC. (2018). *Waste information today: reporting on South African SAWIS*. Retrieved 10 July 2019 from: http://sawic.environment.gov.za/documents/8545.pdf

Schenck, R., Blaauw, D., & Vilijoen, K. (2016). Enabling factors for the existence of waste pickers: A systematic review. *Social Work*, *52*(1), 35–51.

Sentime, K. (2014). The impact of legislative framework governing waste management and collection in South Africa. *African Geographical Review*, *33*(1), 81–93. doi:10.1080/19376812.2013.847253

Simelane, O. (2016). Implementing Alternative Waste Treatment Technologies: A Comparative Study of South Africa and the European Union. *South Africa, 11*(7), 427-430.

Singh, J. (2016). Ranking South African provinces on the basis of MERRA 2D surface incident shortwave flux. *Journal of Energy in Southern Africa*, *27*(3), 50–57. doi:10.17159/2413-3051/2016/v27i3a1555

Singh, J., Laurenti, R., Sinha, R., & Frostell, B. (2014). Progress and challenges to the global waste management system. *Waste Management & Research, 32*(9), 800–812. doi:10.1177/0734242X14537868 PMID:24938296

South African Waste Information Centre (SAWIC). (2013). *Waste classification and management regulations, norms and standards*. Retrieved 13 July 2019 from: http://sawic.environment.gov.za/documents/2177.pdf

Statistics South Africa (StatsSA). (2017a). *Mid-year population estimates (statistical release P0302), Statistics South Africa*. Pretoria: RSA.

Stats, S. A. (2017b). *Living conditions of households in South Africa 2014/2015 (Statistical release P0310) Statistics South Africa*. Pretoria: RSA.

Stubs, K. (2019). *South African industry forced to adapt as waste streams are prohibited from landfill disposal.* Retrieved 14 July 2019 from: https://blog.interwaste.co.za/know-waste/sa-industry-forced-to-adapt-as-waste-streams-are-prohibited-from-landfill-disposal

Sustainable Energy Africa. (2017). *Sustainable energy solutions for South African local government: A practical guide*. Cape Town: RSA.

Tan, S., Ho, W., Hashim, H., Lee, C., Taib, M., & Ho, C. (2015). Energy, economic and environmental (3E) analysis of waste-to-energy (WTE) strategies for municipal solid waste (MSW) management in Malaysia. *Energy Conversion and Management, 102*, 111–120. doi:10.1016/j.enconman.2015.02.010

Thornhill, C. (2012). Improving local government transformation in South Africa. *Administratio Publica, 20*(3), 128–143.

Tsheleza, V., Ndhleve, S., Kabiti, H., Musampa, C., & Nakin, M. (2019). Vulnerability of growing cities to solid waste related environmental hazards: The case of Mthatha. *Jamba: Journal of Disaster Risk Studies, 11*(1), 1–11. PMID:31205617

United Nations. (2015). *World urbanisation prospects: The 2014 revision report (No. ST/ESA/SER-A/366)*. New York: United Nations.

World Bank. (2019). *What a waste 2.0: A global snapshot of solid waste management to 2050*. Retrieved 4 July 2019 from: http://datatopics.worldbank.org/whatwaste/trends_in_solid_waste_management.html

Zhakata, E., Gundani, S. R., Chauke, V., & Odeku, K. O. (2016). A Critic of NEMA: Waste Act 59 of 2008, so Many Promises, Little Implementation and Enforcement. In *SAAPAM Limpopo chapter 5th annual conference proceedings*. Limpopo: RSA.

KEY TERMS AND DEFINITIONS

End-of-Pipe: Practices that promote waste pre-processing prior to disposal through chemical treatment, recycling, and burning.

Environment: The surroundings in which human activities that interact with plants, animals, and the natural world occur.

IWMP: A layout that describes activities to optimise waste management efficiently and concurrently minimize its associated financial costs and environmental impacts.

Landfill: Disposal areas or a system of trashing where waste is buried in engineered facilities.

Solid Waste: Any refuse from discarded materials and facilities of air pollution prevention and waste treatment plants resulting from agriculture, domestic, industrial, mining, and commercial activities.

SWM: Systematic processes involved in the production, collection, transport, disposal, treatment, and reclamation of solid waste to minimize its resultant environmental effects.

Tetratogens and Carcinogens: Agents that cause abnormal formation of unborn babies and are cancer-cancer causatives.

Waste-to-Energy: Processes that lead to power (heat or electricity) generation from waste treatment.

ENDNOTES

1 National Environmental Management: Waste Act

2 Persistent Organic Pollutants

3 Waste Electrical and Electronic Equipment

4 Health Care Risk Waste

5 Danish Cooperation for Environment and Development

6 Polychlorinated Biphenyls

7 Landfills that are strictly designed, operated and monitored

8 Landfills whose operations and management is not strictly monitored

9 Megawatts

10 The public electricity utility in RSA

Chapter 15
Sustainable Waste Management Challenges in Sri Lanka

B. F. A. Basnayake
Faculty of Agriculture, University of Peradeniya, Sri Lanka

R. T. K. Ariyawansha
Postgraduate Institute of Agriculture, University of Peradeniya, Sri Lanka

A. K. Karunarathna
Faculty of Agriculture, University of Peradeniya, Sri Lanka

S. M. Werahera
Ministry of Mahaweli Development and Environment, Sri Lanka

N. Mannapperuma
Waste Management Authority-Western Province, Sri Lanka

ABSTRACT

Sri Lanka faced many catastrophes in managing wastes because of human errors and climate change impacts. Consequently, government administrators with stakeholder participation drafted a comprehensive policy on all aspects of waste management and now requires planning processes in the national, provincial, and local authority levels. Evidenced-Based Interactive Database is under construction. It will be a platform for all actors to participate in managing all types of wastes, inclusive of quantity and quality of wastes. This chapter describes municipal solid waste management, including present status of technology applications with the emphasis on 3R to divert wastes from landfills. It contains reports on e-wastes collection programs and some processing, polythene recycling, hazardous wastes management, including healthcare wastes, disaster and construction and demolition wastes, informal sector and recyclers, economics of waste management, and case studies. Future directives are given with coupling of intended nationally determined contributions for sustainable waste management.

DOI: 10.4018/978-1-7998-0198-6.ch015

INTRODUCTION

The Democratic Socialist Republic of Sri Lanka is an island in the Indian Ocean southwest of the Bay of Bengal, between 5° 55' - 9° 50' North latitude and 79° 42' - 81° 53' East longitude of the equator below the Indian subcontinent. It is separated from the Indian subcontinent by the Gulf of Mannar and the Palk Strait. It is located at a global logistics hub at an intersection with major air and sea routes between Europe and the Far East. Its span is 447 km from north to south and 219 km from east to west. The island has nine provinces and 25 administrative districts as shown in Figure 1 with a total area of land and water of 65,610 km^2 and 62,705 km^2 makes up land area and the rest 2,905 km^2 consists of inland waters. Sri Lanka has a tropical climate and it consists of three major climatic zones namely wet zone, intermediate zone, and dry zone. The main industries consist of processing tea, rubber, coconuts, other agricultural commodities, fishing, telecommunications, insurance, banking, tourism, shipping, clothing, textiles, cement, petroleum refining, information technology services, and construction. The government system consists of a decentralized structure with 9 Provincial Councils (PCs) and 341 Local Authorities (LAs) made up of 24 Municipal Councils (MCs) which corresponds to a city, 41 Urban Councils (UCs) for semi-developed areas, and 276 Pradesheeya Sabhas (PSs) for rural areas. The categorization of LAs is according to its population and size. The LAs are responsible for providing a variety of local public services including sanitation and waste collection.

Figure 1. Provinces and districts of Sri Lanka (Nahallage, 2013)

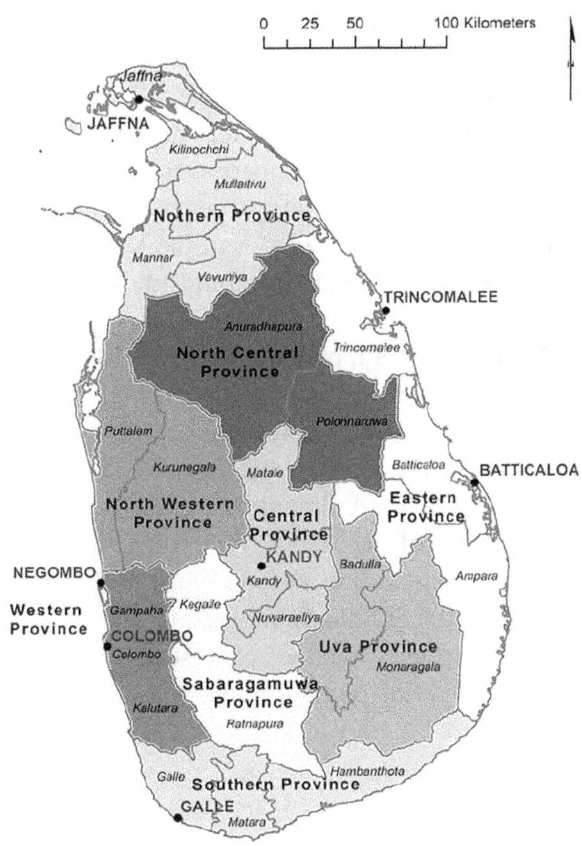

The legal framework of solid waste management in Sri Lanka consists of five major components namely constitution, Acts and Legislations passed by the Parliament, such as National Environmental Act, Pradeshiya Sabha Act, Municipal Council Ordinance, etc. as well as Provincial Statutes, regulation and guidelines and by-laws of the LAs. In Sri Lanka, the basic legal framework required for solid waste management is provided under an umbrella of Government, Provincial Council and LA regulations and legislation. The 13th Amendment to the constitution (1987) and the Provincial Councils Act No. 42 of 1987, Sections 129, 130 and 131 of the Municipal Councils Ordinance (1980), Sections 118, 119 and 120 of the Urban Councils Ordinance, No. 61 of 1989, Sections 41 and 93 to 95 of the PS Act, No. 15 of 1987 and the National Environmental Act (NEA) are the key pieces of legislation governing solid waste management. According to the MC Ordinance, the Urban Councils (UC) Ordinance and the Pradeshiya Sabha (PS) Act, all Municipal Solid Waste (MSW) generated within the boundary of LAs is their property and they are mandated to remove and dispose of such waste materials without causing any nuisance to the public (Vidanaarachchi et al., 2006). At the national level, the responsibility for policies and implementation of plans concerning LAs lies with the Ministry of Internal & Home Affairs and Provincial Councils & Local Government. The formulation of national-level policies for solid waste management is conducted by the Ministry of Mahaweli Development & Environment (MoMD&E), while practical regulatory control and management are undertaken by the Central Environment Authority (CEA). The CEA was established under the National Environmental Act No. 47 of 1980 and is entrusted with the protection, management, and enhancement of the quality of the environment, for the prevention, abatement and control of pollution and with matters connected therewith. This principal Act was subsequently amended in 1988 and in 2000. In 2007, prevention of mosquitos breeding act (Act No. 11 of 2007) was enforced with the aim of ensuring that conditions are unfavorable to the breeding of mosquitoes. The pollution control regulations related to solid waste management that were enforced in the country from 1984 to 2017 are given in Table 1. The Ministry of Megapolis and Western Development (MoMWD) is a new ministry appointed by the Cabinet of Sri Lanka after August 2015 elections and it is responsible for planning for the megapolis development. The Ministry is in charge of planning solutions to resolve garbage management, housing of shanty dwellers and issues related to their needs, as well as drawing new traffic plans to avoid traffic jams in busy towns. And the Ministry of Health, Nutrition, and Indigenous Medicine (MoH) also provides and leads in administrative guidance from the perspective of health and sanitation.

The National Solid Waste Management Support Center (NSWMSC) which was established by the Ministry of Local Government and Provincial Councils in 2007 (JICA & KKC, 2016) assists LAs to improve the solid waste management in administrative areas of LAs through technical, legal, and financial assistance (MoIHA&PCLG, 2019). Waste Management Authority (WMA) of the Western Province (WP) was established in 2004 and The WMA Statute No.1 was formulated in 2007 and it specifies jurisdiction, function and responsibility of the WMA. According to the statute, WMA is responsible for providing technical and financial assistance to all LAs of the WP to build their capacities in SWM, collecting waste data in WP, developing common final disposal sites to LAs and also assisting LAs to publicize and inculcate the public with better methods of waste management discipline (WMA, 2019). This chapter is aimed to provide information on the present status and future directions of waste management in Sri Lanka.

Table 1. Pollution control regulations related to solid waste management

Year	Regulation no	Description
1984	SLS 614 of 1983	Sri Lanka standards specifications for portable water. Prescribes the physical, chemical, and bacteriological requirements, and methods of sampling and test for drinking water.
	SLS 652 of 1984	Tolerance limits for industrial effluents discharged into inland surface waters
1984/1985	SLS 722 of 1984/1985	Tolerance limits for inland surface waters for use as raw water for public water supply
1990	Gazette Extraordinary No. 595/16 of 1990	National environmental protection and quality regulations. Specify content and specifications of EPL, the tolerance limits for wastewater discharge from major industries and activities. Amended by Gazette Extraordinary No. 924/13 of 1996, No. 1159/22 of 2000, No. 1533/16 of 2008. Scheduled waste management license (SWML).
1993	Gazette Extraordinary No. 772/22 of 1993	National environmental regulations (procedure for approval projects). Specify project approval agencies, projects for IEE/EIA approvals. Amended by Gazette Extraordinary No. 859/14 of 1995 and No. 1104/22 of 1999
1994	Gazette Extraordinary No. 850/4 of 1994	National environmental regulations on ambient air quality. Amended by Gazette Extraordinary No. 156/22 of 2008
1996	Gazette Extraordinary No. of 924/12 of 1996	National environmental noise regulations
2000	Gazette Extraordinary No. 1137/35 of 2003	National environmental regulations for mobile air quality
2003	SLSI 1246 of 2003	Sri Lanka standard for compost from municipal solid waste and agricultural waste
2006	SLS 1292 of 2006	Code of practice for design and construction of biogas systems
	Gazette Extraordinary No. 1466/5 of 2006	Order to prohibit the manufacture of polythene or any polythene product of twenty (20) microns or below in thickness for in country use
2008	Gazette Extraordinary No. 1534/18 of 2008	National environmental protection and quality regulations Management of scheduled waste
2009	No. 1627/9 National Environmental (Municipal solid waste) Regulation No 1 of 2009	General rules on SWM discharge and collection of waste (prohibition of waste dumping at national highway and at any place other than places designated for such purpose by the LA
2017	Gazette Extraordinary No. 2034/33 of 2017	Order to prohibit the manufacture of polythene or any polythene product of 20 microns or below in thickness for in country use (may be permitted to be used with the prior written approval of the Authority for specified purposes by the order.
	Gazette Extraordinary No. 2034/34 of 2017	Order to prohibit the manufacture of food wrappers (lunch sheet) etc.
	Gazette Extraordinary No. 2034/35 of 2017	Order to prohibit the manufacture of any bag of high density polyethylene as a raw material for in country use
	Gazette Extraordinary No. 2034/36 of 2017	Order to prohibit burn openly or cause to, allow or permit the open burning of refuse or other combustible matters inclusive of plastics
	Gazette Extraordinary No. 2034/37 of 2017	Order to prohibit the use of all forms of polyethylene, polypropylene, polyethylene products or polypropylene products as decorations
	Gazette Extraordinary No. 2034/38 of 2017	Order to prohibit the manufacture of food containers, plates, cups and spoons from expanded polystyrene for in country use

WASTE DEFINITIONS, POPULATION, WASTE GENERATION, AND CHARACTERISTICS

Waste Definitions

The term wastes (recently given the equivalent term residuals) refers to the materials which are discarded by community activities to the environment and includes solid, liquids, and gases (Barton, 1979). A waste is a material which is thrown away or put aside as worthless. The definition of ''Solid'' waste encompasses all of those wastes which are neither wastewater discharges nor atmospheric emissions. A so-called solid waste may, therefore, be a semi-solid, solid or even a liquid. Refuse normally means ''solid wastes'' but the term is avoided here. Garbage is domestic or household food wastes, rubbish is the domestic non-food waste, and residential or domestic waste is the combination of garbage and rubbish. Urban or MSW is taken as including all waste collected by LAs other than sewage, that is the residential waste, street waste, and other municipal waste, plus commercial waste (Barton, 1979). In Sri Lanka, solid waste is categorized mainly into municipal solid waste, health-care waste (HCW), hazardous waste (HW) (JICA & KKC, 2016), and agricultural waste.

Hazardous wastes can be stable or unstable in terms of changing phase from solids, sludges, liquids, containerized gases and contaminated ambient air with chemicals or biological substances other than radioactive wastes which, by reason of their chemical or biochemical activity or toxic, explosive, corrosive, infectious or other characteristics, cause danger or significantly contribute to an increase in mortality or an increase in serious irreversible, or incapacitating reversible illness, or pose a substantial present or potential hazard to human health or the environment whether alone or when coming in to contact with other waste or organisms and when improperly treated, stored, transported, or disposal of, otherwise managed. In the actual disposal and treatment, HCW is divided into either municipal solid waste (non-hazardous waste) or hazardous waste. One of the hazardous wastes is clinical wastes. Clinical waste means waste that has the potential to cause disease, including the following: animal waste, discarded sharps, human tissue waste, laboratory waste.

Population and Waste Generation

Sri Lanka is a Lower Middle-Income country with a GDP per capita of USD 4,065 in 2017 and a total population of 21.444 million people. The population growth rate was 1.1%. Average population density is 342 persons/km^2 (CB of SL, 2018). The country is developing with increases in population that is generating wastes in the urban centers and rural villages. About 26% of the total population, which is 5.8 million people live in the WP. It is also estimated that the floating population in the Western Province is about one million per day (WMA, 2019).

As reported by MoMD&E, 2016 and Basnayake et al., 2018, the available databases (CEA, 2015; NSWMSC, 2015) of eight provinces were updated by considering reports and publications on collection and generation of wastes. The analysis was based on existing resources for collection, recycling, and disposal of the wastes in each of the LAs. The outliers were corrected by interviewing officials and comparing with the neighbouring LAs and geographical population densities. It entailed an in-depth study on the influence of the transient populations on waste generations. The study included many schools, government institutions, universities, industries, and commercial establishments which produce waste and these were considered and evaluated in the final findings. Total MSW generation per day of

Table 2. Current status of MSW management in relation to population in different districts

Province	District	Land area (km^2)	Inland water area (km^2)	Total Area (km^2)	Population*		MSW Generation			Collection	Recycling		Available resources		
					No	Population density (km^2)	Quantity (tons/day)	Quantity (kg/day/person)	Quantity (kg /capita/year)	Quantity(tons/day)	Recyclables (tons/day)	For composting (tons/day)	Collection crew (No)	Compactors (No)	Tractors (No)
Western**	Colombo[1]	676	23	699	2,419,000	3,578	2,050	0.85	309	1,450	90	21	1,808	135	202
	Gampaha[1]	1,341	46	1,387	2,391,000	1,783	950	0.40	145	450	75	34	1,039	5	185
	Kaluthara[1]	1,576	22	1,598	1,271,000	806	375	0.30	108	170	21	35	589	0	99
Central	Kandy[1,2]	1,917	23	1,940	1,452,000	757	755	0.52	182	262	99	17	690	5	92
	Matale[2,3]	1,952	41	1,993	514,000	263	237	0.46	164	58	23	8	203	1	33
	Nuwara Eliya[1,2]	1,706	35	1,741	756,000	443	362	0.48	167	54	26	5	284	0	27
Southern	Matara[1]	1,270	13	1,283	851,000	670	379	0.45	159	102	110	58	449	0	75
	Galle[1]	1,617	35	1,652	1,113,000	688	505	0.45	169	120	60	13	303	1	101
	Hambantota[2,3]	2,496	113	2,609	647,000	259	276	0.43	184	50	21	32	240	0	49
Northern	Jaffna[3]	929	96	1,025	608,000	654	297	0.49	163	99	27	11	765	1	132
	Kilinochchi[3]	1,205	74	1,279	124,000	103	46	0.37	128	27	1	0	21	0	24
	Mannar[3]	1,880	116	1,996	107,000	57	68	0.64	186	41	3	0	81	1	25
	Mullaitivu[3]	2,491	202	2,693	96,000	39	55	0.57	135	7	0	0	27	0	12
	Vavunia[3]	1,785	106	1,891	184,000	103	118	0.64	191	19	9	0	85	4	41
North Central	Anuradhapura[3]	6,664	515	7,179	918,000	138	438	0.48	160	78	32	46.5	262	0	74
	Polonnaruwa[3]	3,077	216	3,293	431,000	140	180	0.42	143	43	7	12	116	0	33
North Western	Kurunegala[2,3]	4,624	192	4,816	1,694,000	366	772	0.46	189	123	162	88	546	6	86
	Puttalam[2,3]	2,882	190	3,072	814,000	282	396	0.49	169	77	61	30	299	0	63
Eastern	Batticaloa[3]	2,610	244	2,854	560,000	215	296	0.53	175	126	19	9	-	4	76
	Ampara[2,3]	4,222	193	4,415	705,000	167	332	0.47	165	137	27	18	206	5	93
	Trincomalee[3]	2,529	198	2,727	412,000	163	234	0.57	163	69	23	0	197	4	50
Sabaragamuwa	Ratnapura[1,2]	3,236	39	3,275	1,151,000	356	517	0.45	162	120	43	11	185	1	68
	Kegalle[1]	1,685	8	1,693	877,000	520	366	0.42	209	62	31	34	230	0	52
Uva	Badulla[2,3]	2,827	34	2,861	864,000	306	402	0.47	185	92	33	50	320	0	48
	Monaragala[2,3]	5,508	131	5,639	485,000	88	179	0.37	139	36	11	4	160	0	28
Total		62,705	2,905	65,610	21,444,000	342	7,210	0.49	170	3,872	1,014	537	9,105	173	1,768

Note: Adopted from Basnayake et al., 2018; * - CB of SL, 2018; ** - WMA, 2019; 1- Wet zone; 2-Intermediate zone; 3- Dry zone

the country is 7,210 tons/day out of which 3,982 tons/day is being collected as given in Table 2. The average per capita waste generation is 0.49 kg/day as given in Table 2. MCs are generating significantly more wastes than UCs and PSs. Per capita waste generation in MCs in Sri Lanka is around 0.75 kg/day, in UCs is around 0.5 kg/day, whereas in PSs is around 0.4 kg/day. Currently, per capita MSW generation in Colombo Municipal Council (CMC), is 1 kg, which is much higher than the national average (WMA, 2019).

Waste Composition and Characteristics

Accurate information is a pre-requisite of the composition of the generated wastes and other physical and chemical properties for effective management of waste. In Sri Lanka, more than 60% of MSW is biodegradable organic matter (Basnayake & Visvanathan, 2014). The comparison of compositions between cities in the wet and dry zones is given in Table 3. In the Western province, 60% of the composition is bio-degradable waste. Generation of recyclable waste, which contains papers, glass, plastics, and polythene, is also significant and it accounts for about 20 - 21%. The rest of the composition contains 10% - 20% of residual waste (WMA, 2019). Similar compositions to Kandy are found in many of the wet-zone cities and towns. In the dry-zone cities, the percentage of garden wastes are more, but the sand content is very noticeable compared to the wet zone. Moreover, the food waste is given to rearing animals and also decomposes rapidly in these locations. In the recent past, it was made mandatory to use biodegradable plastics of much higher thickness. It has caused considerable impacts on the recycling industry. According to some of the recyclers, the biodegradable plastics amount to 40% of food wrapping (lunch sheets) and day-to-day use bags. It is noteworthy that most of the recyclers are not accepting them (Basnayake et al., 2018). For estimating energy and power generations, the composition studies and energy contents in wastes are essential quality parameters. As reported by Basnayake et al., 2018, proximate analysis of the combustible fraction of MSW (MSW fuel) is given in Table 4 and high heating value (HHV) of 1 kg of waste is given in Table 5.

Table 3. Composition of MSW in LAs

Type of wastes	Percentage (wet basis)															
	Western Province					Central Province		North Western	Northern	Eastern Province					Uva	North Central
	Average[1]	MC[1]	UC[1]	PS[1]	MMC[2]	KMC[3]	NMC[2]	KGMC[2]	JMC[2]	KAMC[2*]	SPS[4*]	KAMC[4#]	SPS[4#]	TUC[4]	KPS[2]	TPS[2]
Combustibles																
Biodegradable*	50	50	61	68	52.5	59.2	74.6	52	68.9	7.1	9.5	0.0	0.0	12	63.3	51.7
Biodegradable**	10	17	2.8	3.2	14.1	24.2	4.8	16	15.1	76.0	68.8	71.7	39.9	37	18.3	15.6
Paper	9	8.5	11	5.6	13.6	2.3	7.8	15.7	2.7	2.8	3.9	1.5	6.3	4	7.1	16.1
Cardboard						3.2				1.3	1.7	1.5	4.8			
Textiles	5				4.3	0.5	1	5.9	1.3	0.7	0.6	1.2	4.4	3	0.6	2.7
Hard plastics	8	16	5.6	5.1	1.4	0.7	0.9	1.4	0.4	3.2	4.3	13.2	9.1	1	0.5	1.4
Soft plastics					9.1	4.7	4.2	4.8	3.6					2	3.9	8.1
Nylon										0.0	0.1	0.0	0.0			
Rubber					0.6	0.8	0.4	0.3	0.2	0.2	0.1	0.5	0.1	1	0.2	1.1
Non-Combustibles																
PVC						0.1										
Glass	4	1.2	3.3	1	1.2	1.6	1.7	0.3	0.4	1.6	1.2	0.8	2.4	2	0.6	0.4
Metal	1	1.8	3.7	2.4	0.7	0.8	0.9	2	0.3	0.4	0.4	1.0	1.0	1	0.2	1.6
Construction & demolition						1.6				5.6	6.7	1.6	26.8			
Stones & ceramic					2.3		0.5	1.1	7.1					15	5.4	1.3
E-wastes						0.0				0.1	0.3	6.8	0.0			
Clinical						0.0				0.0	0.0	0.0	5.1			
Others	13	5.7	13	15	0.1	0.5	3.2	0.6	0.0	1.2	2.3	0.3	0.3	22.0	0.0	0.0
Total	100	100	100	100	100	100	100	100	100	100	100	100	100	100	100	100

Note: MC - MCs of the Western Province in 2019; UC- UCs of the Western Province in 2019 ; PS - PSs of the Western Province in 2019; MMC - Moratuwa MC in 2015; KMC - Kandy MC in 2007; NMC - Nuwara Eliya MC in 2015; KGMC - Kurunagala MC in 2008; JMC - Jaffna MC in 2015; KAMC - Kalmunai MC in 2009; KPS - Sammanthurai PS in 2009, TUC - Trincomalee UC in 2010; KPS Katharagama PS in 2015; TPS - Tamankaduwa PS in 2015; * - Household generated; # - Collected within LA; 1 - WMA, 2019; 2 -JICA & KKC, 2016; 3 - Menikpura et al., 2007; 4 - Ariyawansha et al., 2010a

Table 4. Proximate analysis of MSW fuel (combustible fraction) (Basnayake et al., 2018)

Parameter	Percentage				
	Karadiyana	Colombo	Kandy	Average	SD
Moisture*	7.85	7.91	8.73	8.16	0.49
Volatile matter	83.14	82.44	80.99	82.19	1.10
Ash	9.00	9.65	10.29	9.65	0.64
Fixed ash	4-6	4-6	4-6	5	

*Equilibrium moisture content for biodegradable fraction was 12% and plastics at 0%

Note: Feasibility studies were conducted to ascertain possible power generations from combustible fraction of MSW. In each of the cities, representative samples were obtained to determine the composition and proximate analysis. These results were used along with the published calorific values (Menikpura & Basnayake, 2009) to determine the power generations.

Table 5. High heating value (HHV) of 1kg of waste (Basnayake et al., 2018)

Type of Wastes	Dry Matter (%)				HHV[1,2]	HHV (kJ /kg)
	Karadiyana	Colombo	Kandy	Average	(kJ/kg)	Average
Biodegradable*	31.96	42.50	40.86	38.44	18,400	7,072
Biodegradable**	16.57	15.33	25.12	19.01	14,000	2,661
Coconut shells	20.12	8.46	10.06	12.88	22,335	2,876
Paper	8.08	9.55	4.37	7.33	17,000	1,247
Cardboard	2.19	2.92	6.18	3.76	16,360	616
Textiles	2.98	3.97	0.97	2.64	11,531	304
Plastic 1	13.83	11.58	8.99	11.47	33,300	3,819
Plastic 2	2.77	3.69	1.37	2.61	45,000	1,174
Nylon	1.26	1.68	0.54	1.16	24,200	281
Rubber	0.24	0.32	1.55	0.70	25,500	179
Total	100.00	100.00	100.00	100.00		20,229

Note: * Short term, rapid decomposing volatile biodegradable; ** Long term, lignified biomass of garden wastes, banana stems etc.; [1]Menikpura & Basnayake, 2009; [2]100% per type of waste

WASTE FLOW

At present, resource recovery from waste is growing as a result of various programs launched by the government, non-governmental organizations and also through self-initiatives of LAs. The resource recovery route starts from point of generation where most of readily marketable waste such as metal, cardboard, clean papers, and solid plastics are recovered by generators and channel through formal (municipal) or informal recycling streams. However, on-site composting is limited to rural and semi-urban households and institutes, thus diversion remains low. Most of the low value recyclables are recovered after discharge either at source segregated collection by the LA or waste pickers, collection workers in particular. Once collected separately or as mixed waste, much of the degradable organic waste divert to centralized composting facilities found in many LAs. The waste disposed at dumpsites are mainly low value recyclables, non-recyclable plastics and mix waste that are difficult to be sorted for organic or inorganic resource recovery. Figure 2 shows the waste stream of Kandy Municipality, a typical urban center that encounter a very high floating population in comparison to its residential population. The noticeable feature of the urban waste stream is that a greater portion of waste is generated from commercial and institutional sectors than the residential sector. The waste stream flow shown in Figure 2 can be seen as the general waste flow in Sri Lanka with the exception that waste collection rate remains high in major and secondary (MCs and UCs) cities than PS. Most waste is currently discharged for collection by LAs, yet a major fraction is still being disposed at final disposal sites, often open dumpsites. Onsite disposal accounts for 50% in rural areas; the on-site disposal is much lower in highly urbanized cities (for example, only 6.7% in Kandy Municipality). Illegal disposal includes discharge of waste outside the Local Authority areas and also into someone else's properties. Lower value recyclables (wet or dirt paper, textile, contaminated soft plastic) are present in much larger quantities in mixed waste collections, indicating less of these items are being recycled. Of these materials, paper and plastic have

Figure 2. Waste stream of the Kandy Municipality in 2017 (TPD – tons per day)

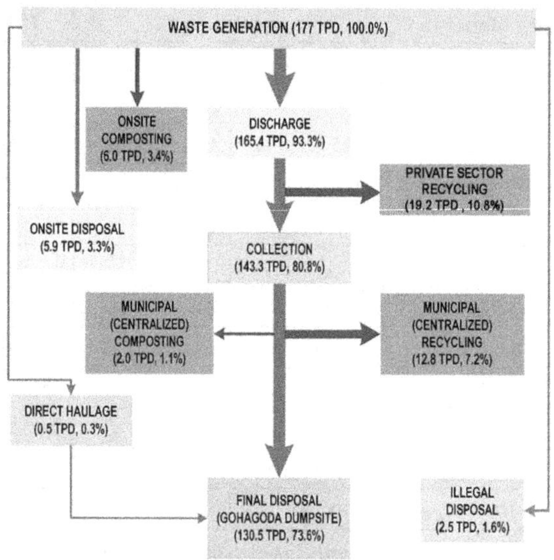

Generation	TPD
Households	35.4
Commercial	45.1
Markets	12.2
Hotels & restaurants	47.4
Privet institutions	10.0
Government institutions	8.1
Schools	2.0
Hospitals	7.7
Religious institutes	6.0
Other	3.5
TOTAL	177

the most (but still limited) potential for increased recycling, particularly if it can be sorted at source and collected separately.

WASTE GENERATION / 3R (REDUCE, REUSE, AND RECYCLE) PRACTICES, WASTE MANAGEMENT TECHNOLOGIES, POLICIES AND REGULATIONS

For creating and improving awareness about the environment, the CEA has been proactive in running numerous educational programs including primary level school programs, eco clubs, awarding of badges, the secondary level environment pioneer program, conduct awareness programs through mass media and also through information centers, exhibitions, celebrating national environmental events, green awards and conduct training and capacity building programs at Nature Field Centers (CEA, 2016). The CEA also assists in effective management and minimization of solid waste through application of different technologies such as the establishment of waste management centers which include composting facilities, plastic/polythene waste collection, storage, and recycling facilities through the 'Pilisaru" national program.

The Ministry of Environment had developed a National Strategy on Solid Waste Management in 2000 and a National Policy on Waste Management in 2007 with a view to facilitating solid waste management in the country with more emphasis on municipal waste. It covers all the three types of waste; solid, liquid and gaseous. Several key strategies of waste management were magnified as detailed policy statements to put the system/s in place first in the right direction and to overcome system deficiencies without deviating from sustainable solutions in 2018 (MoMD&E, 2018).

In the management of waste, the policy document states that the mandatory community involvement in managing waste is recognized as a significant input to ensure that waste managers perform their duties with the highest degree of responsibility and accountability throughout the country. So far individuals and communities are yet to be fully delegated to ensure productive participation with commitment and responsibility. Although, the degree and quantity of waste separations are increasing, but the behavior of the public towards waste disposals is still astonishingly ignorant with communities and individuals resorting to burning wastes and habitually throwing away sizeable quantities of waste and creating waste dumps by roadsides, etc. Inadequate facilities for collections and disposal of wastes are the primary reasons for such indifference towards the wellbeing of the neighbors and communities. On the other hand, officials are yet to implement sound disposal technologies, to facilitate improvements in collections.

However, many of the recent publications on the recycling industry indicate that there is an active informal sector that collects items like old newspaper, bottles, scrap metal, old mattresses, cans, unusable refrigerators and washing machines, etc. and plastic containers and the like. This sector has a very long history dating back to 1960's thus any attempts to regularize is a challenge in itself. According to Jayasinghe et al., 2019, informal waste workers are highly knowledgeable about the recycling industry.

Figure 3. Distribution of recyclable waste collectors and recyclers in Sri Lanka (CEA, 2019a)

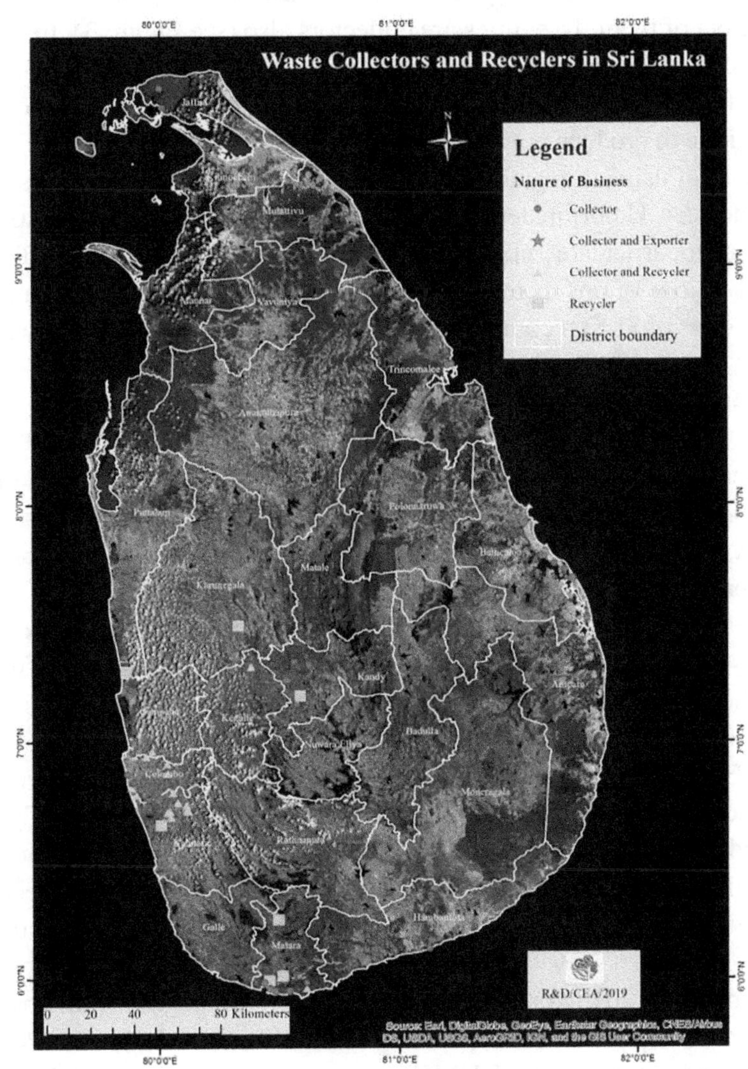

Despite these positive economic and environmental contributions to society, informal waste workers and the knowledge and skills they possess are marginalized. It is indeed difficult to pinpoint who are responsible for their wellbeing. They are involved in door to door collections, dismantling some of the items, grading, sorting, cleaning, and aggregating materials into commercially viable quantities and manufacturing simple consumer products. There are also small businessmen operating at each level of the informal waste economy who buy sorted wastes, pellets, or end products and sell it to interested parties (Jayasinghe et al., 2019). It seems that these businessmen do not like to expose the chain of the informal sector by responding to frequent advertisements in the newspapers for CEA registration. They are responsible for exploiting these workers because these collector/recyclers labor to apply 3R to the best of their knowledge while taking the risk of paying the generators for the wastes. Among these businessmen, there are few entrepreneurs. They are keen registered recyclers, having access to financial incentives provided for small and medium enterprises. Nevertheless, the economic and social structure needs improvement. To this end, these formal recyclers have formed an association called Sri Lanka Recyclers Association (http://www.slr.lk/) well recognized by the government authorities and stakeholder institutions. The CEA correspond, visit regularly and issues environmental protection licenses (EPL) for 21 recyclers island wide out of them, 16 entities are collectors also (see Figure 3). In addition, there are 59 registered recyclable collectors island wide as shown in Figure 3 (CEA, 2019a). The CEA is implementing seven gazette notifications to address the problem of controlling pollution by polythene and plastics. It includes the prohibition to produce polythene sheets less than 20 microns, HDPE for bags, of burning plastics, etc. Exceptions of using thin gauge plastics are given to special categories, like laminating and the pharmaceutical industry. Unfortunately, CEA is promoting biodegradable plastics without knowing the consequences of nanoparticulate entering the food chain (EU, 2018). According to CEA registry, there are 9 registered entities of raw materials importers. Out of them, two are manufacturers, and four finished product importers of SLSI recommended biodegradable products in the country (CEA, 2019b).

E-Waste Management

One of the most challenging aspects of recycling is E-wastes commodities. The consumption of electronic and electrical equipment in Sri Lanka has increased over the years resulting in a significant generation of E-waste. The CEA has introduced a National Corporate E-waste management program with the collaboration of several private sector companies and government institutions. Drop off events and awareness raising campaigns on E-waste have been conducted using mass media, based around various sectors over the years through this program. Declaration of E-waste days and E-waste weeks from time to time has been done to target the general public. E-waste collected through these programs is exported following the Basel procedures for recycling purposes (Suraweera, 2016). Despite these initiatives, Sri Lanka still needs to strengthen E-waste management. As a CSR project Dialog Telekom voluntarily initiated a program to collect and store used mobile phones and accessories in 2008. Specially designed "drop off boxes" to collect used mobile phones and accessories were kept at Dialog Telekom's collection points (UoSJ, 2015). According to Thavalingam and Karunasena, 2016 implemented projects to manage mobile phone waste are malfunctioning due to major gaps such as lack of public contribution and awareness, the illegal mobile phone market, second-hand dealers, and absence of a national policy and regulations.

According to Onuma et al., 2018, a study conducted within Kurunagala Municipal Council (KUMC), the most valuable parts like printed-circuit boards, hard disks, batteries, other PC parts are exported to more than 40 countries. Based on this KUMC export volume, it is estimated that roughly 680 tons/

year of E-Waste 6 item parts are exported from Sri Lanka (Onuma et al., 2018). For the moment, there is no systematic E-Waste recycling program in this country. And Currently, Sri Lanka does not have a comprehensive E-waste recycling facility (UoSJ, 2015). Given the current situation, it is imperative to enact an E-Waste recycling law based on a principle of Extend Producer Responsibility that requires manufacturers and importers to accept responsibility for all stages in a product's lifecycle, including end-of-life management. The important first step in tackling this issue is to introduce a registration system for the informal sector to be listed in an E-Waste national database, and to set the applicable technical standards which define the policy targets of collection, recycling, and the hazardous substances to be treated in a proper way for each product classification in Sri Lanka (Onuma et al., 2018). Sri Lanka exports E-waste to China, India, Pakistan, Hong-Kong, Germany, Belgium, South Korea, Singapore, and the United Kingdom through 19 licensed collectors. These overseas plants are designed so that their emission discharges do not pollute the environment. Sri Lanka lacks such recycling capacity.

According to CEA as of 11.02.2019, there are 11 registered E-waste management facilities in Sri Lanka. Out of them, Asia Recycling (Pvt) Ltd, which is located in Homagama is collecting CFL and Florescence Tube Bulbs and transporting them for storing and recovering materials at its facility. Two companies are located in Colombo and one each in Wennappuwa and Mawanella engaged in the collection, transportation, storage and recovery of E-waste. Apart from them, another 6 facilities do perform the same, but they also are authorized exporters of E-waste (CEA, 2019c).

Composting and Anaerobic Digestion

In the absence of safe disposal facilities, some of the LAs have successfully established home composting systems. It is one of the best solutions to reduce the burden on LAs by diverting wastes from landfills. Improving home composting systems is one of the aspects to be addressed while the collection of made compost should be mandatory for the LAs. There is always a surplus of stabilized compost and garden wastes that need further treatment for marketing the made compost, rather than disposing of valuable materials in landfills.

The LAs are also compelled to stabilize market wastes by composting or by the introduction of anaerobic digestion plants. Composting technologies have been practiced in the country at different locations with varying capacities and levels of success. In the Western Province, 123 tons/day of wastes are converted to 2.23 tons/day of made compost at a very low conversion efficiency of 14.3%, but some reach 54-56%, like Negombo MC and Kattana PS (WMA, 2019). Aerobic composting of static piles was the preferred technology in the past with few Inclined Step Grate composting systems (Ariyawansha et al., 2010b). The latter units still function, producing thermophilic compost of high quality. The quality in both of these systems largely dependent on know-how and motivation of the workers, supervisors, and managers. The know-how came about with research and development (R&D) in the country and there were number of training programs for optimizing process conditions. Unfortunately, management and accountability were not the issues at the beginning of establishing and operating the facilities, because it is necessary to have at least one of the LA officials who are keen and eager to function the system efficiently. However, if there was ever such a person and as soon as that person is transferred or resigns from the post, immediately accountability is neglected and indifference is shown by persons responsible primarily to reduce workload and revert to disposals in the dumpsite. The cost of operating such systems is also a deterrent for the LAs to continue operating such landfill pretreatment technologies. Therefore, profitability is essential for justifying additional expenditure and allocation of the LA workforce for operating the facilities. For example, only three to five systems are able to operate above 60% self-sustainability. The major drawback of increased profitability or breakeven in operating the system is the poor quality of compost. It can be overcome by improving the processing system to retain nitrogen and imbed phosphorous (Gamage et al., 2011; Xiao et al., 2017). In addition to these facilities, nine (9) composting plants with a capacity to process 50 tons/day of compostable waste per unit are being established by the government representing a substantial capacity to meet the need. These systems may not be financially and technically feasible, but like most recycling systems it could be economically justifiable as opposed to dumping. Notably, it was an inappropriate decision without adequate research under tropical conditions. Nevertheless, the systems can be ameliorated to reduce wear and tear and odor nuisance with the incorporation of already known complimentary technologies. The other alternative technology is anaerobic digestion. Lessons have been learned in processing all of the biodegradable wastes in these digesters that need considerable time to convert some of the cellulosic and lignified materials. In these reactors, sludge accumulation is high and thus, removal of sludge is one of the constraining factors and also require large reactors to process fibrous materials. It is then ideal for converting food and non-fibrous kitchen wastes to gaseous end products of methane and carbon dioxide. Anaerobic sludge blanket reactors with liner incorporation for mineralizing inorganic substances is the best solution for macerated high volatile wastes generated from very many eating places and restaurants. It may account for 20-25% of the total wastes from large cities.

Waste to Energy

In fact, the WMA has entered into a Public Private Partnership (PPP) to convert 500 tons/day of waste to energy, having a dual system of bio mechanization and incineration at Karadiyana in the outskirts of Colombo. In the southernmost part of the Western Province in Horana, another PPP project is being materialized to convert 500 tons/day to electricity. These plants are allowed to produce a maximum of 10 MW/plant to supply the national electricity grid. Instead of PPP, a privately owned mass scale of 650 tons/day waste to energy facility is under construction. The company has entered into a feedstock agreement with the CMC.

Dumpsites, Landfills, and Rehabilitation of Dumpsites

The existing and draft national policy on waste management prohibits operation of the dumpsites, thus promoting sustainable waste management practices. The circumstances are such that the policy is not regulated through the judicial process. Naturally, all of the cities and towns continue to dispose of wastes, in most cases in an uncontrolled manner. For example, the Kandy Municipal Council in the absence of support from the central government and due to judicial issues, continue to operate the dump without adequate precautions to prevent dumpsite failures. There are five dumpsites exceeding the disposal rate of 50 tons/day in the Western Province, along with three medium ones between 10-50 tons/day, also five small dumps with the disposal of 1-5 tons/day and eight minor ones less than 5 tons/day.

It was in the year 2003 when the first landfill was constructed with a Japan International Cooperation Agency (JICA) grant in the holiday resort of Nuwera-Eliya city. It is no longer operated efficiently and the leachate is not treated adequately, consequently contaminating the water reservoir which is located downstream of the leachate treatment facility. Another landfill constructed by the Korea International Cooperation Agency (KOICA) in Dompe to dispose of 90 tons/day can only be used for wastes collected from the Dompe PS because of the community protesting about the facility - the not in my backyard (NIMBY) syndrome created and sustained by political influences.

It seems that none of the locations are accepted by communities for constructing, particularly large-scale landfills. Although, 3R programmes are gaining ground and wastes will be diverted from landfills, there will always be residual wastes for final disposal. Besides, landfill sites are needed for disaster waste management and to accommodate waste arising from new cities within cities (developments) proposed by the Ministry of Megapolis and Western development. Therefore, the governments were compelled to find at least one location to construct a sanitary landfill. It happens to be an abundant lime extraction quarry situated in Aruwakkalu, Puttalam which is 170 km from the financial capital Colombo. The transportation of wastes will be by rail. The design engineers opted for several liners, namely; 300 mm clay, HDPE, bentonite layer and several geomembranes (EML, 2015). In the event of a failure, the design criterion is based on natural attenuation of leachate and the landfill is designed for daily disposal of 1,200 tons/day. The project is a fully government-funded project. Proper functioning of all these facilities will reduce the amount of waste to be finally disposed of while compost could be used in agricultural practices.

HAZARDOUS WASTE GENERATION AND MANAGEMENT, POLICY AND REGULATIONS, TRANSBOUNDARY MOVEMENT OF HAZARDOUS WASTE

Sri Lanka has ratified the Basel Convention on Trans-boundary movement of HW and their Disposal on 28th August 1992 and since then, the CEA was designated as the Competent Authority for the Basel Convention. The National Coordinating Committee for the implementation of BC functioning under the Ministry of Environment & Natural Resources has appointed a Technical Expert Committee (TEC) to support the CEA in sound decision making on matters related to technical and legal aspects of managing HW in the country. The TEC consists of the experts from CEA, Ministry of Environment, Sri Lanka Customs Department, Industrial Technology Institute, Universities, Ministry of Health, Registrar of Pesticides, and Board of Investments. The TEC, chaired by the CEA also submits recommendations for the projects that are seeking approval to import waste listed in the Basel List "B" as raw materials. Importing Waste in the List "A" of the Basel Convention (BC) is banned by a decision taken by the Cabinet of Ministers (CEA, 2016). Regulations with respect to the importation of hazardous chemicals and HW materials to the country were gazetted on 05th June of 2013 by the Gazette No. 1813/14 under the Act of import and export control Act No.01 of 1969. New industries depend on the post-consumer waste materials listed in the list "B" of the BC is not allowed. A guidance manual for safe and effective detection and investigation of illegal traffic and transboundary movement of hazardous wastes and other wastes has been prepared. The Environmental Protection License scheme and the Environmental Impact Assessment scheme under the National Environmental Act are in place. Guidelines for locating industries have been formulated (UoSJ, 2015). As reported by CEA, 2016, As part of the implementation of these regulations 686 number of clearances were issued from 2013 to 2015 period. And also, during the period of 2013 to 2015 CEA has given the approvals for 21 transits and 15 exportations (Trans-boundary movement) of waste materials listed in the List "A" of the BC.

Minamata Convention on Mercury was adopted at the Diplomatic Conference held in Kumamoto Prefecture, in Japan in October 2013. Sri Lanka signed the agreement on 8th October 2014. The steering committee on Minamata convention is already established. Measurement of Mercury is not carried out in the national air quality monitoring programmes and a monitoring program is being launched with the collaboration of Asia Pacific Mercury Monitoring Network (APMMN). The Project "Minamata Convention Initial Assessment (MIA) in Sri Lanka" commenced and was implemented by the UNIDO/GEF project. The project is an enabling activity for the Government of Sri Lanka to ratify the Minamata Convention on Mercury through implementing a set of parallel and/or consecutive activities under two major outcomes with assistance from Biodiversity Research Institute to create an enabling environment for the implementation of the Minamata convention and to develop the National Mercury Profile and Mercury Initial Assessment Report.

In 2008, a strategic business unit known from a cement factory obtained the first EPL for HW co-processing in cement kilns to a list of wastes, enabling it to accept a large range of hazardous waste. It also expanded its operations to provide a solution to the waste of 84 companies in the Katunayake Export Processing Zone (KEPZ) and signed a contract with the BOI. Expanding its capacity constructed and opened for operations the first fully-fledged pre-processing plant to process HW in the KEPZ. By 2010 the total customer base consisted of 194 and it was able to co-process more than 40,000 tons of waste.

Healthcare Waste Management

It has been reported that two teaching hospitals in the Southern Province produce hazardous wastes in the range of 0.25-0.47 kg/bed/day (Ministry of Health, 2012), which is close to the average generation rate of 0.37 kg/bed/day. This generation rate has been confirmed by another author of 0.38 kg/bed/day. Nevertheless, peak generation rates should be employed in calculating, selecting and designing for the required treatment system. The separation of wastes is taking place in most Hospitals and other healthcare centers. The infectious waste is highest and pharmaceutical waste is lower for all clinical waste that is generated. The present structure of the Infection Control Unit managing wastes is adequate and effective in most hospitals. The color coded storage containers are available in these Healthcare facilities. It is then possible to weigh daily the wastes and provide mass balances on a weekly or monthly basis based on final disposals. The destinations of the wastes should be given. Absence of an updated, cabinet approved policy and guidelines on HCW management for the Ministry of Health, combined with a lack of an updated database on HCW generation and management, unsafe disposal of HCW, inadequacies in adequate financial and technical resources, as well as commitment of the relevant stakeholders, combined with sustainable methods of waste disposal, and poor research and development programs are some of the most concerning issues related to HCW management.

The Government did allocate for the years 2014, 2015, and 2016 totals of 100, 150, and 250 million LKR, respectively for waste management and the Australian Aid is providing 5 incinerators. These amounts may be adequate to address the issues of solid hazardous waste disposals of government healthcare institutions. Nevertheless, considerable amounts are needed for wastewater treatment. The actual requirements should be calculated on the basis of effluent generation rates for each of the hospitals. Some may require upgrading and others may need new installations. According to Karunasena et al., 2015, the cost-effective disposal strategies were diesel incinerators (28.22 LKR per kg) and dispose in land (12.50 LKR per kg). In general, the cost for disposal of clinical waste in public sector hospitals were 84,084 LKR per day while private sector hospitals were 42,102 LKR per day. A waste incinerator with a processing capacity of 7 tons/day which was constructed in Moratuwa in 2014 under Grant Aid by the Government of Korea, medical waste generated at hospitals in Kandy, Badulla, Galle, and Jaffna is being disposed at this facility. The operation and maintenance of this facility were transferred to a private company under the principle of PPP. The treatment fee (collection and transportation costs are included) for the facility is 67 LKR/kg for general medical waste and 80 LKR/kg for hazardous and infectious medical waste. There are also several small-scale medical organizations with small incinerators (JICA & KKC, 2016).

AGRICULTURAL BIOMASS AND UTILIZATION (WASTE TO ENERGY)

It is estimated that the cattle and buffaloes generate 15,000 tons/day of dung. The calculations were based on 36 kg/day of wastes produced from a bull or cow weighing 454 kg and a buffalo generating 27 kg/day weighing 454 kg and the National Livestock Statistics of the Agriculture and Environment Statistics Division of the Department of Statistics 2019, was used to obtain livestock populations in Sri Lanka. Furthermore, Sri Lankan cattle weigh approximately 160 kg, thus proportionate dung productions were obtained for cattle, cows, and heifers of local and improved breeds. The dry content of the dung excluding total fixed solids are approximately 10% of the generations, thus 300-600 tons of methane can be recuperated from digesters, but most of these animals are free ranged in large extents of land and it is likely that pollution in the form of global warming is arising from 30 tons/day of methane emissions. The maximum quantity of poultry wastes is 700 tons/day. Most of the dung and poultry droppings are used for agriculture lands, deprived of organic matter.

The potential of converting some of the paddy straw presently burnt or left in the field could be used for generating power, since the total paddy straw generations are 1,385,000 tons. It is realistic to suggest that part of the straw generated should be converted to produce biochar and use the syn-gas for waste to energy. The twin benefit could uplift the poor farming communities in Sri Lanka and develop sustainable agricultural systems. Small units of 2.5 MW could be installed within a radius of 5 km for the collection of straw and deliver biochar fertilizer (Ariyawansha et al., 2014).

MARINE AND COASTAL PLASTICS

One of the national surveys indicates (Department of Census and Statistics, 2012) that 46% of the wastes are burnt and destroyed. The remaining ashes which are more than 4% of the daily generations get accumulated and they become vulnerable to washouts during heavy rains apart from flooding. In the most populated lands of the country, the WP, the ashes could be estimated at 320 tons/day, which turns to suspensions in water during heavy rains that are taken away cleansing the lands but polluting low lying lands, river beds and eventually the sea. It could be possible that 4-5 tons/day mostly of polythene and non-degradable wastes arising from flood waters may end up in the sea from the WP. A four-year recurrence flood event due to climate change would then reach 8,000 to 10,000 tons ending up in the sea. The calculation was based on the premise that 7% of plastics dumped on the roadside, streams and low-lying lands (marshes) that is not collected gets washed out. According to Nordrum, (2015), Sri Lanka stands in the first five in the world ranking of polythene wastes washes into the sea, amounting to 1.59 million tons/year. This value may not be accurate, because such large quantities are not generated in Sri Lanka. However, the quantities allowed to be washed into the sea may be considerable as estimated.

CONSTRUCTION AND DEMOLITION WASTE / DISASTER WASTE

Sri Lanka is prone to natural disasters commonly caused by hydro-meteorological (floods, cyclones, landslides, droughts) and coastal erosion for generations with increasing losses of life and property, because of climate change coupled with poor management of ecosystems in watersheds (DMC 2005; Jayawardane, 2006). Over the years, since the establishment of the Disaster Waste Management Center under the National Disaster Management Council, established under the Sri Lanka Disaster Management Act No.13 of 2005 has overcome some of the inadequacies in implementing the Sri Lanka Comprehensive Disaster Management Programme (Ministry of Disaster Management, 2014) but it is still inadequate, compared to most developed and developing countries.

The disastrous event such as the Asian Tsunami in 2004 did engulf almost non-existing waste management systems in Sri Lanka (Basnayake et al., 2005). The Construction Waste Management (COWAM) project which was funded by the European Union from 2005-2009 looked at the most sustainable ways of dealing with Construction & Demolition (C&D) waste after the Tsunami disaster and devised a pilot C&D recycling plant (COWAM Centre) in Galle. The COWAM Centre crushed mix wastes to produce aggregates for use in local roads (Karunasena et al., 2012). A national strategy and a framework for C&D waste management in Sri Lanka is not available (Karunasena et al., 2012). In general, as advocated by many (Karunasena et al., 2009; Brown et al., 2011) the debris generated by most severe disasters overwhelm existing solid waste management facilities or force communities to use disposal options that otherwise would not be acceptable. The major gaps exist in disaster waste and C&D waste management identified (Karunasena et al., 2012; Ranjan et al., 2014; Karunasena and Amaratunga, 2016) as lack of capacity to handle an enormous quantity of waste, difficulty in estimating the quantity and composition of waste, skills and confidence building, organization implementation, linkages and collaborations, continuity and sustainability, investment in infrastructure, research and development, communication and coordination and others like policy issues. They have proposed various measures and approaches to overcome these gaps.

IMPACT OF WASTE MANAGEMENT ON GLOBAL WARMING

The environmental factors affecting global warming are greenhouse gas emissions expressed in terms of global warming potential, acidification potential, eutrophication potential, and fossil fuel consumption in managing solid wastes. For all of these factors, the Intergovernmental Panel on Climate Change (IPCC) has developed models to predict and provide most likely potential values in terms of equivalence. Using this model, Menikpura et al., (2012) found that 29 kg of methane is emitted per ton of wastes disposed in the Gohagoda dumpsite. During the Asian Regional Research Programme on Environmental Technology (ARRPET) study in 2006, actual measurements were made in some of the medium to large dumpsites. For example, the Gohagoda dumpsite with an active filling extent of 2 ha gave a value of 255 $g/m^2/day$, which is approximately 34 kg of methane per ton of wastes at a disposal rate of 150 tons/day. Karadiyana gave a value of 208 $g/m^2/day$. It translates to approximately 18 kg of methane per ton of wastes disposed on an extent of 6 ha at that time. In fact, the disposal at Karadiyana has reduced to 500 tons/day from 700 tons/day. Notably, "hot spots" were identified by Karunarathna et al., 2018, because of disposing fast degradable volatile organic wastes on specific locations as shown in Figure 4.

Figure 4. Surface methane emission concentration map of Karadiyana dumpsite (Karanuarathne et al., 2018)

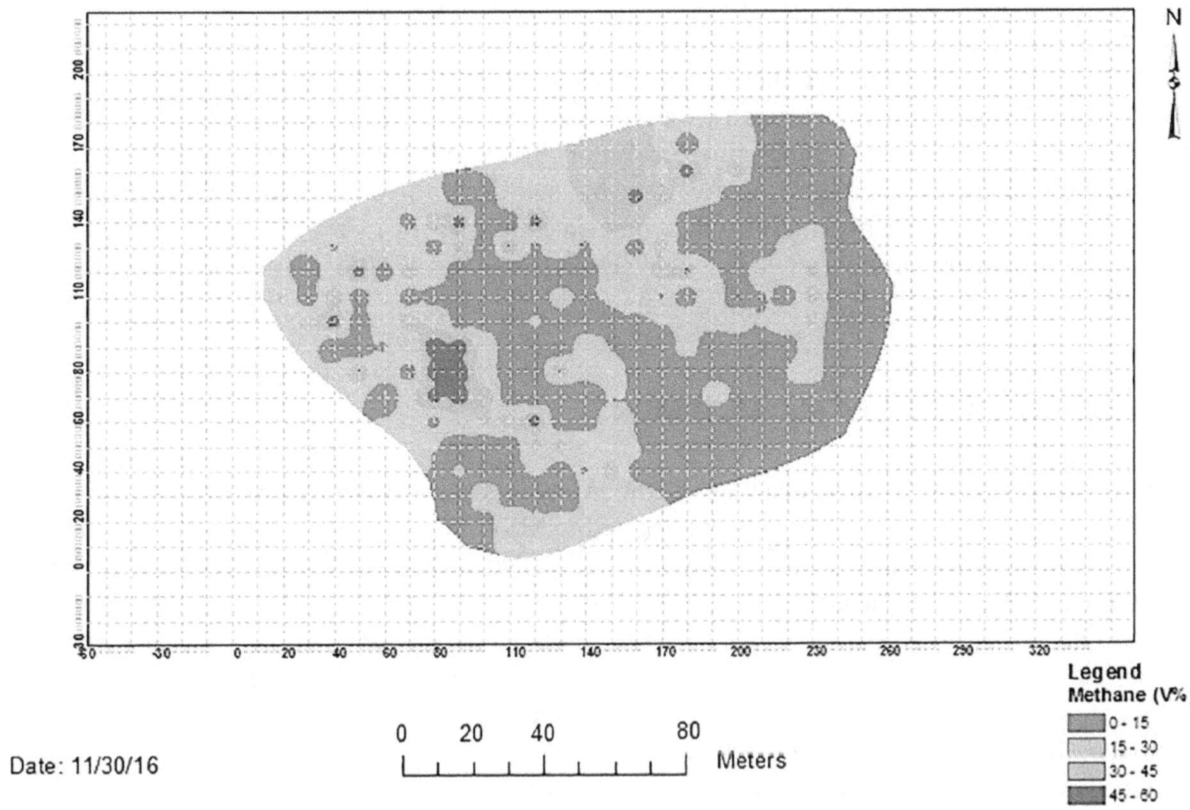

Total methane emissions in the country from MSW can be calculated considering a plausible value of 18-20 kg of methane per ton of wastes. It may amount to 45-50 tons for disposing of 2500 tons/day in dumpsites. It is reported (Ishigaki et al., 2014) that landfills located in the tropical climate zone tended to exhibit higher emission of nitrous oxide and they found that the impact of nitrous oxide on the greenhouse effect at the sites investigated was 20% that of methane when the flux of nitrous oxide was measurable. Interestingly, old dumpsites compared to new ones are generating more nitrous oxide (Nagamoria et al., 2016). Menikpura et al., 2012 also found that the overall acidification potential is 9.71 kg SO_2 equivalents per ton waste, 99.5% of which are from final disposal, and the remainder from fuel production, collection, and transportation. They reported that the overall eutrophication potential per ton of waste is 16.42 kg of NO_3^- equivalents, 99.5% being contributed by final disposal and total fossil fuel consumption from 1 ton of disposed waste is 1.67 kg of crude oil equivalents; of this, 91% comes from transportation and the remaining from manufacture of the collection vehicles.

ISSUES AND CHALLENGES IN WASTE MANAGEMENT/3RS

There is a wide knowledge gap among the professionals, including academia in understanding planning procedures and processers. Most of them are used to developing action plans rather than comprehensive or strategic plans that spell out futuristic approaches with a knowledge base to map out each of the subjects with interaction so as to dovetail with each other. Such a plan was developed by the MMD&E and UN-Habitat. The next step is to formulate and adopt Intended Nationally Determined Contributions to fulfill the obligations of Paris 2015 of sustainable solid waste management practices focusing on proposed sub-plans. For example, CEA will be mandated to issue EPLs to LAs based on diversion of wastes from sustainable landfills while reducing greenhouse gases (GHG) emissions, and acidification and eutrophication potentials.

ECONOMICS OF WASTE MANAGEMENT

A study conducted in 2004, found that high operation and maintenance costs are the major expenses for LAs. Approximately 70-80% of the budget allocated to MSW management is spent for collection and transportation, especially to cover labour and fuel costs, and a much lesser amount covers final disposal costs (Basnayake and Visvanathan, 2014). SWM activities are financed by own-source revenues of LAs. As reported by JICA & KKC, 2016, SWM expenses accounted for 10 to 30% of total expenditure of LAs; and thus, waste collection service is one of the major public services in the LAs. However, the amount of revenue collected by waste fees did not even reach 10% of the total SWM expenses and these were primarily collected from large-scale businesses.

Some LAs collect waste from large waste generators with a collection fee by volume depending on the type of waste. For instance, Kesbewa UC charges (includes tax) 3,420 LKR/ton of MSW, 4,560 LKR/ton of bulky waste and 6,840 LKR/ton of slaughtering and industrial waste from large scale business (JICA & KKC, 2016). The lowest amount as tipping fee is charged by Jaffna MC amounting to 450 LKR/load.

All of these studies reveal poor financial management by LAs in view of the final disposal that causes the most amount of pollution. Out of them, two studies have resulted in similar conclusions. The JICA investigated detailed expenditure of LAs, particularly in two districts (JICA & KKC, 2016). They found that Colombo and Gampaha districts are the most serious as both of the districts have been unable to secure land for waste disposal facilities. Furthermore, they found that the financial capabilities of the LAs in the Colombo and Gampaha districts are relatively low and they conclude that it is important to rely on "polluter pay". On the other hand, the UN-Habitat study relies on acquisition for cluster-based development for the entire district (MoMD&E, 2016) or lands with close proximity for several LAs. The latter study focused on eight of the provinces, hence excluding the WP, which is the worst affected in the country. It did not consider the financial strengths and weaknesses of LAs, but it was based on capital and operational costs of rehabilitation of dumpsites, construction of landfills and compensation. Therefore, the polluter must pay for the base costs of collection and disposal of wastes in landfills. It also considered to increase the collection efficiency, because it was found to be a very low value of 27%. Inevitably, the incentive of recycling wastes is the savings on landfill disposal. In this model, financial stability is assured for all of the LAs. However, five-year plans must be drawn to reduce disposal to 15% of the present collections. It is a transparent and accountable system that was proposed in this unique study (Basnayake et al., 2018).

None of the studies focused on determining the value of the informal sector contributions to resource circulation in the economy by calculating the reduction of wastes ending up in landfills. In other words, the LAs should be paying them equivalent costs of transportation and disposal in engineered or sanitary landfills. Such an incentive can only be realized if a strategic or comprehensive plan for integrated solid waste management is implemented in which the savings on landfilling (Basnayake et al., 2018) could be diverted to these marginalized informal sector workers. These payments could be in the form of machinery and equipment, protective clothing, collection vehicles, etc. They will then become part of the formal sector and the authorities will be able to update databases and revise policies, strategies and plans to reduce consumption while increasing recycling of resources.

One must also consider Life Cycle Costs (LCC) in determining the economics of recycling waste. It is a very important decisive factor for MSW management (Menikpura et al., 2012). It considers total capital expenditure, total operation and maintenance costs and environmental costs. Quite correctly, Menikpura et al., 2012 states that even though LAs or policymakers are not yet concerned with environmental costs, it is important to calculate these externalities, since they may contribute a large share of the LCC as a result of the environmental damage caused by the existing MSW management system. In the absence of a model to calculate these externalities, they based it on the default values of emissions and resources consumption reported in the Swedish Environmental Priority Strategies (EPS) model (Steen, 2000), and expressed as "willingness to pay" (WTP) by society. It is a useful model that considered the potential of global warming, acidification, and eutrophication. They stated that by replacing 50% of open dumps with sanitary landfills with gas recovery systems, the annual GHG emission reduction would be 714,200 tons of CO_2 equivalent, which is 6% reduction of the country's annual GHG emissions and would satisfy 4% of the country's electricity demand, resulting in a reduction of annual importation of crude oil by 35,600 tons, thus a saving amounting 3,378 million LKR in avoidance.

Another or supportive means is to apply Commodity Price Energy Equivalent (CPEE) to divert combustible wastes from landfills (Basnayake et al., 2013). It derives indices based on one unit of energy contents in terms of currency for the production (E_p), residual energy (E_r) and services, including transport (E_s). CPEE was applied as a test on A4 size paper, separating each of the energy flows in terms of indexed to paper production and residual energy, thus an E_s value was deduced. It gave results of 0.12, 0.17 and 0.71 respectively, for the year 2000. In 2013, the values were 0.20, 0.27 and 0.53 respectively. Inevitably, the indexes of production and wastes have increased, while profits have dwindled. Such reductions were due to an exponential increase in diesel prices expressed as $24.93e^{0.123t}$ and paper responded to inflation of $152.7\ e^{0.101t}$. In 2023, the price will hike to 360 LKR/L for diesel and over 1,200 LKR for a A4 paper packet. Instead of LCC calculations, CPEE indexing can provide a host of information, like converting paper wastes to electricity or bio-char in combating inflation, thus a "Hotelling" rule (Gaudet, 2007) can be applied with modifications.

Such benefits can be derived from Residual Derived Fuel (RDF) mined from dumpsites. It will also reduce the impact of polluting the environment. Instead of reaping the benefits, governments, financial institutions and investors viewed rehabilitation of dumpsites as burdens. Within a brief period, all of these stakeholders, particularly financial institutions took interest in rehabilitation of dumpsites with the publication of Senevirathne et al., 2013. It was based on developing a low-cost treatment system to process 9,000 m^3/year of leachate. To sustain rehabilitation works, the private company and the University of Peradeniya (UoP) conducted biochar trials to develop the market to reach 20 tons/week and a similar demand for torrified RDF. The later was used as equivalent to coal in determining the Resource Utility Value (RUV) of the dump. After deducting the rehabilitation costs, the net benefit amounted to US$ 4.6 million, which was sufficient equity for the venture capital to establish a 2.5 MW dual fuel power plant that can be augmented to 10 MW with waste from neighboring LAs. Economic returns of waste to energy can be calculated by comparing with landfilling. The capital and operational costs are virtually similar, but the total operating profit in converting the wastes to electricity is 24,000 LKR/ton of raw wastes, calculated over a period of ten years (Basnayake et al., 2015).

THE WAY FORWARD

The construction of an Evidenced Based Interactive Database (EBID) should be the primary target of the government agencies (MoMD&E, 2016; Basnayake et al., 2018). The databases of the NSWMSC, CEA, and MMD&E will be made into one portrait theater with all of the actors having a specific role to play in the visible interactions. The owner of EBID should be the CEA. Although, most of the actors will have limited access to change the structure of EBID, but all of the actors will have access to information and provide online feedback on the collection of wastes, availability of waste materials and processed products by 3R. It will also have current practices and technological issues, guidelines, information on awareness and training programs, requests, information on hazardous wastes, including E-wastes, registration of the informal sector, R&D on wastes, etc. It is in fact, enacting the legislation on right to information.

Figure 5. Comprehensive integrated solid waste management plans developed by the ministry of Mahaweli development and environment with technical assistance from The United Nations Human Settlements Programme (MoMD&E, 2016; Basnayake et al., 2018)

It is also a tool to implement and make auto changes to the proposed comprehensive integrated solid waste management (ISWM) plans. At the provincial level, it could be strategic ISWM plans. The strategic action plan developed by the WMA could be upgraded to such a plan. All of the plans of strategic, comprehensive ones at district and LAs should be formulated in a cohesive manner that will have all aspects of MSW, Hazardous wastes, HCW, disaster wastes, agricultural wastes, and wastewater. As recommended in the MMD&M/Un-Habitat study of comprehensive plans, strategic and comprehensive ISWM plans should consist of sub-plans as shown in Figure 5. In each of the sub-plans, new approaches are needed to accommodate all of the waste, making cross-reference to individual sub-plans. The considerations to be inclusive and highlighted are given for these plans as;

Institutional Structural Plan

There should be clearly identified institutions and supporting agencies dealing with waste prevention, minimization, application of 3R, application and development of technology, including final disposal. It is expected to formulate coordinating bodies or boards at provincial, district and LA levels. They will function with the participation of citizens, including women as advisory committee or boards to provide the required services of coordinating the activities of different government agencies, private sector and other stakeholders. In all of these committees, regional universities should be involved to support the decision-making process, monitor progress and undertake R&D, including social studies. Publications of the findings will increase and improve communication between stakeholders. This approach will develop effective communication and means of executing directives from the national committee/board to LA level and there should be constructive feedbacks, negative, positive, and neutral responses to those directives.

Infrastructural Development Plan

The capacity development of institutions is one aspect of this plan. The other most importantly is for improvement of existing infrastructure and new facilities for processing of wastes and final disposal.

Research & Development (R&D), and Education & Training (E&T) Plan

All of the institutions should undertake R&D and E&T with support service agencies to provide and facilitate innovations, in other words, marketing aspects of inventions and existing technologies and products. On the other hand, the universities and technical colleges should be made responsible for providing specific levels of E&T in line with Sri Lanka Qualification Framework (SLQF). These E&T programs should strengthen the knowledge base of the officers, managers, supervisors, etc. of public and private sector organizations. They will also deal with targeted R&D outputs for existing problems and encourage them to be creative and inventive to reduce costs, improve working conditions and enhance the environment. They should also focus on social issues and development, business aspects and financial management to bridge the gap between inventions, existing technologies, and products with innovations.

Social Development Plan

The E&T plans will highlight the programs, including awareness programs based on targeted stakeholders to improve social values and status. These plans will forecast participation, information, environmental benefits, rewards, incentives, social justice, empowerment, etc.

Environmental Management Plan

The plans for implementing 3R's with all actors from generation to final products should be well-thought-out with defined improvements to existing and new facilities identified in plan ii. It should also include WTE facilities to reduce systematically over a defined period the waste disposals in landfills. It is best that all of these aspects are included in the initial environmental examination and environmental impact assessment (EIA) study reports and approvals for obtaining environmental protection licenses (EPL). There should be available landfill capacities for unrecyclable disaster wastes.

Resource Allocation Plan

The present and future resources and monetary allocations should be based on "polluter pay" inclusive of the present informal sector. It should give details of the resources needed for existing facilities, including rehabilitation of dumpsites, relocation of venerable populations in existing or new facilities and new infrastructure. Economic and financial feasibilities are within the purview of these plans for different types of wastes and technologies. These plans should have incentives, insurance schemes, commitments of financial institutions, government capital infusions, monitory and technical support of international donor agencies and NGOs.

Monitoring, Risk Assessment, and Disaster Waste Management Plans

It is imperative that not only the physical, chemical, and biological parameters are to be measured and reported, but also the status of implementation programs of waste management, that includes waste from disasters. Methodologies with specific guidelines should be formulated and implemented. The plans should spell out reductions of waste, reports on non-compliance of target reductions and assumed efficiencies, risks assessment with specified time bound risk preventions, vulnerability, and disaster waste management.

Regulatory and Law Enforcement Plan

The establishment of Environmental Court/Tribunal (ECT) is a prerequisite for ensuring good environmental governance in terms of transparency, inclusiveness, and accountability, which are the access rights. As at present, there will be a large number of cases to be resolved. Timely feasibility studies are needed with environmental and social impacts studies for judicial verdicts on site selection, technologies, planning process etc. The rules and regulations between LAs should be consistent and ECT is expected to ensure such cohesive understanding between LAs and stakeholders. The administrative court will function from the beginning and continue to monitor the progress. It is expected that the volume of cases will diminish with time. However, there could be disputes among recyclers and civil or commercial courts too would be involved. Moreover, ECT is needed to improve and bring about new legislation for new and improved technologies and environmental concerns of increased urbanization.

A CASE STUDY: PILOT PROJECT AT GOHAGODA, KANDY

The 30-year-old Gohagoda dumpsite was rehabilitated unlike the one at Meethotamulla in Colombo, killing 36 persons, injuring several others and destroying houses and property. It was a pilot project initiated by the UoP to disseminate the breakthrough research findings, including the patent 15679 (Basnayake, 2011) being an outcome of the ARRPET project. An enhanced & rapid methanification process was a key to this project's takeoff. It was a challenge to curtail one of the major pollution points in the Central Province while relocating the dumpsite tenants. None of the private companies were willing to invest in rehabilitation and relocation activities without a fee and moreover, the dumpsite was not an asset. Like in Colombo, the government and the LA were not willing to provide funding for such activities.

Therefore, the researchers were compelled to incorporate a company to deal with two aspects, namely; making the dump a valuable property with resources and at the same time develop the site by constructing roads to access the dump, undertake further studies like slope stability computations to construct and strengthen the embankments. Some of the research was initiated by the Institute of Fundamental Studies and one of them (Wijesekara et al., 2015) pinpointed the existing problems of perch leachate strata in the body of the dump. This allowed us to precisely dig and install "French drains" and connect them to peripheral pipes, which were also intercepting the seeping leachate flows from the sides of the dump. Then the concept of a landfill bioreactor was used to develop a leachate treatment bioreactor (LTB) (Gnakaran et al., 2017). Instead of raw waste, old waste was used to develop microbial films for treating leachate. The effluent from the LTB was treated in an algal pond and finally in constructed wetlands. It took considerable time and effort to develop these low cost and technically feasible technologies. They were part of the proposed ISWM system illustrated with the project process flow, see Figure 6. The EIA approval was given for the entire project as given in the plan, see Figure 7 while accepting the expected level of pollution from a dumpsite. As predicted by the EIA study report, the results of the leachate treatment were good because all of the polluting parameters were approaching permissible CEA water quality discharge standards (Priyashantha et al., 2015). Improvements were made to the LTB to last much longer. Although funds were limiting, technology was developed to make biochar and compost for diverting some of the wastes from the dump. Also, the first stage of the landfill bioreactor was constructed with the composite liner system and the remaining incoming wastes were disposed as described in the EIA study report (Ecotech Lanka Ltd, 2011).

The social impact assessment of the EIA highlighted the problems facing a significant number of families over a radius covering 500 m from the project site. In fact, out of 43 or 28.7% of the sample work for the Kandy Municipality in various capacities, making the Kandy Municipality the largest single employer. Among them are those who were working at the dump as well. It was also interesting to note that 6.6% of the households were working on site, meaning the dump and they were totally dependent on the dump for income generation. These people and their families were highly exposed to the impact of the dump and it was imperative to relocate 53 houses at the cost of the project (Ecotech Lanka Ltd, 2011). Most of these houses were in a dilapidated state.

The EIA approval and valorization of the dumpsite were the key factors for assured investments. Therefore, armed with a low cost and effective leachate treatment system, researchers resorted to calculate the Resource Utility Value (RUV) of the dumpsite. As a result of the publication, Senevirathne et al., 2013 most of the investors, bankers, accountants, and auditors accepted rehabilitated dumpsites as assets.

The proposed ISWM system had many investment scenarios and one of the leading banks and an investment bank offered the company 70% loan and 30% equity to install plants for 2.5 MW power generation in the integrated system. Unfortunately, the majority shareholder of the company turned down the offer, because he could not raise the capital without forgoing his shares. It was the same fate with a similar offer from the Asian Development Bank through a national leading bank for a higher capacity of minimum 5 MW at a lower interest. One must not get entangled with so-called investors and promoters, because they are opportunists to make a quick buck and get away from the responsibilities (Basnayake et al., 2015).

Figure 6. Proposed layout plan of the project

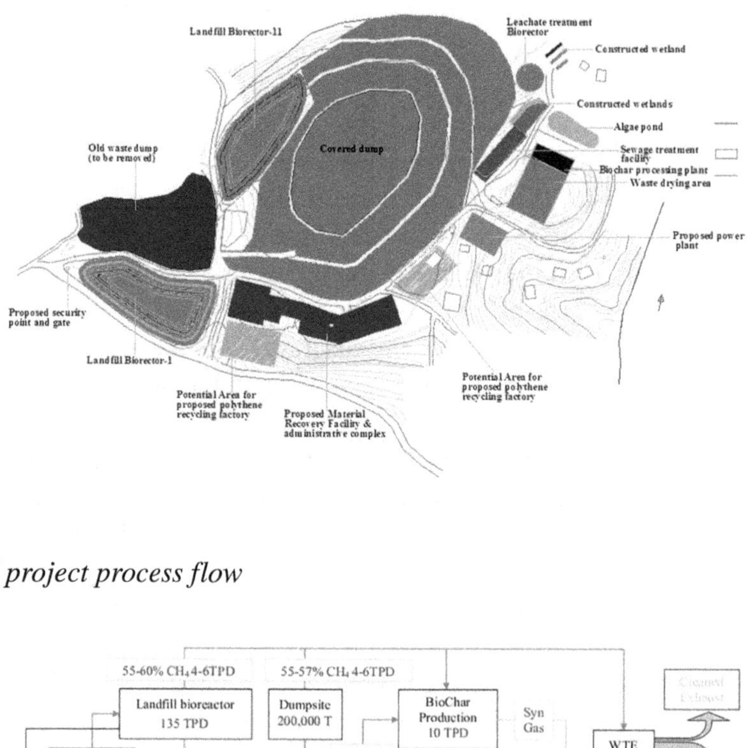

Figure 7. Proposed project process flow

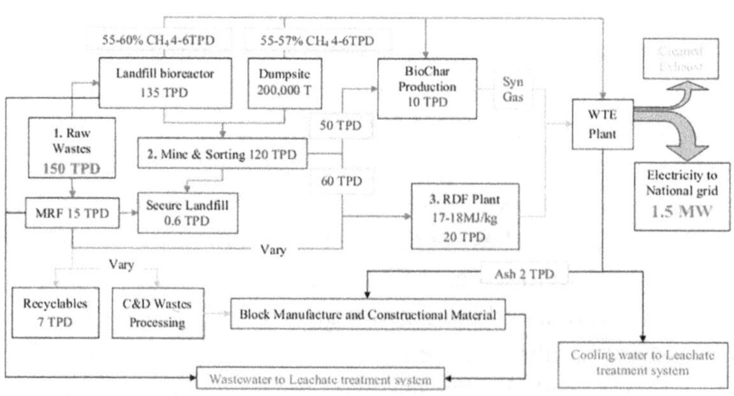

ACKNOWLEDGMENT

Authors are very much grateful to officials of Ministry of Mahaweli Development and Environment, Ministry of Megapolis and Western Development, WMA, UN-Habitat, CEA, NSWMSC, and LAs for their support throughout the study period. We also like to thank Aurora Basnayake for improving the English and writing style.

REFERENCES

Ariyawansha, R. T. K., Chandrasena, A. S. H., Senevirathne, S. A. D. N., & Basnayake, B. F. A. (2010b). Development of a sustainable Environmental Preservation Centre (EPC) at Nawalapitiya for Urban Solid Waste Management. *Proceedings of International Conference on Sustainable Build Environment*, 181-188.

Ariyawansha, R. T. K., Jayakody, L. N., Gunarathne, H. A. Y. R., Senevirathne, S. A. D. N., Kuruparan, P., Chandrasekara, W. W. C. K., ... Basnayake, B. F. A. (2010a). Estimation of quantum of waste generation and solid waste characterization at Kalmunai Municipal Council (KMC) and Sammanthurai Pradeshiya Sabha (SPS). In *Proceedings of third International Symposium*. Sabaragamuwa University of Sri Lanka.

Ariyawansha, R. T. K., Senevirathne, S. A. D. N., & Basnayake, B. F. A. (2014). Potential of Converting Paddy Straw to Bio-char and Electricity in Sri Lanka. *Proceedings of 11th Asia Pacific Roundtable on Sustainable Consumption and Production (APRSCP)*, 207-222.

Barton, F. M. A. (1979). *Resource recovery and recycling*. New York, NY: John Wiley and sons.

Basnayake, B. F. A. (2011). *Composite clay-polyethylene biofilter liner and cover systems for enhancing landfill bioreactors and other bioremediation systems*. Sri Lanka Patent No 15676.

Basnayake, B. F. A., Ariyawansha, R. T. K., Karunarathna, A. K., & Chandrasena, A. S. H. (2013). Influence and counter measures for combating monetary inflation on waste recycling industry in Sri Lanka: Case study on A4 paper. *Proceedings of the International Symposium on Agriculture and Environment at International Symposium on Agriculture and Environment University of Ruhuna*, 87-89.

Basnayake, B. F. A., Ariyawansha, R. T. K., Senevirathne, S. A. D. N., & Chandrasena, A. S. H. (2015). Valorization of municipal solid waste in Sri Lanka. *Economic Review (The Quarterly Economic Journal of People's Bank), 42*(2), 9-13.

Basnayake, B. F. A., Chiemchaisri, C., & Mowjood, M. I. M. (2005). Solid wastes arise from the Asian tsunami disaster and their rehabilitation activities: case study of affected coastal belts in Sri Lanka and Thailand. *Proceedings of the Tenth International Waste Management and Landfill Symposium*.

Basnayake, B. F. A., Popuri, S., Visvanathan, C., Jayatilake, A., Weerasoori, I., & Ariyawansha, R. T. K. (2018). Concerted initiative for planned management of municipal solid waste in target provinces in Sri Lanka. *Journal of Material Cycles and Waste Management*. doi:10.100710163-018-0815-5

Basnayake, B. F. A., & Visvanathan, C. (2014). Solid waste management in Sri Lanka. In Municipal solid waste management in Asia and the Pacific Islands (pp. 299–316). Singapore: Springer Verlag. doi:10.1007/978-981-4451-73-4_15

Brown, C., Milke, M., & Seville, E. (2011). Disaster waste management: A review article. *Waste Management (New York, N.Y.), 31*(6), 1085–1098. doi:10.1016/j.wasman.2011.01.027 PMID:21334871

Central Bank of Sri Lanka (CB of SL). (2018). Annual Report of the Monetary Board 2017. Author.

Central Environmental Authority. (2015). *Data base of solid waste management. 104*. Battaramulla, Sri Lanka: Denzil Kobbekaduwa Mawatha.

Central Environmental Authority (CEA). (2016). *Corporate plan 2016-2020. 104*. Battaramulla, Sri Lanka: Denzil Kobbekaduwa Mawatha.

Central Environmental Authority (CEA). (2019a). *Map of waste collectors and recyclers in Sri Lanka*. Retrieved from http://www.cea.lk/web/?option=com_content&view=article&layout=edit&id=1407

Central Environmental Authority (CEA). (2019b). *Registry of SLSI recommended bio degradable product manufacturers and raw material/ finished product importers*. Retrieved from http://www.cea.lk/web/images/pdf/wm/SLSIRecommendedBio_DegradableproductManufactures.pdf

Central Environmental Authority (CEA). (2019c). *Authorized E - Waste Management Facilities*. Retrieved from http://www.cea.lk/web/en/index-php-option-com-content-view-article-layout-edit-id-983

Department of Census and Statistics. (2012). *Census 2012 results on Population and Housing based on 5% sample, Sri Lanka.* Retrieved from http://www.statistics .gov.lk/PopHo uSat/CPH20 11/Pages / Activities /Repor ts/CPH_2012_5Per_Rpt.pdf

Department of Census and Statistics. Sri Lanka. (2019). *Agriculture and Environment Statistics Division National Livestock Statistics 2007 – 2018*. Retrieved from http://www.statistics.gov.lk/agriculture/Livestock/LivestockStatistics.html

Disaster Management Centre (DMC). (2005). *Towards a Safer Sri Lanka: A Road Map for Disaster Risk Management*. DMC, Ministry of Disaster Management Government of Sri Lanka. Retrieved from http://www.adrc.asia/documents/dm_information/srilanka_plan02.pdf

Ecotech Lanka Limited (ETL). (2011). *Environmental Impact Assessment Study Report: Rehabilitation of Gohagoda dumpsite and establishment of an integrated solid waste management system for Kandy Municipal Council. Business Plan. 629/3, 8th Lane*. Battaramulla, Sri Lanka: Mutthettugoda Road, Thalangama North.

EML Consultants Pvt Ltd. (2015). *Environmental impact assessment report of the proposed project on metro Colombo solid waste management. 6/10 Raja Maha Vihara Rd*. Sri Lanka: Sri Jayawardenepura Kotte.

European Commission (EU). (2018). Report from the Commission to the European Parliament and the Council on the impact of the use of oxo-degradable plastic, including oxo-degradable plastic carrier bags, on the environment. COM (2018) 35 final, Brussels.

Gamage, D. A. S., Wijesekara, S. S. R. M. D. H. R., Sarathchandra, G. W. N. L., Basnayake, B. F. A., & Costa, W. A. J. M. (2011). Estimation of Nitrate and Phosphate Leaching from Lysimeter Simulation of Rice Straw Landfill Bioreactor and Evaluation of Fertilizer Quality of Resulting Compost. *Tropical Agricultural Research*, 22(2), 220–228. doi:10.4038/tar.v22i2.2831

Gaudet, G. (2007). Natural Resource Economics under the Rule of Hotelling. *The Canadian Journal of Economics. Revue Canadienne d'Economique*, *40*(4), 1033–1059. doi:10.1111/j.1365-2966.2007.00441.x

Gnanakaran, M., Ariyawansha, R. T. K., & Basnayake, B. F. A. (2017). Simulation of lab-scale leachate treatment bioreactor with application of logistic growth equation for determining design and operational parameters. *International Journal of Scientific & Engineering Research*, *8*(1).

Ishigaki, T., Nakagawa, M., Nagamori, M., & Yamada, M. (2016). Anaerobic generation and emission of nitrous oxide in waste landfills. *Environmental Earth Sciences*, *75*(9), 750. doi:10.100712665-016-5543-3

Japan International Cooperation Agency (JICA) & Kokusai Kogyo Co., Ltd (KKC). (2016). *Final Report: Data collection survey on solid waste management in Democratic Socialist Republic of Sri Lanka*. Retrieved from http://open_jicareport.jica.go.jp/pdf/12250213.pdf

Jayasinghe, R., Azariadis, M., & Baillie, C. (2019). Waste, Power, and Hegemony: A Critical Analysis of the Wastescape of Sri Lanka. *Journal of Environment & Development*, *0*(0), 1–23. doi:10.1177/1070496518821722

Jayawardane, A. K. W. (2006). *Disaster Mitigation Initiatives in Sri Lanka*. Faculty of Engineering and Professor of Civil Engineering University of Moratuwa, Sri Lanka. Retrieved from http://iiirr.ucalgary.ca/files/iiirr/Kochifullpaper-final.pdf

Karunarathna, A., Gamagedara, Y., Mannapperuma, N., Ariyawansha, R., & Basnayake, B. F. A. (2018). Development of a methane (CH4) surface emission map: a case study at Karadiyana dumpsite, Sri Lanka. *Proceedings of the 10th Asia-Pacific Landfill Symposium*.

Karunasena, D., Amaratunga, R., Haigh, I. L., & Lill, I. (2009). Post disaster waste management strategies in developing countries: Case of Sri Lanka. *International Journal of Strategic Property Management*, *13*(2), 171–190. doi:10.3846/1648-715X.2009.13.171-190

Karunasena, G., & Amaratunga, D. (2016). Capacity building for post disaster construction and demolition waste management: A case of Sri Lanka. *Disaster Prevention and Management*, *25*(2), 137–153. doi:10.1108/DPM-09-2014-0172

Karunasena, G., Jayathilaka, W. M. D. M., & Rathnayake, R. M. N. U. (2015). Comparison on disposal strategies for clinical waste: hospitals in Sri Lanka. *Proceedings of 6th International conference on structural engineering and construction management*.

Karunasena, G., Rameezdeen, R., & Amarathunga, D. (2012). Post-disaster C&D waste management: The case of COWAM project in Sri Lanka. Australasian Journal of Construction Economics and Building. *Conference Series*, *1*(2), 60–71.

Menikpura, S. N. M., & Basnayake, B. F. A. (2009). New applications of 'Hess Law' and comparisons with models for determining calorific values of Municipal Solid Wastes (MSW) in Sri Lankan context. *Renewable Energy*, *34*(6), 1587–1594. doi:10.1016/j.renene.2008.11.005

Menikpura, S. N. M., Basnayake, B. F. A., Boyagoda, P. B., & Kularathne, I. W. (2007). Application of Waste to Energy Concept Based on Experimental and Model Predictions of Calorific Values for Enhancing the Environment of Kandy City.2007. *Journal of Tropical Agricultural Research*, *19*, 389–400.

Menikpura, S. N. M., Gheewala, S. H., & Bonnet, S. (2012). Sustainability assessment of municipal solid waste management in Sri Lanka: Problems and prospects. *Journal of Material Cycles and Waste Management*, *14*(3), 181–192. doi:10.100710163-012-0055-z

Ministry of Disaster Management. (2014). *Sri Lanka Comprehensive Disaster Management Programme (SLCDMP) 2014-2018*. Retrieved from http://www.disastermin.gov.lk/web/images/pdf/slcdmp%20 english.pdf

Ministry of Health. (2012). Environmental management framework for healthcare waste and infrastructure development, Sector development program draft. Sri Lanka: Author.

Ministry of Internal and Home Affairs and Provincial Councils and Local Government (MoIHA&PCLG). (2019). *National Solid Waste Management Support Center of Sri Lanka*. Retrieved from http://www.lgpc.gov.lk/web/index.php?option=com_content&view=article&id=71&Itemid=190&lang=en

Ministry of Mahaweli Development and Environment. (2018). *National Waste Management Policy of Sri Lanka*. Retrieved from http://mmde.gov.lk/web/images/pdf/2018/nationalwastemanagementpolicy-english.pdf

Ministry of Mahaweli Development and Environment (MoMD&E). (2016). *Comprehensive integrated solid waste management plan for target provinces in Sri Lanka, developed with technical assistance from the United Nations Human Settlements Programme. Sobadam Piyasa, No. 416/C/1*. Battaramulla, Sri Lanka: Robert Gunawardana Mawatha.

Nagamori, M., Mowjood, M. I. M., Watanabe, Y., Isobe, Y., Ishigaki, T., & Kawamoto, K. (2016). Characterization of temporal variations in landfill gas components inside an open solid waste dump site in Sri Lanka. *Journal of the Air & Waste Management Association*, *66*(12), 1257–1267. doi:10.1080/10962247.2016.1212746 PMID:27575846

Nahallage, C. A. D., & Huffman, M. A. (2013). Macaque-human interactions in past and present-day Sri Lanka. In S. Radhakrishna, M. Huffman, & A. Sinha (Eds.), *The Macaque connection. Developments in primatology: progress and prospects* (Vol. 43, pp. 135–148). New York, NY: Springer. doi:10.1007/978-1-4614-3967-7_9

National Solid Waste Management Support Centre (NSWMSC). (2015). *Data base of solid waste management-2015*. Sri Lanka: Author.

Nordrum, A. (2015). China, India, US among world's worst marine polluters: country-by-country breakdown of plastics in oceans. *International Business Times*. Retrieved from http://www.ibtimes.com/china-india-us-among-worlds-worst-marine-polluters-country-country-breakdown-plastics-1813476

Onuma, Y., Siriwardana, A. N., & Rajapakse, R. D. D. (2018). Estimation of E-waste flow in Kurunegala Municipal Council, Sri Lanka. In *The 29th Annual Conference of JSMCWM, 2018* (pp. 547-548). Academic Press.

Priyashantha, K. A. S., Ariyawansha, R. T. K., Senevirathne, S. A. D. N., Basnayake, B. F. A., & Chandrasena, A. S. H. (2015). Development and performance Evaluation of the leachate treatment system at Gohagoda municipal solid waste disposal site. In *Proceedings of 6th International Conference on Structural Engineering and Construction Management*, (pp. 31-38). Academic Press.

Ranjan, H., Karunasena, G., & Rathnayake, U. (2014). Construction and Demolition Waste Management Gaps in Construction Industry. In *Proceedings of the 7th FARU International Research Symposium* (pp. 97-104). Academic Press.

Senervirathne, S. A. D. N., Basnayake, B. F. A., Chandrasena, A. S. H., & Ariyawansha, R. T. K. (2013). Municipal solid waste dumpsites; a burden or an obscure buried resource. In *Proceedings of the International Conference on Solid Waste 2013 - Innovation in Technology and Management, Hong Kong SAR, P.R. China* (pp.718-721). Academic Press.

Sri Lanka Recyclers Association (SLRA). No. 19/2A, Parliament Road, Palawatta, Battaramulla, Sri Lanka. Retrieved from http://www.slr.lk/

Steen, B. (2000). A systematic approach to environmental priority strategies in product development (EPS) version 2000-models and data of the default method, CPM report 1999. Chalmers University of Technology, Environmental Systems Analysis, Gotheburg, Sweden.

Suraweera, I. (2016). E-waste issues in Sri Lanka and the Basel Convention. *Reviews on Environmental Health, 31*(1). doi:10.1515/reveh-2015-0069 PMID:26943598

Thavalingam, V., & Karunasena, G. (2016). Mobile phone waste management in developing countries: A case of Sri Lanka. *Resources, Conservation and Recycling, 109*, 34–43. doi:10.1016/j.resconrec.2016.01.017

University of Sri Jawardenapura (UoSJ). (2015). E-waste Management in Sri Lanka. Department of Accounting, Faculty of Management Studies and Commerce, University of Sri Jayewardenepura, Nugegoda, Sri Lanka.

Vidanaarachchi, C. K., Yuen, S. T. S., & Pilapitiya, S. (2005). Municipal solid waste management in the Southern Province of Sri Lanka: Problems, issues challenges. *Waste Management (New York, N.Y.), 26*(8), 920–930. doi:10.1016/j.wasman.2005.09.013 PMID:16298122

Western Province Waste Management Authority (WMA). (2019). *Strategic plan. New Provincial Council Building*. Battaramulla, Sri Lanka: Denzil Kobbekaduwa Mawatha.

Wijesekara, H. R., De Silva, S. N., Wijesundara, D. T., De, S., Basnayake, B. F. A., & Vithanage, M. S. (2015). Leachate plume delineation and lithologic profiling using surface resistivity in an open municipal solid waste dumpsite, Sri Lanka. *Environmental Technology, 36*(23), 2936–2943. doi:10.1080/09593330.2014.963697 PMID:25209886

Xiao, R., Awasthi, M. K., Li, R., Park, J., Pensky, S. M., Wang, Q., ... Zhang, Z. (2017). Recent developments in biochar utilization as an additive in organic solid waste composting: A review. *Bioresource Technology, 246*, 203–213. doi:10.1016/j.biortech.2017.07.090 PMID:28756989

Chapter 16
Sustainable Solid Waste Management in Developing Countries:
Tanzania Outlook

Proscovia Paschal Kamugisha
Mzumbe University, Tanzania

Sebastian Faustin Mhanga
MEDA, Tanzania

ABSTRACT

Resilient economic urbanisation growth in Tanzania signifies national increasing trend of solid waste generation that necessitates proper management. About 50% of Tanzanians suffer from intestinal schistosomiasis and soil transmitted helminths resulting from haphazard solid waste disposal. Various efforts to address solid waste management in Tanzania include establishment of public private partnership and major reforms in policy and regulatory frameworks. Despite these efforts, 60% of waste generated is mismanaged and normally disposed in illegal places along roadsides, sewage canals, and reserved open spaces. Cornerstone of proper solid waste management lies on separation of the waste at source. It is recommended that 'the smart garbage collection bin' be adopted at community collection points and charging higher prices for generators who do not separate the waste.

BACKGROUND

Solid waste has been defined differently. Vaccari and Vitali (2015) described solid waste as items that are useless (unwanted) to initial user which are neither liquid nor atmospheric emissions. Meanwhile, Leblanc (2018) delineated solid waste as garbage arising from animal and human activities that are discarded as unwanted and useless. The solid wastes arise from human activities such as agricultural, industrial, mine wastes and production process from all subsequent steps of converting raw materials to

DOI: 10.4018/978-1-7998-0198-6.ch016

consumption goods along the chain; and the discarded leftovers at consumption point (Tchobanoglous et al., 1993). In Tanzania solid waste is defines as "non-liquid materials arising from domestic, street, commercial, industrial and agricultural activities that includes refuse or garbage, non-liquid materials arising from construction and demolition activities, garden trimmings and mining operations, dead animals and abandoned cars scraps" (URT, 2004). All scholars commonly agree that solid waste refers to non –liquid nor gaseous wastes that have no further use from generator's perspective.

Resilient economic development and growing urban population are key elements that are associated with incremental trends of solid waste generation. This is because as economies urbanize, people's incomes increase due to increased economic activities. In turn the increased income raises consumption of goods and services that enlarges the amount of solid waste generated (IFC, 2014).

With regard to economic development, Khatibu (2015) states that daily generation of solid waste per person rises as the national per capita income increases. In fact, Khatibu (2015) found that low income countries earning per capita income of up to US$ 5,000 generates lower solid waste than higher income countries. While low income countries generate solid waste per capita ranging between 0.3–0.9 kg/day; high income countries generate between 1.4–2.0 kg/day (Khatibu, 2015). Similarly, UNDESA (2010) portrayed direct relationship between solid waste generations and national incomes (Figure 1).

Figure 1. Generation of waste according to income (Source: Kaza et al., 2018)

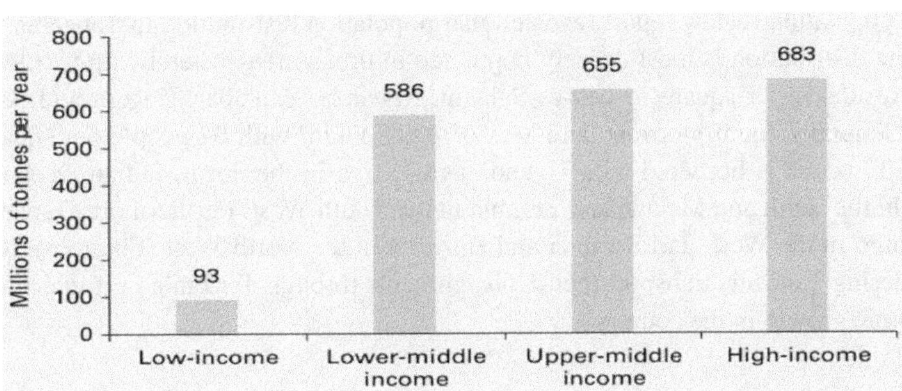

Figure 2. The effect of urbanization on waste generation (Source Kaza et al., 2018)

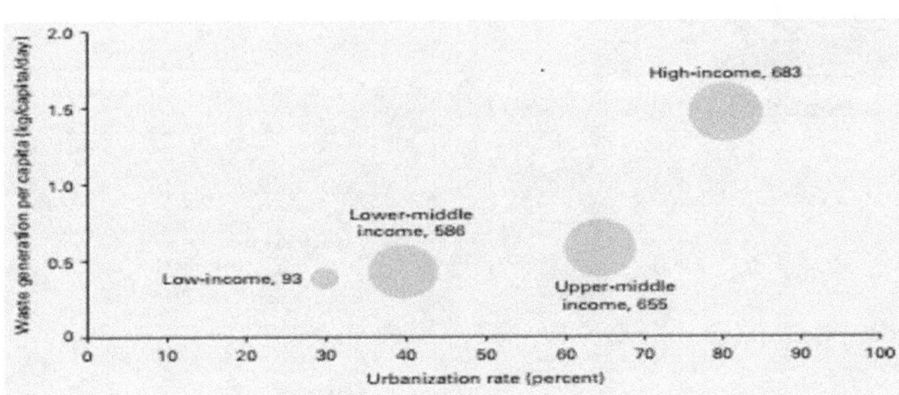

In respect with urbanization, urbanization rate is greater in developing countries than developed countries. Khatibu (2011) revealed that urban growth stood at 23% and 4% between 1990 and 2010 for developing and developed countries respectively (Khatibu, 2011). While Africa's urbanized proportion is 40% that of high-income countries accounts for 75% (UN, 2014). However, UN (2014) projected that by 2050 about 56% of its population will be living in urban areas (Cohen, 2006; UN, 2014). The increasing trend of urbanization and economic development in developing countries that is linked with escalating volumes of solid waste generation (Figure 2) is coupled with waste management challenges in these countries. It is estimated that one to two-thirds of the solid waste generated in developing countries is not properly managed (Firdaus & Ahmad, 2010; Senzige *et al.,* 2014). However, 20-50% of municipalities' total budget is spent on solid waste management (Hoornweg & Bhada-Tata, 2012). The situation is worsened by the fact that the budget spent on solid waste management serves less than 50% of the population (Kingu & Yhdego, 2016).

Tanzania is not exceptional from other developing countries in terms of increasing trend of solid waste generation that is tailored by economic and urban growth. UN (2017) estimates Tanzanian population to stand at 60,913,557 residents in 2019 based on 44,928,923 people estimated from National census survey of 2012 which indicated annual growth rate 3.08%. From arithmetic based on the current population estimates the country's population growth trend is equivalent to an increase of three persons in every minute. With this trend, the national population is expected to reach 72,681,070 and 138,081,621 in 2025 and 2050 respectively (UN, 2017).

The world population review report revealed that population distribution in Tanzania is not evenly distributed. The distribution is most densely populated in urban areas whereby up to 3,920, 330, 593, 110 and 133 residents per square in Dar-es-Salaam, Mwanza, Zanzibar, Kagera and Kilimanjaro is experienced. Generally country covers land area of 885,800km^2 with 64 people per square kilometer (URT, 2017). Tanzania is bordered with Uganda and Kenya in the north, Indian Ocean in the East, Mozambique in the south and Malawi and Zambia in the South West. Furthermore, Tanzania boarders with DRC Congo in the West; and Rwanda and Burundi in the North West (Figure 3). Most of these countries bordering Tanzania transport their Consignments through Tanzania port which increases the level of economic growth in the country.

Figure 3. The map of Tanzania

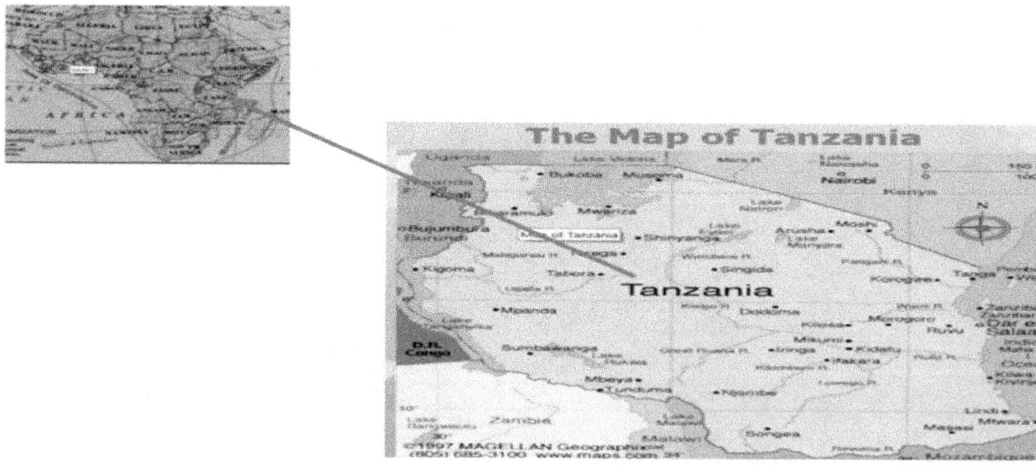

Figure 4. The estimations of population increase in Tanzania

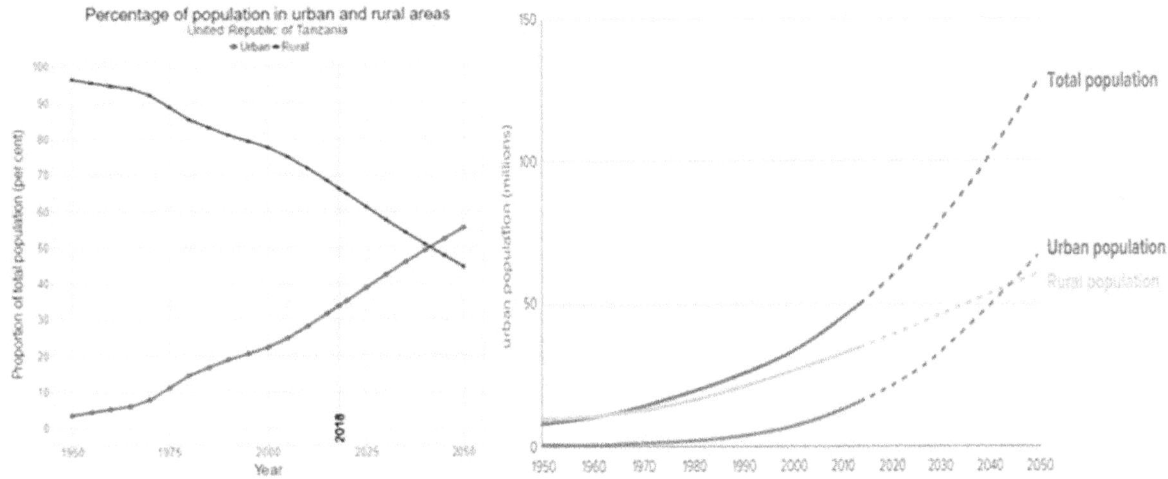

Economic development of Tanzania has sustained Gross Domestic Product (GDP) growth rate of 7% since 2000; while it aims to graduate from lower to middle income country by 2025 as stipulated through Tanzania Vision 2025, The National Strategy for Growth and Reduction of Poverty Phase II (Locally known as MKUKUTA II) and National Development Plan II (2016/17-2021/22) (URT, 1999; 2010; 2016). According to UNDP and ESRF (2017) the country has made significant progress in key indicators of human development. Between 1985 and 2015, Tanzania's Human Development Index score increased from 0.371 to 0.531, representing an increase of 43% over 30-year period that is above Sub-Saharan Africa countries. In addition, Tanzania has recorded a significant reduction in the proportion of the population that is living in multidimensional poverty from 64% in 2010 to 47% in 2016 (UNDP and ESRF, 2017).

As from census conducted in 2012, 30% of the population were based in urban areas with urban growth rate of 5.4% annually; that ranks Tanzania to hold the sixth highest growing urban country in the globe (WDI, 2016). This urbanisation rate is greater than that of the World average (2.1%) and higher than that of Africa (3.5%) annually. Tanzania's urban population will increase by 1.4 million per annum that is projected to reach 68.6 million people by 2050 that is more than fivefold in a 40 years period (IGC, 2014); that is equivalent to 38.8% and 55.4% in 2025 and 2050 respectively (UNDESA, 2018) (Figure 4).

According to NBS (2013) there are 33 urban local government authorities in Tanzanian mainland that include town, municipal, and city councils. In total these urban councils were hubs of 30% of total Tanzanian population. These urban councils are categorized into four main clusters: Dar-es Salaam that is the largest city composed of 4.4million residents (10%) of national population. Other urban councils' populations range between 0.3million and 0.8million residents. These include Mwanza, Arusha, Dodoma, Mbeya and Morogoro with 0.71m, 0.42m, 0.41, 0.38 and 0.31m residents respectively (Worall *et al.*, 2017). The remaining district councils are hubs of about 150,000 inhabitants each. Projections for the population growth in six major urban municipalities are shown in Figure 5 below.

The urbanization and population growth indicate generation of more solid waste in the country. Based on the current per capita generation of solid waste of 0.815kg/day (Huisman et al., 2016) the solid waste shall be more voluminous in the country for some years to come.

Figure 5. The estimated increase in population of six major urban municipalities

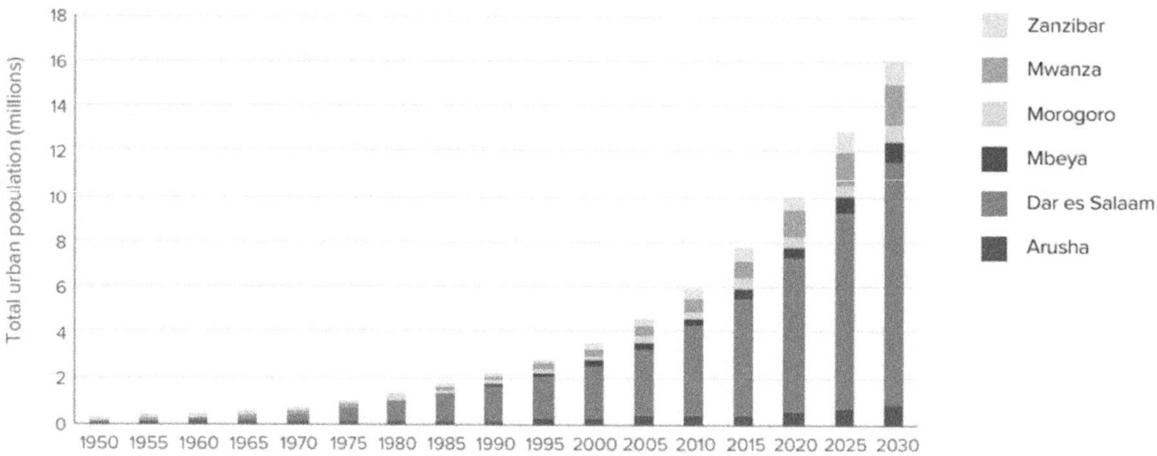

IMPACT OF SOLID WASTE IN LDCS AND TANZANIA IN PARTICULAR

Improper solid waste management experienced lead to generation of pollution problems, potential risks to public human health and the environmental concerns (Medina, 2010). UN- Habitat (2010) estimated the incidence of diarrhoea and acute respiratory infections to be twice and six times higher respectively, for children living in the households where solid wastes are dumped at homesteads compared to children living in the households where wastes are dumped away. Moreover, the safety of waste pickers is endangered as in most cases they perform their duties with bare hands without protective gears. Gwisai et al. (2014) found that major effects of disease symptoms experienced by waste pickers and landfill employees in Botswana was respiratory system related ailments that accounted for 89%; which is aggravated by the fact that scavengers work protective gears. Mothiba et al. (2017) reported similar case in South Africa whereby 89% of waste pickers at landfills had respiratory symptoms related ailments. The general routes through which solid wastes would lead to negative impacts to human being has been well articulated by Oeltzshner and Mutz (1996) (Figure 6).

Negative effects associated with improper waste management in Tanzania is noteworthy that requires attention to all stakeholders. Outwater et al. (2013) revealed that 50% of Tanzanians suffer from intestinal schistosomiasis and soil transmitted helminths resulting from haphazard solid waste dumping. Sakijege et al. (2012) revealed presence of total coliform in drinking water resulting from poor solid waste management to stand at 70 and 23 counts/100ml compared to WHO recommended rate of zero (0) coliform/100ml (coliform free water) (WHO, 2017). Contamination of water results to infectious waterborne diseases such as cholera, dysentery, typhoid fever and schistosomiasis. Moreover, Sakijege et al. (2012) determined incidence of diarrhoea, schistosomiasis and typhoid to stand at 53, 21 and 17 cases respectively at just one dispensary of Keko Machungwa in Dar es Salaam city within three months of March, April and June in 2009.

In response to this situation various efforts have been made to address the issue of solid waste management. Establishment of Public Private Partnership (PPPs) since 1995 has been in place through private companies and Community Based Organisations (CBOs). However, this model in Tanzania has

Figure 6. The route of negative impacts on human health from improper waste management

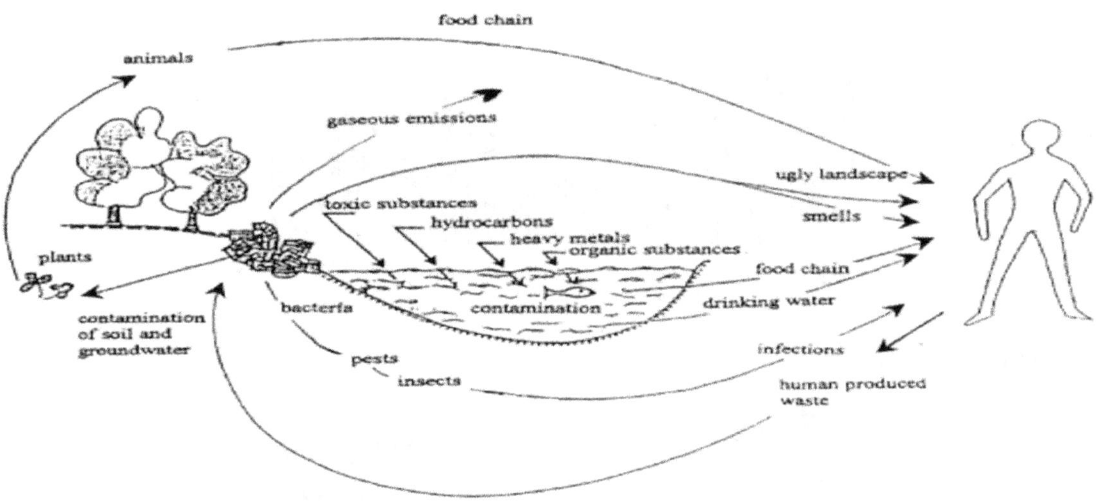

not performed well since its initiation in 1995 when only handful 5.5% of the solid waste (SW) was collected and disposed to the dumping sites (Palfreman, 2014). In Tanzania, the dumping site is an area allocated for disposal of solid waste. Establishment of PPPs has been coupled with major reforms in policy and regulatory framework. Formation of National Environmental Policy (NEP), 1997 and launching of Environmental Management Act 2004 are examples of amendments made in the regulatory framework. Moreover, the Tanzania Strategic cities project among other things has built 317 solid waste collection points and 5 sanitary landfills in eight urbanizing cities of Tanga, Arusha, Mwanza, Kigoma, Dodoma, Ilemela, Mbeya and Mtwara (Kingu and Yhdegho, 2016). Furthermore, the project has built the capacity of community to practice for landfill operators in the mentioned cities.

Despite these initiatives, solid waste has remained high in urban cities. Local authorities in various municipalities have managed to collect, transport and dispose average 40% that ranges between 33% and 90% of the generated waste to open dumping sites (Figure 7) (Breeze, 2012; Palfreman, 2014; The Bartlett UCL & DPU, 2017). The remaining 60% of the waste is disposed in illegal places normally along road sides, sewage canals and reserved open spaces leading to environmental and public health risks (Palfreman, 2014) (Figure 8, 9, 10).

CLASSIFICATION OF SOLID WASTE

Classification of solid waste is mainly done based on its composition and its source/origin. Solid waste composition includes durable goods, non-durable goods, containers and packaging materials, food wastes, yard trimmings, and miscellaneous inorganic wastes. In most developing countries solid wastes are classified as biodegradable (food and kitchen waste, green waste), recyclable (glass, bottles, plastics, metals), electrical (electrical appliances, televisions, computers), hazardous (paints, chemicals, light bulbs), and toxic (pesticides, herbicides) wastes (Karija et al., 2013). Alternatively, classification of solid waste from its sources has been reported to be from households, institutions, market, industries and streets (Membe, 2015).

Figure 7. Open dumping site

Figure 8. Waste littered along the road

Figure 9. Waste littered in water channel

Figure 10. Illegal disposal of waste

Table 1. Sources of solid waste (tons/day)

Authors & location		Household	Commercial	Institution	Market	Industries	Streets	Informal sector waste	Others	Total
Membe 2015 (DSM)	Quantity	3104	223	20	281		5	-	621	4254
	Proportion (%)	73.0	5.2	0.5	6.6	0	0.1	-	14.6	100
Mnyanyi 2014 Kinondoni	Quantity	1619	30.4	12.1	40.5		2	322.2		2026
	Proportion (%)	79.9	1.5	0.6	2.0		0.1	15.9		100
Kingu & Ydegho 2016 -Mbeya	Quantity	218	22	20	30	15		-		305
	Proportion (%)	71.5	7.2	6.6	9.8	4.9		-		100

Table 2. Solid waste proportional composition (%) in Tanzania

	Authors and Locations								
Waste type	**Bubegwa 2012 (DSM)**	**Kingu & Yhdegho 2016 (Arusha)**	**Kingu & Yhdegho 2016 (Mbeya)**	**Kingu & Yhdegho 2016 (Mtwara)**	**Senzige *et al*,. 2014 (DSM)**	**Palfreman 2014 (DSm)**	**Bigirwa 2015 (dsm)**	**Breeze 2012**	**Average**
Organic	39	49	77	71	58	84	70	40	64
Grass/wood	10		3					10	7
Papers	8	8	3	5	10	8	4	8	6.6
Ceramic & stones	6								6
Metals	5	5	3	3	3				3.8
Plastics	16	16	5	9	16	2	19	16	11.9
Glass	2	2	2	7	9				
Leather & rubber	6	6							6
Textiles	5	5	4	1	1	17			3
Brick		6							6
Electronics				1	2				2
Autoparts				3					2
Tetrapak						3			3
Hazardous						0.8			8
Others	3	3	3		1		8	30	4
Total	**100**	**100**	**100**	**100**	**100**	**100**	**100**		**100**

Few studies in Tanzania have classified solid waste based on its origin (Table 1). Households form major source of waste (75%) followed by informal sector (15.9%) while commercial sites, institution, markets and industries account for 7% each (Senzige et al., 2014).

Alternatively, classification of solid waste in Tanzania is classified based on material composition (Magashi and Schluep, 2011; URT, 2018_a). Table 2 indicates that organic wastes dominate (64%) the wastes generated in Tanzanian urban centres that ranges between 39% and 84%. This is followed by plastics with an average composition of 11.9% that ranges between 2% and 19% among Tanzanian cities. The remaining components accounts for at most 7% of waste generated (Table 2).

E - Waste

E-waste is a waste generated from discarded or used up electrical and electronic appliances. An increasing demand of electric and electronic appliances leads to the rising trend of E-waste generation. These electrical and electronic appliances include computers & their accessories; photographic equipment such as cameras; musical instruments like pianos, mobile phones and television sets among others. In 2017, about half of the world's population equivalent to 3.6 billion people were using internet, 4.2 billion active mobile-broadband subscriptions (Baldé et al., 2017). More than four fifth of the world's population is covered by a mobile broadband signal with 54% of households having Internet access at home and 48% having computers (Baldé et al., 2017). This pattern was associated with global E-waste generation of 44.7 million metric tons (Mt) equivalent to 6.1 kilogram per inhabitant (kg/inh). The global amount of world E-waste is expected to increase to 52.2 million metric tons, or 6.8 kg/inh, by 2021 (Baldé et al., 2017).

Globally, only 8.9 Mt (20%) of E-waste generated were documented to be recycled through appropriate channels in 2016 (Baldé et al., 2017). Asia is acknowledged to generate the largest amount of E-waste in the globe (18.2 Mt); that is followed by Europe (12.3 Mt), the Americas (11.3 Mt), Africa (2.2 Mt), and Oceania (0.7 Mt) (Baldé et al., 2017) per annum. Although African continent hosts the least number of direct manufacturers of EEEs, it bears a substantial burden of global E-waste mainly from imports of new and used equipment. Locally generated E-waste accounts for 50% to 85% of the waste (Baldé et al., 2017). Meanwhile the remaining portion is illegally imported from developed countries as far as Americas, Europe and China (Secretariat of the Basel Convention, 2011).

The E-waste management in Tanzania is not well documented. However, several authors have argued that their composition in the municipal solid waste ranges between 1 – 2% (Senzige et al., 2014; Mgimba and Sanga, 2016). Tanzania Communication Regulatory Authority (TCRA) (2013) estimated the country to have 29 million registered mobile phones. Meanwhile, guidelines for Management of Hazardous Waste of June 2013 estimated that E-waste generated in the country was in the range of 18,000–33,000 tons

Figure 11. The import of E-equipment in Tanzania

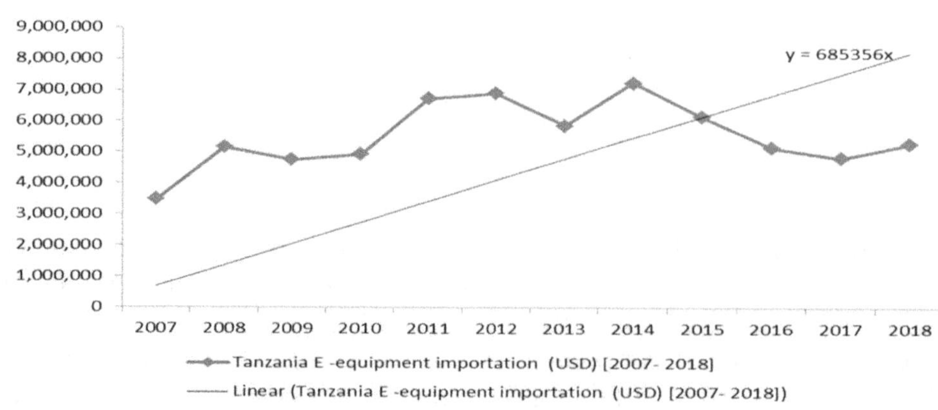

of computers each year (URT, 2018_b). However, the volume of entire E-waste is not documented since it entails dynamics composition that includes mobile phones, computers and television sets. Despite partial documentation of E-waste volumes, the value of electrical and electronic appliances imported is available; and has shown the increasing trend (Figure 11). The imported e – equipment will finally turn into e waste as its economic useful time elapse. Generally, there is potential for increased E-waste generation in the country due to increased use of electric and electronic appliances.

Minerals that are contained in electronic and electrical appliances impair human health status and environment at large. Mercury normally found in fluorescent tubes and tilt switches is known for its sensory impairments, dermatitis, memory loss and muscle weaknesses (Khan, 2017). While Sulphur mostly found in lead acid batteries causes damages on vital body organs like heart, liver and kidney (Sankhla et al., 2016; Khan, 2017); lead normally found in cathode ray tubes; monitor glass and lead acid batteries impair cognitive functions and lower intrigant quotients. Similarly, Brominated Flamed Retardants mostly found in plastics of electronic appliances deter development of nervous systems, thyroid disfunctions and liver problems (Needhidasan & Melvin, 2013; Sankhla et al., 2016). The fact that some components of E-wastes are sources of heavy minerals necessitates proper management options to address health risks caused by these minerals.

Hazardous Waste

Tanzanian Environmental Management Act 2004 defines 'hazardous waste' as any solid, liquid, gaseous or sludge waste which by reason of its chemical reactivity, environmental or human hazardousness, its infectiousness, toxic explosiveness and corrosiveness is harmful to human health, life or environment. However, this chapter deals only with solid hazardous waste in its nature. In this case hazardous waste is considered from health sector since it is widely documented and prevalence of health facilities throughout the country.

About 10% of wastes generated in health care facilities are hazardous. These wastes are classified into infectious waste, pathological waste, sharps, pharmaceutical waste, Geno-toxic waste, chemical waste, heavy metal content waste, pressurised containers, radioactive waste, blood, body fluid waste, and expired/unwanted medicines, drugs and cosmetics (URT, 2018_a). According to URT (2018_a) there are 8,928 health facilities in the country. The facilities are comprised of 6 national hospitals, 6 national super specialized hospitals, 6 Zonal super specialized hospitals, 10 referral hospitals, 25 regional hospitals, 60 other hospitals, 70 district hospitals and 36 designated district hospitals. Other health facilities include 801 health centres and 7,413 dispensaries. According to the Ministry of Health and Social Welfare (2007); 4,745 tons of generated medical waste per year is produced per health facility. Moreover, URT (2014) estimated the rate of generating medical waste to stand at 0.41kg/occupied bed/day in hospitals.

The proportion of Tanzanian health waste managed safely is lower (40%) than that of Sub-Saharan Africa (60%) (WHO, 2015; URT, 2018_a). Failure to safely manage health care waste is associated with non-separation of the waste at source due to lack of storage and transportation facilities. Kuchibanda and Mayo (2015) estimated that average daily solid waste generation per patient at Shinyanga Regional Referral Hospital and Kolandoto District Designated Hospital stood at 1.6kg and 1.3kg respectively. Given the predominant health care facilities in the country standing at 8,928 distributed all over the country (URT, 2018_a) and increasing urban population; projection of hazardous care waste will increase exponentially in future times.

The impact of improper health care waste management includes transmission of diseases and contamination of underground water tables through untreated health care waste landfilling. WHO (1999) estimated that, globally there are about 8 to 16 million new cases of Hepatitis B virus (HBV), 2.3–4.7 million cases of Hepatitis C virus (HCV) and 80,000–160,000 cases of Human Immunodeficiency Virus (HIV) annually due to poor waste management systems.

In most developing countries health care waste is deposited in general waste dumps and neighboring environment without segregation of hazardous and non-hazardous waste (Mgimba and Sanga, 2016). In Tanzania, the health care wastes are disposed onto dumping sites while mixed and managed by municipal authorities (URT, 2018_a). Sharp health care waste management options after use are commonly understood to involve burning, incinerating or burying the same in the land-fills without proper treatment (SBC and WHO, 2004). Manyele and Anicetus (2006) identified that main disposal methods among hospitals involve open pit burning (50%) and burying (30%) of the waste. The rest was transported to the municipal dumping sites. Moreover, a large proportion (71%) of hospitals used dust bins for transporting waste from generation points to incinerator without plastic bags. Mwakanyamale (2017) identified that sharp waste at Makambako health center are disposed and burned in a pit hole or open burning. Moreover, the wastes were not separated from other health care waste such as food left overs and papers surrounding the hospital compound.

Generally, it can be concluded that hazardous waste disposal in the country violates legal and regulatory framework that requires hazardous wastes to be treated through safer and less hazardous routes as stipulated in Environmental Management Act. 2004; and the Solid Waste Management Regulations of 2009 that necessitate landfill sites to be constructed with full leachate and gas management systems (SWMR, 2009).

Construction and Demolition (C&D) Waste in Tanzania

Construction industry entails the use of various resources to construct infrastructure necessary for socio and economic use. It embraces the process by which the said physical infrastructure is planned, designed, procured constructed or produced, altered, repaired, maintained, and demolished (NBS, 2010). However, the construction industry is associated with generation of solid waste that enlarge as construction expands.

Statistics for Construction and Demolition Waste (C&D) in Tanzania is not well documented. However, there is prevalent massive construction projects in the country that include Standard Gauge Railway (SGR), expansion of Mwanza Airport, construction of bridges at TAZARA and Ubungo intersection in

Figure 12. The increasing construction projects from year 2005 to 2017

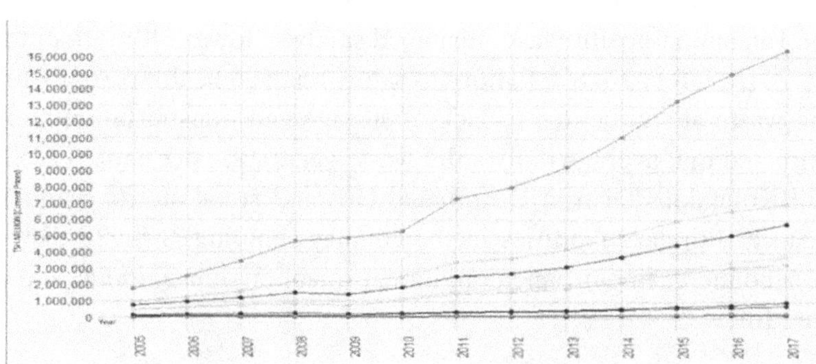

Dar-es-Salaam, the Songosongo natural gas project, the Madimba Processing Centre, the infrastructure for Phase 1 of the Dar es Salaam Rapid Transit Bus, the cement industry and massive road construction. Kikwasi and Escalanta (2018) portrayed growing pattern of values in the construction projects (Figure 12).

The aforementioned construction projects are associated with increasing C&D waste that must be managed well for the future of the nation. Sabai *et al.* (2016) revealed that C&D waste generation increased by more than 100% in Tanzania within 16 years period from 3.03 million tons in 1994 to 7.9 million tons in 2010. Dumping C&D waste puts pressure for acquisition of large portions of land in order to accommodate the disposal of the growing waste generated from construction and demolition sites (Sabai et al., 2016).

Efforts geared turning C&D waste into building materials exists (Sabai et al., 2016; Sabai and Rugudagiza, 2018). Sabai et al. (2016) showed that C&D waste can be recycled into production of building materials in the country while Sabai and Rugudagiza (2018) demonstrated that oil re-refined sludge C&D waste can be recycled into building material in Tanzania. Although C&D waste trend seem to increase in Tanzania; its usage is still huge such that currently it can't put pressure in the environment. Huisman et al. (2016) contended that though construction and demolition waste seems to be generated in substantial quantities in DSM; it is, not considered to be a problem as used by inhabitants for improving dirt roads.

Marine Solid Waste: Litter

Generally, categorization of solid waste to marine waste in Tanzania does not exist. Information on marine litter has been generalized as solid waste generated from urban centers. However, it is acknowledged that marine litter leads to negative effects for marine life and ecosystem if not properly managed. The marine litter is sought to come from illegally deposited wastes in open areas and through water canals especially during the rainy season (Figure 13). Biubwa et al. (2014) found that Zanzibar Islands municipality collect, transport and dispose 45% to 50% of mixed waste to open area without proper treatment. The low disposal rate of municipal authority's wastes is hindered by limited budget, inadequate technical capacity and lack of comprehensive legal and regulatory framework and weak enforcement of the present legal tools. Similarly, Maziku (2014) revealed that low rate of solid waste disposal along Dar-es-Salaam Coastal Belt is influenced by inaccessibility of unplanned settlements, inadequate waste collection facilities, lack of motivation within waste collection companies and weak enforcement of municipal bylaws.

Moreover, Yhdegho and Kingu (2017) revealed that Lindi and Mtwara-Mikindani municipalities along the Indian Ocean coast disposal rate of solid waste stood at 21 tons/day (37.73%) and 59tons/day (60.51%) respectively. The low collection rate in these municipalities was attributed to lack of equipment, lack of staff within the municipal councils, absence of/or weak involvement of community-based organiza-

Figure 13. The abundance of marine debris in water channel

tions to deal with the waste at household levels, failure of municipalities to collect service charges and enforce the solid waste management regulations. It can be concluded that marine litter documentation is mixed up with general solid waste.

SOLID WASTE MANAGEMENT METHODS

The common phenomenon of solid waste management in developing countries entails collection, transportation and disposal to dump sites that mostly are open space areas (Khatib, 2015). This is coupled with burning of some waste in residential areas, or at the dumping sites to reduce its volume (Merina, 2010). Hoornweg and Bhada-Tata (2012) reported various solid waste management approaches practiced by developing countries to include open dumpsites, landfilling, compositing and recycling (Table 3). Accordingly, dumpsite and landfilling dominated other solid waste management practices in developing countries. Compositing and recycling accounted for the lowest proportions of solid waste management options in developing countries. Diaz (2017) estimated that recycling and compositing of solid waste generated in developing countries' municipalities ranges between 10% - 15%.

Solid waste management in Tanzania is limited despite increasing trend of population and urbanization. Most urban center waste collection and disposal rate is averaging at 40% that range between 33% and 90% of the waste generated in Dodoma and Moshi municipalities respectively (Kingu & Yhdego, 2016) (Table 4). Moshi and Tanga cities would save as learning cases for up-scaling in other country's

Table 3. Methods of solid waste management (SWM) by country in Developing countries

Country	Region	Collection Rate (%)	Dumps (%)	Landfills (%)	Composite (%)	Recycled (%)	Others (%)
Ghana	Africa	85					
Haiti	Latin America & Caribbean	11	24				76
Madagascar	Africa	18		97	4		
Mali	Africa	40					
Nepal	South Asia	94					
Niger	Africa			64		4	32
Senegal	Africa	21					
Sierra Leone	Africa	44					
Tanzania	Africa	48					
Uganda	Africa	39		100			
Zambia	Africa	20					
Comoros	Africa	20					
Benin	Africa	23					
Cambodia	East Asia & Pacific	75	100				
Kyrgyz Republic	Europe & Central Asia			100			
Total		**494**	**124**	**361**	**4**	**4**	**108**
Average		**38**	**62**	**90**	**4**	**4**	**54**

Source: Hoornweg and Bhada-Tata. 2012

Table 4. Solid waste management in Tanzanian urban Centers in 2016

Urban centers	Dumping site	Population	Waste generation (tons/day)	Collection (tons/day)	Proportion of waste collected (%)	number of wards in the municipality	Number of wards served by municipality in SWM
Mbeya city council	Nsalaga	385,279	400	140	35	36	8
Mwanza city council	Buhongwa	706,453	338	227	67	18	8
Tanga city council	Mwang'ombe	313,625	185	166	90	27	11
Arusha city council	Murriet	507,903	550	302	55	25	15
Moshi municipality	Kaloleni	210,000	225	203	90	21	12
Dodoma municipality	Municipal	507,350	305	100	33	47	15
Lindi municipality		78,841	23	11	48	18	5
Mtwara-mikindani municipality	Mangamba	108,299	98	59	61	18	5

Sources: Kingu and Yhdego. 2016

town centers. However, even Moshi and Tanga waste management involves collection, transportation and disposals of the waste to open dumping sites.

Generally Tanzanian' solid waste management options entail collection, transportation and disposal to open dumping sites (Palfremen, 2014; Mollel, 2016). Palfreman (2014) revealed that solid waste management practiced by households in Dar-es-Salaam include burning, burying, dumping and re-using through gardening that accounted for 50%, 25%, 20% and 5% respectively in Dar es Salaam city. Mollel (2016) observed that 84% of households disposed their waste to the dumping sites through municipal services while 15.6% had their own pits at home where solid waste is disposed. Accordingly, the waste disposed through municipal services is mainly done by contracted Community based organizations (CBOs). Upon collection from households, CBOs solicit waste by separating plastics for sale and food waste for composting manure and sell plastic bottles to the small-scale factories. It was estimated that approximately 3,000 tons of solid waste per day is buried, burned and/or left somewhere in DSM's streets, open spaces, canals, sewers and shores in Dar es Salaam that in turn contributes to health problems for local residents, flooding and methane emissions (Huisman *et al.,* 2016). Collection and transportation of the waste to disposal sites in Tanzania municipalities employs the following tools: trucks, tractors, trails, pushcarts and wheelbarrows. Transportation of waste is done by mixing various types of waste (Figure 14).

Figure 14. The collection of waste in waste collecting transport

Potential Hazards of Greenhouse Gases (GHGs) From Burning Solid Wastes

GHGs refer to atmosphere gaseous elements that absorb and emit radiation at specific wavelengths within the spectrum of infrared radiation emitted by Earth's surface, atmosphere and clouds (IPCC, 2001). The GHG molecules absorb radiation of which they radiate it back to the atmosphere leading to global warming. The Global warming disastrous effects on the environment and on human health cannot be overstated. Table 5 Illustrates global compositions of GHGs from open waste burning and associated

Table 5. GHGs due to open burning of waste and their potential effects

Pollutant	Global Emission due to Open Burning (Kg/year)	Proportion of Total Global Emission of Pollutants (%)	Potential Health Impacts
GHGs			
Carbon Dioxide (CO_2)	1.4 Trillion	5%	Cardiorespiratory failure Climate change associated risks
Methane (CH_4)	3.6 Billion	1	Respiratory arrest Climate change associated risks
Persistent Organic Pollutant (POP)			
Polycyclic Aromatic Hydrocarbons (PAH)	334 million	39	- Lymphoma - Leukemia - Lung cancer - Reproductive issues - Neurodevelopmental issues - Developmental issues
Polychlorinated Dibenzodioxins/furans (PCDD/F) Toxic Equivalency (TEQ)	206	N/A	
Total Polychlorinated Biphenyls (PCB)	123,000	N/A	
Polybrominated Dibenzodiozins/furans Toxic Equivalency	80	N/A	
Benzene	875 million	25	Chromosomal mutations Acute myeloid leukemia
Particulates matter			
Coarse Particulates (PM_{10})	12 Billion	24	- Lung cancer - Respiratory diseases - Heart failure
Black Carbon (BC)	10 Billion	29	
Organic Carbon (OC)	5.1 Billion	43	

Source: Cogut A. 2016

Figure 15. The emission of greenhouse gases from various sectors

SUB SECTOR	EMISSIONS SUMMARY, Gg			
	Carbon Dioxide, CO_2	Methane, CH_4	Nitrogen Oxides, NO_x	Nitrous Oxide, N_2O
1: Energy				
1A1: Energy Industries (Electricity production)	281.37	0.011527	-	0.002305
1A2: Manufacturing Industries and Construction	482.16	0.018636	-	0.003719
1A3: Transport	1,817.51	0.245014	-	0.119040
1A4: Others (Residential)	368.45	0.015000	-	0.003000
1B: Fugitive Emissions (Coal Mining and Handling)	-	1.085065	-	-
1C: Carbon dioxide Transport and Storage	-	-	-	-
Sub Total	**2,949.49**	**1.375242**	-	**0.128064**
2: Industrial Processes and Product Use				
2A: Mineral Industry (Cement Production)	411.50	-	-	-
3: Agriculture, Forestry and Other Land Use				
3A: Livestock				
3A1: Enteric fermentation	-	616.000000	-	-
3A2: Manure management	-	22.000000	-	1.000000
Sub Total:	-	**638.000000**	-	**1.000000**
3B: Land				
3B2: Agricultural soils	-	-	-	61.000000
3C: Aggregate sources and non-CO_2 emissions sources on land				
3C1: Biomass burning	63,529.83	17.016918	-	2.268922
3C1b: Field burning of agricultural residues	-	89.000000	3,502.00	122.000000
3C1c: Prescribed burning of savannahs	-	4,166.000000	40.00	1.000000
3C7: Rice cultivation	-	54.000000	-	-
Sub Total:	**63,529.83**	**4,326.016918**	**3,542.00**	**125.268922**
4: Waste				
4A: Solid waste disposal	-	1,165.800000	-	-
4B: Biological treatment of solid waste	-	-	-	-
4C: Incineration and open burning of waste	-	-	-	-
4D: Wastewater treatment and discharge	-	10.480000	-	5.210000
Sub Total:	-	**1,176.280000**	-	**5.210000**
GRAND TOTAL:	66,890.82	6,141.672160	3,542.00	192.606986

potential environmental and human health hazards. The amount of emissions differs across countries based on composition of wastes burnt. The open waste burning is not the main source of GHGs globally yet its impact is significant.

In Tanzania, Agriculture, Forest and Other Land Uses (AFOLU) sector emits major portion (94.98%) of CO_2 standing at 63,529.83Gg annually (Figure 15). Although none of emission of CO_2 from waste disposal has been identified, yet AFOLU contribution is from biomass burning (URT, 2014). Moreover, solid waste disposal accounts for 18.98 of Methane (CH_4) emitted in the country. This is equivalent to 1,165.8Gg of methane out of the total 6,141.67 emitted national wide (URT, 2014). This amount is huge given the adverse impact it has on public health. Moreover, the AFOLU sector which composes major part of the methane production is mainly from burning of agricultural residues that would be considered as waste, and burning of Savannah.

Furthermore, waste disposal accounts for 2.7% of the nitrous oxide produced (URT, 2014). It can be concluded that main sources of greenhouse gases in Tanzania are biomass burning, field burning of agricultural residues and burning of savannah grassland that in total account for 94.97%, 69.56%, 100% and 65.03% of carbon-dioxide, Methane, Nitrogen Oxide and Nitrous Oxide respectively (URT, 2014). The solid waste disposal contribution is still minimal in the country. However, the population growth trend of about 5.4% per annum with and the tendency of open burning of solid waste (Figure 16) poses a dilemma for its contribution if its management is not systematically dealt.

Figure 16. Open burning of waste at a disposal site

Policies, Legal and Regulatory Framework by Each Waste Category

Regulatory framework has been amended to address among other things the improvement of solid waste management. One policy and four legislations form the backbone of the legal and regulatory framework for sustainable management of the environment and solid waste management in particular. The policy is the National Environmental Policy (NEP), 1997 while legislation consists of Tanzania & Local Government (Urban Authorities) Act 1982, the Environmental Management Act 2004.

The National Environmental Policy (NEP), 1997, addresses the issue of 'environmental pollution' among six (6) environmental problems it addresses. In this regard NEP (1997) addresses the issue of solid waste management through the issue of environmental pollution that is cited on its Caption 11, 28 and 29 whereby the issue of minimization of solid waste are clearly stipulated through the use of technologies that either lessens associated solid waste in production or recycling of the wastes in a more acceptable manner. The Environmental Management Act 2004 promotes conservation and management of the environment. In this Act some aspects that are directly associated with solid waste management include (i) The legal framework for the overall management of the environment giving power and responsibilities for various organs and institutions enforcement mandate. (ii) Establishment of administrative and institutional framework for the management of the environment. In its Part IX Caption 114 -119, the Act stipulates roles and responsibilities of LGAs in managing the solid waste. The Act gives mandate to the Local Government Authorities to involve the private sector and Non-governmental Organizations (NGOs) in solid waste management activities. Moreover, the Act requires the waste to be separated based on its type. However, separation of the solid waste practices doesn't exist to date. Moreover, the Local government (Urban Authorities) Act 1982 entrusts urban authorities the responsibility to ensure that their areas of jurisdiction and sanitary conditions are kept clean as stipulated in its Caption 55. In addition, the Act delegates the urban authorities with the mandate to make their own by-laws to enable them to execute their responsibility of waste management in their respective areas of administration.

Regulations for E-Waste

The National Environmental Policy 1997 stipulates the overall framework under which environmental management issues in the country are settled. The policy promotes health related programs that require separation of toxic /hazardous wastes and pollution control at household level and proper E-waste management reduction to the environment and associated human health. The National Energy Policy (2003) promotes the use of energy efficient equipment to reduce quantities of E-waste generation. Science and Technology Policy (1996) encourage the use of scientific knowledge and techniques that promotes information sharing on research and development so as to manage E-waste sustainably in the country. The National Information and Communication Technology Policy (2003*)* promotes development of competitive ICT products and services that ensure E-waste recycling activities that could lead to establishment of relationships with manufacturers and designers of ICT equipment and end users of the E-resources.

Currently, Tanzania has no specific E-waste management regulatory framework. The E-waste is regulated through the solid waste and hazardous regulations prescribed under the Environmental Management Act (2004). Part VIII of the Environmental Management (Hazardous Waste Control) regulations, 2009 of the Environmental Management Act (2004) addresses the issue of electrical and electronic waste. In its Regulation 35 (1), it requires every individual who possess or has control of electrical or electronic tools, accessories or equipment to single out the E-waste from other types of waste and deposit separately into different bins as prescribed by entrusted Authorities. Regulation 37 allows manufacturers of E based facilities to set-up and operate individually or collectively voluntary trace-back systems for electrical and electronic waste from customers provided that no fee is chargeable for that service. Moreover, regulation 39 vest powers to Local Government Authorities in ensuring safe handling of e - waste to minimize risks to human health and the environment.

Regulations for Hazardous Waste

Tanzania is a member of the Basel Convention, established in 1989, that prevents shipment and disposal of hazardous waste from industrial to developing countries. This treaty establishes agreements for requirements of trans-boundary movement of hazardous waste. In fact, the convention agreed to adhere to three major aspects. (1) Minimizing generation of hazardous waste. In this aspect the Basel Convention's efforts are to push responsibility to generators of the waste promoting the environmentally sound management of hazardous wastes during the production process. (2) Treat and dispose hazardous wastes as close as possible to where they were generated; and (3) Minimize international movements of hazardous wastes.

To complement Basel Convention, African Nations established Bamako Convention in 1991 that came into force in 1998. The Convention aimed to protect the health of populations and the environment of African countries through a ban on the import of all hazardous and radioactive wastes. It also prohibits the dumping or incineration of hazardous wastes in oceans and inland waters, and promotes the minimization and control of trans-boundary movements of hazardous wastes within the African continent that ensures ecologically friendly management and handling of hazardous waste within Africa, as well as the cooperation between African nations.

Although Tanzania has signed the Basel and Bomako Conventions, it has neither specific policy nor Act for hazardous waste management. The waste is managed through The National Environmental Policy of 1997; and the Environment Management Act. 2004. Part IX of the Act stipulates waste management requirements whereby section 133 clarifies controls mechanisms of governing import, export, transit and internal transport of hazardous wastes. The section empowers the Minister to formulate regulations ensuring proper management of hazardous waste. Hence the Environmental Management (Hazardous Waste Control and Management) Regulations 2009 was formed based on this section whose Part IV (captions 18 – 26) deal with trans-boundary movement of hazardous waste. The regulations vest the powers to the Director of the environment to administer the Basel convention control specifications for monitoring movements of hazardous in the country.

Hazardous waste can be considered from health care wastes regulatory framework that includes National Health Policy (2017). The policy envisages among others to ensure efficient safe management of health care waste in communities; strengthen systems for healthcare waste management at health care delivery settings and improve systems and capacities to adopt and mitigate public health risks from climate change.

Regulations for Construction and Demolition (C&D) Wastes

The construction and Demolition waste are generally regulated under the National Environmental Policy of 1997; and the Environmental Management Act. 2004. However, there are no specific captions dealing with the waste. Other policies related to C&D waste include Tanzania Construction Industry Policy 2003 and National Transport Policy of 2003.

Regulations for Marine Wastes

Marine Litter management in Tanzania is regulated through the National Environmental Policy 1997 and Environmental Management Act 2004. In its part IX, the Act stipulates management of the litter in the country. In its caption 121 and 122 stipulates responsibility of occupier to ensure in public places that he/she provides bins for waste collection, transportation and disposal to the designated areas; and adhere to the guidelines set for effective waste management. Caption 122 vested the power of sanctioning non-compliers to environmental officer to take any reasonable steps to remedy that default; and shall recover the cost of so doing from the occupier as a debt due to the service.

Regulations for Industrial Solid Waste

The Tanzanian industrial solid waste regulation is done through the National Environmental Policy 1997 and Environmental Management Act 2004. Industrial waste'' means waste emanating from processing industries or non-processing industries that is the source of energy, water, treatment plants or communication and includes any other solid waste referred to under Part. IX. In its caption 116. -(1), the Act requires local government authorities to ensure that industries located within their respective areas of jurisdiction provide adequate space and facilities for managing all solid waste generated from such industries before they are collected for disposal at designated places. Moreover, in its section (2) The local government authorities is required to ensure that refuse bays or areas set aside by industries for the collection of solid waste are clean and protected from flies, animals and scavengers. Moreover, Caption 112 of the Act empowers the minister responsible to prescribe rules obliging disclosing of information on status and disposal system of waste to either the council/any other designated person by a person who owns/operate industrial production plants. Moreover, caption 113 vest powers on the council to procure a prohibition notice if the council is of the opinion that the prescribed activity (industry processing plant) involves an imminent risk of serious pollution to the environment.

ECONOMICS OF SOLID WASTE MANAGEMENT

Organic and plastic wastes add up to 75% of waste generated in Tanzania; such that they are of great concern. Fortunately, the markets for plastic waste are well developed; though through informal routes by waste pickers to the plastic recycling factories. It is estimated that 5,500 tons of plastics are recycled monthly (Palfreman, 2014). However, development of markets for organic waste is still at infant stage. Recycling of the wastes generated in Tanzania ranges between 5% and 18% of the total wastes generated in urban cities (Breeze, 2012; Huisman *et al.*, 2016).

Despite its infant stage, economics of organic solid waste management in the country's municipalities seem to be profitable. Mollel (2016) revealed that both organic waste composite and plastic recycling have positive Net Present Value suggesting that both options are profitable in three-year time period using discounted rate of 12%. UCL and DPU (2017) suggested turning of organic waste into chicken feeds through Black Soldier Fly (BSF) technology that can either be fed to chicken directly or be incorporated in chicken feeds as a protein source. Nasilele (2015) estimated that the BSF technology can potentially reduce organic waste by 50% to 78% of all the waste generated.

Moreover, UCL and DPU (2017) suggested transformation of organic waste into energy through the use of Anaerobic Digester (AD) technologies. This is a prototype of bio digesters used in Tanzania rural settings to produce biogas and fertilizer from livestock manure. The study results identified that a bio-digester manufacturer in Dar es Salaam produces modular domestic biogas systems that cost 550,000TSh ($ 236.32) which can be acquired through a financial plan based on credit and instalments for up to 18 months. Clemens *et al.* (2018) estimates that households using this technology could save 0.9 tons and 1.7 tons of fuel wood and charcoal per annum respectively.

UNDP (1994) earlier proposed construction of biogas plant that would treat 1,500 tons of waste produced in Dar-es-Salaam by then; that would produce 41,900 tons of diesel oil and reduce 6,200 tons of carbon dioxide per annum. The plant's life time was projected to last for up to 30 years; that would breakeven in three years life time. Its economic viability was thought to be gained through sale of three major products produced: electricity, biogas to be used as fuel in cars and composite fertilizer. Moreover, (Sabai *et al.* (2013) and Sabai & Rugudagiza (2018) revealed that C&D wastes can be converted into useful building materials.

It can be concluded that solid waste generated in Tanzania can be recycled, reused and reduced for provision of energy and compositing for useful organic fertilizers if separation at source is done. All of technologies suggested to work effectively require waste separation at source. Senzige *et al.* (2014) contend that if wastes are separated at source for composites of organic waste and recycling of recyclables; the wastes to be disposed at the dumping sites would have been reduced to only 2% of all wastes generated. However, in Tanzania separation of waste at source is negligible. In most cases separation is done at the collection points and municipal dumping sites by informal waste scavengers; for processing mainly for plastic industry. Picking waste from mixed ones pose health risks to waste scavengers and the public at large since the separation involves e waste, hazardous waste as those from health care facilities. Moreover, the process becomes time consuming such that labor involved in waste picking would have been employed in other productive sectors.

All the way from organic waste, E-waste, hazardous waste and C&D waste have shown the requirement of separation at source to be a prerequisite condition for proper management of the solid waste. Hence deliberate efforts are needed to take care of separation for effective solid waste management.

WAY FORWARD

The separation of waste at household level requires awareness campaign among community members. Monella and Leyalo (2017) found that enabling factors of household participation in waste separation include levels of household income, education, wards handling status, incentives in terms of monetary and peer influence. Moreover, presence of bins for waste separation at collection points is a prerequisite condition for waste separation to be successful. Coupled with monetary incentives to waste separators (households) would lead to robust results. In fact, the plastic waste management model in the country can be up-scaled to other types of wastes when end markets are available and known. Plastic waste pickers are motivated by earnings they get from collection to the developed market that is further recycled to have different products.

Therefore, we can say that a certain motivation or incentive is required for effective separation of the waste. The waste of those who are separating their waste can be priced higher than those which are mixed up at individual households. From household level after collection, waste pickers can send it to

Figure 17. Waste collection system of smart garbage bins

the nearby community collection waste bins. Various bins for different types of wastes must be in place whereby the waste will be transported further to the final destinations for processing. Smart Garbage Collecting Bins developed by Arunkumar in India (Figure 17) is a prototype that can be adopted in each ward for effective automation of solid waste management in the countries' urban centers. The technology composes of Global System for Mobile Communication (GSM) and ARM (Acorn RISC Machine) 7 controller. Sensors are placed in common garbage bins placed at the public places. When the garbage reaches the level of the sensor, then that indication will be given to ARM 7 Controller. The controller will give indication to the driver of garbage collection truck as to which garbage bin is completely filled and needs urgent attention. ARM 7 will give indication by sending SMS using GSM technology (Arunkumar *et al.,* 2016). Moreover, innovators need to develop technological sensing techniques that would identify type of waste such that if you want to place garbage in inappropriate container, it should not open. However, technical and financial feasibility for those innovations need to be evaluated before promotion of such technologies.

The PPPs in dealing with development and handling of the projects would yield robust results. As such the PPPs will enhance accountability on both sides while government is regulatory authority, the Private sector will monitor the project closely. But when this project is solely handled to government, its incentive to undertake the business solutions will be limited. Therefore, PPPs is recommended to avoid situation like that happened during 1989 when the government sought aid from Japan and was awarded funding to acquire thirty refuse trucks and refuse containers, three skip trucks and six compacter trucks for effective solid waste management. However, by 1992, only three years after receiving the Japanese aid, twenty-six of the thirty refuse trucks, five of the six compacter trucks and two of the three skip trucks had all broken down to scrapers (Solomon, 2011).

Soliciting of soft loans by the government for development agenda from International Development Institutions such as IMF and the World Bank can serve the purpose. Then, collection of waste fee can be traced to pay back the loan. Envisaging the cleanliness agenda is in line with sustainable development goals number 12 that requires responsible consumption and production for management of natural resources, efficiently disposal of toxic waste, and cut food waste by half in 2030 through recycling, reusing and reducing the waste. Moreover, the biogas production project will add to the national energy requirements and contribute achievement of the sustainable development goal number 7 that envisage ensuring access to affordable, reliable and sustainable clean energy requirements for all.

Since the pay of waste management fees have been inefficient, the government can set a certain tax on services that are basic needs that every-body will pay for. Then the organization offering such services becomes an agent of government revenue collection. This service can include mobile phone operators whereby through the purchase of vouchers a certain proportion of the price paid becomes waste collection fee. Alternatively, collection of waste fee can be done through use of electricity whereby people can be charged as they purchase electricity energy for use. In this way solid waste collection fee can be effectively collected.

CONCLUSION

The practical implementation of solid waste separation at source and consecutive solid waste recycling, reusing and reduction can be piloted in one municipal for its trial. Then, upon its successfulness, it can be up-scaled to other parts of Tanzania urban centers.

ACKNOWLEDGMENT

This research received no specific grant from any funding agency in the public, commercial, or not-for-profit sectors.

REFERENCES

Arunkumar, G., Bhanu Priya, G., & Kumar, R. (2016). Smart Garbage Collecting Bin for Municipal Solid Waste. *International Journal of Modern Trends in Engineering and Science*, *3*(3), 9–18.

Baldé, C. P., Forti, V., Gray, V., Kuehr, R., & Stegmann, P. (2017). *The Global E-waste Monitor – 2017, United Nations University (UNU), International Telecommunication Union (ITU) & International Solid Waste Association*. Bonn: ISWA.

Biubwa, A., Norkhadijah, S., Ismail, S., & Rasdi, I. (2014, October). Municipal solid waste management of Zanzibar: Current practice, the challenges and the future. *International Journal of Current Research and Academic Review*, *1*, 5–19.

Breeze, R. (2012). Municipal Solid Waste Management in Dar-es-Salaam. A report prepared for the World Bank.

Clemens, H., Bailis, R., Nyambane, A., & Ndung'u, V. (2018). Africa Biogas Partnership Program: A review of clean cooking implementation through market development in East Africa. *Energy for Sustainable Development*, *46*, 23–31. PMID:31007459

Cogut, A. (2016). *Open Burning of Waste: A Global Health Disaster. The R_{20} Regions of Climate Action.* Retrieved on 26.01.2019 from https://regions20.org/wp-content/uploads/2016/08/OPEN-BURNING-OF-WASTE-A-GLOBAL-HEALTH-DISASTER_R20-Research-Paper_Final_29.05.2017.pdf

Cohen, B. (2006). Urbanization in developing countries: Current trends, future projections, and key challenges for sustainability. *Technology in Society*, *28*, 63–80.

Diaz, L. F. (2017). Waste management in developing countries and the circular economy. *Waste Management & Research, 35*(1), 1–2. PMID:28049391

Firdaus, G., & Ahmad, A. (2010). Management of Urban Solid Waste Pollution in Developing Countries. *International Journal of Environmental of Research, 4*(4), 795–806.

Food and Agriculture Organisation. (2014). World Agriculture: Towards 2015/2030 FAO Perspective. FAO.

Gwisai, R. D., Areola, O., & Segosebe, E. (2014). Respiratory and occupational health problems of scavengers and landfill employees in a municipal landfill site in Lobatse, Botswana. *Journal of Sustainable Development in Africa, 16*, 37–55.

Hoornweg, D., & Bhada-Tata, P. (2012). What a Waste: A Global Review of Solid Waste Management. The World Bank Urban Development Series Knowledge Papers Number 15.

Huisman, H., Breukelman, H., & Keesman, B. (2016). *Expert mission on integrated SWM to DSM, Report of findings November 2016.* Retrieved 15th August 2018 from https://docmh.com/expert-mission-on-integrated-solid-waste-management-iswm-to-dar-es-salaam-mat16tz01-report-of-findings-november-pdf

International Finance Corporation, World Bank Group. (2014). Handshake IFC's Quarterly Journal on Public – Private Partinerships. Author.

International Growth Centre. (2014). Urbanisation in Tanzania: Population Growth, Internal Migration and Urbanisation in Tanzania 1967-2012: A Census-Based Regional Analysis. Author.

International Trade Centre. (n.d.). *The Trade statistics for international business development yearly trade data. Import & export values.* Retrieved 13.03.2019 from https://www.trademap.org/tradestat/Country_SelProduct_TS.aspx?nvpm=1%7c%7c3981%7c%7c%7c35%7c%7c%7c2%7c1%7c1%7c1%7c2%7c1%7c3%7c1%7c1

Karagiannidis, M., Kontogianni, S., Wolff, S., Redwood-Sawyer, J., & Sankoh, S. (2012). Handbook on Waste Management in Developing Countries. Academic Press.

Karija, M. K., Shiua, Q. I., & Lukaw, Y. S. (2013). The Impact of Poor Municipal Solid Waste Management Practices and Sanitation Status on Water Quality and Public Health in Cities of the Least Developed Countries: The Case of Juba, South Sudan. *International Journal of Applied Science and Technology, 3*(4), 87–99.

Kaza, S., Yao, L., Bhada-Tata, P., & Van-Woerden, F. (2018). What a Waste 2.0: A Global Snapshot of Solid Waste Management to 2050. Urban Development Series. Washington, DC: World Bank. doi:-1329-0 doi:10.1596/978-1-4648

Khan, M. (2017). Environment and Health Issues Associated with E-waste. *International Journal of Advanced Research., 5*(1), 1425–1430.

Khatib, I. A. (2011). Municipal Solid Waste Management in Developing Countries: Future Challenges and Possible Opportunities. In *Integrated Waste Management* (vol. 2). InTech, Retrieved 22.03.2018 from: http://www.intechopen.com/books/integrated-waste-management-volumeii/

Kuchibanda, K., & Mayo, A. W. (2015). Public Health Risks from Mismanagement of Healthcare Wastes in Shinyanga Municipality Health Facilities, Tanzania. The Scientific World Journal.

Leblanc, R. (2018). *An Introduction to Solid Waste Management: Integrated Approach Needed for Successful Diversion and Recycling*. Retrieved 22.05.2018 from https://www.thebalancesmb.com/an-introduction-to-solid-waste-management-2878102

Ludete, J. H. (2016). *Factors Influencing Public Private Pertnerships in Solid Waste Management for Local Government Authorities: A Case of Kinondoni Municipality-Dar es Salaam* (Unpublished Master thesis). Sokoine University of Agriculture, Morogoro, Tanzania.

Magashi, A., & Schluep, M. (2011). E-waste Assessment Tanzania. The UNIDO E-waste initiative for Tanzania.

Manyele, S. V., & Anicetus, H. (2006). Management of medical waste in Tanzanian hospitals. *Tanzania Health Research Bulletin*, *8*(3), 177–182. PMID:18254511

McCartney, P. (2014). United Republic of Tanzania: Assessing if notifications, consents, inspections and enforcement of trans-boundary movements of waste and take-back procedures for illegal traffic represent environmentally sound management. Final Version. Retrieved 3.3.2019 from file:///C:/Users/KAMUGISHA/Downloads/UNEP-CHW-IMPL-CLI-ESM-Toolkit-Tanzania-Report-201412.English.pdf

Medina, M. (2010). *Solid waste, Poverty and the Environment in Developing Countries Cities: Challenges and Opportunities*. United Nations University/Institute of Advanced Studies Working Paper 23. Tokyo: UNU.

Mgimba, C., & Sanga, A. (2016). Municipal Solid Waste Composition Characterization for Sustainable Management Systems in Mbeya City, Tanzania. *International Journal of Science. Environmental Technology*, *5*(1), 47–58.

Mnyanyi, N. B. (2014). *Assessment of Solid Waste Management in Kinondoni Municipality* (Unpublished master thesis). Open University of Tanzania.

Mollel, E. L. (2016). *Economic Analysis of the Solid Waste Management Options in Morogoro Municipality* (Unpublished master thesis). Sokoine University of Agriculture, Morogoro, Tanzania.

Monella, J., & Leyaro, V. (2017). Determinants of Households Willingness to Participate In Solid Waste Separation for Reduce, Reuse and Recycle: The Case of Dar es Salaam. *University of Dar es Salaam Journal, 3*(1), 57–82.

Mothiba, M., Moja, S. J., & Loans, C. (2017). A Review of the Working Conditions and Health Status of Waste Pickers at Some Landfill Sites in the of Tshwane Metropolitan Municipality, South Africa. *Advances in Applied City Science Research*, *8*(3), 90–97.

Nasilele, M. R. (2015). *High Value Organic Waste Treatment via Black Soldier Fly Bioconversion* (Unpublished Master Thesis). Royal Institute of Technology, Stockholm, Sweden.

National Bureau of Statistics. (2010). *Integrated Business Survey, 2010: Construction Industry Analytical Report*. Tanzania Mainland.

National Bureau of Statistics. (2013). *Basic Facts and Figures, 2012 on Human Settlements in Tanzania Mainland*. Author.

National Sample Census of Agriculture. (2012). *The (2007/2008) report*. Author.

Needhidasan, S., & Melvin, S. S. (2013). Electronic Waste – Health and Environmental Impact and Recycling Practices. *International Journal of Advance Research, 1*(8), 8–10.

Ntakamulenga, R. (2012, January). *The Status of Solid Waste Management in Tanzania.* A paper presented at the Coastal East Africa solid waste workshop, Flic en Flac, Mauritius.

Oeltzchner, H., & Mutz, D. (1996). *Guidelines for an Appropriate Management of Domestic Sanitary Landfills*. Gesselschaft fur Technische Zusammenarbeit, Eschborn, German.

Outwater, A. H., Pamba, S., & Outwater, A. B. (2013). *Water –Related Diseases of People using Municipal Wasterwater: Risks*. Exposure, Effects on Health and Control Approaches in Tanzania.

Palfreman, J. (2014). *Waste Management and Recycling in Dar es Salaam, Tanzania*. Retrieved 02/05/2016 from https://www.researchgate.net/publication/271441207

Sabai, M. M., Cox, M. G. D. M., & Mato, R. R. A. M., Egmond-de Wilde De Ligny, van, E. L. C., & Lichtenberg, J. J. N. (2013). Concrete block production from construction and demolition waste in Tanzania. *Resources, Conservation and Recycling, 72*, 9–19.

Sabai, S. M. M., Lichtenberg, J. J., Egmond, E. L. C., Florea, M. V. M., & Brouwers, H. J. H. (2016). Construction and Demolition Waste Characteristics in Tanzania. Huria. *Journal of the Open University of Tanzania, 23*(1), 1–19.

Sabai, S. M. M., & Rugudagiza, B. X. (2018). Recycling of Oil Sludge together with Construction and Demolition Waste into Building Materials in Tanzania. *African Journal of Environmental Science and Technology, 12*(2), 84–90.

Sakijege, T., Lupala, J., & Sheuya, S. (2012). Flooding, flood risks and coping strategies in urban informal residential areas: The case of Keko Machungwa, Dar es Salaam, Tanzania. Jamba: Journal of Disaster Risk Studies, 4(1).

Sankhla, M.S., Kumara, M., Nandan, M., Mohril, S., Gaurav Pratap Singh, G.P., Bhaskar Chaturvedi,B., & Kumar, R. (2017). Effect of Electronic waste on Environmental & Human health- A Review. *Journal of Environmental Science, Toxicology and Food Technology, 10*(9), 98-104.

Secretariat of the Basel Convention. (2011). *Where are Weee in Africa?* Geneva: Findings from the Basel Convention E-waste Africa Programme. SBC.

Senzige, J. P., Nkansah-Gyeke, Y., Makinde, D. O., & Njau, K. N. (2016). The Potential for Solid Waste Recycling in Urban Area of Tanzania: The Case of Dar Es Salaam. *International Journal of Environmental Protection and Policy., 2*(5), 147–152.

Solomon, A. O. (2011). *The Role of Households in Solid Waste Management in East Africa Capital Cities* (Unpublished Doctoral Dissertation). Wageningen University.

The Bartlett UCL & Development Planning Unit. (2017). Transforming Solid Waste Management in Dar es Salaam. Final Report. Author.

The Environmental Management Act. (2004). 158pp

United Nations. (2014). *World Urbanization Prospects: The 2014 Revision, Highlights*. New York: United Nations.

United Nations. (2014). *World Urbanization Prospects: The 2014 Revision, Highlights*. New York: United Nations.

United Nations Department of Economic and Social Affairs. (2010). *Trends in solid waste management: issues, challenges, and opportunities*. International Consultative Meeting on Expanding Waste Management Services in Developing Countries (Final Draft), 18-19 March 2010, Tokyo, Japan.

United Nations Department of Economic and Social Affairs, Population Division. (2015). *World Urbanization Prospects: The 2014 Revision*. Author.

United Nations Department of Economic and Social Affairs. (2018). *World Urbanization Prospects 2018*. Retrieved 05.02.2019 from https://population.un.org/wup/Country-Profiles/

United Nations Development Program. (1994). Electericity, Fuel, and Fertilizer from Municipal and Industrial Organic Waste in Tanzania. A Biogas Plant for Africa. The United Nations Development Program Project of the Government of Tanzania Number URT/93/G31/A/1G/99.

United Nations Development Program and Economic and Social Research Foundation. (2017). The Tanzania Human Development Report 2017:Social Policy in the Context of Economic Transformation. Author.

United Nations Environmental Program. (2017). Summary Report: Waste Management in ASEAN Countries. Author.

United Nations Environmental Programm/SBC and World Health Organisations. (2004). *Preparation of National Health-Care Waste Management*. Plans in Sub-Saharan Countries Guidance Manual.

United Nations Human Settlements Programme. (2010). *Solid Waste Management in the World's Cities*. Author.

United Republic of Tanzania. (1997). *The National Environmental Policy*. Author.

United Republic of Tanzania. (1999). The Tanzania Development Vision 2025. Author.

United Republic of Tanzania. (2004). *The Environmental Management Act*. Author.

United Republic of Tanzania. (2010). *National Strategy for Growth and Reduction of Poverty II (NSGRP II)*. Ministry of Finance and Economic Affairs. Retrieved 22.12.2018 from http://www.mof.go.tz/mofdocs/DPE/MKUKUTA%20II%20English.pdf

United Republic of Tanzania. (2014). *The Second National Communication to the United Nations Framework Convention on Climate Change*. Author.

United Republic of Tanzania. (2016). National Five Year Development Plan 2016/17 – 2020/21: "Nurturing Industrialization for Economic Transformation and Human Development". Author.

United Republic of Tanzania. (2017). Tanzania Human Development Report 2017: Social Policy in the Context of Economic Transformation. Author.

United Republic of Tanzania. (2018). Ministry of Health, Community Development, Gender, Elderly and Children]. National Strategic Plan for Health Care Waste Management (2018-2022). Author.

United Republic of Tanzania. (2018). The Performance Audit Report on Electronic Waste Management. A Report of the Controller an Auditor General March 2018. Author.

Vaccari, M., & Vitali, F. (2015). Technical Aspects of Municipal Solid Waste Collection: Case Studies from East Africa: in Case Studies for Developing Globally Responsible Engineers. Global Dimension in Engineering Education.

World Development Indicators. (2016). *World Bank*. Retrieved from http://databank.worldbank.org/data/reports.aspx?source=world-development-indicators

World Health Organization. (2017). Guidelines for Drinking-Water Quality, Fourth Edition Incorporating the First Addendum. World Health Organization.

World Population Prospects. (2017). Retrieved 06.02.2019 from http://worldpopulationreview.com/countries/tanzania-population/

World Population Review. (2019). Retrieved 06.2.2019 from http://worldpopulationreview.com/countries/tanzania-population/

Yhdegho, M., & Kingu, A. (2017). *Solid Waste Management in Coastal Towns of Southern Tanzania Case of Mtwara- Mikindani and Lindi Municipalities*. A Working Paper, Environmental Resource Consultancy. Retrieved 16.03.2019 from https://www.researchgate.net/publication/312604961_SOLID_WASTE_MANAGEMENT_IN_COASTAL_TOWNS_OF_SOUTHERN_TANZANIA_CASE_OF_MTWARA-MIKINDANI_AND_LINDI_MUNICIPALITIES

Yhdego, M., & Kingu, A. (2016). *Solid Waste Management in Urban Centers of Tanzania: Leapfrogging Towards a Circular Economy*. Retrieved 14.5.2019 from https://www.researchgate.net/publication/311846222_SOLID_WASTE_MANAGEMENT_IN_URBAN_CENTERS_OF_TANZANIA_LEAPFROGGING_TOWARDS_A_CIRCULAR_ECONOMY

Chapter 17
Current Situation and Challenges of Waste Management in Thailand

Sandhya Babel
https://orcid.org/0000-0003-2378-4891
Sirindhorn International Institute of Technology, Thammasat University, Pathum Thani, Thailand

Anh Tuan Ta
Thammasat University, Pathum Thani, Thailand

Teshan Udayanga Habarakada Liyanage
Sirindhorn International Institute of Technology, Thammasat University, PathumThani, Thailand

ABSTRACT

Thailand covers an area of 513,120 km^2 and is located in the southern region of the continent of Asia. The total population was around 67.2 million in 2016. With increasing population growth, city expansion and rapid industrialization, generated 27.37 tonnes of waste in 2017. Thailand government has recognized problems associated with solid, hazardous, and electronic waste generation and issued policies/regulations to support the waste management system. Initiatives such as roadmap for municipal and hazardous waste, national master plan on cleaner production and cleaner technology, and national integrated e-waste management strategy have been introduced to improve the waste management system. However, there are many issues preventing the implementation of a successful waste management system in Thailand. Society has suffered from waste mismanagement because of lack of awareness and realistically applicable technology. There are also weaknesses in the processes of public participation, policy implementation, and institutional support.

DOI: 10.4018/978-1-7998-0198-6.ch017

INTRODUCTION

Thailand covers an area of 513,120 km^2 and is located in the southern region of the continent of Asia. The total population was about 67.2 million in 2016, of which 55.4 percent resided in a non-municipal area and 44.6 percent in a municipal area (NSO, 2016). Administratively, Thailand is divided into 76 provinces, 4 regions (northern, northeast, central, southern), and the Bangkok metropolitan area. Structure of Thailand's economy is characterized by the service sector that accounts for 61.5%, following by manufacturing 30.2% and agriculture 8.3%. The Gross National Product (GNP) per capita was 205,339 Baht (6437 USD) for 2016 (BOT, 2018). For 2017, the total gross domestic product (GDP) was 455.3 billion US$, and the GDP per capita was 6,730.2 US$ (NESDC, 2018). Keeping up with the population growth and city expansion, the solid waste generation in Thailand increased rapidly from 2008 to 2017. The Thailand Government in recognizing the problem of solid waste released policies and campaigns to support the waste management system, such as The National 3R Strategic Plan, Roadmap for Municipal and Hazardous Waste in 2014, and the National Solid Waste Management Master Plan (2016 - 2021). However, there are many issues preventing the implementation of a successful solid waste management system in Thailand. In 2017, 7.17 million tonnes of municipal solid waste (MSW) were still improperly disposed of, such as by open dumping or open burning in waste disposal sites (PCD, 2018). This chapter reviews the existing situation and highlights the challenges, with successful case studies of the waste management system in Thailand. Hazardous waste, e-waste issues, and impact of the MSW on Global warming are also discussed in this chapter.

Definition of Solid Waste

In Thai law, waste means residues in a solid, liquid, or gas state generated daily from any sectors, or any other hazardous substances [National Environmental Quality Act, B.E. 2535 (1992)]. Solid waste can be divided into 4 categories, MSW, organic waste, hazardous waste, and electronic waste, which are defined in the Thailand regulations as follows:

MSW means solid waste generated from municipal activities such as from residences, businesses, marketplaces, and institutes. The waste can be organic (food, leaf, grass, etc.), recyclable (glass, paper, metal, etc.), or general waste (wood, fabric), excepting municipal hazardous waste [Pollution Control Department (PCD), Ministry of Natural Resources and Environment (MNRE), B.E. 2550 (2007)].

Electronic waste includes any parts or accessories of electrical and electronic equipment, which reach the end of their useful life or become deteriorated or obsolete [PCD, MNRE, B.E. 2550 (2007)].

Hazardous waste is waste that contains or is contaminated with hazardous substances or exhibits hazardous characteristics, including being flammable, corrosive, reactive, toxic, or having specified components [National Environmental Quality Act, B.E. 2535 (1992)].

Waste Management System in Thailand

In general, the waste management system in Thailand comprises the collection, transportation, treatment, and disposal of waste. Waste is collected directly from households by trucks and indirectly at fixed stations at a frequency of 2 times per week up to every day, relying on the amount of waste generated in each area. The waste collection system in urban areas is generally better and more efficient than in rural areas. As reported by Thailand's PCD in 2016, Nonthaburi, Samut Songkhram, Phuket, and Bangkok are

Figure 1. Current solid waste management system in Thailand

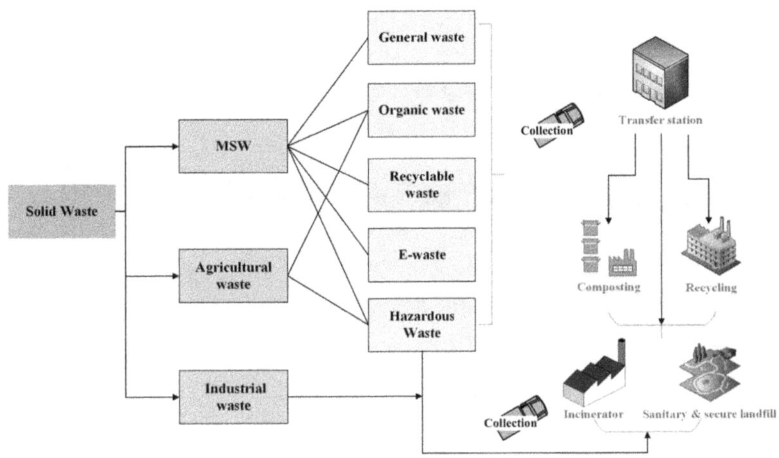

four provinces that could properly collect and treat 100% of the generated waste (PCD, 2017). In many municipalities, there is not enough space for waste disposal or final treatment due to lack of budget or scientific information, and public demonstration of landfill or incineration projects near communities.

Currently, the collection, transportation, and disposal of solid waste have been set up integratively as a cluster for local municipalities. An MSW management (MSWM) system is based on the waste generation rate, distance from the collection area, and disposal technologies. The MSWM system can promote the proper separation for recycling, composting, alternative energy, etc. before final sanitary disposal. A general waste management system is shown in Figure 1.

MUNICIPAL SOLID WASTE MANAGEMENT

Waste Generation and Composition

Like many other developing countries, Thailand also faces rising MSW generation. As reported by PCD (2018), from 2008 to 2017, the amount of waste generation increased by about 3.5 million tonnes, from 23.93 to 27.37 million tonnes (Figure 2). The amount of solid waste increased due to continued economic growth, urbanization, and industrialization. This led to increased consumption of pre-packaged consumer products, resulting in the generation of large amounts of waste from homes, institutions, offices, and business establishments.

For total waste generation, Bangkok has the largest part with 4.86 million tonnes (18%), and the 76 remaining provinces generated around 22.51 million tonnes (82%) in 2017. The efficiency in the operation of MSW disposal and utilization has been improved. However, it still falls behind the total waste generation. In other words, sanitary space for the treatment of the collected waste is not enough.

In 2017, the daily MSW generation rate in Thailand was 1.13 kg/capita. However, waste generation rates are differences in different geographical locations. Factors, including lifestyle, living standards, and population density, have an influence on waste generation rates. In the smallest towns (population

Figure 2. Trends of MSW generation in Thailand (2008 – 2017) (PCD, 2018)

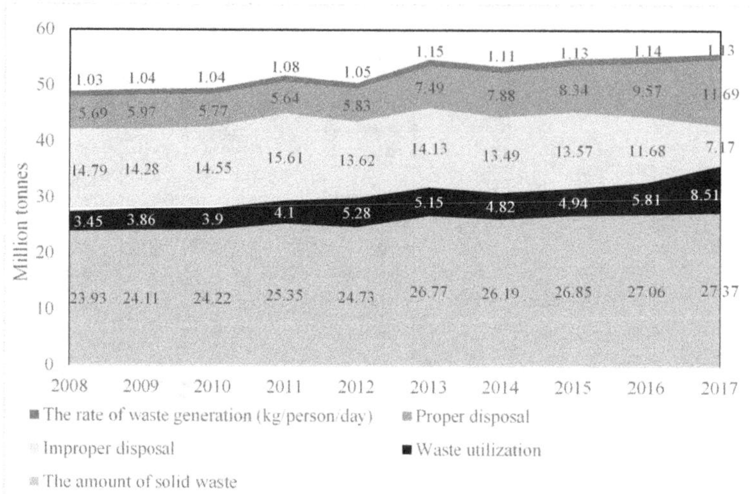

size < 5000) the rate was 1.02 kg/capita, while in fully urbanized towns (population size < 50000) the rate was 1.53 kg/capita (Challcharoenwattana & Pharino, 2016). In tourist zones, waste generation relies on the type of attraction and season. Generally, waste generation is about 0.5 kg per person per day for tourists who do not stay overnight and 1–2.5 kg per person per day for those who do. In national parks, the rate is about 0.5–0.8 kg per person per day during the low season, and 1–2.5 kg per person per day during the high season. (PCD, 2016).

Thailand's MSW characteristics are dominated by organic waste (40 – 66%) which is compostable and has a high moisture content. Plastic, paper, glass, and metals are predominant in non-food waste.

Figure 3. MSW composition from selected provinces in four regions of Thailand: North: Chiang Rai (Dokmaingam & Manomaivibool, 2017), Northeast: Khon Kaen (Hongyon & Hovijit, 2014), Central: Phetchaburi (Suthapanich, 2014), Bangkok (BMA, 2016), and South: Songkhla (REO-16, 2015)

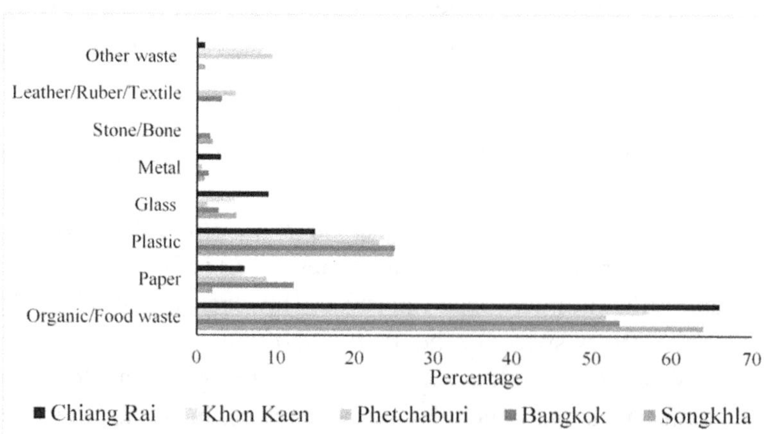

Varieties of waste composition, relying on many factors such as population density, urbanization, and income level. Different regions in Thailand have different waste compositions (Figure 3). In the Bangkok area, the composition of the collected waste shows effects of urbanization and rising income on the MSW composition. As shown in Figure 3, Bangkok residents dispose a lower amount of organic waste, but a higher amount of packaging wastes like paper and plastic. This indicates a high demand of modern society for convenient packaging which contributes a large part of the waste.

Waste Flow and Management Technologies

MSWM in Thailand still faces issues due to a lack of budget and effective integrated management systems. In 2016, the waste collection systems in Thailand only covered 66% of local administrative organizations (5,169 divisions). In the 2,607 remaining divisions of local administrative organizations (34%), the residents had to dispose of waste improperly (PCD, 2017). As a result, in 2016, 11.68 million tonnes of MSW (45% of the nationwide waste) were disposed of by open burning or dumping on owned land and abandoned sites (Figure 4).

As reported by PCD (2018), waste utilization in Thailand has been improved in 9 years, from 3.45 (2008) to 8.51 million tonnes (2017). For instance, in 2016, 5.81 million tonnes or 21.47% of the nationwide waste had been utilized. Most of the utilized waste was recyclable waste such as paper, glass, plastic, metal, and aluminum, with a total amount of about 5.20 million tonnes (89.50% of the total utilized waste). After being sorted, the recyclable wastes could be sold to junkyards, community recycling plants, or waste recycling banks, to be sent back into the manufacturing process. Some may be reused as materials for new products, such as car tires being reused as plant pots. The organic wastes, such as food wastes, vegetables, and fruit, were also utilized as soil supplement compost for agricultural purposes. In Thailand, composting technology varies from simple, like windrow composting which is implemented on small scales at households and communities, to complicated, like in-vessel composting which is applied at municipal levels. In Bangkok, the government has contracted a private company to operate a composting plant located at On Nut transfer station with a capacity of 1200 tonnes of MSW per day.

Figure 4. Flow of MSW in Thailand, 2016 (PCD, 2017)

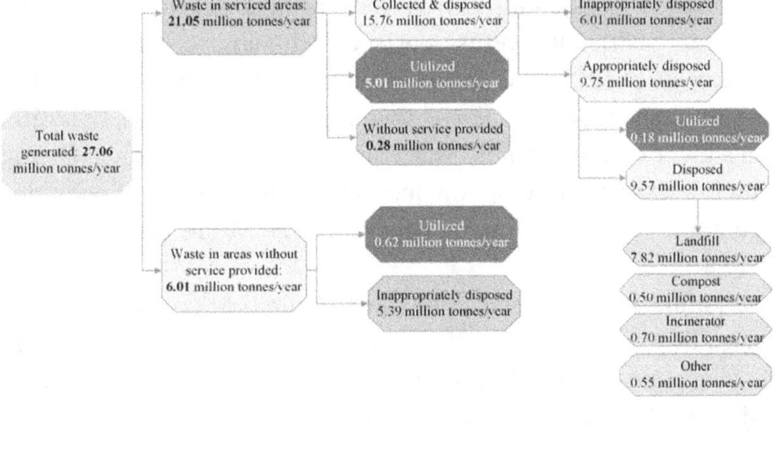

Table 1. MSW treatment sites in Thailand

Quality	MSW Disposal Method	Number of Sites	
		Public	Private
Meet Thailand's MSWM standards (2017) (*)	Sanitary/engineered landfill	96	13
	Controlled dump	379	86
	Incinerators with air pollution control system	25	20
	Mechanical- biological treatment system	19	4
	Segregation system, compost, and correct landfill system	30	5
	Total	549	28
Do not meet Thailand's MSWM standards (2014) (**)	Open dump	1955	
	Controlled dump	24	
	Incinerator does not meet pollution control standard	45	
	Total	2024	

(*)(PCD, 2018)
(**)(ThaiPublica, 2014)

The plant produced 300 tonnes/day of compost. However, the poor of organic waste separation at source results in poor-quality organic fertilizers (BMA, 2016). Recently, the Takakura home composting method has been introduced to 33 districts by the Bangkok government, to convert food waste into compost. Moreover, vermi-composting by earthworms has turned into a popular alternative for making fertilizer from organic waste in Thailand. Some of these organic wastes can also be reused as animal feeds. Approximately 0.60 million tonnes of the organic wastes were utilized (10.33% of the total utilized waste).

In 2017, 2867 MSW disposal sites were operated in Thailand (PCD, 2018). Out of this number, there were 677 disposal systems that properly treated waste, while 2237 systems were improperly operated (one site can include more than one system). There were 18 waste treatment sites that could not start operation after construction because of strong public protests by the nearby residents. Table 1 shows the number of proper and improper disposal sites. As shown in the table, controlled dumping is the preferred treatment method in Thailand compared with engineered or sanitary landfilling. The incineration of MSW is an expensive method that requires less space and produces leachates and gases. Recently, the pollutants from incomplete combustion increase public health concerns (Siriratpiriya, 2014). Furthermore, Thailand government organizations and research institutions have been conducting feasibility studies to produce energy from MSW, such as gasification, refuse-derived fuel, and anaerobic digestion.

Table 2 presents the costs of MSW collection and treatment in Thailand. As shown in the table, the cost of MSWM is mainly contributed by expenses for collecting solid waste. Landfilling and open dumping are popular management options of solid waste in Thailand due to low construction and operation expenses. Incineration is a less preferred option due to its high expenses of operation and maintenance. Construction and operation expenses for more modern MSWM are often higher than the landfilling method (Pharino, 2017).

In Thailand, local administrative organizations are required to collect and discard of MSW under the Thailand Public Health Act B.E. 2535 (1992) (TPHA). However, the TPHA regulation only allows local governments to charge an MSW collection fee from waste generators. The treatment and disposal cost are not indicated in the TPHA because the cost is expected to be covered by the Enhancement and

Table 2. Costs of MSW collection and treatment (ThaiPublica, 2014)

Treatment Method	Collection Fee (USD per tonne)	Treatment Cost (USD per tonne)	Total (USD per tonne)
Sanitary landfill	24.38	9.00	33.38
Incinerator	24.38	31.32	55.69
Refuse-derived fuel (RDF) system	24.38	4.90	29.28
Integrated MSWM	24.38	14.34	38.72

Table 3. MSWM expenses in selected municipalities in Thailand (Challcharoenwattana, 2015)

Parameter	Peri-urban	Urbanized Municipality	Most Urbanized Municipalities
	Thumbon Lamthough, Nakhon Ratchasima	Muang Nong Pru, Chonburi	Nakorn, Nakhon Ratchasima
Population	5950	61198	136,153
MSWM fee per head (USD)	0.91	18.06	23.23
% collected fee covered in MSWM expense	36.36	11.48	0.05

Conservation of the National Environmental Quality Act B.E. 2535 (1992). As a consequence, most local governments use the fees to handle both collection and treatment expenses. However, the collected fees cover only 10-36% of the total cost for MSWM in small municipalities and less than 10% in larger municipalities, as shown in Table 3. The lack of sufficient financing for MSWM is a major problem in finding a sustainable solution.

Current Policies for Municipal Solid Waste Management

The structure of the waste management system in Thailand includes three different levels as specified by the responsible jurisdiction. The central government includes agencies in the MNRE, such as the PCD, Office of Natural Resources and Environmental Policy and Planning (ONEP), Department of Environmental Quality Promotion (DEQP), the Ministry of Interior with the Department of Local Administration (DOLA), and the Ministry of Public Health with the Public Health Department. The central government establishes policies, regulations, guidelines, programs, and standards for waste management and the execution of these regulations. Regional governments are responsible for the connection between the central and local governments. The regional governments help to transfer and implement policies and guidelines from the central government to a local government. Local governments are directly responsible for handling and treating waste. Primarily, national environmental policies are set by the ministries. The departments and organizations under the ministries are responsible for executing the provisions of the law through regulations and technical guidelines. A diagram presenting government organizations related to waste management is shown in Figure 5.

In general, Thailand's national policies on MSWM control the waste generation rate, waste separation, material recovery, and environmentally friendly waste treatment technology that have minimal impacts

Figure 5. Waste management authorities and their responsibilities in Thailand

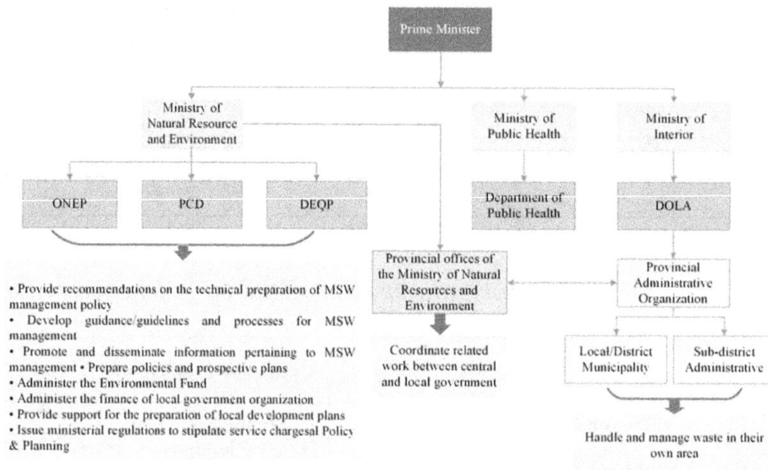

Table 4. Summary of legal framework and policies for SWM in Thailand

Type of Waste	Policies/Plans	Laws/Acts	Key Elements
Overall	National Solid Waste Management Policy Framework for Thailand's Environmental Quality 20-Year Pollution Management Strategy and Pollution Management Plan 2017-2021	The Constitution of the Kingdom of Thailand B.E.2550 (2007) The Enhancement and Conservation of National Environmental Quality Act, 1992	- Affirm the public right to access information and to participate in the prevention and mitigation of public hazards; protecting, conserving, and passing on natural resources and the environment. - Include crucial environmental laws governing environmental standards, including planning, and monitoring environmental quality, and establishing a system for Environmental Impact Assessment - Waste prevention and reduction at source. Increasing efficiency in waste treatment and pollution control at source. Development of waste management system.
MSW	- - National 3R Strategy (draft) - Integrated waste management scheme - National Solid Waste Management Master Plan (2016 - 2021).	Public Health Act, 1992 3R Act (draft)	- Provide a legal role for local administrations to manage MSW by developing and issuing ordinances and regulations for collection, transportation, and disposal of waste. - Promote the segregation of waste at source, encourage collaboration among different stakeholders to implement effective waste management - Encourage the public and the private sector to reduce waste at the source by following the 3R. - Establish proper disposal methods for MSW and household hazardous waste by using centralized facilities for clusters of municipalities, emphasizing waste utilization and waste to energy methods. - All relevant sectors participate in the management of solid and hazardous waste.
Hazardous Waste Management	National Master Plan on Cleaner Production and Cleaner Technology Road Map on Waste and Hazardous Waste Management 2014	Hazardous Substance Act, 1992 Hazardous Substance Act, 2001	- Provide a legal basis to control the import, export, manufacturing, storage, transport, and disposal of hazardous substances. The act governs the methods of managing hazardous materials, hazardous waste, and infectious waste. - Minimize amount of accumulated waste. Encourage appropriate MSW and Hazardous Waste. Specific law on SW Management.

Table 5. Strategies for reduction, reuse, and recycling (PCD, 2011)

Strategies	Reduction Targets (%)		
	1-5 years (2012 – 2016)	5-10 years (2017 – 2021)	> 10 years (2022 – 2026)
1. Waste reduction	1	3	5
2. Utilization of solid waste and recycling materials			
2.1 Material recycling	20	22	25
2.2 Waste to energy	5	10	15
2.3 Utilization organic waste (anaerobic digestion, composting)	5	30	50
Total achievements	31	65	95

on people. Moreover, these policies also intensify the waste treatment capacity of local administrative authorities, and support the participation of the private sector in research and development for the utilizing of raw materials and clean technology. Significant legislation for MSWM is summarized in Table 4.

One of the most important plans for MSWM in Thailand was the National 3R campaign. The draft of the National 3R strategy was finalized by PCD in 2009. The draft aimed to achieve integrated solid waste management, focusing on the 3R (reduce, reuse, and recycle). The objectives of 3R campaign are to reduce the waste quantity, and promote reuse and recycling activities. The concept of this campaign bases on enhancing awareness of the connection between waste generation and resource consumption. The national targets under the 3R campaign are presented in Table 5. The government plans to increase the recovery rate from 22% to 30% by 2016, to 62% by 2021, and to 90% by 2026.

In 2014, an MSWM roadmap issued by Thailand government (PCD, 2014). This roadmap aims to reduce disposal MSW by landfilling without prior treatment. Municipalities much focus on waste segregation and recycling as integrated parts of MSWM. The construction of incinerators applying waste-to-energy technology is also mentioned in the roadmap.

Furthermore, in 2017, a new law also issued to increase the MSW collection fees from 40 to 150 Baht per month. The new law also allows local governments to collect MSW treatment fees directly from the waste generator, which previously was handled by the local government's budget. The new regulation applies the Polluter Pay Principle (PPP) to promote the responsibilities of producers, importers, and consumers. However, many problems still await in executing the new law, including the willingness to pay by local residents, fee collection system, transparency of waste budget management, inspecting, and execution.

Impact of Waste Management on Global Warming

In the past, scientists warned of rising sea levels, shrinkage of glaciers, changes in weather patterns, and extreme weather events in the future. These scenarios are now happening all over the world. They concluded that these extreme events are occurring due to global warming. One of the main reasons for global warming is due to the greenhouse effect, occurring because of Greenhouse Gas (GHG) emission. The burning of fossil fuel, farming, and industrial activities are the major contributors to greenhouse gas emission. Waste management also contributes a significant amount to greenhouse gas emission although it does not contribute as much as farming or the burning of fossil fuel.

According to the second Biennial Update Report of Thailand (ONEP, 2018), the total greenhouse gas emission in 2010 was 226.09 $MtCO_2eq$, and in 2013 it was 318.66 $MtCO_2eq$. This was an increase of 2.68% in the emission per year. The greenhouse gas emission is measured in the form of direct and indirect GHGs. The direct GHGs are carbon dioxide (CO_2), methane (CH_4), and nitrous oxide. (N_2O). Indirect GHGs are carbon monoxide (CO), nitrogen oxides (NO_x), sulfur dioxide (SO_2), and non-methane volatile organic compounds such as hydrofluorocarbon, perfluorocarbon, and sulfur hexafluoride. The major sources of GHGs are the energy sector, industrial processes sector, agricultural sector, land use, land-use change, forestry sector, and waste sector. In 2000, the waste sector generated 9.089 $MtCO_2eq$ of GHGs, and by 2013, it had increased to 11.830 $MtCO_2eq$. The GHG emission increased by 23.17%. In the waste sector, wastewater treatment accounted for 6.376 $MtCO_2eq$, which is 53.90% of the total GHG emission by the waste sector. Landfills accounted for 5.346 $MtCO_2eq$ which is 45.19%, and waste incineration generated 0.108 $MtCO_2eq$ greenhouse gases which is 0.91% (ONEP, 2018).

Thailand adopted the Paris Agreement in December 2015, at the 21st session of the Conference of the Parties (COP21) to the United Nations Framework Convention on Climate Change (UNFCCC). Currently, Thailand has appointed a National Committee on Climate Change to mitigate the greenhouse gas emission. The committee has approved the "Climate Change Master Plan 2015–2050" to tackle climate change problems. The plan was proposed by the Office of Natural Resources and Environmental Policy and Planning. The Climate Change Master Plan 2015–2050 has three strategies. One of the strategies is to reduce the greenhouse gas emission. Thailand has already implemented a nationwide strategy to reduce greenhouse gas emission by 7% to 20% by 2020 (Kamuang, 2017; ONEP, 2018).

According to Thailand's Ministry of Energy, Thailand is promoting waste to energy. The plan is to produce 160 megawatts of power and 100 kilotonnes of oil equivalent (ktoe) thermal power by 2021 using waste under the waste to energy program in the 10-year Alternative Energy Development Plan (2012 to 2021). Currently, Thailand produces 22.23 MW from biogas by utilizing landfill waste. Incineration and gasification produce 20.06 MW of power, and 2.034 MW of power is generated via the anaerobic digestion of waste. As of now a total of 78.59 ktoe of thermal capacity is replaced by the waste to energy program, composed of 77.31 ktoe thermal capacity from Refuse-Derived Fuel (RDF) and 1.28 ktoe thermal capacity by biogas. According to Thailand's Ministry of Energy, waste is being used as a fuel in cement plants, replacing coal (Dawan, 2017; DEDE, 2019). The waste to energy program reduces the greenhouse gas emission in the waste sector, which ultimately helps to control global warming.

Sustainable Development Goals in Waste Management

Sustainable development is a widely discussed topic in many fields including waste management. In 2015, the 2030 Agenda for Sustainable Development was adopted by United Nations member countries including Thailand. This agenda contains 17 goals, known as Sustainable Development Goals (SDGs). These goals are aimed to achieve sustainable development in the economy, society, and environment, simultaneously. Thailand has developed its own pathway to achieve SDGs, known as the "Sufficiency Economy Philosophy". Although there is no specific goal related to waste, several goals address the issue of waste (MFA, 2017), which are:

- SDG 4 Quality education.
- SDG 6 Clean water and sanitation.
- SDG 7 Affordable and clean energy.

- SDG 8 Decent work and economic growth.
- SDG 9 Industry, innovation, and infrastructure.
- SDG 11 Sustainable cities and communities.
- SDG 12 Responsible consumption and production.
- SDG 13 Climate action.

Under these goals, different waste related issues have been identified, and strategies have been introduced to address different types of waste with different management approaches (MFA, 2017). The MNRE's 20-Year Strategy 2017-2036 hopes to achieve some of these SDGs. Under this strategy, the Environmental Quality Management Plan 2017 – 2021 plans to achieve (Karatna, 2017): 7-20% GHG reduction by 2020, 75% waste management, 30% waste recycling.

Case Study: Community-Based Waste Management in Chiang Rai Province

Community-based management (CBM) is a bottom-up approach that involves local members inactive waste management and responsibility, to solve problems in their community. Waste management problems can be resolved effectively by CBM with less burden on the citizen than the traditional waste management system (Cunningham et al., 1995; Liddle et al., 2016).

CBM has been used to manage resources in different fields, for example, water resource management (Dewan et al., 2015; Margerum & Robinson, 2015), forest management (Cagalanan, 2015; Sharma et al., 2015), and community services (Wallerstein & Duran, 2006). For solid waste management, CBM is used for both regular MSW services and specific waste services. In the regular MSW services, CBM members collect garbage from households, as well as sorting and managing waste in their community (Afroz et al., 2009; Wilson et al., 2006). For the specific waste services, CBM focuses on certain types of waste, such as recyclable waste and organic waste for making compost, or bio-methane generation.

CBM was applied in Chiang Rai province through the Chiang Rai Zero Waste (CRZW) program with successes in many sub-districts. As an estimate by the Chiang Rai Provincial Office of Natural Resources and Environment in 2013, the total amount of MSW generated in the province was 434,496 tonnes (Manomaivibool et al., 2018). This equaled to a waste generation rate of 0.93 kg/capita/day. The waste composition in Chiang Rai was predominated by organic waste with a proportion of about 66% in 2015 (Figure 3).

The MSWM system is still a big challenge in Chiang Rai province. Only 43% municipalities in the province could arrange waste collection services to their residents. The failure of the MSWM system in a province is due to two main reasons. First of all, there is a lack of annual revenues for waste disposal. Under the prior fee system, whether all households in Chiang Rai had paid the maximum collection fee of 40 Baht per month, the annual incomes would only reach to 236 million Baht. However, with an average treating cost of 2.62 Baht per kg of waste, the total expense of waste collection and treatment of the province were 1.17 billion Baht in 2013 (Manomaivibool & Dokmaingam, 2017). Thus, local governments have to heavily subsidize the services. Second, there are problems with public opposition when constructing a solid waste disposal site near communities. Currently, the total capacity of three sanitary landfill sites in Chiang Rai is not sufficient for the waste generation. Small incinerators and controlled dumping sites on public and private land were also established. Unfortunately, protests from the community near the new disposal sites resulted in a shutdown.

The CRZW program was an effort to support waste separation in Chiang Rai Province and decrease the pressures on the MSWM system. The CRZW program was started in 2014 with a study to identify the current problems in Chiang Rai's MSWM system and obtain lessons from successful cases of CBM. In 2015, a pilot study was carried out on 18 model municipalities in 18 districts of Chiang Rai (Manomaivibool et al., 2018). A provincial competition of making a "Sa-wian" was held to promote the organic waste separation and home composting in the study areas. The Sa-wian, as shown in Figures 6 (a) and 6 (b), was designed as simple home composters that were made from any available material, such as old roof tiles, nets, woven bamboo, etc. In addition, the local governments also supported villages with training workshops, study visits, awareness campaigns, supporting equipment and materials, appointments of staffs, and marketing of products from recycling programs. As a result, in 2015, 4880 Sa-wians were installed, and they helped to treat 96.7% of the household organic waste in the areas. The total investment cost for all Sa-wians was about 0.98 million Baht (200 Baht /unit) and took only 2.7 months to get a return on the investment. If half of the organic waste generated in Chiang Rai province was utilized at the source by Sa-wians, waste collection and treatment expenses could be saved about 324 million Baht, or more than the total incomes from waste fees in a year.

After the success of the Sa-wians, the separation of recyclable waste was also promoted in the 18 villages. In the same way of organic waste, the encouragement of recycling in CRZW transformed MSWM into a co-production of public services. Households prepared their own containers to store different types of recyclable waste (Figure 6 (c)). Agricultural sacks were the most common choice, used by 60.4% of the households. As a result, 96.7% of households in the areas segregated at least one type of recyclable,

Figure 6. (a) and (b) Two Sa-wians at households in Chiang Rai Province; (c) Separation of recyclables at households in Chiang Rai Province; (d) Hazardous waste collection point at Baan Photanaram, Mae Jan District (Manomaivibool et al., 2018)

and 48% segregated four types or more. Plastic bottles were the most commonly separated recyclables with a sorting rate of 88%, followed by glass bottles at a rate of 73% (Manomaivibool et al., 2018).

Hazardous waste was not covered by the CRZW project. However, with the support of local governments, 83.8% of the households began sorting fluorescent lamps, electronic waste, batteries, and chemical containers. Until 2016, 12 out of 18 municipalities had built a designated point to collect hazardous waste from households, as shown in Fig. 6 (d).

The CRZW project built soft infrastructures to encourage source segregation. The success of the project reduced the burden of the collection and disposal systems in the study areas and even helped to close many of the open dumpsites in some municipalities (14 of 19 village dumpsites were successfully closed in Ngao Tambon Municipality). After the successes in 18 model villages, the number of villages under the program increased to 253 villages within a year.

Case Study: Municipal Solid Waste Management in Rayong City

Rayong City occupies an area of 16.95 km^2 with four sub-districts: Tha Pradu, Pak Nam, Nuen Phra, and Cheang Nuen. The total population of the city in 2013 was 64,406 people with 36,234 households. In addition to the registered residents, Rayong city also accommodates about 50,000 people from nearby areas and non-registered laborers. Waste management in Rayong is an example of medium-scale MSW. Rayong city is a popular tourist destination and a location for many industries, leading to an increased amount of solid waste, which has created serious environmental issues. The quantity of MSW increased from 71 tonnes/day in 2001 to 85 tonnes/day in 2005 (Sharp & Sang-Arun, 2012). The municipal government organized 20 groups with 20 trucks and 218 workers to collect solid waste by the door-to-door method. Like most municipalities in Thailand, the collected fees of MSWM in Rayong city cannot cover the total MSWM expenses. In 2007, annual revenue from waste service fees in the city was only 0.7 million Baht (Kritjaroen, 2011). However, the city government spent 37 million Baht for handling MSW in the year. To solve these problems, Rayong city, in cooperation with private and public organizations from the central government of Thailand, promoted MSWM. Generally, three phases of MSWM including (1) CBM, (2) integrated waste management such as waste-to-fertilizer and energy project (administered by the public sector), and (3) integrated waste management (administered by the private sector) were organized in the city, as shown in Figure 7 (a).

Figure 7. (a) Rayong city waste management model; (b) Waste separation at a school in Rayong city

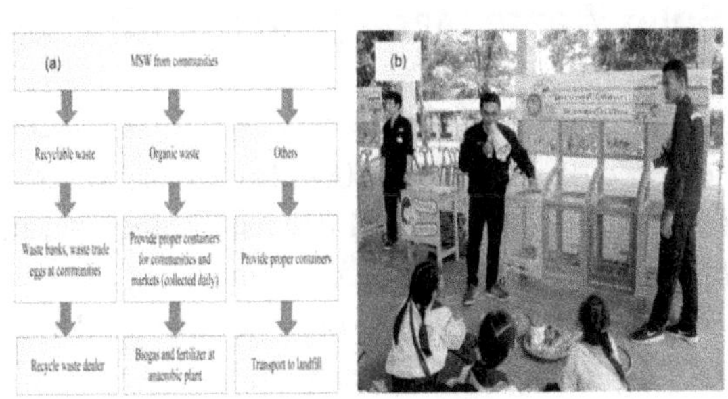

The first phase of waste management started in 1999. The campaign was initiated under the concept of CBM with the participation of 20 volunteers from communities in Rayong city. The team helped to raise awareness about MSW for residents in their communities. In addition to organizing activities about MSWM in the communities, the team helped local governments to connect to the residents. Many activities have been conducted to encourage people to participate in the campaign. 'Garbage for Eggs' was the first program to help local residents separate recyclable wastes from home in exchange for eggs. At schools in the city, waste banks were established to collect recyclable and organic waste, as shown in Figure 7 (b). Small fermenting reactors were provided for schools to produce liquid fertilizer. The city gained good collaboration from residents in the local communities in this project. As a result, during the first two years of the campaign, the quantity of garbage declined from 88 tonnes to 78 tonnes per day (Kritjaroen, 2011).

In the second phase, Rayong city launched a program to produce biogas and organic fertilizer from waste in collaboration with the Development of Environment and Energy Foundation, the International Council for Local Environmental Initiatives, and the Thailand Environment Institute. From December 2001, Rayong city supplies 12- and 100-liter bins for participating households and communities to collect separated organic wastes. Moreover, Rayong city with support from the central government built an anaerobic digestion system with a capacity of about 26,000 tonnes of MSW per year. It is expected to generate 5,100 MWh of electricity annually, along with organic fertilizer. The anaerobic plant was operated in June 2004 by a public organization. During the 16 months of operation, some achievements from selling the produced electricity and fertilizer were observed. Nevertheless, some factors influenced both the quantity and quality of biogas and organic fertilizer due to problems in the separation and collection systems for organic waste at the local communities. Therefore, the anaerobic plant could not reach its maximum capacity with an expenditure per ton of solid waste that was higher than the break-even point (Kritjaroen, 2011).

In October 2006, in order to improve the situation, Rayong city hired a private company to operate both the waste collection system in the city and the anaerobic plant. The company divided Rayong city into seven zones for solid waste collection. The organic waste collection was separated into four zones, which focused on the wet-market and hotel areas. The waste collection, serviced by the private company, operated seven days a week, starting from 7:30 a.m. until 1:30 a.m. the following day. In the first six months of operation, 8,117 tonnes of waste was collected for separating at the plant, containing 5,566 tonnes of solid waste and 2,551 tonnes of organic waste. The amount of sorted organic waste at the plant per day was 30 tonnes higher than the previous operation by the public organization. This showed that the SWM system in Rayong city improved with the cooperation of the private sector (Kritjaroen, 2011).

Case Study: Municipal Solid Waste Management in Bangkok

Bangkok, the capital city of Thailand, is divided into 50 districts and occupies an administrative area of 1568.74 km^2. Bangkok is the home of more than 10 million people (BMA, 2016). Urbanization has enlarged to occupy nearly half of the city area. In 2017, waste collection in Bangkok was 4.86 million tonnes per year, which accounted for 18% of the MSW generated in Thailand (PCD, 2018). As presented in Figure 3, the highest ratios of generated MSW are organic and food waste (54%), recyclable plastic (25%), and recyclable paper (12%). To improve the waste situation in Bangkok, waste management activities were divided into four programs, which were (1) waste reduction and separation, (2) waste collection, (3) waste transfer and transportation, and (4) waste treatment and disposal (Sharp & Sang-Arun, 2012).

Waste Reduction and Separation

The Bangkok government encourages waste reduction and segregation for resource recovery, aiming for public awareness and participation. Waste segregation at the source is encouraged by introducing three waste containers for sorting various types of waste. The containers can be classified into three types: food waste, recyclable waste, and household hazardous waste, as shown in Figure 8 (a). However, a limited number of communities adopted the separation program due to low enforcement of the waste segregation program. According to a survey by BMA, 30% of people in the city are collaborating with the waste segregation program (Sharp & Sang-Arun, 2012).

In the waste reduction campaigns, the Bangkok government organized three main principles (including 3R principles, zero waste management, and CBM) to ensure the suitability in each target group. The 3R principles were used to encourage people to use food waste, sticks, and leaves, and to recycle waste. The campaigns planned for raising awareness, participation, and encouragement of people in waste segregation and waste reduction. This helps to reduce the waste burden transferred to disposal sites, and pollution minimization toward sustainable development. For the zero waste management, the campaign is based on the idea that “waste has economic value where it can be reused”. This program aims to reduce waste and to dispose of the remaining with efficient technologies. Zero-waste management emphasizes the following ideas: to select raw materials that can be mostly recycled, to reduce waste volume, to create environmentally friendly products, to promote using of products made of recycled material. Moreover, the campaign also aims to collect a tax integrated with the raw material price, to improve local business development, and to increase local job opportunities.

The CBM program in Bangkok depended on collaborative activities in discussing and choosing solid waste management techniques for the community, in addition to executing and separating waste at the local municipalities. Generally, in the CBM program, the recyclables were first sorted out and sold. Organic waste such as kitchen and yard waste were collected for making compost, liquid biological fertilization or feeding animals, as shown in Figure 8 (b). There are also some encouragement measures, for example, the preparation of 2 containers for food waste and non-recyclable waste. This helps to keep clean areas in front of houses and clean landscape for communities. As a result of the program, public services can reduce the frequency of waste collection because of a smaller volume and odorless waste. Some communities need waste collection service just once a week (Sharp & Sang-Arun, 2012).

Moreover, to expand the recovery rate of MSW, waste banks were established in communities or organizations (companies or schools), as presented in Figure 8 (c). In the waste bank program, participants get benefits, in the form of money and improving the environmental quality. Participants collect recyclable materials and stored at a community center. Separated wastes are regularly sold to recycle shop. Monetary benefits coming from the selling of recyclables are turned back to the members of the program.

Briefly, the waste reduction and separation program in Bangkok has been organized into 377 communities, 115 educational institutes, 12 business establishments, and 2 markets. The program helped to reduce (by 38 to 80%) the total amount of waste generation in communities. In Chatujak district, the program reduced the solid waste generation by 41% and reduced the MSWM cost by 11.4 million Baht, earning over 6.6 million Baht from the sale of recyclable material. In addition, the program helps to promote waste reduction in schools. In 2014, small anaerobic reactors (2.5 cubic meters) were installed in 40 schools of Bangkok. The reactors can process about 40 kg of organic waste per day and produce about 2.5 cm^3 of biogas per day (BMA, 2014).

Figure 8. (a) Waste separation at a university; (b) Composting and organic fertilizer from organic waste at communities; (c) Waste bank at a school; (d) Flow of MSW in Bangkok (2016)

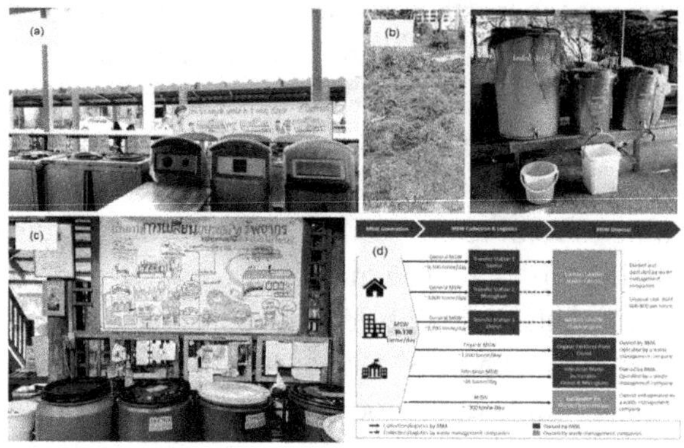

Waste Collection

The Bangkok government aims an effective collection method to minimize uncollected waste. In 2016, there were 1,894 units of waste collection trucks including 2-5 tonne compacting trucks, 8-cubic meter lifting truck, 6-tonne dumping trucks, and 1.5-tonne side-loading trucks. The team is separated and controlled by government districts in the city (BMA, 2016). In Bangkok, general wastes are collected daily at assigned waste dropping areas on main streets, alleys, and markets from 8:00 pm to 03:00 a.m. For areas that are not accessible by collection trucks, wastes are collected at sources by volunteers and transfer them to collection points. Recyclable wastes are collected every Sunday, while hazardous waste is collected on the first and fifteenth of every month.

Waste Transfer and Transportation

The Bangkok government has contracted with private partners to dump off the waste at a sanitary landfill. The landfill is located outside Bangkok. Thus, three waste transfer stations were built up to reduce the expense of transportation and public annoyance due to waste transportation. Each transfer station has the responsibility for solid waste transportation; fixing and maintaining of buildings, vehicles, machinery, and equipment; giving equipment to facilitate and support the cleaning of collection trucks; and other duties.

Waste Treatment and Disposal

At the transfer station, MSW is sorted by its potential utilization such as organic wastes for composting or fermentation, recyclable wastes, and wastes to be disposed of by landfilling or incineration. Currently, the Bangkok government has contracted with private companies to operate a composting plant with a capacity of 1,200 tonnes/day, two organic fertilizer plants with a capacity of 50 and 30 tonnes/day, and two grease disposal and grease transformer plants with a capacity of 300 cm^3/day (each). In addition, demolition waste is transformed at the construction waste processing plant with a capacity of 500 tonnes per day (BMA, 2014). The overall waste in Bangkok is presented in Figure 8 (d). Until 2016, the Bangkok government operated 3 technologies for MSW management, including 11% composting, 3% incinerating for electricity, and 86% landfilling. The government could not recover all of the waste for composting and recycling due to the low efficiency of waste separation at source by residents.

In the "Plan of 20-year Solid Waste Management for Bangkok Metropolis", the Bangkok government planned to increase waste utilization by 28% by 2025, with composting at 17%, waste to energy at 11%, landfilling to 72% (reduced by 17%). Until 2035, the government planned to increase waste utilization by 48%, with composting at 17%, waste to energy at 31% (increased by 20%) and landfilling to 52% (reduced by 20%) (BMA, 2014).

HAZARDOUS WASTE GENERATION AND MANAGEMENT IN THAILAND

The United Nations Environment Program defines hazardous waste as "Wastes (solids, sludge, liquids, and containerized gases) other than radioactive wastes which, by reason of their chemical activity or toxic, explosive, corrosive, or other characteristics, cause danger or likely will cause danger to health or the environment, whether alone or when coming into contact with other wastes" (Pharino, 2017).

Thailand's PCD categorizes hazardous waste into several subcategories which are: municipal hazardous waste or community hazardous waste, industrial hazardous waste, and infectious waste. Municipal hazardous waste contains household hazardous waste and waste electronic and electrical equipment or e-waste (Pharino, 2017). According to the "PCD Thailand State of Pollution Report 2017", the total municipal hazardous waste generation was estimated as 618,700 tonnes, which is around 2.75% of the total municipal solid waste (PCD, 2017).

Thailand cabinet approved the "National Waste Management Master Plan 2016- 2021" on the 3rd of May 2016. The framework of this master plan has the following activities (Siripornprasarn, 2015):

- Following 3R practices at the source to reduce waste.
- Establish proper disposal methods for both hazardous household waste and municipal waste by using centralized facilities.
- Require the participation of all relevant sectors/authorities in the management of solid and hazardous waste.

Thailand's national master plan has 6 goals for different waste sectors which are:

- Properly dispose of 75% of total MSW by 2021.
- Properly dispose of all accumulated waste by 2019.
- Achieve 30% collection and proper disposal of household hazardous waste by 2021.
- Achieve 100% collection and proper disposal of industrial hazardous waste by 2020.
- Achieve 100% collection and proper disposal of infectious waste by 2020.

Household Hazardous Waste

Household hazardous waste is hazardous waste which is generated by households and communities including shops, hotels, schools, etc. There are different types of products that contain hazardous material in households and communities which end up as waste at some point. Here are some common household hazardous materials in different categories:

- **Cleaning Products**: Aerosols, air fresheners, bleaching chemicals.
- **Paints and Solvents**: Wood preservatives, varnishes, paints, and paint thinners.
- **Pesticides**: Insect repellents, insecticides, poisons.
- **Medicine**: Food supplements.
- **Automotive Products**: Lead-acid batteries, different types of oils, gasoline.
- **Other**: Glues and adhesives, used printer cartridges, dry cell batteries.

In 2015, household hazardous waste generation was 206,894 tonnes, and the estimated household waste for 2017 was 216,600 tonnes. This was a 4.6% increase. For both 2015 and 2017, household hazardous waste was 35% of the total household hazardous waste (Chartier, 2014; PCD, 2018; Prüss-Üstün et al., 1999).

There are 80 provincial storage centers for hazardous waste in Thailand. Hazardous waste that is collected from communities at different receiving points of local administrative authorities and related agencies is transported and stored. PCD reported that around 60,600 tonnes of municipal hazardous waste was properly managed in 2017, but this was not adequate (PCD, 2018).

Different programs have been conducted by different government authorities for municipal hazardous waste management, such as used lead-acid battery recycling by the PCD and a special collection day for household hazardous waste.

In Bangkok, the per capita household hazardous waste generation is 2.9 ± 2.2 g/capita/day or 1.033 ± 0.82 kg/capita/year. As a percentage, it is around 0.26% of the total municipal solid waste generation (Pharino, 2017). Table 6 shows the composition of household hazardous waste in Bangkok.

According to a case study by Pharino, 2017, most families in Bangkok regularly purchase self-care products such as nail polish, body wash, shampoo, hairspray, etc., every month. Therefore, they are the most common hazardous waste in Bangkok. Light bulbs can be changed every year, and chemicals are also used regularly (Pharino, 2017).

In 2007, the Bangkok Metropolitan Administration (BMA) and PCD started a household hazardous waste management scheme in Bangkok with different approaches for raising awareness, separation, collection, storage, transportation, recycling, and treatment or disposal. Due to a lack of rules and regulations, these programs were not continued and properly supported. The BMA operates drop-off centers for household hazardous waste. Certified companies can transport the collected waste for disposal or treatment facilities. However, this program is not adequate to properly manage the total household

Table 6. Constituents of household hazardous waste in Bangkok 2013 (Pharino, 2017)

Waste	Amount in Tonnes	Percentage
Self-care products	2796.53	29.83%
Light bulbs	2738.40	29.21%
Chemical containers	2418.72	25.8%
Batteries	982.49	10.48%
Office supplies	219.37	2.34%
Expired cosmetics	152.81	1.63%
Expired medicines	66.56	0.71%
Total	9374.88	100%

hazardous waste in Bangkok. According to the BMA, the amount of collected hazardous waste is much less than the expected amount. In 2011, the hazardous waste collection was 1 tonne per day while the hazardous waste generation per day was 249 tonnes/day. The main reason for this is the lack of segregation of household hazardous waste. The BMA reported that in 2013, 90% of the total household waste in Bangkok was not separated from municipal solid waste (Pharino, 2017).

Industrial Hazardous Waste

In 2015, Thailand's industrial waste generation was 37.4 million tonnes, and 2.8 million tonnes of it was industrial hazardous waste, which was 7.5%. In 2017, 32.95 million tonnes of industrial waste were generated, and 1.95 million tonnes of it were categorized as industrial hazardous waste, which was around 5.9%. This was a 30% reduction in hazardous industrial waste (PCD, 2018).

There are several industrial hazardous waste treatment and disposal facilities in all 6 regions of Thailand. The eastern region has the highest density of industrial plants and facilities. Hence, it has the highest number of treatment and disposal facilities. An important issue in industrial hazardous waste management is illegal dumping (PCD, 2018).

Thailand's national master plan on cleaner production and cleaner technology focuses on industrial waste management, including industrial hazardous waste (Pharino, 2017). The objective of this plan as per PCD:

- Reduce the waste generation and environmentally friendly management of waste.
- Promote appropriate reusing and recycling to avoid landfills.
- Increase the capabilities for collection, storage, and transport from waste generators to waste processors.
- Increase the capabilities for recycling and treatment facilities.

Providing cleaner production and cleaner technology is a process for the reduction of pollution and the reduction of residue generation from industrial processes. The goal is to reduce pollution and residue as much as possible by (first) identifying the sources. Reductions are achieved by changing the materials, technologies, management systems, and products (DIW1; DIW2)

Infectious Waste

The World Health Organization describes infectious waste as waste that is alleged to be contaminated with pathogens such as parasites, bacteria, viruses, and fungi with an adequate concentration or quantity which can cause infections in susceptible hosts. According to PCD, infectious waste includes:

- Laboratory infectious cultures and stocks.
- Waste from surgeries and autopsies on infectious disease patients.
- Waste from patients in isolation wards and waste that has been in contact with infected patients undergoing hemodialysis.
- Laboratory animals with infectious diseases.
- Instruments and materials that were in contact with either infectious humans or animals.

Thailand's Ministry of Public Health (MOPH) describes infectious waste under 4 categories, which are (Pharino, 2017):

1. Body parts or carcasses: surgery, autopsies, and research (for both humans and animals).
2. Sharps: needles, syringes, blades, glassware, etc.
3. Discarded materials: contaminated materials with blood or bodily fluids, live and attenuated vaccines, etc.
4. Wastes from wards: specified waste by the MOPH.

There are more than 26,800 public and private hospitals, private and public clinics, laboratories, and animal hospitals. These are the major sources of infectious waste in Thailand. In 2017, 58,000 tonnes of infectious waste was generated by these facilities, and 65% was generated by public hospitals. Compared to 2015, the total infectious waste generation was a 7% increase. According to the PCD, 51,300 tonnes of infectious waste was properly managed by incineration in 2017. As a percentage, this was 88.5% of the total infectious waste. Compared to 2015, this was a 32% increase. Thailand needs to improve the reporting system for infectious waste (Chartier, 2014; PCD, 2018; Prüss-Üstün et al., 1999).

In Bangkok, infectious waste from hospitals is collected, transported, treated, and disposed of by Krungthep Thanakom Co. Ltd. In 2012 (May), there were 2329 public health facilities using Krungthep Thanakom company services to dispose of about 814 tonnes/month of infectious waste. This waste is collected and transported (using 18 special vehicles) to the On-nuch disposal facility. After incineration, the remaining bottom ash is disposed of in a secure landfill (Jacob et al., 2012).

Current Policies and Regulations

Thailand has taken many steps to control and properly manage hazardous waste. Different government authorities, such as the MNRE, Ministry of Industry, and Ministry of Public Health, regulate different types of hazardous waste.

Household waste is regulated by the public health act and decentralization act for household hazardous waste, but there are no specific laws and regulations on the management of household hazardous waste. The responsibilities of local governments to provide hazardous and infectious management services are specified by the PCD.

Industrial hazardous waste is regulated by several acts which are the Hazardous Substance Act 1992, Factory Act 1992, and Industrial Estates Act 1992. The hazardous substance act regulates chemicals in Thailand. It defines hazardous substances and provides a list of substances available in Thailand. It also guides the authorities to control the transport, handling, storage, and disposal of hazardous substances. The factory act and industrial estates act regulates industrial and factory operations and activities, including waste management. The Ministry of Industry and the Industrial Estate Authority impose the criteria and standards to regulate hazardous waste management. The factory act also administers treatment facilities. The Enhancement and Conservation of National Environmental Quality Act 1992 regulates industrial waste and establishes an environmental impact assessment system for industries with high pollution and central treatment facilities. This act also applies to industrial waste and infectious waste management by setting up quality standards and monitoring (PCD).

Infectious waste management is regulated by several acts, such as the Public Health Act 2007, Determining Planning and Staging of Decentralization Act 1999, and Enhancement and Conservation of National Environmental Quality Act 1992. The Ministry of Public Health's "Notification of Ministry of Public Health B.E. 2546 (2003)" implements the analysis methods for testing the biological standards and labeling of waste containers (PCD).

Although there are several rules and regulations on hazardous waste, there are no specific regulations for the segregation of hazardous waste from municipal solid waste and municipal hazardous waste disposal. Therefore, it is necessary to improve the regulations on municipal hazardous waste management (Kamuang, 2017).

ELECTRONIC WASTE MANAGEMENT IN THAILAND

Electronic and electrical waste or "e-waste" is a rapidly growing stream of waste which is becoming a major problem in many countries, including Thailand. E-waste includes a wide variety of waste that contains electrical and electronic objects and their parts. According to the United Nation's global e-waste monitor report in 2017, there are 6 categories of e-waste: temperature exchange devices, screens and monitors, lamps, large equipment, small equipment, and small IT equipment (Balde et al., 2017).

Situation of E-Waste in Thailand

The PCD categorizes e-waste in Thailand into 10 categories. These categories are electrical appliances and electronic devices in large households, electrical appliances and electronic devices in small households, IT equipment, electrical appliances and consumer electronics, lighting equipment, medical equipment systems, measuring instruments or controls, toys, electrical and electronic tools, and vending machines. Depending on the type and category of electronics or electrical devices or parts, the lifetime varies along with the amount of waste generated and the impact of e-waste on human health and the environment. Table 7 shows the estimated lifespan of different electronic and electrical devices, as per the PCD, Thailand. According to the PCD, in 2017, the total e-waste generation in Thailand was 401,300 tonnes. According to the United Nation's Global E-waste Monitor, the per capita e-waste generation was 7.4 kg. The PCD has forecasted the e-waste generation for different types of e-waste for 2014 to 2020, as shown in Table 8 (Balde et al., 2017; Jiaranaikhajorn, 2013; Piyapanpong).

PCD reported that in Thailand, 51.3% of household e-waste was sold for recycling, 25.3% was kept in homes, 15.6% was disposed of with municipal solid waste, and 7.8% was given or donated (PCD, 2012).

According to the 2017 United Nation's Global E-waste Monitor report, Thailand has several issues in e-waste management. One of the major issues is a lack of awareness of the electronic and electrical waste in the general public. Thailand does not have complete inventories and databases, and Thailand lacks laws and regulations specific to e-waste with proper environmentally friendly e-waste management practices (Balde et al., 2017).

Currently, there are 41 formal e-waste dismantling and recycling factories, and there are around 9000 junk shops (informal). Most e-waste is handled by waste-recyclers, junk shops, waste pickers, buyers, etc. They are considered as informal sectors (Chareonsong, 2014). These informal sectors do not follow environmentally friendly recycling methods, and they lack proper protection measures to control environmental pollution. Invaluable parts and residues generally end up in MSW. Some e-waste that ends up in

Table 7. Estimated lifespan of different electronic and electrical devices by the PCD in 2012 (Pharino, 2017)

Product	Units Per House Hold	Brand New (%)	2nd Hand (%)	Average Use (Years)
CRT TV	1.45	98.98	1.02	6.90
LCD/Plasma TV	0.39	100		3.80
Refrigerator	1.29	98.27	1.73	6.87
Mobile Phone	2.25	99.67	0.33	3.09
DVD Player	0.7	99.47	0.53	3.78
PC/Notebook	0.78	98.57	1.43	3.65
Digital Camera	1.52	99.76	0.24	3.13
AC	0.58	99.15	0.85	5.20
Printer	0.18	100		3.05

Table 8. Thailand's electronic and electrical waste forecast for 2014 to 2020 by Thailand's Pollution Control Department (Chareonsong, 2014)

Product Type	E-Waste Generation x1,000 Units for Year						
	2014	2015	2016	2017	2018	2019	2020
TV	1835	2034	2218	2375	2499	3081	
Media player	3476	3537	3571	3588	3598	3611	3630
Digital camera & VDO camcorder	983	1059	1065	1055	1106	1192	1289
Printer	1520	1532	1542	1546	1547	1545	1543
Mobile phone & land phone	9750	10337	10907	11456	11983	12486	12966
PC	2210	2421	2630	2834	3032	3222	3402
A/C	740	766	796	832	871	911	949
Refrigerator	922	972	1023	1074	1125	11 74	1223
Microwave	313	346	382	419	456	492	527
Washing Machine	467	495	523	551	581	611	644
Fan	1916	1996	2079	2164	2,251	2340	2431

these informal sectors is illegal e-waste imports from other countries. Currently, there is no commercial machinery that is used for e-waste dismantling in Thailand. Therefore, manual labor is used in these recycling facilities. Informal recycling sector workers do not usually have the proper safety equipment, and their working conditions are poor.

E-Waste Management Strategy in Thailand

Thailand has introduced a strategic plan to manage e-waste in an environmentally friendly manner. The MNRE, Ministry of Industry, government, and private authorities/agencies have contributed to the development of the PCD's strategic plan, known as the "National Integrated WEEE (Waste Electrical and Electronic Equipment) Management Strategy (2014-2021) ". It was approved by the cabinet in 2015.

The objectives of this strategic plan are:

- Establish an e-waste collection system.
- Promote eco-design.
- Promote environmentally friendly e-waste dismantling and recycling facilities.
- Prevent Thailand from becoming an e-waste dumpsite.
- Raise awareness among the general public.

There are 6 strategies as follows:

1. Reinforcement of import and export controls.
   Key points of this strategy were: registration of electronic and electrical product importers, promoting the import of better-quality products, and monitoring the impacts of import and export from the Free Trade Area (FTA), ASEAN Economic Community (AEC), and Metropolitan Electricity Authority (MEA).
2. Promoting eco-friendly electronic and electrical products for the general public. In this strategy, the two key points are as follows:
   - Promoting electronic and electrical products to be tested and certificated as eco-friendly products.
   - Proposing a policy on e-product rental.
3. Developing an e-waste database.
   As Thailand lacks an e-waste database, this strategy is to develop a system for data collecting and reporting of e-waste.
4. Developing e-waste segregation, collection, storage, and transport facilities, and laws for local governments.
   The key points are improving the capacity of local governments to manage e-waste and introducing laws for product fees and/or extended producer responsibility.
5. Improving dismantling and recycling facilities.
   This involves supporting research and development in e-waste recycling technologies and supporting investment in environmentally friendly recycling.
6. Raising awareness among the general public by direct and indirect methods.
   In this strategy, the goal is to educate the general public about e-waste by formal or informal education (Kamuang, 2017; PCD, 2018).

Although this is a promising plan, the Thailand e-waste situation has not changed by much. At present, Thailand faces e-waste problems of its own and with other countries due to waste imports. According to Thailand's National News Bureau (16th November 2018), Thailand will deal with e-waste import issues within the next two years, as Thailand is becoming an e-waste dumpsite for other countries.

PLASTIC WASTE MANAGEMENT IN THAILAND

Situation of Plastic Waste in Thailand

Thailand is one of the major plastic producer and contributors in Asia. In 2015, the plastic production capacity of the country was about 6.1 million tonnes. The amount of plastic products consumption was about 4 million tonnes. Plastic packaging is a major part that contributed about 2.1 million tonnes (Wichai-utcha & Chavalparit, 2019). Packaging bags contributed about 0.5 million tonnes, and other packaging such as, plastic boxes and cups contributed about 1.5 million tonnes. In 2016, plastic consumption of Thailand was about 40 kg per year/person.

According to Bureecam et al. (2018), in 2013, Thailand consumed about 6.6 million tonnes of plastic. However, only 3.6 million tonnes of this waste was collected and transported to disposal sites by the local authorities. As shown in Figure 9 (a), 0.77 million tonnes of plastic waste was recycled by local government establishments or utilized as raw material for plastic production by the local industries. About 0.22 million tonnes of plastic waste was incinerated for energy generation, and the remaining 2.08 million tonnes of plastic waste was treated by landfilling. In addition, private manufacturers recycled about 1.1 million tonnes of plastic waste, or 27% of the generated plastic waste.

Plastic Waste Management in Thailand

Currently, the central government of Thailand organizes a plastic waste management network with cooperation between the public and private stakeholders to mitigate plastic waste production and consumption. The network of stakeholders is shown in Figure 9 (b), which depends on the roles of the vital stakeholders. The vertical network includes connections, starting from manufacture to waste disposers such as plastic manufactures, consumers (plastic waste generators), waste collectors, waste recyclers, and waste shops. Plastic wastes were collected and sorted by waste collectors from generators and then provide recyclable plastic waste to waste recyclers and waste shops. Making policies on plastic waste management in Thailand is responsible by a network of organizations includes the MNRE, PCD, Ministry of Energy, and Department of Industrial Work. This network has a significant impact on other stakeholders. The network establishes laws, policies, and plans that affect the plastic waste management of other stakeholders, including plastic producers, waste generators, waste collectors, and waste recyclers. The network also controls the environmental issues that come from each stakeholder, and promotes recycling activities.

Moreover, to enhance the plastic waste management, the network also supports research organizations or universities to carry out research and development of plastic waste management technology, including 3R options, bio-plastic, etc. The research includes many stakeholders, for example, the central government, local government, industries, municipalities, commercial sectors, institutes, waste recyclers, NGOs, and the mass media. Results from the research indicate the important roles of the local government, waste collectors, waste recyclers, and waste shops in the plastic waste management of Thailand (Wichai-utcha & Chavalparit, 2019). Waste generators, research organizations, universities, and plastic producers have minor impact on plastic waste reduction and recycling.

Since 2008, Thailand's government has initiated plans to control plastic pollution such as supporting bio-plastic production and controlling plastic consumption by tax mechanism, as shown in Table 9.

Figure 9 (a) Plastic waste follow of Thailand in 2013; (b) Network of stakeholders on plastic waste management in Thailand (Wichai-utcha & Chavalparit, 2019)

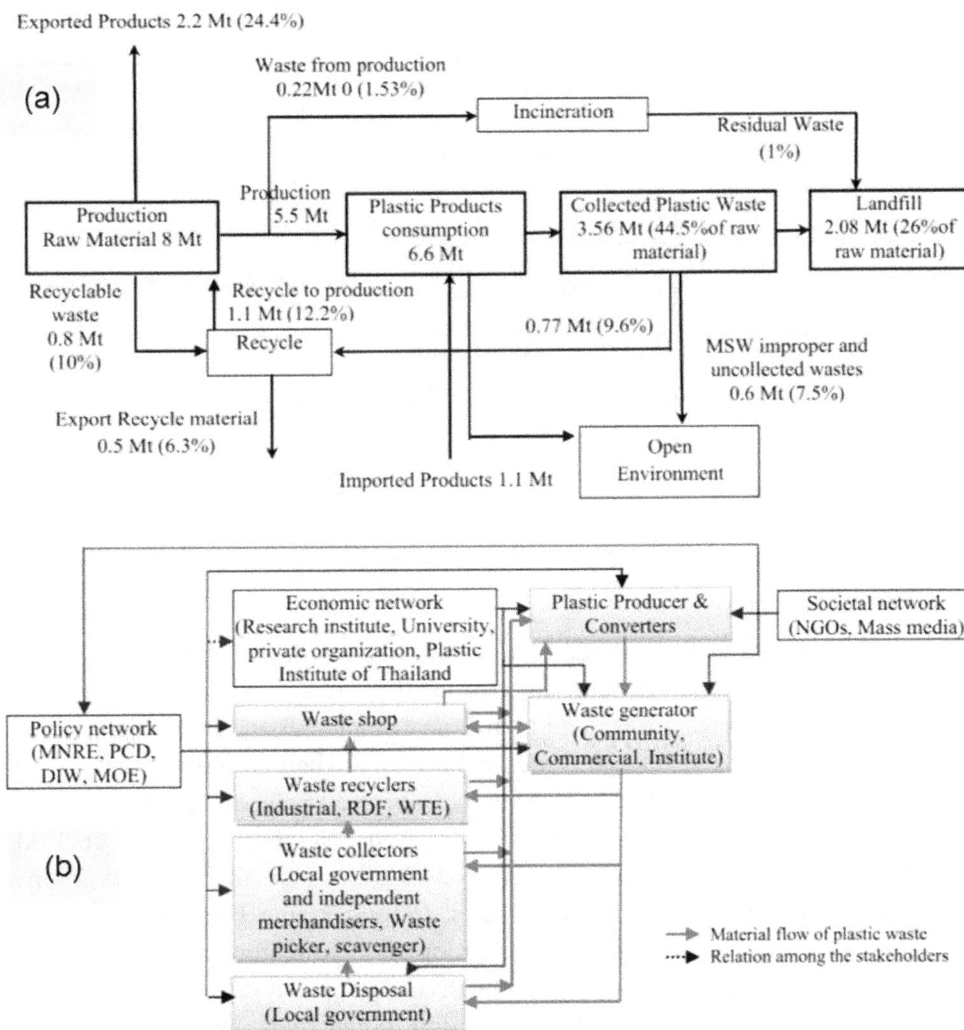

AGRICULTURAL BIOMASS AND UTILIZATION IN THAILAND

Potential Biomass Sources in Thailand

Thailand is heavily dependent on fossil fuels as an energy resource. Fossil fuels account for 80% of the total energy supply. In 2015, Thailand spent 6% of its GDP on energy imports, to fulfill the country's energy requirements. The government set energy security as a top policy objective, to respond to the continued rise in energy demand while draining the domestic energy reserves. In the Paris Agreement 2016, Thailand committed to reducing carbon emissions from its energy, industrial, agricultural, and waste sectors by 20-25% from the business-as-usual scenario by 2030. To achieve this commitment, Thailand has set a new renewable energy target of 30% of the total energy consumption by 2036 (IRENA, 2017).

Table 9. Policies and initiatives on plastic waste management in Thailand

Policies and Initiatives	Stakeholders	Key Element
National roadmap for the development of bio-plastic industry (2008)	Plastic producers, research institutes	Promote research and development of bio-plastic production in Thailand.
Plastic Debris Management Plan (2017–2021)	PCD	Reduce the impacts of plastic waste by involving relevant stakeholders. Increase plastic waste recycling to at least 60%.
Environmental quality management planning (2017- 2021)	MNRE	Control plastic consumption through appropriate tax mechanism, promoting the use of bio-plastic and other alternatives to plastics.
Prohibition of plastic bags and styrofoam containers in national parks (June 2018)	Department of National Parks, Wildlife and Plant Conservation, Thailand	Ban plastic product in all national parks of Thailand.
Campaign on non-plastic bag at hospitals (October 2018)	Department of Medical Services of the Ministry of Public Health	Ban plastic bag at 30 hospitals and reduce the usage of 9 million bags per year.
Campaign on non-single-use plastic at universities	The Sustainable University Network	27 universities nationwide have organized a campaign to reduce single-use plastic on all campuses by 80-90% over the year 2018.
Draft on banning foreign plastic waste (from 2021)	MNRE	Ban all plastic wastes imported from foreign countries.

Biomass is organic material that is considered as a renewable resource. Biomass can be forest and mill residues, wood and wood wastes, agricultural crops and waste animal wastes, aquatic plants, livestock operation residues, fast-growing trees and plants, and municipal and industrial wastes. Thailand is an agricultural country moving to an agro-industrial country. Thailand possesses large amounts of resources in biomass, including agricultural wastes that can be used as feedstocks for renewable energy. In the past, biomass was a vital source of renewable energy in Thailand (Papong et al., 2004). It provides basic energy for cooking and process heating in households and small industries. Main sources of biomass in Thailand include rice, wood wastes, oil palm, and sugar cane. Currently, biomass is used as a fuel for power generation in the country by four technologies: gasification, combustion, pyrolysis, and biogas. Modern applications of biomass energy can be divided into four different technologies, including biomass-fired electric power plants, biogas production, liquid fuels, and improved cooking stove and kiln technologies. In 2015, biomass was used to provide 5990 ktoe of heat for industry, accounting for almost 60% of the renewable energy used in Thailand. A comprehensive list of biomass residues has been compiled by the Ministry of Agriculture and Cooperatives and the Department of Alternative Energy Development and Efficiency, as presented in Table 10.

Biomass Energy Strategy in Thailand

The Government of Thailand has long recognized the significance of alternative and sustainable energy sources, but also realized that there was a need to introduce programs in support of renewable energy development and deployment. A variety of support programs have been set up to support sustainable energy development, ranging from tax exemptions, feed-in premium viper programs, and favorable feed-in tariffs for competitive bidding (IRENA, 2017). The incentives for renewable projects are presented as follows:

Table 10. Biomass residue potential in Thailand (IRENA, 2017)

Biomass type	Available Residues for Energy Purposes (2014)		Plan on Available Residues for Energy Purposes (2036)	
	Tonnes/year	Power potential (MW)	Tonnes/year	Power potential (MW)
Rice husk	432	0.05	432	0.05
Rice straw	4124630	461	4124630	461
Sugar cane and leaf	2928140	411	5265619	738
Bagasse			21280000	1421
Corn cob	80889	7	80889	7
Corn trunk	3369690	300	3369690	300
Cassava rhizome	2838125	141	3372560	168
Cassava trunk	1052636	149	2084755	294
Palm frond	14606671	867	33586191	1993
Palm empty fruit bunch	606541	40	1402455	92
Palm fiber			2944803	304
Palm shell			619959	95
Para wood root	1411834	110	1411834	110
Coconut shell	79678	12	79678	12
Coconut coir	71875	10	71875	10
Coconut bunch and frond	249026	35	249026	35
Total	31420167	2543	79944396	6040

- An eight-year corporate revenue tax exemption without an exemption cap, import duty-free on equipment, and other free tax incentives are available for projects utilizing waste, including refuse-derived fuel, to make electricity or steam.
- The production of solar cells and raw materials for solar cells and the power made from renewable energy sources, e.g., solar energy, wind energy, and biomass or biogas is also eligible for an eight-year revenue tax exemption, import duty-free on equipment, and free tax incentives.
- The production of biofuel production from agricultural products and by-products, for instance, biomass to liquid (BTL) or biogas from wastewater, is also qualified for an eight-year corporate revenue tax exemption, import duty-free on equipment and raw or important materials that were used in manufacturing export products, and other free tax incentives.
- The production of biomass briquettes and pellets is exempt for five years from corporate revenue tax, import duty on equipment, and other free tax incentives.

With the above support, the Thailand government sets a new renewable energy strategy with 30% of the total final energy consumption by 2036, as shown in Table 11. Renewable electricity is expected to contribute 4.2% (5588 ktoe), meeting about 20% of all-out electricity demand in 2036. Renewables for thermal use are expected to contribute for 19% (25088 ktoe). Biofuels represent the remaining 6.7% (8712.43 ktoe). Regarding installed power generating capacity, solar PV is expected to meet 31% of the total with 15% from wind, and the rest will be from biomass-based sources and hydropower plants.

Table 11. Renewable energy use and the Thai government targets (IRENA, 2017)

Energy Type	Real Energy Consumption	Government's Plan		
Power	2015 (ktoe)	Target 2015 (MW)	Target 2021 (MW)	Target 2036 (MW)
Municipal waste	44	131	410	500
Industrial waste	-	-	50	50
Solid biomass	1104	2726	3940	5570
Biogas	92	372	448	600
Biogas (energy crop)	-	-	387	680
Hydropower	314	3078	3165	3282
Wind	28	233	475	3002
Solar	202	1419	2993	6000
Total	1786	7962	11871	19684

ISSUES AND CHALLENGES IN WASTE MANAGEMENT SYSTEM OF THAILAND

Municipal Solid Waste Management

MSWM in Thailand has long been known as one of the most serious environmental issues. Society has suffered from waste mismanagement due to inadequate knowledge and a lack of realistically pertinent technology. There are also weaknesses in the processes of public participation, policy execution, and institutional support. Therefore, the failure to advance waste as a raw material in reusing and recycling processes has happened inside and beyond administrative boundaries. Several existing problems with waste management in Thailand include:

- Accumulating amounts of waste that are not appropriately handled and disposed of.
- Lack of public collaboration in waste separation and reduction.
- Lack of infrastructure, apparatuses, and equipment to gather and disposed waste appropriately from a wide range of various sources.
- Lack of regulations to execute effective waste management, for example, waste separation and waste management expenses.
- Waste collection fees that do exclude the true cost of waste management all through the lifecycle of the waste management system.
- Lack of policy consistency and limited budgets from local government on waste management.
- Lack of a comprehensive approach and cooperation between government agencies in policy, activity planning, and budget planning.
- Expansion of informal recycling sectors and poor working conditions in these facilities.
- Not enough regulations to encourage people to reduce consumption and separate the plastics from waste.
- Not enough support from the government for the stakeholders to encourage recycling activities.

Hazardous and E-Waste Management

The situation of hazardous waste and e-waste management is becoming a matter of high concern in Thailand. With industrialization and economic development, the generation of hazardous waste and e-waste is also increasing. The current management system has many issues to be solved such as:

- Lack of awareness about hazardous waste and e-waste among the general public.
- Lack of segregation of household hazardous waste and e-waste.
- Lack of rules and regulations to enforce the segregation of hazardous waste and e-waste.
- Disposal of household hazardous waste and e-waste with municipal solid waste.
- Lack of effective collection and transport systems for household hazardous waste and e-waste.
- Local governments lack the potential to properly manage wastes.
- Expansion of junk shops and illegal recycling facilities which lack environmentally sound recycling.
- Poor and unsafe working conditions in illegal recycling facilities and junk shops.
- Illegal dumping of industrial hazardous waste in the environment, such as in borrow pits.
- Lack of specific laws and regulations on e-waste.
- Lack of complete e-waste inventories and e-waste databases.

REFERENCES

Afroz, R., Hanaki, K., & Hasegawa-Kurisu, K. (2009). Willingness to pay for waste management improvement in Dhaka city, Bangladesh. *Journal of Environmental Management*, *90*(1), 492–503. doi:10.1016/j.jenvman.2007.12.012 PMID:18242819

Balde, C. P., Forti, V., Gray, V., Kuehr, R., & Stegmann, P. (2017). *The global e-waste monitor 2017: Quantities, flows and resources*. United Nations University, International Telecommunication Union.

BMA. (2014). *Bangkok State of the Environment 2013-2014*. Department of Environment, Bangkok Metropolitan Administration.

BMA. (2016). Bangkok State of Environmental 2015-2016. Department of Environment, Bangkok Metropolitan Administration.

BOT. (2018). *Statistical data EC_EI_027 Thailand's macro economic indicators*. Retrieved March 10, 2019, from http://www2.bot.or.th/statistics/BOTWEBSTAT.aspx?reportID=409&language=ENG

Bureecam, C., Chaisomphob, T., & Sungsomboon, P.-Y. (2018). Material flows analysis of plastic in Thailand. *Thermal Science*, *22*(6).

Cagalanan, D. (2015). Governance challenges in community-based forest management in the Philippines. *Society & Natural Resources*, *28*(6), 609–624. doi:10.1080/08941920.2014.948242

Challcharoenwattana, A. (2015). *Improving municipal solid waste management through recycling in urban settlements of Thailand*. Chulalongkorn University.

Challcharoenwattana, A., & Pharino, C. (2016). Multiple-criteria decision analysis to promote recycling activities at different stages of urbanization. *Journal of Cleaner Production*, *137*, 1118–1128. doi:10.1016/j.jclepro.2016.07.208

Chareonsong, P. (2014). *E-Waste Management in Thailand Pollution*. Paper presented at the Control Department 4 th International E-waste Management Network (IEMN) Workshop, Hanoi, Vietnam.

Chartier, Y. (2014). *Safe management of wastes from health-care activities*. World Health Organization.

Cunningham, C. E., Bremner, R., & Boyle, M. (1995). Large group community-based parenting programs for families of preschoolers at risk for disruptive behaviour disorders: Utilization, cost effectiveness, and outcome. *Journal of Child Psychology and Psychiatry, and Allied Disciplines*, *36*(7), 1141–1159. doi:10.1111/j.1469-7610.1995.tb01362.x PMID:8847377

Dawan, W. (2017). *Thailand's Master Plan for MSW Management*. Paper presented at the Developments of Energy, Environment and Ecosystems (3E) Nexus Initiative for Sustainable Development in Asian Countries, Tokyo, Japan.

DEDE. (2019). *Thailand needs to promote Energy-from-Waste*. Retrieved March 10, 2019, from http://weben.dede.go.th/webmax/content/thailand-needs-promote-energy-waste

Dewan, C., Mukherji, A., & Buisson, M.-C. (2015). Evolution of water management in coastal Bangladesh: From temporary earthen embankments to depoliticized community-managed polders. *Water International*, *40*(3), 401–416. doi:10.1080/02508060.2015.1025196

DIW. (n.d.b). *The concept of Clean Technology*. Retrieved March 19, 2019, from http://php.diw.go.th/ctu/en/index.php

DIW. (n.d.a). *Benefits of Clean Technology*. Retrieved March 19, 2019, from http://php.diw.go.th/ctu/en/benefits_of_ct.php

Dokmaingam, P., & Manomaivibool, P. (2017). Integrated municipal solid waste management for the highland rural area: Case study Patueng sub-district in Chang Rai, Thailand. *Rangsit Journal of Arts and Sciences*, *7*(2), 173–180.

Hongyon, C., & Hovijit, P. (2014). *Evaluation Potential of Oil Production from Khon Kaen Municipality Solid Waste* Paper presented at the The 15th Graduate Research Conferences, Khon Kaen University, Khon Kaen, Thailand.

IRENA. (2017). *Renewable Energy Outlook: Thailand*. Abu Dhabi: International Renewable Energy Agency.

Jacob, P., Kashyap, P., & Visvanathan, C. (2012). *Overview of municipal solid waste - waste to energy in Thailand*. Paper presented at the International brainstorming workshop on waste to energy in India, Mumbai, India.

Jiaranaikhajorn, T. (2013). *WEEE Management Policy Update from Thailand*. Pollution Control Department, Ministry of Natural Resources and Environment Thailand.

Kamuang, T. (2017). *Country Chapter State of the 3Rs in Asia and the Pacific: The kingdom of Thailand*. Secretariat of the Regional 3R Forum in Asia and the Pacific.

Karatna, P. (2017). *Thailand Environmental Quality Management Plan 2017 – 2021*. Paper presented at the National Policy Dialogue on the Urban Nexus in Thailand Conference, Bangkok, Thailand.

Kritjaroen, T. (2011). Understanding urban governance in the context of public-private partnerships: A case study of solid-waste management in Rayong Municipality, Thailand. *Federal Governance, 8*(3).

Liddle, E. S., Mager, S. M., & Nel, E. L. (2016). The importance of community-based informal water supply systems in the developing world and the need for formal sector support. *The Geographical Journal, 182*(1), 85–96. doi:10.1111/geoj.12117

Manomaivibool, P., Srivichai, M., Unroj, P., & Dokmaingam, P. (2018). Chiang Rai Zero Waste: Participatory action research to promote source separation in rural areas. *Resources, Conservation and Recycling, 136*, 142–152. doi:10.1016/j.resconrec.2018.04.002

Margerum, R. D., & Robinson, C. J. (2015). Collaborative partnerships and the challenges for sustainable water management. *Current Opinion in Environmental Sustainability, 12*, 53–58. doi:10.1016/j.cosust.2014.09.003

MFA. (2017). *Suffciency Economy Philosophy: Thailand's Path towards Sustainable Development Goals*. Ministry of Foreign Affairs, Kingdom of Thailand.

NESDC. (2018). Thai Economic Performance in Q3 and Outlook for 2018 - 2019. Office of the National Economic and Social Development Council, Thailand.

NSO. (2016). *Report on Population Characteristics The 2015-2016 Survey of Population Change*. Thailand National Statistical Office.

ONEP. (2018). *Second Biennial Update Report of Thailand*. Office of Natural Resources and Environmental Policy and Planning, Ministry of Natural Resources and Environment.

Papong, S., Yuvaniyama, C., Lohsomboon, P., & Malakul, P. (2004). *Overview of biomass utilization in Thailand*. Paper presented at the ASEAN Biomass Meeting. PCD. An Overview of Hazardous Waste Management & Prevention in Thailand. Retrieved March 15, 2019, from http://www.pcd.go.th/count/mgtdl.cfm?FileName=HWM_111053.pdf&BookName=HWM_111053

PCD. (2011). *Draft of National 3Rs Strategy*. Pollution Control Department, Ministry of Natural Resources and Environment.

PCD. (2012). *Parallel Roundtable 4: Economic Opportunities through 3R ~ Waste Electrical and Electronics Equipment (WEEE)*. Pollution Control Department, Ministry of Natural Resources and Environment.

PCD. (2014). *Roadmap for MSW and hazardous waste management approved by national council for peace and order*. Pollution Control Department, Ministry of Natural Resources and Environment.

PCD. (2016). *Thailand State of Pollution Report 2015*. Pollution Control Department, Ministry of Natural Resources and Environment.

PCD. (2017). *Thailand State of Pollution Report 2016*. Pollution Control Department, Ministry of Natural Resources Bangkok.

PCD. (2018). *Thailand State of Pollution Report 2017*. Pollution Control Department, Ministry of Natural Resources and Environment.

Pharino, C. (2017). *Challenges for sustainable solid waste management: lessons from Thailand: Springer. Piyapanpong, S. Parallel Roundtable 4: Economic Opportunities through 3R ~ Waste Electrical and Electronics Equipment (WEEE)*. Pollution Control Department, Ministry of Natural Resources and Environment.

Prüss-Üstün, A., Giroult, E., Rushbrook, P., & Organization, W. H. (1999). *Safe management of wastes from health-care activities. REO-16. (2015). Performance evaluation of solid waste disposal in Songkhla: Regional Environmental Office 16*. Ministry of Natural Resources and Environment.

Sharma, S. K., Deml, K., Dangal, S., Rana, E., & Madigan, S. (2015). REDD+ framework with integrated measurement, reporting and verification system for Community Based Forest Management Systems (CBFMS) in Nepal. *Current Opinion in Environmental Sustainability*, *14*, 17–27. doi:10.1016/j.cosust.2015.01.003

Sharp, A., & Sang-Arun, J. (2012). *A guide for sustainable urban organic waste management in Thailand: Combining food, energy, and climate co-benefits*. Institute for Global Environmental Strategies (IGES). Available: http://www. iges. or. jp

Sirippornprasarn, A. (2015). *Discovering New E-Waste Management-Initiatives Around the Globe-Thailand*. Pollution Control Department, Ministry of Natural Resources and Environment.

Siriratpiriya, O. (2014). *Municipal solid waste management in Thailand: challenges and strategic solution. In Municipal Solid Waste Management in Asia and the Pacific Islands* (pp. 337–354). Springer. doi:10.1007/978-981-4451-73-4_17

Suthapanich, W. (2014). *Characterization and assessment of municipal solid waste for energy recovery options in Phetchaburi, Thailand*. Degree of Master of Science in Environmental Engineering and Management. Asian Institute of Technology School of Environment, Resources and Development.

ThaiPublica. (2014). National agenda of solid Waste (2 September 2014 ed). ThaiPublica, Bangkok USEPA (2013). Inventory of U.S. greenhouse gas emissions and sinks: 1990–2011 (EPA 430-R-13-001). Washington, DC.

Wallerstein, N. B., & Duran, B. (2006). Using community-based participatory research to address health disparities. *Health Promotion Practice*, *7*(3), 312–323. doi:10.1177/1524839906289376 PMID:16760238

Wichai-utcha, N., & Chavalparit, O. (2019). 3Rs Policy and plastic waste management in Thailand. *Journal of Material Cycles and Waste Management*, *21*(1), 10–22. doi:10.100710163-018-0781-y

Wilson, D. C., Velis, C., & Cheeseman, C. (2006). Role of informal sector recycling in waste management in developing countries. *Habitat International*, *30*(4), 797–808. doi:10.1016/j.habitatint.2005.09.005

Chapter 18
Municipal Waste Management in Turkey

Ahmet Demir
Faculty of Civil Engineering, Department of Environmental Engineering, Yildiz Technical University, Turkey

Adem Basturk
Faculty of Civil Engineering, Department of Environmental Engineering, Yildiz Technical University, Turkey

Bestami Ozkaya
Faculty of Civil Engineering, Department of Environmental Engineering, Yildiz Technical University, Turkey

Mehmet Sinan Bilgili
Faculty of Civil Engineering, Department of Environmental Engineering, Yildiz Technical University, Turkey

ABSTRACT

Due to geographical position at the crossroads between Middle East and Asia and Europe, Turkey has a significant role for neighbor developing countries in terms of economical and social growth. Since solid waste management system mainly depends on these factors, Turkey has been always comprehensive model for many developing countries with its management system. This chapter, consisting of eight sub-sections, evaluates the management of municipal solid waste in Turkey with current data from the Ministry of Environment and Urbanization and provides an overview of the development of waste management practices in Turkey. In addition to the future perspectives and planning, the current status, waste amount, waste characteristics, prevention studies, recycling and re-use applications, collection, treatment and disposal applications of municipal wastes, packaging wastes, and medical wastes are reported in this chapter.

DOI: 10.4018/978-1-7998-0198-6.ch018

INTRODUCTION

Economic development, socio-economic structure, industrialization and climate significantly affect the solid waste generation rates in all over the world. Generally, the amount of solid waste increase with the increase in economic development and the rate of urbanization growth. Hence, the consequence of the unplanned urbanization growth will lead to a huge problem on solid waste management including the collection, transport and disposal in developing countries. Solid waste management is becoming even a more difficult problem of governments to solve due to the scarcity of financial, human, and other critical resources, especially for meeting the increasing demand for proper and healthy municipal service (UNEP, 2005).

It is clear that many developing countries face main problems with proper management of solid waste activities with advanced and expensive technologies due to the lack of adequate administrative and financial resources. An appropriate waste collection, recycling, and disposal methods require a sustainable municipal solid waste management system in a developing country. Therefore, an integrated solid waste management system has to be established a proper system for planning and designing considering technical, political, environmental, socioeconomic, legal, and cultural aspects as well as characteristics of municipal solid waste. Due to differences in waste characteristics, countries may need to use different technologies for waste disposal. Therefore, each country should develop its own waste management system to protect human health and environment, respond to the needs and its situation, also considering its financial situation. However, technology transfer without consideration of solid waste characteristics in developing countries may not be evaluated as an effective option for waste management. Otherwise, this kind of applications may cause more financial debts and adverse impacts on the environment, natural resources, and public health (Japan International Cooperation Agency, 2015).

Over the years, waste management has been a major challenge for Turkey as in whole developing countries. It is even more so with higher population and urbanization as well as socio-economic structure, and therefore more volume of waste is generated. Hence, Turkey has find waste management uneasy because of lack of financial capacity for integrated solid waste management. After European Council gives Turkey the status of candidate country for EU membership in 1999, environmental investments have started to increase, and therefore, waste management has become a pressure point for Turkey as of these years. Turkish Government has implemented several measures in order to protect environmental and public health originated from improper waste disposal. Consequently, a great deal of effort has been taken for appropriate collection, reuse, recycle, and disposal methods for municipal solid wastes during the last decade. This chapter, consisting of eight sub-sections, evaluates the management of municipal solid waste in Turkey with current data from the Ministry of Environment and Urbanization and provides an overview of the development of waste management practices in Turkey.

Section 2 gives a general overview of Turkey, as well as a historical perspective to the background of waste management. Moreover, a summary of population growth and urbanization is given. In addition, the current applications in waste management systems of Turkey are summarized.

Section 3 focuses on legislative development of waste management in Turkey from the 1930s to 2018. Also, the reasons for failure of these regulations to solve the waste management problem are shortly discussed.

Section 4 deals with the existing data on waste classifications, the amount of municipal waste generated and the characteristics of solid wastes in the light of the past and present studies.

Section 5 summarizes the current status of packaging waste management. It gives information about the amount of recyclable waste, the constituents and the collection methods of packaging materials.

Section 6 covers legislative development, collection and disposal of medical waste in Turkey, while Section 7 summaries several milestones of the waste management system in Turkey and gives some information about future perspectives.

COUNTRY SITUATION

Turkey is a country spread to a large part of the territory of Anatolia, while a small part of Thrace that extends to the Balkans (Figure 1). Turkey's neighboring countries are Bulgaria in the northwest, Greece in the west, Georgia in the northeast, Armenia, Iran, Azerbaijan, Nakhchivan in the east, and Iraq and Syria in the southeast. The country is surrounded by important seas on three sides with the Aegean Sea to the west, the Black Sea to the north and the Mediterranean Sea to the south and also contains the Marmara Sea in the northwest. The Marmara Sea separates Anatolia from Thrace, namely Asia from Europe along with the Bosphorus and the Dardanelles. Therefore, Turkey has a significant geo-strategic power

Figure 1. The map of Turkey and its place in the world (from google images)

Figure 2. Turkey's population growth (TurkStat, 2019)

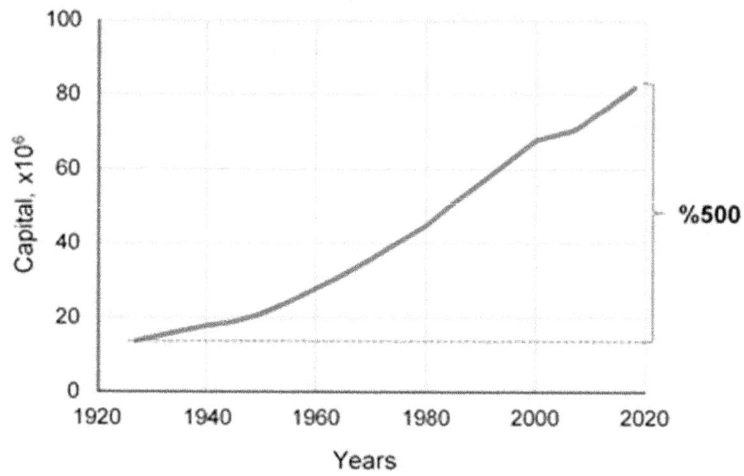

Figure 3. Urban and rural population trends in Turkey (Worldometers, 2019)

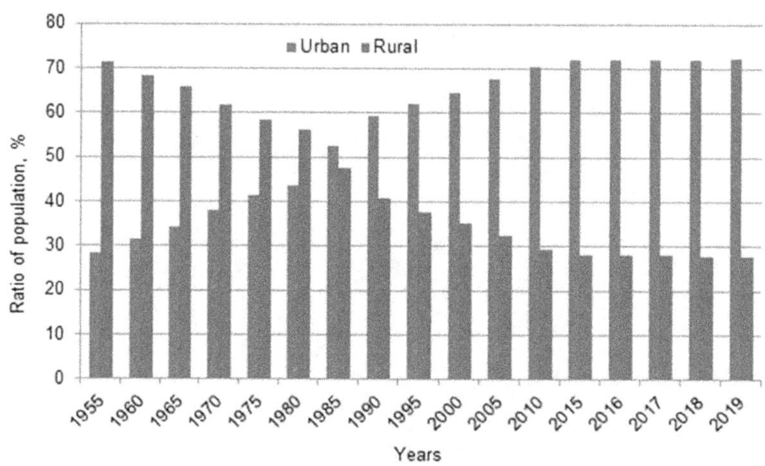

due to its geographical location between Asia and Europe. Turkey lands roughly resemble a rectangular shape, 1600 kilometers long and 800 km wide. It covers an area of 783,562 square kilometers. With these values, it is the 37th largest country in the world in terms of surface area. The current population of Turkey is 82 million as of the end of 2018 the latest Turkish Statistical Institute (Turkstat) estimates. Istanbul, the capital of the imperial period, today is the most populous city with a population of 15.5 million and accommodates 20% of the country's population. The current capital, Ankara, is the second largest city in the country with a population of about 5.5 million. Turkey is forested 13.2%, and 9.6% of this is classified as primary forest, the most biodiverse form of forest. In 2018, the gross domestic product (GDP) in Turkey amounted to around 713.51 billion U.S. dollars.

Turkey is a rapidly developing country as shown in Figure 2. In addition to the growth in population, the urban population is also increasing as can be seen in Figure 3. Accordingly, the amount of solid waste produced increases as the overall consumption increases. Due to geographical location, physical properties' of country and rapid growth, as shown in Figure 1 and Figure 2, waste management must be flexible and suitable for regional planning as well as being convenient with the principles of integrated waste management including waste prevention, waste reduction, reuse, recycling, energy recovery and disposal.

A HISTORICAL PERSPECTIVE TO BACKGROUND OF WASTE MANAGEMENT IN TURKEY

Turkey generated 31.58 million tons of municipal solid waste in 2016; the annual amount of waste generated per capita amounts to 396 kilograms. Currently, landfilling is mainly used for disposal of municipal solid waste. In this Section, the history of solid waste management from 1950 to 2000s was summarized in Istanbul, which has almost 20% of the country population, has an even more than some European countries' population. This mega city currently has a social mosaic that lives from all geographical parts of the country.

For this reason, this section mainly gives examples from Istanbul. Until 1950s, most of the solid waste disposed in un-sanitary dumping sites in inland areas, and into the sea in the coastal regions of Turkey. However, this activity was ended due to environmental problems especially in shores after 1953. Particularly, since the 1970s, solid waste has become a common problem with a major element for environmental pollution depending on rapid urbanization, the increase in production and consumption. After 1953, the solid waste was dumped in uncontrolled dump sites around Istanbul and it was continued to store in the valleys in the following years uncontrollably. From these years until 1993, it continued to be stored in uncontrolled dumped some sites on the Asian and European sides. The year 1994 was a turning point for the establishment of the sanitary landfill for the first time in Turkey. Odayeri and Komurcuoda Sanitary Landfills located in Asian and European sides of Istanbul were firstly established, which were the first sanitary landfills of the country established in accordance with technical standards including landfill gas management and leachate treatment.

The first academic research on solid wastes for Istanbul was carried out by Basturk, 1979. The same researcher realized the first systematic study in these years on solid waste for Istanbul. In this study, Istanbul is divided into 5 sub-regions considering the type of fuel and socio-economic structure it uses; then the first waste characterization study was carried out. Basturk (1979) was firstly recorded unit volume weight as 328 to 690 kg/m^3, water content as 17-58%, organics as 30-60%, calorific value as 3000-7000 kJ/kg in this study.

The second noteworthy work was realized on waste amount and characterization by the same researcher (Basturk, 1980) with the support of "Scientific and Technological Research Council of Turkey" (TUBITAK). This work recorded daily amount of waste of 2800-3000 tons/day and waste per capita of 246 kg/person-year in Istanbul and recorded calorific value varied between 2000 and 6500 kJ/kg.

Metropolitan Istanbul Solid Waste Management Study was carried out by CH2M Hill/Antel company in 1991-1993 with the support of the American TDA (Trade Developing Agency) (CH2M Hill/Antel) Project, 1992). In this study, the methodology and sample regions used in the previous studies were used.

Istanbul Solid Waste Recycling and Composting Plant Input Analysis (TBU, 2000) and in 2000, the Lurgi / Biat consortium building the composting plant built an operation for solid waste to feed the plant to the German TBU company. This study is mainly aimed at determining the properties of solid

Figure 4. The relationship between municipal waste and capita (TurkStat, 2016)

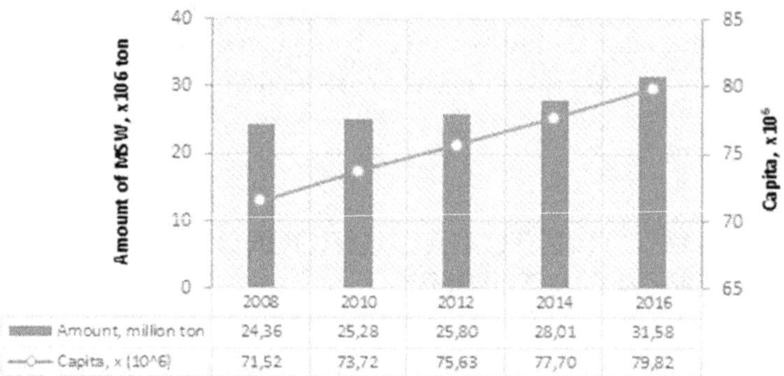

waste. Report on Training and Consultancy Services for the Operation of İstanbul Composting Facility (IGA, 2002) was realized during the trial run of the composting plant and the current collection system was evaluated.

Municipal Waste Statistics

The basis for integrated waste management is based on a hierarchy covering waste management, waste prevention, waste reduction, reuse, recycling, energy recovery and disposal. In Turkey, the efforts related to integrated waste management rapidly grow and successfully continues in recent years. Turkey has 81 provinces that in the status of the 30 metropolitan provinces with the spread to wide geography, which is one of the difficulties of waste management. This section contains statistical information on the amount of waste and disposal technologies in Turkey.

In this section, the values used for the waste statistics are the last data recorded in the "National Waste Management Action Plan – 2023" report published in 2017, which is taking into account the values in this report and Turkish Statistical Institution (TurkStat, 2016). Figure 4 shows the relationship between capita and the amount of waste. The amount of collected municipal solid waste reached to 31.58 million tons in 2016, while this amount was 24.36 million tons in 2008 (TurkStat, 2016). Population and economic growth affect the increase in the amount of waste, which reveals the importance of waste reduction and recovery. When the data of population and municipal waste between the years of 2008-2016 are examined in Figure 4, it is seen that both data increase in direct proportion to each other.

The two most important parameters in waste management are the waste amount and its characteristic. These two parameters guide all planning processes from waste collection to transport, recovery and disposal. Waste characterization data is discussed in detail in Section 4.

Disposal of municipal solid waste without recovery and recycling of valuable materials causes serious loss of raw material and energy. Waste generation in Turkey continues to increase and this increase reveals the need for the sustainable management of waste from the past to the present (European Environment Agency, 2013). Sustainable waste management requires a focus on waste prevention, recycling, recovery and reuse. According to 2016 data, the total amount of municipal waste is 31,584,000 tons in Turkey, corresponding to the value of 1.17 kg/capita-day (TurkStat, 2016). When the waste statistics in Figure 4 are analyzed, it is seen that the population and urbanization continue to increase compared to

Figure 5. Waste management methods in Turkey (MoEU, 2017)

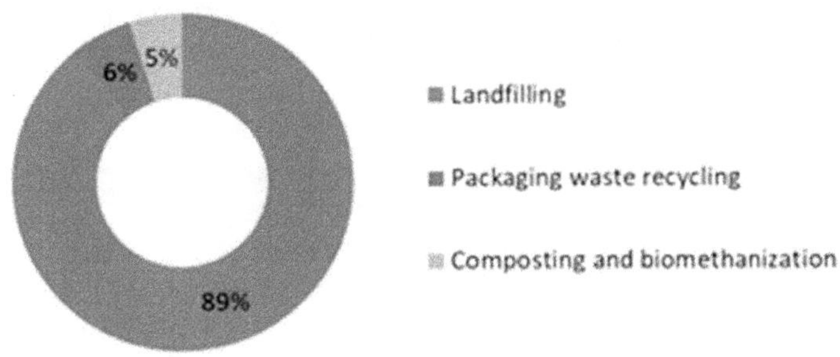

the previous years, and correspondingly the amount of waste is growing which requires the need and development of a sustainable waste management system.

Disposal of municipal solid waste without recovery and recycling of valuable materials causes serious loss of raw material and energy. Waste generation in Turkey continues to increase and this increase reveals the need for the sustainable management of waste from the past to the present. Sustainable waste management requires a focus on waste prevention, reuse, recycling and disposal.

Figure 5 shows the percentage distribution of the methods used for the disposal of wastes produced in 2014. It is observed that 6% of the wastes are recovered from packaging waste recovery, 5% are recovered by compost and biomethanization, and 89% are disposed by landfill method (Figure 5).

Waste characterization is the basis of Integrated Waste Management Planning. It is essential to know the waste composition in terms of planning and implementation of the waste management system. The quantity and characteristics of waste vary from country to country and vary from region to region in the same country, even from the district to the district in the same city. Although this change depends on the socio-economic structure of society, it depends mostly on income levels and consumption and usage patterns. Figure 6 gives the change of waste components since 1979 in Istanbul.

WASTE MANAGEMENT LEGISLATIONS

The environmental legislation in Turkey is mainly based on Article 56 of the Constitution, which relates directly to the environment in the 1982 Constitution and, indirectly, of various other articles of the same. In the part of the 1982 Constitution relating to "Fundamental Rights and Duties", in Article 56, which is under the heading "Social and Economic Rights and Duties", the rights of the individual in terms of the environment are defined as "Everyone has the right to live in a healthy, balanced environment. It is the duty of the state and citizens to improve the natural environment, and to prevent environmental pollution."

In addition to the 1982 Constitution, other main legislations are Environmental Law No:2872, related By-laws and other legal arrangements. The laws and by-laws of the environmental legislation are listed in Figure 7 which also includes the chronological structure of waste management legislations in Turkey.

After the right to enjoy a healthy and balanced environment had been defined by an edict in the 1982 Constitution, the Environmental Law (No: 2872) came into force on 11 August 1983. It was inevitable that amendments would be made over time to this Law, which had appeared to be sufficient to meet

Figure 6. Historical changes in municipal solid waste composition in Istanbul

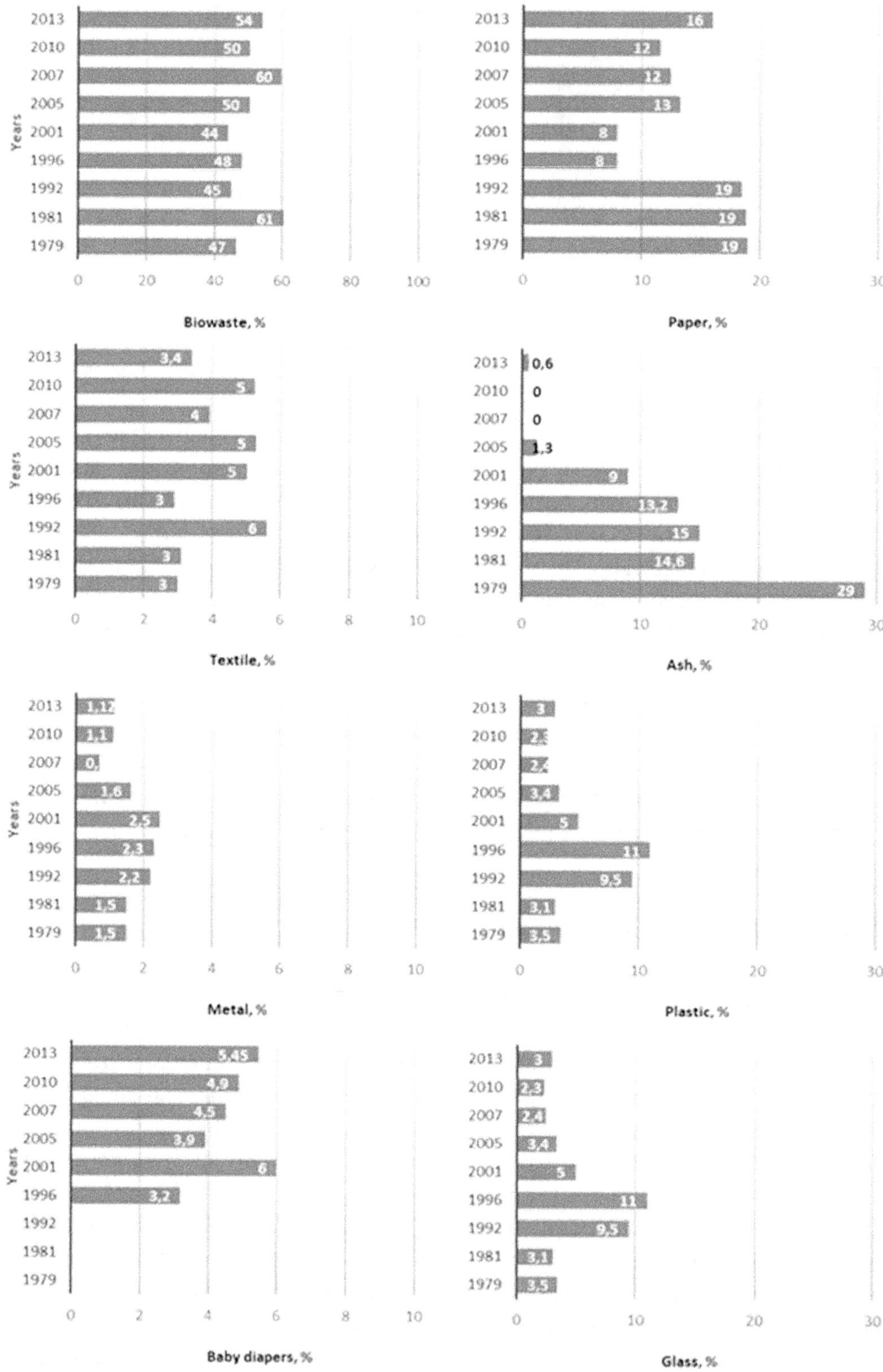

Figure 7. Development of waste management regulations in Turkey

Municipality Law	(1930, 2005)[1]
General Health Law	(1930)[2]
Environmental Law	(1983)[2]
Regulation on Control of Solid Wastes	(1991)[3]
Regulation on Control of Soil Pollution	(2001)[3]
Regulation on Control of Medical Wastes	(1993, 2005, 2017)[1]
Regulation on Control of Hazardous Wastes	(1995, 2005)[3]
Regulation on Control of Packaging Wastes	(2004, 2007, 2011, 2017)[1]
Regulation on Control of Used Batteries and Accumulators	(2004)[2]
Regulation on Collection and Control of Wastes from Ships	(2004)[2]
Regulation on Control of Excavation Soils, Construction and Demolition Wastes	(2004)[2]
Regulation on Control of Waste Electrical and Electronic Equipment	(2012)[2]
Regulation on Control of Vegetable Waste Oils	(2005, 2015)[1]
Regulation on Control of End-of-Life Tires	(2006)[2]
Regulation on PCBs and PCTs Control	(2007)[2]
Waste Oil Control Regulation	(2008)[2]
Regulation on General Principles of Waste Management	(2008)[3]
Regulation on the Control of End-of-Life Vehicles	(2009)[2]
Regulation on Control of Soil Pollution and Point Source Contaminated Sites	(2010)[2]
Regulation on Landfilling of Wastes	(2010)[2]
Regulation on Incineration of Wastes	(2010)[2]
Regulation on Mining Wastes	(2015)[2]
Regulation on Waste Management	(2015)[2]

1 Changed 2 In force 3 Repealed

requirements at the time it was enacted, and major changes were made in 2006. After these amendments had been made, the aim of the Law as defined in Article 1 was "to ensure the protection of the environment, which is the common asset of all living creatures, in accordance with the principles of a sustainable environment and sustainable development".

Another regulation on waste management is Metropolitan Municipality Law (No: 5216) with the aim of this Law is to regulate the legal status of metropolitan municipality administrations and to ensure that services are performed in a planned, programmed, effective, efficient and harmonious manner. The part of the Law relating to solid waste is Article 7 regarding the duties and responsibilities of metropolitan municipalities in keeping with the principles of sustainable development...

Municipality Law (No: 5393) is another regulation and the aim of this law is to arrange the establishment, organs, administration, duties, authorities and responsibilities of municipalities, as well as the way and principles according to which their work is performed.

Figure 8. Waste flow diagram (left) and waste hierarchy (right) in accordance with the EU directives (MoEU, 2017)

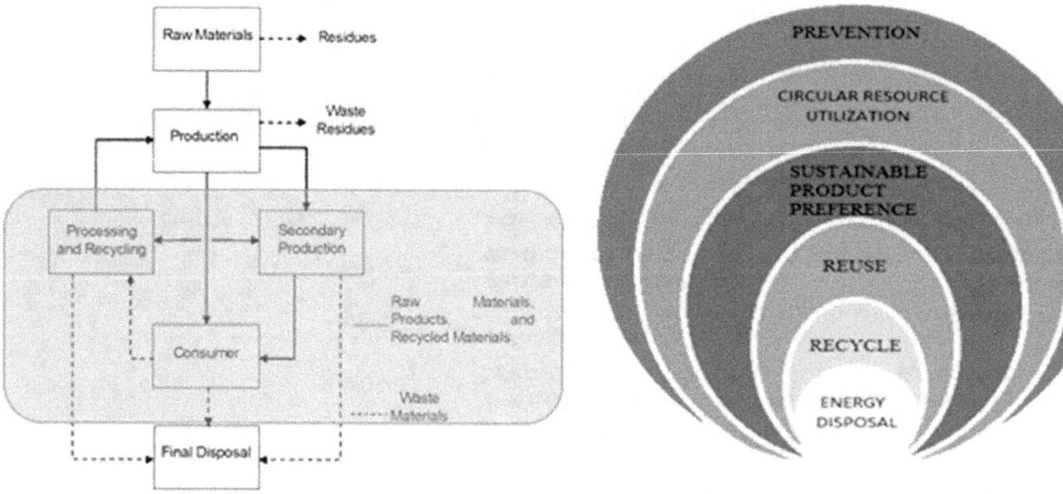

The Turkish Penal Code (Law Number 5237), the scope of which is to protect the rights and freedoms of the individual, public order and security, the rule of law in the state, public health and the environment and social peace, and to prevent crime, came into force on 26.09.2004.

Regulation on Control of Solid Wastes enacted in 1991, is the first secondary level legislation specialized on solid waste management. Initially, the regulation covered the minimization, collection, transport, recycling, and disposal of all types of solid wastes including household, hazardous, special, and packaging wastes. Due to the difficulties in the management of all types of wastes in one regulation, the number of secondary regulations is increased according to the waste characteristics and types. Regulation on Control of Medical Wastes (1993) and Regulation on Control of Hazardous Wastes (1995) are enacted in accordance with the Regulation on Control of Solid Wastes, to provide the proper separate collection, transport, and disposal of these special waste groups.

Environmental policies represent one of the most complex and costly points of the EU accession process including the need to adapt almost 300 legislations and around € 60 billion for complete implementation. Among the directives that constitute the EU acquis, 12 legislation titles are predicted to be harmonized within the scope of waste management. The required regulative changes have been mostly completed with the rapid legislative developments in the last decade.

The EU Waste Framework Directive (2008/98/EC) constitutes the basis of the waste management strategy for all types of wastes including household, hazardous, and special waste groups. The predicted waste flow and waste hierarchy in accordance with the Waste Framework Directive are given in Figure 8.

In accordance with the EU Waste Framework Directive (2008/98/EC), wastes are coded and classified according to the Regulation on Waste Management (2015). This regulation aims to ensure the management of wastes from source to disposal without any impact on the environment and human health, and the reduction, recycle, reuse, and recovery of wastes by means of natural resource utilization. Upon the entry of the force of this regulation, some of the previous regulations such as Regulation on Control of Solid Wastes, Regulation on Control of Hazardous Wastes, and Regulation on General Principles of Waste Management have been repealed.

The first particular regulation on packaging waste control came into force in 2004 with the Regulation on Control of Packaging Wastes and this regulation was revised in 2007, 2011, and finally in 2017 in order to increase the recycled waste amount and to minimize the amount of packaging waste in municipal waste. Recycling targets are identified for authorized institutions and suppliers by this regulation. While the existing recycling rate is 5.3%, the estimated target for 2023 is to reach to 12% of the total generated waste (60% of the recyclable wastes) in accordance with the EU directives.

Separate control regulations were established for medical wastes, construction and demolition wastes, waste electrical and electronic equipment, used batteries and accumulators, end of life vehicles, end of life tires, mining wastes, PCBs and PCTs, and waste oils. The shortcomings will be completed and all regulations will be harmonized with the EU directives during the accession process.

According to the summarized regulatory inventory, it is clear that the Turkish Government has implemented several measures in order to prevent the environment and citizen's health from improper waste disposal. The appropriate collection, reuse, recycle, and disposal methods for wastes and the provision of penalties for non-compliance with the regulations were emphasized even in the earlier legislations. A great deal of effort has been taken to harmonize regulations in the EU accession process during the last decade. The most important factor that could have contributed to the failure of the implementation of the regulations is the huge amount of investment cost of the required facilities for recycling, recover, and disposal of wastes.

A new action plan entitled "Waste Management and Action Plan – 2023" had been prepared by the Ministry of Environment and Urbanization in 2017. This action plan includes more reliable projections and investment costs than the previous plans for the waste sector. Thus, it is expected that the all predicted regulations, most of which are today harmonized to EU directives, and investments will be completed until 2023 in accordance with the Waste Management Action Plan.

MUNICIPAL WASTE MANAGEMENT

The high population and population growth rate causes some difficulties in the creation of modern waste management systems due to the continuous increase in the waste amount and municipal budgets for the management of these waste (Guerrero et al., 2013). Therefore, waste generation rates and characterization is a key point for the perspectives of future projects and applications. In this section, collection/transfer and disposal of municipal solid waste are given in Turkey.

The history of waste characterization studies lies to early 1980s in Turkey. However, the first researches were realized only for several municipalities and the results from various sources differ from each other. The reliable data about waste amounts and characteristics can only be reached from the beginning of the 2000s. A circular for the standardization of the waste characterization studies (Solid Waste Characterization Analysis Method Booklet) has been prepared in 2007 by the Ministry of Environment and Urbanization. All municipalities were held responsible to realizing the characterization studies for at least twice a year according to this circular. Additionally, Turkish Standards Institution (TurkStat) is collecting and reporting the whole statistics including economic, social, industrial, technological, agricultural, and environmental issues of the country regularly since 2008, and the reliable waste data depends on these statistics.

This section summarizes the variations of the waste amount and characteristics for municipal solid waste (MSW) with the existing available data for Istanbul and Turkey.

Municipal Wastes

Municipal wastes are defined as wastes that are classified as non-hazardous wastes and originate from houses or are in similar content from different sources. The amount and characterization of MSW is a key parameter for the planning of the ultimate waste management system including collection, transport, recycle, and disposal methods.

Solid waste disposal without both material and/or energy recycle operations causes serious economic and natural resource losses. The increase in the waste generation reveals the need for the sustainable management of these wastes. The first step for the sustainable management of waste depends on the identification of the waste with its amount and characteristics. The rate of wastes from different sources is classified in Figure 9. Municipal wastes constitute the largest group with its 87.18% in the total waste stream. The remaining 12.82% consists of packaging wastes, hazardous wastes, medical waste, and special wastes. Construction and demolition wastes and wastes from mining industry are not included in this figure.

Figure 9. Waste distribution in Turkey (MoEU, 2017)

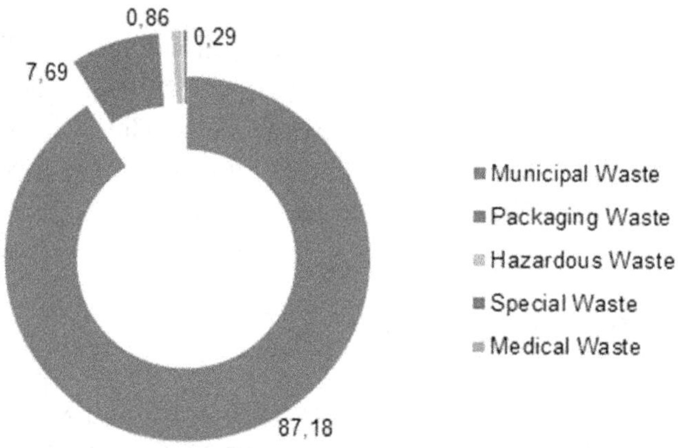

Figure 10. Total amount of collected municipal solid waste and waste amount per capita (TurkStat, 2016)

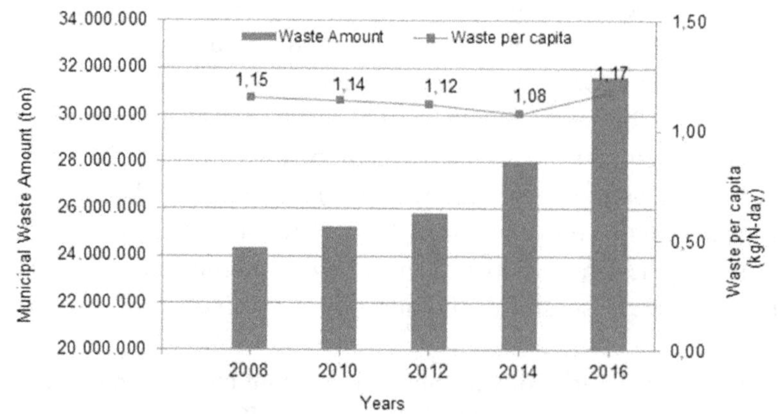

Figure 10 gives the amount of collected MSW and the relevant waste generation rate per capita for the period between 2008 and 2016. The per capita values range from 1.08 to 1.17 kg per person per day with an average of 1.13 kg/capita/day, according to the existing data. The amounts of the generated waste per capita decreased due to the increase in the recycling and reuse operations. Another factor that is effective on the decrease of the waste amount per capita is the increase in the number of the citizens that receive waste services from municipalities. The rate of the citizens living in urban areas and receiving services from a municipality increased from 70.5% in 2007 to 92.3% in 2016. Therefore, despite the increase in the amount of generated waste the amount per capita decreased gradually.

MSW amount is increasing rapidly as can be seen from Figure 10. In addition to the increase in the population, the increase in the waste amount is also related for the collection of real data due to the increased awareness to waste management operations in the last decade. As a result of the socio-economic development in the country and the population growth, it is estimated that the generated waste amount will also continue to increase in the next years.

Municipal Waste Characterization

The characteristics of MSW vary not only from city to city but also within the same city depending on several factors including living standards, socio-economic development and seasonal variations. The average results of the waste characterization studies are given in Figure 11. The organic waste or biowaste constitute the majority of the total waste. Biowaste and packaging wastes constitute 55.54% and 18.72%, respectively, of the total municipal wastes.

Istanbul is the largest city in Turkey and the fifth largest city in the world in terms of population within the city boundaries. Almost 20% of the population which corresponds to around 16 million people lives in Istanbul. Therefore, almost all of the first waste management and disposal applications are realized in Istanbul during the last decades. The success of these operations and applications depends on reliable waste characterizations. Table 1 gives the results of the previous waste characterization studies of Istanbul. Depending on the socio-economic development and the variations in the consumption habits, characteristics of the solid waste varies substantially.

Figure 11. Average composition of MSW in Turkey, % (MoEU, 2017)

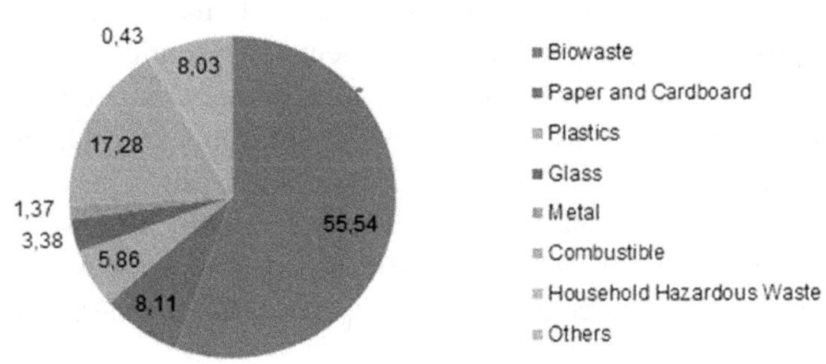

Table 1. Composition of municipal waste in Istanbul, 1979-2013 (Bilgili, 2006; Yildiz, et al. 2013)

Parameter	Basturk 1979	WHO/UNDP, (1981)	CH2M Hill, (1992)	Ozturk et al, (1997)	Y.T.U, (2003)	Yildiz et al., (2013)
Ash	29	14.6	15	13.2	7	0.63
Organic Waste	46.5	60.6	45	48	53.8	54.09
Paper	19	18.8	14.5	8.4	7.5	15.57
Plastics	3.5	3.1	9.5	11	14.4	13.71
Glass	3	0.7	3.8	4.6	4.3	3.03
Textile	3	3.1	5.6	2.9	4.4	3.42
Metal	1.5	1.5	2.2	2.3	1.7	1.12
Diaper	-	-	-	3.2	6.9	5.45
Others	1.5	6.9	4.4	6.3	-	2.99

Collection and Transfer of Municipal Solid Waste

The first step in a successful waste management system is to separately collect in a sustainable manner at the source of waste. Accordingly, the legislations on waste management in Turkey aims to promote recycling/recovery of this recyclable fraction within the municipal solid waste. Municipalities collect and transfer municipal wastes by themselves or by private companies in the current situation.

The municipal waste collection system is affected by population density, characteristics of settlements and physical structure of street (narrow and cross streets, etc.). In case of the streets are narrow and the buildings are very close to each other, the wastes are kept in garbage bags on the sidewalk, while garbage containers are generally preferred on large streets. The most common surface container capacity used in Turkey is 300 lt-1100 lt, and the underground container capacity is 2200 to 5000 lt. In order to reduce waste transfer costs and to ensure transfer optimization, municipal solid waste carries by means of transfer stations. There are 104 transfer stations in Turkey that 59 of these numbers operate in metropolis and 45 in other cities (MoEU, 2017).

Management and Disposal of Municipal Solid Waste

Municipal solid wastes constitute the largest proportion of the waste produced in Turkey. It is known that the packaging waste, which has to be collected separately at the source, is sent to the landfill sites where there is no dual collection system. Municipal solid waste management systems including dual collection system are progressing but further developments are required in Turkey. Only 64% of the wastes generated by Turkey's population are stored in landfills according to technical standard and the other 30% of the population receives waste removal services that are not within regulations and stored in un-controlled landfill sites. The percentage of organic domestic waste received to landfills is projected to decrease in connection to waste management regulation in Turkey. In EU-27 countries, the amount of waste disposed of by means of landfill was significantly decreased between 1995 and 2014, while the amount of waste disposed by material recovery, biological and incineration methods increased in European countries.

According to the 2014 data, the amount of municipal waste sent to recycling in Turkey is about 1.5 million tons. There are 6 compost facilities in Amasya, Aydin, Balikesir, Canakkale, Denizli, Kutahya, where municipal wastes have been collected separately. 2 biomethanization plants are operating in Kocaeli and Sakarya provinces. There are 6 mechanical and biological treatment plants in which mixed wastes are processed. In Istanbul, 2 mechanical and biological treatment plants are operated as composting and bio-drying process.

In 29 provinces in Turkey, landfill gas to energy plant is currently active. The total amount of electricity produced from these facilities is over 900.000 MWh per year. With the increase in the number of landfills, it is expected that the amount of electricity produced will increase with the increase in the number of landfill gas to energy facilities. In addition, the gas produced from biomethanization plant in Ankara, Adana and Samsun provinces is burned together with the landfill gas from the landfills and electricity is produced.

The final disposal of municipal wastes is performed by landfill method. The collected amount of municipal waste has reached to 27.13 million tons with an increase of 11.4% since 2008. While 17.5 million tons of municipal waste is disposed of with sanitary landfill method, approximately 8 million tons of waste is left to store with un-controlled landfill sites. As of 2014, the number of landfill facilities is 79 and in these 1073 municipalities, 47.4 million people are served. As of the end of 2015, the number of landfills has increased to 81 and the number of serving population has reached 48.9 million in 1091 municipalities. In 2016, the number of provinces establishing and operating landfill sites reached 59, and the number of sanitary landfills reached 82. There are about 800 un-controlled landfills over the casting in Turkey. Some of these sites have been rehabilitated and some of them are still active. Figure 12 shows the percentages of municipal waste amounts in 2008 and 2014 according to the disposal method. In 2014, the amount of municipal waste sent to un-controlled landfills decreased by 44%, while the amount of municipal waste sent to sanitary landfill increased by 42%.

Figure 12. Municipal waste rates sent to sanitary landfill and un-controlled landfill in 2008 and 2014

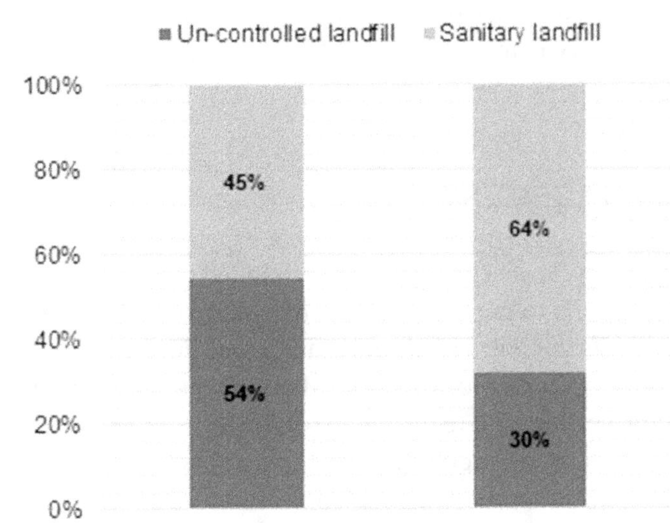

Figure 13. Waste amounts in Istanbul in terms of management activities

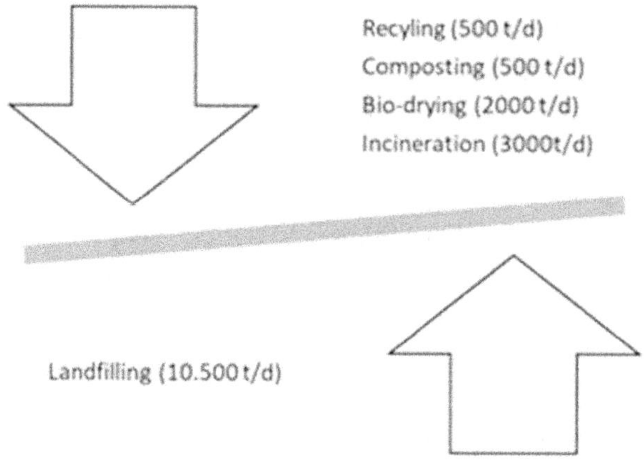

Municipal Solid Waste Management of Istanbul

A huge amount of municipal solid waste, nearly 16.500 tons per day is handled by Istanbul Metropolitan Municipality. Therefore, several challenges present regarding landfill space, collection and transport. This problem is actually becoming a driving force in finding a different method instead of landfilling for waste management in Istanbul. Thermal process is an alternate method overcoming these challenges and waste to energy facility has currently been started to establish with the capacity of 3,000 tons per day taking the first step (Yildiz et al., 2014). Figure 13 shows waste amounts in Istanbul in terms of management activities including incineration plant of 3000 tons/day capacity.

Currently, Composting and Waste Recovery Facility with the capacity of 1,000 tons per day in the European side and a biodrying facility of 2,000 tons per day in the Asian side. There are two landfill gas to energy facilities located in European (Odayeri) and Asian sides (Komurcuoda). Approximately 32,000,000 tons of solid wastes has the potential use to recovery energy from landfill gas in Odayeri sanitary landfill and has a landfill gas to energy plant with the installed capacity of 25 MW. Komurcuoda landfill has an energy recovery potential of 15,000,000 tons of solid wastes and 10 MW energy recovery from this storage.

PACKAGING WASTE MANAGEMENT

Real and legal entities aiming to recycle packaging waste are obliged to obtain a license from the MoEU. The reason for this implementation is to impose a certain discipline to the facilities where such waste is collected, sorted and recovered/recycled. The licenses are given in two categories – one for collection and sorting of waste, and the other one for recycling. This process has progressed quickly in three separate operations – collection, sorting, and recycling in Turkey.

In the existing situation, the reuse, recovery and recycling of packaging waste can be carried out on a limited scale. The recovery and recycling of packaging waste have not reached the desired level. Furthermore, the number of licensed collection and sorting companies is still insufficient.

The By-law on Packaging Waste was published in the Official Gazette of 24.06.2007 and thus came into force. This By-law includes all the packages and package wastes offered in the domestic market regardless of the materials used (plastic, metal, glass, paper-cardboard, composite etc.) and the resource of the waste (domestic, commercial, industrial and all kinds of offices).

In Turkey, the amount of packaging waste introduced to the market in 2014 is 4,200,000 tons. The municipalities are responsible for collecting and recycling the packaging wastes generated at the source. Municipalities conduct the packaging waste management either through authorized institutions and/or licensed companies (MoEU, 2017).

In 2014, 4,200,000 tons of packaging waste was launched. The amount of packaging waste collected separately by the municipalities at the source is 1,527,960 tons, the amount of packaging waste collected from the industry is 586,690 tons, the amount of packaging waste collected from the airports is 14,362 tons and the amount of packaging waste collected from outside the area is 263,361 tons. The total amount of packaging waste collected at the total source is 2,392,374 tons/year (MoEU, 2017).

Figure 14. Number of packaging waste management facilities (MoEU, 2017)

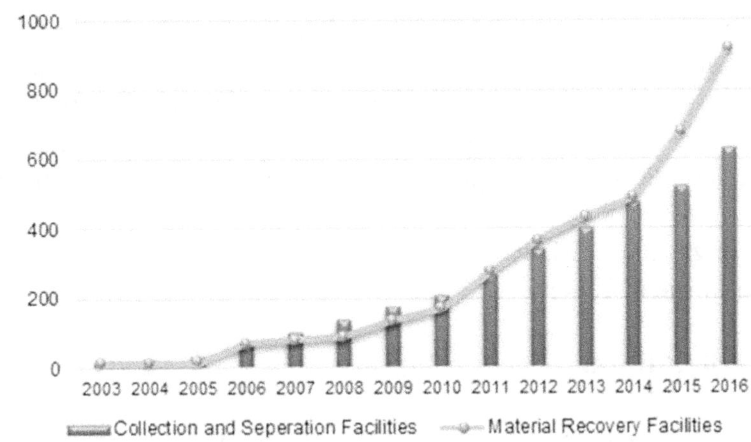

Figure 15. Packaging waste characteristics (MoEU, 2018)

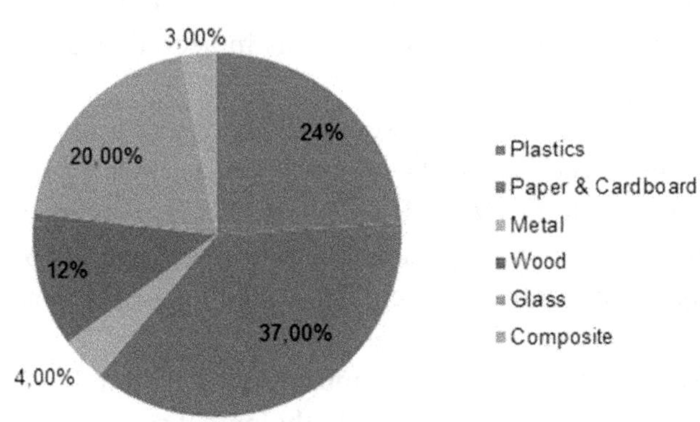

Many municipalities provide environmental trainings in schools and workplaces, especially in schools, in order to increase the efficiency of the separate collection at the source. The most common collection method in our country is collecting with 58% (door and door), 33% container and 9% fetching center. Mixed packaging wastes collected separately at their source are separated according to their type in collection facilities and sent to licensed recycling facilities. The data for licensed facilities in Turkey is given in Figure 14 (MoEU, 2017).

The characteristics of the packaging wastes collected from municipalities are given in Figure 15. The highest constituent of packaging waste is paper and cardboard according to the results of the characterization studies.

MEDICAL WASTE MANAGEMENT

The health sector is one of the significant sectors that have improved all over the world in the last decades. However, medical waste which can be defined as the fraction of waste generated at healthcare services has not attracted the same level of attention as other types of wastes, especially in developing countries. Hospitals, clinics, laboratories, veterinary clinics and many more establishments have to dispose of waste materials that have been generated in the process of medical care and treatment (Abdulla et al., 2008). Medical wastes are potentially dangerous since they may contain pathogenic agents.

Medical waste management and disposal is an emerging issue that is magnified by a lack of training, awareness, and financial resources to support solutions. Due to the direct impact on both public and environmental health, the collection and disposal of medical waste are of great importance (Birpinar et al., 2009).

This section summarizes the legislative development of medical waste management, collection and disposal of this type of waste in Turkey.

Medical Waste Legislations

The first legislation that deals with the control of medical wastes was entered into force by the Ministry of Environment initially in May 1993. This regulation changed in July 2005 and a new Medical Waste Control Regulation (MWCR) was published concordant with the EU Environmental Directives. This regulation determined the principles of collection, transportation, temporary storage in the healthcare facilities, and removing of medical wastes. Medical waste control regulation was changed finally in 2017 in accordance with the harmonization of the regulations in accordance with the EU directives during the accession process.

Wastes generated from healthcare services were classified in four main groups named as municipal wastes (general and packaging waste), medical wastes (infectious, pathological, sharps), hazardous waste, and radioactive waste according to the MWCR (MWCR, 2017). This regulation covers the principles of separate collection, transport, temporary storage and disposal options of medical waste. A detailed organization of the management system from separate collection to a disposal of medical waste is given in the regulation which is summarized below.

According to MWCR, municipal solid wastes must be collected in black plastic bags, while the recyclable materials must be collected in blue bags. Pathological medical wastes, non-pathological, and infectious wastes must be collected in red bags with a minimum thickness of 100 μm and a capacity of 10 kg. Additionally, sharp objects must be collected in yellow plastic containers. All of the bags or containers must include international "Biohazard" emblem and indicate the type of waste covered. Moreover, hazardous chemicals, pharmaceutical wastes, and heavy metal containing wastes are removed according to the Hazardous Waste Management Regulation. Radioactive wastes are also removed according to the Turkey Atomic Energy Council Act (Ministry of Environment and Urbanization, 2017).

The temporary storage depots must be covered with a solid, impermeable, microorganism and dirt-repellent material that is easy to clean and disinfect. The depots must have a storage capacity of medical wastes up two days. Special vehicles must be used for the transport of medical waste to the disposal plant. These vehicles cannot be used for the transport of other types of wastes. The vehicles must be orange colored and must include the international "Biohazard" emblem in each side.

Sterilization can be used for the infectious and sharp materials, and these wastes can be disposed of in Class II landfills after sterilization. A biological indicator must be used in the sterilization unit in order to control the success of the sterilization process. Incineration is mandatory for the disposal of chemically treated pathological wastes. The incineration temperature must be at least 1100 °C if the waste stream includes a large quantity of genotoxic wastes.

Medical Waste Quantities

Medical waste quantities are generally depicted as a function of the bed capacity of the healthcare services. The generation rates differ by type of establishment, the proportion of reusable items used and proportion treated on an outpatient basis (Eker & Bilgili, 2011). Similar to the trend of MSW generation, hospital waste quantity is also directly proportional to the income level (Chaerul et al., 2008). Medical waste generation rates of different countries are given in Table 2. Healthcare wastes are one of the most dangerous causes of the pollution of the environment and natural resources due to their infectious nature. The amount of hospital wastes is increasing due to the use of more disposable products as a result of developing healthcare technologies in healthcare services (Sabour et al., 2007; Patwary et al., 2009).

Table 2. Healthcare waste generation in different countries (Eker & Bilgili, 2011)

Country	Waste Generation Rate kg/bed/day
Bangladesh	0.55-1.67
China	0.68
India	1.60
Iran	1.25
Italy[a]	3-5
Japan	2.05
Jordan	1.88-3.49
Portugal	2.5-4.5
Tanzania	0.84
Thailand	1.75
Thailand	1
Turkey	2.35
USA[a]	5-7

Information on medical waste generation should be obtained for the development of proper waste management strategies. Medical waste generation depends on several factors including the size of the healthcare facility, occupancy rate of hospital beds, medical waste segregation system, the location of the healthcare facility, and type of the services provided. A detailed statistical study is conducted by Eker and Bilgili (2011) for the determination of medical waste quantities in Istanbul. According to the results of the survey totally 43,411.20 kg of regulated medical wastes (infectious and pathological wastes) which corresponds to 28.8% of the total waste streams, are collected from the healthcare services in Istanbul in 2010. Also, the generation rate for regulated medical waste was calculated as 2.11 ± 3.83 kg/bed day, while private hospitals and private medical centers are the major producers (Eker & Bilgili, 2011).

Medical Waste Management in Istanbul

Almost 20% of Turkey's population which corresponds to around 16 million people lives in Istanbul. Correspondingly, around 20% of the total bed capacity (40000 beds from state and private hospitals) and 25% of the hospitals are located in Istanbul (Demir et al., 2002). Medical Waste Control Regulation gives the responsibility of collection, transport, and disposal duties to the municipalities. The task of medical waste management organization is carried out by Istanbul Environmental Management Industry and Trade Inc. (ISTAC), a semi-private subsidiary of Istanbul Metropolitan Municipality.

The amount of medical waste collected annually in Istanbul is given in Figure 16. According to the available data for 2016, the collected medical waste amount is about 65 ton per day, which corresponds to around 23,700 tons of waste collected with around six million bags, annually. The average waste generation rate, from healthcare services with bed capacity higher than 20, is between 1.5 and 2 kg per bed.

The first and the only medical waste incineration plant with its 24 tons/day capacity is serving Istanbul since 1995 in order to protect the human and environmental health. The risks of disease transmission are eliminated by incineration of pathological and infectious wastes. Medical waste decreased by 95% by volume, and 75% by mass after incineration between 1000 and 1200 °C in a rotary kiln.

Figure 16. The annual medical waste amount in Istanbul

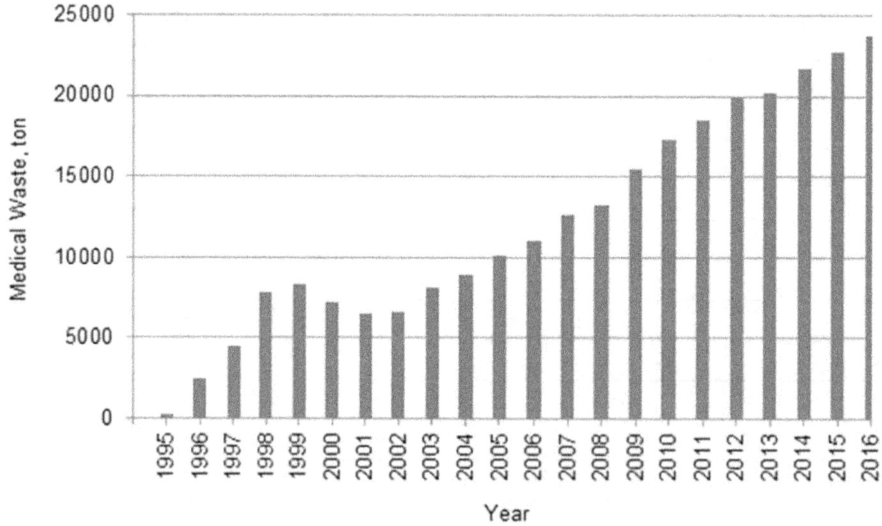

Istanbul Metropolitan Municipality built Turkey's biggest sterilization unit with 40,000 tons/year capacity for the infectious medical wastes in 2013. The sterilization plant is equipped with the latest technology including full automatic conveyor systems, remote monitoring and intervention systems. Medical wastes are being sterilized at 145 °C 3 bar pressurized steam in the autoclaves, each having a capacity of 1.5 tons/hour. Total reaction time is about 45 min for each autoclave. The annual amount of sterilized wastes was 17,300 tons in 2015. Sterilized wastes are disposed of at landfills after shredding. The plant is designed to serve Istanbul for the next 30 years.

MILESTONES OF WASTE MANAGEMENT IN TURKEY AND FUTURE PERSPECTIVES

Environmental protection has become one of the major priorities of all countries due to the increasing environmental awareness and waste management is one of the major fields of environmental protection policies. Therefore, waste management strategies have become the basis of "sustainable development" and "circular economy" strategies, aiming to prevent waste generation and protection of natural resources for almost all countries.

Environment is one of the most challenging sections in the accession negotiations with EU. Environmental protection is regulated by nearly 300 Directives and regulations. Turkey realized significant progress in the development of waste management since the beginning of the 1990s, which is further accelerated during the EU accession process after the year 2005. Besides the regulatory harmonization, the number of waste treatment, recycling and disposal facilities are increasing with the aim of the environmental and human health protection. Table 3 summarizes the most significant activities of waste management applications since the 1990s.

Action Required

In accordance with Turkey's compliance with EU environmental legislations several plans, which are listed below, have been prepared:

- Planning of High Cost Environmental Investments.

Table 3. Milestones of waste management services in Turkey

Year	Application
1995	The first sanitary landfill site commissioned. Today, 87% of the municipal wastes are being disposed in landfills.
1995	The first medical waste incineration plant commissioned. Due to the regulation change in 2005, sterilization becomes the most preferred method for medical wastes today. Around 90% of medical wastes are disposed into landfills after sterilization.
1997	The first hazardous and industrial waste treatment and disposal plant commissioned.
2001	The first composting and recycling plant commissioned. Composting plant is one of the biggest facilities in Europe with its capacity of 1000 t/day waste processing capacity.
2007	The first landfill gas to energy facility commissioned. The number of the plants is still increasing as a result of the increasing number of landfill sites.

Table 4. Required facility capacities compatible with national strategies and estimated initial investment costs

Required Facilities	Required Capacity (ton/day)	Unit Investment Cost (€/ton)	Investment Range (million €)
Recycling (Packaging Waste)	12.509	25 – 50	114,1 – 228,3
Biological Processes	4.050	75 – 200	110,9 – 295,6
Mechanical-Biological Processes	7.250	100 – 150	264,6 – 396,9
Thermal Processes	8.046	300 – 450	881,0 – 1.321,6
Landfilling	67.732	15 – 25	370,8 – 618,1
TOTAL INVESTMENT COST			**1.741,5 – 2.860,5**

- National Environment Integrated Adaptation Strategy.
- Solid Waste Master Plan.
- Waste Management Action Plan.
- National Recycling Strategy Document and Action Plan.
- National Waste Management and Action Plan – 2023.

National Waste Management and Action Plan (2016-2023) prepared in line with both the implementation of national legislations and harmonization with the EU acquis. The current situation of waste management in 81 municipalities, source separation, recycling, energy and/or material recovery, and disposal methods are aimed to be revealed within the scope of this plan. The types of facilities, capacities and investment costs for waste management activities planned to be made until 2023 are presented on a regional basis with compliant to regulations.

The amount of waste, demographic structure, geographical characteristics, socio-economic status, tourism sector, agriculture and forest areas and livestock activities were taken into account when preparing the waste generation models. In accordance with these models, middle and long term targets can be listed as below.

- Considering the amount of waste that will be produced in 2023; 35% of the waste will be recycled and 65% will be disposed in landfills. For this purpose:
    - The amount of source separated packaging waste must be increased from 5.3% to 12% in 2023.
    - The recovery rate of municipal wastes by biological methods must be increased from 0,2% to 4% by the end of 2023.
    - The rate of mechanical-biological treatment of municipal wastes must be increased from 5,4% to 11%.
    - The rate of thermal treatment processes must be increased from 0,3% to 8%.
    - The amount of the landfilled wastes must be decreased from 88,7% to 65% by the end of 2023.
- All of the open dumps will be closed and rehabilitated.
- Construction and demolition waste and excavation soil management will be ensured throughout the country.

- The efficiency of collection and recycling of special wastes will be increased.
- Additional facility investments will be increased for the recovery and disposal of hazardous wastes.

The facility capacities and the estimated initial investment costs required complying with the targets given in the National Waste Management and Action Plan - 2023 strategies are summarized in Table 4.

In line with these strategies, new sanitary landfill facilities are being constructed nationwide. Additionally, compilation with the EU Landfill Directive (99/31/EC) requires the reduction of the rate of biodegradable wastes that disposed into the landfill sites. For this purpose, mechanical biological treatment (MBT) facilities and biogas production facilities are being constructed in metropolitan municipalities. However, the success of these facilities depends on the source separated organic waste amount, while separate collection is not well established yet in Turkey. Therefore, further effort should be made to ensure the source separation of organic wastes.

Thermal processes have not yet been used for MSW in Turkey, due to the low calorific value and high investment and operation costs. However, the number of the incineration plants is expected to increase in the near future in accordance with the developed strategies in the National Plans. Thermal processes will also help to achieve the reduction target of the organic waste amount going to the landfills.

Turkey has gained acceleration in improving both the legal infrastructure and application of waste management systems in a short time. Therefore, it is believed that there will be a sound effective waste management system and the establishment of many solid waste management facilities will be provided in Turkey soon.

FUTURE REMARKS

Municipal solid waste is still almost completely disposed of in landfills in Turkey. However, the number of un-controlled landfills remains in high percentage. In order to establish a sustainable recycling economy or a circular economy in the country, the Turkish Ministry of the Environment and Urbanization has initiated the Zero Waste Project. The initiative calls for the infrastructure of a nationwide recycling system to be established in public and private institutions. In addition, the project aims to raise public awareness of environmental issues, particularly recycling. Specifically, it provides for training measures at businesses, modification of the system to fit the respective establishments, and assessment of its implementation. The pilot phase at the ministry for the environment began in June 2017. Starting in April 2018, the project will be expanded to another 230 state institutions in the capital of Ankara. Turkey wants to have the recycling system established nationwide by 2023.

In the last decade, MSW management in Turkey has improved. The well-funded municipalities effectively organized the planning and implementation of the steps of Integrated Solid-Waste-Management Plan. The Metropolitan Municipality of Istanbul has currently intensified its efforts in the separation of MSW at the source. However, there are still problems with MSW management. Landfilling, the last choice in the hierarchy of waste management remains the main method adopted for MSW treatment in Istanbul. Biogas energy recovery is also not well organized. Although there have been many published studies showing the potential value and use of biogas, biogas energy recovery is still low and inefficient in Istanbul and Turkey, probably due to high investment cost and operational problems. However, the disposal or treatment methods such as thermal methods, bio-methanization, mechanical and biological treatment, composting will be the main actor for the municipal solid waste management instead of the landfill on coming years.

REFERENCES

Abdulla, F., Qdais, H. A., & Rabi, A. (2008). Site Investigation on Medical Waste Management Practices In Northern Jordan. *Waste Management (New York, N.Y.)*, *28*(2), 450–458. doi:10.1016/j.wasman.2007.02.035 PMID:17507209

Atienza, V. (2011). Review of waste management system in Philippines: Initiatives to promote waste segregation and recycling through good governance. In *Economic Integration and Recycling in Asia: An Interim Report*. Academic Press.

Basturk, A. (1979). *A Research Model on Solid Wastes and Its Application for Istanbul*. Academy Associate Professor Thesis.

Basturk, A. (1980). *Remedies for the Elimination of Solid Waste Hazards in Istanbul*. TUBITAK, Environmental Research Group, Project No: CAG-11.

Basturk, A. (1994). *Solid Waste Course Notes*. İstanbul, Turkey: Yildiz Technical University, Environmental Engineering Department.

Bilgili, M. S. (2006). *Effect of leachate recirculation on degradation of muicipal solid wastes in aerobic and anaerobic landfills* (Ph.D Thesis). Yildiz Technical University, İstanbul, Turkey.

Birpinar, M.E., Bilgili, M.S., & Erdogan, T. (2009). Medical waste management in Turkey: A case study of Istanbul. *Waste Management, 29*, 445-448.

CH2M Hill-Antel. (1992). *Solid Waste Management Study for Istanbul Main City*. İstanbul: Author.

Chaerul, M., Tanaka, M., & Shekdar, A. V. (2008). A system dynamics approach for hospital waste management. *Waste Management (New York, N.Y.)*, *28*(2), 442–449. doi:10.1016/j.wasman.2007.01.007 PMID:17368013

Demir, A., Sengun, R. B., & Ozkaya, B. (2002). The Management of Medical Wastes in Istanbul. ISWA-2002 Appropriate Environmental and Solid Waste Management and Technologies for Developing Countries, Istanbul, Turkey.

Eker, H. H., & Bilgili, M. S. (2011). Statistical analysis of waste generation in healthcare services: A case study. *Waste Management & Research*, *29*(8), 791–796. doi:10.1177/0734242X10396755 PMID:21382873

European Environment Agency (EEA). (2013). *Municipal waste management in Turkey*. ETC/SCP Working Paper.

Guerrero, L. A., Maas, G., & Hogland, W. (2013). Solid waste management challenges for cities in developing countries. *Waste Management, 33*(1), 220-232. Retrieved from http://www.worldometers.info/world-population/turkey-population/

Japan International Cooperation Agency (JICA). (2015). *Data Collection Survey on Solid Waste Management in Turkey*. Final Report. Author.

Ministry of Environment and Urbanization. (2017a). *Medical Waste Control Regulation*. MWCR.

Ministry of Environment and Urbanization. (2017b). National Waste Management and Action Plan 2023. Ankara, Turkey: Author.

Ministry of Environment and Urbanization. (2018). Packaging Waste Statistics 2016. Ankara, Turkey: Author.

Patwary, M. A., O'Hare, W. T., Street, G., Elahi, K. M., Hossain, S. S., & Sarker, M. H. (2009). Quantitative assessment of medical waste generation in the capital city of Bangladesh. *Waste Management (New York, N.Y.)*, *29*(8), 2392–2397. doi:10.1016/j.wasman.2009.03.021 PMID:19375297

Sabour, M. R., Mohamedifard, A., & Kamalan, H. (2007). Mathematical model to predict the composition and generation of hospital wastes in Iran. *Waste Management (New York, N.Y.)*, *27*(4), 584–587. doi:10.1016/j.wasman.2006.05.010 PMID:17239577

Turkish Statistical Institution. (2018). Retrieved from http://www.turkstat.gov.tr

Yildiz, S., Balahorli, V., Hosoglu, F., & Sezer, K. (2014). *Waste Management in the Metropolis of Istanbul – Waste Incineration and Energy Production Facility in Istanbul.* Retrieved from http://www.vivis.de/phocadownload/Download/2014_wm/2014_wm_265_276_yildiz.pdf

Yildiz, S., Yaman, C., Demir, G., Ozcan, H. K., Coban, A., Okten, H. E., ... Goren, S. (2013). Characterization of Municipal Solid Waste in Istanbul, Turkey. *Environmental Progress & Sustainable Energy*, *32*(3), 734–739. doi:10.1002/ep.11640

Chapter 19
Solid Waste Management in Vietnam:
State and Institutional Framework and Challenges

Thang Trung Nguyen
Institute of Strategy and Policy on Natural Resources and Environment, Vietnam

Hanh Hoang Hong
Institute of Strategy and Policy on Natural Resources and Environment, Vietnam

Phuong Anh Thi Duong
Institute of Strategy and Policy on Natural Resources and Environment, Vietnam

Tu Ngoc Nguyen
Institute of Strategy and Policy on Natural Resources and Environment, Vietnam

ABSTRACT

This chapter will provide an overview of waste management in Vietnam, including drivers, the current waste management situation, impacts, an institutional framework, issues, and challenges. There have been improvements in the last 10 years. However, waste management in Vietnam continues to face issues and constraints related to inefficient waste reduction, a national level efforts to implement separation at the source, environmental pollution caused by informal sector recycling efforts, and 70% to 75% of municipal solid waste in landfills. These issues are caused by incomplete legislation, inadequate financial resources, weak enforcement and implementation efforts, and low stakeholder awareness. Based on the analysis, several recommendations are provided for future improvement.

DOI: 10.4018/978-1-7998-0198-6.ch019

INTRODUCTION

Waste management plays an important role in environmental protection and sustainable development of a country. Globally, waste is often well managed in developed countries. However, many issues face developing countries. Poor waste management leads to environmental pollution, health impacts, and global warming. The United Nations (UN) has appealed to countries to implement measures to promote green growth, green economy, and circular economy. These efforts include waste management.

After over 30 years of renovation, Vietnam has achieved remarkable economic development success. Its economic growth has been maintained, poverty reduction has achieved remarkable progress, and the income per capita has increased. However, the country has experienced environmental pollution and degradation. A key environmental issue is inadequate waste management. In collaboration with the international community, Vietnam has committed to implement the 2030 Agenda for Sustainable Development, which will implement 17 sustainable development goals (SDGs), and the Paris Agreement on climate change.

The purpose of this chapter is to provide an overview of waste management in Vietnam by analyzing drivers of waste generation, current waste generation, collection, recycling, and treatment/disposal. It will also review the impacts of waste management and the institutional framework. The chapter will use identified issues, reasons, and challenges to provide recommendations and measures for improvement of waste management. It aims to achieve the SDGs and climate change commitments.

COUNTRY CONTEXT

Located in south-east Asia, Vietnam is the third most populated country in the region, ranking 14th in the world by population (see Figure 1). The country population increased from 86.95 million in 2010 to 94.67 million in 2018. Its average of 1 million people per year is expected to increase. Vietnam has also experienced a rapid urbanization process with the urban population increasing from 26.51 million (~30% of total population) in 2010 to 33.83 million (~ 36%) in 2018 (see Figure 2).

Figure 1. Administration map of Vietnam (Ministry of Natural Resources and Environment [MONRE], 2018a)

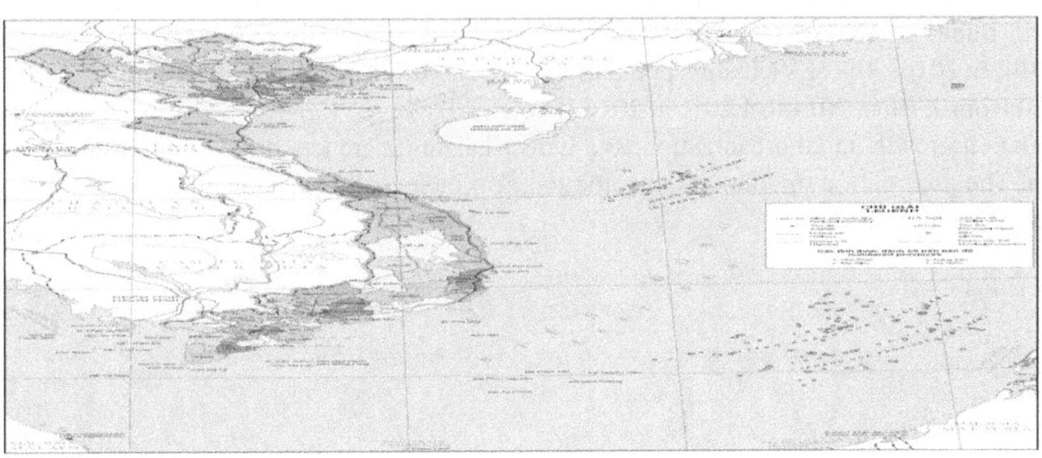

Figure 2. Total, rural and urban population of Vietnam 2010-2018 (adapted from General Statistics Office of Vietnam [GSO], 2018)

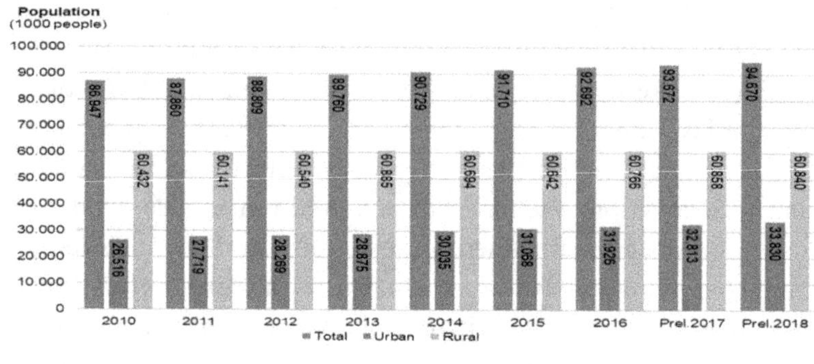

Figure 3. Vietnam GDP growth 1986-2018 (adapted from GSO, 2018)

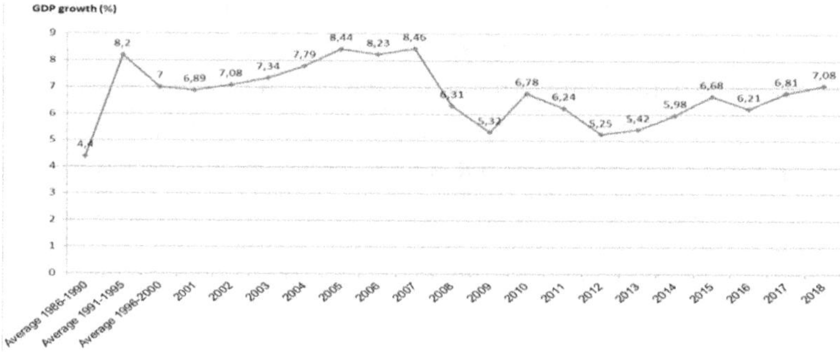

The adoption of the renovation policy shifted the country's focus from centrally planning into a market economy. In turn, Vietnam has achieved high economic growth in the last 30 years. The average gross domestic product (GDP) growth rate is about 6.63% per year from 1986-2017. It reached 7.08% in 2018 (see Figure 3). One of the reasons for this is the high industrialization process. The industrial production value increased sharply from ~9,500 billion Vietnam dong (VND) (408.5 million USD) in 1990 to 1,700,000 billion VND (73.1 billion USD) in 2017 (GSO, 2018).

The middle-income population has increased along with the economic development. According to the World Bank, an average of 1.5 million Vietnamese has joined the global middle-class each year since 2014. In 2016, 70% of Vietnam's population were classified as financially secure; the middle class increased from 7.7% in 2014 to 13.3% in 2016 (Dione, 2018).

Population growth, rapid urbanization, and industrialization are key drivers for waste generation in Vietnam. The growth of a middle-class population also increases unsustainable consumption, leading to higher generation of waste.

SOLID WASTE MANAGEMENT IN VIETNAM

Waste Definition and Classification

In Vietnam, waste and waste-related definitions have been regulated by the environmental legislations. According to this, waste is generally classified into two categories: (i) non-hazardous and; (ii) hazardous waste (HW). Besides, waste also can be classified as household waste, industrial waste, agricultural waste, medical waste and construction waste (National Assembly of Vietnam [NAV], 2014; Government of Vietnam [GOV], 2015). However, there are still no clear definitions for municipal solid waste (MSW) and e-waste. In the national state of environment reports 2011 and 2016, however, MSW has been addressed as waste generated from urban areas. According to these reports, the MSW includes domestic/household waste, street waste, construction and demolition waste (CDW), and waste generated from offices, hospitals, industries, and urban markets. More often used is the concept of urban domestic/household waste (UDW) (MONRE, 2011, 2016). (see Figure 4).

Figure 4. Solid waste classification in Vietnam (adapted from National Assembly of Vietnam, 2014; Government of Vietnam [GOV], 2015)

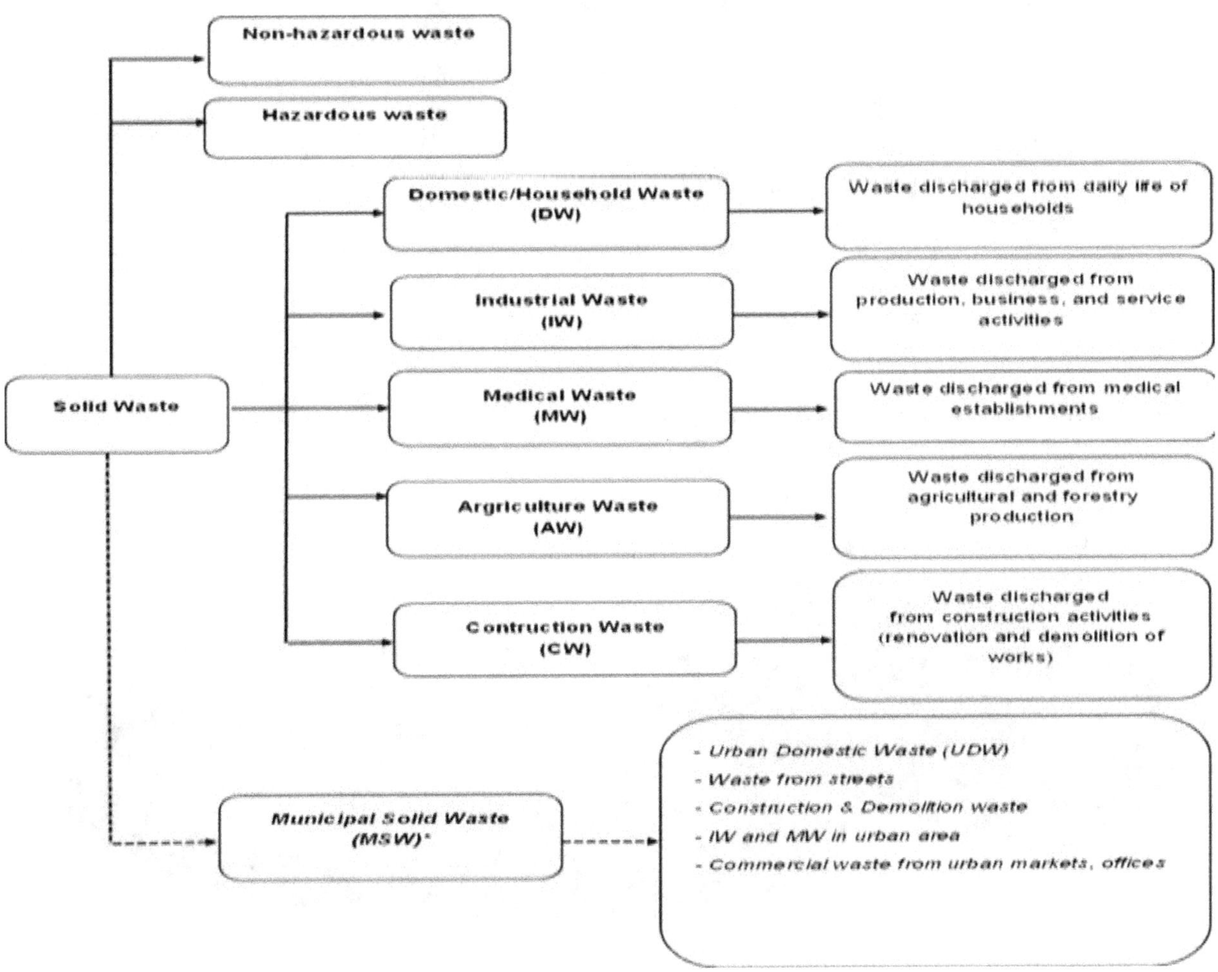

Waste Generation

According to MONRE, the generated amount of household solid waste in 2017 was about 25.5 million tons, in which UDW was 38,000 tons/day and rural household waste was 32,000 tons/day. The UDW increased by 10% to 16% per year, accounting for 60% to 70% of the total MSW. In some cities, the share of domestic waste is as high as 90% of the total MSW (MONRE, 2018b). The UDW has increased from 30,000 tons/day in 2013 to 38,000 tons/day in 2017 while the rural household waste increased from 22,000 tons/day in 2013 to 30,000 tons/day in 2017. It is observed a correlation between the growth of urban population and UDW generation (Fig 5).

Figure 5. Trend of domestic waste generation in urban and rural area in correlation with urban population growth (adapted from GSO, 2018; MONRE, 2018c)

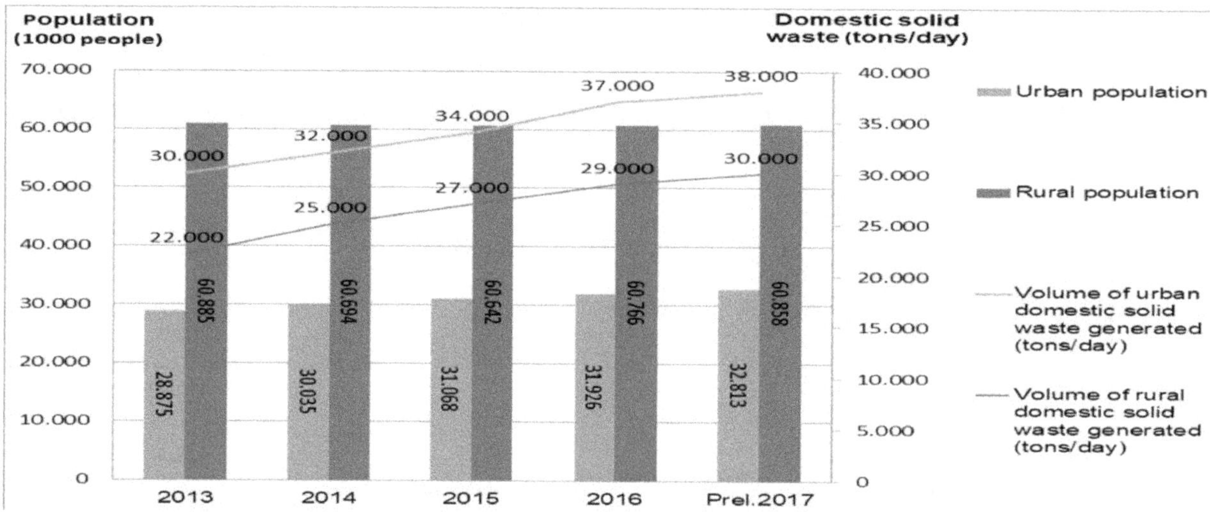

Figure 6. Solid waste composition in Hanoi (adapted from MONRE, 2015)

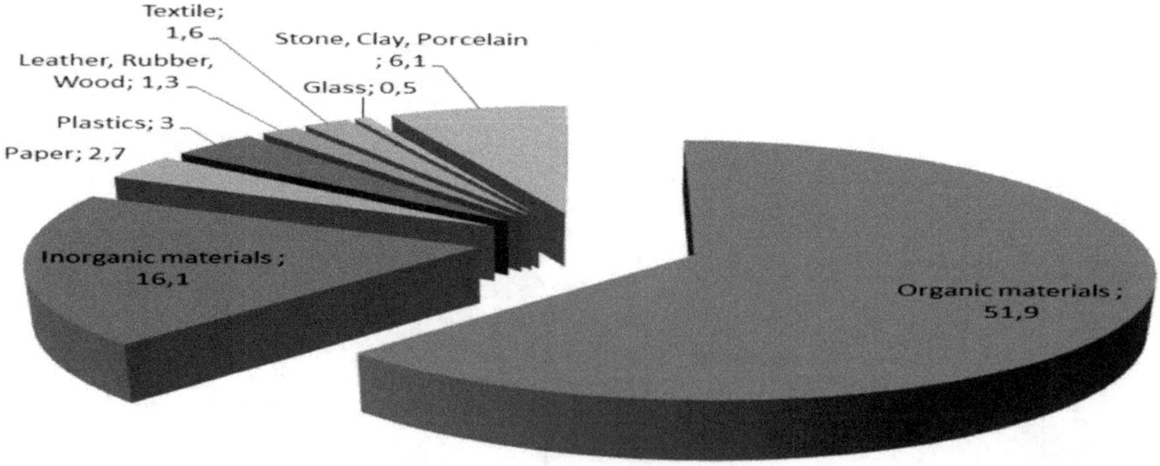

Table 1. Forecast of solid waste generation in some regions (adapted from Prime Minister [PM], 2013, 2015a, 2015b)

Type of Waste	Dong Nai River Basin (ton/day)		Nhue-Day River Basin (ton/day)		Northern Key Economic Region (ton/day)	
	2020	2030	2020	2030	2020	2030
Domestic waste	28,700	35,800	9,100	11,200	17,460	26,390
Construction waste	4,400	5,600	2,100	4,000	3,900	6,420
Craft villages and industry ordinary waste	15,000	26,800	2,200	3,700	9,590	16,300
Hazardous solid waste (industry, craft villages)	3,400	6,000	750	1,250	2,940	4,310
Total	**51,500**	**74,200**	**14,150**	**20,150**	**33,890**	**53,420**

With regards to waste composition, according to World Bank research, food and green materials vary 53% to 56% in middle- to low-income countries, respectively. Organic component is increasing as the development level decreases (Kaza, Yao, Bhada-Tata, & Woerden, 2018). Vietnam is not an exception of this as a survey in Hanoi showed that organic material accounts for 51.9% of waste (see Figure 6).

Waste generation in Vietnam has been forecast to increase continuously. According to master plans for waste management in the Dong Nai river basin, Nhue-Day river basin, and the northern key economic region, solid waste generation is forecast to increase in all categories. The domestic (household) waste accounts for the largest share in all regions and the Dong Nai river basin will be the region with the highest waste generation in the next 10 years (Table 1).

WASTE REDUCTION, COLLECTION, RECYCLING, TREATMENT, AND DISPOSAL

Waste Prevention and Reduction: Waste prevention and reduction have not been applied widely or efficiently in Vietnam in regards to both production and daily life. Although the National Strategy on Cleaner Industrial Production to 2020 has been issued since 2009, cleaner production, waste audit, and ISO 14000 standards have not been actively implemented by industries. In fact, only around 24% of enterprises have applied cleaner production measures (Vietnam Cleaner Production Center, 2016). In daily life, sustainable consumption has not been popular with the public. There are no specific incentives or measures for households to reduce their solid waste generation. The solid waste fee system is based on a flat rate rather than a volume based or pay-as-you-throw mechanism like in other countries (Thang, 2016).

Waste Separation at Source: Although regulated by law, household waste separation at source has not been applied widely yet due to lack of treatment facilities for separated wastes. Recently, Ho Chi Minh City issued a regulation on waste separation at source for household waste, which went into effect in November 2018. According to the regulation, waste should be separated into: (1) organic; (2) recyclable; and (3) remaining parts. Separated waste is collected and treated by different facilities (City People Committee of Ho Chi Minh City, 2018). In the industry sector, most recyclables are either sorted for reuse and recycling in the same company or sold to other companies or recycling facilities. In the health sector, around 95% of central hospitals nationwide have implemented waste segregation (MONRE, 2011).

Collection and Transportation: The average collection rate of MSW has increased from 82-84% in 2010 to 85.5% in 2018 (Government Party Committee [GPC], 2018). This rate is higher in large cities and lower in small towns. The lowest collection rate of MSW, which is found in small towns, is about 65%. Hanoi has the highest rate at 98% (MONRE, 2015). However, the collection rate is still low in sub-urban areas (~60%) and rural areas (~40-55%, MONRE, 2018b). Waste collection is often implemented by the state-owned urban and environment companies (URENCO). However, in some provinces/cities (i.e., Hanoi, Ho Chi Minh City), the private sector also participates.

Waste Recycling: Although there is no official data, it is estimated that the recycling rate of MSW is about 8% to 12% of the total generated volume (MONRE, 2011). Recyclables (i.e., paper, plastics, metal, discarded electronics) are usually collected by garbage collectors and shipped to craft villages. Recycling in craft villages (i.e. in Hung Yen, Bac Ninh, Hai Duong, other provinces) uses informal, primitive technologies at household scale, which pollutes the environment and impacts public health. It is reported that there are at least 35 composting facilities nationwide. In addition, compost products have not been applied widely throughout the country but mainly in the South (MONRE, 2018b).

Waste Treatment and Disposal: Although there have been some improvements, landfilling is still the main method for treatment/disposal of MSW in Vietnam. The rate of MSW landfilling has slightly decreased from 76%-82% in 2011 (MONRE 2011) to 70%-75% in 2018 (MONRE 2018b). In 2016, there were around 660 landfills, excluding small dump sites at commune level. Of these, 203 (~30%) were considered sanitary (MONRE, 2017a; Thang, 2016). Open burn is popular in some rural and mountainous provinces. There are about 300 incinerating furnaces (MONRE, 2018b) and most of them are small with capacity about 400-500kg/hour. Waste-to-energy (WtE) is at the starting stage; 03 projects of waste incineration for electricity are developed in Hanoi, Ha Nam, and Can Tho (see Case Study 2).

MANAGEMENT OF SPECIAL WASTE

Hazardous Waste: HW in domestic waste, which varies from 0.02% to 0.82%, is neither collected nor treated separately. Both HW and MWS are usually discharged to landfills (MONRE, 2016). In the health sector, the amount of HW generated is about 300 tons/day, accounting for about 20% of total medical solid waste. It is mainly treated by incineration. Nearly 200 specialized incinerators serve 453 (or 40%) hospitals and health facilities. Hospitals without specialized incinerators treat HW in primitive incinerators or illegally discharge it to landfills (MONRE, 2017a). The master plan for medical hazardous waste treatment through 2025 has been approved and promulgated (PM, 2012).

Generated industrial HW accounts for 15% to 20% of total industrial solid waste. It is higher in mechanical, electrical, electronic, and chemical industries (MONRE, 2017a). It was estimated to be about 874,588 tons in 2016 (GPC, 2018). In agriculture, there are about 14,000 tons of HW generated annually. This includes packaging and containers of fertilizers and pesticides, which are collected but not fully treated (MOC, 2017a; MONRE, 2015).

According to regulations, the treatment of HW must be carried out by licensed enterprises. By March 2018, there were 114 HW treatment facilities nationwide licensed by MONRE and 97% of them were private enterprises (GPC, 2018). The collected and treated amount of industrial HW has increased from

Table 2. Collected and treated HW 2012-2016

Year	Collected and Treated HW (tons/year)
2012	165,624
2013	186,657
2014	320, 275
2015	400,000
2016	752,181

(adapted from GPC, 2018; Hien et al., 2015; MONRE, 2016)

165,624 tons in 2012 to 752,181 tons in 2016 (Table 2). The estimated collection and treatment rate of total generated HW is about 85%-90% through the years.

Agricultural Waste: Agricultural solid waste includes biodegradable components like straw, post-harvest byproducts, animal waste, and HW like pesticide and fertilizer packaging. It is estimated that over 76 million tons of straw and 47 million tons of livestock waste are generated each year (MONRE, 2017a). Most of the straw and post-harvest byproducts are used as fertilizer, animal feed, or mushroom cultivation. Livestock waste includes manure, uneaten feed, and cattle and poultry carcasses. Approximately 40% to 50% of the livestock waste is treated (MONRE, 2017a). The most common treatment method, anaerobic digestion, is applied in 31.79% of the total number of livestock farms nationwide (MONRE, 2014). It has been reported that there are 500,000 biogas tanks across the country, mainly at household scale with volume below 10m^3 (Vietnam Energy Efficiency Program [VNEEP], 2016).

Plastic Waste: The plastics sector has experienced a high growth rate of 16% to 18% per year in Vietnam (2010-2015). There were 5-6 million tons of plastic products produced in 2015, including 39% packaging, 32% domestic, 14% construction, and 9% technical products. The plastic consumption per capita has increased from 3.9 kg in 1990 to 41 kg in 2015 (Vietnam Plastics Association [VPA], 2016). However, 80% of raw material (~4.4 million tons in 2016) must be imported from countries like Arab Saudi, Korea, Thailand, and China (Truong, 2016; VPA, 2016).

There is no official data on plastic waste in Vietnam. According to a recent international publication, Vietnam ranked fourth in the top five countries in marine plastic waste generation (following China, Indonesia, and the Philippines). Vietnam discharges around 0.28-0.73 tons of marine plastic debris annually (Jambeck et al., 2015). It has been estimated that the percentage of plastic waste in lower-middle income countries like Vietnam is 12% of total MSW (Hoornweg & Bhada-Tata, 2012). This figure varies from 8% to 16% according to a MONRE estimation (MONRE, 2011).

Waste from plastic bags is a big concern in Vietnam. According to Thanh, Matsui, and Fujiwara (2011), the average plastic waste generation rate in Can Tho City was 17.24 g/cap/day, of which plastic packaging and plastic containers dominated with 95.64%. Plastic shopping bags were identified as the major component, accounting for 45.72% of total plastic waste (Thanh et al., 2011). It is estimated that the number of plastic bags in Vietnam is over 30 billion bags per year; 17% of the bags are reused and others become waste after a single use (MONRE, 2012). Hanoi and Ho Chi Minh City generate an average of 80 tons of plastic waste per day. Discarded plastic bags account for 7% to 8% of the waste in Hanoi and 10% in Ho Chi Minh City (MONRE, 2017a).

In October 2017, China banned the import of 24 categories of solid waste, including plastics, textiles, and unsorted paper (Washington Refuse & Recycling Association [WRRA], 2018). This ban led to a

Figure 7. Production of mobile phones and air conditioners 2006-2016 (adapted from GSO, 2018)

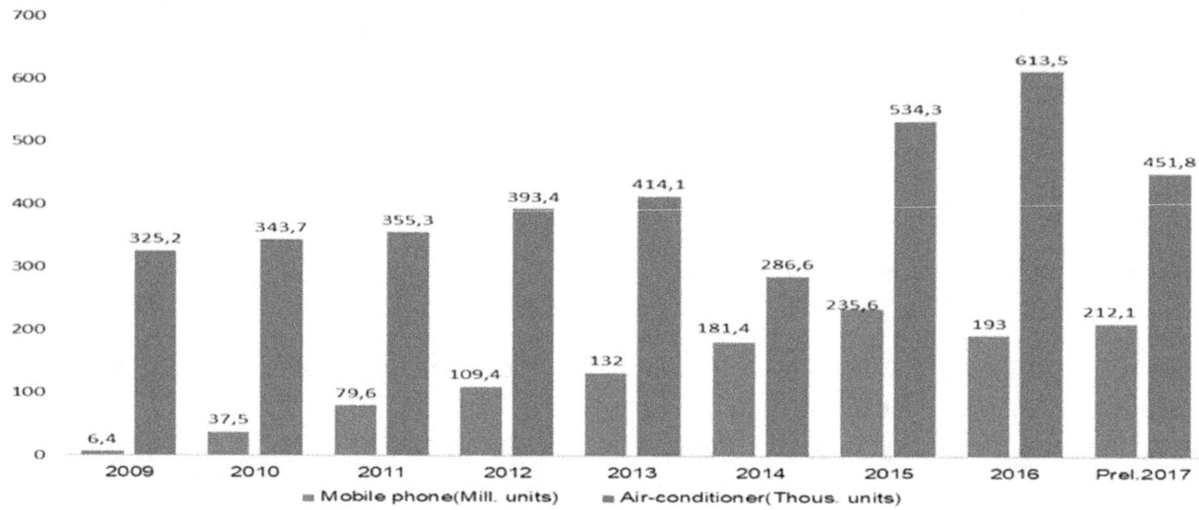

sharp increase of the plastic recyclable waste import to Vietnam, from 18,548 tons in 2016 to 90,839 tons in 2017 and 157,880 tons between January and May 2018 (MONRE, 2018d). Responding to the situation, the Vietnamese Government has issued the Resolution 27/CT-TTg to impose strict control of scrap import (PM, 2018b). However, this policy has also reduce the imported scrap and thus impacted to recycling businesses (see Case study 1).

Plastic waste has not been controlled separately in Vietnam but together with MSW. Discarded nylon bags are usually mixed with MSW and often sent to landfills. A few waste treatment facilities in Lam Dong and Nghe An provinces recycle nylon waste into diesel oil (Vietnam Environment Administration [VEA], 2017). Plastic recycling is still implemented by household facilities in craft villages like the Minh Khai craft village (Hung Yen province, see Case Study 1).

E-Waste: The production and consumption of electronic and electrical goods in Vietnam have grown steadily. In 2018, 72% of the population uses smartphones and the average phone ownership rate was 1.7 phones/person (GSO, 2018). In seven years, the number of produced mobile phones increased from 6.4 million units (2009) to 193 million (2016). Air conditioners increased sharply from 189,000 to 613,500 units in the same period (see Figure 7).

There is no official data on e-waste in Vietnam. International studies show that the country generated 116,000 tons of e-waste (approximately 1.3 kg/inhabitant) in 2014 (Baldé, Wang, Kuehr, & Huisman, 2015) and increased to 141,000 tons in 2016 (Baldé, Forti, Gray, Kuehr, & Stegmann, 2017). Hai, Hung, and Quang (2015) in their research have estimated the generated e-waste ranged from 61,100 to 113,500 tons in 2010. Quang (2017) forecasted that e-waste from Vietnam's discarded televisions, refrigerators, washing machines, and air conditioners will increase steadily in the coming years. Specifically, by 2025, there will be discarded about 215,000 tons TVs, 192,000 tons refrigerators, 130,000 tons washing machines, and 23,000 tons air conditioners (Quang, Okumura, Yamasue, & Ishihara 2009).

Formally, e-waste is classified and managed as HW in Vietnam. The mechanism of extended producers' responsibility (EPR), which was applied in 2016, regulates that discarded electronic and electrical equipment must be retrieved and treated by their producers and importers. However, there is still a large gap between regulation and implementation, causing the implementation to be inefficient.

Informally, after being discarded, e-waste is usually sold by households to private collectors. The e-waste is then sorted into three categories: (1) devices for reuse; (2) devices for dismantling and refurbishment; and (3) devices for recycling. After that, e-waste is transported to refurbishment shops and recycling facilities in craft villages for the recovery of iron, copper, aluminum, and plastic (MONRE, 2011; Thang & Anh, 2016).

Construction and Demolition Waste (CDW): CDW is composed of sand, broken bricks, glass, concrete, and metal. It normally accounts for about 10% to 15% of MSW. In big cities like Hanoi and Ho Chi Minh City, this rate is 25%. In some provinces/city such as Bac Giang, Hai Phong, and An Giang, CDW accounts for 12% to 13% of MSW (MONRE, 2017a).Tuan et. al (2018) in their research showed that the daily generation of CDW in Hanoi and Ho Chi Minh City reached 3,000 tons/day in 2013 (Tuan, NV et.al, 2018). It is forecasted that the amount of CDW will increase sharply in the coming years because of planned demolition and renovation of old apartment buildings in big cities. Until recently, there has been limited attention on CDW management as it is usually mixed with MSW to make ground or landfill. Illegal dumping of CDW still exists, which pollutes the environment and landscape and clogs the drainage system.

IMPACTS OF WASTE MANAGEMENT

Impacts to the Environment: Improper waste management causes pollution and impacts the soil, water, air, and biota. Once discharged to the land, solid waste can remain in soils for an extended period. Illegal dumping of CDW greatly impacts soil quality (MONRE, 2011, 2015, 2016). Landfill leachate with heavy metals also causes pollution to soil and water (MONRE, 2016, 2017a). For example, soils in surrounding landfill in Dong Nai province contain a higher level of arsenic than applicable soil standards (MONRE, 2016).

Illegal littering into urban water bodies (e.g., canals, lakes, ponds, rivers) causes serious surface water pollution. In Ho Chi Minh City, the amount of collected floating trash in main canals (i.e., Nhieu Loc - Thi Nghe, Doi - Te, Ben Nghe - Tau Hu) was 12,820 tons in 2015 (Kieu, Strady, & Perset, 2016). In Hanoi, high levels of fecal, inorganic, and organic nitrogen in the wells could have been partly attributed by inappropriate disposal of solid waste (Asian Productivity Organization [APO], 2007). Also, high concentrations of the total dissolved solids (TDS), chemical oxygen demand (COD), arsenic, cadmium, lead, copper, and iron were found in underground water in the vicinity of the Nam Son landfill. Of this, the TDS and some heavy metals concentrations greatly exceeded the standards (Giang, Kochanek, Vu, & Duan, 2018).

Regarding air pollution, landfill odors are problematic in Vietnam (Luong, Giang, Thanh, & Hung, 2013). One example is the bad odor from the Da Phuoc landfill in Ho Chi Minh City (Ngan, 2018). In addition, uncontrolled burning of MSW and improper incineration contribute significantly to air pollution (MOC, 2016; MONRE, 2015, 2017a). For example, combustion of one ton of MSW can release more than 7 kg of hydrochloric acid (HCl) (Kafando et al., 2013). This is a quite serious impact as approximately 40% to 50% of MSW transported to open dumping sites are uncontrolled burns (Luong et al., 2013). In the health sector, about two-thirds of medical waste incinerators are without exhaust treatment systems (MONRE, 2016).

Figure 8. Trends of GHG emission in waste sector (adapted from MONRE, 2003, 2010, 2017b)

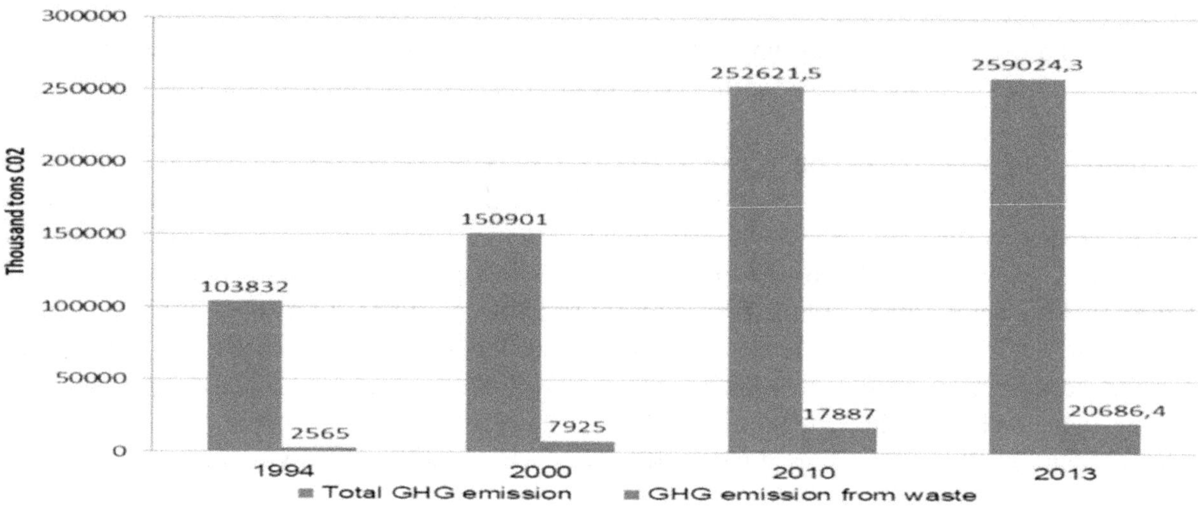

Impact to Global Warming: Increasing greenhouse gas (GHG) emissions from the waste sector contributes to global warming. For 20 years, GHG emissions in Vietnam (along with land use, land-use change, and forestry - LULUCF) increased about 2.5 times from 103.8 million tons in 1994 to 259 million tons of CO_2-eq in 2013. The GHG emission, from the waste sector increased faster, from 2.56 to 20.7 million tons of CO_2-eq, respectively (see Figure 8). It accounted for about 7% of the total GHG emission in 2013. Although the percentage of GHG emissions from the waste sector is relatively small compared to other sectors, it is increasing quickly. Methane (CH_4) emission from solid waste disposal sites was the second largest source of GHG emission from the waste sector. It accounted for 35.9%, with 7.4 million tons of CO_2-eq in 2013 (MONRE, 2017b).

Impact on Human Health: In Vietnam, poor waste management has negatively impacted to human health. Waste collectors/workers are considered vulnerable to different diseases as most of them are women (more than 60%) and children (~9%, Alam & Ahmade, 2013; Luong et al., 2013; Netherlands Enterprise Agency, 2018). Residents who live near open dumping sites, unsanitary landfills, and recycling villages have various ailments, including cancer, diarrhea, and dermatological and respiratory diseases (Luong et al., 2013; MONRE, 2011). Furthermore, emission of dioxin and furan from open MSW burning is a significant threat to public health (Luong et al., 2013; MONRE, 2015). According to Daniel et al. (2015), all examined children in the Dong Mai lead recycling village have high blood lead levels and require urgent remediation. Before operating the Nam Son landfill in Hanoi City, no people in the nearby Lai Son village suffered from oncological diseases. However, 40 people are reported to have died in the last three years from cancer. A dozen were diagnosed with a terminal stage of cancer (Giang et al., 2018).

Table 3. Discarded products to be retrieved and treated

Discarded Products	Effective Date
Batteries	July 1, 2016
Electric and electronic equipment *Compact and fluorescent lamps; computers; computer display; CPU; printing machines; fax, scanner machines; camera, video camera; mobile phones, ipad; DVD; VCD; CD and other players; photocopy machines; TV; fridge; air conditioners; washing machines*	July 1, 2016
Lubricant oils	July 1, 2016
Tires	July 1, 2016
End-of-life vehicles	January 1, 2018

(adapted from PM, 2015a)

INSTITUTIONAL FRAMEWORK ON WASTE MANAGEMENT

Policies and Legislation

Policies and legislation on waste management in Vietnam comprise of laws, by-law documents, strategies, plans and programs, and economic and information instruments (Box 1). It is regulated that all kinds of waste must be separated at source, collected, transported, and treated in an environmentally friendly manner. Waste generators are legislated to reduce, reuse, recycle (3Rs), and recover energy from waste. Waste is classified into hazardous and non-hazardous. HW must be strictly managed by special regulations and implemented by licensed entities. The EPR mechanism, which was applied in 2016, requires that certain discarded products be retrieved and treated by producers and importers (see Table 3).

Policy and Legislation on Waste Management in Vietnam

Laws, By-Law Documents, and Strategies

- Law on Environmental Protection (2014)
- Decree 19/2015/NĐ-CP on guidance to implementation of Law on Environmental Protection
- Decree 59/2007/ND-CP on solid waste management
- Decree 38/2015/ND-CP on waste and scrap management
- Decision 16/2015/QD-TTg on take-back and treatment of discarded products
- Circular 36/2015/TT-BTNMT on management of HW (2015)
- Inter-ministerial circular 58/2015/BYT-BTNMT on medical waste management (2015)
- Circular 08/2017/TT-BXD on management of construction waste (2017)
- Decision 73/QD-TTg stipulating the list of allowed scraps for import
- Circular 41/2015/TT-BTNMT dated September 9, 2015 on environmental protection in scraps import
- National Strategy for Environment Protection Until 2020 and Vision Toward 2030 (Decision No. 1216/QD-TTg dated September 5, 2012)
- Revised National Strategy for Integrated Management of Solid Waste Until 2025, and Vision Toward 2050 (Decision 419/QD-TTg dated May 7, 2018)

Master Plans and Programs

- Master plan for the construction of solid waste treatment sites in three key economic areas in the North, Center, and South of Vietnam to 2020 (Decision 1440/QD-TTg dated October 6, 2008)
- Master plan for the construction of the solid waste treatment site in the key economic area of Mekong Delta to 2020 (Decision 1873/QD-TTg dated October 11, 2010)
- Master plan for solid waste management in Cau river basin to 2020 (Decision 2211/QD-TTg dated November 14, 2013)
- Master plan for solid waste management in Nhue-Day river basin to 2020 (Decision 223/QD-TTg dated February 12, 2015)
- Master plan for solid waste management in Dong Nai river basin to 2030 (Decision 07/QD-TTg dated January 6, 2015)
- Master plan on Hazardous Medical Solid Waste Treatment System to 2025 (Decision No. 170/QĐ-TTg February 8, 2012)
- Program on strengthening control on environmental pollution by nylon bags to 2020 (Decision 582/QĐ-TTg dated November 4, 2013)
- National Action Program on sustainable production and consumption to 2020, vision to 2030 (Decision 76/QĐ-TTg dated January 11, 2016)
- Solid waste treatment investment program for the period of 2011-2020 (Decision 798/QD-TTg dated May 25, 2011)
- Program on mobilization of investment on water supply and drainage and solid waste treatment (Decision 1196/QĐ-TTg dated July 23, 2014)

Economic Instruments

- Provincial/city regulations on waste collection and transportation fee
- MOC Decision 1354/QD-BXD on cost norms of waste treatment
- Law on Environmental Tax (2010)
- Regulations on support mechanism to WtE projects development in Vietnam (Decision 31/2014/QĐ-TTg dated May 5, 2014)
- Vietnam Environmental Protection Fund

Information Instruments

- Circular 07/2012/TT-BTNMT stipulating criteria, procedure on certification of environmentally friendly nylon bags
- Decision 253/QD-BTNMT dated March 5, 2009 on program on eco-labeling

To provide direction for waste management, the government set targets and tasks to be performed in the national waste management strategy (see Table 4). The core concepts of the strategy consider waste as a resource, strongly promote 3R measures, and limit landfilling. It also enhances waste recycling and energy recovery for a circular economy. It is worth noting that there is a certain difference between strategies for solid waste management in urban and rural area. Particularly, target for collection and treatment of UDW (90%) is higher than that in rural area (80%) (Table 4). The UDW is required to be

Table 4. Key targets of solid waste management to 2025

Indicators	Targets to 2025
Rate of HW generated from production, businesses, services, medical facilities and craft villages collected, transported and treated	100%
Rate of UDW collected, transported and treated	90%
Rate of domestic solid waste generated from rural areas collected, transported, treated	80%
Rate of CDW from urban areas collected, transported and treated	90%
Rate of medical solid waste generated from medical facilities or hospitals sorted, collected, stored, transported and treated	100%
Rate of electronic equipment producers establish facilities for the recovery of discarded products	100%
Rate of grade-I urban areas have solid waste recycling facilities appropriate to waste sorting at households	100%
Ratio of domestic solid waste treated by landfills to total amount of domestic solid waste collected.	<30%
Rate of environmentally-friendly plastic bags replacing non-biodegradable plastic bags in trade centers and supermarkets.	100%

(adapted from PM, 2018a)

recycled, treated with energy recovery or organic fertilizer production while rural household waste to be composted or treated at households (PM, 2018a).

Institutional Arrangement

In Vietnam, waste management policies are developed and issued by the national government; waste management activities are implemented by local (province or municipality) agencies. At the national level, the Law on Environmental Protection 2014 has decentralized waste management by the principle that line ministries are responsible for waste management in their sector, specifically:

- **MONRE:** This ministry serves as the focal ministry on environmental protection in general. It is also responsible for management of hazardous and ordinary industrial solid waste.
- **MOC**: This ministry is responsible for the management of household waste and construction waste. This includes sludge from septic tanks and urban drainage systems.
- **Ministry of Agriculture and Rural Development (MARD)**: This ministry is responsible for the management of agricultural waste.
- **Ministry of Health (MOH):** MOH is responsible for classification, storage, management of medical waste on the campus of medical facilities, and waste generated from burials and cremation.
- **Ministry of Transport (MOT):** This ministry manages hazardous and ordinary wastes generated from transport activities.
- **Ministry of Science and Technology (MOST):** MOST appraises new domestic solid waste treatment technologies.

This decentralization involved different sectors. However, this led to overlaps and gaps in waste management. In February 2019, after recognizing these problems, the government assigned MONRE to be the focal point of solid waste management (GOV, 2019). It transferred the waste management functions back from the ministries to MONRE. Currently, MONRE is reviewing and revising relevant legal documents, including the Law on Environmental Protection 2014.

At the local level, the Provincial/City People's Committees (P/CPC) are taking responsibility of waste management within their individual territory. Based on national policy and regulations, P/CPC issued guidance documents and master plans for waste management. Focal agencies on waste management include the Department of Construction (DOC). However, in some cities, including Da Nang and Ho Chi Minh City, the Department of Natural Resources and Environment (DONRE) are the focal points. Collection, transportation, and treatment of waste are usually implemented by URENCO. In some cities/ provinces, these issues are handled by private companies.

Waste Management Planning

Vietnam's solid waste planning focused on regional and provincial levels. The regional level includes the following master plans approved by the prime minister: (1) three northern, central, and southern key economic regions; (2) key economic areas of the Mekong Delta; (3) Cau river basin; (4) Dong Nai river basin; (5) Nhue-Day river basin; and (6) the northern key economic region. These master plans provide guidelines for selection of treatment technologies, spatial positions of 86 waste treatment facilities with an area of around 8,600 ha, and estimated investment capital. This totals around US$3.5 billion through 2030 (see Table 5).

At the provincial level, 58/63 provinces have approved solid waste management planning through 2018 (MOC, 2019). Based on the approval, programs and projects on solid waste treatment have been implemented in localities.

Table 5. Number of facilities, areas, and estimated investment capitals in regional waste management master plans

Master Plan	Number of Treatment Facilities	Area *(ha)*	Estimated Investment Capital
Planning on construction of solid waste treatment facilities in the northern, central, and southern key economic regions through 2020	8	2280-2300	US$598 million
Master plan for the construction of a solid waste treatment site in the key economic area of Mekong Delta through 2020	5	367	*not specified*
Planning on solid waste management in Cau river basin through 2020	15	376.7	US$ 430 million
Planning on solid waste management in Nhue-Day river basin through 2020	18	300.95	US$ 327 million
Planning on solid waste management in Dong Nai river basin through 2030	34	4724.4	US$ 1,505 million
Planning on solid waste management in the northern economic region through 2030	6	488	US$ 575 million
Total	**86**	**~ 8,537-8,737**	**US$3,465 million**

(adapted from PM, 2008, 2010, 2013, 2015b, 2015c, 2016b)
Note: 1 USD ~ 23,250 VND in 2019.

Waste Treatment Technology

Currently, main technologies used in domestic solid waste treatment in Vietnam are landfilling, composting and incineration (with recovering and without energy recovery) or combining these 03 technologies. The most common method is open dumping (accounted for 2/3 of landfills) while controlled landfills is still not so popular. Incinerations have also been applied in many provinces but mainly without energy recovery. There are 3 projects for incineration with electricity generation.

Through the years, research and development of solid waste treatment technologies are managed by MOST, MOC and carried out by different institutes and enterprises. Until now, there have been 05 domestic technologies certified by the MOC, including: 02 composting technologies (SERAPHIN and An Sinh - ASC); 01 RDF technology MBT-CD.08; 02 incineration technologies (ENVIC and BD-ANPHA Technology) (MONRE, 2017). However, application of these technologies is still limited due to unsynchronized equipment and their incompleteness.

Along with the domestic technologies, foreign solid waste treatment technologies have also been applied such as technologies from Spain, France, Belgium, Denmark, Korea, etc. However, these technologies are not suitable for Vietnam as the waste is not separated at-source and climate weather is different. In promulgated master plans on solid waste management, 04 technologies are recommended including sanitary landfilling, recycling, composting, and incineration with energy recovery. Prioritization is given to domestic and energy-efficient, environmentally friendly technologies (PM, 2008, 2010, 2013, 2015b, 2015c). In recent years, interests on incineration for electricity generation are increasing due to the policy on development of renewable energy and high diversion rate from landfilling.

Economic Instruments in Waste Management

Waste management in Vietnam is mainly subsidized by the provincial/city budget. Based on the province/city's socio-economic development, the Provincial/City People Council decides the environmental sanitation fee. The fee, which covers part of the solid waste collection cost, varies by localities, the types of households (street side are higher than those in an alley), and waste discharge volume for business households, offices, and schools. In the five biggest cities (i.e., Can Tho, Hai Phong, Da Nang,

Table 6. Waste collection and transportation service fee in the five biggest cities (adapted from CPC Can Tho, 2016; CPC Hai Phong, 2016; CPC Da Nang, 2017; CPC Ho Chi Minh City, 2008; CPC Hanoi, 2016)

City	Waste Collection and Transportation Service Fee (VND/Month)		
	Household	Business, Household, Office, School, etc. (with ≤ 1m3 waste/month)	Business, Household, Office, School, etc. (with ≥ 1m3 waste/month)
Can Tho	10,000 – 20,000	20,000 – 100,000	160,000/m3
Da Nang	15,000 – 30,000	50,000 – 165,000	205,000 – 265,000/m3
Ha Noi	3,000 – 6,000 *(per person)*	90,000 – 130,000	208,000/m3 or 500,000/ton
Hai Phong	20,000 – 40,000	90,000 – 324,000	324,000/m3
Ho Chi Minh	15,000 – 20,000	60,000 – 110,000	176.800/m3

Table 7. Standard cost norms of treatment of MSW based on technology

Technology	Capacity (ton/ day)	Standard MSW Treatment Cost Norms (million VND/ton.day)	
		Foreign Technology and Equipment	Domestic Technology and Equipment
Bio-fertilizer processing technology	100 - 1,000	0.22 – 0.34	0.20 – 0.30
Incineration technology without power generation and energy recovery	50 - 800	0.34 – 0.50	0.32 – 0.47
Combined bio-fertilizer processing technology and incineration	100 -1,500	0.25 – 0.41	0.22 – 0.38
		With Depreciation Cost	**Without Depreciation Cost**
Sanitary landfill technology	100 - 800	0.10 – 0.14	0.09 – 0.13

(adapted from MOC Minister Decision, 2017b)

Hanoi, Ho Chi Minh City), the fee varies from 10,000-40,000 VND (~0.43-1.74 USD) per household per month. For businesses, offices, and schools, it varies from 20,000-324,000 VND (0.87-14.08 USD) per entity per month with waste discharge no higher than 1m3 waste/month. For establishments with discharge higher than 1m3 waste/month, the fee is based on volume and varies from 160,000-324,000 VND (~7.00-14.00 USD) per m3 of waste. From these five cities, Hai Phong imposes the highest fees to households and businesses (see Table 6).

Regarding the cost of solid waste treatment, it is the responsibility of the Provincial/City People Council to determine the price for contracting with waste companies. The cost norms framework, as promulgated by the MOC, varies by treatment capacity, type, and origin of technologies and equipment. Specifically, the treatment cost varies by capacities and the lowest cost is sanitary landfilling with 0.09-0.14 million VND (~ 4.0-6.0 USD) per ton.day. The highest cost is incineration without energy recovery, with 0.32-0.50 million VND (~14.0-22.0 USD) per ton.day (see Table 7). To promote WtE projects, the buying price of electricity from waste has also promulgated with 10.05 US cents/kWh for incineration and 7.28 US cents/kWh for landfill gas projects (PM, 2014).

Other economic instruments in Vietnam's waste sector include the Law on Environmental Tax 2010 and the Vietnam Environment Protection Fund (VEPF). Environmental taxes are imposed on eight groups of products, including: (1) petrol, oil, and grease; (2) coal; (3) hydro-chloro-fluoro-carbon (HCFC) solution; (4) taxable plastic bags; (5) herbicides restricted from use; (6) termiticides restricted from use; (7) forest product preservatives restricted from use; and (8) warehouse disinfectant restricted from use. The significance of these products is the tax on plastics bags, which is 40,000 VND/kg. The VEPF was established in 2005 with a capital of 1,000 billion VND (~45 million USD). It provides incentive support (i.e., soft loan, interest-rate support) to solid waste recycling/treatment projects.

CASE STUDIES

Case Study 1. Informal Recycling of Plastics at the Minh Khai Craft Village

Located at Nhu Quynh town, Hung Yen province, the Minh Khai craft village was established in 1980. It began with a small number of households that processed plastics recycling activities (Figure 9). At present, the number of households is 1,236, with a recycling capacity of approximately 1,000 tons of recycled plastics per day. The main final products are plastic pellets, nylon bags, and household appliances (UNIDO, 2018). The recyclable materials include both domestic scraps (~30%) and imported scraps (~70%). With China's ban of scraps import in 2017, the scrap import to Vietnam has increased sharply in 2018. However, after tightened control of scrap import by the Vietnamese government started in Sep 2018, the village faced a shortage of material for recycling.

The recycling technology includes three processes: (1) washing and cutting plastic scraps into small pieces; (2) thermal processing; and (3) producing pellet. The equipment is usually backward and rudimentary. There is no control of wastewater and air emission.

While recycling activities bring economic benefits to the village, they also adversely impact the environment and human health (Hoang, 2018; Tham, Huong, Phuong, & Trinh, 2018). Since 2003, Minh Khai village has been classified as 1 of 439 "hot spots" with serious environmental pollution nationwide (PM, 2003). However, it has not been thoroughly cleaned up (Huong, 2018). Besides wastewater and the odor, solid waste remains the biggest problem.

The unrecycled part of the scraps can total up to 30% of the recyclables accumulated in villages. It is estimated that 60 to 65 tons/day are accumulated. More than 10,000 tons of solid waste from purchasing and recycling plastic scraps are dumped at adjacent vacant lands (Figure 10) and/or along sidewalks (Thang & Hanh, discussion with Minh Khai village authority, February 26, 2019). In addition, about

Figure 9 Plastic recycling technology in a household in Minh Khai village (Photo by Thang, 2019)

Figure 10 Open dump site at Minh Khai craft village (Photo by Hanh, 2019)

7,000 m^3 of wastewater is discharged directly into the environment every day. Some pollutants are observed in ambient air in the village, including SO_2 and NO_2 (MONRE, 2014).

Environmental pollution has impacted people in the village with diseases related to the respiratory system, digestive system, skin, throat, and eyes (Huong, 2018; Trung, Mahanty, & Van, 2012). 60% of the cases are respiratory disease.

As an effort to deal with pollution, the provincial, district, and commune authorities have planned an industrial cluster of more than 28ha through a two-phase project. In 2015, the first phase (using 10ha) was completed, with 143 households added to the industrial cluster. As of 2016, the number of households in the industrial cluster rose to 200 (UNIDO, 2018) and the second phase continues. However, the industrial cluster does not yet have centralized wastewater or solid waste treatment facilities due to financial constraints. Therefore, environmental issues remain unresolved (MONRE, 2014, 2017; Thang & Hanh, 2019).

Case Study 2. WtE Projects in Vietnam: A New Approach in Waste Management

Waste-to-Energy (WtE) has become a popular treatment technology in developed countries due to its high diversion rate (90% to 95%) from landfilling and energy recovery in the form of electricity and/or heat (British Business Group Vietnam [BBGV], 2018; Quang, 2017). As a low middle-income country, WtE in Vietnam is drawing new attention. It is estimated that the energy potential from waste in Vietnam could reach 320MW (Das, 2018). The waste could generate up to 1 billion kWh in 2020 and 6 billion kWh in 2050 (Hien, 2017; Das 2018; BBGV, 2018).

In the national electricity master plan and national strategy on renewable energy, the government aims to increase electricity from renewable sources to 7% in 2020, more than 10% in 2030, and 70% of MSW to be used for energy purpose by 2030 (PM, 2015d, 2016a). The mechanism of buying electricity from WtE projects in 20 years at a price of 10.05 US cents/kWh (for incineration) and 7.28 US cents/kWh (for landfill gas collection) has been promulgated with incentives in tax and land (PM, 2014).

These policies include WtE projects that started with the new approach to waste management in Vietnam:

- In October 2016, the Hydraulic Machine Company successfully completed a pilot with 208 tons of non-classified solid waste to generate enough power to light the Dong Van 2 Industrial Park, Ha Nam province for 10 days.
- In April 2017, the Go Cat WtE Plant in Ho Chi Minh City started operation and generated 168,000 kWh/day from industrial waste. The plant is operated by a system of preprocessed garbage equipment, including primary and secondary cutting equipment, industrial garbage pressing equipment, gasification furnaces, synthesis gas containers, and three synthesis gas generators with a capacity of 550 kVA each.
- In April 2017, the first industrial WtE plant was set up in Ha Noi at the Nam Sơn Waste Treatment Complex. Equipment came from Japan's Hitachi Zosen Corporation. The facility processes 75 tons of waste and generates 1.93 MW of energy per day.
- In December 2018, the Thoi Lai WtE plant in Can Tho City started operation with a capacity of 400 tons of waste per day. It generated 150,000 kWh of electricity per day and consumed up to 77% of the city's total waste.

New WtE projects have been approved in Hanoi, Phu Tho, and Thanh Hoa provinces. Construction will begin soon.

ISSUES AND CHALLENGES OF WASTE MANAGEMENT

Issues, Constraints, and Reasons

There has been some improvement in Vietnam's waste collection and treatment in the last 10 years. However, waste management in Vietnam still faces the following issues, ranked by pressing level:

1. Uncontrolled landfills/dumping sites are the main disposal method for 70% to 75% of MSW.
2. Collection rate of household waste in rural areas have no improvements but remained with the rate 40% to 55% in the last 10 years.
3. Waste recycling is still carried out by an informal sector with primitive technology and caused pollution and health impacts.
4. Waste reduction has not been effectively implemented. There are just 24% enterprises which have applied cleaner production while there has not been any measure for reduction of MSW generation.
5. Waste at-source separation for MSW is not applied nationwide but just started in Ho Chi Minh City.

There are many reasons for these issues. First, waste management legislation is incomplete. The definitions on certain kinds of waste, including MSW and e-waste, are not regulated. They lack specific regulations on types of waste, including e-waste, packaging waste, food waste, and plastic waste. National standards and technical regulations for vehicles, equipment, and technological lines of solid waste collection, transportation, and treatment are neither sufficient nor synchronous. Standards for recycled materials and products have not been developed and issued. There is still the need for a comprehensive, systematic, and regularly updated database on waste management in the country.

Second, resources for waste management are inadequate. Being a low middle-income country, Vietnam faces a shortage of financial resources for the development of waste treatment facilities, especially in rural areas. State budget allocations are limited regarding the implementation of approved master plans. Participation from private and foreign investment sectors in waste management has not been active due to a low treatment price by municipalities. Mobilization of financial resources for solid waste management based on the principle of "polluter pays" is difficult due to the low living standard of the population. The environmental sanitation fee only meets a part of the collection and transportation cost.

Third, legislation enforcement and implementation is weak. Waste collection and treatment in rural areas are not fully implemented. Capacity of authority at district and commune levels is low. Although waste separation at-source has been regulated, it is not implemented nationwide. The EPR mechanism has been set for four years for many products. However, its implementation is not as active as expected. Although master plans have been approved, many cases are not implemented.

Fourth, coordination/collaboration between ministries and stakeholders is poor. There are overlaps between ministries like MONRE and MARD regarding non-HW in agriculture or MONRE and MOC on environmental planning and waste planning. There is a gap in management responsibility of rural household waste. In addition, there is a lack of effective collaboration between provinces/cities in the implementation of regional master plans. There is an inconsistency in assignment of a focal agency for waste management at local levels. For example, DONRE is used in some cities but DOC is used in most provinces. Additionally, there is a lack of coordination among relevant ministries, sectors, and localities in reporting, database building, and information sharing.

Finally, there is limited awareness of stakeholders on integrated waste management and circular economy. The awareness of people and communities on a sustainable lifestyle is low. Littering and illegal dumping continues in many places, especially in rural areas. Nylon bags are used extensively. These discharge directly into the environment. The concept of a circular economy is neither elaborated nor understood by policymakers.

CHALLENGES

In the next 10 years, Vietnam will continue to face challenges in waste management. First, waste generation is increasing due to population growth and steady urbanization. With economic development, the middle-income population is increasing. However, their consumption pattern is unsustainable. Second, the economic development model is unsustainable, which leads to high material consumption and waste generation. Although Vietnam advocates for industrialization processes, the production technologies are backward. Productivity is low and there is high material intensity. Third, emerging waste issues include e-waste, marine plastic waste, and CDW. Due to the rapid development of the information, communication, and technology (ICT) industry and urbanization needs, waste streams will be more pressing while infrastructure systems to manage waste has been slow to develop.

RECOMMENDATIONS

In the next decade, the implementation of the 2030 Agenda for Sustainable Development and the SDGs will be a high priority for many countries. Reduction of GHG emissions to meet the commitment of the Paris Agreement is also a compulsory responsibility. Like many countries, Vietnam will have to promote waste management to meet the SDGs and climate change mitigation commitment. Based on the analyzed constraints, reasons, and challenges, there are several measures suggested in this chapter for improvement of future waste management.

First, the institutional framework for waste management must be strengthened. Incentive policies for waste reduction, like waste audit and ISO 14000, should be developed. Specific regulations on types of waste (i.e., e-waste, packaging waste, food waste, plastic waste) should be issued. The standards system on waste management must be reviewed to promote 3R practices, including recycled materials, standards for fly ash, and slags for reuse.

The Law on Environmental Protection (LEP) and regulations should be revised to transfer waste management responsibilities from other ministries to MONRE, as required by government resolution 09/NQ-CP. Consistency should be ensured through institutional arrangements between national- and local-level waste management systems. Efficient collaboration between line ministries and provinces/cities should be strengthened in the development of waste management infrastructures and a national database.

Second, financial resources for waste management should be diversified. Policy incentives should be developed and issued to mobilize financial resources. Participation is necessary from both private and foreign investment sectors. The state budget on waste management should be increased to promote investments in waste treatment facilities, specifically in the regional and provincial master plans. Economic instruments for waste management would enhance effectiveness. Therefore, the country should gradually increase the environmental sanitation fee, especially in big cities like Hanoi and Ho Chi Minh.

Third, policy enforcement and implementation should be strengthened. This includes enhancing control and inspection of waste management activities and increasing fines for legislative violations of littering and illegal dumping. Waste separation at source should be conducted, especially in bigger cities. Monitoring systems should be developed and applied to push efficient implementation of EPR mechanisms. Local authorities, especially in rural areas, should implement effectively their responsibilities in waste collection and treatment.

Fourth, science and technology development, as well as application in waste management, should be promoted. Measures should be taken to develop the environmental industry to resolve issues surrounding waste treatment. State-of-the-art technologies, including WtE, should be used to increase the diversion rate to landfills. Recycling activities should be formalized in craft villages through industrial clusters and proper environmental infrastructures. The development and implementation of technological waste plans should be applied to different regions and provinces. Vietnam should enhance interprovincial collaboration. Additionally, measures should promote the application of Industry 4.0 in the waste management sector.

Finally, the country should continue to raise awareness and responsibilities of relevant stakeholders. Communication campaigns can raise awareness on individual and community sustainable life styles, especially in the reduction of litter and use of plastic products. Disseminating knowledge on the circular economy to policymakers and enterprises will promote its practice. Responsibilities of local authorities on waste management should be strengthened.

To realize these recommendations, many actions and plans must be implemented in the next 10 years: (i) By 2020, to complete revising the Law on Environmental Protection; (ii) By 2022, to issue National Environmental Master-Plan; (iii) By 2025, to successfully implement the National Strategy on Integrated Solid Waste Management and master-plans; (iv) By 2030, to achieve Vietnam Sustainable Development Goals (VSDGs).

CONCLUSION

In the past 30 years, Vietnam has achieved remarkable success in socio-economic development. However, the environment is now polluted and degraded. There are still problems on waste reduction, separation, collection, recycling, and treatment/disposal. Vietnam will have to improve waste management to achieve the SDGs and to realize its GHG reduction commitment by 2030. Strengthening the institutional framework, mobilizing investment resources, enhancing legislative enforcement, promoting technology, and raising awareness on waste management are the recommended measures that need to be realized in the next 10 years.

REFERENCES

Alam, P., & Ahmade, K. (2013). Impact of solid waste on health and the environment. *International Journal of Sustainable Development and Green Economics*, *2*, I-1.

Asian Productivity Organization (APO). (2007). Solid waste management: Issues and challenges in Asia. Tokyo: Author.

Baldé, C., Forti, V., Gray, V., Kuehr, R., & Stegmann, P. (2017). *The global e-waste monitor – 2017. United Nations University (UNU), International Telecommunication Union (ITU) & International Solid Waste Association*. Bonn, Geneva, Vienna: ISWA.

Baldé, C. P., Wang, F., Kuehr, R., & Huisman, J. (2015). The global e-waste monitor-2014. United Nations University, IAS – SCYCLE, Bonn, Germany.

British Business Group Vietnam (BBGV). (2018, May). *Renewable energy: Waste-to-energy*. Retrieved from https://bbgv.org/wp-content/uploads/2018/05/Waste-to-energy.pdf

City People Committee (CPC) of Can Tho. (2016). *Decision 40/2016/QĐ-UBND dated 30 December 2016 approved service fee*. Author.

City People Committee (CPC) of Da Nang. (2017). *Decision 33 /2017/QĐ-UBND dated 18 November 2017 promulgated maximal prices for collection, transportation of household waste paid by city budget*. Author.

City People Committee (CPC) of Hai Phong. (2016). *Decision 3257/2016/QĐ-UBND dated 22 December 2016 for the price of collection, transportation and treatment of household waste management in the urban areas of the city*. Author.

City People Committee (CPC) of Hanoi. (2016). *Decision No. 54/2016/QĐ-UBND dated 31 December 2016 promulgated price of household waste collection and transportation, prices of environmental sanitary service on non-hazardous industrial waste*. Author.

City People Committee (CPC) of Ho Chi Minh City. (2008). *Decision 88/2008/NQ-UBND dated 20 December 2008 on sanitary and environmental protection fee on non-hazardous solid waste*. Author.

City People Committee (CPC) of Ho Chi Minh City. (2018). *Decision 44/2018/QD-UBND dated 14 November 2018 on issuance of regulation on sorting at source the domestic waste*. Author.

Daniel, W., Van Tung, L., Wallace, R., Havens, D., Karr, C., & Bich Diep, N., … Duy Bao, N. (2015). Childhood lead exposure from battery recycling in Vietnam. *BioMed Research International*. doi:10.1155/2015/193715

Das, K. (2018). *Vietnam: Opportunities in the Waste-to-Energy Sector, Vietnam Briefing, 2018*. Retrieved 15th March 2019 from https://www.vietnam-briefing.com/news/vietnam-opportunities-waste-energy-sector.html/

Dione, O. (2018, April 5). *Climbing the ladder: Poverty reduction and shared prosperity in Vietnam* [Speeches & Transcripts]. The World Bank. Retrieved from https://www.worldbank.org/en/news/speech/2018/04/05/poverty-reduction-and-shared-prosperity-in-vietnam

General Statistics Office of Vietnam (GSO). (2018). Retrieved 25th February 2019 from https://www.gso.gov.vn/default.aspx?tabid=714

Giang, N. V., Kochanek, K., Vu, N. T., & Duan, N. B. (2018). Landfill leachate assessment by hydrological and geophysical data: Case study Nam Son, Hanoi, Vietnam. *Journal of Material Cycles and Waste Management*, *20*(3), 1648–1662. doi:10.100710163-018-0732-7

Government of Vietnam (GOV). (2015). *Decree 38/2015/ND-CP dated April 2015 on waste and scrap management*. Author.

Government of Vietnam (GOV). (2019). *Resolution 09/NQ-CP dated 03 February 2019 on Government Regular Meeting of Jannuary 2019*. Author.

Government Party Committee (GPC). (2018). *Report on assessment of 05 years implementing Resolution 24-NQ/TW on proactive response to climate change, strengthening resource management and environmental protection*. Author.

Hai, H. T., Hung, H. V., & Quang, N. D. (2015, December). An overview of electronic waste recycling in Vietnam. *Journal of Material Cycles and Waste Management*.

Hien, L. (2017, March 21). Energy figures. *Vietnam Electricity*. Retrieved from http://www.tietkiemnangluong.vn

Hien, N. T., Yen, N. T., Doan, D. T., Lam, N. T., Thuan, L. T. M., Giang, P. T., & Lam, L. N. (2015). Assessment of application of hazardous waste treatment and domestic solid waste in Vietnam. *Proceedings of the IV National Environment Conference*.

Hoang, H. (2018, June 1). There should be solutions to promptly handle environmental pollution in Minh Khai plastic craft village, Hung Yen province. *Vietnam Environment Administration Magazine*. Retrieved from http://tapchimoitruong.vn

Hoornweg, D., & Bhada-Tata, P. (2012). *What a waste: A global review of solid waste management* [urban development series; knowledge papers no. 15]. World Bank.

Huong, N. T. (2018). *The study on development of a procedure to identify polybrominated diphenyl ether (PBDEs) in tissue of the Anodonta cygnea and Pomacea canaliculata and pilot application to evaluate accumulation of PBDEs in the environment in the plastic recycling village of Minh Khai, Nhu Quynh town, Hung Yen province* (Unpublished master thesis). Hanoi University of Natural Resources & Environment.

Jambeck, J. R., Andrady, A., Geyer, R., Narayan, R., Perryman, M., Siegler, T., ... Lavender Law, K. (2015). Plastic waste inputs from land into the ocean. *Science*, *347*(6223), 768–771. doi:10.1126cience.1260352 PMID:25678662

Kafando, P., Segda, B. G., Nzihou, J. F., & Koulidiati, J. (2013). Environmental Impacts of Waste Management Deficiencies and Health Issues: A Case Study in the City of Kaya, Burkina Faso. *Journal of Environmental Protection*, *4*(10), 1080–1087. doi:10.4236/jep.2013.410124

Kaza, S., Yao, L., Bhada-Tata, P., & Woerden, F. V. (2018). *What a waste 2.0: A global snapshot of solid waste management to 2050* [overview booklet]. Washington, DC: World Bank. doi:10.1596/978-1-4648-1329-0

Kieu, L. T. C., Strady, E., & Perset, M. (2016). *Life cycle of floating debris in the canals of Ho Chi Minh*. Working paper of the Urban Development Management Support Centre (PADDI). Retrieved from http://horizon.documentation.ird.fr/exl-doc/pleins_textes/divers17-07/010070479.pdf

Luong, N. D., Giang, H. M., Thanh, B. X., & Hung, N. T. (2013). Challenges for municipal solid waste management practices in Vietnam. *Water Technology*, *1*(1), 17–21.

Ministry of Construction (MOC). (2016). *Current situation of technical infrastructure system in Vietnam urban areas*. Author.

Ministry of Construction (MOC). (2017a). *Report on implementation of the National Strategy on Integrated Management of Solid Waste to 2025, vision to 2050*. Author.

Ministry of Construction (MOC). (2017b). *Decision 1354/QD-BXD dated 29 December 2017 promulgated investment and treatment cost norms for household solid waste*. Author.

Ministry of Construction (MOC). (2019). *Annual Report for Reviewing Activities in 2018 and Implement Plans in 2019 of Construction Sector*. Author.

Ministry of Natural Resources and Environment (MONRE). (2003). *First National Communication to the United Nations Framework Convention on Climate Change (UNFCCC)*. Author.

Ministry of Natural Resources and Environment (MONRE). (2010). *Second National Communication to the United Nations Framework Convention on Climate Change (UNFCCC)*. Author.

Ministry of Natural Resources and Environment (MONRE). (2011). *National state of environment report 2011 - Solid waste*. Vietnam Publishing House of Natural Resources, Environment and Cartography.

Ministry of Natural Resources and Environment (MONRE). (2012). *Report on development of Program of strengthening control of environmental pollution caused by using non-biodegradable plastic bag in daily life up to 2020*. Author.

Ministry of Natural Resources and Environment (MONRE). (2014). *National State of Environment Report 2014 - Rural Environment*. Vietnam Publishing House of Natural Resources, Environment and Cartography.

Ministry of Natural Resources and Environment (MONRE). (2015). *National State of Environment Report 2011-2015*. Author.

Ministry of Natural Resources and Environment (MONRE). (2016). *National State of Environment Report 2016 - Urban Environment*. Vietnam Publishing House of Natural Resources, Environment and Cartography.

Ministry of Natural Resources and Environment (MONRE). (2017a). *National State of Environment Report 2017 - Waste Management*. Vietnam Publishing House of Natural Resources, Environment and Cartography.

Ministry of Natural Resources and Environment (MONRE). (2017b). *The second Biennial Updated Report of Vietnam to the UNFCCC on Climate Change*. Vietnam Publishing House of Natural Resources, Environment and Cartography.

Ministry of Natural Resources and Environment (MONRE). (2018a). *Administrative map of Socialist Republic of Vietnam*. Author.

Ministry of Natural Resources and Environment (MONRE). (2018b, July). *Report on solid waste management in Vietnam*. Workshop with Ministry of Environment of Japan.

Ministry of Natural Resources and Environment (MONRE). (2018c). *Report on management of household waste, 5/2018*. Author.

Ministry of Natural Resources and Environment (MONRE). (2018d). *Report on environmental management in scrap import and usage for production, 7/2018*. Author.

Ministry of Natural Resources and Environment (MONRE) & Ministry of Construction (MOC). (2009). *Report on development of the National Strategy of Integrated Solid Waste Management to 2025, vision to 2050*. Author.

National Assembly of Vietnam (NAV). (2014). *Law No. 55/2014/QH13 on Environmental Protection*. Author.

Netherlands Enterprise Agency. (2018). *Circular economy Vietnam waste management & recycling*. Retrieved from https://www.rvo.nl/sites/default/files/2018/09/Waste-management-recycling.pdf

Ngan, T. (2018). *Solid waste management in Vietnam: Current situation, challenges and strategies for development* (Unpublished undergraduate dissertation). Metropolia University of Applied Sciences.

Prime Minister (PM). (2003). *Decision 64/2003/QD-TTg dated 22 April 2003 on approval of Plan for treatment of seriously polluting establishment*. Author.

Prime Minister (PM). (2008). *Decision 1440/QĐ-TTg dated 06 October 2008 approved Master plan on construction of solid waste treatment facilities in three Northern, Central and Southern key economic regions up to 2020*. Author.

Prime Minister (PM). (2010). *Decision 1873/QD-TTg dated 11 October 2010 approved Master plan for the construction of the solid waste treatment site in the key economic area of Mekong Delta to 2020*. Author.

Prime Minister (PM). (2012). *Decision 170/QD-TTg dated 08 February 2012 approved master plan for comprehensive system for treatment of hazardous medical waste to 2025*. Author.

Prime Minister (PM). (2013). *Decision 2211/QĐ-TTg dated 14 November 2013 approved Master-plan for solid waste management in Cau river basin to 2020*. Author.

Prime Minister (PM). (2014). *Decision 31/2014/QĐ-TTg dated 05 May 2014 approved Regulations on support mechanism to waste-to-energy projects development in Vietnam*. Author.

Prime Minister (PM). (2015a). *Decision 16/2015/QD-TTg dated 22 May 2015 on take-back and treatment of discarded products*. Author.

Prime Minister (PM). (2015b). *Decision 07/QĐ-TTg dated 06 January 2015 approved Master plan for solid waste management in Dong Nai river basin to 2030*. Author.

Prime Minister (PM). (2015c). *Decision 223/QĐ-TTg dated 12 February 2015 approved Master-plan for solid waste management in Nhue-Day river basin to 2020*. Author.

Prime Minister (PM). (2015d). *Decision 2068/QĐ-TTg dated 25 November 2015 approved Vietnam strategy on development of renewable energy*. Author.

Prime Minister (PM). (2016a). *Decision 428/QD-TTg dated 18 March, 2016, approved Revision of national master-plan on electricity development 2011-2020 vision to 2030*. Author.

Prime Minister (PM). (2016b). *Decision 1979/QD-TTg dated October 14, 2016, on approval for solid waste management planning in Northern Key economic region by 2030*. Author.

Prime Minister (PM). (2018a). *Decision 491/QD-TTg dated 07 May 2018 on approval of Adjusted National strategy on Integrated solid waste management to 2025, vision to 2050*. Author.

Prime Minister (PM). (2018b). *Resolution 27/CT-TTg dated 17 Sep 2018 on immediate measures in management of import and usage of scraps for production materials*. Author.

Quang, N. D., Okumura, H., Yamasue, E., & Ishihara, K. (2009). Use and disposal of large home electronic appliances in Vietnam. *Journal of Material Cycles and Waste Management*, 364.

Quang, N. X. (2017). Waste-to-Energy: Potential and Reality for Vietnam. *Vietnam Energy Online*. Retrieved from http://nangluongvietnam.vn/news/vn/bao-ton-nang-luong/dot-rac-phat-dien-tiem-nang-va-hien-thuc-cho-viet-nam.html

Tham, T. T., Huong, N. T., Phuong, B. T., & Trinh, L. T. (2018). Assessment of the distribution and accumulation of polybrominated diphenyl ethers in the environment in the plastic recycling village of Minh Khai, Nhu Quynh town, Hung Yen province. *VNU Journal of Science: Earth and Environmental Sciences*, *34*(2). doi:10.25073/2588-1094/vnuees.4243

Thang, N. T. (2016). *State of 3R I Asia and Pacific – Country report for Vietnam*. 8th Regional 3R Forum in Asia and the Pacific.

Thang, N. T., & Anh, D. T. P. (2016). e-Waste management and resource recovery in Vietnam. In *Proceedings of the 6th International Waste Management Conference*. Jadavpur University.

Thang, N. T., & Hanh, H. H. (2019, February). *Discussion result with Nhu Quynh Town People's Committee about the environmental status in Minh Khai plastic craft village*. Academic Press.

Thanh, N. P., Matsui, Y., & Fujiwara, T. (2011). Assessment of plastic waste generation and its potential recycling of household solid waste in Can Tho City, Vietnam. *Environmental Monitoring and Assessment*, *175*(1-4), 23–35. doi:10.100710661-010-1490-8 PMID:20490914

Trung, D. D., Mahanty, S., & Van, N. T. (2012). *Vietnam's craft villages and water pollution: A review of previous research*. Australian National University. Retrieved from https://crawford.anu.edu.au/rmap/pdf/_docs/water_pollution_craft/dang_et_al_2010.pdf

Truong, T. X. (2016). *Report on plastic sector*. FPT Security Company. Retrieved from http://images1.cafef.vn/Images/Uploaded/DuLieuDownload/PhanTichBaoCao/Baocaonganhnhua_0317_FPTS.pdf

Tuan, N. V., Kien, T. K., Huyen, D. T. T., Nga, T. T. V., Giang, N. H., Dung, N. T., ... Kawamoto, K. (2018). Current Status of construction and demolition waste management in Vietnam: Challenges and Opportunities. *International Journal of GEOMATE, 15*(52), 24.

United Nations Industrial Development Organization (UNIDO). (2018). *Terms of Reference for the Provision of services and equipment for two pilot facilities to be located at Phan Boi and Minh Khai craft villages in Vietnam*. Retrieved from https://www.unido.org/sites/default/files/files/2018-03/Appendix%20 1%20-%20Terms%20of%20reference%20dated%2013%20February%202018_1.pdf

Vietnam Energy Efficiency Program (VNEEP). (2016). *Biogas in Vietnam: Potential in a hand.* Retrieved March 7, 2019 from http://tietkiemnangluong.com.vn/tin-tuc/pho-bien-kien-thuc/t13304/biogas-o-viet-nam-trien-vong-trong-tam-tay.html

Vietnam Cleaner Production Centre (VNCPC). (2016). *Cleaner production nationwide in Vietnam.* Retrieved from http://www.sxsh.vn/vi-VN/Home/tongquansanxuatsachhon-14/2011/San-xuat-sach-hon-tren-toan-quoc-874.aspx

Vietnam Plastics Association (VPA). (2016). *Overview on Plastic Sector in Vietnam.* Retrieved October 20, 2018 from http://www.vpas.vn/gioi-thieu/tong-quan-nganh.html

Washington Refuse & Recycling Association (WRRA). (2018). *Comprehensive list of banned materials under National Sword.* Retrieved from https://www.wrra.org/wp-content/uploads/2018/04/WRRA-Comprehensive-China-Ban-List.pdf

Chapter 20
Waste Management in Developing Countries:
An Overview

Agamuthu Pariatamby
https://orcid.org/0000-0001-9346-6843
Sunway University, Malaysia

Mehran Sanam Bhatti
https://orcid.org/0000-0001-9165-7195
University of Malaya, Malaysia

Fauziah Shahul Hamid
https://orcid.org/0000-0002-4083-0602
University of Malaya, Malaysia

ABSTRACT

This chapter summarizes and compares the current situation of waste management practices in developing countries as reported in other chapters of this book. Due to increasing rate of urbanization, population, and economic growth, generation of waste is also on the rise. It further makes the waste management situation trickier for developing countries as effective regulations and policies are either lacking or their stringent enforcement is absent. Thus, it leads to major issues of inefficient waste collection, open dumping, open burning, and over-reliance on landfilling. It was also discovered that holistic approach of all relevant stakeholders from governmental level to waste generators level is glaringly missing in most of the developing countries. At the end of the chapter, several recommendations are presented to overcome the challenges.

DOI: 10.4018/978-1-7998-0198-6.ch020

INTRODUCTION

Waste management has always been a challenging task for local authorities. Nevertheless, it is a sector which should not be taken lightly, as improper waste management can lead to disastrous consequences. The weight of the burden for waste management differs from one country to the other. Different countries face different challenges, while one country is challenged by one social factor, same factor could be the strength for waste management in another country. This chapter summarizes and compares the current situation of waste management practices in some of the global developing countries. With the underlying foundation, this chapter synthesizes present and future challenges faced by waste management sector of developing countries and discusse the means of overcoming these challenges.

WASTE DEFINITIONS IN DEVELOPING COUNTRIES

Definition of waste is important in order to regulate the waste material while imposing its appropriate management. Table 1 defines waste in accordance to the respective developing countries.

A clear definition is very important in authorizing the responsibility and jurisdiction of the various waste to the appropriate party. This not only ensures a more effective and efficient management of the waste, but also prevents overlapped jurisdiction which are not economical and interest conflicting. On the other hand, the clear definition can avoid misconception and confusion particularly in legal execution. Nevertheless, all stakeholders should be made aware of the existence of such jurisdictions and proper management of respective waste streams should be carried out, as well as collecting the data on generation of each type of waste. Following section discusses the various types of solid waste generated in developing countries.

TRENDS IN SOLID WASTE GENERATION IN SELECTED DEVELOPING COUNTRIES

The waste generation of different countries of the same region may vary from one another as it can be seen in previous chapters. However, it is evident that different cultures, living styles, income level, and urbanization, among others, dictate the trend in waste generation in all developing countries. Table 2 gives the Municipal Solid Waste (MSW) generation of developing countries from previous chapters and it is worth noting that all of the developing countries focused on management of MSW but not on the reduction of waste generation. Unlike developed countries where the main focus is on waste reduction and resource circulation.

When studying the factors of MSW generation in developing countries, it is seen that certain factors influence greatly in some countries than the others. For instance, in 2012, Bangladesh, with a population of 163 million inhabitants generated 15 million tons of waste with per capita generation of 0.35 kg/day. South Africa, on the other hand, generated 54 million tons of waste from the whole country populated with only 57.8 million people due to higher gross domestic product (GDP).

On the contrary, Cambodia's population (16 million) was half of the population of Ghana (30 million), produced an annual waste generation of 3 million tons of waste while Ghana has almost double the amount of waste generation with 5 million tons. This implies that population size has a significant

Table 1. Definition of waste according the different developing countries in the world

Countries	Definition of Waste
Argentina	Wastes are substances or objects which are disposed of or are intended to be disposed of or are required to be disposed of by the provisions of national law
Bangladesh	Any solid, liquid, gaseous, radioactive substance, the discharge, disposal, and dumping of which may cause harmful change to the environment
Cambodia	Solid waste refers to hard objects, hard substances, products or refuse which are useless, disposed of, are intended to be disposed of, or required to be disposed of
India	Solid Waste includes solid or semi-solid domestic waste, sanitary waste, commercial waste, institutional waste, catering and market waste and other nonresidential wastes, street sweepings, silt removed or collected from the surface drains, horticulture waste, agriculture and dairy waste, treated bio-medical waste excluding industrial waste, bio-medical waste and e-waste, battery waste, radio-active waste generated in the area under the local authorities and other entities
Malaysia	The laws of Malaysia (Act 672 – Solid Waste and Public Cleansing Management Act 2007) define "solid waste" as o Any scrap material or other unwanted surplus substance or rejected products rising from the application of any process; o Any substance required to be disposed of as being broken, worn out, contaminated or otherwise spoiled; or o Any other material that according to this Act or any other written law is required by the authority to be disposed of, but does not include scheduled wastes as prescribed under the Environmental Quality Act 1974 (Act 127), sewage as defined in the Water Services Industry Act 2006 (Act 655) or radioactive waste as defined in the Atomic Energy Licensing Act 1984 (Act 304)
Pakistan	Hazardous waste is defined as waste which is or which contains a hazardous substance, and includes hospital waste and nuclear waste
Philippines	Solid waste - refers to all discarded household, commercial waste, non-hazardous institutional and industrial waste, street sweepings, construction debris, agricultural waste, and other non-hazardous/non-toxic solid waste. Unless specifically noted otherwise, the term "solid waste" as used in this Act shall not include: (1) Waste identified or listed as hazardous waste of a solid, liquid, contained gaseous or semisolid form which may cause or contribute to an increase in mortality or in serious or incapacitating reversible illness, or acute/chronic effect on the health of persons and other organisms; (2) Infectious waste from hospitals such as equipment, instruments, utensils, and fomites of a disposable nature from patients who are suspected to have or have been diagnosed as having communicable diseases and must therefore be isolated as required by public health agencies, laboratory wastes such as pathological specimens (i.e. all tissues, specimens of blood elements, excreta, and secretions obtained from patients or laboratory animals) and disposable fomites that may harbor or transmit pathogenic organisms, and surgical operating room pathologic materials from outpatient areas and emergency rooms; and (3) Waste resulting from mining activities, including contaminated soil and debris. Municipal waste - refers to waste produced from activities within local government units which include a combination of domestic, commercial, institutional, and industrial wastes and street litters;
Rwanda	A product or a substance that is no longer suited for its intended use
South Africa	waste as superfluous and unwanted in nature or applies the precautionary principle to describe waste as toxic and hazardous unless proven otherwise
Tanzania	non-liquid materials arising from domestic, street, commercial, industrial and agricultural activities that includes refuse or garbage, non-liquid materials arising from construction and demolition activities, garden trimmings and mining operations, dead animals and abandoned cars scraps
Thailand	Waste means residues either in a solid, liquid, or gas state generated in daily life from any sectors, or any other hazardous substances
Turkey	Municipal wastes are defined as wastes that are classified as non-hazardous wastes and originate from houses or are in similar content from different sources.
Viet Nam	Waste is generally classified into two categories: (i) non-hazardous and; (ii) hazardous waste (HW). Waste also can be classified as household waste, industrial waste, agricultural waste, medical waste and construction waste.

Table 2. Municiapal Solid Waste Generation of Selected Developing Countries

Country	Year	Waste Generation (million tons)	Population (million)
Argentina	2015	16.4	43
Bangladesh	2012	14.8	155
Cambodia	2017	3	16
China	2017	215.2	1,390
Ghana	2018	5.2	30
India	2018	51.5	1,250
Lebanon	2017	2.7	5
Malaysia	2016	13.7	32
Myanmar	2017	7.3	54
Pakistan	2018	48.5	204.6
Philippines	2018	13.1	106
Rwanda	2019	0.2	12
Serbia	2018	2.5	7
South Africa	2018	54	57.8
Sri Lanka	2018	2.6	21.4
Tanzania	2019	1.1	61
Thailand	2017	27.4	67.2
Turkey	2018	31.6	82
Viet Nam	2017	25.5	94.7

contribution towards the generation of waste within a country, that countries with larger population generally would generate more waste as compared to a smaller populated country but depending on the per capita generation.

While, population size undeniably plays a significant factor in influencing the generation of waste in a country, the opulation size alone is not the only contributing factor that may influence the waste generation in the country. This was observed in Sri Lanka and Republic of Serbia, where their annual waste generation was approximately 2.4 – 2.6 million tons, but the latter's population was only one third of the former. As established before, GDP and urbanization also impacts the waste generation in a country. Therefore, in the case of Sri Lanka and Republic of Serbia, it highly important to note that GDP played a more significant role than population size. Hence, countries with higher GDP tend to generate higher volume of waste because waste generation is greatly influenced by the ability of the population to consume products and dispose them. Thus, countries with higher purchasing parity create a more effluent society which results in generation of more waste. As a result, the waste per capita generation may differ within a country in accordance to the standard of living and the income level of the community.

Bangladesh being one developing country with low income as listed by World Bank has per capita waste generation of 0.35 kg/day, while in Nigeria and Ghana, it is 0.43 and 0.47 kg/day, respectively. Moving towards the lower middle-income countries like Cambodia and the Philippines, per capita generation for both was 0.53 kg/day. As for the high middle-income countries like China, Argentina, Thailand,

Figure 1. GDP and per capita waste generation of selected developing countries in the world

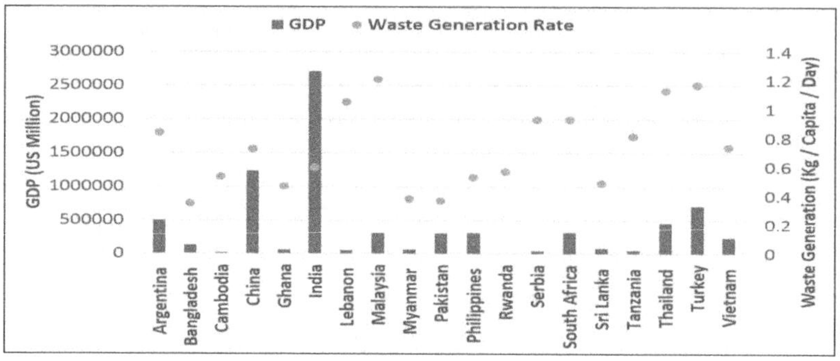

and Malaysia, the per capita waste generation was 0.75, 0.84, 1.13 and 1.21 kg/day, respectively. Figure 1 illustrates the per capita waste generation and the GDP of selected developing countries in the world.

Per capita waste generation varies according to the GDP of a country, particularly among the developing economies. Unfortunately, generation of excessive volume of waste makes its management a huge challenge and burdensome to many local authorities of the developing nations. On the contrary, this is not necessarily the case for developed nations where their existing integrated waste management can be a profitable sector with reputable income. This is only applicable if effective waste management is in place to cater for the need of the sector. Thus, it is highly imperative that proper strategies are implemented to ensure that the selected waste management is highly effective and sustainable.

Another major and rapidly emerging solid waste stream in developing countries is electronic waste (E-waste) or waste of electrical and electronic equipment (WEEE). The availability of data on generation of E-waste in developing countries is not always consistent and for some countries, it has not been reported in country chapters. Figure 2 shows the amount of generation of E-waste in selected developing countries.

Although, Bangladesh is reported to have generated 2.7 million tons of E-waste in the country, the majority of E-waste is ship breaking yards. The studies in E-waste generation in Myanmar has produced two different values of 29,000 tons and 5,920 tons, respectively and author cited that no national database exists for generation of E-waste in the country. When it comes to African countries, Rwanda generates E-waste in the range of 10,000 – 15,000 tons per year. The exact amount of E-waste generated in Tanzania is not known, while some authors suggest that E-waste consititute 1 to 2% of total MSW generated, one source reported in the Tanzania chapter states that there may be 18,000 to 33,000 tons of only E-waste comprised of computers.

Figure 2. Amount of E-waste Generated in Selected Developed Countries

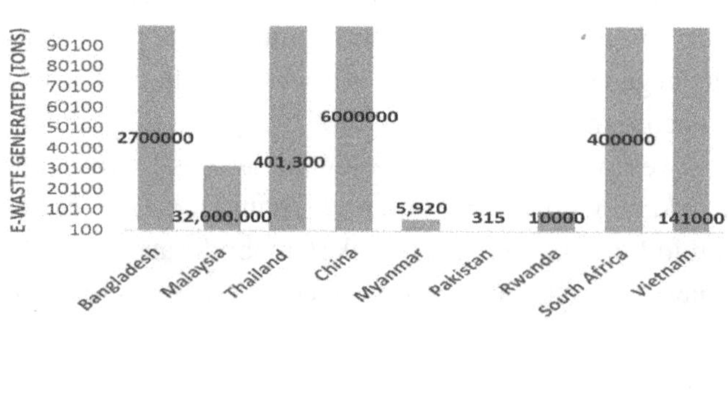

Another type of solid waste generated in developing countries is hazardous waste. However, hazardous waste generated may also include hazardous waste in aqeous state. Figure 3 demonstrates the hazardous waste generated in developing countries that reported the amount of genearated hazardous waste. It must be noted that 250,000 tons of hazardous waste generated in Pakistan is medical waste and the quantity of industrial hazardous waste is not available. Whereas, in Viet Nam, 874,588 tons of hazardous waste was generated from industries in 2016, remaining 14,000 tons of hazardous waste is produced from agriculture. For instance, containers and packaging of pesticides and fertilizers are examples of hazardous waste generated from agriculture.

Successful implementation of a sustainable waste management strategy can only be executed with sufficient and reliable background data. Reliable database on waste management among developing countries can be very scarce. The following section discusses the reliability of data reported, collected and published from the waste management sector of selected developing countries in the world.

Waste Data Reliability

Waste generation data can be very scarce among developing nations due to various factors. Among others is the absence of dedicated policy pertaining to waste management, hence local authorities tend to carry out their responsibilities according to their convenience. As a result, waste is managed with minimum effort, with the sole objective to get rid of the waste, by applying the concept of "out of sight and out of mind". Therefore, local authorities are not concerned with waste data collection while managing waste but are focused on ensuring successful disposal of waste and on avoiding complaints from the public. The absence of necessary legal commitment on appropriate data collection has resulted in failure to produce database on waste management.

However, Argentina is one example of a developing country with a standard on collection of waste generation rates. Nevertheless, despite standardized methods of gathering waste generation rates at source, the volume of waste is calculated at the final disposal sites in Argentina. As pointed out by the authors, such numbers represent the amount of waste collected and disposed at final disposal sites but not the actual amount that was generated at sources. Since, inconsistent waste collection rates and waste recovery by informal waste recyclers may reduce the total amount of waste reaching final disposal sites. Therefore, while majority of developing countries do not have explicit policies or regulations on obtaining waste generation rates, those that have such legislations, fail to implement it.

Figure 3. Amount of Hazardous Waste Generated in Selected Developed Countries

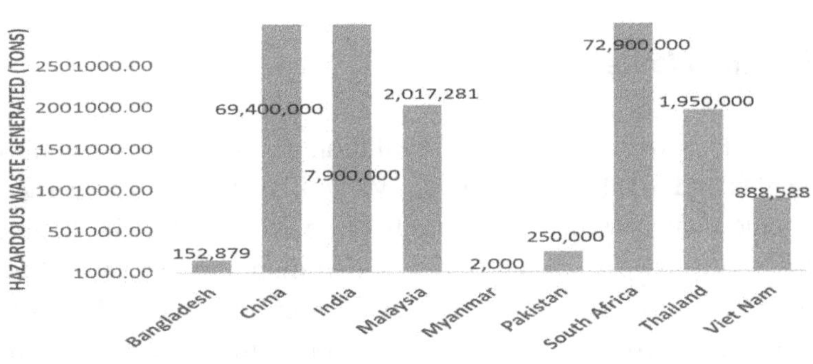

While top-down commitment can be highly effective such that its absence in many developing countries leads to insufficient data collection in waste management, the bottom-up approach can be useful to fill the missing gaps. This is evident in many developed nations, that when public demands for an improved waste management system, reliable and sufficient data collection benefits the subsequent planning and strategizing a viable system. Unfortunately, such movement from the public that demands for improved quality of waste management services are rarely observed among the developing nations and the bottom-up approach is also non-effective. This is strongly observed in most of the developing countries where absence of public awareness on waste management issues was discovered in such as Pakistan, Bangladesh, Serbia and others. However, in case of Rwanda despite the involvement of public in solid waste management system, reliable data on national waste generation is unavailable.

Although most of the earlier chatpers on developing countries reported waste generation data clearly, suggesting easier access to waste generation data in respective countries butsome countries such as Lebanon, Bangladesh, Malaysia and Cambodia had highlighted issues related to availability of waste generation data. For instance, waste generation data reported in Lebanon chapter is based on extrapolation and interpolation. Whereas, for Bangladesh and Malaysia, different source provide different amount of waste generated. In case of Bangladesh, Shams et al, (2017) and Islam, (2017) reported urban waste generation for Bangladesh and Kaza et al, (2018) reported waste generation for the entire country. As for Malaysia, one value of waste generation is estimated (13 million) and the other value is reported by a waste sector related source (Pauze, 2016).

The lack of reliable and consistent data on waste management can be further aggravated due to financial constraint faced by the country. As the task of compiling data can be significantly expensive if proper data collecting system is not in placed. Also, many waste managers would avoid to commit to such activities particularly when it is not legally required by the local authorities. As a result, data collection which was normally done only involved basic information which may not be reliable and consistent, or even accurate to represent the actual situation. Nevertheless, many waste managers in developing countries have become aware of the critical need of reliable and consistent data. Frequent data collection is crucially important to understand the evolution in waste management, its issues and challenges, and the options to deal with the obstacles.

The responsibility of collecting and collating data on waste generation, waste composition, the chemical compounds and calorific values of the waste, is the task to be accomplished by the waste managers. This information is not only necessary for the sake of easing and smoothening the management, but is also crucial for planning the future strategies. The obvious trend in majority of developing nations is the transition of waste composition from organic nature to waste with highly complex composition due to the presence of various plastics compounds (PE, PET, PP, PS etc.) and other synthetic materials. The following section discussed the composition of waste in selected developing countries in the world.

Composition of Solid Waste

Many countries in the world have experienced the evolution of waste composition from natural products mainly to a more complex and highly heterogenous synthetic materials. The earlier waste management system dealt with organic matters which were highly putrescible and easily degraded. Due to the fast decomposition of this waste, proper waste collection and disposal was crucial to prevent unsanitary environment and pest. The composition of waste evolved over the years and after the industrial revolution, synthetic waste, namely plastics has become one of the significant fraction in the waste stream.

The increase in synthetic materials within the waste stream has complicated the conventional waste disposal technologies, particularly those that are highly dependent on open-landfilling. This is because, open dumping will result in escapement of un-degradable wastes away from its disposal site by animals, wind and/or water. The seriousness of this situation is highlighted in vast amount of reports on plastics and other synthetics combinations accumulating in the ocean and threatening the survival of human and other organisms as summarized in the Malaysian chapter.

The understanding on the waste composition is important in order to identify the best possible option to manage the waste, with minimum impact to the environment and human health. Figure 4 illustrates the percentage of plastic waste and organic portion in selected developing nations in the world.

The drastic increase in plastic waste conponent has been an alarming scene for many developing nations whose resource recovery activities are not effective. Resulting from the absence of resource recovery, disposal ground will be rapidly exhausted. This not only disrupt the operation of disposal facilities, but also required more land to be allocated for the construction of new disposal sites. Situation as this is not wanted by any waste managers since locating and siting a landfill site can be very difficult and expensive.

The waste generators are encouraged to separate their waste via various recycling campaign to reduce the waste which can be retrieved as "resources". With the retrieval of these "resources", waste volume can be reduced and lead to the reduction of waste generation in a country. On the other hand, retrieval of the "resources" will result with the composition to be less complex with more portion of the remaining waste be easier to manage. Yet, the scenario in the change and evolution of waste generation and composition is highly influenced by several factors. The consecutive section discussed the parameters that affected waste generation and composition of a country.

Parameters Affecting The Waste Generation and Composition

Though it is inevitable to prevent waste generation from any anthropogenic activity, waste generation is highly influenced by human behavior. The fact that humans can utilize the easily available resources, had created a consumer society which generally has very low regards to materials they consider as waste. Humans tend to generate more waste when they can afford to do so. Thus, one factor which has high influence in waste generation is the affordability of the community.

Figure 4. Percentage of plastic and organic waste in municipal solid waste stream in selected developing countries in the world

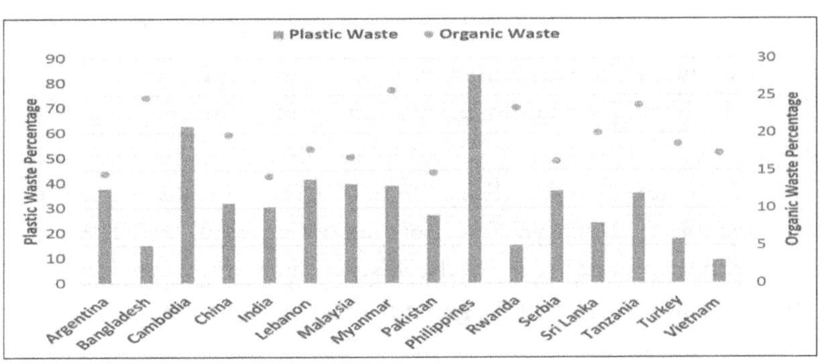

Other factors that influenced the waste generation and composition include improved standard of living, urbanization, and local anthropogenic activities. Improved standard of living dictates the living style that waste generation can changed from the norm. This is seen in change in the purchasing of goods to materials which is more luxurious to suit the need of the society. This is evident in many cities of developing countries where economic prosperity also transforms the community into a consumer society which demand for various goods not only to cater their "needs" but also their "wants".

Urbanization also affects the waste generation and composition of a country where more processed goods are being consumed to cater to the needs for fast and speedy lifestyle. Food which are prepared from fresh products would generate waste of organic nature that can be easily decomposed without much difficulty. Yet, the fact that this approach of food preparation needs more time and attention, makes it less convenient to the urbanites, who prefer conveniently processed food products that is less time consuming in its preparation. On the other hand, more packaging materials will be produced along the process. Similar observation can be found in the consumption of numerous products which create conveniences to human life but with the production of more packaging waste to be dealt with.

The influence of the existing anthropogenic activities on waste generation is also observed among the developing nations. Areas that have more activities that relate to natural resources tend to produce waste which mainly composed of organic matters. Whist area with more downstream processes tend to generate waste which are highly complex due to the complexity of the input and goods. Regardless of the situation, composition and generation, waste needs to be properly managed to prevent detrimental impacts to the environment and human health.

THE ELEMENTS OF SUSTAINABLE WASTE MANAGEMENT

In ensuring an effective and successful implementation of a waste management strategy, a sustainable system is highly important. Among the developed nations that have been known to be highly efficient in managing their waste sector are such as Germany, Japan, Denmark and Korea. The sustainable waste management adopts the inverted pyramid of the waste management hierarchy. This is also the trend that most developing countries are heading towards to achieve a sustainable waste management. In order to achieve a sustainable waste management system, many countries in the world strategized their waste management plan based on the waste management hierarchy. The elements of sustainable waste management are discussed in the following sections.

Implementation of 3R

3R, most commonly referred to reuse, recycle and recover is the practice that most developing countries are aiming to incorporate in their solid waste management systems. However, as observed in the country chapters earlier, most of the developing countries are struggling to implement 3R practices.

However, before discussing 3R in developing countries, the first and most crucial strategy must be waste prevention. It can be achieved by instilling the appropriate attitude of waste prevention amongst the public. Malaysia for example, campaigned on the concept of "think before you throw", to encourage public to consider their need to dispose their goods. Also, in many other developing countries including India and the Philippines, non-governmental agencies promote the public to donate their waste rather than disposing it.

Creating a viable platform for waste prevention is an effective diversion option for waste management as the volume will not accumulate within the waste stream until they are discarded intentionally when the value diminished. Thailand for example, encourages the establishment of waste bank, as a strategy to prevent the generation of waste particularly of non-putrescible nature in the country. Some cities in India, on the other hand, promote the donation of uneaten food products to reduce their food waste generation.

The option following waste prevention is waste reduction. Should prevention of waste be inevitable, waste reduction must be considered. Earlier chapters have highlighted the effective top-down approach implemented by several governments among the developing countries, particularly for plastic waste products. This includes the banning or restricting the use of single-use plastic bags. Such an implementation may reduce the generation and disposal of waste bags into the waste stream. Not only it can extend the lifespan of landfills to a certain extent, but can also reduce the risk of plastic contamination to the marine environment. Another strategy for waste reduction involves the polluter pay principle that requires waste generators to pay some fee for the volume of waste they generate. The pay-as-you-throw scheme has been popular strategy in waste reduction as seen in many developing countries of the Eastern Europe, including Republic of Serbia.

The recycling rate can be highly variable in developing countries. Countries with transitory economy generally have lower recycling rate as compared to that of other developing nations. This is because the main driver of recycling i.e. economic incentives does not exist among the nations of transitory countries, that are experiencing improvement in standards of living. However, the challenges in implementation of recycling can also range from absence of regulations or tax incentives from the government to irregular market prices as observed in Ghana (lower middle-income country). In Ghana, inconsistent supply of energy and high operation costs negatively impact the recycling of waste in established recycling centers.

On the contrary, the economic encouragement to recycling is not usually appealing to the community, such that people do not want to be involved in recycling if there is a slightest inconvenience in their daily life. This scenario is evident in transitory countries like Malaysia and Thailand. Plently of awareness initiatives have been taken in Myanmar to improve and promote 3R but the participation of public is still underwhelming to the least. Moreover, there are also established reuse and recycling centers in the country. However, informal sector in Myanmar is responsible for some form of recycling and waste recovery. Similar involvement of informal sector in 3R is reported in Ghana, Viet Nam and Pakistan. (Detailed information on informal sector in developing countries is discussed in later section of this chapter). Due to the efforts of informal sector and other stakholders, recycling rate can reach up to 20-30% among the lower income countries, since the income obtained by the group of people that are involved in recycling business, is substantially significant to improve their livelihood. India, Indonesia, Nigeria, Rwanda and Ghana are among the developing countries that have relatively high rate of recycling due to the drivers from the economic sector. Figure 5 illustrates the percentage of recyclables in the waste stream of selected developing countries.

Waste segregation is paramount to successful implementation of 3R. Argentina, despite waste segregation (into organics and inorganics), does not practice resource circulation nationwide. However, there are some technologies that are locally deployed for executing 3R including, biogas capture, composting, and mechanical biological treatment. Similarly, in China composting and anerobic digestion is mainly carried out in rural areas. Nevertheless, the rate of composting and anerobic digestion is still minimal. These two technologies are funded and supported by Chinese government. Similarly, in Lebanon, with the help of top-down approach, a four year plan is recycling and composting the MSW generated in Beirut and Mount Lebanon. However, the rest of Lebanon does not have similar plans of resource recovery and recycling.

Figure 5. Percentage of organics, recyclables and non-recyclables in the waste stream of selected developing countries in the world

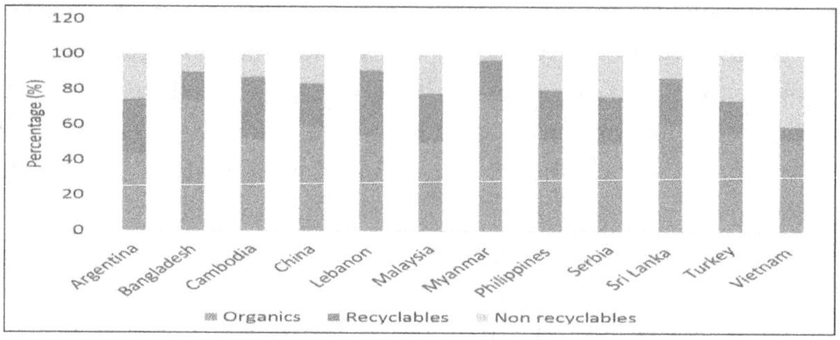

In addition to recycling of dry recyclables, 3R for organic portion of MSW is common in developing countries. But use of such technologies may not be nationwide in respective countries. For instance, Viet Nam has functional composting facilities and composts that are mainly utilized in the southern parts of the country. Cambodia, on the other hand, only carries out composting at industrial scale but exports inorganic waste to neighbouring countries such as Thailand and Viet Nam. Similarly, composting is commonly deployed in South Africa and anaerobic digestion has not been successful.

There are some success stories of 3R in some parts of the Philippines such as Quezon city in Metro Manila. There is material recovery facility, and had successful implementation of biogas capture and refuse derived fuel manufacturing projects in the city in 2013. Whereas, San Carlos City, Negros Occidental in Philippines was also reported to have operational material recovery facility. Similarly, waste recycling only occurs in cities of Tanzania as well, where rural areas do not contribute as much. ON the other hand, few companies in Rwanda are responsible for manufacturing of briquettes and composting. Whereas, informal sector also plays a role in recycling of dry recyclables.

Furthermore, the resource recovery in Sri Lanka is on the rise. However, it is still not substantial. Majority of recycling is performed by informal sector who also collect recyclables from the waste generators. Nevertheless, there are also formal waste collectors and recyclers in Sri lanka that are lisenced by Central Environment Authority, who recycle. The implementation of 3R in form of recycling and composting is also growing over the years in Thailand. Similarly, Turkey has been focusing on execution of 3R as well and has a number of operational composting facilities, biomethanization plants, and mechanical and biological treatment plants. Moreover, reyclcing is also carried out in Turkey.

Recycling is highly influenced by the market price which can affect the recycling of goods based on the market value. Materials with no market value may not be collected at all by waste collectors, while those with greater monitary value will be extensively sought after for recycling purpose. This is evident in many developing nations in Asia which had experienced the impact of the banning of importation of recycled plastic by China.

The absence of market demand will result in inappropriate management of the unwanted materials and illegal dumping. Therefore, it is important for a country to strategize the utilization of the resource from the waste stream to prevent its loss, when it was not in demand in the recycling business. All stakeholders need to play their role in managing the waste to ensure that retrievable resources are not lost to waste disposal. For example, recycling in South Africa is carried out by non-governmental organisations, industries and volunteers, however it is still underdeployed. Another mean of reducing the loss of resources in form of final disposal is active participation of the manufacturers by creating channels to enable "their waste" to be collected for reuse or recycling purposes. Thus, implementation of the extended producer responsibility (EPR) needs to be encouraged, particularly by the local government. Following section discusses the practice and involvement of industries and manufacturing sector in EPR among the developing countries.

Extended Producer Responsibility

Extended Producer Responsibility (EPR) has been reported to be the most effective tool in reducing waste generation by enabling various options for waste generators to divert their waste for reutilization and/or recycling. The involvement of manufacturer and industrial players are essential to allow the diversion, by creating values to the waste. As a result, this can generate a close-loop system where materials will continue to be utilized with a prolonged lifespan. EPR is closely related to the concept of circular economy where discarded materials will not be exiting the system but continue to be utilized within for similar intended purpose or channeled to cater for other purposes.

Majority of the developing countries in this book do not have EPR in practice. Those that has integrated EPR, has focused on materials that are most commonly channeled back into the system are electronic items. Due to their hazardous nature, E-wastes need to be diverted away from other municipal wastes. Otherwise, the heavy-metal and persistent organic compound may contaminate via leachate pollution to the environment. Ghana ratified Hazardous and Electronic Waste Control and Management Act, Act 917 in 2016 which has a component on EPR for the management of WEEE. Similarly, Electronic waste (Management and Handling) Rules 2011 was enacted in India in 2012 with the focus on EPR. EPR for E-waste has been fully implemented in India since 2018 and has targets that must be achieved for each calender year starting from October 2017. Whereas, Thailand aims to assimilate EPR in the respective regulations for the management of E-waste. While EPR is part of legislations on the management of E-wate in Viet Nam, it has not been fully implemented in the country.

EPR is also applicable to other waste streams in developing countries. While EPR is part of South Africa's waste management system, EPR charge system is applicable to beverage bottles, batteries, cars and tyres in the country and aims to extend to WEEE, lighting, paper and packaging industries. Moreover, the government is imposing EPR tax on the producers as well, to reduce waste generation. Philippines, on the other hand, does not have EPR in the regulations related to waste management.

Argentina is formulating a law on management of waste tyres based on EPR. On the contrary, Malaysia had incorporated EPR in their earliest policies for waste management but it has not been part of succeeding policies until Solid Waste and Public Cleansing Management Act in 2007 and Malaysian National Plans. In Malaysia, several communication industries allow trade-back system where discount to purchase a new product will be given when the old merchandise was returned. This is applicable to mobile phones, computers and other electronic gadgets. This will reduce the disposal of e-waste into the waste stream. Whereas, Thailand has several locations of facilities owned by specific brand-name to carry out resource recovery. Retrieved materials are channeled back into the input stream for production of suitable products.

Lastly, EPR is implemented in Serbia in form of consumers paying for recycling fee when purchasing products such as electronic items, tires and batteries. The management of packaging waste is funded by similar mode of EPR in Serbia.

Therefore, it is undeniable that EPR plays a huge role in enabling a circular economy within a country. With the encouragement by government for industries to carry out EPR, resource recovery can be effectively implemented for the improvement of a country's waste management system. However, there are various issues and challenges among the developing that need to be addressed before a successful integrated waste management can be realized. The next section elaborates the issues and challenges in the implementation of an sustainable and integrated waste management.

CHALLENGES TO SUSTANABLE WASTE MANAGEMENT

The implementation of an effective integrated waste management has always been challenging in developing countries. This is mainly due to various issues pertaining to waste management system. Among the most crucial factors that influence the efficiency of an implemented waste system, is the policy and regulations that may set the stage for further challenges to arise. Such challenges are elucidated more in detail in following sections.

National Policies and Regulations on Waste Management

Among the developing countries in the world, challenges in implementing an effective waste management system differ from one country to another. Nevertheless, due to the similarity in the financial ability and social-perception among the public, some countries share similar challenges. Thus, implementation of appropriate policies and regulations are necessary to overcome some to these challenges. Table 3 lists the regulations and policies of respective developing countries in the world that directly or indirectly assist the waste management sector.

As highlighted in the Table 3, some of these developing countries have an extensive set of regulations and policy in order to assist the implementation of an integrated waste management system. However, many of these regulations are yet to be implemented or their enforcement has not been stringent or effective. Several developing countries have a holistic regulation and policy which cater not only the typical municipal waste but also the hazardous waste and clinical waste. These separate Acts and policies are necessary to ease the implementation, and the enforcement of the regulations to achieve effective management of the waste.

Another issue with the policy and regulations in developing countries is that they can be a decade old such as in Sri Lanka, Rwanda, Philippines, and Malaysia among others. However, it is also observed that some developing countries have just established their policy on waste management recently. Yet, regardless of the maturity of the policy in a country, majority of the developing countries struggle to implement these policies. As a matter of a fact, enforcement can be a huge challenge to any developing nations. This is an absolute contradiction with the developed countries where policy and regulations ease the waste management system and create wide range of opportunities to improve the system further.

The absence of appropriate policy and regulations halts any possibility of the improvement. This is because, in order for a system to work effectively, a proper regulation is necessary to provide guidelines as how to go about achieving the objectives and targets. This scenario is clearly absent in most of the developing countries when it comes to waste management sector. The waste management is of the lower priority in the national agenda among developing countries as their main focus has been on stabilizing and developing strong economy to a point that environment is often compromised. Therefore, even

Table 3. Regulations and policies pertaining to waste management sector of selected developing countries

Countries	Regulations and Policies
Bangladesh	National Environmental Management Action Plan (NEMAP) 1995 Dhaka Declaration on Waste Management by SAARC countries 2004 Lead Acid Battery Recycling and Management Rules 2006 Constitution of Bangladesh 1972 National Environment Policy 1992 Environmental Conservation Act 1995 Bangladesh Environmental Conservation Act (ECA) 1995 National Environmental Management Action Plan (NEMAP) 1995 Bangladesh Environmental Conservation Rules (ECR) 1997 Urban Management Policy Statement 1998 National Agriculture Policy 1999 National Industrial Policy 2005 Environment Management Plan 2005 Poverty Reduction Strategy Paper (PRSP) 2005 Solid Waste Management Action Plan for Eight Secondary Towns in Bangladesh 2005 Private Sector Housing Development Guideline 2005 Lead Acid Battery (LAB) Recycling and Management Rules 2006 Biomedical Waste Management Rules 2008 National Solid Waste Management Handling Rules 2010 National 3R (Reduce, Reuse and Recycle) Strategy for Waste Management 2010 Hazardous Waste and Ship Breaking Waste Management Rules 2011
Cambodia	**Municipal Solid Waste:** 1. Sub-Decree on Solid Waste Management, No. 36, 1999 2. Inter-Ministerial Declaration of Ministry of Interior-Ministry of Environment on Waste and Solid Waste Management in Province / Municipalities of Cambodia, No. 80, 2003 3. Sub-Decree on Urban Solid Waste Management, No. 113, 2015 4. Inter-Ministerial Prakas on Maximum Fee Setting of Urban Solid Waste Management Services, No. 195, 2018 5. Sub-Decree on Plastic Waste Management, No. 168, 2017 6. Sub-Decree on Encouragement of Eco-friendly Practice, No. 60, 2018 **E-Waste:** 1. Sub-Decree on E-Waste and Electronic Equipment, 2016 **Hazardous Waste:** 1. Sub-Decree No 466 on the Organization and Function of the Department of Hazardous Substance Management, 2015 2. Declaration No. 387 on Enforcement to Standard Level the Amount of Toxic or Hazardous Substance that Allow to Abandoned, 2015 **Medical Waste:** 1. The Declaration on Waste Management from Health Care Services in the Kingdom of Cambodia (Ministry Health, 2008) 2. Decision on Creation of Medical Waste Management Unit, No. 96 of Red Cross Cambodia, 2009 3. National Guide on Waste Management from Health Care Services, 2012 **Industrial Waste:** 1. Guideline on Solid Waste Management at Factories, Enterprises and Companies, No. 11, 2003 2. Guideline on Sludge Waste Management at Factories – Enterprises, 2000 3. Guideline on manufacturing Hazardous Waste Management, No. 87, 2000
China	13th Five-year Plan for China's National Economic and Social Development (2016 – 2020) Notice on Promoting the Classification of Wastes in Public Institutions such as Party and Government Organs, 2017 – No. 180 Opinions on the Implementation of Waste Classification System by Military Units, 2017 – No. 485 Notice on Promoting the Classification Management of Waste in Medical Institutions, 2017 – No. 30 Notice on Accelerating the Classification of Waste in Some Key Cities, 2017 – No. 253 Standard for pollution control on the landfill site of municipal solid waste (GB16889-2008) Standard for pollution control on co-processing of solid wastes in cement kiln" (GB30485-2013) Standard for the Management of Rural Household Waste Classification Regulations on the Management of Rural Household Waste Classification in Jinhua City
Ghana	Hazardous and Electronic Waste Control and Management Act, Act 917 – 2016
India	Factories Act 1948 The Environment (Protection) Act, 1986 Hazardous Wastes (Management and Handling) Rules, 1989 Manufacture, Storage and Import of Hazardous Chemical Rules, 1989 Biomedical Waste Management and Handling Rules, 1998 Plastic Manufacture, Sale, Usage Rules, 1999 MSW (Management and Handling) Rules, 2000 Batteries (Management and Handling) Rules, 2001 E-waste (Management and Handling) Rules, 2011 Biomedical waste management and handling rules, 2011 Hazardous and other wastes (Management and Transboundary Movement) Rules, 2016 Construction and Demolition Waste Management Rules, 2016 Biomedical Waste Management Rules, 2016 E-Waste (Management) Rules, 2016 Plastic Waste Management Rules, 2016 Solid Waste (Management) Rules, 2016
Lebanon	Decree 8735/1974 - Assigns to Municipalities the roles, of public cleanliness and of collection and disposal of waste Law 64/1988 - Preservation against harmful and hazardous waste pollution Decree 9093/2002 - Provides municipalities with incentives to host solid waste facilities Decree 8006/2002 and Decree 13389/2004 - Identifying the categories of health care waste and means their disposal Decree 1117/2008 - Provides municipalities with incentives to host solid waste facilities COM Decision 45/2018 - Integrated solid waste management policy, provides provisions related to solid waste management Law 80/2018 - Integrated solid waste law provides frameworks and provisions for solid waste management

continued on following page

Table 3. Continued

Countries	Regulations and Policies
Malaysia	**Municipal Solid Waste:** 1. Environmental Quality Act, 1974 2. Solid Waste and Public Cleansing Management Act, 2007 3. Solid Waste Management and Public Cleansing Corporation, 2007 4. Eleventh Malaysian Plan (2016 – 2020) **Hazardous Waste:** 1. Environmental Quality Act 2. Environmental Quality (Prescribed Premises) (Scheduled Waste Treatment and Disposal Facilities) Regulations, 1989 3. Environmental Quality (Scheduled Waste) Regulations, 2005
Myanmar	National Environmental Policy – 2018 Myanmar Sustainable Development Plan (2018 – 2030) National Waste Management Strategy and Master Plan (2018 – 2030) Myanmar Climate Change Strategy and Action Plan (2017 – 2030) Mater Plan for Hazardous Waste Management in Myanmar – 2019
Pakistan	Pakistan Environmental Protection Act – 1997 National Environmental Policy – 2005 The hazardous substance rules of Pakistan – 2014
Philippines	Toxic Substances and Hazardous and Nuclear Waste Act of 1990 (RA 6969) Local Government Code of 1991 (RA 7160) Philippine Clean Air Act of 1999 (RA 8749) Philippine Clean Water Act of 2004 (RA 9275) National Environmental Awareness and Education Act of 2008 (RA 9512) Renewable Energy Act of 2008 (RA 9513) The Ecological Solid Waste Management Act of 2000 (RA 9003)
Rwanda	**Regulations:** 1. Law N° 04/2005 of 08/04/2005 on the protection, conservation and promotion of the environment in Rwanda 2. Law N° 39/2001 of 13 September 2001 establishing as utility regulator the Rwanda Utilities Regulatory Authority (RURA) 3. LAW N° 57/2008 OF 10/09/2008 prohibiting the manufacturing, importation, use and sale of polythene bags in Rwanda 4. Regulations Governing of Solid Wastes Recycling in Rwanda, April 2015 (RURA, 2015) 5. Governing Liquid Waste Collection and Transportation, November 10, 2016. No: 005/R/SANEWS/RURA/2016 (GoR, 2016) 6. Law Relating to the Prohibition of Manufacturing, Importation, Use and Sale of Polythene Bags in Rwanda, October 9, 2008. No: 57/2008 (Ministry of Natural Resources, 2008) 7. Ministerial Order Establishing Special Regulations Relating to Burying Toxic Wastes, July 1 2010 No: 006/16.01 8. Draft Regulations Governing E-waste Management in Rwanda, Republic of Rwanda, Rwanda Utilities Regulatory Authority, 2016 (RURA, 2016) **Strategies and Policies:** 1. Economic Development and Poverty Reduction Strategy – The Republic of Rwanda (MINECOFIN, 2007). 2. National Strategic Plan, and five-year operational plan, for the Management of Healthcare Waste (Ministry of Health, 2012) 3. National Industrial Policy (Ministry of Trade and Industry,2011) 4. National Sanitation Policy (MININFRA, 2016). 5. National Policy and Strategy for Water Supply and Sanitation, February 2010 (MININFRA, 2010) 6. National E-Waste Management Policy for Rwanda, September 2015 (Ministry of Natural Resources, 2015)
Serbia	Regulation on Waste Landfilling (No 36/2009 i 88/2010) – The Reclamation and Closure of Landfills Requirements Law on Waste Management ("RS Official Gazette", No. 36/09, 88/2010) – Collection of Waste Generation and Composition Data
South Africa	1989 – Environmental conservation Act (ECA) 1996 – RSA constitution (Act 108): Green paper environmental policy 1998 – National environmental management (NEMA) Act 1999 – First national waste management strategy (NWMS) 2003 – Flat bag and plastic bag regulations 2004 – National waste management strategy implementation (NWMSI) 2004 – Gauteng waste regulations 2005 – Glass memorandum of agreement 2006 – Draft policy on economic instruments (EIs) 2007 – Draft waste bill 2007 – RSA waste information system 2008 – Draft policy: waste incineration 2008 – Asbestos regulation 2009 – Waste tyre regulation 2009 – Policy on thermal waste treatment 2010 – Policy on EIs 2011 – By-law on waste model 2012 – Redisa tyre Industry waste management plan (IndWMP) 2012 – Tariff strategy for municipal waste 2012 – Waste information regulations 2012 – Second NWMS 2013 – Norms and standards on extraction and flaring of landfill gas 2013 – Norms and standards on scrapping motor vehicles 2013 – Waste classification and management regulations 2013 – Regulations on assessment of waste for landfilling 2013 – List of SWM activities regulation 2014 – Regulations on remediation of contaminated land 2015 – Regulation on the list of waste management activities 2015 – Regulations on admission of guilt fines 2015 – Residue deposits and stockpiles regulation 2016 – Standards of dangerous goods 2017 – Standards and norms on general waste bailing, screening, crushing, grinding, shredding and sorting

continued on following page

Table 3. Continued

Countries	Regulations and Policies
Sri Lanka	SLSI 1246 of 2003 – Sri Lanka standard for compost from municipal solid waste and agricultural waste SLS 1292 of 2006 – Code of practice for design and construction of biogas systems Gazette Extraordinary No. 1534/18 of 2008 – National environmental protection and quality regulations Management of scheduled waste No. 1627/9 National Environmental (Municipal solid waste) Regulation No 1 of 2009 – General rules on SWM discharge and collection of waste (prohibition of waste dumping at national highway and at any place other than places designated for such purpose by the Local Authority) Gazette Extraordinary No. 2034/36 of 2017 – Order to prohibit burn openly or cause to, allow or permit the open burning of refuse or other combustible matters inclusive of plastics
Thailand	**Municipal Solid Waste:** 1. Public Health Act, 1992 2. 3R Act (draft) **Hazardous Waste:** 1. National Solid Waste Management Master Plan (2016 - 2021) 2. National Master Plan on Cleaner Production and Cleaner Technology 3. Road Map on Waste and Hazardous Waste Management 2014 4. Hazardous Substance Act, 1992 5. Hazardous Substance Act, 2001
Turkey	General Health Law – 1930 Environmental Law – 1983 Regulation on Control of Medical Wastes – 1993 Regulation on Control of Packaging Wastes – 2004 Regulation on Control of Used Batteries and Accumulators – 2004 Regulation on Collection and Control of Wastes from Ships – 2004 Regulations on Control of Excavation Soils, Construction and Demolition Wastes – 2004 Regulation on Control of Waste Electrical and Electronic Equipment – 2012 Regulation on Control of Vegetabe Waste Oils – 2005 Regulation on Control of End-of-Life Tires – 2006 Regulation on PCBs and PCTs Control – 2007 Waste Oil Control Regulation – 2008 Regulation on the Control of End-of-Life Vehicles – 2009 Regulation on Control of Soil Pollution and Point Source Contaminated Sites – 2010 Regulation on Landfilling of Wastes – 2010 Regulation on Incineration of Wastes – 2010 Regulation on Mining Wastes – 2015 Regulation on Waste Management – 2015
Viet Nam	Law on Environmental Protection – 2014 Decree 19/2015/ND-CP on guidance to implementation of Law on Environmental Protection Decree 59/2007/ND-CP on solid waste management Decree 38/2015/ND-CP on waste and scrap management Decision 16/2015/QD-TTg on take-back and treatment of discarded products Circular 36/2015/TT-BTNMT on management of HW – 2015 Inter-ministerial circular 58/2015/BYT-BTNMT on medical waste management – 2015 Circular 08/2017/TT-BXD on management of construction waste – 2017 Decision 73/QD-TTg stipulating the list of allowed scraps for import Circular 41/2015/TT-BTNMT dated September 9, 2015 on environmental protection in scraps import National Strategy for Environment Protection Until 2020 and Vision Toward 2030 (Decision No. 1216/QD-TTg dated September 5, 2012 Revised National Strategy for Integrated Management of Solid Waste Until 2025, and Vision Toward 2050 (Decision 419/QD-TTg dated May 7, 2018)

when policy and regulations are in place, it has no effect in improving the waste management system since in most developing countries, they are not properly implemented or enforced. Such scenarios are observed in China and Malaysia where both countries have introduced the regulation pertaining to waste separation to improve the recycling potential in the country while aiming to divert and reduce waste from disposal options. Yet, not much improvement has been recorded in the two countries where waste collected for disposal is still highly commingled with recyclables. The system seems to fail due to the lack of participation from the general public, who are the main stakeholder for being the largest waste generator of any country.

Participation from the relevant stakeholders in waste sector is crucially important to ensure the success of the implemented policy and regulations. Participation from the stakeholders can be anticipated should the policy and regulations are made compulsory. As far as any waste management scheme in developing countries are concerned, the major failures are due to the absence of strict mandatory requirement from the stakeholders. To compare, stakeholders of waste management sector among the developed nations generally are more abiding and open to regulatory system implemented by their government. This probably has begun with strict regulation with appropriate incentive, or disincentive. Once the system is already in place and the proposed scheme has begun to give positive results, subsequent improvements would come as a voluntary participation.

Voluntary participations among stakeholders in many developing nations has been at a very low level. This is unsurprising since participation in voluntary actions particularly for the improvement of waste management system in a country may require change of habit which can be very difficult in most society. It is extremely challenging to convert human habit and altering the "comfort zone" of a society if it not made mandatory. The lack of interest in participating in changes and improvement of waste management scheme is applicable to both the general public, as well as, the industry. Being in their comfort zone would hinder any possibility to cause them to change, particularly if it caused additional inconveniences. Thus, voluntary participation is unlikely to be effective in developing nations for many years to come.

While incentives can create more interesting opportunities in waste management improvement, the option can only be considered as bonus to accessorize an existing effective system. Incentives alone may not be sufficient to alter the whole waste management system in a country. Thus, mandatory requirement would serve a better solution in ensuring the success of a waste management scheme in any developing country.

Furthermore, the most crucial steps towards an integrated waste management is an effective waste collection system. The following section elaborates the collection rate of selected developing countries in the world, and the challenges they face.

Collection Rate

Waste collection is the first and foremost step in realizing an effective waste management system. An appropriate waste collection system eliminates illegal waste disposal which can cause adverse environmental impact to the local water, soil and air quality. Waste collection is very crucial, particularly in densely populated area so that sanitation and hygiene are not jeopardized. Among the developing countries, higher collection rate is recorded mainly in the urban areas, while sub-urban and rural areas experience a very low waste collection activity or are completely absent. As a matter of a fact, the waste collection rate highly differs from one country to another. Figure 6 illustrates the waste collection percentage in selected developing countries in the world.

Figure 6. Percentage of waste collected in selected developing countries in the world

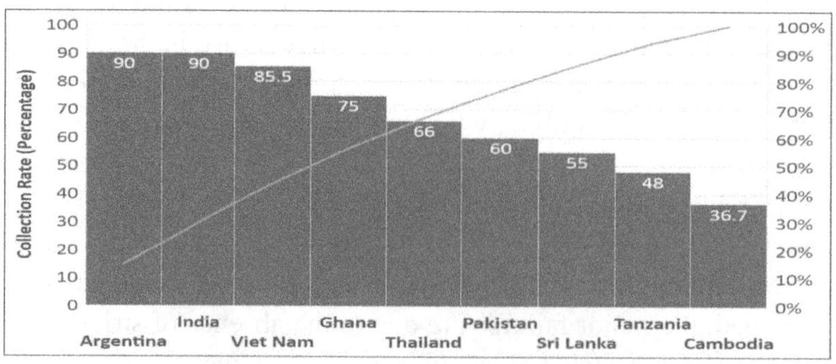

While some developing countries have high collection rate such as that in Argentina (90%) and Viet Nam (86%), many others have a much lower rate. The main obstacles in achieving a high collection rate among the developing countries are the lack of facilities and substantial financial constraints. Countries like Cambodia and Tanzania has waste collection rate of less than 40%. Nevertheless, a positive trend has been observed in majority of the developing countries that percentage of waste collected are increasing every year. This highlights the continuous improvement in the waste management system in developing nations. High percentage of waste collected is essential to ensure that the waste will be appropriately managed. The remaining balance of the waste which has not being collected by the authorized agency might indicate the improper waste management in the form of illegal dumping. Illegal dumping in developing countries is very rampant.

Illegal dumping are frequently reported in countries like Bangladesh, Pakistan, Cambodia, Nigeria and Rwanda.. This is due to the absence of proper collection system that many communities are forced to find alternative solution to dispose their waste. As a result, waste is dumped illegally, usually away from the local community to avoid accumulation of waste and other impacts such as horrid stench, and unsightly scene. Similarly, in above mentioned African countries, households often bury MSW in their backyard as well due to lack of waste collection. Additionally, illegal dumping is often accompanied by open burning as well in above mentioned countries. Aside from the absence of proper waste collection services, illegal dumping also occurred due to the ignorance and indifferent attitude of the waste generators. For example, while in Malaysia waste collection system covers more than 90% of the population, illegal dumping still occurs as waste collectors try to avoid paying the tipping fee for landfills.

Having a proper waste collection system enables a more efficient monitoring by the authority. This is even more crucial in the developing countries whereby data on waste management are very scarce. Data collection particularly on waste generation and its composition is important to create a database for future planning in the waste management sector, including the planning of waste disposal facilities.

Eliminating Open Dumping/ Burning

In order to achieve a sustainable integrated waste management, an appropriate waste disposal facility such as landfill is necessary. Unfortunately, many of the developing countries still dispose their waste into open dumps which has no precautionary measures, but only natural liners, such as clay layer.

The lack of proper landfill liners to prevent leachate from penetrating and contaminating the groundwater system poses huge environmental risks to human beings and environment. Therefore, open dumping needs to be avoided at all costs. Nevertheless, countries that have no expertise and/or funds may look to open dumping as establishing and managing a landfill can be very expensive and might require some technical expertise during its operation. Open dumping in developing countries normally involved an area which is dedicated by the local authority as dumping site. Without proper fence, perimeter drain, leachate accumulation pond, landfill gas treatment facilities and others, result in the migration of various contaminants such as leachate to surface and groundwater, and soil. Landfill gas is emitted directly to the atmosphere, contributing more greenhouse gases (GHGs) and impacting the global warming phenomenon. In fact, open dumps in developing countries has been identified as one major source of GHG emission from the waste sector. Figure 7 shows the percentage of waste sent to open dump for disposal in selected developing countries. It must be noted that only three country chapters have reported rate of open dumping and while other countries have reported practice of open dumping, the rate of open dumping in not available in those countries.

Figure 7. Percentage of waste disposed into open dumps in selected developing countries in the world

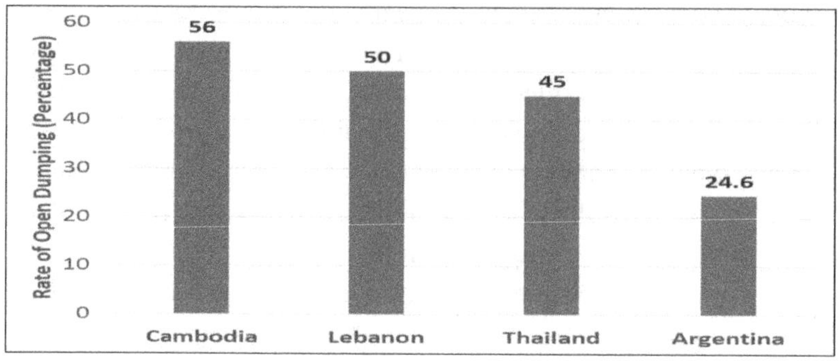

The dependency of open dumps by many developing countries need to be changed, which may require huge assistance from other countries. Some developing countries are undergoing a transition whereby open dumps are upgraded to controlled landfill in a fashion to reduce the risk of pollution from landfilling. Malaysia for example upgraded more than 100 dumps into controlled dumps in order to comply with Solid Waste and Public Cleansing Management Act 2007. Similarly, this is seen in other countries such as Viet Nam and Turkey. Figure 8 depicts the percentage of waste disposed into controlled landfill in Argentina, Turkey and Viet Nam (countries that have clearly reported landfilling rate).

Although open dumping is prevalent in majority of developing countries, several developing nations have moved towards establishing sanitary landfill as final solid waste disposal site. Countries like Turkey, South Africa have send more than 50% of their MSW to sanitary landfills. Similar trend is also noted among many other developing nations such that the percentage of waste sent for disposal into sanitary landfill has increased at a positive rate. Figure 9 shows the percentage of waste sent for disposal in sanitary landfills in selected developing countries in the world.

In majority of the developing countries, public generally have no regards as to how their household waste will be managed as long as it is collected and cleared from their bins. This is due to the fact that the public are ignorant of anything that does not affect them directly, particularly from the financial aspect. Survey on awareness among public in many developing countries showed that majority of the people have no serious regards on waste management and many more are not keen to be involved in such "unhygienic" activities. In fact, many have no awareness of what would happen to their waste once it is disposed of into their bin, while having absolutely no knowledge on the various option of waste disposal such as landfills, and incineration.

Figure 8. Percentage of waste send to controlled landfills in selected developing country in the world

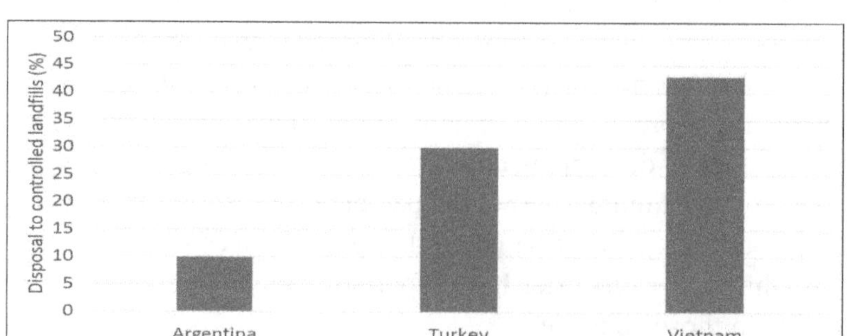

Figure 9. Percentage of waste disposed into sanitary landfills in selected developing countries of the world

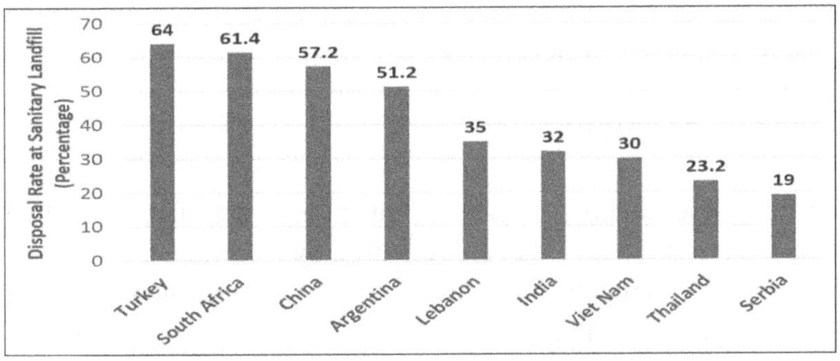

Reducing Landfilling

Since many other options and alternatives have been proposed and effectively implemented in developed nations, dependency on landfilling for disposal purpose needs to be significantly reduced. It has been proven that diverting waste away from landfilling can be an economical solution in the long run. The reduction of landfilling can be achieved by diverting waste towards resource circulation option, including waste-to-eneryg. The status of resource circulation has been discussed earlier in the chapter. As it is established that biological treatment of waste can be applied to organic component of the waste stream to generate value added products. In Bangladesh, huge amount of organic waste (85%) were channeled into composting stream to produce compost. The produced compost is distributed to farmers, to be utilized as organic fertilizer or soil conditioner. This practice has successfully diverted the organics from the waste stream while the volume of waste that need disposal has been significantly reduced.

Other options to divert the organic waste include the treatment via anaerobic digestion and conversion to animal feed. While anaerobic digestion produced biogas, which can be utilized to generate energy, conversion to animal feed allows the production of high-quality animal feed in high-protein forms of maggots and worms. This is gaining popularity in many developing nations particularly in Asia, as reported in Thailand, Malaysia and Indonesia.

Production of briquette from combustible component of the waste too has been gaining popularity. This is repeatedly reported in many developing countries in the African continent. Not only the conversion of waste into briquette enables the reduction of waste for disposal, it also provides source of energy to householders for cooking and heating purposes. In addition, the production and utilization of briquette has improved the environmental quality with the reduce dependency on forest resources namely woody trees as firewood. This practice is gaining popularity in Rwanda, Bangladesh, and India.

The conversion of waste to energy is also popular in China when the government began to encourage the establishment of incineration plants to divert waste from landfilling. As a result incineration plants deal with more volume of waste than those sent for landfilling. Waste is burned to generate heat energy to produce steam and turn turbine for electricity production. However, incineration is yet to gain popularity in other developing nations in the world such as Nigeria, India, Malaysia, and Thailand.

Regardless of the numerous options available to improve the waste management system in a country, the role of the government is crucial and essentially important. As reported in many literatures, many policy and regulations fail to fruit any improvement in waste management of developing country. This is mainly due to the lack of commitment from the government to execute the policy rigorously. The commitment of the local country is undeniably necessary to ensure the effectiveness of any stipulated and implement policy and regulation. Following section discusses the existing top-down approaches in selected developing countries in the world.

Increasing Awareness of Public

While the stakeholders are experimenting with compliance to the set rules and policy imposed on them, the government would need to strengthen this participation by instilling and creating the awareness and consciousness on the issues pertaining to waste management in the country. Creating awareness among public and the lack of it has been highlighted in almost all the chapters of the book. The public in most developing countries do not have information on waste management aspects, including waste management of complex and new waste streams such as WEEE, household hazardous waste and others.

Intensive campaigning is necessary to create awareness and instill the sense of responsibility among the stakeholders. It can be carried out at various level in order to produce effective results according to the target group. To compare between generations, many reports on environmental awareness from the developing countries highlighted that the younger generations are more conscious on the environmental issues as compared to their adults. This can be translated that campaigns conducted by many entities have budded some inspiration. However, it can also see some visible results in years to come. Therefore, in order to improve the current system, the main waste generators which are adults should be made aware of the issues and consequences of improper waste management and unsustainable waste disposal.

In general, the awareness of public in developing nations have been on positively increasing trend. Several years ago, the majority of the public in the developing nations were generally not aware of many environmental issues. However, this trend has evolved positively when many global issues particularly those related to environmental pollution and anthropogenic environmental disasters changed the perception of the world that it is not an option to live more sustainably, but a must for survival. Thus, more campaigns have been initiated by various agencies including the non-governmental agencies which relate impacts of improper waste management with the survival of humankind and the quality of the environment. To improve the awareness among the public can be very challenging especially among the adults. Yet, solution and effective approaches have been identified to overcome this and other challenges in improving towards an effective integrated waste management in developing countries. The following section particularizes the various initiates taken by developing countries in the world to overcome their challenges in moving towards a more sustainable waste management system.

Informal Sector

In majority of the developing countries, informal sector plays an invaluable role in the waste management from collection of waste to waste segregation and their contribution to the national recycling rates. However, it is also the stakeholder in waste management which is at the most risk. For instance, in Argentina, sometimes healthcare waste is mixed with MSW and waste pickers are then prone to health risks due to infectious needles, glass or other materials.

Informal sector in Ghana is also responsible for recovering recyclables from dumping sites and selling it to respective recycling companies in the country. Hence they play a significant role in overall national recycling rate in the absence of top-down approaches. Similar role is also played by informal sector in Myanmar, Cambodia, India, Bangladesh, Rwanda, Serbia, Thailand, Viet Nam and Pakistan who are solely or partly responsible for recycling and recovery. A staggering estimate of 90,000 informal waste collectors are reported in South Africa who also play their part in recycling of waste. Similarly, waste pickers in Argentina also contribute to waste recycling by recovering recyclables. Whereas, these waste pickers also constitute around 30% of workers in composting plant in Bella Vista, Argentina.

Perhaps, Phillipines is one of the fewest developing countries that has taken iniative to incorporate informal sector in their national solid waste management system by launching "National Framework Plan for the Informal Sector in Solid Waste Management".

The formalization of waste pickers or informal sector has been recommended in all of the country chapters that have reported the role of informal sector in waste management of the country.

DRIVERS OF SUSTAINABLE WASTE MANAGEMENT IN SELECTED DEVELOPING COUNTRIES

As established throughout this chapter, it is evident that the institutional drivers are critical in creating a sustainable waste management system in a country. Especially among the developing countries, the existence of appropriate rules and policy enable the government to effectively regulate the system. Also, rules and regulations are necessary to ensure that stakeholders abide by the requirements stipulated in the acts. It is also necessary for regulating and monitoring purposes. Since waste management sector is a sensitive area that involves the sanitation and hygiene, no mismanagement can be allowed to happen. Otherwise, the impact can be really disastrous to the environment and human health.

The regulation and monitoring of stakeholders and players in the waste management sector can be easily managed with the provision and licensing procedures. This awarding and revoking of license can be an effective tool to ensure competitiveness among all players. This incentive and disincentive approaches are also valuable to promote improvement in technologies in waste management system of a country.

The institutional drivers are found to be an effective tool in developing countries for managing the waste sector to a certain extent. This is because the existence of various regulations and policies on waste management provide basic guidelines on conforming to the stipulated requirements. Thus, to a certain extent, it aids towards the betterment and a more sustainable waste management. This is seen in several countries including China, which had elaborated the strict regulation on waste disposal options and emission from waste incinerators, while Malaysia sets certain key performance index for the waste managers to achieve in ensuring improvement in the waste management system.

Beside the institutional drivers, economic drivers are also crucial in improving the waste management in developing countries towards a more sustainable system. As observed in developed nations of the European Union, EU Landfill Directive strictly imposes the promotion of sustainable waste management by its member countries, which has led to drastic changes in the standard of waste management in these developed nations. Such strong directives together with the incentives and disincentives have pushed the members in the developing nations to strive towards the improvement of their waste management. Republic of Serbia for example has taken numerous initiatives by implementing programs to improve the waste management towards a more sustainable system.

Incentives and disincentive can be an excellent economic driver towards a more sustainable waste management among the developing countries, as it promotes the active participations of the main players and stakeholders to continuously improve their system. Awarding system also aids the improvement of waste management in developing nations by encouraging good practices in waste management. On the other hand, tax and levy can also be utilized as an economic driver towards the development of a sustainable waste management among the developing countries.

In the promotion of 3R, economic drivers are the most effective tools for the developing countries. This is because resource recovery can be carried out more effectively if the market price of certain materials is profitable. This is seen in many countries in Asia such as Malaysia where recycling practices fluctuate with the changes in the market price. Therefore, with careful studies, developing nations may be able to regulate the price of the recovered resources with some governmental intervention, such as having stock piling and other system.

Nevertheless, the improvement of a waste management system among the developing nations needs a strong social drive from the main players and stakeholders in the country. Human driver is necessary since human is the main player in waste generation, and having the right attitudes would serve a more prominent purpose in waste management. Human drivers relate to the social drive in order to obtain a better quality of life. Thus, participation towards the improvement of waste management system would come naturally as the need to ensure less risk and peril from improper waste management system. These human drivers have been seen to emerge from many non-governmental movements in developing countries like Malaysia, India, the Philippines and Indonesia, that compels towards improvement of the waste management system in order to reduce risk to human life.

Human drivers are also closely related to the environmental drivers. Yet, the depth of the two differ very much. While human drivers basically lead to improve the waste management system in order to reduce and hinder risk and hazard from improper management to human, environmental driver meant to serve the conservation and preservation of the environment, for the sake of environmental protection.

Environmental drivers require the strong awareness and consciousness to protect the environment from contamination and pollution due to inappropriate waste management system. Thus, having the environmental drivers would lead a country towards options that can really benefit the environment, regardless of the impacts to other aspect, economy, institutional, and human drivers. Environmental drivers towards a sustainable waste management in developing countries are very scarce. Yet, if governments of the developing nation are able to contemplate the environmental drive into the society, improvement towards sustainable waste management can be easily and efficiently achieved.

SIMILARITIES AND DIFFERENCES IN WASTE MANAGEMENT AMONG THE DEVELOPING COUNTRIES

Waste management system practiced among the developing nations can be grouped into two major systems. Table 4 summarizes the main waste management system commonly observed in developing countries in the world.

While there are two main groups in the practiced waste management among the developing countries, many of these countries share similar issues and challenges in the implementation of an integrated waste management. Table 5 summarizes the issues and challenges commonly shared by the developing countries in the implementation of an effective waste management system.

Table 4 Practices in waste management system among the developing countries in the world

Practices	System 1	System 2
Percentage of waste collection	Less than 40%	More than 40%
Waste separation	Not available	Occurs at waste treatment facility (transfer station, landfill, composting plant etc.)
Resource recovery	Achieve higher rate due to scavenging activities by informal sector	Limited to the separation activities at selected facilities
Contribution of informal sector in 3R	High	Low
Final waste disposal sites	Majorly open dumps with few controlled landfills	Majorly controlled landfills and few sanitary landfills
Illegal dumping	Rampant	Less frequent due to enforcement
State of controlled landfills	Minimum protection on emissions from landfill	Sufficient protection on emissions from landfill
Waste treatment options	Diversion of organic component for biological treatment	Diversion of organics and combustibles for value added products and waste to energy, respectively.

Table 5. Common issues and challenges shared by the developing countries in the world

Issue	Challenges
Policy and regulations on waste management	i) Absence of effective policy and regulations ii) Policy and regulations are present but not implemented or properly enforced. iii) Lack of enforcement by the local authority iv) Stringent regulation is not in place
Financial constraint in waste management sector	i) Insufficient fund to invest in new technology ii) Lack of skilled manpower and experts iii) Limited funding to promote environmental awareness among the stakeholders
Public participation	i) Low level of participation in activities to promote good waste management practices. ii) Low awareness and consciousness on waste management issues and challenges iii) Indifferent attitude among the general public. iv) Lack of interest among public in waste management issue, which is considered as 'low-class' issue
Involvement of private sector	i) Low level due to the low awareness level among the industrial players ii) Very lacking due to lack of incentive by the government iii) Occur at the minimal level if driven by institutional factor e.g. laws and regulations, but will occur extensively due to monetary drive rather than environmental drive
Informal sector involvement	i) Pose serious risk to the health and well-being of the individual involved in waste picking activities ii) Prohibiting the involvement of this sector will lessen the recycling rate in a country iii) Occur due to the dependency on monetary value generated from the recovered resources
Technology and skill	i) Insufficient fund to invest in new technology ii) Lack of skilled manpower and experts iii) Lack of financial support by local government to invest in unfamiliar practices iv) Adaption of technologies from developed country always failed to accommodate the need of the local condition v) Adopted technology failed to address the significant differences in local waste management sector
NIMBY syndrome	i) Highly embedded among the general public ii) Resistance to any changes in technology if they will be affected by the changed system. iii) Difficulties to local government and waste managers to evolve to an improved technology. iv) Refusal to accept changes if they are directly affected by it.

RECOMMENDATIONS

Institutional initiative is very essential in ensuring the establishment of an effective and sustainable waste management system in the developing countries. Thus, it is imperative that institutional foundation can be consolidated to assist the formulation of effective and practical legislations, regulations and policies. Similarly, in establishing standard and strategies to improve the waste management system, viable and reasonable approaches can be proposed to serve as guidance towards a more sustainable and improved options.

Relevant stakeholders and main players in the waste sectors must have clear roles and responsibilities, as well as explicit aims and objectives with the best approaches to improve the waste management system. Key performance criteria and indexes must be made available as to guide the relevant players in selecting the technologies to be implemented in the waste management sector to ensure efficiency and prevent pollution. Capacity building is crucial to ensure the sustainability of a chosen technology. Thus, it is imperative that development of the needed skills is in tandem with the application of the technology.

Reducing gaps between communities of different socio-economic groups is also essential in order to make the improvement of waste management system a national agenda which will benefit the country as a whole. This can be addressed by ensuring each community receives the adequate and appropriate services and technologies which suit their local needs. While working on reducing existing gaps, financial stability needs to be looked into as to ensure sustainability of the sector. This can be achieved by promoting the participation of private sector as main players in the waste management system of the country.

By addressing all the areas of concern, the implementation of a sustainable waste management system can be realized in majority of the developing nation. Though it might seem unrealistic to implement and address all the areas simultaneously, developing countries can begin by probing one issue at a time and invest in high priority ones first, and followed by the rest with decreasing priority. The strongest factor may be prioritized as this may lead to the ease of implementation and improvement of the other areas. Yet, this may not be applicable to all countries. Thus, this necessitates wise decision making by the authority. In the case of countries that have an obstacle or major weakness that hinder the implementation of effective waste management, perhaps elimination of this challenge must be prioritized, regardless of the sacrifice needed. This will ensure that once the obstacles are removed, other improvements can be implemented smoothly and efficiently. In conclusion, regardless of the situations faced by the developing nations in the world, there are rooms for improvement. Therefore, bold and determined steps need to be taken to allow the improvement towards a sustainable waste management for the betterment of environment and mankind.

REFERENCES

Islam, K. M. N. (2017). Greenhouse gas footprint and the carbon flow associated with different solid waste management strategy for urban metabolism in Bangladesh. *The Science of the Total Environment*, *580*, 755–769. doi:10.1016/j.scitotenv.2016.12.022 PMID:28024747

Kaza, S., Yao, L., Bhada-Tata, P., & Woerden, F. V. (2018). *What a waste 2.0: A global snapshot of solid waste management to 2050. Urban Development Series*. Washington, DC: World Bank. doi:10.1596/978-1-4648-1329-0

Pauze. (2016). *Integrated Solid Waste Management: Challenge and Future. NEHAP Conference 2016.* 26 September 2016. Putrajaya International Convention Centre. Retrieved from http://nehapmalaysia.moh.gov.my/wp-content/uploads/2016/03/Paper-2-Solid-Waste.pdf

Shams, S., Sahu, J. N., Rahman, S. S., & Ahsan, A. (2017). Sustainable waste management policy in Bangladesh for reduction of greenhouse gases. *Sustainable Cities and Society*, *33*, 18–26. doi:10.1016/j.scs.2017.05.008

Compilation of References

Abdel-Shafy, H., & Mansour, M. (2018). Solid waste issue: Sources, composition, disposal, recycling and valorisation. *Egyptian Journal of Petroleum*, *27*(4), 1275–1290. doi:10.1016/j.ejpe.2018.07.003

Abdul, G. W. (2014). *USDA Global Agricultural Information Network (GAIN) Report: Malaysia Grain and Feed Annual 2014. Foreign Agricultural Service, MY400.* Retrieved from http://www.ers.usda.gov/media/978994/rcs-121-01.pdf

Abdulla, F., Qdais, H. A., & Rabi, A. (2008). Site Investigation on Medical Waste Management Practices In Northern Jordan. *Waste Management (New York, N.Y.)*, *28*(2), 450–458. doi:10.1016/j.wasman.2007.02.035 PMID:17507209

Abedin, M. A., & Jahiruddin, M. (2015). Waste generation and management in Bangladesh: An overview. *Asian Journal of Medical and Biological Research*, *1*(1), 114–120. doi:10.3329/ajmbr.v1i1.25507

Abor, P. A. (2007). *Medical Waste Management at Tygerberg Hospital in the Western Cape, South Africa* (Unpublished doctoral dissertation). Cape Peninsular University of Technology.

Abor, P. A., & Anton, B. (2008). Medical Waste Management Practices in a Southern African Hospital. *International Journal of Health Care Quality Assurance*, *4*(4), 356–364. doi:10.1108/09526860810880153 PMID:18785461

Abushammala, M. F. M., Basri, N. E. A., Basri, H., El-Shafie, A. H., & Kadhum, A. A. H. (2011). Regional landfills methane emission inventory in Malaysia. *Waste Management & Research: The Journal of the International Solid Wastes and Public Cleansing Association, ISWA*, *29*(8), 863–873. doi:10.1177/0734242X10382064 PMID:20858637

Achankeng, E. (2003). *Globalization, Urbanization and Municipal Solid Waste Management in Africa*. African Studies Association of Australasia and the Pacific. Retrieved from http://www.inclusivecities.org/pdfs/achankeng Globalization Urbanization and MSW Mgmt in Africa.pdf

Acosta, V., Paul, J., & Hanuschke, K. (2013). Eco-center: Integrated solid waste management facilities with sanitary landfill and resource recovery technologies. Version 1. Manila: Deutsche Gesellchaft fur Internationale Zusammenarbeit (GIZ) GmbH.

ADB. (2017). *Urban Development and Water Sector Assessment, Strategy and Road Map*. Manila, Philippines: ADB.

ADB. (2018). *Myanmar Economy*. Retrieved from https://www.adb.org/countries/myanmar/economy

Afroz, R., Hanaki, K., & Hasegawa-Kurisu, K. (2009). Willingness to pay for waste management improvement in Dhaka city, Bangladesh. *Journal of Environmental Management*, *90*(1), 492–503. doi:10.1016/j.jenvman.2007.12.012 PMID:18242819

Agamuthu, P. (2001). *Solid Waste: Principles and Management: With Malaysian Case Studies*. Kuala Lumpur: Institute of Biological Sciences, University of Malaya Press.

Agamuthu, P., & Tanaka, M. (Eds.). (2014). *Municipal Solid Waste Management in Asia and the Pacific Islands*. Singapore: Springer.

Agamuthu, P., & Victor, D. (2011). Policy trends of extended producer responsibility in Malaysia. *Waste Management & Research*, *29*(9), 945–953. doi:10.1177/0734242X11413332 PMID:21771873

Aguinaldo, E. (2009). *Philippines solid waste management*. In Regional 3R Forum in Asia, Tokyo City.

Ahmed, S. A., & Ali, S. M. (2006). People as partners: Facilitating people's participation in public-private partnerships for solid waste management. *Habitat International*, *30*(4), 781–796. doi:10.1016/j.habitatint.2005.09.004

Ahsan, A., Daud, N. N. N., El-Sergany, M. M., Alamgir, M., Shams, S., & Rowshon, M. K. (2014). Assessment of municipal solid waste management system in a developing country. *Chinese Journal of Engineering*, 1–11. Retrieved from /561935 doi:10.1155/2014

Ahsan, A., Alamgir, M., El-Sergany, M. M., Shams, S., Rowshon, M. K., & Daud, N. N. N. (2014). Assessment of Municipal Solid Waste Management System in a Developing Country. *Chinese Journal of Engineering*, *14*, 1–11. doi:10.1155/2014/561935

Ahsan, A., Alamgir, M., Imteaz, M., Daud, N. N., & Islam, R. (2012). Role of NGOs and CBOs in waste management. *Iranian Journal of Public Health*, *41*(6), 27. PMID:23113191

Aich, A., & Ghosh, S. K. (2016). Organic fraction of municipal solid waste – a valuable source of green energy in India. *International Journal of Energy Sector Management*, *10*(4), 526–545. doi:10.1108/IJESM-06-2015-0001

Akese, G. A. (2014). *Price realization for electronic waste (e-waste) in Accra*. Memorial University of Newfoundland.

Akese, G. A., & Little, P. C. (2018). Electronic Waste and the Environmental Justice Challenge in Agbogbloshie. *Environmental Justice*, *11*(2), 77–83. doi:10.1089/env.2017.0039

Akhund, M. A., Memon, N. A., Ali, T. H., Memon, A. H., & Imad, H. U. (2018). A Comprehensive Review on Waste Generating Attributes: Way Forward for Pakistan's Construction Industry. *Engineering Science & Technology International Research Journal*, *2*(3), 1–7.

Al Mamun, M. R. (2007). *Utilization pattern of biomass for rural energy supply in Bangladesh*. Bangladesh Agricultural University.

Al Masha'an, M., & Mahrous, F. (1999). Environmental Strategy for Managing the Kuwait's Solid Wastes until 2020. *Proceedings of Arab and Middle East International Conference and Fair on Solid Waste Management*.

Alamgir, M., Bidlingmaier, W., & Cossu, R. (2012). Successful waste management strategies in developing countries require meaningful involvement of the concerned stakeholders. *Waste Management (New York, N.Y.)*, *32*(11), 2007–2008. doi:10.1016/j.wasman.2012.07.028 PMID:22959155

Alam, P., & Ahmade, K. (2013). Impact of solid waste on health and the environment. *International Journal of Sustainable Development and Green Economics*, *2*, I-1.

Ali, A. (2010). Wasting time on solid waste in developing countries. *Waste Management (New York, N.Y.)*, *30*(8-9), 1437–1438. doi:10.1016/j.wasman.2010.04.025 PMID:20493678

Amin, A. N. (2017). *State of the 3Rs in Asia and the Pacific country chapter: The people's republic of Bangladesh*. doi:10.1111/eth.12371

Amoyaw-Osei, Y., Agyekum, O. O., Pwamang, J. A., Mueller, E., Fasko, R., & Schluep, M. (2011). Ghana e-waste country assessment. *SBC e-waste Africa Project.* Available online: http://www.basel.int/portals/4/basel%20convention/docs/ewaste/e-wasteassessmentghana.pdf

Amuzu, D. (2018). Environmental injustice of informal e-waste recycling in Agbogbloshie-Accra: Urban political ecology perspective. *Local Environment*, *23*(6), 603–618. doi:10.1080/13549839.2018.1456515

Ansar, M., Ani, F., & Kader, A. (2014). The Energy Potential of Municipal Solid Waste for Power Generation in Indonesia. *Jurnal Mekanikal*, *37*, 42–54.

Anton, E., & Raine, N. (2016). The South African renewable energy independent power producer procurement programme: A review and lessons learned. *Journal of Energy in Southern Africa*, *27*(4), 1–14. doi:10.17159/2413-3051/2016/v27i4a1483

Ariyawansha, R. T. K., Chandrasena, A. S. H., Senevirathne, S. A. D. N., & Basnayake, B. F. A. (2010b). Development of a sustainable Environmental Preservation Centre (EPC) at Nawalapitiya for Urban Solid Waste Management. *Proceedings of International Conference on Sustainable Build Environment*, 181-188.

Ariyawansha, R. T. K., Jayakody, L. N., Gunarathne, H. A. Y. R., Senevirathne, S. A. D. N., Kuruparan, P., Chandrasekara, W. W. C. K., ... Basnayake, B. F. A. (2010a). Estimation of quantum of waste generation and solid waste characterization at Kalmunai Municipal Council (KMC) and Sammanthurai Pradeshiya Sabha (SPS). In *Proceedings of third International Symposium.* Sabaragamuwa University of Sri Lanka.

Ariyawansha, R. T. K., Senevirathne, S. A. D. N., & Basnayake, B. F. A. (2014). Potential of Converting Paddy Straw to Bio-char and Electricity in Sri Lanka. *Proceedings of 11th Asia Pacific Roundtable on Sustainable Consumption and Production (APRSCP)*, 207-222.

Arshad, H., Qasim, M., Thaheem, M. J., & Gabriel, H. F. (2017). Quantification of Material Wastage in Construction Industry of Pakistan: An Analytical Relationship between Building Types and Waste Generation. *Journal of Construction in Developing Countries*, *22*(2), 19–34. doi:10.21315/jcdc2017.22.2.2

Arunkumar, G., Bhanu Priya, G., & Kumar, R. (2016). Smart Garbage Collecting Bin for Municipal Solid Waste. *International Journal of Modern Trends in Engineering and Science*, *3*(3), 9–18.

Asante, B., Yanful, E., & Yaokumah, B. (2014). Healthcare Waste Management; Its Impact: A Case Study Of The Greater Accra Region, Ghana. *International Journal of Scientific & Technology Research, 3*(3).

Asase, M., Yanful, E. K., Mensah, M., Stanford, J., & Amponsah, S. (2009). Comparison of municipal solid waste management systems in Canada and Ghana: A case study of the cities of London, Ontario, and Kumasi, Ghana. *Waste Management (New York, N.Y.)*, *29*(10), 2779–2786. doi:10.1016/j.wasman.2009.06.019 PMID:19615883

Asian Development Bank. (2018). *Asian Development Bank member fact sheet: Bangladesh.* Retrieved from https://www.adb.org/sites/default/files/publication/27753/ban-2017. pdf

Asian Productivity Organization (APO). (2007). Solid waste management: Issues and challenges in Asia. Tokyo: Author.

Asumadu-Sarkodie, S., Owusu, P. A., & Rufangura, P. (2015). Impact analysis of flood in Accra, Ghana. *Advances in Applied Science Research*, *6*(9), 53–78.

Atiemo, S., Faabeluon, L., Manhart, A., Nyaaba, L., & Schleicher, T. (2018). *Baseline Assessment of E-Waste Management in Ghana.* Academic Press.

Atiemo, S. M., Ofosu, F. G., Aboh, I., & Kuranchie-Mensah, H. (2012). Assessing the heavy metals contamination of surface dust from waste electrical and electronic equipment (e-waste) recycling site in Accra, Ghana. *Res J Environ Earth Sci*, *4*(5), 605–611.

Atienza, V. (2008). Breakthroughs in solid waste management: Lessons from selected municipality and barangay in the Philippines. *Asian Review of Public Administration*, *XX*(1-2), 82–98.

Atienza, V. (2011). Review of waste management system in Philippines: Initiatives to promote waste segregation and recycling through good governance. In *Economic Integration and Recycling in Asia: An Interim Report*. Academic Press.

Atienza, V., Sison, E., & Cardenas, L. (2011). *Review of the Philippines' waste segregation and collection system and the trading of recyclables. Joint Research Program Series 158*. Institute of Developing Economies, Japan External Trade Organization.

Awuah, H. K. (2017). *Waste reduction strategies on construction projects: the case of Accra metropolis*. Academic Press.

Ayoub, G., Acra, A., Abdallah, R., & Merhebi, F. (1994). *Fundamental Aspects of Municipal Refuse Generated in Beirut and Tripoli. Technical Report*. Beirut: Department of Civil and Environmental Engineering, American University of Beirut.

Ayoub, G., Acra, A., Abdallah, R., & Merhebi, F. (1996). *Fundamental Aspects of Municipal Refuse Generated in Beirut-Phase II. Technical Report*. Beirut: Department of Civil and Environmental Engineering, American University of Beirut.

Azizin, S. M. (2010). *Isolation of hemicellulose from sugarcane bagasse by alkaline extraction*. Faculty of Applied Sciences, Universiti Teknologi MARA. Retrieved from http://eprints.uitm.edu.my/2366/

Bahafun, K. A. (2007). *Analysis of renewable energy potential in Malaysia*. Postgraduate Coursework Student Faculty of Electrical Engineering and Telecommunications (Eet), University of New South Wales.

Baker, N. (2016). *Myanmar's air pollution among the worst in the world*. WHO. Retrieved from http://www.mmtimes.com/index.php/national-news/22840-m

Baldé, C. P., Wang, F., Kuehr, R., & Huisman, J. (2015). The global e-waste monitor-2014. United Nations University, IAS – SCYCLE, Bonn, Germany.

Baldé, C. P., Forti, V., Gray, V., Kuehr, R., & Stegmann, P. (2017). *The Global E-waste Monitor – 2017, United Nations University (UNU), International Telecommunication Union (ITU) & International Solid Waste Association*. Bonn: ISWA.

Balde, C. P., Forti, V., Gray, V., Kuehr, R., & Stegmann, P. (2017). *The global e-waste monitor 2017: Quantities, flows and resources*. United Nations University, International Telecommunication Union.

Balde, C. P., Wang, F., Kuehr, R., & Huisman, J. (2015). *The global e-waste monitor – 2014, United Nations University*. IAS – SCYCLE.

Baldé, C., Forti, V., Gray, V., Kuehr, R., & Stegmann, P. (2017). *The global e-waste monitor – 2017. United Nations University (UNU), International Telecommunication Union (ITU) & International Solid Waste Association*. Bonn, Geneva, Vienna: ISWA.

Bangladesh Bureau of Statistics (BBS). (2010). Compendium of environment statistics of Bangladesh 2009. Dhaka: Bangladesh: Bureau of Statistics (BBS), Ministry of Planning, Government of Bangladesh.

Bangladesh Bureau of Statistics (BBS). (2017). *Bangladesh statistics 2017*. Statistics and Informatics Division (SID), Ministry of Planning. Dhaka: Bangladesh. Retrieved from http://bbs.portal.gov.bd/sites/default/files/files/bbs.portal.gov.bd/ page/a1d32f13_8553_ 44f1_92e6_8ff80a4ff82e/Bangladesh%20%20Statistics-2017.pdf

Bangladesh Bureau of Statistics (BBS). (2018a). *Bangladesh economic review 2018*. Finance Division, Ministry of Finance, Government of the People's Republic of Bangladesh. Retrieved from https://mof.portal.gov.bd/sites/default/files/files/mof.portal.gov.bd/ page/e8bc0eaa_463d_4cf9_b3be_26ab70a32a47/01.%20BER-2018_Cover%20page.pdf

Bangladesh Bureau of Statistics (BBS). (2018b). Statistical pocket book, Bangladesh 2017. Dhaka: Bangladesh Bureau of Statistics (BBS), Ministry of Planning, Government of Bangladesh.

Barker, J.C., & Walls, F.R. (2011). *North Carolina agricultural chemicals manual: livestock manure production rates and nutrient content December 2001*. Academic Press.

Bartone C. (1988). The Value in Wastes. *Decade Watch,* 3-4.

Barton, F. M. A. (1979). *Resource recovery and recycling*. New York, NY: John Wiley and sons.

Basnayake, B. F. A. (2011). *Composite clay-polyethylene biofilter liner and cover systems for enhancing landfill bioreactors and other bioremediation systems*. Sri Lanka Patent No 15676.

Basnayake, B. F. A., & Visvanathan, C. (2014). Solid waste management in Sri Lanka. In Municipal solid waste management in Asia and the Pacific Islands (pp. 299–316). Singapore: Springer Verlag. doi:10.1007/978-981-4451-73-4_15

Basnayake, B. F. A., Ariyawansha, R. T. K., Senevirathne, S. A. D. N., & Chandrasena, A. S. H. (2015). Valorization of municipal solid waste in Sri Lanka. *Economic Review (The Quarterly Economic Journal of People's Bank), 42*(2), 9-13.

Basnayake, B. F. A., Ariyawansha, R. T. K., Karunarathna, A. K., & Chandrasena, A. S. H. (2013). Influence and counter measures for combating monetary inflation on waste recycling industry in Sri Lanka: Case study on A4 paper. *Proceedings of the International Symposium on Agriculture and Environment at International Symposium on Agriculture and Environment University of Ruhuna*, 87-89.

Basnayake, B. F. A., Chiemchaisri, C., & Mowjood, M. I. M. (2005). Solid wastes arise from the Asian tsunami disaster and their rehabilitation activities: case study of affected coastal belts in Sri Lanka and Thailand. *Proceedings of the Tenth International Waste Management and Landfill Symposium.*

Basnayake, B. F. A., Popuri, S., Visvanathan, C., Jayatilake, A., Weerasoori, I., & Ariyawansha, R. T. K. (2018). Concerted initiative for planned management of municipal solid waste in target provinces in Sri Lanka. *Journal of Material Cycles and Waste Management.* doi:10.100710163-018-0815-5

Basturk, A. (1979). *A Research Model on Solid Wastes and Its Application for Istanbul.* Academy Associate Professor Thesis.

Basturk, A. (1980). *Remedies for the Elimination of Solid Waste Hazards in Istanbul.* TUBITAK, Environmental Research Group, Project No: CAG-11.

Basturk, A. (1994). *Solid Waste Course Notes.* İstanbul, Turkey: Yildiz Technical University, Environmental Engineering Department.

Batinić, B., Vujić, G., Zivancev, M., Ubavin, D., & Stanisavljević, N. (2015). Knowledge of the MSW quantity and composition as a basic prerequisite for the successful implementation of the EU Directives. *Proceedings of the Symposium "The Role of Communication in Waste Management".*

Batool, S. A., Chaudhry, N., & Majeed, K. (2008). Economic potential of recycling business in Lahore, Pakistan. *Waste Management, 28*(2), 294-298.

Batteries Rules. (2001). *India.* Retrieved from http://www.pccdaman.info/pdf/Batteries_Rules.pdf

Bazimenyera, J. D., Qiang, F., & Karangwa, A. (2012). Solid waste management in Kigali City, Rwanda. *East African Journal of Science and Technology, 1*, 46–58.

BBC Monitoring. (2019). *Bangladesh country profile.* Retrieved from https://www.bbc.com/news/world-south-asia-12650940

Beigl, P., Lebersorger, S., & Salhofer, S. (2008). Modelling municipal solid waste generation: A review. *Waste Management (New York, N.Y.)*, *28*(1), 200–214. doi:10.1016/j.wasman.2006.12.011 PMID:17336051

Bello, I., Ismail, M., & Kabbashi, N. (2016). Solid waste management in Africa: A review. *International Journal of Waste Resources*, *2*, 216–221.

Bell, R. G., & Russell, C. (2018). Environmental Policy for Developing Countries. In G. E. Helfand & P. Berck (Eds.), *The Theory and Practice of Command and Control in Environmental Policy* (1st ed.; pp. 239–246). By P. Berck. doi:10.4324/9781315197296-10

Bhuiyan, S. H. (2010). A crisis in governance: Urban solid waste management in Bangladesh. *Habitat International, 34*(1), 125–133. Retrieved from . 002 doi:10.1016/j.habitatint.2009.08

BIGD. (2015). *The state of cities: Solid waste management of Dhaka city–towards decentralized governance*. Dhaka, Bangladesh: BRAC Institute of Governance and Development, BRAC University.

Bilgili, M. S. (2006). *Effect of leachate recirculation on degradation of muicipal solid wastes in aerobic and anaerobic landfills* (Ph.D Thesis). Yildiz Technical University, İstanbul, Turkey.

Bingham Centre for the Rule of Law. (2013). *Constitutional Awareness Myanmar*. Retrieved from http://binghamcentre.biicl.org/binghamcentre/projects/burmamyanmar

Biomass-SP. (2014). *Overview of biomass industry in Malaysia. EU-Malaysia Biomass Sustainable Production Initiative*. Retrieved from http://biomass-sp.net/about/overview/

Birpinar, M.E., Bilgili, M.S., & Erdogan, T. (2009). Medical waste management in Turkey: A case study of Istanbul. *Waste Management, 29*, 445-448.

Birpinar, M. E., Bilgili, M. S., & Erdorgan, T. (2008). Medical waste management in Turkey: A case study of Istanbul. *Waste Management (New York, N.Y.)*, *29*(1), 445–448. doi:10.1016/j.wasman.2008.03.015 PMID:18550354

Biubwa, A., Norkhadijah, S., Ismail, S., & Rasdi, I. (2014, October). Municipal solid waste management of Zanzibar: Current practice, the challenges and the future. *International Journal of Current Research and Academic Review*, *1*, 5–19.

Blaauw, P. (2017). Informal employment in South Africa: Still missing pieces in the vulnerability puzzle. *Southern African Business Review*, *21*, 339–360.

BMA. (2014). *Bangkok State of the Environment 2013-2014*. Department of Environment, Bangkok Metropolitan Administration.

BMA. (2016). Bangkok State of Environmental 2015-2016. Department of Environment, Bangkok Metropolitan Administration.

BMWM Rules. (2016). *India*. Retrieved from http://www.indiaenvironmentportal.org.in/files/file/BMW%20Rules,%20 2016.pdf

Boadi, K. O., & Kuitunen, M. (2003). Municipal solid waste management in the Accra Metropolitan Area, Ghana. *The Environmentalist*, *23*(3), 211–218. doi:10.1023/B:ENVR.0000017283.09117.20

Boadi, K. O., & Kuitunen, M. (2005). Environmental and health impacts of household solid waste handling and disposal practices in third world cities: The case of the Accra Metropolitan Area, Ghana. *Journal of Environmental Health*, *68*(4). PMID:16334095

Bogner, J., Abdelrafie Ahmed, M., Diaz, C., Faaij, A., Gao, Q., Hashimoto, S., Pipatti, R., ... Zhang, T. (2007). Waste Management, In Climate Change 2007: Mitigation. Contribution of Working Group III to the Fourth Assessment Report of the Intergovernmental Panel on Climate Change. Cambridge University Press.

BOT. (2018). *Statistical data EC_EI_027 Thailand's macro economic indicators*. Retrieved March 10, 2019, from http://www2.bot.or.th/statistics/BOTWEBSTAT.aspx?reportID=409&language=ENG

Breeze, R. (2012). Municipal Solid Waste Management in Dar-es-Salaam. A report prepared for the World Bank.

British Business Group Vietnam (BBGV). (2018, May). *Renewable energy: Waste-to-energy*. Retrieved from https://bbgv.org/wp-content/uploads/2018/05/Waste-to-energy.pdf

Brown, C., Milke, M., & Seville, E. (2011). Disaster waste management: A review article. *Waste Management (New York, N.Y.)*, *31*(6), 1085–1098. doi:10.1016/j.wasman.2011.01.027 PMID:21334871

Bual, H. (2018). *Waste Management*. Retrieved April 12, 2019, from https://www.khmertimeskh.com/548828/waste-management/

Buffalo City News. (2018). *Partnering for sustainable waste management*. Retrieved 8 July 2019 from: https://infra-structurenews.co.za/2018/06/27/partnering-for-sustainable-waste-management/

Bundhoo, Z. M. (2018). Solid waste management in least developed countries: Current status and challenges faced. *Journal of Material Cycles and Waste Management*, 1–11.

Bunrith, S., Ochiai, S., & Kaneko, H. (2012). Resident willingness of food waste separation in Phnom Penh, Cambodia. In *Proceedings of the 23rd Annual Conference of Japan Society of Material Cycles and Waste Management* (p. 615). Japan Society of Material Cycles and Waste Management. Retrieved from https://www.jstage.jst.go.jp/article/jsmcwm/23/0/23_615/_pdf/-char/ja

Bureecam, C., Chaisomphob, T., & Sungsomboon, P.-Y. (2018). Material flows analysis of plastic in Thailand. *Thermal Science*, *22*(6).

Burns, K. N., Sayler, S. K., & Neitzel, R. L. (2019). Stress, health, noise exposures, and injuries among electronic waste recycling workers in Ghana. *Journal of Occupational Medicine and Toxicology (London, England)*, *14*(1), 1. doi:10.118612995-018-0222-9 PMID:30647766

Business Sweden. (2017). *High potential opportunity waste to energy South Africa: Energy for Africa*. Retrieved 5 May 2019 from https://www.business-sweden.se/globalassets/international-markets1/reports-and-facts---markets/waste-to-energy-in-south-africa.pdf

Cagalanan, D. (2015). Governance challenges in community-based forest management in the Philippines. *Society & Natural Resources*, *28*(6), 609–624. doi:10.1080/08941920.2014.948242

Cambodia Tribune. (2018). Sub-decree to Discourage Too Much Use of Plastic Bag. *Cambodian Tribune*. Retrieved April 7, 2019, from https://cambodiatribune.com/sub-decree-to-discourage-too-much-use-of-plastic-bag/

Caravanos, J., Clark, E., Fuller, R., & Lambertson, C. (2011). Assessing worker and environmental chemical exposure risks at an e-waste recycling and disposal site in Accra, Ghana. *Journal of Health and Pollution, 1*(1), 16-25.

Caravanos, J., Clarke, E. E., Osei, C. S., & Amoyaw-Osei, Y. (2013). Exploratory health assessment of chemical exposures at e-waste recycling and scrapyard facility in Ghana. *Journal of Health and Pollution, 3*(4), 11-22.

Carto G. I. S. Services. (2018). *Cambodia base*. Retrieved April 14, 2019, from http://asiapacific.anu.edu.au/mapsonline/base-maps/cambodia-base

CCAC. (2014). *The Climate and Clean Air Coalition to* (J. Maréchal, S. Bonnard, & N. BorgfordParnell, Eds.). Paris: Climate and Clean Air Coalition to Reduce Short-Lived Climate Pollutants. Retrieved from www.ccacoalition.org

CDR. (2006). *Municipal Solid Waste Management in Lebanon*. Beirut: Council for Development and Reconstruction.

CEDARE/ Blue Plan/ UNEP. (2000). *Policy and institutional assessment of solid waste management in five countries Cyprus, Egypt, Lebanon, Syria, Tunisia*. Blue Plan- Regional Activity Centre.

Central Bank of Sri Lanka (CB of SL). (2018). Annual Report of the Monetary Board 2017. Author.

Central Committee of the Communist Party of China and State Council. (2019). *Announcement on Accelerating the Construction of Ecological Civilization*. Retrieved July 1, 2019, from: http://www.xinhuanet.com/politics/2015-05/05/c_1115187518_3.htm

Central Environmental Authority (CEA). (2016). *Corporate plan 2016-2020. 104*. Battaramulla, Sri Lanka: Denzil Kobbekaduwa Mawatha.

Central Environmental Authority (CEA). (2019a). *Map of waste collectors and recyclers in Sri Lanka*. Retrieved from http://www.cea.lk/web/?option=com_content&view=article&layout=edit&id=1407

Central Environmental Authority (CEA). (2019b). *Registry of SLSI recommended bio degradable product manufacturers and raw material/finished product importers*. Retrieved from http://www.cea.lk/web/images/pdf/wm/SLSIRecommendedBio_DegradableproductManufactures.pdf

Central Environmental Authority (CEA). (2019c). *Authorized E - Waste Management Facilities*. Retrieved from http://www.cea.lk/web/en/index-php-option-com-content-view-article-layout-edit-id-983

Central Environmental Authority. (2015). *Data base of solid waste management. 104*. Battaramulla, Sri Lanka: Denzil Kobbekaduwa Mawatha.

CEPIS/PAHO. (2003). Gestión de residuos sólidos en situación de desastres [Management of solid waste in disasters]. Serie salud ambiental y desastres.

CH2M Hill-Antel. (1992). *Solid Waste Management Study for Istanbul Main City*. İstanbul: Author.

Chaerul, M., Tanaka, M., & Shekdar, A. V. (2008). A System dynamic approach for hospital waste management. *Waste Management (New York, N.Y.)*, *28*(2), 442–449. doi:10.1016/j.wasman.2007.01.007 PMID:17368013

Challcharoenwattana, A. (2015). *Improving municipal solid waste management through recycling in urban settlements of Thailand*. Chulalongkorn University.

Challcharoenwattana, A., & Pharino, C. (2016). Multiple-criteria decision analysis to promote recycling activities at different stages of urbanization. *Journal of Cleaner Production*, *137*, 1118–1128. doi:10.1016/j.jclepro.2016.07.208

Chareonsong, P. (2014). *E-Waste Management in Thailand Pollution*. Paper presented at the Control Department 4 th International E-waste Management Network (IEMN) Workshop, Hanoi, Vietnam.

Chartier, Y. (2014). *Safe management of wastes from health-care activities*. World Health Organization.

Chen, T., & Lin, C. (2008). Greenhouse gases emissions from waste management practices using Life Cycle Inventory Model. *Journal of Hazardous Materials*, *155*(1-2), 23–31. doi:10.1016/j.jhazmat.2007.11.050 PMID:18164811

Chen, Y. (2018). Effects of urbanisation of municipal solid waste composition. *Waste Management (New York, N.Y.)*, *79*, 828–836. doi:10.1016/j.wasman.2018.04.017 PMID:30078496

China Electricity Council. (2018). *Annual Report of Electricity 2018*. Retrieved July 1, 2019, from: http://www.cec.org.cn/guihuayutongji/tongjxinxi/niandushuju/

Chowdhury, F. H., Raihan, M. T., Islam, G. M. S., & Ramiz, F. (2016). Construction waste management practice: Bangladesh perception. *Proceedings of 3rd International Conference on Advances in Civil Engineering*.

Chung, S. S., & Poon, C. S. (2001). Characterization of municipal solid waste and its recyclable contents of Guangzhou. *Waste Management & Research*, *19*(6), 473–485. doi:10.1177/0734242X0101900603 PMID:12201677

City of Kitakyushu & Phnom Penh Capital Administration. (2017). *Phnom Penh Capital City Strategic Action Plan for Climate Change*. Phnom Penh: Authors.

City People Committee (CPC) of Can Tho. (2016). *Decision 40/2016/QĐ-UBND dated 30 December 2016 approved service fee*. Author.

City People Committee (CPC) of Da Nang. (2017). *Decision 33 /2017/QĐ-UBND dated 18 November 2017 promulgated maximal prices for collection, transportation of household waste paid by city budget*. Author.

City People Committee (CPC) of Hai Phong. (2016). *Decision 3257/2016/QĐ-UBND dated 22 December 2016 for the price of collection, transportation and treatment of household waste management in the urban areas of the city*. Author.

City People Committee (CPC) of Hanoi. (2016). *Decision No. 54/2016/QĐ-UBND dated 31 December 2016 promulgated price of household waste collection and transportation, prices of environmental sanitary service on non-hazardous industrial waste*. Author.

City People Committee (CPC) of Ho Chi Minh City. (2008). *Decision 88/2008/NQ-UBND dated 20 December 2008 on sanitary and environmental protection fee on non-hazardous solid waste*. Author.

City People Committee (CPC) of Ho Chi Minh City. (2018). *Decision 44/2018/QD-UBND dated 14 November 2018 on issuance of regulation on sorting at source the domestic waste*. Author.

Clemens, H., Bailis, R., Nyambane, A., & Ndung'u, V. (2018). Africa Biogas Partnership Program: A review of clean cooking implementation through market development in East Africa. *Energy for Sustainable Development*, *46*, 23–31. PMID:31007459

Cogut, A. (2016). *Open Burning of Waste: A Global Health Disaster. The R_{20} Regions of Climate Action*. Retrieved on 26.01.2019 from https://regions20.org/wp-content/uploads/2016/08/OPEN-BURNING-OF-WASTE-A-GLOBAL-HEALTH-DISASTER_R20-Research-Paper_Final_29.05.2017.pdf

Cohen, B. (2006). Urbanization in developing countries: Current trends, future projections, and key challenges for sustainability. *Technology in Society*, *28*, 63–80.

COMPED. (2012). *Multi Fuel Policy Development for Solid Biomass Utilization and Capacity Building Plan for EEP Mekong Countries Overview of solid biomasses volumes and availability in Cambodia*. Phnom Penh: Author.

COMPED. (2018). Retrieved April 9, 2019, from https://comped-cam.org/

Concern, W. (2014). *Bangladesh waste database 2014*. Retrieved from www.wasteconcern.org

Construction and Demolition Waste Management Rules. (2016). *India*. Retrieved from http://sscgj.in/wp-content/uploads/2016/07/Waste-Management-Rules-2016.pdf

Cossu, R., Grossule, V., & Lavagnolo, M. C. (2019). Sustainable low cost waste management: Learning from airlines. *Detritus*, *6*, 1–3.

CP Bangladesh, . (2010). *2nd Meeting of the regional 3R forum in Asia.* Kuala Lumpur, Malaysia: Author.

CPCB Report. (n.d.). *LATS/21/2015-16, Plastic Waste Generation in India.* Retrieved from http://www.cpcbenvis.nic.in/publication_list.html

Crouse, D.A., Smyth, T.J., & Crozier, C.R. (2014). North Carolina Agricultural Chemicals Manual 39: Livestock & Poultry Manure Production Rates and Nutrient Content. *Soil Science, Biological & Agricultural Engineering.*

CSARO. (2018). *CSARO Organization Information.* Retrieved April 9, 2019, from http://www.bigpond.com.kh/users/csaro/organization_info.htm

CTI Engineering International, Nippon Koei, & Kitakyushu Water Service. (2016). *The Study on Drainage and Sewerage Improvement Project in Phnom Penh Metropolitan Area.* Authors. Retrieved from http://open_jicareport.jica.go.jp/pdf/12270294_01.pdf

Cunningham, C. E., Bremner, R., & Boyle, M. (1995). Large group community-based parenting programs for families of preschoolers at risk for disruptive behaviour disorders: Utilization, cost effectiveness, and outcome. *Journal of Child Psychology and Psychiatry, and Allied Disciplines, 36*(7), 1141–1159. doi:10.1111/j.1469-7610.1995.tb01362.x PMID:8847377

Daniel, W., Van Tung, L., Wallace, R., Havens, D., Karr, C., & Bich Diep, N., … Duy Bao, N. (2015). Childhood lead exposure from battery recycling in Vietnam. *BioMed Research International.* doi:10.1155/2015/193715

Das, K. (2018). *Vietnam: Opportunities in the Waste-to-Energy Sector, Vietnam Briefing, 2018.* Retrieved 15th March 2019 from https://www.vietnam-briefing.com/news/vietnam-opportunities-waste-energy-sector.html/

Daskalopoulos, E., Badr, O., & Probert, S. (1998). An integrated approach to municipal solid waste management. *Resources, Conservation and Recycling, 24*(1), 33–50. doi:10.1016/S0921-3449(98)00031-7

Daum, K., Stoler, J., & Grant, R. (2017). Toward a more sustainable trajectory for E-waste Policy: A review of a decade of E-waste research in Accra, Ghana. *International Journal of Environmental Research and Public Health, 14*(2), 135. doi:10.3390/ijerph14020135 PMID:28146075

Davis, J.-M., Akese, G., & Garb, Y. (2019). Beyond the pollution haven hypothesis: Where and why do e-waste hubs emerge and what does this mean for policies and interventions? *Geoforum, 98*, 36–45. doi:10.1016/j.geoforum.2018.09.020

Dawan, W. (2017). *Thailand's Master Plan for MSW Management.* Paper presented at the Developments of Energy, Environment and Ecosystems (3E) Nexus Initiative for Sustainable Development in Asian Countries, Tokyo, Japan.

de Titto, E., Montecchia, M. F., Brunstein, L., & Chesini, F. (2015). Normativas para la gestión de residuos biopatogénicos en Argentina [Regulations on Bio-Pathogenic Waste Management in Argentina]. *Rev Argent Salud Pública, 6*(24), 7–14.

DEA. (2011). *South Africa state of waste report, first draft. DEA.* Pretoria: RSA.

DEA. (2013). *National norms and standards for disposal of waste to landfill.* Retrieved from https://www.environment.gov.za/sites/default/files/gazetted_notices/nemwa59of2008_norms_standards_fordisposa_0.pdf

DEA. (2016a). *Policy brief 8: Transitioning South Africa to a green economy: opportunities for green jobs in the waste sector.* Retrieved 15 July 2019 from: https://www.sagreenfund.org.za/wordpress/wp-content/uploads/2016/11/Policy-Brief-No-8.pdf

DEA. (2016b). *National pricing strategy for waste management.* Retrieved 7 July 2019 from: https://cer.org.za/wp-content/uploads/2010/03/National-Pricing-Strategy-for-Waste-Management.pdf

DEA. (2018). *South Africa state of waste. A report on the state of the environment. First draft report.* Pretoria: RSA.

DEA. (2019). *National waste management strategy*. Retrieved on 6 July 2019 from: https://www.environment.gov.za/documents/strategicdocuments/wastemanagement

DEDE. (2019). *Thailand needs to promote Energy-from-Waste*. Retrieved March 10, 2019, from http://weben.dede.go.th/webmax/content/thailand-needs-promote-energy-waste

Dela Rueda, J. B. (2017). *The effects of change in political leadership on service delivery of a program: the case of municipality of los baños' solid waste management program (SWM)* (Unpublished masteral field study). College of Public Affairs and Development, University of the Philippines Los Baños. Los Banos, Laguna, Philippines.

Demir, A., Sengun, R. B., & Ozkaya, B. (2002). The Management of Medical Wastes in Istanbul. ISWA-2002 Appropriate Environmental and Solid Waste Management and Technologies for Developing Countries, Istanbul, Turkey.

Denney, L. (2016). *Reforming solid waste management in Phnom Penh* (WORKING POLITICALLY IN PRACTICE SERIES-CASE STUDY No. 8). Retrieved from https://asiafoundation.org/wp-content/uploads/2016/06/Working-Politically-and-Flexibly-to-Reform-Solid-Waste-Management-in-Phnom-Penh.pdf

Department of Census and Statistics. (2012). *Census 2012 results on Population and Housing based on 5% sample, Sri Lanka.* Retrieved from http://www.statistics .gov.lk/PopHo uSat/CPH20 11/Pages /Activities /Repor ts/CPH_2012_5Per_Rpt.pdf

Department of Census and Statistics. Sri Lanka. (2019). *Agriculture and Environment Statistics Division National Livestock Statistics 2007 – 2018*. Retrieved from http://www.statistics.gov.lk/agriculture/Livestock/LivestockStatistics.html

Department of Cooperative Governance and Traditional Affairs (COGTA). (2015). *Integrated urban development framework: Draft for discussion. COGTA*. Pretoria: RSA.

Department of Environment (DoE). (2010). *National 3R strategy for waste management*. Ministry of Environment and Forests, GOB, Dhaka. Retrieved from http://doe.portal.gov.bd/sites/default/files/files/doe.portal.gov.bd/publications/7dc258b2_4501_400d_b066_5e56d84f439f/National_3R_Strategy.pdf

Department of Environmental Affairs. (2011). *National waste management strategy. DEA*. Pretoria: RSA.

Department of Science and Technology (DST). (2014). *A National Waste R&D and Innovation Roadmap for South Africa: Phase 2 Waste RDI Roadmap. Trends in waste management and priority waste streams for the Waste RDI Roadmap*. Department of Science and Technology.

Department of Science and Technology. (2012). *A National Waste R&D and Innovation Roadmap for South Africa: Phase 1 Status Quo Assessment. Current and required institutional mechanisms to support waste innovation*. Department of Science and Technology.

Department of Statistics Malaysia Official Portal. (2018). Retrieved from https://www.dosm.gov.my/v1/index.php?r=column/cthemeByCat&cat=155&bul_id=c1pqTnFjb29HSnNYNUpiTmNWZHArdz09&menu_id=L0pheU43NWJwRWVSZklWdzQ4TlhUUT09

Department of the Interior and Local Government (DILG) Calabarzon. (2019). *Local government performance management system*. Retrieved August2, 2019, from http://calabarzon.dilg.gov.ph/programsprojects/265-lgpms

Department of Water Affairs (DWA). (1998). *National Water Act*. Retrieved 8 August 2019 from: http://www.dwa.gov.za/Documents/Legislature/nw_act/NWA.pdf

Derraik, J. G. (2002). The pollution of the marine environment by plastic debris: A review. *Marine Pollution Bulletin*, *44*(9), 842–852. doi:10.1016/S0025-326X(02)00220-5 PMID:12405208

Dewan, C., Mukherji, A., & Buisson, M.-C. (2015). Evolution of water management in coastal Bangladesh: From temporary earthen embankments to depoliticized community-managed polders. *Water International*, *40*(3), 401–416. doi:10.1080/02508060.2015.1025196

Diaz, L. F. (2017). Waste management in developing countries and the circular economy. SAGE Publications.

Diaz, L. F. (2017). Waste management in developing countries and the circular economy. *Waste Management & Research*, *35*(1), 1–2. PMID:28049391

Dione, O. (2018, April 5). *Climbing the ladder: Poverty reduction and shared prosperity in Vietnam* [Speeches & Transcripts]. The World Bank. Retrieved from https://www.worldbank.org/en/news/speech/2018/04/05/poverty-reduction-and-shared-prosperity-in-vietnam

Disaster Management Centre (DMC). (2005). *Towards a Safer Sri Lanka: A Road Map for Disaster Risk Management*. DMC, Ministry of Disaster Management Government of Sri Lanka. Retrieved from http://www.adrc.asia/documents/dm_information/srilanka_plan02.pdf

DIW. (n.d.a). *Benefits of Clean Technology*. Retrieved March 19, 2019, from http://php.diw.go.th/ctu/en/benefits_of_ct.php

DIW. (n.d.b). *The concept of Clean Technology*. Retrieved March 19, 2019, from http://php.diw.go.th/ctu/en/index.php

Dlamini, S. (2016). *Solid waste management in South Africa: Exploring the role of the informal sector in solid waste recycling in Johannesburg* (MSc. Thesis). University of Witwatersrand, RSA.

Dlamini, S., Simatele, M., & Kubanza, M. (2019). Municipal solid waste management in South Africa: From waste-to-energy recovery through waste-to-energy technologies in Johannesburg. *Local Environment*, *24*(3), 249–257. doi:10.1080/13549839.2018.1561656

DoE, Waste Concern, & ITN-BUET. (2004). *Country paper, Bangladesh*. In SAARC Workshop on Solid Waste Management, Dhaka, Bangladesh. Retrieved from http://wasteconcern.org/wp-content/uploads/2016/04/SAARC-Country-Paper_Banglaadesh.pdf

Dokmaingam, P., & Manomaivibool, P. (2017). Integrated municipal solid waste management for the highland rural area: Case study Patueng sub-district in Chang Rai, Thailand. *Rangsit Journal of Arts and Sciences*, *7*(2), 173–180.

DST. (2016). *A waste RDI roadmap for South Africa*. Retrieved from https://www.wasteroadmap.co.za/download/presentation_06.pdf

Duku, M. H., Gu, S., & Hagan, E. B. (2011). A comprehensive review of biomass resources and biofuels potential in Ghana. *Renewable & Sustainable Energy Reviews*, *15*(1), 404–415. doi:10.1016/j.rser.2010.09.033

Džolev, N. M., & Vujić, G. V. (2016). Impact assessment of concentrate re-circulation on the landfill gas production. *Thermal Science*, *20*(4), 1283–1294. doi:10.2298/TSCI160401137D

Earth Day Network. (2018). *Fact Sheet – Single Use Plastic*. Retrieved from https://www.earthday.org/2018/03/29/fact-sheet-single-use-plastics/

ECD, MONREC. (2018). *National Waste Management Strategy and Master Plan for Myanmar, Nay Pyi Taw* (Waiting for official approvals). Author.

ECD, MONREC. (2019). *Master Plan for Hazardous Waste Management in Myanmar, 2019, Nay Pyi Taw* (Waiting for official approvals). Author.

Ecotech Lanka Limited (ETL). (2011). *Environmental Impact Assessment Study Report: Rehabilitation of Gohagoda dumpsite and establishment of an integrated solid waste management system for Kandy Municipal Council. Business Plan. 629/3, 8th Lane*. Battaramulla, Sri Lanka: Mutthettugoda Road, Thalangama North.

Ejaz, N., Hashmi, H.N., & Ghumman, A.R. (2011). Water quality assessment of effluent receiving streams in Pakistan: A case study of Ravi River. *Mehran University Research Journal of Engineering & Technology, 30*(3), 383-396.

Eker, H. H., & Bilgili, M. S. (2011). Statistical analysis of waste generation in healthcare services: A case study. *Waste Management & Research, 29*(8), 791–796. doi:10.1177/0734242X10396755 PMID:21382873

El Fadel, M., & Sbayti, H. (2000). Economics of mitigation greenhouse gas emissions from solid waste in Lebanon. *Waste Management & Research, 18*(4), 329–340. doi:10.1034/j.1399-3070.2000.00140.x

Electronic Reporting System of the Basel Convention. (2017). Retrieved from http://ers.basel.int/ERS-Extended/FeedbackServer/fsadmin.aspx?fscontrol=respondentReport&surveyid=75&voterid=49405&readonly=1&nomenu=1

Embassy of Japan in Myanmar. (2019). *Exchange of Notes on Grant Aid to Myanmar "the Project for the Urgent Improvement of Solid Waste Management in Yangon City"*. Retrieved from https://www.mm.emb-japan.go.jp/profile/english/press/20190218.html

EML Consultants Pvt Ltd. (2015). *Environmental impact assessment report of the proposed project on metro Colombo solid waste management. 6/10 Raja Maha Vihara Rd.* Sri Lanka: Sri Jayawardenepura Kotte.

Enayetullah, I. (2019). *IRRC Lessons Learned: Contribution to Urban Resource Efficiency, Circular Economy and Sustainability in the Asia Pacific Region*. Dhaka: Waste Concern.

Enayetullah, I., Sinha, A. H. M. M., & Khan, S. S. A. (2005). *Urban solid waste management scenario of Bangladesh: Problem and prospects*. Dhaka, Bangladesh: Waste Concern.

Environment and Social Development Organization-ESDO. (2012). *Guidelines for e-waste management in Bangladesh*. Dhaka, Bangladesh: Author.

Environmental Protection Agency. (2005). *Guidelines for solid waste management; present status of SWM in Pakistan and strategy for its improvement*. Author.

Environmental Protection agency. (2014). *The Hazardous Substances Rules. Sindh environment protection agency*. Government of Sindh.

Environmental Protection Agency. (2017). *Ghana State of the Environment (2016) Report, Environmental Protection Agency, Ministry of Environment, Science, Technology and Innovation, Accra*. Retrieved from http://www.epa.gov.gh/epa/publications/state-environment-report

ENVIS Newsletter. (2010). Retrieved from http://mahenvis.nic.in/Pdf/Environemnt%20Literacy_Eng2011.PDF

E-Thekwini Municipality. (2016). *E-Thekwini municipality integrated waste management plan 2016-2021 (*Project no. 4024-022*)*. Durban, RSA: Author.

European Commission (EU). (2018). Report from the Commission to the European Parliament and the Council on the impact of the use of oxo-degradable plastic, including oxo-degradable plastic carrier bags, on the environment. COM (2018) 35 final, Brussels.

European Environment Agency (EEA). (2013). *Municipal waste management in Turkey*. ETC/SCP Working Paper.

Eurostat. (2016). *Municipal Waste Statistics*. Retrieved from http:// http://ec.europa.eu/eurostat/statistics-explained/index.php/Municipal_waste_statistics

E-waste Management (Amended) Rules. (2018). *India*. Retrieved from http://sscgj.in/wp-content/uploads/2016/07/Waste-Management-Rules-2016.pdf

Ezeah, C., & Roberts, C. L. (2012). Analysis of barriers and success factors affecting the adoption of sustainable management of municipal solid waste in Nigeria. *Journal of Environmental Management*, *103*, 9–14.

Fakoya, M. (2014). Institutional challenges to municipal waste management service delivery in South Africa. *Journal of Human Ecology (Delhi, India)*, *45*(2), 119–125. doi:10.1080/09709274.2014.11906685

FAO, Food and Agricultural Organisation of United States,FAO Statistics Division. (2013). Available: http://faostat.fao.org/site/567/DesktopDefault.aspx?PageI D=567#ancor

FAO. (2018). *Myanmar Economy and the Role of Agriculture*. Retrieved from http://www.fao.org/myanmar/fao-in-myanmar/myanmar/en/

Fatemi, M. N. (2012). Strategies to reduce construction and demolition (C&D) waste for sustainable building design in Dhaka: Role of architects. *Proceedings of International Seminar on Architecture: Education, Practice and Research.*

Fauziah, S. H., Liyana, I. A., & Agamuthu, P. (2015). Plastic debris in the coastal environment: The invincible threat? Abundance of buried plastic debris on Malaysian beaches. *Waste Management & Research*, *33*(9), 812–821. doi:10.1177/0734242X15588587 PMID:26092255

Feldt, T., Fobil, J. N., Wittsiepe, J., Wilhelm, M., Till, H., Zoufaly, A., ... Göen, T. (2014). High levels of PAH-metabolites in urine of e-waste recycling workers from Agbogbloshie, Ghana. *The Science of the Total Environment*, *466*, 369–376. doi:10.1016/j.scitotenv.2013.06.097 PMID:23921367

Firdaus, G., & Ahmad, A. (2010). Management of Urban Solid Waste Pollution in Developing Countries. *International Journal of Environmental of Research*, *4*(4), 795–806.

Fobil, J. N., Armah, N. A., Hogarh, J. N., & Carboo, D. (2008). The influence of institutions and organizations on urban waste collection systems: An analysis of waste collection system in Accra, Ghana (1985-2000)'. *Journal of Environmental Management*, *86*(1), 262–271. doi:10.1016/j.jenvman.2006.12.038 PMID:17291677

Fondazione, A. C. R. A. (2015). *Cambodia's plastic bag ecosystem and usage summary of research report excess baggage: Reducing plastic bag waste in major cities of Cambodia*. Retrieved from www.excessbaggage-cambodia.org

Food and Agriculture Organisation. (2014). World Agriculture: Towards 2015/2030 FAO Perspective. FAO.

Forman, E., & Peniwati, K. (1998). Aggregating individual judgments and priorities with the analytic hierarchy process. *European Journal of Operational Research*, *108*(1), 165–169. doi:10.1016/S0377-2217(97)00244-0

FTN – Faculty of Technical Sciences. (2009). *Determination of waste composition and quantity estimation in order to define a strategy for the management of secondary raw materials within the sustainable development of the Republic of Serbia*. FTN - Department of Environmental Engineering and Occupational Safety and Health, University of Novi Sad.

Gaisie, E., Kim, H. M., & Han, S. S. (2019). Accra towards a city-region: Devolution, spatial development and urban challenges. *Cities (London, England)*, *95*, 102398. doi:10.1016/j.cities.2019.102398

Gamage, D. A. S., Wijesekara, S. S. R. M. D. H. R., Sarathchandra, G. W. N. L., Basnayake, B. F. A., & Costa, W. A. J. M. (2011). Estimation of Nitrate and Phosphate Leaching from Lysimeter Simulation of Rice Straw Landfill Bioreactor and Evaluation of Fertilizer Quality of Resulting Compost. *Tropical Agricultural Research*, *22*(2), 220–228. doi:10.4038/tar.v22i2.2831

Garces-Ayerbe, C., Rivera-Torres, P., Ines, S., & Dante, L. (2019). Is it possible to change from a linear to a circular economy? An overview of opportunities and barriers for European small and medium sized enterprise companies. *International Journal of Environmental Research and Public Health, 16*(5), 851. doi:10.3390/ijerph16050851 PMID:30857193

Gaudet, G. (2007). Natural Resource Economics under the Rule of Hotelling. *The Canadian Journal of Economics. Revue Canadienne d'Economique, 40*(4), 1033–1059. doi:10.1111/j.1365-2966.2007.00441.x

General Statistics Office of Vietnam (GSO). (2018). Retrieved 25th February 2019 from https://www.gso.gov.vn/default.aspx?tabid=714

Getahun, T., Mengistie, E., Haddis, A., Wasie, F., Alemayehu, E., Dadi, D., ... Van der Bruggen, B. (2012). Municipal solid waste generation in growing urban areas in Africa: Current practices and relation to socioeconomic factors in Jimma, Ethiopia. *Environmental Monitoring and Assessment, 184*(10), 6337–6345. doi:10.100710661-011-2423-x PMID:22042042

GFA (Consulting Group GmbH). (2017). *Support to Reforms – Environmental Governance, Beirut, Lebanon. Assessment of Solid Waste Management Practices in Lebanon in 2015.* Hamburg, Germany: European Union.

Ghana Statistical Services. (2019). *Ghana Population and Housing Census: Projection_Districts_2015_2020.* Retrieved from http://www.statsghana.gov.gh/

Ghosh, S. K. (2016). Sustainable SWM in developing countries focusing on faster growing economies, India and China. *Procedia Environmental Sciences, 35*, 176–184. doi:10.1016/j.proenv.2016.07.073

Ghosh, S. K., Debnath, B., Baidya, R., De, D., Li, J., Ghosh, S. K., ... Tavares, A. N. (2016). Waste electrical and electronic equipment management and Basel Convention compliance in Brazil, Russia, India, China and South Africa (BRICS) nations. *Waste Management & Research, 34*(8), 693–707. doi:10.1177/0734242X16652956 PMID:27422616

Giang, N. V., Kochanek, K., Vu, N. T., & Duan, N. B. (2018). Landfill leachate assessment by hydrological and geophysical data: Case study Nam Son, Hanoi, Vietnam. *Journal of Material Cycles and Waste Management, 20*(3), 1648–1662. doi:10.100710163-018-0732-7

Ginindza, B., & Muzenda, E. (2016). *Waste management challenges to opportunities in the west rand district municipality, Gauteng, South Africa: initiatives.* Johannesburg: RSA.

GIZ - Deutsche Gesellschaft für Internationale Zusammenarbeit - German Corporation for International Cooperation. (2016). *Solid Waste Data Collection in South East Europe.* Belgrade, Serbia: Author.

GIZ/SWEEP-Net. (2014). *Country profile on the solid waste management situation in Lebanon.* Retrieved from http://www.sweep-net.org/sites/default/files/LEBANON%20PROFIL.pdf

Gnanakaran, M., Ariyawansha, R. T. K., & Basnayake, B. F. A. (2017). Simulation of lab-scale leachate treatment bioreactor with application of logistic growth equation for determining design and operational parameters. *International Journal of Scientific & Engineering Research, 8*(1).

Godfrey, L., Strydom, W., & Phukubye, R. (2016). *Integrating the informal sector into the South African waste and recycling economy in the context of extended producer responsibility.* Retrieved on 12 July 2019 from: https://www.wasteroadmap.co.za/download/informal_sector_2016.pdf

Godfrey, L. (2019). Waste plastic, the challenge facing developing countries-ban it, change it, collect it? *Recycling, 4*(3), 1–6.

Godfrey, L., & Oelofse, S. (2017). Historical Review of Waste Management and Recycling in South Africa. *Resources, 6*(4), 57. doi:10.3390/resources6040057

Government Gazette. (2014). *Act No. 26: National Environmental Management: Waste Act (2014).* Retrieved 6 July 2017 from: https://cer.org.za/wp-content/uploads/2010/03/Waste-Amenment-Act-2014.pdf

Government of Bangladesh (GoB). (1995). *National environmental management action plan final report.* Dhaka, Bangladesh: Author.

Government of Bangladesh (GoB). (1998). *Urban management policy statement.* Dhaka, Bangladesh: Author.

Government of Ghana. (2010). *National Environmental Sanitation Strategic Action Plan.* Retrieved from https://www.ircwash.org/sites/default/files/MLGRD-2010-National.pdf

Government of Ghana. (2017). *Enhancing urban resilience in the Greater Accra Metropolitan Area. Global Practice on Social, Urban, Rural and Resilience.* The World Bank Group. Retrieved from http://documents.worldbank.org/curated/en/949241495793834492/Enhancing-urban-resilience-in-the-Greater-Accra-Metropolitan-Area

Government of Ghana. (2018). *The Health Sector in Ghana: Facts and Figures.* Retrieved from https://www.ghanahealthservice.org/

Government of Lebanon/ Council of Ministers. (2015). *Amendment of the Council of Ministers' decision No. 1 dated 12/01/2015.* Retrieved from http://www.pcm.gov.lb/arabic/landing.aspx?pageid=9

Government of Lebanon/Council of Ministers. (1997). *COM Decision No.58 (dated 1/2/1997).* Retrieved from http://www.pcm.gov.lb/arabic/landing.aspx?pageid=9

Government of Lebanon/Council of Ministers. (2005). *COM Decision No.1/4952 (dated 8/18/2005).* Retrieved from http://www.pcm.gov.lb/arabic/landing.aspx?pageid=9

Government of Lebanon/Council of Ministers. (2010). *COM Decision No.55 (dated 9/1/2010).* Retrieved from http://www.pcm.gov.lb/arabic/landing.aspx?pageid=9

Government of Lebanon/Council of Ministers. (2014). *COM Decision No.46 (dated 10/30/2014).* Retrieved from http://www.pcm.gov.lb/arabic/landing.aspx?pageid=9

Government of Myanmar and UNFPA. (2014). *Population and Housing Census of Myanmar – Data Sheet.* Retrieved from http://myanmar.unfpa.org/sites/asiapacific/files/pub-pdf/Census%20Data%20Sheet%20-%20ENGLISH_0.pdf

Government of Rwanda. (2016). *Governing Liquid Waste Collection and Transportation, November 10, 2016 No: 005/R/SANEWS/RURA/2016.* Retrieved from http://www.rura.rw/uploads/media/Regulations_Governing__liquid_waste_collection_and_transportation.pdf

Government of the Punjab (GoP). (2019). *Climate and Clean Air Coalition Municipal Solid Waste Initiative.* Retrieved from http://www.waste.ccacoalition.org/

Government of Vietnam (GOV). (2015). *Decree 38/2015/ND-CP dated April 2015 on waste and scrap management.* Author.

Government of Vietnam (GOV). (2019). *Resolution 09/NQ-CP dated 03 February 2019 on Government Regular Meeting of Jannuary 2019.* Author.

Government Party Committee (GPC). (2018). *Report on assessment of 05 years implementing Resolution 24-NQ/TW on proactive response to climate change, strengthening resource management and environmental protection.* Author.

Grant, R., & Oteng-Ababio, M. (2019). Electronic-Waste Circuitry and Value Creation in Accra, Ghana. In Value Chains in Sub-Saharan Africa (pp. 115-131). Springer.

Grant, R., & Oteng-Ababio, M. (2012). Mapping the invisible and real" African" economy: Urban e-waste circuitry. *Urban Geography*, *33*(1), 1–21.

Green, D. W. J. (2010). Technical assistance consultant's report: Managing hazardous wastes. Hong Kong, China: Academic Press.

GreenCape. (2017). *Waste economy-2017: A market Intelligence Report. GreenCape*. Cape Town: RSA.

Greenpeace Report. (2008). *Hazardous other wastes management and Transboundary movement rules 2016, India*. Retrieved from https://www.greenpeace.org/international/publication/19714/annual-report-2008/

Grossule, V., & Lavagnolo, M. C. (2018). From 3R to 3S: An appropriate strategy for developing countries. *Detritus*, *4*, 1–3. doi:10.31025/2611-4135/2018.13703

Gu, B., Zhu, W., Wang, H., Zhang, R., Liu, M., Chen, Y., ... Bi, J. (2014). Household hazardous waste quantification, characterization and management in China's cities: A case study of Suzhou. *Waste Management (New York, N.Y.)*, *34*(11), 2414–2423. doi:10.1016/j.wasman.2014.06.002 PMID:25022547

Guerrero, L. A., Maas, G., & Hogland, W. (2013). Solid waste management challenges for cities in developing countries. *Waste Management, 33*(1), 220-232. Retrieved from http://www.worldometers.info/world-population/turkey-population/

Guerrero, L. A., Maas, G., & Hogland, W. (2013). Solid waste management challenges for cities in developing countries. *Waste Management (New York, N.Y.)*, *33*(1), 220–232. doi:10.1016/j.wasman.2012.09.008 PMID:23098815

Guilberto, B., & Okumura, S. (2010). *Municipal Waste Management Report: Status-quo and Issues in Southeast and East Asian Countries*. AIT/UNEP Regional Resource Center for Asia and the Pacific (RRC.AP). Thematic Working Group on Solid and Hazardous Waste. Retrieved from http://www.environment-health.asia/twg.cfm?themeid=3

Gullett, B. K., Wyrzykowska, B., Grandesso, E., Touati, A., Tabor, D. G., & Ochoa, G. S. (2010). PCDD/F, PBDD/F, and PBDE Emissions from Open Burning of a Residential Waste Dump. *Environmental Science & Technology*, *44*(1), 394–399. doi:10.1021/es902676w PMID:19950929

Gumbi, S. (2016). *Current waste management and minimisation patterns and practices: an exploratory study on the Ekurhuleni metropolitan municipality in South Africa* (MSc Thesis). University of South Africa, RSA.

Gupta, N., Yadav, K. K., & Kumar, V. (2015). A review on current status of municipal solid waste management in India. *Journal of Environmental Sciences (China). Elsevier B.*, *37*, 206–217. doi:10.1016/j.jes.2015.01.034

Gutberlet, J. (2018). *Waste in the city: Challenges and opportunities for urban agglomerations, urban agglomeration Mustafa Ergen*. IntechOpen. Retrieved 09 July 2019 from: https://www.intechopen.com/books/urban-agglomera

Gu, Y., Wu, Y., Xu, M., Mu, X., & Zuo, T. (2016). Waste Electrical and Electronic Equipment (WEEE) Recycling for a Sustainable Resource Supply in The Electronics Industry in China. *Journal of Cleaner Production*, *27*, 331–338. doi:10.1016/j.jclepro.2016.04.041

Gwisai, R. D., Areola, O., & Segosebe, E. (2014). Respiratory and occupational health problems of scavengers and landfill employees in a municipal landfill site in Lobatse, Botswana. *Journal of Sustainable Development in Africa*, *16*, 37–55.

Hai, H. T., Hung, H. V., & Quang, N. D. (2015, December). An overview of electronic waste recycling in Vietnam. *Journal of Material Cycles and Waste Management*.

Halder, P. K., Paul, N., & Beg, M. R. A. (2014). Assessment of biomass energy resources and related technologies practice in Bangladesh. *Renewable & Sustainable Energy Reviews*, *39*, 444–460. doi:10.1016/j.rser.2014.07.071

Harshwardhan & Upadhyay. (2017). Effective Utilization of Agricultural Waste [Review]. *J Fundam Renewable Energy Appl, 7*, 5. doi:10.4172/2090-4541.1000237

Health-care waste. (2018). Retrieved from https://www.who.int/news-room/fact-sheets/detail/health-care-waste

Henry, R. K., Yongsheng, Z., & Jun, D. (2006). Municipal solid waste management challenges in developing countries–Kenyan case study. *Waste Management (New York, N.Y.), 26*(1), 92–100. doi:10.1016/j.wasman.2005.03.007 PMID:16006111

Hettiarachchi, H., Meegoda, J., & Ryu, S. (2018). Organic waste buyback as a viable method to enhance sustainable municipal solid waste management in developing countries. *International Journal of Environmental Research and Public Health, 15*(11), 1–15. doi:10.3390/ijerph15112483 PMID:30405058

Hien, L. (2017, March 21). Energy figures. *Vietnam Electricity*. Retrieved from http://www.tietkiemnangluong.vn

Hien, N. T., Yen, N. T., Doan, D. T., Lam, N. T., Thuan, L. T. M., Giang, P. T., & Lam, L. N. (2015). Assessment of application of hazardous waste treatment and domestic solid waste in Vietnam. *Proceedings of the IV National Environment Conference*.

Hoang, H. (2018, June 1). There should be solutions to promptly handle environmental pollution in Minh Khai plastic craft village, Hung Yen province. *Vietnam Environment Administration Magazine*. Retrieved from http://tapchimoitruong.vn

Hongyon, C., & Hovijit, P. (2014). *Evaluation Potential of Oil Production from Khon Kaen Municipality Solid Waste* Paper presented at the The 15th Graduate Research Conferences, Khon Kaen University, Khon Kaen, Thailand.

Hoornweg, D., & Bhada-Tata, P. (2012). *What a waste: a global review of solid waste management* (Vol. 15). World Bank.

Hoornweg, D., & Bhada-Tata, P. (2012). What a Waste: A Global Review of Solid Waste Management. The World Bank Urban Development Series Knowledge Papers Number 15.

Hoornweg, D., & Bhada-Tata, P. (2012). *What a waste: A global review of solid waste management* [urban development series; knowledge papers no. 15]. World Bank.

Hoornweg, D., & Bhada-Tata, P. (2012). *What a Waste: A Global Review of Solid Waste Management, Urban development series, knowledge papers no. 15*. Washington, DC: World Bank.

Hoornweg, D., & Bhada-Tata, P. (2012). *What a Waste: A Global Review of Solid Waste Management*. Washington, DC: World Bank.

Hosseini, S. E., & Abdul Wahid, M. (2014). The role of renewable and sustainable energy in the energy mix of Malaysia: A review. *International Journal of Energy Research, 38*(14), 1769–1792. doi:10.1002/er.3190

Hossen, M. M., Rahman, A. H. M. S., Kabir, A. S., Hasan, M. M. F., & Ahmed, S. (2017). Systematic assessment of the availability and utilization potential of biomass in Bangladesh. *Renewable and Sustainable Energy Reviews, 67*, 94–105. .2016.09.008 doi:10.1016/j.rser

Hristovski, K., Olson, L., Hild, N., Peterson, D., & Burge, S. (2007). The municipal solid waste system and solid waste characterization at the municipality of Veles, Macedonia. *Waste Management (New York, N.Y.), 27*(11), 1680–1689. doi:10.1016/j.wasman.2006.09.003 PMID:17098409

Huang, J., Nkrumah, P. N., Anim, D. O., & Mensah, E. (2014). E-waste disposal effects on the aquatic environment: Accra, Ghana. *Reviews of Environmental Contamination and Toxicology, 229*, 19–34. doi:10.1007/978-3-319-03777-6_2 PMID:24515808

Huisman, H., Breukelman, H., & Keesman, B. (2016). *Expert mission on integrated SWM to DSM, Report of findings November 2016*. Retrieved 15th August 2018 from https://docmh.com/expert-mission-on-integrated-solid-waste-management-iswm-to-dar-es-salaam-mat16tz01-report-of-findings-november-pdf

Huong, N. T. (2018). *The study on development of a procedure to identify polybrominated diphenyl ether (PBDEs) in tissue of the Anodonta cygnea and Pomacea canaliculata and pilot application to evaluate accumulation of PBDEs in the environment in the plastic recycling village of Minh Khai, Nhu Quynh town, Hung Yen province* (Unpublished master thesis). Hanoi University of Natural Resources & Environment.

Ian, S. (2015). *Coastal communities dumping 8m tons of plastic in oceans every year*. Academic Press.

Ibrahim, C. A. (2013). *Policy Update from Malaysia*. Department of Environment. Retrieved from https://19january2017snapshot.epa.gov/sites/production/files/2014-05/documents/malaysia.pdf

ICC. (2010). *Trash travels. International Coastal Cleanup, Ocean Conservancy*. Retrieved from https://oceanconservancy.org/wp-content/uploads/2017/04/2010-Ocean-Conservancy-ICC-Report.pdf

ICC. (2012). *The ocean trash index*. International Coastal Cleanup, Ocean Conservancy. Retrieved from https://oceanconservancy.org/wp-content/uploads/2017/04/2012-Ocean-Conservancy-ICC-Report.pdf

ICC. (2014). *Turning the tide on trash*. International Coastal Cleanup, Ocean Conservancy. Retrieved from https://oceanconservancy.org/wp-content/uploads/2017/04/2014-Ocean-Conservancy-ICC-Report.pdf

IDB. (2013). *Guidebook for the Application of Waste to Energy Technologies in Latin America and the Caribbean*. IDB.

IDB-AIDIS-PAHO. (2011). *Regional Evaluation on Urban Solid Waste Management in Latin America and the Caribbean - 2010 Report*. Author.

IMG - International Management Group. (2014). *Municipal Solid Waste Information – Final Report*. Belgrade. Serbia: Author.

INDEC-Argentina. (2010). *Censo Nacional de Población, Hogares y Viviendas*. National Census of Population, Homes and Housing.

INFORM. (2019). *Information Report (Canadian Wood Fibre Centre)*. Retrieved from http://www.inform-index.org/Portals/0/InfoRM/2019/Inform%202019%20WEB%20spreads%20(3).pdf?ver=2019-02-07-113610-123

Infrastructure News. (2015). *First bottle-2-bottle plant in Africa*. Retrieved 14 July 2019 from: https://infrastructurenews.co.za/2015/05/12/first-bottle-2-bottle-plant-in-africa/

Instituto de Ingeniería Sanitaria, Facultad de Ingeniería, Universidad de Buenos Aires- CEAMSE. (2015). *Estudio de Calidad de los Residuos Sólidos Urbanos de la CABA* [Study on the quality of Urban Solid Waste in the CABA]. Author.

International Committee of the Red Cross. (2011). *Medical Waste Management*. International Committee of the Red Cross, Geneva, Switzerland. Retrieved from https://www.icrc.org/en/doc/assets/files/publications/icrc-002-4032.pdf

International Finance Corporation, World Bank Group. (2014). Handshake IFC's Quarterly Journal on Public – Private Partinerships. Author.

International Growth Centre. (2014). Urbanisation in Tanzania: Population Growth, Internal Migration and Urbanisation in Tanzania 1967-2012: A Census-Based Regional Analysis. Author.

International Trade Centre. (n.d.). *The Trade statistics for international business development yearly trade data. Import & export values*. Retrieved 13.03.2019 from https://www.trademap.org/tradestat/Country_SelProduct_TS.aspx?nvpm=1%7c%7c3981%7c%7c%7c35%7c%7c%7c2%7c1%7c1%7c1%7c2%7c1%7c3%7c1%7c1

International Union for Conservation of Nature (IUCN). (2004). *AbbottabadfState of the Environment and Development.* Karachi: IUCN Pakistan. Available at http://cmsdata.iucn.org/downloads/abbottabad_soed.pdf

Iqbal, M., Breivik, K., Syed, J. H., Malik, R. N., Li, J., Zhang, G., & Jones, K. C. (2015). Emerging issue of e-waste in Pakistan: A review of status, research needs and data gaps. *Environmental Pollution*, *207*, 308–318. doi:10.1016/j.envpol.2015.09.002 PMID:26433180

IRENA. (2017). *Renewable Energy Outlook: Thailand.* Abu Dhabi: International Renewable Energy Agency.

Ishak, M. B., & Samah, M. A. A. (2010). Strict liability versus policy and regulation for environmental protection and agricultural waste management in Malaysia. *Environment Asia*, *3*, 11–19.

Ishigaki, T., Nakagawa, M., Nagamori, M., & Yamada, M. (2016). Anaerobic generation and emission of nitrous oxide in waste landfills. *Environmental Earth Sciences*, *75*(9), 750. doi:10.100712665-016-5543-3

Islam, M. S., & Emon, M. M. H. (2016). *Current status with facing challenges by entrepreneurs and future prospect of plastic sector in Bangladesh.* Paper presented at the International Plastic Summit, Bangladesh Plastic Goods Manufacturers and Exporters Association (BPGMEA), Dhaka, Bangladesh.

Islam, K. M. (2016). Municipal solid waste to energy generation in Bangladesh: Possible scenarios to generate renewable electricity in Dhaka and Chittagong City. *Journal of Renewable Energy.*

Islam, K. M. N. (2017). Greenhouse gas footprint and the carbon flow associated with different solid waste management strategy for urban metabolism in Bangladesh. *The Science of the Total Environment*, *580*, 755–769. doi:10.1016/j.scitotenv.2016.12.022 PMID:28024747

Isugi, J., & Niu, D. (2016). Research on Landfill and Composting Guidelines in Kigali City, Rwanda Based on China's Experience. *International Proceedings of Chemical, Biological & Environmental Engineering*, *94*, 62–68. doi:10.7763/IPCBEE

ITU. (2015). *Gestión Sostenible de Residuos de Aparatos Eléctricos y Electrónicos en América Latina* [Sustainable Management of waste from electrical and electronic equipment in Latin America]. ITU.

Jacob, P., Kashyap, P., & Visvanathan, C. (2012). *Overview of municipal solid waste - waste to energy in Thailand.* Paper presented at the International brainstorming workshop on waste to energy in India, Mumbai, India.

JacobsenS. (2005). Plastic Bag Pollution. Retrieved from http://dpw.lacounty.gov/epd/PlasticBags/Articles/Googobits_07-21-05.pdf

Jaffar, I. (2019). *Current Scenario of Solid Waste in Pakistan.* Retrieved from https://www.technologytimes.pk/current-solid-waste-pakistan/

Jamal, M. S., & Sobuz, M. H. R. (2017). A sustainable approach of demolition, reconstruction, reuses and recycling construction materials in Bangladesh. *BAUET Journal, 1*(1). Retrieved from https://bauet.ac.bd/uploads/0.57107300 1505019927Article-9.pdf

Jambeck, J. R. (2015). *Plastic waste inputs from land into the ocean.* Retrieved, June 2016 http://www.iswa.org/fileadmin/user_upload/Calendar_2011_03_AMERICANA/Science-2015-Jambeck-768-71__2_.pdf

Jambeck, J. R., Geyer, R., Wilcox, C., Siegler, T. R., Perryman, M., Andrady, A., . . . Lavender Law, K. (2015). *Plastic waste inputs from land into the ocean.* Retrieved from https://www.iswa.org/fileadmin/user_upload/Calendar_2011_03_AMERICANA/Science-2015-Jambeck-768-71__2_.pdf

Jambeck, J. R., Geyer, R., Wilcox, C., Siegler, T. R., Perryman, M., Andrady, A., ... Law, K. L. (2015). Plastic waste inputs from land into the ocean. *Science*, *347*(6223), 768–771. doi:10.1126cience.1260352 PMID:25678662

Japan International Cooperation Agency (JICA) & Kokusai Kogyo Co., Ltd (KKC). (2016). *Final Report: Data collection survey on solid waste management in Democratic Socialist Republic of Sri Lanka*. Retrieved from http://open_jicareport.jica.go.jp/pdf/12250213.pdf

Japan International Cooperation Agency (JICA). (2015). *Data Collection Survey on Solid Waste Management in Turkey*. Final Report. Author.

Javied, S., Hanan, F., Munawar, S., Qasim, M., Anees, M. M., Ghani, M. U., ... Ansar, A. (2014). Management of municipal solid waste generated in eight cities of Pakistan. *International Journal of Scientific & Engineering Research*, *5*(12), 1186–1192.

Jayasinghe, R., Azariadis, M., & Baillie, C. (2019). Waste, Power, and Hegemony: A Critical Analysis of the Wastescape of Sri Lanka. *Journal of Environment & Development*, *0*(0), 1–23. doi:10.1177/1070496518821722

Jayawardane, A. K. W. (2006). *Disaster Mitigation Initiatives in Sri Lanka*. Faculty of Engineering and Professor of Civil Engineering University of Moratuwa, Sri Lanka. Retrieved from http://iiirr.ucalgary.ca/files/iiirr/Kochifullpaper-final.pdf

Jiaranaikhajorn, T. (2013). *WEEE Management Policy Update from Thailand*. Pollution Control Department, Ministry of Natural Resources and Environment Thailand.

Joshi & Ahmed. (2016). Status and challenges of municipal solid waste management in India: A review. *Cogent Environmental Science*, *2*, 1139434. doi:10.1080/23311843.2016.1139434

Kabera, T., Wilson, D. C., & Nishimwe, H. (2019). Benchmarking performance of solid waste management and recycling systems in East Africa: Comparing Kigali Rwanda with other major cities. *Waste Management & Research*, *37*(1_suppl), 58-72.

Kabera, T., Wilson, D. C., & Nishimwe, H. (2019). Benchmarking performance of solid waste management and recycling systems in East Africa: Comparing Kigali Rwanda with other major cities. *Waste Management & Research*, *37*(1), 58–72. doi:10.1177/0734242X18819752 PMID:30761955

Kabir, M. R. (2015). Municipal solid waste management system: A study on Dhaka North and South city corporations. *Journal of Bangladesh Institute of Planners*, *8*, 35–48.

Kafando, P., Segda, B. G., Nzihou, J. F., & Koulidiati, J. (2013). Environmental Impacts of Waste Management Deficiencies and Health Issues: A Case Study in the City of Kaya, Burkina Faso. *Journal of Environmental Protection*, *4*(10), 1080–1087. doi:10.4236/jep.2013.410124

Kamuang, T. (2017). *Country Chapter State of the 3Rs in Asia and the Pacific: The kingdom of Thailand*. Secretariat of the Regional 3R Forum in Asia and the Pacific.

Kandil, S. (2013). *World e-waste map reveals national volumes. Int. Flows*. Retrieved from http://i. unu.edu/media/unu.edu/news/41225/World-E-Waste-Map-Reveals-National-Volumes-International-Flows.pdf

Karagiannidis, M., Kontogianni, S., Wolff, S., Redwood-Sawyer, J., & Sankoh, S. (2012). Handbook on Waste Management in Developing Countries. Academic Press.

Karatna, P. (2017). *Thailand Environmental Quality Management Plan 2017 – 2021*. Paper presented at the National Policy Dialogue on the Urban Nexus in Thailand Conference, Bangkok, Thailand.

Karija, M. K., Shiua, Q. I., & Lukaw, Y. S. (2013). The Impact of Poor Municipal Solid Waste Management Practices and Sanitation Status on Water Quality and Public Health in Cities of the Least Developed Countries: The Case of Juba, South Sudan. *International Journal of Applied Science and Technology*, *3*(4), 87–99.

Kartam, N., Al-Mutairi, N., Al-Ghusain, I., & Al-Humoud, J. (2004). Environmental management of construction and demolition waste in Kuwait. *Waste Management (New York, N.Y.)*, *24*(10), 1049–1059. doi:10.1016/j.wasman.2004.06.003 PMID:15567670

Karunarathna, A., Gamagedara, Y., Mannapperuma, N., Ariyawansha, R., & Basnayake, B. F. A. (2018). Development of a methane (CH4) surface emission map: a case study at Karadiyana dumpsite, Sri Lanka. *Proceedings of the 10th Asia-Pacific Landfill Symposium.*

Karunasena, G., Jayathilaka, W. M. D. M., & Rathnayake, R. M. N. U. (2015). Comparison on disposal strategies for clinical waste: hospitals in Sri Lanka. *Proceedings of 6th International conference on structural engineering and construction management.*

Karunasena, D., Amaratunga, R., Haigh, I. L., & Lill, I. (2009). Post disaster waste management strategies in developing countries: Case of Sri Lanka. *International Journal of Strategic Property Management*, *13*(2), 171–190. doi:10.3846/1648-715X.2009.13.171-190

Karunasena, G., & Amaratunga, D. (2016). Capacity building for post disaster construction and demolition waste management: A case of Sri Lanka. *Disaster Prevention and Management*, *25*(2), 137–153. doi:10.1108/DPM-09-2014-0172

Karunasena, G., Rameezdeen, R., & Amarathunga, D. (2012). Post-disaster C&D waste management: The case of COWAM project in Sri Lanka. Australasian Journal of Construction Economics and Building. *Conference Series*, *1*(2), 60–71.

Kaza, S., Yao, L. C., & Bhada-Tata, P. Van, Woerden, F. (2018). What a Waste 2.0: A Global Snapshot of Solid Waste Management to 2050. World Bank, Washington, DC.

Kaza, S., Yao, L., Bhada-Tata, P., & Van-Woerden, F. (2018). What a Waste 2.0: A Global Snapshot of Solid Waste Management to 2050. Urban Development Series. Washington, DC: World Bank. doi:-1329-0 doi:10.1596/978-1-4648

Kaza, S., Yao, L., Bhada-Tata, P., & Woerden, F. V. (2018). *What a waste 2.0: A global snapshot of solid waste management to 2050. Urban Development Series.* Washington, DC: World Bank. doi:10.1596/978-1-4648-1329-0

Kemausuor, F., Kamp, A., Thomsen, S. T., Bensah, E. C., & Østergård, H. (2014). Assessment of biomass residue availability and bioenergy yields in Ghana. *Resources, Conservation and Recycling*, *86*, 28–37. doi:10.1016/j.resconrec.2014.01.007

Khan, A. A., Ahmed, Z., & Siddiqui, M. A. (2012). Issues with solid waste management in South Asian countries: A situational analysis of Pakistan. *J. Environ. Occup. Sci.*, *1*(2), 129–131. doi:10.5455/jeos.20120731084356

Khan, M. (2017). Environment and Health Issues Associated with E-waste. *International Journal of Advanced Research.*, *5*(1), 1425–1430.

Khatib, I. A. (2011). Municipal Solid Waste Management in Developing Countries: Future Challenges and Possible Opportunities. In *Integrated Waste Management* (vol. 2). InTech, Retrieved 22.03.2018 from: http://www.intechopen.com/books/integrated-waste-management-volumeii/

Kieu, L. T. C., Strady, E., & Perset, M. (2016). *Life cycle of floating debris in the canals of Ho Chi Minh.* Working paper of the Urban Development Management Support Centre (PADDI). Retrieved from http://horizon.documentation.ird.fr/exl-doc/pleins_textes/divers17-07/010070479.pdf

Kokusai Kogyo. (2005). *The study on solid waste management in the municipality of Phnom Penh in the Kingdom of Cambodia Final Report.* Retrieved from http://open_jicareport.jica.go.jp/pdf/11784469_01.pdf

Kol Mardi, M., Sam, N., Vathanak Thida, K., Channarith, K., Yagasa, R., Akhmedkhodjaeva, N., … Totoki, Y. (2018). *Phnom Penh Waste Management Strategy and Action Plan 2018-2035*. Phnom Penh Capital Administration. Retrieved from https://www.ccet.jp/publication/phnom_penh_waste_management_strategy_and_action_plan

Korai, M. S., Mahar, R. B., & Uqaili, M. A. (2017). The feasibility of municipal solid waste for energy generation and its existing management practices in Pakistan. *Renewable & Sustainable Energy Reviews*, *72*, 338–353. doi:10.1016/j.rser.2017.01.051

Kormoker, T., Proshad, R., & Khan, M. M. (2017). Analysis of water quality in urban water supply system of Bangladesh. *Journal of Environmental & Analytical Toxicology*, *7*(4), 2161–0525. doi:10.4172/2161-0525.1000492

Kritjaroen, T. (2011). Understanding urban governance in the context of public-private partnerships: A case study of solid-waste management in Rayong Municipality, Thailand. *Federal Governance, 8*(3).

Kuchibanda, K., & Mayo, A. W. (2015). Public Health Risks from Mismanagement of Healthcare Wastes in Shinyanga Municipality Health Facilities, Tanzania. The Scientific World Journal.

Kumar, S. (2016). *Municipal solid waste management in developing countries*. CRC Press. doi:10.1201/9781315369457

Laceco. (2018). *Supervision of Greater Beirut Solid Waste Treatment Plants Contract N 19942*. Beirut, Lebanon: CDR.

Laceco/Ramboll. (2012). *Preparation of Pre-qualification documents and Tender Documents for Solid Waste Management in Lebanon, Sub Report 1, Baseline Study*. Beirut, Lebanon: CDR.

Laurent, A., Ioannis, B., & Julie, C. (2014). *Review of LCA studies of solid waste management systems - Part I: Lessons learned and perspectives*. Waste Management. doi:10.1016/j.wasman.2013.10.045

Lauritzen, E. K. (2018). *Construction, Demolition and Disaster Waste Management: An Integrated and Sustainable Approach*. CRC Press. doi:10.1201/b20145

Leal Filho, W., Doni, F., Vargas, V. R., Wall, T., Hindley, A., Rayman-Bacchus, L., ... Avila, L. V. (2019). The integration of social responsibility and sustainability in practice: Exploring attitudes and practices in Higher Education Institutions. *Journal of Cleaner Production*, *220*, 152–166. doi:10.1016/j.jclepro.2019.02.139

Leal Filho, W., Shiel, C., Paço, A., Mifsud, M., Ávila, L. V., Brandli, L. L., ... Vargas, V. R. (2019). Sustainable Development Goals and sustainability teaching at universities: Falling behind or getting ahead of the pack? *Journal of Cleaner Production*, *232*, 285–294. doi:10.1016/j.jclepro.2019.05.309

Leal Filho, W., Vargas, V. R., Salvia, A. L., Brandli, L. L., Pallant, E., Klavins, M., ... Conticelli, E. (2019). The role of higher education institutions in sustainability initiatives at the local level. *Journal of Cleaner Production*, *233*, 1004–1015. doi:10.1016/j.jclepro.2019.06.059

Lebersorger, S., & Beigl, P. (2011). Municipal solid waste generation in municipalities: Quantifying impacts of household structure, commercial waste and domestic fuel. *Waste Management (New York, N.Y.)*, *31*(9–10), 1907–1915. doi:10.1016/j.wasman.2011.05.016 PMID:21689921

Leblanc, R. (2018). *An Introduction to Solid Waste Management: Integrated Approach Needed for Successful Diversion and Recycling*. Retrieved 22.05.2018 from https://www.thebalancesmb.com/an-introduction-to-solid-waste-management-2878102

Lebreton, L. C. M., van der Zwet, J., Damsteeg, J.-W., Slat, B., Andrady, A., & Reisser, J. (2017). River Plastic Emissions to the World's Oceans. *Nature Communications*, *8*, 15611. doi:10.1038/ncomms15611 PMID:28589961

Lepawsky, J., Akese, G., Billah, M., Conolly, C., & McNabb, C. (2015). Composing urban orders from rubbish electronics: Cityness and the site multiple. *International Journal of Urban and Regional Research, 39*(2), 185–199. doi:10.1111/1468-2427.12142

Liddle, E. S., Mager, S. M., & Nel, E. L. (2016). The importance of community-based informal water supply systems in the developing world and the need for formal sector support. *The Geographical Journal, 182*(1), 85–96. doi:10.1111/geoj.12117

Li, J., Yang, L., & Liu, L. (2015). Development Potential of E-waste Recycling Industry in China. *Waste Management & Research, 33*(6), 533–542. doi:10.1177/0734242X15584839 PMID:25990983

Lipinski, B., Hanson, C., Lomax, J., Kitinoja, L., Waite, R., & Searchinger, T. (2013). *Reducing Food Loss and Waste.* Working Paper, Installment 2 of Creating a Sustainable Food Future. Washington, DC: World Resources Institute. Available online at: http://www.worldresourcesreport.org

Liu, C., Hotta, Y., Totoki, Y., Yagasa, R., Aoki-Suzuki, C., Hengesbaugh, M., & Inoue, M. (2018). *State of the 3Rs in Asia and the Pacific-Experts' Assessment of Progress in Ha Noi 3R Goals.* Retrieved from http://www.uncrd.or.jp/content/documents/6777[full document] State of the 3Rs in Asia and the Pacific.pdf

Li, W., Ma, Z., Huang, Q., & Jiang, X. (2018). Distribution and Leaching Characteristics of Heavy Metals in a Hazardous Waste Incinerator. *Fuel.* doi:10.1016/j.fuel.2018.06.041

Ludete, J. H. (2016). *Factors Influencing Public Private Pertnerships in Solid Waste Management for Local Government Authorities: A Case of Kinondoni Municipality-Dar es Salaam* (Unpublished Master thesis). Sokoine University of Agriculture, Morogoro, Tanzania.

Luken, R., & Clarence-Smith, E. (2019). *Green industrialisation in sub-Saharan Africa: A guide for policy makers. The Institute of African Leadership for Sustainable Development.* Tanzania: Uongozi Institute.

Lundgren, K. (2012). *The global impact of e-waste: addressing the challenge.* International Labor Organization. Retrieved from https://www.ilo.org/wcmsp5/groups/public/---ed_dialogue/---sector/documents/publication/wcms_196105.pdf

Luong, N. D., Giang, H. M., Thanh, B. X., & Hung, N. T. (2013). Challenges for municipal solid waste management practices in Vietnam. *Water Technology, 1*(1), 17–21.

Lu, W., Yuan, H., Li, J., Hao, J. J., Mi, X., & Ding, Z. (2011). An empirical investigation of construction and demolition waste generation rates in Shenzhen city, South China. *Waste Management (New York, N.Y.), 31*(4), 680–687. doi:10.1016/j.wasman.2010.12.004 PMID:21208794

MAFF. (2016). *Policy On Biodigester Development in Cambodia.* MAFF.

Magashi, A., & Schluep, M. (2011). E-waste Assessment Tanzania. The UNIDO E-waste initiative for Tanzania.

Maisiri, W. (2016). *A techno-economic evaluation of a waste-to-energy grate incineration power plant for a small South African city* (MSc. Thesis). North West University, RSA.

MalaysiaP. C. R. 2016. (2017). Retrieved from http://cpmaindia.com/pdf/apic-country-2017/apic2017-malaysia-report.pdf

Manhart, A., Meinel, J., & Walgenbach, S. (2014). *Legal and institutional requirements for e-waste recycling in Ghana: Work package 3.5of the Bo2W project.* Retrieved from http://www.oeko/de/oekodoc/2142/2014-695-en.pdf

Manila Bulletin. (2017). *San Carlos City honors eco-friendly schools.* Retrieved August 22, 2019, from https://news.mb.com.ph/2017/02/01/san-carlos-city-honors-eco-friendly-schools/

Mannie, N. M., & Bowers, A. (2014). Challenges in Determining the Correct Waste Disposal Solutions for Local Municipalities - A South African Overview. *Proceedings of the 20th wastecon conference*.

Manomaivibool, P., Srivichai, M., Unroj, P., & Dokmaingam, P. (2018). Chiang Rai Zero Waste: Participatory action research to promote source separation in rural areas. *Resources, Conservation and Recycling*, *136*, 142–152. doi:10.1016/j.resconrec.2018.04.002

Manyele, S. V., & Anicetus, H. (2006). Management of medical waste in Tanzanian hospitals. *Tanzania Health Research Bulletin*, *8*(3), 177–182. PMID:18254511

Mar, W. W. (2018). *The 2018 World Environmental Day – Steps to beat plastic pollution*, Retrieved from http://www.globalnewlightofmyanmar.com/2018-world-environment-day-steps-beat-plastic-pollution/

Margerum, R. D., & Robinson, C. J. (2015). Collaborative partnerships and the challenges for sustainable water management. *Current Opinion in Environmental Sustainability*, *12*, 53–58. doi:10.1016/j.cosust.2014.09.003

Marshall, R. E., & Farahbakhsh, K. (2013). Systems approaches to integrated solid waste management in developing countries. *Waste Management (New York, N.Y.)*, *33*(4), 988–1003. doi:10.1016/j.wasman.2012.12.023 PMID:23360772

Matschullat, J. (2014). Save Cambodia's wildlife: Atlas of Cambodia. Maps on socio-economic development and environment. *Environmental Earth Sciences*, *72*(4), 1295–1298. doi:10.100712665-014-3325-3

McCartney, P. (2014). United Republic of Tanzania: Assessing if notifications, consents, inspections and enforcement of trans-boundary movements of waste and take-back procedures for illegal traffic represent environmentally sound management. Final Version. Retrieved 3.3.2019 from file:///C:/Users/KAMUGISHA/Downloads/UNEP-CHW-IMPL-CLI-ESM-Toolkit-Tanzania-Report-201412.English.pdf

MCD and ECD. (2017). *City Waste Management Strategy and Action Plan for Mandalay, the Republic of the Union of Myanmar*. Author.

McDougall, F. R., White, P. R., Franke, M., & Hindle, P. (2008). *Integrated solid waste management: a life cycle inventory*. John Wiley & Sons.

McLeod, F., & Cherrett, T. (2013). *Reverse logistics for sustainable waste management processes, Sustainable Practices: Concepts*. Methodologies, Tools, and Applications. doi:10.4018/978-1-4666-4852-4.ch084

Medina, M. (2010). *Solid waste, Poverty and the Environment in Developing Countries Cities: Challenges and Opportunities*. United Nations University/Institute of Advanced Studies Working Paper 23. Tokyo: UNU.

Menikpura, N., & Premakumara, D. G. J. (2018). *User's Manual - Emission Quantification Tool (EQT) for Estimating Short Lived Climate Pollutants (SLCPs) and Other Greenhouse Gases (GHGs) from Waste Sector Institute for Global Environmental Strategies (IGES) Prepared on behalf: Climate and Clean Air Coalition (CCAC)*. Hayama, Kanagawa: Initiative. Retrieved from https://www.ccet.jp/sites/default/files/2018-12/User Manual_EQT version II_June 2018.pdf

Menikpura, S. N. M., & Basnayake, B. F. A. (2009). New applications of 'Hess Law' and comparisons with models for determining calorific values of Municipal Solid Wastes (MSW) in Sri Lankan context. *Renewable Energy*, *34*(6), 1587–1594. doi:10.1016/j.renene.2008.11.005

Menikpura, S. N. M., Basnayake, B. F. A., Boyagoda, P. B., & Kularathne, I. W. (2007). Application of Waste to Energy Concept Based on Experimental and Model Predictions of Calorific Values for Enhancing the Environment of Kandy City.2007. *Journal of Tropical Agricultural Research*, *19*, 389–400.

Menikpura, S. N. M., Gheewala, S. H., & Bonnet, S. (2012). Sustainability assessment of municipal solid waste management in Sri Lanka: Problems and prospects. *Journal of Material Cycles and Waste Management*, *14*(3), 181–192. doi:10.100710163-012-0055-z

MFA. (2017). *Suffciency Economy Philosophy: Thailand's Path towards Sustainable Development Goals*. Ministry of Foreign Affairs, Kingdom of Thailand.

Mgimba, C., & Sanga, A. (2016). Municipal Solid Waste Composition Characterization for Sustainable Management Systems in Mbeya City, Tanzania. *International Journal of Science. Environmental Technology*, *5*(1), 47–58.

Miezah, K., Obiri-Danso, K., Kádár, Z., Fei-Baffoe, B., & Mensah, M. Y. (2015). Municipal solid waste characterization and quantification as a measure towards effective waste management in Ghana. *Waste Management (New York, N.Y.)*, *46*, 15–27. doi:10.1016/j.wasman.2015.09.009 PMID:26421480

MININFRA – Ministry of Infrastructure. (2016). National Sanitation Policy. Kigali, Rwanda: Retrieved from http://www.mininfra.gov.rw/fileadmin/user_upload/new_upload/NATIONAL_SANITATION_POLICY__DECEMBER_2016.pdf

MININFRA. (2018). *Solid Waste Management*. Retrieved from http://www.mininfra.gov.rw/index.php?id=90

Ministry of Construction (MOC). (2016). *Current situation of technical infrastructure system in Vietnam urban areas*. Author.

Ministry of Construction (MOC). (2017a). *Report on implementation of the National Strategy on Integrated Management of Solid Waste to 2025, vision to 2050*. Author.

Ministry of Construction (MOC). (2017b). *Decision 1354/QD-BXD dated 29 December 2017 promulgated investment and treatment cost norms for household solid waste*. Author.

Ministry of Construction (MOC). (2019). *Annual Report for Reviewing Activities in 2018 and Implement Plans in 2019 of Construction Sector*. Author.

Ministry of Disaster Management and Relief. (2015). *Bangladesh national debris management guidelines*. Dhaka, Bangladesh: Author.

Ministry of Disaster Management. (2014). *Sri Lanka Comprehensive Disaster Management Programme (SLCDMP) 2014-2018*. Retrieved from http://www.disastermin.gov.lk/web/images/pdf/slcdmp%20english.pdf

Ministry of Ecology and Environment. (2018). *Environmental Access Conditions for Waste to Energy Construction Project*. Retrieved July 1, 2019, from: http://www.sdfgw.gov.cn/art/2018/3/20/art_ 4996_817510.html

Ministry of Environment and Urbanization. (2017a). *Medical Waste Control Regulation*. MWCR.

Ministry of Environment and Urbanization. (2017b). National Waste Management and Action Plan 2023. Ankara, Turkey: Author.

Ministry of Environment and Urbanization. (2018). Packaging Waste Statistics 2016. Ankara, Turkey: Author.

Ministry of Environment Cambodia. (2019a). *Draft National Waste Management Strategy and Action Plan*. Phnom Penh: Ministry of Environment Cambodia.

Ministry of Environment Cambodia. (2019b). *Waste Management Data*. doi:10.1007/978-1-62703-176-9_8

Ministry of Environment Cambodia. Sub-Decree No. 168 on the management of plastic bags, Pub. L. No. No. 168 GNKR.BK, Ministry of Environment Cambodia 11. (2018). *OD Mekong Datahub*. Retrieved from https://data.opendevelopmentmekong.net/lo/dataset/sub-decree-no-168-on-the-management-of-plastic-bags/resource/ea9eeff5-78db-4ba9-b9c3-ad5135ba1d47?type=laws_record

Ministry of Environment. (2019). *Regional 3Rs Forum in Asia and the Pacific Cambodia - Country Report*. UNCRD. Retrieved from http://www.uncrd.or.jp/?page=view&nr=1174&type=13&menu=198

Ministry of Environmental Protection. (2018). *Standard for Pollution Control on the Landfill Site of Municipal Solid Waste*. Retrieved July 1, 2019, from: http://codeofchina.com/standard/GB16889-2008.html

Ministry of Health. (2012). Environmental management framework for healthcare waste and infrastructure development, Sector development program draft. Sri Lanka: Author.

Ministry of Health. (2017). *Medical Waste Management Plan (MWMP)*. Retrieved from http://www.rbc.gov.rw/fileadmin/user_upload/SPRP_MWMP_November_21_2017.pdf

Ministry of Housing and Urban-Rural Development. (2016). *13th Five-year Plan for Housing and Urban-rural Construction*. Retrieved July 1, 2019, from: https://max.book118.com/html/2017/1219/ 145056289.shtm

Ministry of Housing and Urban-Rural Development. (2018). *Notice on Promoting the Experience of Waste Classification and Resource Utilization in Rural Areas of Jinhua City*. Retrieved July 1, 2019, from: http://www.mohurd.gov.cn/wjfb/201612/t20161228_230118.html

Ministry of Internal and Home Affairs and Provincial Councils and Local Government (MoIHA&PCLG). (2019). *National Solid Waste Management Support Center of Sri Lanka*. Retrieved from http://www.lgpc.gov.lk/web/index.php?option=com_content&view=article&id=71&Itemid=190&lang=en

Ministry of Mahaweli Development and Environment (MoMD&E). (2016). *Comprehensive integrated solid waste management plan for target provinces in Sri Lanka, developed with technical assistance from the United Nations Human Settlements Programme. Sobadam Piyasa, No. 416/C/1*. Battaramulla, Sri Lanka: Robert Gunawardana Mawatha.

Ministry of Mahaweli Development and Environment. (2018). *National Waste Management Policy of Sri Lanka*. Retrieved from http://mmde.gov.lk/web/images/pdf/2018/nationalwastemanagementpolicy-english.pdf

Ministry of Natural Resources and Environment (MONRE) & Ministry of Construction (MOC). (2009). *Report on development of the National Strategy of Integrated Solid Waste Management to 2025, vision to 2050*. Author.

Ministry of Natural Resources and Environment (MONRE). (2003). *First National Communication to the United Nations Framework Convention on Climate Change (UNFCCC)*. Author.

Ministry of Natural Resources and Environment (MONRE). (2010). *Second National Communication to the United Nations Framework Convention on Climate Change (UNFCCC)*. Author.

Ministry of Natural Resources and Environment (MONRE). (2011). *National state of environment report 2011 - Solid waste*. Vietnam Publishing House of Natural Resources, Environment and Cartography.

Ministry of Natural Resources and Environment (MONRE). (2012). *Report on development of Program of strengthening control of environmental pollution caused by using non-biodegradable plastic bag in daily life up to 2020*. Author.

Ministry of Natural Resources and Environment (MONRE). (2014). *National State of Environment Report 2014 - Rural Environment*. Vietnam Publishing House of Natural Resources, Environment and Cartography.

Ministry of Natural Resources and Environment (MONRE). (2015). *National State of Environment Report 2011-2015*. Author.

Ministry of Natural Resources and Environment (MONRE). (2016). *National State of Environment Report 2016 - Urban Environment*. Vietnam Publishing House of Natural Resources, Environment and Cartography.

Ministry of Natural Resources and Environment (MONRE). (2017a). *National State of Environment Report 2017 - Waste Management*. Vietnam Publishing House of Natural Resources, Environment and Cartography.

Ministry of Natural Resources and Environment (MONRE). (2017b). *The second Biennial Updated Report of Vietnam to the UNFCCC on Climate Change*. Vietnam Publishing House of Natural Resources, Environment and Cartography.

Ministry of Natural Resources and Environment (MONRE). (2018a). *Administrative map of Socialist Republic of Vietnam*. Author.

Ministry of Natural Resources and Environment (MONRE). (2018b, July). *Report on solid waste management in Vietnam*. Workshop with Ministry of Environment of Japan.

Ministry of Natural Resources and Environment (MONRE). (2018c). *Report on management of household waste, 5/2018*. Author.

Ministry of Natural Resources and Environment (MONRE). (2018d). *Report on environmental management in scrap import and usage for production, 7/2018*. Author.

Ministry of Natural Resources. (2015). *National e-Waste Policy for Rwanda*. Retrieved from http://www.minirena.gov.rw/fileadmin/Environment_Subsector/Laws__Policies_and_Programmes/Laws/E-waste_policy_-_FINAL.pdf

Ministry of New and Renewable Energy Sadhan Kumar Ghosh. (2017). *India Report on 3R Status*. UNCRD. Retrieved from http://www.uncrd.or.jp/content/documents/5688[1121]%20India.pdf

Ministry of Planning and Finance. (2018). *Myanmar Sustainable Development Plan (2018-2030), the Government of the Republic of the Union of Myanmar*. Nay Pyi Taw.

Ministry of Trade and Industry. (2011). *National Industrial Policy*. Retrieved from http://www.minicom.gov.rw/fileadmin/minicom_publications/policies/Industrial_Policy-2.pdf

Misra, V., & Pandey, S. D. (2005). Hazardous waste, impact on health and environment for development of better waste management strategies in future in India. *Environment International*, *31*(3), 417–431. doi:10.1016/j.envint.2004.08.005 PMID:15734194

Mnyanyi, N. B. (2014). *Assessment of Solid Waste Management in Kinondoni Municipality* (Unpublished master thesis). Open University of Tanzania.

MOE. (2018). *Policy Summary on Integrated Solid Waste Management*. Beirut: Republic of Lebanon – Ministry of Environment.

MOE/EU/UNDP. (2014). *Lebanon Environmental Assessment of the Syrian Conflict & Priority Interventions*. Retrieved from http://www.undp.org/content/dam/lebanon/docs/Energy%20and%20Environment/Publications/EASC-WEB.pdf

MOE/GEF/UNDP. (2011). *Lebanon's Second National Communication to the UNFCCC*. Retrieved from http://www.lb.undp.org/content/lebanon/en/home/library/environment_energy/lebanon_s-second-national-communication-to-the-unfccc.html

MOE/GEF/UNDP. (2015). *National Greenhouse Gas Inventory Report and Mitigation Analysis for the Waste Sector in Lebanon*. Retrieved from http://climatechange.moe.gov.lb/viewfile.aspx?id=222

MOE/UNDP. (2017). *Nationally Appropriate Mitigation Action in Lebanon's Municipal Solid Waste Sector NAMA Proposal and Design Document*. Beirut: Republic of Lebanon – Ministry of Environment.

MOE/UNDP. (2017). *Updated master plan for the closure and rehabilitation of uncontrolled dumpsites throughout the country of Lebanon*. Beirut: Republic of Lebanon – Ministry of Environment.

MOE/UNDP/ELARD. (2011). *Provision of consultancy services for the Preparation of a Master plan for the Closure and Rehabilitation of Uncontrolled Dumps throughout the country of Lebanon*. Retrieved from http://www.moe.gov.lb/The-Ministry/Reports/Master-plan-for-the-closure-and-rehabilitation-of.aspx

MOE/UNDP/GEF. (2016). Lebanon's third national communication to the UNFCCC. Beirut, Lebanon: Author.

MOECAF. (2012). *Myanmar's Initial National Communication under the United Nations Framework Convention on Climate Change*. Retrieved from https://unfccc.int/resource/docs/natc/mmrnc1.pdf

MOF. (2011). *Lebanon Country Profile 2011*. Beirut: Ministry of Finance. Retrieved from http://www.finance.gov.lb/en-us/Finance/Rep-Pub/DRI-MOF/Sovereign%20and%20Investment%20Reports/Lebanon%20Country%20Profile%202011.pdf

Mollel, E. L. (2016). *Economic Analysis of the Solid Waste Management Options in Morogoro Municipality* (Unpublished master thesis). Sokoine University of Agriculture, Morogoro, Tanzania.

Monella, J., & Leyaro, V. (2017). Determinants of Households Willingness to Participate In Solid Waste Separation for Reduce, Reuse and Recycle: The Case of Dar es Salaam. *University of Dar es Salaam Journal, 3*(1), 57–82.

Mongtoeun, Y., Fujiwara, T., & Spoann, V. (2019). Household Solid Waste Generation and Socioeconomic Factors in the Capital City of Cambodia. *International Journal of Environmental Sciences & Natural Resources, 20*(1), 4. doi:10.19080/IJESNR.2019.20.556029

Mothiba, M., Moja, S. J., & Loans, C. (2017). A Review of the Working Conditions and Health Status of Waste Pickers at Some Landfill Sites in the of Tshwane Metropolitan Municipality, South Africa. *Advances in Applied City Science Research, 8*(3), 90–97.

Mrkajić, V., Stanisavljevic, N., Wang, X., Tomas, L., & Haro, P. (2018). Efficiency of packaging waste management in a European Union candidate countries. *Resources, Conservation and Recycling, 1*(136), 130-141.

Mun, V. (2016). *Progress and Challenges of Deconcentration in Cambodia: The Case of Urban Solid Waste Management* (CDRI Working Paper Series No. 110). Retrieved from www.cdri.org.kh

Municipality of Los Banos, Laguna. (2001). Los Banos Municipal Ordinance No. 2001-08.

Muny, M. (2016). *Survey Report on Waste Management Practices At Municipality/District Level*. Retrieved from https://www.kas.de/c/document_library/get_file?uuid=9e9be00d-4355-3a39-a006-9e856bb5eb58&groupId=252038

Mutezo, G. (2016). *Challenges impeding South African municipalities from adopting waste-to-energy schemes: an exploratory approach* (Mphil Thesis). University of Cape Town, RSA.

Muzenda, E. (2014). A Discussion on Waste Generation and Management Trends in South Africa. International Journal of Chemical. *Environmental and Biological Sciences, 2*(2), 105–113.

Muzenda, E., & Popa, C. (2015). Waste tyre management in Gauteng, South Africa: Government, industry and community perceptions. *International Journal of Environmental Sciences and Development, 6*(4), 311–315. doi:10.7763/IJESD.2015.V6.609

MYANMORE. (2018). *Myanmar targets plastic waste in growing green movement*. Retrieved from https://www.myanmore.com/yangon/2018/06/myanmar-targets-plastic-waste-in-growing-green-movement/

Nagamori, M., Mowjood, M. I. M., Watanabe, Y., Isobe, Y., Ishigaki, T., & Kawamoto, K. (2016). Characterization of temporal variations in landfill gas components inside an open solid waste dump site in Sri Lanka. *Journal of the Air & Waste Management Association, 66*(12), 1257–1267. doi:10.1080/10962247.2016.1212746 PMID:27575846

Nahallage, C. A. D., & Huffman, M. A. (2013). Macaque-human interactions in past and present-day Sri Lanka. In S. Radhakrishna, M. Huffman, & A. Sinha (Eds.), *The Macaque connection. Developments in primatology: progress and prospects* (Vol. 43, pp. 135–148). New York, NY: Springer. doi:10.1007/978-1-4614-3967-7_9

Nahar, N. (1998). *Handling, storage, use and disposal of toxic and hazardous materials in Bangladesh.* Bangladesh University of Engineering and Technology.

Nasilele, M. R. (2015). *High Value Organic Waste Treatment via Black Soldier Fly Bioconversion* (Unpublished Master Thesis). Royal Institute of Technology, Stockholm, Sweden.

Nasrin, F. (2016). Waste management in Bangladesh: Current situation and suggestions for action. *International Research Journal of Social Sciences, 5*(10), 36–42. Retrieved from http://www.isca.in/IJSS/Archive/v5/i10/7.ISCA-IRJSS-2016-113.pdf

National Assembly of Vietnam (NAV). (2014). *Law No. 55/2014/QH13 on Environmental Protection.* Author.

National Biomass Strategy. (2020). *New wealth creation for Malaysia's palm oil industry.* Agensi Inovasi Malaysia.

National Bureau of Statistics of China. (2018). *China Statistical Yearbook 2018.* Beijing: China Statistics Press.

National Bureau of Statistics of China. (2018). *National Data.* Retrieved may 28, 2019, from: http://data.stats.gov.cn/

National Bureau of Statistics. (2010). *Integrated Business Survey, 2010: Construction Industry Analytical Report.* Tanzania Mainland.

National Bureau of Statistics. (2013). *Basic Facts and Figures, 2012 on Human Settlements in Tanzania Mainland.* Author.

National Development and Reform Commission, Ministry of Housing and Urban-Rural Development. (2016). *13th Five-year Plan for Construction of MSW Treatment Facilities.* Retrieved April 20, 2019, from: http://www.ndrc.gov.cn/zcfb/zcfbtz/201701/t20170122_836016.html

National Development and Reform Commission, Ministry of Housing and Urban-Rural Development. (2017). *Implementation Plan for Municipal Solid Waste Classification System.* Retrieved June 20, 2019, from: http://www.gov.cn/zhengce/content/2017-03/30/content_5182124.htm

National Development and Reform Commission. (2016). *Notice on The Publication of the 13th Five Year Plan for The Development of Renewable Energy by The National Development and Reform.* Retrieved June 18, 2019, from: https://chinaenergyportal.org/en/13th-fyp-development-plan-renewable-energy/

National Development and Reform Commission. (2017). *13th Five-year Plan for Biomass Energy.* Retrieved June 20, 2019, from: http://www.ndrc.gov.cn/fzgggz/fzgh/ghwb/gjjgh/201708/20170809_857320.html

National Development and Reform Commission. (2017). *13th Five-year Plan of Biogas Development in Villages.* Retrieved June 20, 2019, from: http://www.ndrc.gov.cn/fzgggz/fzgh/ghwb/gjjgh/201706/ t20170607_850195.html

National Development and Reform Commission. (2017). *Notice on Further Planning and Site Selection of MSW Power Plants.* Retrieved June 19, 2019, from: http://www.ndrc.gov.cn/ zcfb/gfxwj/201712/ t20171220_ 871129.html

National Development and Reform Commssion. (2015). *Notice on Further Accelerating the Comprehensive Utilization of Crop Straw and Prohibition of Burning.* Retrieved June 25, 2019, from: http://www.ndrc.gov.cn/zcfb/zcfbtz/201511/t20151125_759523.html

National Institute of Statistics of Rwanda. (2012a). *Size of the resident population.* Retrieved from http://www.statistics.gov.rw/publication/size-resident-population

National Institute of Statistics of Rwanda. (2012b). *The Fourth Population and Housing Census in Rwanda.* Retrieved from http://microdata.statistics.gov.rw/index.php/catalog/65

National Sample Census of Agriculture. (2012). *The (2007/2008) report.* Author.

National Solid Management Commission. (2016). *National Solid Waste Management Strategy 2012-2016.* Retrieved from nswmc.emb.gov.ph/wp-content/uploads/2016/07/NSWM-Strategy-2012-2016.pdf

National Solid Waste Management Support Centre (NSWMSC). (2015). *Data base of solid waste management-2015.* Sri Lanka: Author.

Needhidasan, S., & Melvin, S. S. (2013). Electronic Waste – Health and Environmental Impact and Recycling Practices. *International Journal of Advance Research, 1*(8), 8–10.

NESDC. (2018). Thai Economic Performance in Q3 and Outlook for 2018 - 2019. Office of the National Economic and Social Development Council, Thailand.

Netherland Enterprise Agency (2017). *Scoping Mission on Integrated Solid Waste Management (ISWM) to Myanmar.* Author.

Netherlands Enterprise Agency. (2018). *Circular economy Vietnam waste management & recycling.* Retrieved from https://www.rvo.nl/sites/default/files/2018/09/Waste-management-recycling.pdf

Ngan, T. (2018). *Solid waste management in Vietnam: Current situation, challenges and strategies for development* (Unpublished undergraduate dissertation). Metropolia University of Applied Sciences.

Niekerk, S., & Weghmann, V. (2019). *Municipal solid waste management services in Africa.* Working Paper, Public Service International, Johannesburg, RSA.

Nie, Y. (2008). Development and Prospect of Municipal Solid Waste (MSW) Incineration in China. *Frontiers of Environmental Science & Engineering, 2*(1), 1–7. doi:10.100711783-008-0028-6

Nilsson, M., Eklund, M., & Tyskeng, S. (2009). Environmental integration and policy implementation: Competing governance modes in waste management decision making. *Environment and Planning. C, Government & Policy, 27*(1), 1–18. doi:10.1068/c0794j

Nixon, H., Joelene, K. C., Chit Saw, P., Aung Lynn, T., & Arnold, M. (2013). *State and Region Governments in Myanmar.* Yangon: MDRI-CESD and the Asia Foundation.

Njoku, P., Odiyo, J., Durowoju, O., & Edokpayi, J. (2018). A review of landfill gas generation and utilisation in Africa. *Open Environmental Sciences, 10*(1), 1–15. doi:10.2174/1876325101810010001

Nordrum, A. (2015). China, India, US among world's worst marine polluters: country-by-country breakdown of plastics in oceans. *International Business Times.* Retrieved from http://www.ibtimes.com/china-india-us-among-worlds-worst-marine-polluters-country-country-breakdown-plastics-1813476

Nshimiyimana, V. (2015). *Assessment of municipal solid waste management practices case study: Nyarugenge district-Rwanda (Master's Thesis).* University of Makelele. Retrieved from http://makir.mak.ac.ug /handle/10570/5863

NSO. (2016). *Report on Population Characteristics The 2015-2016 Survey of Population Change.* Thailand National Statistical Office.

NSWMC/Environment Management Bureau/ Department of Environment and Natural Resources. (2015). *National solid waste management status report (2008-2014).* Author.

NSWMD. (2012). *Solid Waste Management in Malaysia: The Way Forward.* Retrieved from http://ensearch.org/wp-content/uploads/2012/07/Paper-13.pdf

Ntakamulenga, R. (2012, January). *The Status of Solid Waste Management in Tanzania.* A paper presented at the Coastal East Africa solid waste workshop, Flic en Flac, Mauritius.

Oduro-Appiah, K., & Aggrey, B. E. (2013). Determinants of source separation of municipal solid waste in developing countries: The case of Ghana. *Journal of Sustainable Development in Africa, 15*, 47-60.

Oduro-Appiah, K., Scheinberg, A., Afful, A., & de Vries, N. (2019). *The contribution of participatory planning processes to reliable data gathering and inclusive policy development in municipal solid waste management in the greater Accra metropolitan area of Ghana.* Manuscript submitted for publication.

Oduro-Appiah, K., Scheinberg, A., Mensah, A., Kotey, V., Afful, A., & de Vries, N. (2017). *Locally responsive intervention to improve municipal solid waste collection coverage in Accra, Ghana.* Paper presented at the 40th WEDC International Conference, Loughborough, UK.

Oduro-Appiah, K., Afful, A., Kotey, V., & de Vries, N. (2019). Working with the Informal Service Chain as a Locally Appropriate Strategy for Sustainable Modernization of Municipal Solid Waste Management Systems in Lower-Middle Income Cities: Lessons from Accra, Ghana. *Resources*, *8*(1), 12. doi:10.3390/resources8010012

Oduro-Appiah, K., Scheinberg, A., Mensah, A., Afful, A., Boadu, H. K., & de Vries, N. (2017). Assessment of the municipal solid waste management system in Accra, Ghana: A'' Wasteaware' benchmark indicator approach. *Waste Management & Research*, *35*(11), 1149–1158. doi:10.1177/0734242X17727066 PMID:28931348

Oduro-Kwarteng, S., & van Dijk, M. P. (2013). The effect of increased private sector involvement in solid waste collection in five cities in Ghana. *Waste Management & Research, 31*(10_suppl), 81-92.

OECD. (2008). *Measuring Material Flows and Resource Productivity, Volume 1.* The OECD Guide. Retrieved from http://www.oecd.org/environment/indicators-modellingoutlooks/MFA - Guide. pdf#page=1&zoom=auto,-82

Oelofse, S. (2018). Opportunities and challenges in waste management. *CSIR*. Retrieved 9 July 2019 from: http://sustainabilityweek.co.za/assets/files/Day%203%20-%20CSIR%20-%20Waste.pdf

Oeltzchner, H., & Mutz, D. (1996). *Guidelines for an Appropriate Management of Domestic Sanitary Landfills.* Gesselschaft fur Technische Zusammenarbeit, Eschborn, German.

Off. Gazette RS, No 36/2009 i 88/2010. Law on Waste Management

Off. Gazette RS, No 61/2010. The Rulebook on the Methodology for Collection of Data on Composition and Quantities of Municipal Waste at the Territory of Local Self-Government Units

Office of the Auditor General of State Finances. (2016). *Performance audit report on management of solid and liquid (sewage) waste in city of Kigali, Kigali-Rwanda.* Retrieved from http://www.oag.gov.rw/fileadmin/user_upload/Performance_Reports/MANAGEMENT_OF_SOLID_AND_LIQUID

Office of the Auditor General of State Finances. (2016). Performance audit report on management of solid and liquid (sewage) waste in city of Kigali, Kigali-Rwanda. Retrieved from http://www.oag.gov.rw/fileadmin/user_upload/Performance_Reports/MANAGEMENT_OF_SOLID_AND_LIQUID __SEWAGE__IN_CITY_OF_KIGALI.pdf

Office of the State Council. (2016). *Pilot Work Plan for None Waste City Construction.* Retrieved June 18, 2019, from: http://www.gov.cn/xinwen/2019-01/21/content_5359705.htm

Office of the State Council. (2017). *China to Boost Recycling of Agriculture Waste.* Retrieved June 18, 2019, from: http://english.gov.cn/policies/latest_releases/2017/06/12/content_281475684141592.htm

Office of the State Council. (2017). *Prohibition of Foreign Garbage Entry and Promotion of Solid Waste Import Management System Reform Implementation Plan.* Retrieved June 20, 2019, from: https://www.isri.org/docs/default-source/int'l-trade/170727chinaannouncement-en.pdf

Omar, D., Razali, S. N., & Karunpannan, S. (2012). Clinical Waste Management in District Hospital of Tumpat, Batu Pahat and Taiping. Asia Pacific International Conference on Environment-Behavior Studies. *Procedia: Social and Behavioral Sciences*, *68*, 134–135. doi:10.1016/j.sbspro.2012.12.213

One Earth One Ocean. (2018). *PROJECTS - One Earth One Ocean.* Retrieved April 5, 2019, from https://oneearth-oneocean.com/en/the-association/projects/

ONEP. (2018). *Second Biennial Update Report of Thailand.* Office of Natural Resources and Environmental Policy and Planning, Ministry of Natural Resources and Environment.

Onuma, Y., Siriwardana, A. N., & Rajapakse, R. D. D. (2018). Estimation of E-waste flow in Kurunegala Municipal Council, Sri Lanka. In *The 29th Annual Conference of JSMCWM, 2018* (pp. 547-548). Academic Press.

Oteng-Ababio, M. (2011). Beyond technical details: The stalled Kwabenya engineered sanitary landfill project in Accra, Ghana. *Geografisk Tidsskrift-Danish Journal of Geography*, *111*(2), 169–179. doi:10.1080/00167223.2011.10669531

Oteng-Ababio, M. (2014). "Guilty with explanation": Rethinking the destiny of landfills in a Millennium City in Ghana. *Management of Environmental Quality*, *25*(2), 200–215. doi:10.1108/MEQ-11-2012-0074

Oteng-Ababio, M. (2018). Crossing conceptual boundaries: Re-envisioning coordination and collaboration among women for sustainable livelihoods in Ghana. *Local Environment*, *23*(3), 316–334. doi:10.1080/13549839.2017.1418847

Oteng-Ababio, M., Amankwaa, E. F., & Chama, M. A. (2014). The local contours of scavenging for e-waste and higher-valued constituent parts in Accra, Ghana. *Habitat International*, *43*, 163–171. doi:10.1016/j.habitatint.2014.03.003

Oteng-Ababio, M., Arguello, J. E. M., & Gabbay, O. (2013). Solid waste management in African cities: Sorting the facts from the fads in Accra, Ghana. *Habitat International*, *39*, 96–104. doi:10.1016/j.habitatint.2012.10.010

Oteng-Ababio, M., Owusu-Sekyere, E., & Amoah, S. T. (2017). Thinking globally, acting locally: Formalizing informal solid waste management practices in Ghana. *Journal of Developing Societies*, *33*(1), 75–98. doi:10.1177/0169796X17694517

Outwater, A. H., Pamba, S., & Outwater, A. B. (2013). *Water –Related Diseases of People using Municipal Wasterwater: Risks.* Exposure, Effects on Health and Control Approaches in Tanzania.

PAHO-Argentine Ministry of Health. (2018). *Argentina Indicadores Básicos 2017* (Argentine, Basic Indicators 2017). Author.

Pak EPA Pakistan. (2019). *Waste Management.* Available at file:///G:/Sir%20arshad/book%202/Pakistan%20-%20Waste%20Management%20_%20export.gov.html

Pakistan - Waste Management. (n.d.). Retrieved from https://www.export.gov/article?id=Pakistan-Waste-Management

Pakistan Map and Satellite Image. (2019). Assessed on 29 March 2019. https://geology.com/world/pakistan-satellite-image.shtml

Pakistan National Report submitted to The United Nations Conference on Environment and Development. (1991). Environmental and Urban Division, Government of Pakistan (GOP).

Palfreman, J. (2014). *Waste Management and Recycling in Dar es Salaam, Tanzania*. Retrieved 02/05/2016 from https://www.researchgate.net/publication/271441207

Papong, S., Yuvaniyama, C., Lohsomboon, P., & Malakul, P. (2004). *Overview of biomass utilization in Thailand*. Paper presented at the ASEAN Biomass Meeting. PCD. An Overview of Hazardous Waste Management & Prevention in Thailand. Retrieved March 15, 2019, from http://www.pcd.go.th/count/mgtdl.cfm?FileName=HWM_111053.pdf&BookName=HWM_111053

Patwary, M. A., O'Hare, W. T., Street, G., Elahi, K. M., Hossain, S. S., & Sarker, M. H. (2009). Quantitative assessment of medical waste generation in the capital city of Bangladesh. *Waste Management (New York, N.Y.)*, *29*(8), 2392–2397. doi:10.1016/j.wasman.2009.03.021 PMID:19375297

Pauze. (2016). *Integrated Solid Waste Management: Challenge and Future*. NEHAP Conference 2016. 26 September 2016. Putrajaya International Convention Centre. Retrieved from http://nehapmalaysia.moh.gov.my/wp-content/uploads/2016/03/Paper-2-Solid-Waste.pdf

PCD. (2011). *Draft of National 3Rs Strategy*. Pollution Control Department, Ministry of Natural Resources and Environment.

PCD. (2012). *Parallel Roundtable 4: Economic Opportunities through 3R ~ Waste Electrical and Electronics Equipment (WEEE)*. Pollution Control Department, Ministry of Natural Resources and Environment.

PCD. (2014). *Roadmap for MSW and hazardous waste management approved by national council for peace and order*. Pollution Control Department, Ministry of Natural Resources and Environment.

PCD. (2016). *Thailand State of Pollution Report 2015*. Pollution Control Department, Ministry of Natural Resources and Environment.

PCD. (2017). *Thailand State of Pollution Report 2016*. Pollution Control Department, Ministry of Natural Resources Bangkok.

PCD. (2018). *Thailand State of Pollution Report 2017*. Pollution Control Department, Ministry of Natural Resources and Environment.

Perez, C. (2006). *Local ecological governance: Viable solutions to the solid waste management problem of Los Banos*. Academic Press.

Perez, C., Faylon, P., Pantua, L., & Valdez. (2002). *Managing town-community solid waste: Malinis at maayos na kapaligiran malusog at matinong pamayanan*. Paper presented at the 2002 PACS Seminar Series of the Philippine Association of Career Scientists, Philippines.

Perrin, D. (2018). *Key results from a baseline study on waste generation of all the provinces*. Phnom Penh: UNDP.

Pharino, C. (2017). *Challenges for sustainable solid waste management: lessons from Thailand: Springer. Piyapanpong, S. Parallel Roundtable 4: Economic Opportunities through 3R ~ Waste Electrical and Electronics Equipment (WEEE)*. Pollution Control Department, Ministry of Natural Resources and Environment.

Philippine Statistics Authority. (2017). Retrieved June 2019, from https://psa.gov.ph/population-and-housing/node/120080

Phnom Penh Capital Administration. (2017). *Phnom Penh Green City Strategic Plan 2017-2026*. Author.

Pichtel, J. (2014). *Waste management practices: municipal, hazardous, and industrial* (2nd ed.). Taylor & Francis Group. doi:10.1201/b16576

Pirsaheb, M., Khosravi, T., & Sharafi, K. (2013). Domestic scale vermicomposting for solid waste management. *International Journal of Recycling of Organic Waste in Agriculture*, *2*(1), 4. doi:10.1186/2251-7715-2-4

Planning Commission Report. (2014). *Reports of the task force on waste to energy (Vol-I) (in the context of Integrated MSW management).* Retrieved from http:// planningcommission.nic.in/reports/genrep/rep_wte1205.Pdf

Plastic Free Cambodia. (2018). *Plastic Free Cambodia – Reducing plastic pollution in SEA*. Retrieved April 13, 2019, from https://plasticfreecambodia.com/

Plastics Waste Management Rules. (2016). *India*. Retrieved from http://sscgj.in/wp-content/uploads/2016/07/Waste-Management-Rules-2016.pdf

Po, A. V. A. (n.d.). *Technology interventions and 3R strategies to address Quezon City's solid waste generation*. Academic Press.

PPCA. (2015). *Sixth Regional 3R Forum in Asia and the Pacific - City Report - Phnom Penh*. Retrieved from http://www.uncrd.or.jp/content/documents/25936-3R_City-Report_PhnomPenh.pdf

Präger, F., Paczkowski, S., Sailer, G., Derkyi, N. S. A., & Pelz, S. (2019). Biomass sources for a sustainable energy supply in Ghana–A case study for Sunyani. *Renewable & Sustainable Energy Reviews*, *107*, 413–424. doi:10.1016/j.rser.2019.03.016

Prairie Village, K. (1998). *Characterization of building-related construction and demolition debris in the United States*. The US Environmental Protection Agency Municipal and Industrial Solid Waste Division Office of Solid Waste.

Premakumara, D. G. J., & May Tin, H. O. (2019). Results of Quick Survey on Solid Waste Management in Mandalay Region. A Presentation make at the Regional Workshop for Improving Municipal Solid Waste Management in Mandalay, Myanmar.

Premakumara, D. G. J., Kataoka, Y., & Chowdhury, M. (2016a). *Development of Environmental Learning Programme for Establishing a Sustainable Solid Waste Management System in Mandalay City, Myanmar*. The HDCA 2016 Conference: Capability and Diversity in a Global Society, 1-3 September 2016, IGES, Japan.

Premakumara, D.G.J., Hengseshaugh, M., Onogawa, K., & May Tin, H. O. (2017). *Waste Management in Myanmar, Current status, key challenges and recommendations for National and City Waste Management Strategies*. IGES and UN Environment.

Premakumara, D. G. J., & Hengesbaugh, M. (2016). *Quick Study on Waste Management in Myanmar: Current Situation and Key Challenges*. IGES.

Press Information Bureau, Ministry of New and Renewable Energy, Government of India. (2014). *Production of Energy from Agricultural Waste*. Retrieved from https://www.progressivegardening.com/agricultural-wastes/energy-production-from-agricultural-waste.html

Prime Minister (PM). (2003). *Decision 64/2003/QD-TTg dated 22 April 2003 on approval of Plan for treatment of seriously polluting establishment*. Author.

Prime Minister (PM). (2008). *Decision 1440/QĐ-TTg dated 06 October 2008 approved Master plan on construction of solid waste treatment facilities in three Northern, Central and Southern key economic regions up to 2020*. Author.

Prime Minister (PM). (2010). *Decision 1873/QD-TTg dated 11 October 2010 approved Master plan for the construction of the solid waste treatment site in the key economic area of Mekong Delta to 2020*. Author.

Prime Minister (PM). (2012). *Decision 170/QD-TTg dated 08 February 2012 approved master plan for comprehensive system for treatment of hazardous medical waste to 2025*. Author.

Prime Minister (PM). (2013). *Decision 2211/QĐ-TTg dated 14 November 2013 approved Master-plan for solid waste management in Cau river basin to 2020*. Author.

Prime Minister (PM). (2014). *Decision 31/2014/QĐ-TTg dated 05 May 2014 approved Regulations on support mechanism to waste-to-energy projects development in Vietnam*. Author.

Prime Minister (PM). (2015a). *Decision 16/2015/QD-TTg dated 22 May 2015 on take-back and treatment of discarded products*. Author.

Prime Minister (PM). (2015b). *Decision 07/QĐ-TTg dated 06 January 2015 approved Master plan for solid waste management in Dong Nai river basin to 2030*. Author.

Prime Minister (PM). (2015c). *Decision 223/QĐ-TTg dated 12 February 2015 approved Master-plan for solid waste management in Nhue-Day river basin to 2020*. Author.

Prime Minister (PM). (2015d). *Decision 2068/QĐ-TTg dated 25 November 2015 approved Vietnam strategy on development of renewable energy*. Author.

Prime Minister (PM). (2016a). *Decision 428/QD-TTg dated 18 March, 2016, approved Revision of national master-plan on electricity development 2011-2020 vision to 2030*. Author.

Prime Minister (PM). (2016b). *Decision 1979/QD-TTg dated October 14, 2016, on approval for solid waste management planning in Northern Key economic region by 2030*. Author.

Prime Minister (PM). (2018a). *Decision 491/QD-TTg dated 07 May 2018 on approval of Adjusted National strategy on Integrated solid waste management to 2025, vision to 2050*. Author.

Prime Minister (PM). (2018b). *Resolution 27/CT-TTg dated 17 Sep 2018 on immediate measures in management of import and usage of scraps for production materials*. Author.

Priyashantha, K. A. S., Ariyawansha, R. T. K., Senevirathne, S. A. D. N., Basnayake, B. F. A., & Chandrasena, A. S. H. (2015). Development and performance Evaluation of the leachate treatment system at Gohagoda municipal solid waste disposal site. In *Proceedings of 6th International Conference on Structural Engineering and Construction Management*, (pp. 31-38). Academic Press.

Prüss-Üstün, A., Giroult, E., Rushbrook, P., & Organization, W. H. (1999). *Safe management of wastes from health-care activities. REO-16. (2015). Performance evaluation of solid waste disposal in Songkhla: Regional Environmental Office 16*. Ministry of Natural Resources and Environment.

Quang, N. X. (2017). Waste-to-Energy: Potential and Reality for Vietnam. *Vietnam Energy Online*. Retrieved from http://nangluongvietnam.vn/news/vn/bao-ton-nang-luong/dot-rac-phat-dien-tiem-nang-va-hien-thuc-cho-viet-nam.html

Quang, N. D., Okumura, H., Yamasue, E., & Ishihara, K. (2009). Use and disposal of large home electronic appliances in Vietnam. *Journal of Material Cycles and Waste Management*, 364.

Rachael, L. (2018). *Solid Waste Management in Pakistan*. Retrieved from https://www.bioenergyconsult.com/solid-waste-management-in-pakistan/

Rahman, M. A. (2017). *E-waste management: A study on legal framework and institutional preparedness in Bangladesh*. North South University. Retrieved from http://www.northsouth. edu/newassets/files/ppg-research/PPG_5th_Batch/14._Anis_E-waste_Management. pdf

Ramanathan, R., & Ganesh, L. S. (1994). Group preference aggregation methods employed in AHP: An evaluation and an intrinsic process for deriving members' weightages. *European Journal of Operational Research, 79*(2), 249–265. doi:10.1016/0377-2217(94)90356-5

Ranjan, H., Karunasena, G., & Rathnayake, U. (2014). Construction and Demolition Waste Management Gaps in Construction Industry. In *Proceedings of the 7th FARU International Research Symposium* (pp. 97-104). Academic Press.

REMA. (2009). *Rwanda State of Environment and Outlook*. Ministry of Natural Resources. doi:10.13140/RG.2.1.5148.6328

Republic of Rwanda Ministry of Infrastructure. (2010). *National Policy & Strategy for Water Supply and Sanitation Services*. Retrieved from http://www.rura.rw/fileadmin/docs/Board_Decisions/WATSAN_Policy_Strategy.pdf

Republic of the Philippines. (2019). *Official Website of the Republic of the Philippines*. Retrieved August 2019, from https://www.gov.ph/about-the-philippines

Republic of the Philippines. RA 6969. Toxic Substances and Hazardous and Nuclear Waste Act of 1990.

Rode, S. (2012). E-waste management in Mumbai metropolitan region: Constraints and opportunities. *Theoretical and Empirical Researches in Urban Management, 7*(2), 89-103. Retrieved from http://um.ase.ro/no72/5.pdf

Rodić, L., & Wilson, D. C. (2017). Resolving governance issues to achieve priority sustainable development goals related to solid waste management in developing countries. *Sustainability, 9*(3), 404. doi:10.3390u9030404

RURA. (2009). *Guidelines on the management of waste disposal site (landfill)*. Retrieved from http://www.gicumbi.gov.rw/fileadmin/templates/document/GUIDELINES_Landfill_.pdf

RURA. (2015). *Regulations for the Governing of Solid Wastes Recycling in Rwanda, April 24, 2015. No: 002/ESWASTAN/SW/RURA/2015*. Retrieved from http://www.rura.rw/uploads/media/CLEANED_FINAL_SOLID_WASTES_RECYCLING_REGULATIONS.pdf

RURA. (2016). *Draft Regulations Governing E-waste Management in Rwanda*. Retrieved from http://www.rura.rw/fileadmin/laws/Ewaste_regullations_Rwanda_with_technical_requirements_.pdf

Saaty, T. L. (1980). Multicriteria Decision Making: The Analytic Hierarchy Process. RWS Publications.

Saaty, T. L., & Kearns, K. P. (1985). *Analytical Planning: The organization of system* (1st ed.). Pergamon Press.

Sabai, M. M., Cox, M. G. D. M., & Mato, R. R. A. M., Egmond-de Wilde De Ligny, van, E. L. C., & Lichtenberg, J. J. N. (2013). Concrete block production from construction and demolition waste in Tanzania. *Resources, Conservation and Recycling, 72*, 9–19.

Sabai, S. M. M., Lichtenberg, J. J., Egmond, E. L. C., Florea, M. V. M., & Brouwers, H. J. H. (2016). Construction and Demolition Waste Characteristics in Tanzania. Huria. *Journal of the Open University of Tanzania, 23*(1), 1–19.

Sabai, S. M. M., & Rugudagiza, B. X. (2018). Recycling of Oil Sludge together with Construction and Demolition Waste into Building Materials in Tanzania. *African Journal of Environmental Science and Technology, 12*(2), 84–90.

Sabour, M. R., Mohamedifard, A., & Kamalan, H. (2007). Mathematical model to predict the composition and generation of hospital wastes in Iran. *Waste Management (New York, N.Y.), 27*(4), 584–587. doi:10.1016/j.wasman.2006.05.010 PMID:17239577

Saeed, A. (2013). *Dangerous labour: the little known e-waste phenomenon*. Retrieved from http://www.dawn.com/news/1032967

Saeed, M. A., Irshad, A., & Sattar, H. (2015). Agricultural Waste Biomass Energy Potential In Pakistan. *Proceedings of the International Conference.*

Sahmakum Teang Tnaut. (2019). *Urban Governance Waste Management in Phnom Penh A review of waste management from the streets.* Retrieved from http://teangtnaut.org/wp-content/uploads/2019/01/20190129_STT_Final_Annual_Report_English_version-compressed.pdf

Saifullah, I., Uddin, S. M. K., Alam, A., & Quazi, B. (2009). Practice of demolition waste management. *Proceedings of the International Conference on Solid Waste Management - WasteSafe.* .192510.13140/RG.2.1.4222

Sajid, M., Syed, J. H., Iqbal, M., Abbas, Z., Hussain, I., & Baig, M. A. (2018). *Assessing the generation, recycling and disposal practices of electronic/electrical-waste (E-waste) from major cities in Pakistan.* Waste Mangement. doi:10.1016/j.wasman.2018.11.026

Sakijege, T., Lupala, J., & Sheuya, S. (2012). Flooding, flood risks and coping strategies in urban informal residential areas: The case of Keko Machungwa, Dar es Salaam, Tanzania. Jamba: Journal of Disaster Risk Studies, 4(1).

Saleh, H. M. (2016). *Introduction to hazardous waste management.* INTECH. doi:10.5772/61668

San Carlos City. (2017). *San Carlos City wins anew in SGLG. Tribute: Official publication of San Carlos City, Negros Occidental,* 11, 4.

Sang-Arun, J., Chau, K. H., Uch, R., & Sam, P. (2012). *A guide for technology selection and implementation of urban organic waste utilisation projects in Cambodia* (Vol. 2011–06). Institute for Global Environmental Strategies. Retrieved from https://pub.iges.or.jp/pub/guide-technology-selection-and-implementation

Sankhla, M.S., Kumara, M., Nandan, M., Mohril, S., Gaurav Pratap Singh, G.P., Bhaskar Chaturvedi,B., & Kumar, R. (2017). Effect of Electronic waste on Environmental & Human health- A Review. *Journal of Environmental Science, Toxicology and Food Technology, 10*(9), 98-104.

Sassoon, A. M., & Chheng, N. (2018). *Lack of control at landfill fire.* Retrieved April 12, 2019, from https://www.phnompenhpost.com/national/lack-control-landfill-fire

SAWIC. (2015). *Waste policy and regulations.* Retrieved 12 July 2019 from: http://sawic.environment.gov.za/?menu=13

SAWIC. (2016). *Approach to waste in South Africa.* Retrieved 5 July 2019 from: http://sawic.environment.gov.za/?menu=60

SAWIC. (2018). *Waste information today: reporting on South African SAWIS.* Retrieved 10 July 2019 from: http://sawic.environment.gov.za/documents/8545.pdf

Scheinberg, A. (2011). *Value added: modes of sustainable recycling in the modernisation of waste management systems* (PhD thesis). Wageningen University.

Scheinberg, A., Wilson, D. C., & Rodic-Wiersma, L. (2010). *Solid waste management in the world's cities. In UN-Habitat's State of Water and Sanitation in the World's Cities Series.* Earthscan.

Schenck, R., Blaauw, D., & Vilijoen, K. (2016). Enabling factors for the existence of waste pickers: A systematic review. *Social Work, 52*(1), 35–51.

Schübeler, P., Wehrle, K., & Christen, J. (1996). *Conceptual framework for municipal solid waste management in low-income countries.* Academic Press.

Scopus. (2019). *Advanced Search.* Retrieved July 20, 2019, from: http://www.scopus.com/

Secretariat of the Basel Convention. (2011). *Where are Weee in Africa?* Geneva: Findings from the Basel Convention E-waste Africa Programme. SBC.

Secretary of Environment and Natural Resources (SEMARNAT) -IPN. (2009). *Estudio de análisis, evaluación y definición de estrategias de solución de la corriente de residuos generada por los vehículos usados al final de su vida útil.* Author.

Secretary of Environment and Sustainable Development (SAyDS), Argentina. (2005). *Estrategia Nacional para la Gestión Integral de Residuos Sólidos Urbanos* [National Strategy for Integral Municipal Solid Waste Management, ENGIRSU]. Author.

Secretary of Environment and Sustainable Development (SAyDS). (2011). *Economic and Financial Matrix for calculating the costs of the integrated municipal solid waste management in Argentina.* Argentina: Author.

Seingheng, H., Chiemchaisri, C., Tanchuling, M., Xaypanya, P., & Takemura, J. (2018). Characterization of Landfill Leachates and Sediments in Major Cities of Indochina Peninsular Countries—Heavy Metal Partitioning in Municipal Solid Waste Leachate. *Environments*, *5*(6), 65. doi:10.3390/environments5060065

Seldman, N., & Morris, D. (2019). *China's New Recycling import Policy and the Future of US Municipal Recycling.* Retrieved July 2, 2019, from: https://ilsr.org/china-recycling-policy-opportunity/

Senate Economic Planning Office (SEPO). (2017). *Philippine solid waste at glance.* Retrieved August 20, 2019, from https://www.senate.gov.ph

Senervirathne, S. A. D. N., Basnayake, B. F. A., Chandrasena, A. S. H., & Ariyawansha, R. T. K. (2013). Municipal solid waste dumpsites; a burden or an obscure buried resource. In *Proceedings of the International Conference on Solid Waste 2013 - Innovation in Technology and Management, Hong Kong SAR, P.R. China* (pp.718-721). Academic Press.

Seng, K., Kouk, F., Hul, S., & Penh, P. (2015). *Solid Waste Composition, Forecasting of Waste Generation and Methane generation in Phnom.* Academic Press.

Seng, B., Hirayama, K., Katayama-Hirayama, K., Ochiai, S., & Kaneko, H. (2013). Scenario analysis of the benefit of municipal organic-waste composting over landfill, Cambodia. *Journal of Environmental Management*, *114*, 216–224. doi:10.1016/j.jenvman.2012.10.002 PMID:23168253

Seng, K. (2015). *L'analyse de la composition des dechets solide et la prevision de la production de dechets en Phnom Penh avec la production du methane des sites d'enfouissement Dangkao au Cambodge.* Institut de Technologie du Cambodge.

Seng, K. (2016). *L'analyse de la composition des dechets solide et la prevision de la production de dechets en Phnom Penh avec la prodution du methane des sites d'enfouissement Dangkao au Cambodge.* Institute of Technology of Cambodia.

Sentime, K. (2014). The impact of legislative framework governing waste management and collection in South Africa. *African Geographical Review*, *33*(1), 81–93. doi:10.1080/19376812.2013.847253

Senzige, J. P., Nkansah-Gyeke, Y., Makinde, D. O., & Njau, K. N. (2016). The Potential for Solid Waste Recycling in Urban Area of Tanzania: The Case of Dar Es Salaam. *International Journal of Environmental Protection and Policy.*, *2*(5), 147–152.

Sethy, S. (2017). *State of the 3Rs in Asiat and the Pacific Country Chapter - The Kingdom of Cambodia.* Retrieved from http://www.uncrd.or.jp/content/documents/5686[Nov 2017] Cambodia.pdf

Sethy, S., Sothun, C., & Wildblood, R. (2014). Municipal Solid Waste Management in Cambodia. In A. Pariatamby & M. Tanaka (Eds.), *Municipal Solid Waste Management in Asia and the Pacific Islands* (pp. 77–94). Singapore: Springer Singapore. doi:10.1007/978-981-4451-73-4_5

Shafie, S. M., Mahlia, T. M. I., Masjuki, H. H., & Ahmad-Yazid, A. (2012). A review on electricity generation based on biomass residue in Malaysia. *Renewable & Sustainable Energy Reviews*, *16*(8), 5879–5889. doi:10.1016/j.rser.2012.06.031

Shams, S., Sahu, J. N., Rahman, S. S., & Ahsan, A. (2017). Sustainable waste management policy in Bangladesh for reduction of greenhouse gases. *Sustainable Cities and Society*, *33*, 18–26. doi:10.1016/j.scs.2017.05.008

Shanghai City Government. (2019). *The Municipal Solid Waste Management Regulation of Shanghai Municipality*. Retrieved June 20, 2019, from: http://www.shanghai.gov.cn/nw2/nw2314/nw2319/nw12344/u26aw58275. html

Sharma, M., & McBean, E. (2007). A methodology for solid waste characterization based on diminishing marginal returns. *Waste Management (New York, N.Y.)*, *27*(3), 337–344. doi:10.1016/j.wasman.2006.02.007 PMID:16600585

Sharma, S. K., Deml, K., Dangal, S., Rana, E., & Madigan, S. (2015). REDD+ framework with integrated measurement, reporting and verification system for Community Based Forest Management Systems (CBFMS) in Nepal. *Current Opinion in Environmental Sustainability*, *14*, 17–27. doi:10.1016/j.cosust.2015.01.003

Sharp, A., & Sang-Arun, J. (2012). *A guide for sustainable urban organic waste management in Thailand: Combining food, energy, and climate co-benefits*. Institute for Global Environmental Strategies (IGES). Available: http://www. iges. or. jp

Simelane, O. (2016). Implementing Alternative Waste Treatment Technologies: A Comparative Study of South Africa and the European Union. *South Africa, 11*(7), 427-430.

Singh, R. K., Gamaralalage, P. J. D., Yagasa, R., & Onogawa, K. (2018). *State of Waste Management in Phnom Penh, Cambodia*. Retrieved from https://pub.iges.or.jp/pub/state-waste-management-phnom-penh-cambodia

Singh, J. (2016). Ranking South African provinces on the basis of MERRA 2D surface incident shortwave flux. *Journal of Energy in Southern Africa*, *27*(3), 50–57. doi:10.17159/2413-3051/2016/v27i3a1555

Singh, J., Laurenti, R., Sinha, R., & Frostell, B. (2014). Progress and challenges to the global waste management system. *Waste Management & Research*, *32*(9), 800–812. doi:10.1177/0734242X14537868 PMID:24938296

Siripornprasarn, A. (2015). *Discovering New E-Waste Management-Initiatives Around the Globe-Thailand*. Pollution Control Department, Ministry of Natural Resources and Environment.

Siriratpiriya, O. (2014). *Municipal solid waste management in Thailand: challenges and strategic solution. In Municipal Solid Waste Management in Asia and the Pacific Islands* (pp. 337–354). Springer. doi:10.1007/978-981-4451-73-4_17

Sixteenth, L. S. F. (2015). *Standing committee on energy (2015-16)*. Government of India. Retrieved from http://www. wwfenvis.nic.in/Database/ConsulativeCommittee_5204.aspx

SolidW. M. R. (2016). *India*. Retrieved from http://sscgj.in/wp-content/uploads/2016/07/Waste-Management-Rules-2016. pdf

Solomon, A. O. (2011). *The Role of Households in Solid Waste Management in East Africa Capital Cities* (Unpublished Doctoral Dissertation). Wageningen University.

Song, X., Lu, B., & Wu, W. (2019). Environmental Management of E-waste in China. *Electronic Waste Management and Treatment Technology*, 285–310.

Soós, R., Whiteman, A. D., Wilson, D. C., Briciu, C., Nürnberger, S., Oelz, B., ... Schwehn, E. (2017). Operator models for delivering municipal solid waste management services in developing countries: Part B: Decision support. *Waste Management & Research*, *35*(8), 842–862.

South African Waste Information Centre (SAWIC). (2013). *Waste classification and management regulations, norms and standards*. Retrieved 13 July 2019 from: http://sawic.environment.gov.za/documents/2177.pdf

Spiess, R., & Sokhorng, C. (2017). *Cintri chided by former exec*. Retrieved April 12, 2019, from https://www.phnom-penhpost.com/business/cintri-chided-former-exec

Sri Lanka Recyclers Association (SLRA). No. 19/2A, Parliament Road, Palawatta, Battaramulla, Sri Lanka. Retrieved from http://www.slr.lk/

Stanisavljević, N., Ubavin, D., Batinić, B., Fellner, J., & Vujić, G. (2012). Methane emissions from landfills in Serbia and potential mitigation strategies: A case study. *Waste Management and Research, 30*(10), 1095-1103. doi:10.1177/0734242X12451867

State Council. (2016a). *13th Five-year Plan for China's National Economic and Social Development*. Retrieved June 20, 2019, from: https://wenku.baidu.com/view/aafe5d3549d7c1c708a1284ac850ad02de8007 bf.html

State Council. (2016b). *13th Five-Year Plan for Ecological Environmental Protection Plan*. Retrieved June 20, 2019, from: http://www.gov.cn/xinwen/2016-12/05/content_5143464.htm

State Council. (2016c). *13th Five-year Plan for Energy Saving and Emission Reduction*. Retrieved June 21, 2019, from: http://www.gov.cn/zhengce/content/2017-01/05/content_5156789.htm

Statistics South Africa (StatsSA). (2017a). *Mid-year population estimates (statistical release P0302), Statistics South Africa*. Pretoria: RSA.

Stats, S. A. (2017b). *Living conditions of households in South Africa 2014/2015 (Statistical release P0310) Statistics South Africa*. Pretoria: RSA.

Steen, B. (2000). A systematic approach to environmental priority strategies in product development (EPS) version 2000-models and data of the default method, CPM report 1999. Chalmers University of Technology, Environmental Systems Analysis, Gotheburg, Sweden.

Stickel, B. H., Jahn, A., & Kier, W. (2012). *The Cost to West Coast Communities of Dealing with trash, reducing marine debris*. Retrieved from https://www.coastal.ca.gov/publiced/coordinators/WestCoastCommsCost-MngMarineDebris.pdf

Stubs, K. (2019). *South African industry forced to adapt as waste streams are prohibited from landfill disposal*. Retrieved 14 July 2019 from: https://blog.interwaste.co.za/know-waste/sa-industry-forced-to-adapt-as-waste-streams-are-prohibited-from-landfill-disposal

Suraweera, I. (2016). E-waste issues in Sri Lanka and the Basel Convention. *Reviews on Environmental Health, 31*(1). doi:10.1515/reveh-2015-0069 PMID:26943598

Sustainable Energy Africa. (2017). *Sustainable energy solutions for South African local government: A practical guide*. Cape Town: RSA.

Suthapanich, W. (2014). *Characterization and assessment of municipal solid waste for energy recovery options in Phetchaburi, Thailand.* Degree of Master of Science in Environmental Engineering and Management. Asian Institute of Technology School of Environment, Resources and Development.

Suy, P. (2018). *Svay Rieng to build recycling plant - Khmer Times*. Retrieved April 13, 2019, from https://www.khmer-timeskh.com/536529/svay-rieng-to-build-recycling-plant/

Swachh, S. G. A. (2018). *Press Information Bureau*. Retrieved from http://pib.nic.in/newsite/PrintRelease.aspx?relid=183949

SWEEP-Net. (2010). *Country Report on the Solid Waste Management in Lebanon*. Retrieved from http://www.sweep-net.org/ckfinder/userfiles/files/country-profiles/CountryreportLebanon-En-mai2011,pdf

Takele. (2004). *Solid waste and liquid waste management.* Addis Ababa: Ethiopia Public Health Training Initiative.

Tang, K. M. (2014). *Towards Environmental & Economic Sustainability in Malaysia via Biomass Industry.* Retrieved from https://www.oeaw.ac.at/forebiom/WS2lectures/02-01-TKMUN.pdf

Tan, S., Ho, W., Hashim, H., Lee, C., Taib, M., & Ho, C. (2015). Energy, economic and environmental (3E) analysis of waste-to-energy (WTE) strategies for municipal solid waste (MSW) management in Malaysia. *Energy Conversion and Management*, *102*, 111–120. doi:10.1016/j.enconman.2015.02.010

Tanskanen, J.-H. (2000). Strategic planning of municipal solid waste management. *Resources, Conservation and Recycling*, *30*(2), 111–133. doi:10.1016/S0921-3449(00)00056-2

Tchobanoglous, G., Theisen, H., & Vigil, S. A. (1993). *Integrated Solid Waste Management: Engineering: Principles and Management Issues*. New York: McGraw Hill Inc.

ThaiPublica. (2014). National agenda of solid Waste (2 September 2014 ed). ThaiPublica, Bangkok USEPA (2013). Inventory of U.S. greenhouse gas emissions and sinks: 1990–2011 (EPA 430-R-13-001). Washington, DC.

Tham, T. T., Huong, N. T., Phuong, B. T., & Trinh, L. T. (2018). Assessment of the distribution and accumulation of polybrominated diphenyl ethers in the environment in the plastic recycling village of Minh Khai, Nhu Quynh town, Hung Yen province. *VNU Journal of Science: Earth and Environmental Sciences*, *34*(2). doi:10.25073/2588-1094/vnuees.4243

Thang, N. T. (2016). *State of 3R I Asia and Pacific – Country report for Vietnam.* 8th Regional 3R Forum in Asia and the Pacific.

Thang, N. T., & Hanh, H. H. (2019, February). *Discussion result with Nhu Quynh Town People's Committee about the environmental status in Minh Khai plastic craft village*. Academic Press.

Thang, N. T., & Anh, D. T. P. (2016). e-Waste management and resource recovery in Vietnam. In *Proceedings of the 6th International Waste Management Conference*. Jadavpur University.

Thanh, N. P., Matsui, Y., & Fujiwara, T. (2011). Assessment of plastic waste generation and its potential recycling of household solid waste in Can Tho City, Vietnam. *Environmental Monitoring and Assessment*, *175*(1-4), 23–35. doi:10.100710661-010-1490-8 PMID:20490914

Thavalingam, V., & Karunasena, G. (2016). Mobile phone waste management in developing countries: A case of Sri Lanka. *Resources, Conservation and Recycling*, *109*, 34–43. doi:10.1016/j.resconrec.2016.01.017

The Bartlett UCL & Development Planning Unit. (2017). Transforming Solid Waste Management in Dar es Salaam. Final Report. Author.

The Environmental Management Act. (2004). 158pp

The Guardian. (2017). *A million bottles a minute: world's plastic binge 'as dangerous as climate change.* Retrieved from https://www.theguardian.com/environment/2017/jun/28/a-million-a-minute-worlds-plastic-bottle-binge-as-dangerous-as-climate-change

The Royal Government of Cambodia. Sub-decree on Management of Urban Garbage and Solid Waste (English Translation), Pub. L. No. No. 113 NKR.PR, 23 (2015). Cambodia.

The Telegraph. (2017). *Seafood eaters ingest up to 11,000 tiny pieces of plastic every year, study shows.* Retrieved from https://www.telegraph.co.uk/science/2017/01/24/seafood-eaters-ingest-11000-tiny-pieces-plastic-every-year-study/

The World Bank. (2015). *Diagnóstico de la Gestión Integral de Residuos Sólidos Urbanos en la Argentina*. Author.

The World Bank. (2019). *World Bank Open Data*. Retrieved March 11, 2019, from https://data.worldbank.org/

Thornhill, C. (2012). Improving local government transformation in South Africa. *Administratio Publica*, *20*(3), 128–143.

Tiong, C. S., Latiff, P. A., & Karupannan, S. (2012). Medical waste management in private medical clinics Taiping. *Perak. International Conference on Ecological, Environmental and Bio-Sciences (ICEEBS'2012)*, Pattaya.

Toolkit, S. S. (2019). *Ministry of Housing and Urban Affairs, Government of India*. Retrieved from http://164.100.228.143:8080/sbm/content/writereaddata/Survekshan%20Survey%202019%20Toolkit%2013.09.2018.pdf (Accessed on 4 October 2018)

Tot, B., Srđević, B., Vujić, B., Tavares Russo, M. A., & Vujić, G. (2016). Evaluation of key driver categories influencing sustainable waste management development with the Analytic Hierarchy Process (AHP): Serbia example. *Waste Management & Research*, *34*(8), 740–377. doi:10.1177/0734242X16652963 PMID:27288689

Tot, B., Vujić, G., Srđević, Z., Ubavin, D., & Russo, M. A. T. (2017). Group assessment of key indicators of sustainable waste management in developing countries. *Waste Management & Research*, *35*(9), 913–922. doi:10.1177/0734242X17709911 PMID:28606005

Trung, D. D., Mahanty, S., & Van, N. T. (2012). *Vietnam's craft villages and water pollution: A review of previous research*. Australian National University. Retrieved from https://crawford.anu.edu.au/rmap/pdf/_docs/water_pollution_craft/dang_et_al_2010.pdf

Truong, T. X. (2016). *Report on plastic sector*. FPT Security Company. Retrieved from http://images1.cafef.vn/Images/Uploaded/DuLieuDownload/PhanTichBaoCao/Baocaonganhnhua_0317_FPTS.pdf

Tsheleza, V., Ndhleve, S., Kabiti, H., Musampa, C., & Nakin, M. (2019). Vulnerability of growing cities to solid waste related environmental hazards: The case of Mthatha. *Jamba: Journal of Disaster Risk Studies*, *11*(1), 1–11. PMID:31205617

Tuan, N. V., Kien, T. K., Huyen, D. T. T., Nga, T. T. V., Giang, N. H., Dung, N. T., ... Kawamoto, K. (2018). Current Status of construction and demolition waste management in Vietnam: Challenges and Opportunities. *International Journal of GEOMATE*, *15*(52), 24.

Tue, N. M., Goto, A., Takahashi, S., Itai, T., Asante, K. A., Kunisue, T., & Tanabe, S. (2016). Release of chlorinated, brominated and mixed halogenated dioxin-related compounds to soils from open burning of e-waste in Agbogbloshie (Accra, Ghana). *Journal of Hazardous Materials*, *302*, 151–157. doi:10.1016/j.jhazmat.2015.09.062 PMID:26474377

Tue, N. M., Matsushita, T., Goto, A., Itai, T., Asante, K. A., Obiri, S., ... Kunisue, T. (2019). Complex Mixtures of Brominated/Chlorinated Diphenyl Ethers and Dibenzofurans in Soils from the Agbogbloshie e-Waste Site (Ghana): Occurrence, Formation, and Exposure Implications. *Environmental Science & Technology*, *53*(6), 3010–3017. doi:10.1021/acs.est.8b06929 PMID:30793891

Turkish Statistical Institution. (2018). Retrieved from http://www.turkstat.gov.tr

UN Habitat. (2010). *WTE Industry in Latin America 2010*. Author.

UNCRD (United Nations Centre for Regional Development). (2017). *State of 3Rs in Asia and the Pacific Country Report: Malaysia*. Secretariat of the Regional 3R Forum in Asia and the Pacific, UNCRD, Nagono, Japan, and Institute for Global Environmental Strategies, Kamiyamaguchi, Japan.

UNDP. (2017). *UNDP – Myanmar Annual Report 2017*. Retrieved from http://www.mm.undp.org/content/myanmar/en/home/library/poverty/2017-annual-report.html

UNEP & WMO. (2011). *Integrated Assessment of Black Carbon and Tropospheric Ozone | Climate & Clean Air Coalition*. Retrieved from http://www.ccacoalition.org/ar/resources/integrated-assessment-black-carbon-and-tropospheric-ozone

UNEP. (2011). *Near-term Climate Protection and Clean Air Benefits: Actions for Controlling Short-Lived Climate Forcers*. Retrieved from http://www.unep.org/publications/ebooks/SLCF/

UNEP. (2011). *Revised technical guidelines for the environmentally sound management of used and waste pneumatic tires*. Document UNEP/CHW.10/6/Add.1/Rev.1. 2011. UNEP.

UNEP. (2015). *Plastics in Cosmetics*. UNEP.

UNEP. (2016). *GEO-6 Regional Assessment for Latin America and the Caribbean*. Nairobi, Kenya: United Nations Environment Programme.

UNEP. (2017). *Regional Waste Management Outlook in Latin America and the Caribbean*. UNEP.

UNEP. (2017). *The Emissions Gap Report 2017 A UN Environment Synthesis Report*. Retrieved from www.unenvironment.org/resources/emissions-gap-report

UNEP. (2018). *Africa Waste Management Outlook*. United Nations Environment Programme. Retrieved from http://hdl.handle.net/20.500.11822/25514

UNEP. (2019). Retrieved from http://www.unep.org/

UNEP-IETC. (2010). *Waste and Climate Change - Global Trends and Strategy Framework*. Retrieved from http://www.unep.or.jp/ietc/Publications/spc/Waste&ClimateChange/Waste&ClimateChange.pdf

UNEP-ISWA (2015). *Global Waste Management Outlook*. Author.

UNFCCC (United Nations Framework Convention on Climate Change). (2012). *Background on the UNFCCC: The international response to climate change*. Retrieved from https://unfccc.int/essential_background/items/6031.php

UNFCCC (United Nations Framework Convention on Climate Change). (2015). *Adoption of the Paris Agreement*. Retrieved from https://unfccc.int/resource/docs/2015/cop21/eng/l09r01.pdf

UNFCCC. (2005). *Sixth compilation and synthesis of initial national communications from Parties not included in Annex I to the Convention*. Retrieved from http://unfccc.int/resource/docs/2005/sbi/eng/18a02.pdf

UNFPA Myanmar. (2017). *Census report projects population growth over 35 years, and supports long-term planning*. Retrieved from https://myanmar.unfpa.org/en/news/census-report-projects-population-growth-over-35-years-and-supports-long-term-planning

United Nation Industrial Development Organization. (2016). *Policy on biomass energy technology*. Assessed 5 May 2019. https://www.aedb.org/images/UNIDOBiomassPolicyDocumentSecondDraftV1.pdf

United Nations Department of Economic and Social Affairs, Population Division. (2015). *World Urbanization Prospects: The 2014 Revision*. Author.

United Nations Department of Economic and Social Affairs. (2010). *Trends in solid waste management: issues, challenges, and opportunities*. International Consultative Meeting on Expanding Waste Management Services in Developing Countries (Final Draft), 18-19 March 2010, Tokyo, Japan.

United Nations Department of Economic and Social Affairs. (2018). *World Urbanization Prospects 2018*. Retrieved 05.02.2019 from https://population.un.org/wup/Country-Profiles/

United Nations Development Program and Economic and Social Research Foundation. (2017). The Tanzania Human Development Report 2017:Social Policy in the Context of Economic Transformation. Author.

United Nations Development Program. (1994). Electericity, Fuel, and Fertilizer from Municipal and Industrial Organic Waste in Tanzania. A Biogas Plant for Africa. The United Nations Development Program Project of the Government of Tanzania Number URT/93/G31/A/1G/99.

United Nations Environmental Program. (2017). Summary Report: Waste Management in ASEAN Countries. Author.

United Nations Environmental Programm/SBC and World Health Organisations. (2004). *Preparation ofNational Health-Care Waste Management*. Plans in Sub-Saharan Countries Guidance Manual.

United Nations Environmental Programme- International Environmental Technology Centre (UNEP-IETC). (1996). *International Source Book on Environmentally Sound Technologies for Municipal Solid Waste Management*. Technical Publication Series. Available at UNEP website: http://www.unep.or.jp

United Nations Human Settlements Programme. (2010). *Solid Waste Management in the World's Cities*. Author.

United Nations Industrial Development Organization (UNIDO). (2018). *Terms of Reference for the Provision of services and equipment for two pilot facilities to be located at Phan Boi and Minh Khai craft villages in Vietnam*. Retrieved from https://www.unido.org/sites/default/files/files/2018-03/Appendix%201%20-%20Terms%20of%20reference%20dated%2013%20February%202018_1.pdf

United Nations. (2012). *Department of Field Support, Cartographic Section, Map No. 4365 Rev. 1*. Author.

United Nations. (2014). *World Urbanization Prospects: The 2014 Revision, Highlights*. New York: United Nations.

United Nations. (2015). *World urbanisation prospects: The 2014 revision report (No. ST/ESA/SER-A/366)*. New York: United Nations.

United Nations. (2019). *World Countries by Population Density 2019*. Retrieved from http://worldpopulationreview.com/countries/countries-by-density/

United Republic of Tanzania. (1997). *The National Environmental Policy*. Author.

United Republic of Tanzania. (1999). The Tanzania Development Vision 2025. Author.

United Republic of Tanzania. (2004). *The Environmental Management Act*. Author.

United Republic of Tanzania. (2010). *National Strategy for Growth and Reduction of Poverty II (NSGRP II)*. Ministry of Finance and Economic Affairs. Retrieved 22.12.2018 from http://www.mof.go.tz/mofdocs/DPE/MKUKUTA%20II%20English.pdf

United Republic of Tanzania. (2014). *The Second National Communication to the United Nations Framework Convention on Climate Change*. Author.

United Republic of Tanzania. (2016). National Five Year Development Plan 2016/17 – 2020/21: "Nurturing Industrialization for Economic Transformation and Human Development". Author.

United Republic of Tanzania. (2017). Tanzania Human Development Report 2017: Social Policy in the Context of Economic Transformation. Author.

United Republic of Tanzania. (2018). Ministry of Health, Community Development, Gender, Elderly and Children]. National Strategic Plan for Health Care Waste Management (2018-2022). Author.

United Republic of Tanzania. (2018). The Performance Audit Report on Electronic Waste Management. A Report of the Controller an Auditor General March 2018. Author.

University of Sri Jawardenapura (UoSJ). (2015). E-waste Management in Sri Lanka. Department of Accounting, Faculty of Management Studies and Commerce, University of Sri Jayewardenepura, Nugegoda, Sri Lanka.

Unnisa, S. A., & Rav, S. B. (Eds.). (2012). *Sustainable solid waste management*. CRC Press. doi:10.1201/b13116

Urban Voice Cambodia. (2018). *Flooding and Drainage Infrastructure Development in Phnom Penh - Urban Voice Cambodia*. Retrieved March 28, 2019, from https://urbanvoicecambodia.net/flooding-and-drainage-infrastructure-development-in-phnom-penh/?lang=en

USAID. (2013). *Experiencias internacionales en el composteo de residuos sólidos orgánicos*. USAID.

Vaccari, M., & Vitali, F. (2015). Technical Aspects of Municipal Solid Waste Collection: Case Studies from East Africa: in Case Studies for Developing Globally Responsible Engineers. Global Dimension in Engineering Education.

Vidanaarachchi, C. K., Yuen, S. T. S., & Pilapitiya, S. (2005). Municipal solid waste management in the Southern Province of Sri Lanka: Problems, issues challenges. *Waste Management (New York, N.Y.)*, *26*(8), 920–930. doi:10.1016/j.wasman.2005.09.013 PMID:16298122

Vietnam Cleaner Production Centre (VNCPC). (2016). *Cleaner production nationwide in Vietnam*. Retrieved from http://www.sxsh.vn/vi-VN/Home/tongquansanxuatsachhon-14/2011/San-xuat-sach-hon-tren-toan-quoc-874.aspx

Vietnam Energy Efficiency Program (VNEEP). (2016). *Biogas in Vietnam: Potential in a hand.* Retrieved March 7, 2019 from http://tietkiemnangluong.com.vn/tin-tuc/pho-bien-kien-thuc/t13304/biogas-o-viet-nam-trien-vong-trong-tam-tay.html

Vietnam Plastics Association (VPA). (2016). *Overview on Plastic Sector in Vietnam.* Retrieved October 20, 2018 from http://www.vpas.vn/gioi-thieu/tong-quan-nganh.html

Vujić, G., Stanisavljević, N., Batinić, B., Jurakić, Ž., & Ubavin, D. (2017). Barriers for implementation of ''waste to energy'' in developing and transition countries: a case study of Serbia. *Journal of Material Cycles and Waste Management, 19*(1), 55-69. DOI doi:10.100710163-015-0377-8

Vujić, G.V., Jovičić, N.M., Babić, M.J., Stanisavljević, N.Z., Batinić, B.M., & Pavlović, A.R. (2010). Assessment of plastic flows and stocks in Serbia using material flow analysis. *Thermal Science, 14*, S89-S95. Doi:10.2298/TSCI100621031V

Wallerstein, N. B., & Duran, B. (2006). Using community-based participatory research to address health disparities. *Health Promotion Practice*, *7*(3), 312–323. doi:10.1177/1524839906289376 PMID:16760238

Wang, C., Wu, G., Reniers, P., Huang, L., Kang, L., & Zhang, L. (2018). The Future of Hazardous Chemical Safety in China: Opportunities, Problems, Challenges and Tasks. *The Science of the Total Environment*, *643*(12), 1–11. doi:10.1016/j.scitotenv.2018.06.174 PMID:29935358

Wang, Z., Zhang, B., & Guan, D. (2016). Take Responsibility for Electronic-Waste Disposal. *Nature*, *536*(7614), 23–25. doi:10.1038/536023a PMID:27488785

Washington Refuse & Recycling Association (WRRA). (2018). *Comprehensive list of banned materials under National Sword*. Retrieved from https://www.wrra.org/wp-content/uploads/2018/04/WRRA-Comprehensive-China-Ban-List.pdf

WAT. (2012). Report on consulting services for detailed design and supervision for the construction works of a new landfill site and recycling centre. Kigali, Rwanda: Report to City of Kigali, funded by the United Nations Development Programme.

Western Province Waste Management Authority (WMA). (2019). *Strategic plan. New Provincial Council Building.* Battaramulla, Sri Lanka: Denzil Kobbekaduwa Mawatha.

White, P., Dranke, M., & Hindle, P. (2012). *Integrated solid waste management: a lifecycle inventory*. Springer Science & Business Media.

WHO. (2016). *Ambient air pollution: A global assessment of exposure and burden of disease*. Geneva: World Health Organisation.

Wichai-utcha, N., & Chavalparit, O. (2019). 3Rs Policy and plastic waste management in Thailand. *Journal of Material Cycles and Waste Management*, *21*(1), 10–22. doi:10.100710163-018-0781-y

Wijesekara, H. R., De Silva, S. N., Wijesundara, D. T., De, S., Basnayake, B. F. A., & Vithanage, M. S. (2015). Leachate plume delineation and lithologic profiling using surface resistivity in an open municipal solid waste dumpsite, Sri Lanka. *Environmental Technology*, *36*(23), 2936–2943. doi:10.1080/09593330.2014.963697 PMID:25209886

Wilson, D. C. (2007). Development drivers for waste management. *Waste Management & Research*, *25*(3), 198–207. doi:10.1177/0734242X07079149 PMID:17612318

Wilson, D. C., Araba, A. O., Chinwah, K., & Cheeseman, C. R. (2009). Building recycling rates through the informal sector. *Waste Management (New York, N.Y.)*, *29*(2), 629–635. doi:10.1016/j.wasman.2008.06.016 PMID:18701272

Wilson, D. C., Kanjogera, J. B., Soós, R., Briciu, C., Smith, S. R., Whiteman, A. D., ... Oelz, B. (2017). Operator models for delivering municipal solid waste management services in developing countries. Part A: The evidence base. *Waste Management & Research*, *35*(8), 820–841. doi:10.1177/0734242X17705723 PMID:28617175

Wilson, D. C., Rodic, L., Cowing, M. J., Velis, C. A., Whiteman, A. D., Scheinberg, A., ... Oelz, B. (2015). 'Wasteaware'benchmark indicators for integrated sustainable waste management in cities. *Waste Management (New York, N.Y.)*, *35*, 329–342. doi:10.1016/j.wasman.2014.10.006 PMID:25458855

Wilson, D. C., Velis, C. A., & Rodic, L. (2013). Integrated sustainable waste management in developing countries. *Proceedings of the Institution of Civil Engineers-Waste and Resource Management*, *166*(2), 52–68. doi:10.1680/warm.12.00005

Wilson, D. C., Velis, C., & Cheeseman, C. (2006). Role of informal sector recycling in waste management in developing countries. *Habitat International*, *30*(4), 797–808. doi:10.1016/j.habitatint.2005.09.005

World Bank. (1999). *What a Waste: Solid Waste Management in Asia*. Washington, DC: The World Bank.

World Bank. (2012). *What a Waste*. Retrieved from http://siteresources.worldbank.org/INTURBANDEVELOPMENT/Resources/336387-1334852610766/What_a_Waste2012_Final.pdf

World Bank. (2014). *World Bank Development Indicators*. Retrieved from http://data.worldbank.org/country/malaysia

World Bank. (2018). Lebanon's *Economic Outlook*. Retrieved from https://www.worldbank.org/en/country/lebanon/publication/economic-outlook-october-2018

World Bank. (2018). *Overview of Myanmar*. Retrieved from http://www.worldbank.org/en/country/myanmar/overview

World Bank. (2018). *What a Waste 2.0*. Washington, DC: World Bank.

World Bank. (2019). *Cambodia Overview*. Retrieved April 8, 2019, from https://www.worldbank.org/en/country/cambodia/overview

World Bank. (2019). *GNI per capita, Atlas methods (current US$)*. Retrieved from https://data.worldbank.org/indicator/ny.gnp.pcap.cd

World Bank. (2019). Retrieved August 21, 2019, from https://data.worldbank.org/country/philippines

World Bank. (2019). *The World Bank In Lebanon*. Retrieved from https://data.worldbank.org/country/lebanon

World Bank. (2019). *What a waste 2.0: A global snapshot of solid waste management to 2050.* Retrieved 4 July 2019 from: http://datatopics.worldbank.org/whatwaste/trends_in_solid_waste_management.html

World Development Indicators. (2016). *World Bank.* Retrieved from http://databank.worldbank.org/data/reports.aspx?source=world-development-indicators

World Economic Forum. (2016). *The New Plastic Economy – Rethinking the Future of Plastic.* Retrieved from http://www3.weforum.org/docs/WEF_The_New_Plastics_Economy.pdf

World Health Organization (WHO). (2019). *Definition and characterization of health-care waste.* Retrieved from https://www.who.int/water_sanitation_health/medicalwaste/002to019.pdf?ua=1

World Health Organization. (2017). Guidelines for Drinking-Water Quality, Fourth Edition Incorporating the First Addendum. World Health Organization.

World Population Prospects. (2017). Retrieved 06.02.2019 from http://worldpopulationreview.com/countries/tanzania-population/

World Population Review. (2019). Retrieved 06.2.2019 from http://worldpopulationreview.com/countries/tanzania-population/

World Resource Institute. (2011). *Climate Analysis Indicators Tool (CAIT) Version 9.0.* Washington, DC: WRI.

WRI-CAIT. (2017). *Myanmar.* Retrieved from https://www.climatewatchdata.org/ndcs/country/MMR/full

Wuhu Ecology Center. (2018). *WtE Location Map.* Retrieved April 1, 2019, from: http://www.waste-cwin.org/

Xiao, R., Awasthi, M. K., Li, R., Park, J., Pensky, S. M., Wang, Q., ... Zhang, Z. (2017). Recent developments in biochar utilization as an additive in organic solid waste composting: A review. *Bioresource Technology*, *246*, 203–213. doi:10.1016/j.biortech.2017.07.090 PMID:28756989

Yagasa, R., Sang-Arun, J., Uch, R., Sam, P., & Gamaralalage, P. J. D. (2018). *Participatory Waste Management Approach for Climate Change Mitigation: the Case of Battambang City*. Hayama, Kanagawa: Institute for Global Environmental Strategies. Retrieved from https://pub.iges.or.jp/pub/participatory-waste-management-approach

Yan, M., Li, X., Lu, S., Chen, C., Chi, Y., & Yan, J. (2011). Persistent Organic Pollutant Emissions from Medical Waste Incinerators in China. *Journal of Material Cycles and Waste Management*, *13*(3), 213–218. doi:10.100710163-011-0020-2

Yan, M., Xiong, Z., Li, X., & Hu, Y. (2015). Development Trend and Present Status of Municipal Solid Waste Incineration in China and USA. *Environmental Engineering*, *3*, 87–91.

Yhdegho, M., & Kingu, A. (2017). *Solid Waste Management in Coastal Towns of Southern Tanzania Case of Mtwara-Mikindani and Lindi Municipalities*. A Working Paper, Environmental Resource Consultancy. Retrieved 16.03.2019 from https://www.researchgate.net/publication/312604961_SOLID_WASTE_MANAGEMENT_IN_COASTAL_TOWNS_OF_SOUTHERN_TANZANIA_CASE_OF_MTWARA-MIKINDANI_AND_LINDI_MUNICIPALITIES

Yhdego, M., & Kingu, A. (2016). *Solid Waste Management in Urban Centers of Tanzania: Leapfrogging Towards a Circular Economy*. Retrieved 14.5.2019 from https://www.researchgate.net/publication/311846222_SOLID_WASTE_MANAGEMENT_IN_URBAN_CENTERS_OF_TANZANIA_LEAPFROGGING_TOWARDS_A_CIRCULAR_ECONOMY

Yildiz, S., Balahorli, V., Hosoglu, F., & Sezer, K. (2014). *Waste Management in the Metropolis of Istanbul – Waste Incineration and Energy Production Facility in Istanbul.* Retrieved from http://www.vivis.de/phocadownload/Download/2014_wm/2014_wm_265_276_yildiz.pdf

Yildiz, S., Yaman, C., Demir, G., Ozcan, H. K., Coban, A., Okten, H. E., ... Goren, S. (2013). Characterization of Municipal Solid Waste in Istanbul, Turkey. *Environmental Progress & Sustainable Energy*, *32*(3), 734–739. doi:10.1002/ep.11640

Yim, M., Fujiwara, T., & Sethy, S. (2013). *A Study on Household Solid Waste Characteristics in Phnom Penh City*. Academic Press.

Yim, M., Takeshi, F., & Sour, S. (2014). Current status of commercial solid waste generation, composition and management in Phnom Penh city, Cambodia. *Environment and Waste Management*, *1*(3), 31–38.

Yoada, R. M., Chirawurah, D., & Adongo, P. B. (2014). Domestic waste disposal practice and perceptions of private sector waste management in urban Accra'. *BMC Public Health*, *14*(1), 697. doi:10.1186/1471-2458-14-697 PMID:25005728

Yusop, M. H. (2006). *Production of bio-ethanol from biomass waste(sugarcane)*. Faculty of Chemical and Natural Resources Engineering, University College of Engineering & Technology Malaysia. Retrieved from http://umpir.ump.edu.my/611/

Zaman, A. U. (2014). Measuring waste management performance using the "Zero Waste Index": The case of Adelaide, Australia. *Journal of Cleaner Production*, *66*, 407–419. doi:10.1016/j.jclepro.2013.10.032

Zeng, X., Duan, H., Wang, F., & Li, F. (2017). Examining Environmental Management of E-Waste: China's Experience and Lessons. *Renewable and Sustainable Energy Review*, *72*, 1076-1082.

Zhakata, E., Gundani, S. R., Chauke, V., & Odeku, K. O. (2016). A Critic of NEMA: Waste Act 59 of 2008, so Many Promises, Little Implementation and Enforcement. In *SAAPAM Limpopo chapter 5th annual conference proceedings*. Limpopo: RSA.

Zhang, D., Huang, G., Zu, G., & Gong, Q. (2015). Waste to Energy in China: Key Challenges and Opportunities. *Energies*, *8*(12), 14182–14196. doi:10.3390/en81212422

Zhang, L., & Sun, X. (2018). Influence of Sugar Beet Pulp and Paper Waste as Bulking Agents on Physical, Chemical, and Microbial Properties During Green Waste Composting. *Bioresource Technology*, *267*, 182–191. doi:10.1016/j.biortech.2018.07.040 PMID:30021150

Zhang, X. (2016). *Emission Standards and Control of PM2.5 From Coal-Fired Power Plants*. London: IEA Clean Coal Center.

Zhou, C., Gong, Z., Hu, J., Cao, A., & Liang, H. (2015). A Cost-Benefit Analysis of Landfill Mining and Material Recycling in China. *Waste Management (New York, N.Y.)*, *35*, 191–198. doi:10.1016/j.wasman.2014.09.029 PMID:25453315

Zhou, C., Liu, X., Wang, Z., Yang, T., Shi, L., Wang, L., ... Zhang, C. (2016). Assessment of Marine Debris in Beaches or Seawaters Around the China Seas and Coastal Provinces. *Waste Management (New York, N.Y.)*, *48*, 625–660. doi:10.1016/j.wasman.2015.11.010 PMID:26589868

Zhou, Z., Tang, Y., Dong, J., Chi, Y., Ni, M., Li, N., & Zhang, Y. (2018). Environmental Performance Evolution of Municipal Solid Waste Management by Life Cycle Assessment in Hangzhou, China. *Journal of Environmental Management*, *227*, 23–33. doi:10.1016/j.jenvman.2018.08.083 PMID:30172156

About the Contributors

Agamuthu Pariatamby, Professor Agamuthu FASc, is currently at Sunway University. Previously he was Professor in University of Malaya, Kuala Lumpur for 44 years. He is appointed as the High Level End Foreign Expert position with the Ministry of Science and Technology, China, and is also visiting professor to Zhejiang University of Technology, Hangzhou, China. The other affiliations of Professor Agamuthu are Chairman of Organization for Climate Change (OFCC), Vice-President/International Advisory Board Member of Society of Solid Waste Management Experts in Asia and Pacific Islands (SWAPI), Fellow of Academy of Sciences Malaysia, Senior Editor in Chief of Waste Management & Research, and Editor of Journal of Material Cycles and Waste Management. Professor Agamuthu has written more than 19 books on waste management and has published more than 450 papers . His research interests include plastics and microplastics, marine debris, plastic biodegradation, municipal solid waste, hazardous waste, industrial waste, agriculture waste, electronic waste, circular Economy, 3R, composting, and anaerobic digestion.

Mehran Sanam Bhatti acquired BSc Hons in Applied Geology in 2014 and Master's degree in Environmental Management in 2019 from University of Malaya, Kuala Lumpur, Malaysia. The research in Master's program was focused on bacterial biodegradation of microplastics. In the last one year, the focus of research and academic writings have been on waste management, marine debris and microplastics.

Fauziah Shahul Hamid obtained her PhD in 2010 in Waste Management, and has been affiliated with University of Malaya since 2007, as the academic staff. Her research among others, covers municipal solid waste management, bioremediation, plastic and marine debris, and microplastics. She has been actively publishing many articles in various aspects of waste management in peer-reviewed journals, as well as, contributing her expertise as reviewers to those prestigious journals. Her publications include three chapters in books on waste management and more than 80 papers. Her current research interest are bioremediation of contaminated soil (heavy metal, persistent organic pollutants, microplastics, etc.), studies on abundance, distribution and patterns of marine debris, conversion of waste to value added products, environmental awareness creation and community engagement on waste management issues.

* * *

Renuka Ariyawansha is currently a doctoral candidate at the Postgraduate Institute of Agriculture, University of Peradeniya. B. Sc in Agriculture in 2007 and the Master Degree in Agricultural Biosystems Engineering in 2009, University of Peradeniya. Serving as an active researcher over a period of 12 years on solid waste management and biochemical transformation kinetics and actively involved in designing and developing bioreactors, leachate and wastewater treatment systems, solid waste management systems, renewable energy generation systems, and organic fertilizer production systems from biochar biocatalyst and from active phase composting with biochar and municipal solid waste dumpsites rehabilitations.

Muhammad Arshad is associate professor head of Environmental Sciences Department at Institute of Environmental Sciences and Engineering (IESE), National University of Sciences and Technology (NUST) Islamabad Pakistan. He got PhD in Ecology from Institut National Polytechnique de Toulouse (INPT), University of Toulouse, France in 2009. His area of research is environmental biotechnology and waste management. He has supervised more then 30 MS thesis and produced 4 PhD's. Up till now, he has won and completed more then 6 national and international research projects in the area of environmental sciences.

Md. Ashikuzzaman is a full-time faculty at Khulna University. Formerly he worked in national NGOs in Bangladesh. He completed his bachelor in Urban and Rural Planning Discipline from Khulna University, Bangladesh and accomplished Master of Urban Engineering from University of Tokyo, Japan. His research interests are urban planning and management.

Vella Atienza is an Assistant Professor at the Institute for Governance and Rural Development (IGRD), College of Public Affairs and Development (CPAf), University of the Philippines Los Baños (UPLB), Philippines. She obtained her Doctor of Philosophy (Ph.D.) in Asia Pacific Studies from the Ritsumeikan Asia Pacific University (APU), Beppu, Japan; and her Master of Management and Bachelor of Science in Forestry both from UP Los Baños. She used to work as a Research Fellow at the Institute of Developing Economies (IDE), Japan External Trade Organization (JETRO), Japan from April 2009 to March 2012. Her research focuses on environmental governance in waste management in the Philippines and other Asian countries. She is the Country Chapter author for the Philippines and a drafting committee member of the State of the 3Rs in Asia and the Pacific. She has also authored and edited various publications related to waste management and recycling, forestry and the environment.

Sandhya Babel is a full time Professor at the Sirindhorn International Institute of Technology (SIIT), Thammasat University, Thailand. She received her M.S. and PhD from Asian Institute of Technology, Thailand. Dr. Sandhya is an expert in environmental technology and management with extensive professional experience in teaching and research over 25 years. Her current research interest includes solid waste management, conversion of waste to energy and adsorbents, recovery of valuable materials from waste, low-cost technologies for wastewater treatment, photocatalysis, phytoremediation, and adsorption. She has carried out many research and sponsored projects in collaboration with international organizations. Dr. Sandhya has published more than 100 publications in both International and National journals.

Benedict Basnayake is a Senior Professor of Energy and Waste Management at the University of Peradeniya. He obtained B.Sc in Agricultural Engineering from Cranfield University, UK., and DEA and D.Eng from the University of Pierre et Marine Curie, France on Applied Kinetic Chemistry. He has 5 recognized patents. He has 40 years of teaching and research experience and 24 years of research in Municipal Solid Waste as the principal investigator. His primary research interests are energy and waste management, and biochemical transformation kinetics. Secondary research interests are effluent treatment and inventions in the field of Agricultural Engineering.

Adem Basturk was born in 1950 in Kayseri. He graduated from Istanbul Technical University (ITU), Faculty of Civil Engineering in 1972 as a M.Sc. in Civil Engineering. In the same year, he started to work as an assistant at Yildiz Technical University (YTU). Between 1974 and 1976, he realized his doctoral studies at Stuttgart University / Germany and he received PhD degree in 1976. He appointed Associate Professor position in 1981 and he became a Professor at YTU Environmental Engineering Department in 1988. Prof.Dr.Adem BASTURK has founded Environmental Engineering Department, and served as the Head of this department at YTU. Between 1994 and 1999, he served as Deputy Secretary General responsible for investments in Istanbul Metropolitan Municipality (IBB) and as the Secretary General of the IBB between 1999-2002. Dr. Basturk was elected as a member of Turkish Parliament in 2002. During his term as a Member of Parliament, he served as the President of the Commission on Public Works, Transportation, Reconstruction and Tourism, as a member of the Environmental Commission, as a member of the International Parliamentary Commission Turkish Group, and as the Chair of the Global Warming Research Commission until 2007. In 2009, he was again appointed as the Secretary General of IBB and this service continued until 2014. He has several national and international publications and books on solid wastes. Dr. Basturk is regarded as the pioneer of solid waste activities in Turkey. In addition to his academic experience, he has also pioneered in setting up and implementing regulations on solid wastes, as well as the establishment of regular landfill sites in Istanbul, the establishment of a composting facility, and the establishment of a medical waste incineration plant in IBB. Prof. Dr. Adem Başturk currently continue to works on environmental engineering field, especially on solid waste technologics, as well as various foundation activities and training services.

Mehmet Sinan Bilgili is a professor at Yildiz Technical University (YTU) and a member of the International Waste Working Group (IWWG). Since 2006, he has supervised over 30 Bachelor and 5 Master of Science Degree students as well as 5 doctorate students. He is currently supervising 3 Master's degree research projects and one doctoral degree research project. Prof. Bilgili is the Editor of an International Journal, Environmental Research & Technology, and the organizer of EurAsia Waste Management Symposium which is being organized bi-annually. He has authored more than 40 scientific publications and more than 30 international symposium presentations. His research interests focus on Sustainable landfill gas management and treatment, waste characterization, municipal waste management, hazardous waste management, biological treatment of waste, hybrid methods for leachate treatment, water and wastewater treatment, hospital wastewater, micropollutant treatment and control, sewage sludge gasification, and environmental effects of nanomaterials.

Ernesto De Titto Former Researcher of the National Research Council (CONICET) and National Director of Health Determinants and Research (2007-18). Previously Responsible of the Health and Environment Coordination Unit (2004-7); Director of Health Promotion and Protection (1996-2004), Director of the National Institute of Microbiology "Dr. Carlos Malbrán" (1995-6), and Chief of the Research Department (1991-6), all in the Argentinean Ministry of Health; research fellow of the WHO and the Palo Alto Medical Foundation (USA). Author or coauthor of more than 50 original papers. Director of several Ph D and Magister thesis in various Universities. Academic Coordinator of the Magister in Environmental Health Management at the ISALUD University.

Nanne de Vries is a social psychologist, professor of health promotion and the Vice Dean of the Faculty of Health, Medicine and Life Sciences of Maastricht University, the Netherlands, with expertise at mapping interventions towards behavioural change and system improvements.

Ahmet Demir was born in 1959 in Uşak, Turkey. He was graduated from Istanbul Technical University (ITU) as a civil engineer in 1981. He started his academic career as a research assistant at Yildiz Technical University, Civil Engineering Department in 1983. He received, M.Sc. degree in 1985, and PhD degree in 1990 from the same university. In 1989, Dr. Demir graduated MSc from I.H.E. (Sanitary Engineering) in Delft, the Netherlands. He appointed professor in 2002 and worked as the Head of YTU Environmental Engineering Department between 2006 and 2008 and as the Dean of YTU Civil Engineering Faculty between 2008-2010 and 2014-2016. He was the General Director of Istanbul Water and Sewerage Administration (ISKI) between 2010-2014. Prof. Dr. Demir supervised over 100 BSc and more than 20 Ms.C. thesis as well as 10 doctorate students since 1995. He has edited two boks and he is currently Editor in Chief of Environmental Research & Technology, a quarterely published international journal. He is also serving as the Chair of EurAsia Waste Management Symposium which is organized biannually in Turkey. Prof. Dr. Demir published more than 100 scientific publications in the refereed technical engineering journals and symposium proceedings published in Turkish and English. Dr. Demir currently works at Yildiz Technical University as a full professor with the research topic of municipal waste management, aste to energy processes, water and wastewater treatment.

Premakumara Dickella Gamaralalage is a development planner, specialising in participatory environmental planning and sustainable urban development. He holds a PhD in Management Development from the Nihon Fukushi University, Japan in 2006. He has over 25 years' experiences in working with academic, government, non-governmental organisations, bilateral and international development programmes. After serving as a postdoctoral fellow at the Meijo Asian Center, Meijo University, he started working for IGES, Kitakyushu Urban Centre in Japan in 2010. At the Kitakyushu Urban Centre, he focused on participatory planning and community-based development towards achieving sustainable cities. Currently, he works as a Senior Researcher/ Programme Manager of the IGES Centre Collaborating with UNEP on Environmental Technologies (CCET) and provides technical support for improving waste and resource management in developing countries. His work focuses on developing integrated/ holistic waste management strategies at national and local levels, application of participatory learning and action methods to promote 3Rs (reduce, reuse and recycling) and circular economy/ resource efficiency societies, integration of informal sector and women participation in waste management, and linkages between waste and climate change as well as sustainable development goals (SDGs). He is a member of academic societies and regional networks and written regularly for academic and policy journals.

Megersa Dinka is a graduate with PhD from University of Natural Resources and Applied Life Science (Vienna) in 2010. He has expert knowledge in water resource engineering discipline specific to hydrology, hydraulics and water management aspects. He has about 14 years of experience as academician and 18 years of experience as a researcher. He has taught various courses and modules at undergraduate and postgraduate levels successfully. Most of the courses he taught are related to hydrology, hydraulics and watershed (water) management aspects. Currently he is teaching Hydrology, Hydraulics and Water Treatment Technology Modules at University of Johannesburg. Moreover, he also supervised a number of postgraduate students (>23 MSc and 2 PhD) successfully and still supervising a number of undergraduate and postgraduate students at University of Johannesburg. He has published a number of peer reviewed scientific articles and book chapters in accredited international journals and book publishers. He has reviewed a number of articles for various international journals.

Phuong Anh Duong has been working as a researcher at the Institute of Strategy and Policy on Natural Resources and Environment (ISPONRE) which belongs to the Ministry of Natural Resources and Environment (MONRE) of Viet Nam for over 10 years. She is now Deputy Head of Department of Environment and Sustainable under ISPONRE, who is responsible for research field on waste management. She joined in a lot of projects in environment including waste management such as projects in composting, marine waste, 3R, etc. at ISPONRE.

Mutasem El-Fadel is currently Professor and Chairperson of the Department of Civil and Environmental Engineering at the American University of Beirut (AUB) where he also holds the Dar Al-Handasah (Shair & Partners) Chair in Engineering. A Civil Engineer by background, Dr. El-Fadel holds an MS in Environmental Engineering, an MS in Water Resources Engineering, and a PhD in Environmental Engineering, all from Stanford University, California. Prof. El-Fadel has a distinguished record of high impact international and regional publications and is the recipient of the award of Scientific Research Excellence by the National Council for Scientific Research for outstanding contributions to research and distinguished scientific publication record in Environmental & Water Resources Engineering. Dr. El-Fadel has been a registered Professional Engineer in the States of California, Wisconsin, Nevada, Arizona, Utah, and Michigan as well as Lebanon. he has been for over a decade a Senior Consultant to the World Bank on environmental safeguards in the Middle East and North Africa (MENA) region. He has also worked for nearly a decade in the US where he was designated as an expert witness in several US Federal Courts to provide testimony related to cases in Environmental litigation and risk assessment of industrial facilities. He equally served as an expert on projects funded by the World Bank, the European Commission, the United Nations Development Program, the United Nations Environment Program, the United Nations Office of Project Services, the Economic and Social Commission of Western Asia, the US Information Agency, the US Geological Survey, the US Agency for International Development, the International Development and Research Center of Canada, the World Health Organization, the Swedish International Development Agency, the Swedish International Water Institute, the German Society for International Cooperation, the International Renewable Energy Agency, the International Center for Biosaline Agriculture, Centre International des Technologies de l'Environnement de Tunis, ministries, councils, municipalities, non-governmental organizations, and regional / international firms.

Sadhan Ghosh Professor & Former Head of Mechanical Engineering Department, being the founder coordinator of the Centre for QMS at Jadavpur University, Kolkata, India, is a renowned personality in the field on Waste Management, QEES Management systems, Circular Economy, Green Manufacturing, Supply Chain and Sustainable Development. He was the Director, CBWE, Ministry of Labour and Employment, Govt. of India with starting career in Larsen & Toubro Ltd. He is the Chair Elect of 9th IconSWM 2019, at Odisha, Nov 27-30, '19 by KIIT. He specializes in waste management, Green Manufacturing, supply chain management, circular economy and ISO standards on QMS, EMS, EnMS and OHSAS. He has published more than 50 research papers in leading international refereed journals and 120 papers in conference proceedings, edited more than 30 books and proceedings. He has accomplished and has been involved in several impactful interdisciplinary research projects on sustainable supply chain of small and medium sized enterprises (SMEs) across the globe, Circular Economy in 34 countries (2018-2021) and waste to energy. His projects have been funded by British Council (2011-2016), Royal Society (2015), Royal academy of Engineering (2012 & 2019-2021), EU horizon 2020 (2018-2022), Jute Technology Mission (2008-2011), and Central Pollution Control Board (1999-2002), Govt. of India, UNCRD/DESA (2016-2018),, APO (2016-2019), and Shota Rustaveli National Science Foundation of Georgia (2019 -2021) and a few others. He is currently facilitating SMEs in India to adopt low carbon initiatives and achieve competitiveness. He received two patents. He is the Associate editor of Journal of Japan Society of Materials Cycles and Waste Management. He has been invited for key note and plenary speeches in many countries across the globe. He was the International Expert/ Consultant of UN DESA/UNCRD IN 2016-18. He works as expert in many government committees in India, Japan an in South Asian. Received Indian patent on "Eco-friendly plastics recycling machine and the process there of" [Patent no.202532 dated 02/03/2007]; Bangladesh patent "Automatic High Speed Jute Ribboning Machine" [Patent no. 1005146 dated 17/02/2014]. He wrote nine books and a number of book chapters.

Iram Gul got PhD in Environmental Sciences from National University of Sciences and Technology (NUST), Islamabad, Pakistan in 2019. Her areas of research are environmental biotechnology, soil contamination and remediation, phytoremediation, toxicology and nanotechnology. She has working experience at national and international research organizations.

Teshan Habarakada Liyanage is a young engineer from Sri Lanka, graduated in Bio-systems Engineering from Asian Institute of Technology, Thailand in 2017. Currently he is a full time master student in Environmental Engineering at Sirindhorn International Institute of Technology and Thailand since January 2018. His undergraduate research has focused on photosensitization. At present he is focusing on pretreatment methods in anaerobic digestion.

Matthew Hengesbaugh is a Communications Coordinator supporting IGES's Sustainability Governance Centre with action research on the 2030 Development Agenda. Mr. Hengesbaugh has worked extensively in the Asia-Pacific region advocating and delivering sustainable development policy objectives since 2006, over which time he has advised and consulted with the United Nations, including the International Labour Organization and the United Nations Economic and Social Commission for Asia and the Pacific (UNESCAP) on issues including climate change, green jobs, green economy, and sustainable consumption and production. He received his M.A. from University of Trento, Italy, where as an Erasmus Mundus scholar he studied Local Economic Development. In addition, he also holds a M.A. in International Relations from the University of Wollongong, Australia, as well as B.A.s in Psychology and Anthropology from the University of California, Santa Cruz.

Hanh Hong has been working at Institute for Strategy, Policy on Natural Resources and Environment (ISPONRE) under Ministry of Natural Resources and Environment (MONRE). She received Doctor of Environment and Sustainable Development in 2018 at the Hanoi National University. Before working at ISPONRE, she was an auditor on Environmental Management System-ISO 14000 in Vietnam Certification Centre- Directorate for Standard, Metrology and Quality. Currently, she is Head, Department of Environment and Sustainable Development, ISPONRE. Beside regular researching activities such as advising and supporting the General Director of ISPONRE in conducting research studies, proposing mechanisms and policies for environmental protection and sustainable development, she has participated in some main projects, researches such as main member in the Project "Investigation and proposed orientation for implementing environmental related (Sustainable Development Goals) SDGs of Viet Nam post 2015" (2016-2018). The main output of the project is MONRE's Action Plan on implementation of (Vietnam Sustainable Development Goals) VSDGs in natural resources and environment sector approved by MONRE's Minister; participated in the mision "Assessment of 5 year implementation of Resolution No.24-NQ/TW of the Central Excutive Committee on the active in response to climate change, improvement of natural resource management and environmental protection" and "Assessment of implementation of National Strategy on Environmental protection to 2010 and development of the Environment Strategy 2011-2020"; manager for the Project "Phasing out Incandescent Lamps through Lighting Market Transformation in Viet Nam" (2012-2018); and some reports on Strategic Environment Assessment (SEA) for some Master Plans in Vietnam.

Md. Howlader completed the Bachelor of Social Science (Hons.) and Master of Social Science in Sociology from Sociology Discipline, Khulna University, Bangladesh. Currently he is the Assistant professor of Development Studies Discipline of the same University. The research interest of Howlader are Education and Development; Disaster and Development; Gender and Development; Health and Development, Environment and Development.

Irum Iqrar, PhD student of Biotechnology at Quaid-i-Azam University Islamabad, involved in research related to plant associated bacteria and their application in agriculture, skilled in Microbiological Research, Biotechnology, Plant Breeding and Genetics and Life Sciences. Irum's research interest includes analyzing the potential of bacterial endophytes for biological control using bioassays and identifying secondary metabolites responsible for antagonism of target phyto-pathogens, potentially will lead to a broad spectrum biopesticides for our agriculturally important crops. Irum has demonstrated history of working in the Microbiology and Biotechnology institutes, skilled in high throughput techniques for performing microbiological and biotechnological experiments.

Telesphore Kabera awarded a Distinguished International Students Scholarship by the China Scholarship Council to study for his MSc and PhD at the prestigious China University of Geosciences in Wuhan-Hubei. He returned to his Alma mater in Kigali in 2012, where his research interests cover a wide range of environmental fields. These include solid waste management (SWM); biogas; indoor air quality; CO_2 sequestration and climate change; environmental impact assessment (EIA) and strategic environmental assessment (SEA); and water and wastewater, specifically water harvesting and constructed wetland. Dr Kabera is currently Principal Investigator for an International Council for Science project: Health effects of indoor air pollution from cooking stoves in Kigali and Dar es Salaam cities: An assessment and solution. He is also a team leader for SIDA (Swedish International Development Cooperation Agency)'s: Strengthening capacity for sustainable bioenergy, waste management and air pollution control in Rwanda. He has also led several SWM projects in Rwanda and recently published a paper in the first open access issue of Waste Management & Research, benchmarking performance of the SWM and recycling system in Kigali and comparing that with four other East African capital cities (Dar-es-Salaam, Kampala, Nairobi and Maputo). Dr Kabera is a regular peer reviewer for research proposals submitted to the International Council for Science, and for reputable journals including Environment, Development and Sustainability; Environmental Monitoring and Assessment Journal; and International Journal of Environmental Studies. He is currently a member of two Editorial Boards: for the Waste Management and Research journal and forthe only Rwanda Journal of Engineering, Science, Technology and Environment (RJESTE); and of the technical committee for international conferences on Renewable Energy and Environmental Engineering (REEE).

Proscovia Kamugisha is a trainer at Mzumbe University as Lecturer in the Department of Economics under the Faculty of Social Science since 2008 to date. she is working in three major domains: training, research and consultancy. In the course of her responsibility Dr. Kamugisha has attained research skills in various social disciplines. She has published one book and several papers. She is expert in agricultural, natural resources and developmental economics.

Anurudda Karunarathna is a Senior Lecturer in Environmental Engineering. His professional career is been focused on understanding the waste management issues in regional and national levels, development and evaluation of waste treatment methods, environmental remediation, fate and transport of contaminants in the environment, environmental risk assessment, and devising effective measures for risk mitigation. He is a national expert on technical and policy dialogs associated with the municipal solid waste assessment and remediation of waste dumpsites. He has authored and coauthored many technical reports in the field of environmental science and engineering, and has authored more than 25 refereed technical papers.

Qike Liu born in Zhejiang on November 18th, 1999. She is a student in Institute of Energy and Thermal Engineering, Zhejiang University of Technology, PR China. Her current area of interest is waste to clean energy.

Amani Maalouf received her Ph.D. degree in Environmental and Water Resources Engineering from the American University of Beirut (AUB) in 2019. Her general research interest is in environmental engineering with emphasis on waste management. Currently, her research focuses on life cycle assessment (LCA) modeling of waste and wastewater management systems and the development of decision-making tools with emphasis on developing countries. During her studies, she was awarded a competitive scholarship from the Lebanese National Council for Scientific Research (CNRS-L) and AUB for her excellent academic record and quality of her doctoral research. She equally won an award to attend intensive courses on advanced waste management processes at the Technical University of Denmark. Besides her Ph.D. research topic, Dr. Maalouf collaborated on other research projects with the Department of Industrial Engineering at the University of Perugia, Italy. She published her work in numerous prestigious and highly ranked international peer reviewed journals and presented her work at equally renowned international conferences. In her last participation in the 8th International Conference on Sustainable Waste Management, she received the IconSWM Springer Excellence Paper Award 2018 for best paper. Dr. Maalouf managed to engage herself with the wider local community by contributing to many newspaper articles. Besides her MS and PhD theses, to date, she published 8 articles in refereed international scientific journals with 1 other under peer-review process, 2 book chapters, 1 monogram, 7 peer-reviewed conference proceedings, 5 conference presentations (invited speaker), and 10 local newspaper articles. She has worked as a technical consultant to the Regional External Programs (REP) Office at AUB; the United Nations Development Programmes, Lebanon; and the International Solid Waste Association (ISWA), Wien, Austria. She equally served as an expert on projects funded by the World Bank, the European Commission, the United Nations Environment Programme, the Swedish

International Development Agency, the Stockholm International Water Institute, and the Friedrich Ebert German foundation.

Nalin Mannapperuma is the Director General of the Waste Management Authority of the Western province (WMA) he joined to the WMA as a senior officer of the authority in October 2005. Mr. Nalin Mannapperuma obtained his fist degree in the field of Agriculture in 1996 from the Faculty of Agriculture, University of Peradeniya. Further he is a Master Degree holder in the field of Environmental Engineering from the University of SAITHAMA, Japan. Before, He joining to Waste Management Authority he has experience in working with different Research Developments institutes who, are working on the Municipal Solid Waste Management. He had experience in handling different types of Municipal Solid Waste composting techniques in Sri Lanka and other Asian countries such as China, Thailand, Japan, India, Malaysia, Singapore, Germany, UK, and Korea. And also he has a special research experience on waste to Energy, Bio merthanization of Municipal solid waste. He has published many Local and international research papers on MSW composting and MSW management. After, Mr Nalin Mannapperuma joining to the Waste Management Authority (W.P.) he took the responsibility of the technical division of the Authority. During past few years he has extensity involved in the areas of designing and construction of composting facilities, operation, maintenance and treble shouting of compost facilities. He has been trained on using of different types of composting technique through several local and overseer training programs. Further Mr Nalin Mannapperuma has been delicately working on establishing Waste Energy, recycling bio-Gass Facilities of the province for last few years. Presently he is holding the topmost position of the Authority and as CEO he is driving authority to achieve objectives of the Waste management statute of the Western Province.

Maria Manzoor PhD in Environmental science. And working experience at national and international research organizations in the area of waste management, phytoremediation and toxicology.

Aung Maw is Deputy Head of Department, Pollution Control and Cleansing Department (PCCD), Yangon City Development Committee (YCDC), Myanmar. He holds a PhD in Environment in 2013. Dr. Aung also serves as a Visiting Professor to the Yangon Technological University, Myanmar.

Kodwo Meizah is a lecturer and researcher at the Environmental Science Department of the Kwame Nkrumah University of Science and Technology (KNUST). He has previously worked with the research department of the biggest private waste management company in Ghana. Miezah in 2015 comprehensively characterized the solid waste management of Ghana (Miezah, Obiri-Danso, Kádár, Fei-Baffoe, & Mensah, 2015).

Anthony Mensah is the Director of Sanitation at the Ministry of Sanitation and Water Resources of Ghana. He has over 20 years working experience as the Director of Waste Management Department in Accra, the capital city, and Kumasi, the second city of Ghana. He has also worked as a consultant for UNICEF and ILO in Liberia.

Sebastian Mhanga was born in Tanzania in 1975. He received his first degree Bsc. Agricultural Economics and Agribusiness from Sokoine University of Agriculture, Tanzania, in 2002. He joined Private Agriculture Sector Support Trust for his first appointment in July, 2002 as a Business Development

Specialist up to April 2004. Thereafter he moved to Technoserve, a US based NGO where he worked as a High Value Opportunities Advisor until August 2005. In September 2005 he joined SNV- Netherlands Development Organization where he worked as an Economic Development Advisor until August 2013. Thereafter he left for self-employment in farming and trading general merchandise. He has recently finished writing his thesis for a Masters degree program in Management of Natural Resources for Sustainable Agriculture at Sokoine University of Agriculture. He is currently working for MEDA in the Feed the Future - ENGINE program in Tanzania as a BDS Coordinator to support MSMEs growth through Investment and Enterprise. While working at SNV Sebastian subscribed some articles to internal SNV Newletters and attended various pro-poor development short courses such as Critical Writing for Change and Impact Investment Management within and outside Tanzania. Over his service period Sebastian has developed competences in areas of market linkage facilitation, value chain analysis, investment analysis, multi-stakeholder process facilitation and monitoring & evaluation. His main areas of research interest are Economic Development and Natural Resources Management. Sebastian is also a member of World Economics Association (WEA), Sustainable Aquaculture & Research Network for Information Sharing in Africa (SARNISA) and The Platform for African-European Partnership in Agricultural Research for Development Phase II (PAEPARD II).

Shivani Mishra is working as Full Professor at Nanotechnology and water sustainability unit, University of South Africa. She works in the field of Nano material and nanocomposites development and its application in the field of water research and waste minimizations. She has published over 150 articles and book chapters in peer reviewed journals and has edited 3 books. Prof Mishra has supervised over 25 postgraduate researchers. She is reviewer for many prestigious journals and panel member for numerous academic programs. She is Fellow of Royal Society of Chemistry and recipient of prestigious distinguished woman in science award from DST South Africa and Women in leadership award from Unisa.

Thang Nguyen is the Deputy Director General of the Institute of Strategy and Policy on Natural Resources and Environment (ISPONRE), a think tank that belongs to the Ministry of Natural Resources and Environment (MONRE) of Viet Nam. He has 15 year experience in study/research and development of strategies/policies on environmental protection and climate change. His expertise focuses on pollution control, waste management, sustainable production and consumption. He participated actively in development of the National Strategy on Environmental Protection to 2020, Party Resolution 24-NQ/TW on proactive responding to climate change, strengthening natural resources management and environmental protection… He was also a Coordinating Lead Author (CLA) of the UNEP Global Environmental Outlook 6, Regional Assessment for Asia-Pacific and participated in some joint researches such as The State of 3R in Asia and the Pacific, 2018 (UNCRD, IGES) and Eco-innovation and Sustainable Consumption and Production in Vietnam (Hanns Seidel Foundation, ASEIC), GEO 6 – Industry (UNEP and IGES), ect… Dr. Nguyen Trung Thang got his PhD degree in Chemical Technology from Sofia University of Chemical Technology, Bulgaria.

Tu Nguyen is Deputy Head of Department of Environment and Sustainable Development, Institute of Strategy and Policy on Natural Resources and Environment, Ministry of Natural Resources and Environment. He has over 10 year experience in study/research of policies on environment and sustainable development. His expertise focuses on development economics, environmental economics and manangement. He participated in a variety of projects on environment and sustainable development such as compost production, 3R, waste audit, SDGs, etc. Mr. Nguyen Ngoc Tu got his Master Degree in Environmental Economics and Management from National Economics University.

Joan Nyika is currently a PhD student at the Department of Civil and Chemical Engineering conducting her research in solid waste management. She has also worked for the past 5 years at the Technical University of Kenya, where she served both as a graduate and research assistant. Her research interests include solid waste management, environmental management, water and land resources management. She has written several publications relating to these fields in peer reviewed journals.

Kwaku Oduro-Appiah is a lecturer at the Water and Sanitation Department of the University of Cape Coast, Ghana, and a PhD fellow at Maastricht University, the Netherlands, with research interest in sustainable solid waste management systems in developing countries.

Edna Onyari is currently a senior lecturer working at the University of South Africa in the civil engineering department. She has a PhD in Civil & Environmental engineering from the University of the Witwatersrand and a MSc degree in Water Engineering from UNESCO-IHE in the Netherlands. Dr Onyari has contributed to knowledge and has several publications in the field of water resources and environmental engineering.

Bestami Ozkaya was born in Elazığ/Turkey on 02 March 1976. He earned a BS in Environmental Engineering field at Firat University placed at the birth city. Dr. Özkaya left at the age of 22 to complete his MSc and pHD education at Yildiz Technical University (YTU), İstanbul. In 2001 he graduated MSc from YTU. In 2004 the same university awarded his a phD in waste management field. Dr. Özkaya then was appointed as an assistant professor in the Yildiz Technical University. He moved to Tampere (Finland) to perform his post-doctoral researches from 2006 to 2008 at Tampere University of Technology. There he was promoted to associated professor position. After returning to İstanbul in 2010 he was promoted to associate professor position in YTU. In 2013 he was awarded to full professor position in 2013 at the same university. He is a Turkish researcher who has specialization in Environmental Engineering, Environmental Biotechnology and Bioenergy. Dr. Ozkaya's main objective of research is to develop the energy harnessing methodologies that can easily be adapted by the effluent treatment plants so as to produce energy besides treatment. Dr. Ozkaya currently works in Yildiz Technical University as a full professor with the research topic of municipal waste management, biohydrogen (acidogenic fermentation), bioelectricity (microbial fuel cell (MFC)), and biodiesel production from waste matrix.

Dicka Rahim was born in Jakarta, 5 May 1986. He is a Ph.D student at Institute of Energy and Power Engineering, Zhejiang University of Technology, China. He is also a lecturer at Chemical Engineering Department, Bandung Institute of Technology, Indonesia. He has scientific background in process engineering and thermochemical conversion of biomass. His current area of interest are waste to clean energy and syngas desulfurization.

Phalla Sam is a co-founder and a deputy director of Cambodian Education and Waste Management Organization (COMPED). He supervises Waste Management Project Groups (COMPED) and manages a composting project in Battambang province, Cambodia. Over the last twenty years, Mr. Phalla has coordinated and implemented several projects, including capacity building, public awareness raising, and waste, energy and climate change, and conducted studies such as analysis of institutional and legal framework on solid waste management and of waste characteristics for composting and energy. In 2000s, he involved in developing guidelines for solid waste management and urban organic waste utilization in Cambodia and a manual on installation and maintenance of renewable energy technologies. Prior to COMPED, he had been with Ministry of Public Work and Transport of Cambodia before working as Manager of Transportation at one waste collection company in Phnom Penh. Mr. Phalla holds master's degree in Public Administration from Chamroeun University of Polytechnology, Phnom Penh, Cambodia and is diploma engineer in Mechanical Engineering at Rostock University in Germany.

Atilio Savino is vice-president of ARS-Association for Solid Waste Studies, National Member of ISWA in Argentina, and Honorary Member of ISWA-International Solid Waste Association. Previously he has been president of ISWA, Secretary of Sanitary Health Determinants-Ministry of Health of Argentina, and secretary of Environment and Sustainable Development of Argentina-National Ministry of Health and Environment. Chief Editor and author of the Waste Management Outlook for Latin America and the Caribbean – UNEP. Designated author of the Cross-cutting issue Waste in the Global Environment Outlook (GEO-6) assessment, UNEP. Regional Editor and Author of "Municipal Solid Waste Energy Conversion in Emerging Countries" Elsevier.

Anne Scheinberg is an action researcher and recycling specialist; her current areas of focus are inclusive urban waste systems and the application of participatory research appraisal in planning and evaluating integrated sustainable waste management and recycling in emerging economies.

Anh Ta is pursuing Ph.D. program in Environmental Technology at Sirindhorn International Institute of Technology, Thammasat University, Thailand. He graduated from the same institute, with a Master Degree in Engineering and Technology in 2018. His research interest includes solid waste management, recovery of valuable materials from waste, and wastewater treatment.

Ohnmar Tin Hlaing Environmental Health Consultant has been working as a local consultant for the IGES - Centre Collaborating with the UN Environment on Environmental Technologies (CCET) since 2016 in the development of National and City Level Waste Management Strategy and Action Plans.

Bojana Tot is Teaching Assistant at the Faculty of Technical Sciences, Department of Environmental Engineering and Occupational Safety and Health, University of Novi Sad in Republic of Serbia. She graduated 2009 in Environmental Engineering, Faculty of Technical Science at University of Novi Sad and became a graduated engineer of environmental protection-master. Currently she is General Secretary of SeSWA (Serbian Solid Waste Association). The present scientific and research work was focused on systems of waste management and multi criteria decision making (MCDM, AHP, Borda count), the development of waste management systems in developing and transition countries, impact assessments on the environment.

Rithy Uch is the Director of Waste Management Projects Group of Cambodian Education and Waste Management Organization (COMPED-WM), and has 12 years of experience in managing various projects on waste management in Cambodia, including research, analysis and consulting on waste management, development of environmental and waste management guidelines, and organizing diverse public engagement programs. Mr. Uch is also recently engaged in a waste cleansing operation in local water bodies using customized catamaran called Sea Hamster while managing composting facilities to promote recycling, with the aim of closing the loop in Cambodia in long-term.

Goran Vujic is Full Professor at the Faculty of Technical Sciences, Department of Environmental Engineering and Occupational Safety and Health, University of Novi Sad in Republic of Serbia. He holds his M.Sc. (2003) and Ph.D. (2007) degrees in Environmental Engineering from Faculty of Technical Science at University of Novi Sad. In general his research work is focused on landfilling technologies and management, landfill gas management, development of waste management systems in developing and transition countries, transfer of environmental technologies from developed to developing countries and environmental impact assessments. He served as ISWA board member for South East Europe. Currently he is Executive Director of the Serbian Solid Waste Association and guest lecturer at Vienna University of Technology.

S. Werahera currently the Deputy Director/Environmental Pollution Control and Chemical Management of Ministry of Mahaweli Development and Environment, Sri Lanka. B.Sc. in Agriculture and M.Sc. in Agricultural Economics, University of Peradeniya and M.Sc. in Environment and Resource Economics from Australian National University. During his 19 years carrier in the Ministry, he represented Sri Lanka in several international meetings including conferences of parties. Made many technical contributions to the waste and chemical management sector that include initiation of the second National waste management of Sri Lanka, first public-private sector partnership projects and on Extended Producer Responsibility.

Haryo Wibowo was born in Jakarta, January 5th 1983. He is PhD candidate in Institute of Energy and Thermal Engineering, Zhejiang University of Technology, PR China. He has scientific background in process engineering and materials engineering. His current area of interest are waste to clean energy and acid gas removal.

Ran Yagasa is a researcher at IGES Centre Collaborating with UNEP on Environmental Technologies (CCET), a unit established within Sustainable Consumption and Production Group, Institute for Global Environmental Strategies, with the mandate of assisting countries to strengthen waste management systems and governance. Since 2015, he has been working as the chief coordinator for the center's project in Cambodia for the development of national and city-level waste management strategies, while facilitating implementation of pilot project in Phnom Penh for promoting environmental education in primary schools. He has also worked with the Government of Maldives as a member of center's country project team. Before joining IGES, he received MA in Public Management and Policy Analysis from the International University of Japan (IUJ), and has previously worked in MICE industry. His research interests are various topics under waste management governance and strategies including public participation, environmental education, and selection of appropriate technologies.

Mi Yan was born in Zhejiang on July 6th 1985. He is an associate professor in Institute of Energy and Thermal Engineering, Zhejiang University of Technology, PR China. As an academic staff, he was involved in teaching, research, and community services. He has scientific background in thermal engineering and thermochemical conversion of biomass. His current area of interest is waste to clean energy

Hu Yanjun an associate professor in the Institute of Energy and Thermal Engineering, Zhejiang University of Technology. her research focuses on pyrolysis and fly ash treatment.

Cai Yi currently works in research and development for Zhejiang Energy Group, PR China. His current area of work is waste-to-clean energy.

Index

P

Q

R

S

T

U

W

Y

Z

www.ingramcontent.com/pod-product-compliance
Lightning Source LLC
LaVergne TN
LVHW080404011026
849161LV00009B/2345